PROBABILITY AND RANDOM PROCESSES FOR ELECTRICAL AND COMPUTER ENGINEERS

The theory of probability is a powerful tool that helps electrical and computer engineers explain, model, analyze, and design the technology they develop. The text begins at the advanced undergraduate level, assuming only a modest knowledge of probability, and progresses through more complex topics mastered at the graduate level. The first five chapters cover the basics of probability and both discrete and continuous random variables. The later chapters have a more specialized coverage, including random vectors, Gaussian random vectors, random processes, Markov Chains, and convergence. Describing tools and results that are used extensively in the field, this is more than a textbook: it is also a reference for researchers working in communications, signal processing, and computer network traffic analysis. With over 300 worked examples, some 800 homework problems, and sections for exam preparation, this is an essential companion for advanced undergraduate and graduate students.

Further resources for this title, including solutions, are available online at www.cambridge.org/9780521864701.

JOHN A. GUBNER has been on the Faculty of Electrical and Computer Engineering at the University of Wisconsin-Madison since receiving his Ph.D. in 1988, from the University of Maryland at College Park. His research interests include ultra-wideband communications; point processes and shot noise; subspace methods in statistical processing; and information theory. A member of the IEEE, he has authored or co-authored many papers in the *IEEE Transactions*, including those on Information Theory, Signal Processing, and Communications.

Discrete random variables

Bernoulli(p)

$P(X=1) = p$, $\quad P(X=0) = 1-p$.
$E[X] = p$, $\quad \text{var}(X) = p(1-p)$, $\quad G_X(z) = (1-p) + pz$.

Poisson(λ)

$$P(X=k) = \frac{\lambda^k e^{-\lambda}}{k!}, \quad k = 0, 1, \ldots.$$

$E[X] = \lambda$, $\quad \text{var}(X) = \lambda$, $\quad G_X(z) = e^{\lambda(z-1)}$.

geometric$_0$(p)

$P(X=k) = (1-p)p^k$, $\quad k = 0, 1, 2, \ldots$.

$E[X] = \dfrac{p}{1-p}$, $\quad \text{var}(X) = \dfrac{p}{(1-p)^2}$, $\quad G_X(z) = \dfrac{1-p}{1-pz}$.

geometric$_1$(p)

$P(X=k) = (1-p)p^{k-1}$, $\quad k = 1, 2, 3, \ldots$.

$E[X] = \dfrac{1}{1-p}$, $\quad \text{var}(X) = \dfrac{p}{(1-p)^2}$, $\quad G_X(z) = \dfrac{(1-p)z}{1-pz}$.

binomial(n, p)

$P(X=k) = \binom{n}{k} p^k (1-p)^{n-k}$, $\quad k = 0, \ldots, n$.

$E[X] = np$, $\quad \text{var}(X) = np(1-p)$, $\quad G_X(z) = [(1-p) + pz]^n$.

negative binomial or Pascal(m, p)

$P(X=k) = \binom{k-1}{m-1}(1-p)^m p^{k-m}$, $\quad k = m, m+1, \ldots$.

$E[X] = \dfrac{m}{1-p}$, $\quad \text{var}(X) = \dfrac{mp}{(1-p)^2}$, $\quad G_X(z) = \left[\dfrac{(1-p)z}{1-pz}\right]^m$.

Note that Pascal($1, p$) is the same as geometric$_1$(p).

Note. To obtain the moment generating function from $G_X(z)$, let $z = e^s$. To obtain the characteristic function, let $z = e^{jv}$. In other words, $M_X(s) = G_X(e^s)$ and $\varphi_X(v) = G_X(e^{jv})$.

Fourier transforms

Fourier transform
$$H(f) = \int_{-\infty}^{\infty} h(t) e^{-j2\pi ft}\, dt$$

Inversion formula
$$h(t) = \int_{-\infty}^{\infty} H(f) e^{j2\pi ft}\, df$$

$h(t)$	$H(f)$
$I_{[-T,T]}(t)$	$2T\dfrac{\sin(2\pi Tf)}{2\pi Tf}$
$2W\dfrac{\sin(2\pi Wt)}{2\pi Wt}$	$I_{[-W,W]}(f)$
$(1-|t|/T)I_{[-T,T]}(t)$	$T\left[\dfrac{\sin(\pi Tf)}{\pi Tf}\right]^2$
$W\left[\dfrac{\sin(\pi Wt)}{\pi Wt}\right]^2$	$(1-|f|/W)I_{[-W,W]}(f)$
$e^{-\lambda t}u(t)$	$\dfrac{1}{\lambda + j2\pi f}$
$e^{-\lambda|t|}$	$\dfrac{2\lambda}{\lambda^2 + (2\pi f)^2}$
$\dfrac{\lambda}{\lambda^2 + t^2}$	$\pi e^{-2\pi\lambda|f|}$
$e^{-(t/\sigma)^2/2}$	$\sqrt{2\pi}\,\sigma\, e^{-\sigma^2(2\pi f)^2/2}$

Note. The indicator function $I_{[a,b]}(t) := 1$ for $a \le t \le b$ and $I_{[a,b]}(t) := 0$ otherwise. In particular, $u(t) := I_{[0,\infty)}(t)$ is the unit step function.

Series formulas

$$\sum_{k=0}^{N-1} z^k = \frac{1-z^N}{1-z},\ z \ne 1 \qquad e^z := \sum_{k=0}^{\infty} \frac{z^k}{k!} \qquad (a+b)^n = \sum_{k=0}^{n} \binom{n}{k} a^k b^{n-k}$$

$$\sum_{k=0}^{\infty} z^k = \frac{1}{1-z},\ |z| < 1 \qquad \lim_{n\to\infty}\left(1+\frac{z}{n}\right)^n = e^z \qquad \sum_{k=1}^{n} k = \frac{n(n+1)}{2}$$

PROBABILITY AND RANDOM PROCESSES FOR ELECTRICAL AND COMPUTER ENGINEERS

JOHN A. GUBNER
University of Wisconsin-Madison

CAMBRIDGE
UNIVERSITY PRESS

University Printing House, Cambridge CB2 8BS, United Kingdom

Cambridge University Press is part of the University of Cambridge.

It furthers the University's mission by disseminating knowledge in the pursuit of education, learning and research at the highest international levels of excellence.

www.cambridge.org
Information on this title: www.cambridge.org/9780521864701

© Cambridge University Press 2006

This publication is in copyright. Subject to statutory exception and to the provisions of relevant collective licensing agreements, no reproduction of any part may take place without the written permission of Cambridge University Press.

First published 2006
6th printing 2014

A catalogue record for this publication is available from the British Library

ISBN 978-0-521-86470-1 Hardback

Cambridge University Press has no responsibility for the persistence or accuracy of URLs for external or third-party internet websites referred to in this publication, and does not guarantee that any content on such websites is, or will remain, accurate or appropriate.

To Sue and Joe

Contents

			page	
	Chapter dependencies			xii
	Preface			xiii
1	**Introduction to probability**			**1**
	1.1	Sample spaces, outcomes, and events		6
	1.2	Review of set notation		8
	1.3	Probability models		17
	1.4	Axioms and properties of probability		22
	1.5	Conditional probability		26
	1.6	Independence		30
	1.7	Combinatorics and probability		34
		Notes		43
		Problems		48
		Exam preparation		62
2	**Introduction to discrete random variables**			**63**
	2.1	Probabilities involving random variables		63
	2.2	Discrete random variables		66
	2.3	Multiple random variables		70
	2.4	Expectation		80
		Notes		96
		Problems		99
		Exam preparation		106
3	**More about discrete random variables**			**108**
	3.1	Probability generating functions		108
	3.2	The binomial random variable		111
	3.3	The weak law of large numbers		115
	3.4	Conditional probability		117
	3.5	Conditional expectation		127
		Notes		130
		Problems		132
		Exam preparation		137
4	**Continuous random variables**			**138**
	4.1	Densities and probabilities		138
	4.2	Expectation of a single random variable		149
	4.3	Transform methods		156
	4.4	Expectation of multiple random variables		162
	4.5	Probability bounds		164
		Notes		167
		Problems		170
		Exam preparation		183
5	**Cumulative distribution functions and their applications**			**184**
	5.1	Continuous random variables		185
	5.2	Discrete random variables		194
	5.3	Mixed random variables		197
	5.4	Functions of random variables and their cdfs		200
	5.5	Properties of cdfs		205
	5.6	The central limit theorem		207
	5.7	Reliability		215

	Notes	219
	Problems	222
	Exam preparation	238
6	**Statistics**	**240**
6.1	Parameter estimators and their properties	240
6.2	Histograms	244
6.3	Confidence intervals for the mean – known variance	250
6.4	Confidence intervals for the mean – unknown variance	253
6.5	Confidence intervals for Gaussian data	256
6.6	Hypothesis tests for the mean	262
6.7	Regression and curve fitting	267
6.8	Monte Carlo estimation	271
	Notes	273
	Problems	276
	Exam preparation	285
7	**Bivariate random variables**	**287**
7.1	Joint and marginal probabilities	287
7.2	Jointly continuous random variables	295
7.3	Conditional probability and expectation	302
7.4	The bivariate normal	309
7.5	Extension to three or more random variables	314
	Notes	317
	Problems	319
	Exam preparation	328
8	**Introduction to random vectors**	**330**
8.1	Review of matrix operations	330
8.2	Random vectors and random matrices	333
8.3	Transformations of random vectors	340
8.4	Linear estimation of random vectors (Wiener filters)	344
8.5	Estimation of covariance matrices	348
8.6	Nonlinear estimation of random vectors	350
	Notes	354
	Problems	354
	Exam preparation	360
9	**Gaussian random vectors**	**362**
9.1	Introduction	362
9.2	Definition of the multivariate Gaussian	363
9.3	Characteristic function	365
9.4	Density function	367
9.5	Conditional expectation and conditional probability	369
9.6	Complex random variables and vectors	371
	Notes	373
	Problems	375
	Exam preparation	382
10	**Introduction to random processes**	**383**
10.1	Definition and examples	383
10.2	Characterization of random processes	388
10.3	Strict-sense and wide-sense stationary processes	393
10.4	WSS processes through LTI systems	401
10.5	Power spectral densities for WSS processes	403
10.6	Characterization of correlation functions	410
10.7	The matched filter	412
10.8	The Wiener filter	417

10.9	The Wiener–Khinchin theorem	421
10.10	Mean-square ergodic theorem for WSS processes	423
10.11	Power spectral densities for non-WSS processes	425
	Notes	427
	Problems	429
	Exam preparation	440

11 Advanced concepts in random processes — 443
11.1	The Poisson process	443
11.2	Renewal processes	452
11.3	The Wiener process	453
11.4	Specification of random processes	459
	Notes	466
	Problems	466
	Exam preparation	475

12 Introduction to Markov chains — 476
12.1	Preliminary results	476
12.2	Discrete-time Markov chains	477
12.3	Recurrent and transient states	488
12.4	Limiting n-step transition probabilities	496
12.5	Continuous-time Markov chains	502
	Notes	507
	Problems	509
	Exam preparation	515

13 Mean convergence and applications — 517
13.1	Convergence in mean of order p	518
13.2	Normed vector spaces of random variables	522
13.3	The Karhunen–Loève expansion	527
13.4	The Wiener integral (again)	532
13.5	Projections, orthogonality principle, projection theorem	534
13.6	Conditional expectation and probability	537
13.7	The spectral representation	545
	Notes	549
	Problems	550
	Exam preparation	562

14 Other modes of convergence — 564
14.1	Convergence in probability	564
14.2	Convergence in distribution	566
14.3	Almost-sure convergence	572
	Notes	579
	Problems	580
	Exam preparation	589

15 Self similarity and long-range dependence — 591
15.1	Self similarity in continuous time	591
15.2	Self similarity in discrete time	595
15.3	Asymptotic second-order self similarity	601
15.4	Long-range dependence	604
15.5	ARMA processes	606
15.6	ARIMA processes	608
	Problems	610
	Exam preparation	613

Bibliography — 615

Index — 618

Chapter dependencies

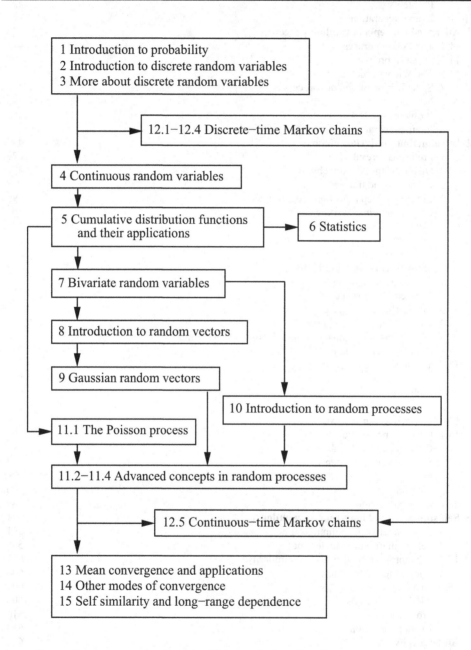

Preface

Intended audience

This book is a primary text for **graduate-level courses** in probability and random processes that are typically offered in electrical and computer engineering departments. The text starts from first principles and contains more than enough material for a two-semester sequence. The **level of the text** varies from advanced undergraduate to graduate as the material progresses. The principal **prerequisite** is the usual undergraduate electrical and computer engineering course on signals and systems, e.g., Haykin and Van Veen [25] or Oppenheim and Willsky [39] (see the Bibliography at the end of the book). However, later chapters that deal with random vectors assume some familiarity with linear algebra; e.g., determinants and matrix inverses.

How to use the book

A first course. In a course that assumes at most a modest background in probability, the core of the offering would include Chapters 1–5 and 7. These cover the basics of probability and discrete and continuous random variables. As the chapter dependencies graph on the preceding page indicates, there is considerable flexibility in the selection and ordering of additional material as the instructor sees fit.

A second course. In a course that assumes a solid background in the basics of probability and discrete and continuous random variables, the material in Chapters 1–5 and 7 can be reviewed quickly. In such a review, the instructor may want include **sections and problems marked with a ⋆, as these indicate more challenging material that might not be appropriate in a first course**. Following the review, the core of the offering would include Chapters 8, 9, 10 (Sections 10.1–10.6), and Chapter 11. Additional material from Chapters 12–15 can be included to meet course goals and objectives.

Level of course offerings. In any course offering, the level can be adapted to the background of the class by omitting or including the **more advanced sections, remarks, and problems that are marked with a ⋆**. In addition, discussions of a highly technical nature are placed in a Notes section at the end of the chapter in which they occur. Pointers to these discussions are indicated by **boldface numerical superscripts** in the text. These notes can be omitted or included as the instructor sees fit.

Chapter features

- Key equations are boxed:

$$\boxed{\mathsf{P}(A|B) := \frac{\mathsf{P}(A \cap B)}{\mathsf{P}(B)}.}$$

- Important text passages are highlighted:

  Two events A and B are said to be independent if $\mathsf{P}(A \cap B) = \mathsf{P}(A)\mathsf{P}(B)$.

- Tables of discrete random variables and of Fourier transform pairs are found inside the front cover. A table of continuous random variables is found inside the back cover.
- The index was compiled as the book was written. Hence, there are many cross-references to related information. For example, see "chi-squared random variable."
- When cumulative distribution functions or other functions are encountered that do not have a closed form, MATLAB commands are given for computing them; see "Matlab commands" in the index for a list. The use of many commands is illustrated in the examples and the problems throughout most of the text. Although some commands require the MATLAB Statistics Toolbox, alternative methods are also suggested; e.g., the use of `erf` and `erfinv` for `normcdf` and `norminv`.
- Each chapter contains a **Notes** section. Throughout each chapter, numerical superscripts refer to discussions in the Notes section. These notes are usually rather technical and address subtleties of the theory.
- Each chapter contains a **Problems** section. There are more than 800 problems throughout the book. Problems are grouped according to the section they are based on, and this is clearly indicated. This enables the student to refer to the appropriate part of the text for background relating to particular problems, and it enables the instructor to make up assignments more quickly. In chapters intended for a first course, **the more challenging problems are marked with a ***. Problems requiring MATLAB are indicated by the label **MATLAB**.
- Each chapter contains an **Exam preparation** section. This serves as a chapter summary, drawing attention to key concepts and formulas.

Acknowledgements

The writing of this book has been greatly improved by the suggestions of many people.

At the University of Wisconsin–Madison, the sharp eyes of the students in my classes on probability and random processes, my research students, and my postdocs have helped me fix countless typos and improve explanations of several topics. My colleagues here have been generous with their comments and suggestions. Professor Rajeev Agrawal, now with Motorola, convinced me to treat discrete random variables before continuous random variables. Discussions with Professor Bob Barmish on robustness of rational transfer functions led to Problems 38–40 in Chapter 5. I am especially grateful to Professors Jim Bucklew, Yu Hen Hu, and Akbar Sayeed, who taught from early, unpolished versions of the manuscript.

Colleagues at other universities and students in their classes have also been generous with their support. I thank Professors Toby Berger, Edwin Chong, and Dave Neuhoff, who have used recent manuscripts in teaching classes on probability and random processes and have provided me with detailed reviews. Special thanks go to Professor Tom Denney for his multiple careful reviews of each chapter.

Since writing is a solitary process, I am grateful to be surrounded by many supportive family members. I especially thank my wife and son for their endless patience and faith in me and this book, and I thank my parents for their encouragement and help when I was preoccupied with writing.

1
Introduction to probability

Why do electrical and computer engineers need to study probability?

Probability theory provides powerful tools to explain, model, analyze, and design technology developed by electrical and computer engineers. Here are a few applications.

Signal processing. My own interest in the subject arose when I was an undergraduate taking the required course in probability for electrical engineers. We considered the situation shown in Figure 1.1. To determine the presence of an aircraft, a known radar pulse $v(t)$

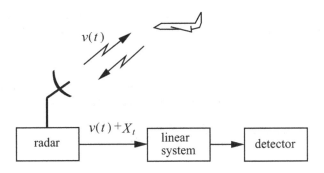

Figure **1.1.** Block diagram of radar detection system.

is sent out. If there are no objects in range of the radar, the radar's amplifiers produce only a noise waveform, denoted by X_t. If there is an object in range, the reflected radar pulse plus noise is produced. The overall goal is to decide whether the received waveform is noise only or signal plus noise. To get an idea of how difficult this can be, consider the signal plus noise waveform shown at the top in Figure 1.2. Our class addressed the subproblem of designing an optimal linear system to process the received waveform so as to make the presence of the signal more obvious. We learned that the optimal transfer function is given by the matched filter. If the signal at the top in Figure 1.2 is processed by the appropriate matched filter, we get the output shown at the bottom in Figure 1.2. You will study the matched filter in Chapter 10.

Computer memories. Suppose you are designing a computer memory to hold k-bit words. To increase system reliability, you employ an error-correcting-code system. With this system, instead of storing just the k data bits, you store an additional l bits (which are functions of the data bits). When reading back the $(k+l)$-bit word, if at least m bits are read out correctly, then all k data bits can be recovered (the value of m depends on the code). To characterize the quality of the computer memory, we compute the probability that at least m bits are correctly read back. You will be able to do this after you study the binomial random variable in Chapter 3.

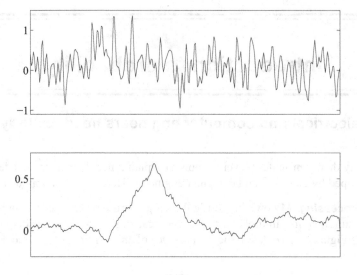

***Figure* 1.2.** Matched filter input (top) in which the signal is hidden by noise. Matched filter output (bottom) in which the signal presence is obvious.

Optical communication systems. Optical communication systems use photodetectors (see Figure 1.3) to interface between optical and electronic subsystems. When these sys-

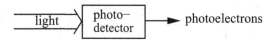

***Figure* 1.3.** Block diagram of a photodetector. The rate at which photoelectrons are produced is proportional to the intensity of the light.

tems are at the limits of their operating capabilities, the number of photoelectrons produced by the photodetector is well-modeled by the Poisson[a] random variable you will study in Chapter 2 (see also the Poisson process in Chapter 11). In deciding whether a transmitted bit is a zero or a one, the receiver counts the number of photoelectrons and compares it to a threshold. System performance is determined by computing the probability that the threshold is exceeded.

Wireless communication systems. In order to enhance weak signals and maximize the range of communication systems, it is necessary to use amplifiers. Unfortunately, amplifiers always generate thermal noise, which is added to the desired signal. As a consequence of the underlying physics, the noise is Gaussian. Hence, the Gaussian density function, which you will meet in Chapter 4, plays a prominent role in the analysis and design of communication systems. When noncoherent receivers are used, e.g., noncoherent frequency shift keying,

---

[a]Many important quantities in probability and statistics are named after famous mathematicians and statisticians. You can use an Internet search engine to find pictures and biographies of them on the web. At the time of this writing, numerous biographies of famous mathematicians and statisticians can be found at http://turnbull.mcs.st-and.ac.uk/history/BiogIndex.html and at http://www.york.ac.uk/depts/maths/histstat/people/welcome.htm. Pictures on stamps and currency can be found at http://jeff560.tripod.com/.

this naturally leads to the Rayleigh, chi-squared, noncentral chi-squared, and Rice density functions that you will meet in the problems in Chapters 4, 5, 7, and 9.

Variability in electronic circuits. Although circuit manufacturing processes attempt to ensure that all items have nominal parameter values, there is always some variation among items. How can we estimate the average values in a batch of items without testing all of them? How good is our estimate? You will learn how to do this in Chapter 6 when you study parameter estimation and confidence intervals. Incidentally, the same concepts apply to the prediction of presidential elections by surveying only a few voters.

Computer network traffic. Prior to the 1990s, network analysis and design was carried out using long-established Markovian models [41, p. 1]. You will study Markov chains in Chapter 12. As self similarity was observed in the traffic of local-area networks [35], wide-area networks [43], and in World Wide Web traffic [13], a great research effort began to examine the impact of self similarity on network analysis and design. This research has yielded some surprising insights into questions about buffer size vs. bandwidth, multiple-time-scale congestion control, connection duration prediction, and other issues [41, pp. 9–11]. In Chapter 15 you will be introduced to self similarity and related concepts.

In spite of the foregoing applications, probability was not originally developed to handle problems in electrical and computer engineering. The first applications of probability were to questions about gambling posed to Pascal in 1654 by the Chevalier de Mere. Later, probability theory was applied to the determination of life expectancies and life-insurance premiums, the theory of measurement errors, and to statistical mechanics. Today, the theory of probability and statistics is used in many other fields, such as economics, finance, medical treatment and drug studies, manufacturing quality control, public opinion surveys, etc.

Relative frequency

Consider an experiment that can result in M possible outcomes, $O_1, \ldots, O_M$. For example, in tossing a die, one of the six sides will land facing up. We could let O_i denote the outcome that the ith side faces up, $i = 1, \ldots, 6$. Alternatively, we might have a computer with six processors, and O_i could denote the outcome that a program or thread is assigned to the ith processor. As another example, there are $M = 52$ possible outcomes if we draw one card from a deck of playing cards. Similarly, there are $M = 52$ outcomes if we ask which week during the next year the stock market will go up the most. The simplest example we consider is the flipping of a coin. In this case there are two possible outcomes, "heads" and "tails." Similarly, there are two outcomes when we ask whether or not a bit was correctly received over a digital communication system. No matter what the experiment, suppose we perform it n times and make a note of how many times each outcome occurred. Each performance of the experiment is called a **trial**.[b] Let $N_n(O_i)$ denote the number of times O_i occurred in n trials. The **relative frequency** of outcome O_i,

$$\frac{N_n(O_i)}{n},$$

is the fraction of times O_i occurred.

---

[b]When there are only two outcomes, the repeated experiments are called **Bernoulli trials**.

Here are some simple computations using relative frequency. First,

$$N_n(O_1) + \cdots + N_n(O_M) = n,$$

and so

$$\frac{N_n(O_1)}{n} + \cdots + \frac{N_n(O_M)}{n} = 1. \quad (1.1)$$

Second, we can group outcomes together. For example, if the experiment is tossing a die, let E denote the event that the outcome of a toss is a face with an even number of dots; i.e., E is the event that the outcome is O_2, O_4, or O_6. If we let $N_n(E)$ denote the number of times E occurred in n tosses, it is easy to see that

$$N_n(E) = N_n(O_2) + N_n(O_4) + N_n(O_6),$$

and so the relative frequency of E is

$$\frac{N_n(E)}{n} = \frac{N_n(O_2)}{n} + \frac{N_n(O_4)}{n} + \frac{N_n(O_6)}{n}. \quad (1.2)$$

Practical experience has shown us that as the number of trials n becomes large, the relative frequencies settle down and appear to converge to some limiting value. This behavior is known as **statistical regularity**.

Example 1.1. Suppose we toss a fair coin 100 times and note the relative frequency of heads. Experience tells us that the relative frequency should be about $1/2$. When we did this,[c] we got 0.47 and were not disappointed.

The tossing of a coin 100 times and recording the relative frequency of heads out of 100 tosses can be considered an experiment in itself. Since the number of heads can range from 0 to 100, there are 101 possible outcomes, which we denote by $S_0, \ldots, S_{100}$. In the preceding example, this experiment yielded S_{47}.

Example 1.2. We performed the experiment with outcomes $S_0, \ldots, S_{100}$ 1000 times and counted the number of occurrences of each outcome. All trials produced between 33 and 68 heads. Rather than list $N_{1000}(S_k)$ for the remaining values of k, we summarize as follows:

$$N_{1000}(S_{33}) + N_{1000}(S_{34}) + N_{1000}(S_{35}) = 4$$
$$N_{1000}(S_{36}) + N_{1000}(S_{37}) + N_{1000}(S_{38}) = 6$$
$$N_{1000}(S_{39}) + N_{1000}(S_{40}) + N_{1000}(S_{41}) = 32$$
$$N_{1000}(S_{42}) + N_{1000}(S_{43}) + N_{1000}(S_{44}) = 98$$
$$N_{1000}(S_{45}) + N_{1000}(S_{46}) + N_{1000}(S_{47}) = 165$$
$$N_{1000}(S_{48}) + N_{1000}(S_{49}) + N_{1000}(S_{50}) = 230$$
$$N_{1000}(S_{51}) + N_{1000}(S_{52}) + N_{1000}(S_{53}) = 214$$
$$N_{1000}(S_{54}) + N_{1000}(S_{55}) + N_{1000}(S_{56}) = 144$$

---

[c]We did not actually toss a coin. We used a random number generator to simulate the toss of a fair coin. Simulation is discussed in Chapters 5 and 6.

$$N_{1000}(S_{57}) + N_{1000}(S_{58}) + N_{1000}(S_{59}) = 76$$
$$N_{1000}(S_{60}) + N_{1000}(S_{61}) + N_{1000}(S_{62}) = 21$$
$$N_{1000}(S_{63}) + N_{1000}(S_{64}) + N_{1000}(S_{65}) = 9$$
$$N_{1000}(S_{66}) + N_{1000}(S_{67}) + N_{1000}(S_{68}) = 1.$$

This summary is illustrated in the histogram shown in Figure 1.4. (The bars are centered over values of the form $k/100$; e.g., the bar of height 230 is centered over 0.49.)

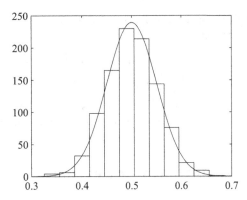

Figure **1.4.** Histogram of Example 1.2 with overlay of a Gaussian density.

Below we give an indication of why most of the time the relative frequency of heads is close to one half and why the bell-shaped curve fits so well over the histogram. For now we point out that the foregoing methods allow us to determine the bit-error rate of a digital communication system, whether it is a wireless phone or a cable modem connection. In principle, we simply send a large number of bits over the channel and find out what fraction were received incorrectly. This gives an estimate of the bit-error rate. To see how good an estimate it is, we repeat the procedure many times and make a histogram of our estimates.

What is probability theory?

Axiomatic probability theory, which is the subject of this book, was developed by **A. N. Kolmogorov**[d] in 1933. This theory specifies a set of axioms for a well-defined mathematical model of physical experiments whose outcomes exhibit random variability each time they are performed. The advantage of using a model rather than performing an experiment itself is that it is usually much more efficient in terms of time and money to analyze a mathematical model. This is a sensible approach only if the model correctly predicts the behavior of actual experiments. This is indeed the case for Kolmogorov's theory.

A simple prediction of Kolmogorov's theory arises in the mathematical model for the relative frequency of heads in n tosses of a fair coin that we considered in Example 1.1. In the model of this experiment, the relative frequency converges to $1/2$ as n tends to infinity;

---

[d]The website http://kolmogorov.com/ is devoted to Kolmogorov.

this is a special case of the the **strong law of large numbers**, which is derived in Chapter 14. (A related result, known as the **weak law of large numbers**, is derived in Chapter 3.)

Another prediction of Kolmogorov's theory arises in modeling the situation in Example 1.2. The theory explains why the histogram in Figure 1.4 agrees with the bell-shaped curve overlaying it. In the model, the strong law tells us that for each k, the relative frequency of having exactly k heads in 100 tosses should be close to

$$\frac{100!}{k!(100-k)!} \frac{1}{2^{100}}.$$

Then, by the **central limit theorem**, which is derived in Chapter 5, the above expression is approximately equal to (see Example 5.19)

$$\frac{1}{5\sqrt{2\pi}} \exp\left[-\frac{1}{2}\left(\frac{k-50}{5}\right)^2\right].$$

(You should convince yourself that the graph of e^{-x^2} is indeed a bell-shaped curve.)

Because Kolmogorov's theory makes predictions that agree with physical experiments, it has enjoyed great success in the analysis and design of real-world systems.

1.1 Sample spaces, outcomes, and events

Sample spaces

To model systems that yield uncertain or random measurements, we let Ω denote the set of all possible distinct, indecomposable measurements that could be observed. The set Ω is called the **sample space**. Here are some examples corresponding to the applications discussed at the beginning of the chapter.

Signal processing. In a radar system, the voltage of a noise waveform at time t can be viewed as possibly being any real number. The first step in *modeling* such a noise voltage is to consider the sample space consisting of all real numbers, i.e., $\Omega = (-\infty, \infty)$.

Computer memories. Suppose we store an n-bit word consisting of all 0s at a particular location. When we read it back, we may not get all 0s. In fact, any n-bit word may be read out if the memory location is faulty. The set of all possible n-bit words can be modeled by the sample space

$$\Omega = \{(b_1, \ldots, b_n) : b_i = 0 \text{ or } 1\}.$$

Optical communication systems. Since the output of a photodetector is a random number of photoelectrons. The logical sample space here is the nonnegative integers,

$$\Omega = \{0, 1, 2, \ldots\}.$$

Notice that we include 0 to account for the possibility that no photoelectrons are observed.

Wireless communication systems. Noncoherent receivers measure the energy of the incoming waveform. Since energy is a nonnegative quantity, we model it with the sample space consisting of the nonnegative real numbers, $\Omega = [0, \infty)$.

Variability in electronic circuits. Consider the lowpass RC filter shown in Figure 1.5(a). Suppose that the exact values of R and C are not perfectly controlled by the manufacturing process, but are known to satisfy

$$95 \text{ ohms} \leq R \leq 105 \text{ ohms} \quad \text{and} \quad 300 \text{ } \mu\text{F} \leq C \leq 340 \text{ } \mu\text{F}.$$

1.1 Sample spaces, outcomes, and events

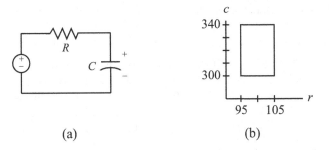

Figure 1.5. (a) Lowpass *RC* filter. (b) Sample space for possible values of *R* and *C*.

This suggests that we use the sample space of ordered pairs of real numbers, (r,c), where $95 \leq r \leq 105$ and $300 \leq c \leq 340$. Symbolically, we write

$$\Omega = \{(r,c) : 95 \leq r \leq 105 \text{ and } 300 \leq c \leq 340\},$$

which is the rectangular region in Figure 1.5(b).

Computer network traffic. If a router has a buffer that can store up to 70 packets, and we want to model the actual number of packets waiting for transmission, we use the sample space

$$\Omega = \{0, 1, 2, \ldots, 70\}.$$

Notice that we include 0 to account for the possibility that there are no packets waiting to be sent.

Outcomes and events

Elements or points in the sample space Ω are called **outcomes**. Collections of outcomes are called **events**. In other words, an event is a subset of the sample space. Here are some examples.

If the sample space is the real line, as in modeling a noise voltage, the individual numbers such as 1.5, -8, and π are outcomes. Subsets such as the interval

$$[0,5] = \{v : 0 \leq v \leq 5\}$$

are events. Another event would be $\{2, 4, 7.13\}$. Notice that singleton sets, that is sets consisting of a single point, are also events; e.g., $\{1.5\}, \{-8\}, \{\pi\}$. Be sure you understand the difference between the outcome -8 and the event $\{-8\}$, which is the set consisting of the single outcome -8.

If the sample space is the set of all triples (b_1, b_2, b_3), where the b_i are 0 or 1, then any particular triple, say $(0,0,0)$ or $(1,0,1)$ would be an outcome. An event would be a subset such as the set of all triples with exactly one 1; i.e.,

$$\{(0,0,1), (0,1,0), (1,0,0)\}.$$

An example of a singleton event would be $\{(1,0,1)\}$.

In modeling the resistance and capacitance of the *RC* filter above, we suggested the sample space
$$\Omega = \{(r,c) : 95 \leq r \leq 105 \text{ and } 300 \leq c \leq 340\},$$
which was shown in Figure 1.5(b). If a particular circuit has $R = 101$ ohms and $C = 327$ μF, this would correspond to the outcome $(101, 327)$, which is indicated by the dot in Figure 1.6. If we observed a particular circuit with $R \leq 97$ ohms and $C \geq 313$ μF, this would correspond to the event
$$\{(r,c) : 95 \leq r \leq 97 \text{ and } 313 \leq c \leq 340\},$$
which is the shaded region in Figure 1.6.

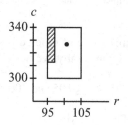

Figure 1.6. The dot is the outcome $(101, 327)$. The shaded region is the event $\{(r,c) : 95 \leq r \leq 97 \text{ and } 313 \leq c \leq 340\}$.

1.2 Review of set notation

Since sample spaces and events use the language of sets, we recall in this section some basic definitions, notation, and properties of sets.

Let Ω be a set of points. If ω is a point in Ω, we write $\omega \in \Omega$. Let A and B be two collections of points in Ω. If every point in A also belongs to B, we say that A is a **subset** of B, and we denote this by writing $A \subset B$. If $A \subset B$ and $B \subset A$, then we write $A = B$; i.e., two sets are equal if they contain exactly the same points. If $A \subset B$ but $A \neq B$, we say that A is a **proper subset** of B.

Set relationships can be represented graphically in **Venn diagrams**. In these pictures, the whole space Ω is represented by a rectangular region, and subsets of Ω are represented by disks or oval-shaped regions. For example, in Figure 1.7(a), the disk A is completely contained in the oval-shaped region B, thus depicting the relation $A \subset B$.

Set operations

If $A \subset \Omega$, and $\omega \in \Omega$ does not belong to A, we write $\omega \notin A$. The set of all such ω is called the **complement** of A in Ω; i.e.,
$$A^c := \{\omega \in \Omega : \omega \notin A\}.$$
This is illustrated in Figure 1.7(b), in which the shaded region is the complement of the disk A.

The **empty set** or **null set** is denoted by $\emptyset$; it contains no points of Ω. Note that for any $A \subset \Omega$, $\emptyset \subset A$. Also, $\Omega^c = \emptyset$.

1.2 Review of set notation

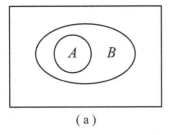 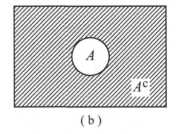

Figure 1.7. (a) Venn diagram of $A \subset B$. (b) The complement of the disk A, denoted by A^c, is the shaded part of the diagram.

The **union** of two subsets A and B is

$$A \cup B := \{\omega \in \Omega : \omega \in A \text{ or } \omega \in B\}.$$

Here "or" is inclusive; i.e., if $\omega \in A \cup B$, we permit ω to belong either to A or to B or to both. This is illustrated in Figure 1.8(a), in which the shaded region is the union of the disk A and the oval-shaped region B.

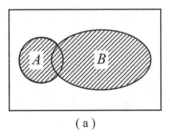 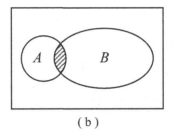

Figure 1.8. (a) The shaded region is $A \cup B$. (b) The shaded region is $A \cap B$.

The **intersection** of two subsets A and B is

$$A \cap B := \{\omega \in \Omega : \omega \in A \text{ and } \omega \in B\};$$

hence, $\omega \in A \cap B$ if and only if ω belongs to both A and B. This is illustrated in Figure 1.8(b), in which the shaded area is the intersection of the disk A and the oval-shaped region B. The reader should also note the following special case. If $A \subset B$ (recall Figure 1.7(a)), then $A \cap B = A$. In particular, we always have $A \cap \Omega = A$ and $\emptyset \cap B = \emptyset$.

The **set difference** operation is defined by

$$B \setminus A := B \cap A^c,$$

i.e., $B \setminus A$ is the set of $\omega \in B$ that do not belong to A. In Figure 1.9(a), $B \setminus A$ is the shaded part of the oval-shaped region B. Thus, $B \setminus A$ is found by starting with all the points in B and then removing those that belong to A.

Two subsets A and B are **disjoint** or **mutually exclusive** if $A \cap B = \emptyset$; i.e., there is no point in Ω that belongs to both A and B. This condition is depicted in Figure 1.9(b).

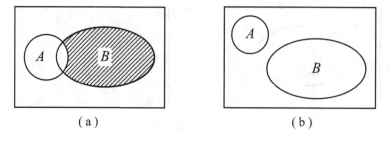

Figure 1.9. (a) The shaded region is $B \setminus A$. (b) Venn diagram of disjoint sets A and B.

Example 1.3. Let $\Omega := \{0,1,2,3,4,5,6,7\}$, and put

$$A := \{1,2,3,4\}, \quad B := \{3,4,5,6\}, \quad \text{and} \quad C := \{5,6\}.$$

Evaluate $A \cup B, A \cap B, A \cap C, A^c$, and $B \setminus A$.

Solution. It is easy to see that $A \cup B = \{1,2,3,4,5,6\}$, $A \cap B = \{3,4\}$, and $A \cap C = \emptyset$. Since $A^c = \{0,5,6,7\}$,

$$B \setminus A = B \cap A^c = \{5,6\} = C.$$

Set identities

Set operations are easily seen to obey the following relations. Some of these relations are analogous to the familiar ones that apply to ordinary numbers if we think of union as the set analog of addition and intersection as the set analog of multiplication. Let A, B, and C be subsets of Ω. The **commutative laws** are

$$A \cup B = B \cup A \quad \text{and} \quad A \cap B = B \cap A. \tag{1.3}$$

The **associative laws** are

$$A \cup (B \cup C) = (A \cup B) \cup C \quad \text{and} \quad A \cap (B \cap C) = (A \cap B) \cap C. \tag{1.4}$$

The **distributive laws** are

$$A \cap (B \cup C) = (A \cap B) \cup (A \cap C) \tag{1.5}$$

and

$$A \cup (B \cap C) = (A \cup B) \cap (A \cup C). \tag{1.6}$$

De Morgan's laws are

$$(A \cap B)^c = A^c \cup B^c \quad \text{and} \quad (A \cup B)^c = A^c \cap B^c. \tag{1.7}$$

Formulas (1.3)–(1.5) are exactly analogous to their numerical counterparts. Formulas (1.6) and (1.7) do not have numerical counterparts. We also recall that $A \cap \Omega = A$ and $\emptyset \cap B = \emptyset$; hence, we can think of Ω as the analog of the number one and $\emptyset$ as the analog of the number zero. Another analog is the formula $A \cup \emptyset = A$.

1.2 Review of set notation

We next consider infinite collections of subsets of Ω. It is important to understand how to work with unions and intersections of infinitely many subsets. Infinite unions allow us to formulate questions about some event ever happening if we wait long enough. Infinite intersections allow us to formulate questions about some event never happening no matter how long we wait.

Suppose $A_n \subset \Omega$, $n = 1, 2, \ldots$. Then

$$\bigcup_{n=1}^{\infty} A_n := \{\omega \in \Omega : \omega \in A_n \text{ for some } 1 \leq n < \infty\}.$$

In other words, $\omega \in \bigcup_{n=1}^{\infty} A_n$ if and only if for at least one integer n satisfying $1 \leq n < \infty$, $\omega \in A_n$. This definition admits the possibility that $\omega \in A_n$ for more than one value of n. Next, we define

$$\bigcap_{n=1}^{\infty} A_n := \{\omega \in \Omega : \omega \in A_n \text{ for all } 1 \leq n < \infty\}.$$

In other words, $\omega \in \bigcap_{n=1}^{\infty} A_n$ if and only if $\omega \in A_n$ for every positive integer n.

Many examples of infinite unions and intersections can be given using intervals of real numbers such as (a,b), $(a,b]$, $[a,b)$, and $[a,b]$. (This notation is reviewed in Problem 5.)

***Example* 1.4.** Let Ω denote the real numbers, $\Omega = \mathbb{R} := (-\infty, \infty)$. Then the following infinite intersections and unions can be simplified. Consider the intersection

$$\bigcap_{n=1}^{\infty} (-\infty, 1/n) = \{\omega : \omega < 1/n \text{ for all } 1 \leq n < \infty\}.$$

Now, if $\omega < 1/n$ for all $1 \leq n < \infty$, then ω cannot be positive; i.e., we must have $\omega \leq 0$. Conversely, if $\omega \leq 0$, then for all $1 \leq n < \infty$, $\omega \leq 0 < 1/n$. It follows that

$$\bigcap_{n=1}^{\infty} (-\infty, 1/n) = (-\infty, 0].$$

Consider the infinite union,

$$\bigcup_{n=1}^{\infty} (-\infty, -1/n] = \{\omega : \omega \leq -1/n \text{ for some } 1 \leq n < \infty\}.$$

Now, if $\omega \leq -1/n$ for some n with $1 \leq n < \infty$, then we must have $\omega < 0$. Conversely, if $\omega < 0$, then for large enough n, $\omega \leq -1/n$. Thus,

$$\bigcup_{n=1}^{\infty} (-\infty, -1/n] = (-\infty, 0).$$

In a similar way, one can show that

$$\bigcap_{n=1}^{\infty} [0, 1/n) = \{0\},$$

as well as

$$\bigcup_{n=1}^{\infty} (-\infty, n] = (-\infty, \infty) \quad \text{and} \quad \bigcap_{n=1}^{\infty} (-\infty, -n] = \emptyset.$$

The following **generalized distributive laws** also hold,

$$B \cap \left(\bigcup_{n=1}^{\infty} A_n \right) = \bigcup_{n=1}^{\infty} (B \cap A_n),$$

and

$$B \cup \left(\bigcap_{n=1}^{\infty} A_n \right) = \bigcap_{n=1}^{\infty} (B \cup A_n).$$

We also have the **generalized De Morgan's laws**,

$$\left(\bigcap_{n=1}^{\infty} A_n \right)^c = \bigcup_{n=1}^{\infty} A_n^c,$$

and

$$\left(\bigcup_{n=1}^{\infty} A_n \right)^c = \bigcap_{n=1}^{\infty} A_n^c.$$

Finally, we will need the following definition. We say that subsets $A_n, n = 1, 2, \ldots$, are **pairwise disjoint** if $A_n \cap A_m = \emptyset$ for all $n \neq m$.

Partitions

A family of nonempty sets B_n is called a **partition** if the sets are pairwise disjoint and their union is the whole space Ω. A partition of three sets B_1, B_2, and B_3 is illustrated in Figure 1.10(a). Partitions are useful for chopping up sets into manageable, disjoint pieces. Given a set A, write

$$\begin{aligned} A &= A \cap \Omega \\ &= A \cap \left(\bigcup_n B_n \right) \\ &= \bigcup_n (A \cap B_n). \end{aligned}$$

Since the B_n are pairwise disjoint, so are the pieces $(A \cap B_n)$. This is illustrated in Figure 1.10(b), in which a disk is broken up into three disjoint pieces.

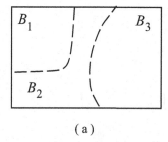
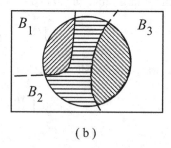

(a) (b)

Figure 1.10. (a) The partition B_1, B_2, B_3. (b) Using the partition to break up a disk into three disjoint pieces (the shaded regions).

1.2 Review of set notation

If a family of sets B_n is disjoint but their union is not equal to the whole space, we can always add the remainder set

$$R := \left(\bigcup_n B_n\right)^c \tag{1.8}$$

to the family to create a partition. Writing

$$\Omega = R^c \cup R$$
$$= \left(\bigcup_n B_n\right) \cup R,$$

we see that the union of the augmented family is the whole space. It only remains to show that $B_k \cap R = \emptyset$. Write

$$B_k \cap R = B_k \cap \left(\bigcup_n B_n\right)^c$$
$$= B_k \cap \left(\bigcap_n B_n^c\right)$$
$$= B_k \cap B_k^c \cap \left(\bigcap_{n \neq k} B_n^c\right)$$
$$= \emptyset.$$

*Functions

A **function** consists of a set X of admissible inputs called the **domain** and a **rule** or **mapping** f that associates to each $x \in X$ a value $f(x)$ that belongs to a set Y called the **co-domain**. We indicate this symbolically by writing $f: X \to Y$, and we say, "f maps X into Y." Two functions are the same if and only if they have the same domain, co-domain, and rule. If $f: X \to Y$ and $g: X \to Y$, then the mappings f and g are the same if and only if $f(x) = g(x)$ for all $x \in X$.

The set of all possible values of $f(x)$ is called the **range**. In symbols, the range is the set $\{f(x) : x \in X\}$. Since $f(x) \in Y$ for each x, it is clear that the range is a subset of Y. However, the range may or may not be equal to Y. The case in which the range is a proper subset of Y is illustrated in Figure 1.11.

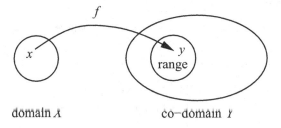

Figure 1.11. The mapping f associates each x in the domain X to a point y in the co-domain Y. The range is the subset of Y consisting of those y that are associated by f to at least one $x \in X$. In general, the range is a proper subset of the co-domain.

A function is said to be **onto** if its range is equal to its co-domain. In other words, every value $y \in Y$ "comes from somewhere" in the sense that for every $y \in Y$, there is at least one $x \in X$ with $y = f(x)$.

A function is said to be **one-to-one** if the condition $f(x_1) = f(x_2)$ implies $x_1 = x_2$.

Another way of thinking about the concepts of onto and one-to-one is the following. A function is onto if for every $y \in Y$, the equation $f(x) = y$ has a solution. This does not rule out the possibility that there may be more than one solution. A function is one-to-one if for every $y \in Y$, the equation $f(x) = y$ can have at most one solution. This does not rule out the possibility that for some values of $y \in Y$, there may be *no* solution.

A function is said to be **invertible** if for every $y \in Y$ there is a unique $x \in X$ with $f(x) = y$. Hence, a function is invertible if and only if it is both one-to-one and onto; i.e., for every $y \in Y$, the equation $f(x) = y$ has a unique solution.

***Example* 1.5.** For any real number x, put $f(x) := x^2$. Then

$$f:(-\infty,\infty) \to (-\infty,\infty)$$
$$f:(-\infty,\infty) \to [0,\infty)$$
$$f:[0,\infty) \to (-\infty,\infty)$$
$$f:[0,\infty) \to [0,\infty)$$

specifies four different functions. In the first case, the function is not one-to-one because $f(2) = f(-2)$, but $2 \neq -2$; the function is not onto because there is no $x \in (-\infty,\infty)$ with $f(x) = -1$. In the second case, the function is onto since for every $y \in [0,\infty)$, $f(\sqrt{y}) = y$. However, since $f(-\sqrt{y}) = y$ also, the function is not one-to-one. In the third case, the function fails to be onto, but is one-to-one. In the fourth case, the function is onto and one-to-one and therefore invertible.

The last concept we introduce concerning functions is that of inverse image. If $f: X \to Y$, and if $B \subset Y$, then the **inverse image** of B is

$$f^{-1}(B) := \{x \in X : f(x) \in B\},$$

which we emphasize is a subset of X. This concept applies to any function whether or not it is invertible. When the set X is understood, we sometimes write

$$f^{-1}(B) := \{x : f(x) \in B\}$$

to simplify the notation.

***Example* 1.6.** Suppose that $f:(-\infty,\infty) \to (-\infty,\infty)$, where $f(x) = x^2$. Find $f^{-1}([4,9])$ and $f^{-1}([-9,-4])$.

Solution. In the first case, write

$$\begin{aligned} f^{-1}([4,9]) &= \{x : f(x) \in [4,9]\} \\ &= \{x : 4 \leq f(x) \leq 9\} \\ &= \{x : 4 \leq x^2 \leq 9\} \\ &= \{x : 2 \leq x \leq 3 \text{ or } -3 \leq x \leq -2\} \\ &= [2,3] \cup [-3,-2]. \end{aligned}$$

In the second case, we need to find
$$f^{-1}([-9,-4]) = \{x : -9 \leq x^2 \leq -4\}.$$
Since there is no $x \in (-\infty, \infty)$ with $x^2 < 0$, $f^{-1}([-9,-4]) = \emptyset$.

***

Remark. If we modify the function in the preceding example to be $f:[0,\infty) \to (-\infty,\infty)$, then $f^{-1}([4,9]) = [2,3]$ instead.

*Countable and uncountable sets

The number of points in a set A is denoted by $|A|$. We call $|A|$ the **cardinality** of A. The cardinality of a set may be finite or infinite. A little reflection should convince you that if A and B are two disjoint sets, then
$$|A \cup B| = |A| + |B|;$$
use the **infinity conventions** that if x is a real number, then
$$x + \infty = \infty \quad \text{and} \quad \infty + \infty = \infty,$$
and be sure to consider the three cases: (*i*) A and B both have finite cardinality, (*ii*) one has finite cardinality and one has infinite cardinality, and (*iii*) both have infinite cardinality.

A nonempty set A is said to be **countable** if the elements of A can be enumerated or listed in a sequence: $a_1, a_2, \ldots$. In other words, a set A is countable if it can be written in the form
$$A = \bigcup_{k=1}^{\infty} \{a_k\},$$
where we emphasize that the union is over the positive integers, $k = 1, 2, \ldots$. *The empty set is also said to be countable.*

Remark. Since there is no requirement that the a_k be distinct, every finite set is countable by our definition. For example, you should verify that the set $A = \{1,2,3\}$ can be written in the above form by taking $a_1 = 1, a_2 = 2, a_3 = 3$, and $a_k = 3$ for $k = 4, 5, \ldots$. By a **countably infinite set**, we mean a countable set that is not finite.

Example 1.7. Show that a set of the form
$$B = \bigcup_{i,j=1}^{\infty} \{b_{ij}\}$$
is countable.

Solution. The point here is that a sequence that is doubly indexed by positive integers forms a countable set. To see this, consider the array

$$b_{11} \quad b_{12} \quad b_{13} \quad b_{14}$$
$$b_{21} \quad b_{22} \quad b_{23}$$
$$b_{31} \quad b_{32}$$
$$b_{41} \qquad \qquad \ddots$$

Now list the array elements along antidiagonals from lower left to upper right defining

$$a_1 := b_{11}$$
$$a_2 := b_{21}, \ a_3 := b_{12}$$
$$a_4 := b_{31}, \ a_5 := b_{22}, \ a_6 := b_{13}$$
$$a_7 := b_{41}, \ a_8 := b_{32}, \ a_9 := b_{23}, \ a_{10} := b_{14}$$
$$\vdots$$

This shows that

$$B = \bigcup_{k=1}^{\infty} \{a_k\},$$

and so B is a countable set.

***Example* 1.8.** Show that the positive rational numbers form a countable subset.

Solution. Recall that a rational number is of the form i/j where i and j are integers with $j \neq 0$. Hence, the set of positive rational numbers is equal to

$$\bigcup_{i,j=1}^{\infty} \{i/j\}.$$

By the previous example, this is a countable set.

You will show in Problem 16 that the union of two countable sets is a countable set. It then easily follows that the set of all rational numbers is countable.

A set is **uncountable** or **uncountably infinite** if it is not countable.

***Example* 1.9.** Show that the set S of unending row vectors of zeros and ones is uncountable.

Solution. We give a proof by contradiction. In such a proof, we assume that what we are trying to prove is false, and then we show that this leads to a contradiction. Once a contradiction is obtained, the proof is complete.

In this example, we are trying to prove S is uncountable. So, we assume this is false; i.e., we assume S is countable. Now, the assumption that S is countable means we can write $S = \bigcup_{i=1}^{\infty} \{\mathbf{a}_i\}$ for some sequence $\mathbf{a}_i$, where each $\mathbf{a}_i$ is an unending row vector of zeros and ones. We next show that there is a row vector $\mathbf{a}$ that does not belong to

$$\bigcup_{i=1}^{\infty} \{\mathbf{a}_i\}.$$

To show how to construct this special row vector, suppose

$$\begin{aligned}
\mathbf{a}_1 &:= \boxed{1}\ 0\ 1\ 1\ 0\ 1\ 0\ 1\ 1\ \cdots \\
\mathbf{a}_2 &:= 0\ \boxed{0}\ 1\ 0\ 1\ 1\ 0\ 0\ 0\ \cdots \\
\mathbf{a}_3 &:= 1\ 1\ \boxed{1}\ 0\ 1\ 0\ 1\ 0\ 1\ \cdots \\
\mathbf{a}_4 &:= 1\ 1\ 0\ \boxed{1}\ 0\ 0\ 1\ 1\ 0\ \cdots \\
\mathbf{a}_5 &:= 0\ 1\ 1\ 0\ \boxed{0}\ 0\ 0\ 0\ 0\ \cdots \\
&\ \vdots
\end{aligned}$$

where we have boxed the diagonal elements to highlight them. Now use the following **diagonal argument**. Take $\mathbf{a} := 0\,1\,0\,0\,1\,\cdots$ to be such that kth bit of $\mathbf{a}$ is the complement of the kth bit of $\mathbf{a}_k$. In other words, viewing the above row vectors as an infinite matrix, go along the diagonal and flip all the bits to construct $\mathbf{a}$. Then $\mathbf{a} \neq \mathbf{a}_1$ because they differ in the first bit. Similarly, $\mathbf{a} \neq \mathbf{a}_2$ because they differ in the second bit. And so on. Thus,

$$\mathbf{a} \notin \bigcup_{i=1}^{\infty} \{\mathbf{a}_i\} = S.$$

However, by definition, S is the set of all unending row vectors of zeros and ones. Since $\mathbf{a}$ is such a vector, $\mathbf{a} \in S$. We have a contradiction.

The same argument shows that the interval of real numbers $[0,1)$ is not countable. To see this, write each such real number in its binary expansion, e.g., $0.11010101110\ldots$ and identify the expansion with the corresponding row vector of zeros and ones in the example.

1.3 Probability models

In Section 1.1, we suggested sample spaces to model the results of various uncertain measurements. We then said that events are subsets of the sample space. In this section, we add probability to sample space models of some simple systems and compute probabilities of various events.

The goal of probability theory is to provide mathematical machinery to analyze complicated problems in which answers are not obvious. However, for any such theory to be accepted, it should provide answers to simple problems that agree with our intuition. In this section we consider several simple problems for which intuitive answers are apparent, but we solve them using the machinery of probability.

Consider the experiment of tossing a fair die and measuring, i.e., noting, the face turned up. Our intuition tells us that the "probability" of the ith face turning up is $1/6$, and that the "probability" of a face with an even number of dots turning up is $1/2$.

Here is a *mathematical model* for this experiment and measurement. Let the sample space Ω be any set containing six points. Each sample point or outcome $\omega \in \Omega$ corresponds to, or models, a possible result of the experiment. For simplicity, let

$$\Omega := \{1,2,3,4,5,6\}.$$

Now define the events
$$F_i := \{i\}, \quad i = 1,2,3,4,5,6,$$
and
$$E := \{2,4,6\}.$$

The event F_i corresponds to, or models, the die's turning up showing the ith face. Similarly, the event E models the die's showing a face with an even number of dots. Next, for every subset A of Ω, we denote the number of points in A by $|A|$. We call $|A|$ the **cardinality** of A. We *define* the **probability** of any event A by

$$\mathsf{P}(A) := |A|/|\Omega|.$$

In other words, for the model we are constructing for this problem, the probability of an event A is defined to be the number of outcomes in A divided by the total number of possible outcomes. With this definition, it follows that $\mathsf{P}(F_i) = 1/6$ and $\mathsf{P}(E) = 3/6 = 1/2$, which agrees with our intuition. You can also compare this with MATLAB simulations in Problem 21.

We now make four observations about our model.

(i) $\mathsf{P}(\emptyset) = |\emptyset|/|\Omega| = 0/|\Omega| = 0$.
(ii) $\mathsf{P}(A) \geq 0$ for every event A.
(iii) If A and B are mutually exclusive events, i.e., $A \cap B = \emptyset$, then $\mathsf{P}(A \cup B) = \mathsf{P}(A) + \mathsf{P}(B)$; for example, $F_3 \cap E = \emptyset$, and it is easy to check that

$$\mathsf{P}(F_3 \cup E) = \mathsf{P}(\{2,3,4,6\}) = \mathsf{P}(F_3) + \mathsf{P}(E).$$

(iv) When the die is tossed, *something* happens; this is modeled mathematically by the easily verified fact that $\mathsf{P}(\Omega) = 1$.

As we shall see, these four properties hold for all the models discussed in this section.

We next modify our model to accommodate an unfair die as follows. Observe that for a fair die,[e]

$$\mathsf{P}(A) = \frac{|A|}{|\Omega|} = \sum_{\omega \in A} \frac{1}{|\Omega|} = \sum_{\omega \in A} p(\omega),$$

where $p(\omega) := 1/|\Omega|$. For example,

$$\mathsf{P}(E) = \sum_{\omega \in \{2,4,6\}} 1/6 = 1/6 + 1/6 + 1/6 = 1/2.$$

For an *unfair* die, we simply change the definition of the function $p(\omega)$ to reflect the likelihood of occurrence of the various faces. This new definition of P still satisfies (i) and (iii); however, to guarantee that (ii) and (iv) still hold, we must require that p be nonnegative and sum to one, or, in symbols, $p(\omega) \geq 0$ and $\sum_{\omega \in \Omega} p(\omega) = 1$.

Example 1.10. Construct a sample space Ω and probability P to model an unfair die in which faces 1–5 are equally likely, but face 6 has probability $1/3$. Using this model, compute the probability that a toss results in a face showing an even number of dots.

---
[e]If $A = \emptyset$, the summation is taken to be zero.

1.3 Probability models

Solution. We again take $\Omega = \{1,2,3,4,5,6\}$. To make face 6 have probability $1/3$, we take $p(6) = 1/3$. Since the other faces are equally likely, for $\omega = 1,\ldots,5$, we take $p(\omega) = c$, where c is a constant to be determined. To find c we use the fact that

$$1 = \mathsf{P}(\Omega) = \sum_{\omega \in \Omega} p(\omega) = \sum_{\omega=1}^{6} p(\omega) = 5c + \frac{1}{3}.$$

It follows that $c = 2/15$. Now that $p(\omega)$ has been specified for all ω, we define the probability of any event A by

$$\mathsf{P}(A) := \sum_{\omega \in A} p(\omega).$$

Letting $E = \{2,4,6\}$ model the result of a toss showing a face with an even number of dots, we compute

$$\mathsf{P}(E) = \sum_{\omega \in E} p(\omega) = p(2) + p(4) + p(6) = \frac{2}{15} + \frac{2}{15} + \frac{1}{3} = \frac{3}{5}.$$

This unfair die has a greater probability of showing an even numbered face than the fair die.

This problem is typical of the kinds of "word problems" to which probability theory is applied to analyze well-defined physical experiments. The application of probability theory requires the modeler to take the following steps.

- Select a suitable sample space Ω.
- Define $\mathsf{P}(A)$ for all events A. For example, if Ω is a finite set and all outcomes ω are equally likely, we usually take $\mathsf{P}(A) = |A|/|\Omega|$. If it is not the case that all outcomes are equally likely, e.g., as in the previous example, then $\mathsf{P}(A)$ would be given by some other formula that must be determined based on the problem statement.
- Translate the given "word problem" into a problem requiring the calculation of $\mathsf{P}(E)$ for some specific event E.

The following example gives a family of constructions that can be used to model experiments having a finite number of possible outcomes.

Example 1.11. Let M be a positive integer, and put $\Omega := \{1,2,\ldots,M\}$. Next, let $p(1)$, $\ldots, p(M)$ be nonnegative real numbers such that $\sum_{\omega=1}^{M} p(\omega) = 1$. For any subset $A \subset \Omega$, put

$$\mathsf{P}(A) := \sum_{\omega \in A} p(\omega).$$

In particular, to model equally likely outcomes, or equivalently, outcomes that occur "at random," we take $p(\omega) = 1/M$. In this case, $\mathsf{P}(A)$ reduces to $|A|/|\Omega|$.

Example 1.12. A single card is drawn at random from a well-shuffled deck of playing cards. Find the probability of drawing an ace. Also find the probability of drawing a face card.

Solution. The first step in the solution is to specify the sample space Ω and the probability P. Since there are 52 possible outcomes, we take $\Omega := \{1,\ldots,52\}$. Each integer corresponds to one of the cards in the deck. To specify P, we must define $P(E)$ for all events $E \subset \Omega$. Since all cards are equally likely to be drawn, we put $P(E) := |E|/|\Omega|$.

To find the desired probabilities, let $1,2,3,4$ correspond to the four aces, and let $41,\ldots,52$ correspond to the 12 face cards. We identify the drawing of an ace with the event $A := \{1,2,3,4\}$, and we identify the drawing of a face card with the event $F := \{41,\ldots,52\}$. It then follows that $P(A) = |A|/52 = 4/52 = 1/13$ and $P(F) = |F|/52 = 12/52 = 3/13$. You can compare this with MATLAB simulations in Problem 25.

While the sample spaces Ω in Example 1.11 can model any experiment with a finite number of outcomes, it is often convenient to use alternative sample spaces.

Example 1.13. Suppose that we have two well-shuffled decks of cards, and we draw one card at random from each deck. What is the probability of drawing the ace of spades followed by the jack of hearts? What is the probability of drawing an ace and a jack (in either order)?

Solution. The first step in the solution is to specify the sample space Ω and the probability P. Since there are 52 possibilities for each draw, there are $52^2 = 2704$ possible outcomes when drawing two cards. Let $D := \{1,\ldots,52\}$, and put

$$\Omega := \{(i,j) : i,j \in D\}.$$

Then $|\Omega| = |D|^2 = 52^2 = 2704$ as required. Since all pairs are equally likely, we put $P(E) := |E|/|\Omega|$ for arbitrary events $E \subset \Omega$.

As in the preceding example, we denote the aces by $1,2,3,4$. We let 1 denote the ace of spades. We also denote the jacks by $41,42,43,44$, and the jack of hearts by 42. The drawing of the ace of spades followed by the jack of hearts is identified with the event

$$A := \{(1,42)\},$$

and so $P(A) = 1/2704 \approx 0.000370$. The drawing of an ace and a jack is identified with $B := B_{aj} \cup B_{ja}$, where

$$B_{aj} := \{(i,j) : i \in \{1,2,3,4\} \text{ and } j \in \{41,42,43,44\}\}$$

corresponds to the drawing of an ace followed by a jack, and

$$B_{ja} := \{(i,j) : i \in \{41,42,43,44\} \text{ and } j \in \{1,2,3,4\}\}$$

corresponds to the drawing of a jack followed by an ace. Since B_{aj} and B_{ja} are disjoint, $P(B) = P(B_{aj}) + P(B_{ja}) = (|B_{aj}| + |B_{ja}|)/|\Omega|$. Since $|B_{aj}| = |B_{ja}| = 16$, $P(B) = 2 \cdot 16/2704 = 2/169 \approx 0.0118$.

Example 1.14. Two cards are drawn at random from a *single* well-shuffled deck of playing cards. What is the probability of drawing the ace of spades followed by the jack of hearts? What is the probability of drawing an ace and a jack (in either order)?

1.3 Probability models

Solution. The first step in the solution is to specify the sample space Ω and the probability P. There are 52 possibilities for the first draw and 51 possibilities for the second. Hence, the sample space should contain $52 \cdot 51 = 2652$ elements. Using the notation of the preceding example, we take

$$\Omega := \{(i,j) : i,j \in D \text{ with } i \neq j\},$$

Note that $|\Omega| = 52^2 - 52 = 2652$ as required. Again, all such pairs are equally likely, and so we take $\mathsf{P}(E) := |E|/|\Omega|$ for arbitrary events $E \subset \Omega$. The events A and B are defined as before, and the calculation is the same except that $|\Omega| = 2652$ instead of 2704. Hence, $\mathsf{P}(A) = 1/2652 \approx 0.000377$, and $\mathsf{P}(B) = 2 \cdot 16/2652 = 8/663 \approx 0.012$.

In some experiments, the number of possible outcomes is countably infinite. For example, consider the tossing of a coin until the first heads appears. Here is a model for such situations. Let Ω denote the set of all positive integers, $\Omega := \{1, 2, \ldots\}$. For $\omega \in \Omega$, let $p(\omega)$ be nonnegative, and suppose that $\sum_{\omega=1}^{\infty} p(\omega) = 1$. For any subset $A \subset \Omega$, put

$$\mathsf{P}(A) := \sum_{\omega \in A} p(\omega).$$

This construction can be used to model the coin tossing experiment by identifying $\omega = i$ with the outcome that the first heads appears on the ith toss. If the probability of tails on a single toss is α ($0 \leq \alpha < 1$), it can be shown that we should take $p(\omega) = \alpha^{\omega-1}(1-\alpha)$ (cf. Example 2.12). To find the probability that the first head occurs before the fourth toss, we compute $\mathsf{P}(A)$, where $A = \{1, 2, 3\}$. Then

$$\mathsf{P}(A) = p(1) + p(2) + p(3) = (1 + \alpha + \alpha^2)(1 - \alpha).$$

If $\alpha = 1/2$, $\mathsf{P}(A) = (1 + 1/2 + 1/4)/2 = 7/8$.

For some experiments, the number of possible outcomes is more than countably infinite. Examples include the duration of a cell-phone call, a noise voltage in a communication receiver, and the time at which an Internet connection is initiated. In these cases, P is usually defined as an integral,

$$\mathsf{P}(A) := \int_A f(\omega) \, d\omega, \quad A \subset \Omega,$$

for some nonnegative function f. Note that f must also satisfy $\int_\Omega f(\omega) \, d\omega = 1$.

***Example* 1.15.** Consider the following model for the duration of a cell-phone call. For the sample space we take the nonnegative half line, $\Omega := [0, \infty)$, and we put

$$\mathsf{P}(A) := \int_A f(\omega) \, d\omega,$$

where, for example, $f(\omega) := e^{-\omega}$. Then the probability that the call duration is between 5 and 7 time units is

$$\mathsf{P}([5,7]) = \int_5^7 e^{-\omega} \, d\omega = e^{-5} - e^{-7} \approx 0.0058.$$

Example 1.16. An on-line probability seminar is scheduled to start at 9:15. However, the seminar actually starts randomly in the 20-minute interval between 9:05 and 9:25. Find the probability that the seminar begins at or after its scheduled start time.

Solution. Let $\Omega := [5,25]$, and put

$$P(A) := \int_A f(\omega)\,d\omega.$$

The term "randomly" in the problem statement is usually taken to mean $f(\omega) \equiv$ constant. In order that $P(\Omega) = 1$, we must choose the constant to be $1/\text{length}(\Omega) = 1/20$. We represent the seminar starting at or after 9:15 with the event $L := [15,25]$. Then

$$P(L) = \int_{[15,25]} \frac{1}{20}\,d\omega = \int_{15}^{25} \frac{1}{20}\,d\omega = \frac{25-15}{20} = \frac{1}{2}.$$

Example 1.17. A cell-phone tower has a circular coverage area of radius 10 km. If a call is initiated from a random point in the coverage area, find the probability that the call comes from within 2 km of the tower.

Solution. Let $\Omega := \{(x,y) : x^2 + y^2 \le 100\}$, and for any $A \subset \Omega$, put

$$P(A) := \frac{\text{area}(A)}{\text{area}(\Omega)} = \frac{\text{area}(A)}{100\pi}.$$

We then identify the event $A := \{(x,y) : x^2 + y^2 \le 4\}$ with the call coming from within 2 km of the tower. Hence,

$$P(A) = \frac{4\pi}{100\pi} = 0.04.$$

1.4 Axioms and properties of probability

In this section, we present Kolmogorov's axioms and derive some of their consequences.

The probability models of the preceding section suggest the following axioms that we now require of any probability model.

Given a nonempty set Ω, called the **sample space**, and a function P defined on the subsets[1] of Ω, we say P is a **probability measure** if the following four axioms are satisfied.[2]

(i) The empty set $\emptyset$ is called the **impossible event**. The probability of the impossible event is zero; i.e., $P(\emptyset) = 0$.

(ii) Probabilities are nonnegative; i.e., for any event A, $P(A) \ge 0$.

(iii) If $A_1, A_2, \ldots$ are events that are mutually exclusive or pairwise disjoint, i.e., $A_n \cap A_m = \emptyset$ for $n \ne m$, then

$$P\left(\bigcup_{n=1}^{\infty} A_n\right) = \sum_{n=1}^{\infty} P(A_n). \quad (1.9)$$

1.4 Axiomes and properties of probability

> The technical term for this property is **countable additivity**. However, all it says is that the probability of a union of disjoint events is the sum of the probabilities of the individual events, or more briefly, "the probabilities of disjoint events add."
>
> (*iv*) The entire sample space Ω is called the **sure event** or the **certain event**, and its probability is one; i.e., $P(\Omega) = 1$. If an event $A \neq \Omega$ satisfies $P(A) = 1$, we say that A is an **almost-sure event**.

We can view $P(A)$ as a function whose argument is an event, A, and whose value, $P(A)$, is greater than or equal to zero. The foregoing axioms imply many other properties. In particular, we show later that $P(A)$ satisfies $0 \leq P(A) \leq 1$.

We now give an interpretation of how Ω and P model randomness. We view the sample space Ω as being the set of all possible "states of nature." First, Mother Nature chooses a state $\omega_0 \in \Omega$. We do not know which state has been chosen. We then conduct an experiment, and based on some physical measurement, we are able to determine that $\omega_0 \in A$ for some event $A \subset \Omega$. In some cases, $A = \{\omega_0\}$, that is, our measurement reveals exactly which state ω_0 was chosen by Mother Nature. (This is the case for the events F_i defined at the beginning of Section 1.3). In other cases, the set A contains ω_0 as well as other points of the sample space. (This is the case for the event E defined at the beginning of Section 1.3). In either case, we do not know before making the measurement what measurement value we will get, and so we do not know what event A Mother Nature's ω_0 will belong to. Hence, in many applications, e.g., gambling, weather prediction, computer message traffic, etc., it is useful to compute $P(A)$ for various events to determine which ones are most probable.

Consequences of the axioms

Axioms (*i*)–(*iv*) that characterize a probability measure have several important implications as discussed below.

Finite disjoint unions. We have the finite version of axiom (*iii*):

$$\boxed{P\left(\bigcup_{n=1}^{N} A_n\right) = \sum_{n=1}^{N} P(A_n), \quad A_n \text{ pairwise disjoint.}}$$

To derive this, put $A_n := \emptyset$ for $n > N$, and then write

$$P\left(\bigcup_{n=1}^{N} A_n\right) = P\left(\bigcup_{n=1}^{\infty} A_n\right)$$

$$= \sum_{n=1}^{\infty} P(A_n), \quad \text{by axiom } (iii),$$

$$= \sum_{n=1}^{N} P(A_n), \quad \text{since } P(\emptyset) = 0 \text{ by axiom } (i).$$

Remark. It is not possible to go backwards and use this special case to derive axiom (*iii*).

Example **1.18.** If A is an event consisting of a finite number of sample points, say $A = \{\omega_1, \ldots, \omega_N\}$, then[3] $P(A) = \sum_{n=1}^{N} P(\{\omega_n\})$. Similarly, if A consists of countably many sample points, say $A = \{\omega_1, \omega_2, \ldots\}$, then directly from axiom (iii), $P(A) = \sum_{n=1}^{\infty} P(\{\omega_n\})$.

Probability of a complement. Given an event A, we can always write $\Omega = A \cup A^c$, which is a finite disjoint union. Hence, $P(\Omega) = P(A) + P(A^c)$. Since $P(\Omega) = 1$, we find that

$$\boxed{P(A^c) = 1 - P(A).} \tag{1.10}$$

Monotonicity. If A and B are events, then

$$\boxed{A \subset B \quad \text{implies} \quad P(A) \leq P(B).} \tag{1.11}$$

To see this, first note that $A \subset B$ implies

$$B = A \cup (B \cap A^c).$$

This relation is depicted in Figure 1.12, in which the disk A is a subset of the oval-shaped

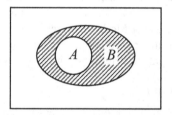

Figure 1.12. In this diagram, the disk A is a subset of the oval-shaped region B; the shaded region is $B \cap A^c$, and $B = A \cup (B \cap A^c)$.

region B; the shaded region is $B \cap A^c$. The figure shows that B is the disjoint union of the disk A together with the shaded region $B \cap A^c$. Since $B = A \cup (B \cap A^c)$ is a disjoint union, and since probabilities are nonnegative,

$$\begin{aligned} P(B) &= P(A) + P(B \cap A^c) \\ &\geq P(A). \end{aligned}$$

Note that the special case $B = \Omega$ results in $P(A) \leq 1$ for every event A. In other words, *probabilities are always less than or equal to one*.

Inclusion–exclusion. Given any two events A and B, we always have

$$\boxed{P(A \cup B) = P(A) + P(B) - P(A \cap B).} \tag{1.12}$$

This formula says that if we add the entire shaded disk of Figure 1.13(a) to the entire shaded ellipse of Figure 1.13(b), then we have counted the intersection twice and must subtract off a copy of it. The curious reader can find a set-theoretic derivation of (1.12) in the Notes.[4]

1.4 Axioms and properties of probability

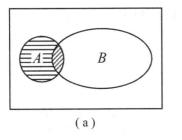 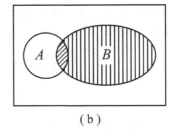

(a) (b)

Figure 1.13. (a) Decomposition $A = (A \cap B^c) \cup (A \cap B)$. (b) Decomposition $B = (A \cap B) \cup (A^c \cap B)$.

Limit properties. The following limit properties of probability are essential to answer questions about the probability that something ever happens or never happens. Using axioms (i)–(iv), the following formulas can be derived (see Problems 33–35). For *any* sequence of events A_n,

$$\mathsf{P}\left(\bigcup_{n=1}^{\infty} A_n\right) = \lim_{N \to \infty} \mathsf{P}\left(\bigcup_{n=1}^{N} A_n\right), \qquad (1.13)$$

and

$$\mathsf{P}\left(\bigcap_{n=1}^{\infty} A_n\right) = \lim_{N \to \infty} \mathsf{P}\left(\bigcap_{n=1}^{N} A_n\right). \qquad (1.14)$$

In particular, notice that if the A_n are increasing in the sense that $A_n \subset A_{n+1}$ for all n, then the finite union in (1.13) reduces to A_N (see Figure 1.14(a)). Thus, (1.13) becomes

$$\mathsf{P}\left(\bigcup_{n=1}^{\infty} A_n\right) = \lim_{N \to \infty} \mathsf{P}(A_N), \quad \text{if } A_n \subset A_{n+1}. \qquad (1.15)$$

Similarly, if the A_n are decreasing in the sense that $A_{n+1} \subset A_n$ for all n, then the finite intersection in (1.14) reduces to A_N (see Figure 1.14(b)). Thus, (1.14) becomes

$$\mathsf{P}\left(\bigcap_{n=1}^{\infty} A_n\right) = \lim_{N \to \infty} \mathsf{P}(A_N), \quad \text{if } A_{n+1} \subset A_n. \qquad (1.16)$$

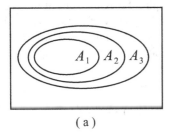 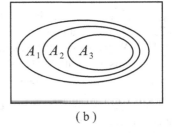

(a) (b)

Figure 1.14. (a) For increasing events $A_1 \subset A_2 \subset A_3$, the union $A_1 \cup A_2 \cup A_3 = A_3$. (b) For decreasing events $A_1 \supset A_2 \supset A_3$, the intersection $A_1 \cap A_2 \cap A_3 = A_3$.

Formulas (1.15) and (1.16) are called **sequential continuity** properties. Formulas (1.12) and (1.13) together imply that for any sequence of events A_n,

$$\boxed{\mathsf{P}\!\left(\bigcup_{n=1}^{\infty} A_n\right) \leq \sum_{n=1}^{\infty} \mathsf{P}(A_n).} \qquad (1.17)$$

This formula is known as the **union bound** in engineering and as **countable subadditivity** in mathematics. It is derived in Problems 36 and 37 at the end of the chapter.

1.5 Conditional probability

A computer maker buys the same chips from two different suppliers, S1 and S2, in order to reduce the risk of supply interruption. However, now the computer maker wants to find out if one of the suppliers provides more reliable devices than the other. To make this determination, the computer maker examines a collection of n chips. For each one, there are four possible outcomes, depending on whether the chip comes from supplier S1 or supplier S2 and on whether the chip works (w) or is defective (d). We denote these outcomes by $O_{w,S1}$, $O_{d,S1}$, $O_{w,S2}$, and $O_{d,S2}$. The numbers of each outcome can be arranged in the matrix

$$\begin{bmatrix} N(O_{w,S1}) & N(O_{w,S2}) \\ N(O_{d,S1}) & N(O_{d,S2}) \end{bmatrix}. \qquad (1.18)$$

The sum of the first column is the number of chips from supplier S1, which we denote by $N(O_{S1})$. The sum of the second column is the number of chips from supplier S2, which we denote by $N(O_{S2})$.

The relative frequency of working chips from supplier S1 is $N(O_{w,S1})/N(O_{S1})$. Similarly, the relative frequency of working chips from supplier S2 is $N(O_{w,S2})/N(O_{S2})$. If $N(O_{w,S1})/N(O_{S1})$ is substantially greater than $N(O_{w,S2})/N(O_{S2})$, this would suggest that supplier S1 might be providing more reliable chips than supplier S2.

Example **1.19.** Suppose that (1.18) is equal to

$$\begin{bmatrix} 754 & 499 \\ 221 & 214 \end{bmatrix}.$$

Determine which supplier provides more reliable chips.

Solution. The number of chips from supplier S1 is the sum of the first column, $N(O_{S1}) = 754 + 221 = 975$. The number of chips from supplier S2 is the sum of the second column, $N(O_{S2}) = 499 + 214 = 713$. Hence, the relative frequency of working chips from supplier S1 is $754/975 \approx 0.77$, and the relative frequency of working chips form supplier S2 is $499/713 \approx 0.70$. We conclude that supplier S1 provides more reliable chips. You can run your own simulations using the MATLAB script in Problem 51.

Notice that the relative frequency of working chips from supplier S1 can also be written as the quotient of relative frequencies,

$$\frac{N(O_{w,S1})}{N(O_{S1})} = \frac{N(O_{w,S1})/n}{N(O_{S1})/n}. \qquad (1.19)$$

1.5 Conditional probability

This suggests the following definition of conditional probability. Let Ω be a sample space. Let the event S_1 model a chip's being from supplier S1, and let the event W model a chip's working. In our model, the **conditional probability** that a chip works given that the chip comes from supplier S1 is defined by

$$P(W|S_1) := \frac{P(W \cap S_1)}{P(S_1)},$$

where the probabilities model the relative frequencies on the right-hand side of (1.19). This definition makes sense only if $P(S_1) > 0$. If $P(S_1) = 0$, $P(W|S_1)$ is not defined.

Given any two events A and B of positive probability,

$$P(A|B) = \frac{P(A \cap B)}{P(B)} \tag{1.20}$$

and

$$P(B|A) = \frac{P(A \cap B)}{P(A)}.$$

From (1.20), we see that

$$P(A \cap B) = P(A|B) P(B). \tag{1.21}$$

Substituting this into the numerator above yields

$$P(B|A) = \frac{P(A|B) P(B)}{P(A)}. \tag{1.22}$$

We next turn to the problem of computing the denominator $P(A)$.

The law of total probability and Bayes' rule

The law of total probability is a formula for computing the probability of an event that can occur in different ways. For example, the probability that a cell-phone call goes through depends on which tower handles the call. The probability of Internet packets being dropped depends on which route they take through the network.

When an event A can occur in two ways, the law of total probability is derived as follows (the general case is derived later in the section). We begin with the identity

$$A = (A \cap B) \cup (A \cap B^c)$$

(recall Figure 1.13(a)). Since this is a disjoint union,

$$P(A) = P(A \cap B) + P(A \cap B^c).$$

In terms of Figure 1.13(a), this formula says that the area of the disk A is the sum of the areas of the two shaded regions. Using (1.21), we have

$$P(A) = P(A|B) P(B) + P(A|B^c) P(B^c). \tag{1.23}$$

This formula is the simplest version of the **law of total probability**.

***Example* 1.20.** Due to an Internet configuration error, packets sent from New York to Los Angeles are routed through El Paso, Texas with probability 3/4. Given that a packet is routed through El Paso, suppose it has conditional probability 1/3 of being dropped. Given that a packet is *not* routed through El Paso, suppose it has conditional probability 1/4 of being dropped. Find the probability that a packet is dropped.

Solution. To solve this problem, we use the notation[f]

$$E = \{\text{routed through El Paso}\} \quad \text{and} \quad D = \{\text{packet is dropped}\}.$$

With this notation, it is easy to interpret the problem as telling us that

$$P(D|E) = 1/3, \quad P(D|E^c) = 1/4, \quad \text{and} \quad P(E) = 3/4. \tag{1.24}$$

We must now compute $P(D)$. By the law of total probability,

$$\begin{aligned} P(D) &= P(D|E)P(E) + P(D|E^c)P(E^c) \\ &= (1/3)(3/4) + (1/4)(1-3/4) \\ &= 1/4 + 1/16 \\ &= 5/16. \end{aligned} \tag{1.25}$$

To derive the simplest form of **Bayes' rule**, substitute (1.23) into (1.22) to get

$$P(B|A) = \frac{P(A|B)P(B)}{P(A|B)P(B) + P(A|B^c)P(B^c)}. \tag{1.26}$$

As illustrated in the following example, it is not necessary to remember Bayes' rule as long as you know the definition of conditional probability and the law of total probability.

***Example* 1.21** (continuation of Internet Example 1.20). Find the conditional probability that a packet is routed through El Paso given that it is not dropped.

Solution. With the notation of the previous example, we are being asked to find $P(E|D^c)$. Write

$$\begin{aligned} P(E|D^c) &= \frac{P(E \cap D^c)}{P(D^c)} \\ &= \frac{P(D^c|E)P(E)}{P(D^c)}. \end{aligned}$$

From (1.24) we have $P(E) = 3/4$ and $P(D^c|E) = 1 - P(D|E) = 1 - 1/3$. From (1.25), $P(D^c) = 1 - P(D) = 1 - 5/16$. Hence,

$$P(E|D^c) = \frac{(2/3)(3/4)}{11/16} = \frac{8}{11}.$$

---

[f]In working this example, we follow common practice and do not explicitly specify the sample space Ω or the probability measure P. Hence, the expression "let $E = \{\text{routed through El Paso}\}$" is shorthand for "let E be the subset of Ω that models being routed through El Paso." The curious reader may find one possible choice for Ω and P, along with precise mathematical definitions of the events E and D, in Note **5**.

1.5 Conditional probability

If we had not already computed $P(D)$ in the previous example, we would have computed $P(D^c)$ directly using the law of total probability.

We now generalize the law of total probability. Let B_n be a sequence of pairwise disjoint events such that $\sum_n P(B_n) = 1$. Then for any event A,

$$P(A) = \sum_n P(A|B_n)P(B_n).$$

To derive this result, put $B := \bigcup_n B_n$, and observe that[g]

$$P(B) = \sum_n P(B_n) = 1.$$

It follows that $P(B^c) = 1 - P(B) = 0$. Next, for any event A, $A \cap B^c \subset B^c$, and so

$$0 \leq P(A \cap B^c) \leq P(B^c) = 0.$$

Hence, $P(A \cap B^c) = 0$. Writing (recall Figure 1.13(a))

$$A = (A \cap B) \cup (A \cap B^c),$$

it follows that

$$\begin{aligned} P(A) &= P(A \cap B) + P(A \cap B^c) \\ &= P(A \cap B) \\ &= P\left(A \cap \left[\bigcup_n B_n\right]\right) \\ &= P\left(\bigcup_n [A \cap B_n]\right) \\ &= \sum_n P(A \cap B_n). \end{aligned} \quad (1.27)$$

This formula is illustrated in Figure 1.10(b), where the area of the disk is the sum of the areas of the different shaded parts.

To compute $P(B_k|A)$, write

$$P(B_k|A) = \frac{P(A \cap B_k)}{P(A)} = \frac{P(A|B_k)P(B_k)}{P(A)}.$$

In terms of Figure 1.10(b), this formula says that $P(B_k|A)$ is the ratio of the area of the kth shaded part to the area of the whole disk. Applying the law of total probability to $P(A)$ in the denominator yields the general form of Bayes' rule,

$$\boxed{P(B_k|A) = \frac{P(A|B_k)P(B_k)}{\sum_n P(A|B_n)P(B_n)}.}$$

---

[g] Notice that since we do not require $\bigcup_n B_n = \Omega$, the B_n do not, strictly speaking, form a partition. However, since $P(B) = 1$ (that is, B is an almost sure event), the remainder set (cf. (1.8)), which in this case is B^c, has probability zero.

In formulas like this, A is an event that we observe, while the B_n are events that we cannot observe but would like to make some inference about. Before making any observations, we know the **prior probabilities** $P(B_n)$, and we know the conditional probabilities $P(A|B_n)$. After we observe A, we compute the **posterior probabilities** $P(B_k|A)$ for each k.

***Example* 1.22.** In Example 1.21, before we learn any information about a packet, that packet's prior probability of being routed through El Paso is $P(E) = 3/4 = 0.75$. After we observe that the packet is not dropped, the posterior probability that the packet was routed through El Paso is $P(E|D^c) = 8/11 \approx 0.73$, which is different from the prior probability.

1.6 Independence

In the previous section, we discussed how a computer maker might determine if one of its suppliers provides more reliable devices than the other. We said that if the relative frequency of working chips from supplier S1 is substantially different from the relative frequency of working chips from supplier S2, we would conclude that one supplier is better than the other. On the other hand, if the relative frequencies of working chips from both suppliers are about the same, we would say that whether a chip works not does not depend on the supplier.

In probability theory, if events A and B satisfy $P(A|B) = P(A|B^c)$, we say A does not depend on B. This condition says that

$$\frac{P(A \cap B)}{P(B)} = \frac{P(A \cap B^c)}{P(B^c)}. \tag{1.28}$$

Applying the formulas $P(B^c) = 1 - P(B)$ and

$$P(A) = P(A \cap B) + P(A \cap B^c)$$

to the right-hand side yields

$$\frac{P(A \cap B)}{P(B)} = \frac{P(A) - P(A \cap B)}{1 - P(B)}.$$

Cross multiplying to eliminate the denominators gives

$$P(A \cap B)[1 - P(B)] = P(B)[P(A) - P(A \cap B)].$$

Subtracting common terms from both sides shows that $P(A \cap B) = P(A)P(B)$. Since this sequence of calculations is reversible, and since the condition $P(A \cap B) = P(A)P(B)$ is symmetric in A and B, it follows that A does not depend on B if and only if B does not depend on A.

When events A and B satisfy

$$P(A \cap B) = P(A)P(B), \tag{1.29}$$

we say they are **statistically independent**, or just **independent**.

1.6 Independence

Caution. The reader is warned to make sure he or she understands the difference between disjoint sets and independent events. Recall that A and B are disjoint if $A \cap B = \emptyset$. This concept does not involve P in any way; to determine if A and B are disjoint requires only knowledge of A and B themselves. On the other hand, (1.29) implies that independence *does* depend on P and not just on A and B. To determine if A and B are independent requires not only knowledge of A and B, but also knowledge of P. See Problem 61.

In arriving at (1.29) as the definition of independent events, we noted that (1.29) is equivalent to (1.28). Hence, if A and B are independent, $\mathsf{P}(A|B) = \mathsf{P}(A|B^c)$. What is this common value? Write

$$\mathsf{P}(A|B) = \frac{\mathsf{P}(A \cap B)}{\mathsf{P}(B)} = \frac{\mathsf{P}(A)\mathsf{P}(B)}{\mathsf{P}(B)} = \mathsf{P}(A).$$

We now make some further observations about independence. First, it is a simple exercise to show that if A and B are independent events, then so are A and B^c, A^c and B, and A^c and B^c. For example, writing

$$\begin{aligned}\mathsf{P}(A) &= \mathsf{P}(A \cap B) + \mathsf{P}(A \cap B^c) \\ &= \mathsf{P}(A)\mathsf{P}(B) + \mathsf{P}(A \cap B^c),\end{aligned}$$

we have

$$\begin{aligned}\mathsf{P}(A \cap B^c) &= \mathsf{P}(A) - \mathsf{P}(A)\mathsf{P}(B) \\ &= \mathsf{P}(A)[1 - \mathsf{P}(B)] \\ &= \mathsf{P}(A)\mathsf{P}(B^c).\end{aligned}$$

By interchanging the roles of A and A^c and/or B and B^c, it follows that if any one of the four pairs is independent, then so are the other three.

Example 1.23. An Internet packet travels from its source to router 1, from router 1 to router 2, and from router 2 to its destination. If routers drop packets independently with probability p, what is the probability that a packet is successfully transmitted from its source to its destination?

Solution. A packet is successfully transmitted if and only if neither router drops it. To put this into the language of events, for $i = 1, 2$, let D_i denote the event that the packet is dropped by router i. Let S denote the event that the packet is successfully transmitted. Then S occurs if and only if the packet is not dropped by router 1 and it is not dropped by router 2. We can write this symbolically as

$$S = D_1^c \cap D_2^c.$$

Since the problem tells us that D_1 and D_2 are independent events, so are D_1^c and D_2^c. Hence,

$$\begin{aligned}\mathsf{P}(S) &= \mathsf{P}(D_1^c \cap D_2^c) \\ &= \mathsf{P}(D_1^c)\mathsf{P}(D_2^c) \\ &= [1 - \mathsf{P}(D_1)][1 - \mathsf{P}(D_2)] \\ &= (1-p)^2.\end{aligned}$$

Now suppose that A and B are any two events. If $\mathsf{P}(B) = 0$, then we claim that A and B are independent. We must show that

$$\mathsf{P}(A \cap B) = \mathsf{P}(A)\mathsf{P}(B) = 0.$$

To show that the left-hand side is zero, observe that since probabilities are nonnegative, and since $A \cap B \subset B$,

$$0 \leq \mathsf{P}(A \cap B) \leq \mathsf{P}(B) = 0. \tag{1.30}$$

We now show that if $\mathsf{P}(B) = 1$, then A and B are independent. Since $\mathsf{P}(B) = 1$, $\mathsf{P}(B^c) = 1 - \mathsf{P}(B) = 0$, and it follows that A and B^c are independent. But then so are A and B.

Independence for more than two events

Suppose that for $j = 1, 2, \ldots$, A_j is an event. When we say that the A_j are independent, we certainly want that for any $i \neq j$,

$$\mathsf{P}(A_i \cap A_j) = \mathsf{P}(A_i)\mathsf{P}(A_j).$$

And for any distinct i, j, k, we want

$$\mathsf{P}(A_i \cap A_j \cap A_k) = \mathsf{P}(A_i)\mathsf{P}(A_j)\mathsf{P}(A_k).$$

We want analogous equations to hold for any four events, five events, and so on. In general, we want that for every *finite* subset J containing two or more positive integers,

$$\mathsf{P}\left(\bigcap_{j \in J} A_j\right) = \prod_{j \in J} \mathsf{P}(A_j).$$

In other words, we want the probability of every intersection involving finitely many of the A_j to be equal to the product of the probabilities of the individual events. If the above equation holds for all *finite* subsets of two or more positive integers, then we say that the A_j are **mutually independent**, or just independent. If the above equation holds for all subsets J containing exactly two positive integers but not necessarily for all finite subsets of 3 or more positive integers, we say that the A_j are **pairwise independent**.

Example 1.24. Given three events, say A, B, and C, they are mutually independent if and only if the following equations *all* hold,

$$\mathsf{P}(A \cap B \cap C) = \mathsf{P}(A)\mathsf{P}(B)\mathsf{P}(C)$$
$$\mathsf{P}(A \cap B) = \mathsf{P}(A)\mathsf{P}(B)$$
$$\mathsf{P}(A \cap C) = \mathsf{P}(A)\mathsf{P}(C)$$
$$\mathsf{P}(B \cap C) = \mathsf{P}(B)\mathsf{P}(C).$$

It is possible to construct events A, B, and C such that the last three equations hold (pairwise independence), but the first one does not.[6] It is also possible for the first equation to hold while the last three fail.[7]

1.6 Independence

***Example* 1.25.** Three bits are transmitted across a noisy channel and the number of correct receptions is noted. Find the probability that the number of correctly received bits is two, assuming bit errors are mutually independent and that on each bit transmission the probability of correct reception is λ for some fixed $0 \leq \lambda \leq 1$.

Solution. When the problem talks about the event that two bits are correctly received, we interpret this as meaning exactly two bits are received correctly; i.e., the other bit is received in error. Hence, there are three ways this can happen: the single error can be in the first bit, the second bit, or the third bit. To put this into the language of events, let C_i denote the event that the ith bit is received correctly (so $\mathsf{P}(C_i) = \lambda$), and let S_2 denote the event that two of the three bits sent are correctly received.[h] Then

$$S_2 = (C_1^c \cap C_2 \cap C_3) \cup (C_1 \cap C_2^c \cap C_3) \cup (C_1 \cap C_2 \cap C_3^c).$$

This is a disjoint union, and so $\mathsf{P}(S_2)$ is equal to

$$\mathsf{P}(C_1^c \cap C_2 \cap C_3) + \mathsf{P}(C_1 \cap C_2^c \cap C_3) + \mathsf{P}(C_1 \cap C_2 \cap C_3^c). \tag{1.31}$$

Next, since C_1, C_2, and C_3 are mutually independent, so are C_1 and $(C_2 \cap C_3)$. Hence, C_1^c and $(C_1 \cap C_2)$ are also independent. Thus,

$$\begin{aligned} \mathsf{P}(C_1^c \cap C_2 \cap C_3) &= \mathsf{P}(C_1^c)\mathsf{P}(C_2 \cap C_3) \\ &= \mathsf{P}(C_1^c)\mathsf{P}(C_2)\mathsf{P}(C_3) \\ &= (1-\lambda)\lambda^2. \end{aligned}$$

Treating the last two terms in (1.31) similarly, we have $\mathsf{P}(S_2) = 3(1-\lambda)\lambda^2$. If bits are as likely to be received correctly as incorrectly, i.e., $\lambda = 1/2$, then $\mathsf{P}(S_2) = 3/8$.

***Example* 1.26.** If $A_1, A_2, \ldots$ are mutually independent, show that

$$\mathsf{P}\left(\bigcap_{n=1}^{\infty} A_n\right) = \prod_{n=1}^{\infty} \mathsf{P}(A_n).$$

Solution. Write

$$\begin{aligned} \mathsf{P}\left(\bigcap_{n=1}^{\infty} A_n\right) &= \lim_{N \to \infty} \mathsf{P}\left(\bigcap_{n=1}^{N} A_n\right), \quad \text{by limit property (1.14),} \\ &= \lim_{N \to \infty} \prod_{n=1}^{N} \mathsf{P}(A_n), \quad \text{by independence,} \\ &= \prod_{n=1}^{\infty} \mathsf{P}(A_n), \end{aligned}$$

where the last step is just the definition of the infinite product.

---

[h]In working this example, we again do not explicitly specify the sample space Ω or the probability measure P. The interested reader can find one possible choice for Ω and P in Note **8**.

Example 1.27. Consider the transmission of an unending sequence of bits over a noisy channel. Suppose that a bit is received in error with probability $0 < p < 1$. Assuming errors occur independently, what is the probability that every bit is received in error? What is the probability of ever having a bit received in error?

Solution. We use the result of the preceding example as follows. Let Ω be a sample space equipped with a probability measure P and events $A_n, n = 1, 2, \ldots$, with $\mathsf{P}(A_n) = p$, where the A_n are mutually independent.[9] Thus, A_n corresponds to, or models, the event that the nth bit is received in error. The event that all bits are received in error corresponds to $\bigcap_{n=1}^{\infty} A_n$, and its probability is

$$\mathsf{P}\left(\bigcap_{n=1}^{\infty} A_n\right) = \lim_{N \to \infty} \prod_{n=1}^{N} \mathsf{P}(A_n) = \lim_{N \to \infty} p^N = 0.$$

The event of ever having a bit received in error corresponds to $A := \bigcup_{n=1}^{\infty} A_n$. Since $\mathsf{P}(A) = 1 - \mathsf{P}(A^c)$, it suffices to compute the probability of $A^c = \bigcap_{n=1}^{\infty} A_n^c$. Arguing exactly as above, we have

$$\mathsf{P}\left(\bigcap_{n=1}^{\infty} A_n^c\right) = \lim_{N \to \infty} \prod_{n=1}^{N} \mathsf{P}(A_n^c) = \lim_{N \to \infty} (1-p)^N = 0.$$

Thus, $\mathsf{P}(A) = 1 - 0 = 1$.

1.7 Combinatorics and probability

There are many probability problems, especially those concerned with gambling, that can ultimately be reduced to questions about cardinalities of various sets. We saw several examples in Section 1.3. Those examples were simple, and they were chosen so that it was easy to determine the cardinalities of the required sets. However, in more complicated problems, it is extremely helpful to have some systematic methods for finding cardinalities of sets. **Combinatorics** is the study of systematic counting methods, which we will be using to find the cardinalities of various sets that arise in probability. The four kinds of counting problems we discuss are:

(i) ordered sampling with replacement;
(ii) ordered sampling without replacement;
(iii) unordered sampling without replacement; and
(iv) unordered sampling with replacement.

Of these, the first two are rather straightforward, and the last two are somewhat complicated.

Ordered sampling with replacement

Before stating the problem, we begin with some examples to illustrate the concepts to be used.

Example 1.28. Let A, B, and C be finite sets. How many triples are there of the form (a, b, c), where $a \in A$, $b \in B$, and $c \in C$?

1.7 Combinatorics and probability

Solution. Since there are $|A|$ choices for a, $|B|$ choices for b, and $|C|$ choices for c, the total number of triples is $|A| \cdot |B| \cdot |C|$.

Similar reasoning shows that for k finite sets $A_1, \ldots, A_k$, there are $|A_1| \cdots |A_k|$ k-tuples of the form $(a_1, \ldots, a_k)$ where each $a_i \in A_i$.

Example **1.29.** Suppose that to send an Internet packet from the east coast of the United States to the west coast, a packet must go through a major east-coast city (Boston, New York, Washington, D.C., or Atlanta), a major mid-west city (Chicago, St. Louis, or New Orleans), and a major west-coast city (San Francisco or Los Angeles). How many possible routes are there?

Solution. Since there are four east-coast cities, three mid-west cities, and two west-coast cities, there are $4 \cdot 3 \cdot 2 = 24$ possible routes.

Example **1.30** (ordered sampling with replacement). From a deck of n cards, we draw k cards with replacement; i.e., we draw each card, make a note of it, put the card back in the deck and re-shuffle the deck before choosing the next card. How many different sequences of k cards can be drawn in this way?

Solution. Each time we draw a card, there are n possibilities. Hence, the number of possible sequences is

$$\underbrace{n \cdot n \cdots n}_{k \text{ times}} = n^k.$$

Ordered sampling without replacement

In Example 1.28, we formed triples (a, b, c) where no matter which $a \in A$ we chose, it did not affect which elements we were allowed to choose from the sets B or C. We next consider the construction of k-tuples in which our choice for the each entry affects the choices available for the remaining entries.

Example **1.31.** From a deck of 52 cards, we draw a hand of 5 cards without replacement. How many hands can be drawn in this way?

Solution. There are 52 cards for the first draw, 51 cards for the second draw, and so on. Hence, there are

$$52 \cdot 51 \cdot 50 \cdot 49 \cdot 48 = 311\,875\,200.$$

different hands

Example **1.32** (ordered sampling without replacement). A computer virus erases files from a disk drive in random order. If there are n files on the disk, in how many different orders can $k \leq n$ files be erased from the drive?

Solution. There are n choices for the first file to be erased, $n-1$ for the second, and so on. Hence, there are

$$n(n-1)\cdots(n-[k-1]) = \frac{n!}{(n-k)!}$$

different orders in which files can be erased from the disk.

Example 1.33. Let A be a finite set of n elements. How may k-tuples $(a_1,\ldots,a_k)$ of *distinct* entries $a_i \in A$ can be formed?

Solution. There are n choices for a_1, but only $n-1$ choices for a_2 since repeated entries are not allowed. Similarly, there are only $n-2$ choices for a_3, and so on. This is the same argument used in the previous example. Hence, there are $n!/(n-k)!$ k-tuples with distinct elements of A.

Given a set A, we let A^k denote the set of all k-tuples $(a_1,\ldots,a_k)$ where each $a_i \in A$. We denote by A_*^k the subset of all k-tuples with *distinct* entries. If $|A|=n$, then $|A^k|=|A|^k=n^k$, and $|A_*^k|=n!/(n-k)!$.

Example 1.34 (the birthday problem). In a group of k people, what is the probability that two or more people have the same birthday?

Solution. The first step in the solution is to specify the sample space Ω and the probability P. Let $D := \{1,\ldots,365\}$ denote the days of the year, and let

$$\Omega := \{(d_1,\ldots,d_k) : d_i \in D\}$$

denote the set of all possible sequences of k birthdays. Then $|\Omega|=|D|^k$. Assuming all sequences are equally likely, we take $\mathsf{P}(E) := |E|/|\Omega|$ for arbitrary events $E \subset \Omega$.

Let Q denote the set of sequences $(d_1,\ldots,d_k)$ that have at least one pair of repeated entries. For example, if $k=9$, one of the sequences in Q would be

$$(364,17,201,17,51,171,51,33,51).$$

Notice that 17 appears twice and 51 appears 3 times. The set Q is complicated. On the other hand, consider Q^c, which is the set of sequences $(d_1,\ldots,d_k)$ that have *no* repeated entries. Then

$$|Q^c| = \frac{|D|!}{(|D|-k)!},$$

and

$$\mathsf{P}(Q^c) = \frac{|Q^c|}{|\Omega|} = \frac{|D|!}{|D|^k(|D|-k)!},$$

where $|D|=365$. A plot of $\mathsf{P}(Q) = 1-\mathsf{P}(Q^c)$ as a function of k is shown in Figure 1.15. As the dashed line indicates, for $k \geq 23$, the probability of two more more people having the same birthday is greater than $1/2$.

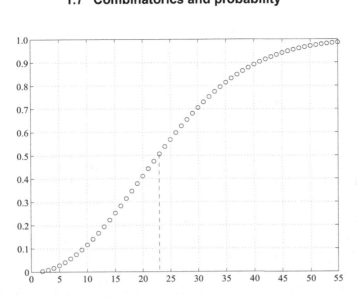

Figure 1.15. A plot of P(Q) as a function of k. For $k \geq 23$, the probability of two or more people having the same birthday is greater than $1/2$.

Unordered sampling without replacement

Before stating the problem, we begin with a simple example to illustrate the concept to be used.

Example 1.35. Let $A = \{1,2,3,4,5\}$. Then A^3 contains $5^3 = 125$ triples. The set of triples with distinct entries, A_*^3, contains $5!/2! = 60$ triples. We can write A_*^3 as the disjoint union

$$A_*^3 = G_{123} \cup G_{124} \cup G_{125} \cup G_{134} \cup G_{135}$$
$$\cup G_{145} \cup G_{234} \cup G_{235} \cup G_{245} \cup G_{345},$$

where for distinct i,j,k,

$$G_{ijk} := \{(i,j,k),(i,k,j),(j,i,k),(j,k,i),(k,i,j),(k,j,i)\}.$$

Each triple in G_{ijk} is a rearrangement, or **permutation**, of the same three elements.

The above decomposition works in general. Write A_*^k as the union of disjoint sets,

$$A_*^k = \bigcup G, \qquad (1.32)$$

where each subset G consists of k-tuples that contain the same elements. In general, for a k-tuple built from k distinct elements, there are k choices for the first entry, $k-1$ choices for the second entry, and so on. Hence, there are $k!$ k-tuples that can be built. In other words, each G in (1.32) has $|G| = k!$. It follows from (1.32) that

$$|A_*^k| = (\text{number of different sets } G) \cdot k!, \qquad (1.33)$$

and so the number of different subsets G is

$$\frac{|A_*^k|}{k!} = \frac{n!}{k!(n-k)!}.$$

The standard notation for the above right-hand side is

$$\binom{n}{k} := \frac{n!}{k!(n-k)!}$$

and is read "n choose k." In MATLAB, $\binom{n}{k} = \text{nchoosek}(n,k)$. The symbol $\binom{n}{k}$ is also called the **binomial coefficient** because it arises in the **binomial theorem**, which is discussed in Chapter 3.

Example **1.36** (unordered sampling without replacement). In many card games, we are dealt a hand of k cards, but the order in which the cards are dealt is not important. From a deck of n cards, how many k-card hands are possible?

Solution. First think about ordered hands corresponding to k-tuples with distinct entries. The set of all such hands corresponds to A_*^k. Now group together k-tuples composed of the same elements into sets G as in (1.32). All the ordered k-tuples in a particular G represent rearrangements of a single hand. So it is really the number of different sets G that corresponds to the number of unordered hands. Thus, the number of k-card hands is $\binom{n}{k}$.

Example **1.37.** A new computer chip has n pins that must be tested with all patterns in which k of the pins are set high and the rest low. How many test patterns must be checked?

Solution. This is exactly analogous to dealing k-card hands from a deck of n cards. The cards you are dealt tell you which pins to set high. Hence, there are $\binom{n}{k}$ patterns that must be tested.

Example **1.38.** A 12-person jury is to be selected from a group of 20 potential jurors. How many different juries are possible?

Solution. There are

$$\binom{20}{12} = \frac{20!}{12!\,8!} = 125\,970$$

different juries.

Example **1.39.** A 12-person jury is to be selected from a group of 20 potential jurors of which 11 are men and nine are women. How many 12-person juries are there with five men and seven women?

Solution. There are $\binom{11}{5}$ ways to choose the five men, and there are $\binom{9}{7}$ ways to choose the seven women. Hence, there are

$$\binom{11}{5}\binom{9}{7} = \frac{11!}{5!\,6!} \cdot \frac{9!}{7!\,2!} = 16\,632$$

possible juries with five men and seven women.

1.7 Combinatorics and probability

Example **1.40.** An urn contains 11 green balls and nine red balls. If 12 balls are chosen at random, what is the probability of choosing exactly five green balls and seven red balls?

Solution. Since balls are chosen at random, the desired probability is

$$\frac{\text{number of ways to choose five green balls and seven red balls}}{\text{number of ways to choose 12 balls}}.$$

In the numerator, the five green balls must be chosen from the 11 available green balls, and the seven red balls must be chosen from the nine available red balls. In the denominator, the total of $5+7=12$ balls must be chosen from the $11+9=20$ available balls. So the required probability is

$$\frac{\binom{11}{5}\binom{9}{7}}{\binom{20}{12}} = \frac{16632}{125970} \approx 0.132.$$

Example **1.41.** Consider a collection of N items, of which d are defective (and $N-d$ work properly). Suppose we test $n \leq N$ items at random. Show that the probability that k of the n tested items are defective is

$$\frac{\binom{d}{k}\binom{N-d}{n-k}}{\binom{N}{n}}. \tag{1.34}$$

Solution. Since items are chosen at random, the desired probability is

$$\frac{\text{number of ways to choose } k \text{ defective and } n-k \text{ working items}}{\text{number of ways to choose } n \text{ items}}.$$

In the numerator, the k defective items are chosen from the total of d defective ones, and the $n-k$ working items are chosen from the total of $N-d$ ones that work. In the denominator, the n items to be tested are chosen from the total of N items. Hence, the desired numerator is $\binom{d}{k}\binom{N-d}{n-k}$, and the desired denominator is $\binom{N}{n}$.

Example **1.42** (lottery). In some state lottery games, a player chooses n distinct numbers from the set $\{1,\ldots,N\}$. At the lottery drawing, balls numbered from 1 to N are mixed, and n balls withdrawn. What is the probability that k of the n balls drawn match the player's choices?

Solution. Let D denote the subset of n numbers chosen by the player. Then $\{1,\ldots,N\} = D \cup D^c$. We need to find the probability that the lottery drawing chooses k numbers from D and $n-k$ numbers from D^c. Since $|D| = n$, this probability is

$$\frac{\binom{n}{k}\binom{N-n}{n-k}}{\binom{N}{n}}.$$

Notice that this is just (1.34) with $d = n$. In other words, we regard the numbers chosen by the player as "defective," and we are finding the probability that the lottery drawing chooses k defective and $n - k$ nondefective numbers.

Example 1.43 (binomial probabilities). A certain coin has probability p of turning up heads. If the coin is tossed n times, what is the probability that k of the n tosses result in heads? Assume tosses are independent.

Solution. Let H_i denote the event that the ith toss is heads. We call i the toss index, which takes values $1, \ldots, n$. A typical sequence of n tosses would be

$$H_1 \cap H_2^c \cap H_3 \cap \cdots \cap H_{n-1} \cap H_n^c,$$

where H_i^c is the event that the ith toss is tails. The probability that n tosses result in k heads and $n - k$ tails is

$$\mathsf{P}\left(\bigcup \widetilde{H}_1 \cap \cdots \cap \widetilde{H}_n\right),$$

where $\widetilde{H}_i$ is either H_i or H_i^c, and the union is over all such intersections for which $\widetilde{H}_i = H_i$ occurs k times and $\widetilde{H}_i = H_i^c$ occurs $n - k$ times. Since this is a disjoint union,

$$\mathsf{P}\left(\bigcup \widetilde{H}_1 \cap \cdots \cap \widetilde{H}_n\right) = \sum \mathsf{P}(\widetilde{H}_1 \cap \cdots \cap \widetilde{H}_n).$$

By independence,

$$\mathsf{P}(\widetilde{H}_1 \cap \cdots \cap \widetilde{H}_n) = \mathsf{P}(\widetilde{H}_1) \cdots \mathsf{P}(\widetilde{H}_n)$$
$$= p^k (1-p)^{n-k}$$

is the same for every term in the sum. The number of terms in the sum is the number of ways of selecting k out of n toss indexes to assign to heads. Since this number is $\binom{n}{k}$, the probability that k of n tosses result in heads is

$$\binom{n}{k} p^k (1-p)^{n-k}.$$

Example 1.44 (bridge). In bridge, 52 cards are dealt to four players; hence, each player has 13 cards. The order in which the cards are dealt is not important, just the final 13 cards each player ends up with. How many different bridge games can be dealt?

Solution. There are $\binom{52}{13}$ ways to choose the 13 cards of the first player. Now there are only $52 - 13 = 39$ cards left. Hence, there are $\binom{39}{13}$ ways to choose the 13 cards for the second player. Similarly, there are $\binom{26}{13}$ ways to choose the second player's cards, and $\binom{13}{13} = 1$ way to choose the fourth player's cards. It follows that there are

$$\binom{52}{13}\binom{39}{13}\binom{26}{13}\binom{13}{13} = \frac{52!}{13!\,39!} \cdot \frac{39!}{13!\,26!} \cdot \frac{26!}{13!\,13!} \cdot \frac{13!}{13!\,0!}$$
$$= \frac{52!}{(13!)^4} \approx 5.36 \times 10^{28}$$

1.7 Combinatorics and probability

games that can be dealt.

Example 1.45. Traditionally, computers use binary arithmetic, and store n-bit words composed of zeros and ones. The new m–Computer uses m-ary arithmetic, and stores n-symbol words in which the symbols (m-ary digits) come from the set $\{0, 1, \ldots, m-1\}$. How many n-symbol words are there with k_0 zeros, k_1 ones, k_2 twos, $\ldots$, and k_{m-1} copies of symbol $m-1$, where $k_0 + k_1 + k_2 + \cdots + k_{m-1} = n$?

Solution. To answer this question, we build a typical n-symbol word of the required form as follows. We begin with an empty word,

$$\underbrace{(\ ,\ ,\ldots,\ )}_{n\text{ empty positions}}.$$

From these n available positions, there are $\binom{n}{k_0}$ ways to select positions to put the k_0 zeros. For example, if $k_0 = 3$, we might have

$$\underbrace{(\ ,0,\ ,0,\ ,\ldots,\ ,0)}_{n-3\text{ empty positions}}.$$

Now there are only $n - k_0$ empty positions. From these, there are $\binom{n-k_0}{k_1}$ ways to select positions to put the k_1 ones. For example, if $k_1 = 1$, we might have

$$\underbrace{(\ ,0,1,0,\ ,\ldots,\ ,0)}_{n-4\text{ empty positions}}.$$

Now there are only $n - k_0 - k_1$ empty positions. From these, there are $\binom{n-k_0-k_1}{k_2}$ ways to select positions to put the k_2 twos. Continuing in this way, we find that the number of n-symbol words with the required numbers of zeros, ones, twos, etc., is

$$\binom{n}{k_0}\binom{n-k_0}{k_1}\binom{n-k_0-k_1}{k_2}\cdots\binom{n-k_0-k_1-\cdots-k_{m-2}}{k_{m-1}},$$

which expands to

$$\frac{n!}{k_0!(n-k_0)!} \cdot \frac{(n-k_0)!}{k_1!(n-k_0-k_1)!} \cdot \frac{(n-k_0-k_1)!}{k_2!(n-k_0-k_1-k_2)!} \cdots$$

$$\cdots \frac{(n-k_0-k_1-\cdots-k_{m-2})!}{k_{m-1}!(n-k_0-k_1-\cdots-k_{m-1})!}.$$

Canceling common factors and noting that $(n - k_0 - k_1 - \cdots - k_{m-1})! = 0! = 1$, we obtain

$$\frac{n!}{k_0!k_1!\cdots k_{m-1}!}$$

as the number of n-symbol words with k_0 zeros, k_1 ones, etc.

We call
$$\binom{n}{k_0,\ldots,k_{m-1}} := \frac{n!}{k_0!k_1!\cdots k_{m-1}!}$$
the **multinomial coefficient**. When $m=2$,
$$\binom{n}{k_0,k_1} = \binom{n}{k_0,n-k_0} = \frac{n!}{k_0!(n-k_0)!} = \binom{n}{k_0}$$
becomes the binomial coefficient.

Unordered sampling with replacement

Before stating the problem, we begin with a simple example to illustrate the concepts involved.

Example **1.46.** An automated snack machine dispenses apples, bananas, and carrots. For a fixed price, the customer gets five items from among the three possible choices. For example, a customer could choose one apple, two bananas, and two carrots. To record the customer's choices electronically, 7-bit sequences are used. For example, the sequence $(0,1,0,0,1,0,0)$ means one apple, two bananas, and two carrots. The first group of zeros tells how many apples, the second group of zeros tells how many bananas, and the third group of zeros tells how many carrots. The ones are used to separate the groups of zeros. As another example, $(0,0,0,1,0,1,0)$ means three apples, one banana, and one carrot. How many customer choices are there?

Solution. The question is equivalent to asking how many 7-bit sequences there are with five zeros and two ones. From Example 1.45, the answer is $\binom{7}{5,2} = \binom{7}{5} = \binom{7}{2}$.

Example **1.47** (unordered sampling with replacement). Suppose k numbers are drawn with replacement from the set $A = \{1,2,\ldots,n\}$. How many different sets of k numbers can be obtained in this way?

Solution. Think of the numbers $1,\ldots,n$ as different kinds of fruit as in the previous example. To count the different ways of drawing k "fruits," we use the bit-sequence method. The bit sequences will have $n-1$ ones as separators, and the total number of zeros must be k. So the sequences have a total of $N := n-1+k$ bits. How many ways can we choose $n-1$ positions out of N in which to place the separators? The answer is
$$\binom{N}{n-1} = \binom{n-1+k}{n-1} = \binom{k+n-1}{k}.$$

Just as we partitioned A_*^k in (1.32), we can partition A^k using
$$A^k = \bigcup G,$$
where each G contains all k-tuples with the same elements. Unfortunately, different Gs may contain different numbers of k-tuples. For example, if $n=3$ and $k=3$, one of the sets G would be
$$\{(1,2,3),(1,3,2),(2,1,3),(2,3,1),(3,1,2),(3,2,1)\},$$

while another G would be
$$\{(1,2,2),(2,1,2),(2,2,1)\}.$$
How many different sets G are there? Although we cannot find the answer by using an equation like (1.33), we see from the above analysis that there are $\binom{k+n-1}{k}$ sets G.

Notes

1.4: Axioms and properties of probability

Note **1.** When the sample space Ω is finite or countably infinite, $P(A)$ is usually defined for all subsets of Ω by taking
$$P(A) := \sum_{\omega \in A} p(\omega)$$
for some nonnegative function p that sums to one; i.e., $p(\omega) \geq 0$ and $\sum_{\omega \in \Omega} p(\omega) = 1$. (It is easy to check that if P is defined in this way, then it satisfies the axioms of a probability measure.) However, for larger sample spaces, such as when Ω is an interval of the real line, e.g., Example 1.16, and we want the probability of an interval to be proportional to its length, it is not possible to define $P(A)$ for all subsets and still have P satisfy all four axioms. (A proof of this fact can be found in advanced texts, e.g., [3, p. 45].) The way around this difficulty is to define $P(A)$ only for some subsets of Ω, but not all subsets of Ω. It is indeed fortunate that this can be done in such a way that $P(A)$ is defined for all subsets of interest that occur in practice. A set A for which $P(A)$ is defined is called an **event**, and the collection of all events is denoted by $\mathscr{A}$. The triple $(\Omega, \mathscr{A}, P)$ is called a **probability space**. For technical reasons discussed below, the collection $\mathscr{A}$ is always taken to be a σ-field.

If $\mathscr{A}$ is a collection of subsets of Ω with the following properties, then $\mathscr{A}$ is called a **σ-field** or a **σ-algebra**.

(i) The empty set $\varnothing$ belongs to $\mathscr{A}$, i.e., $\varnothing \in \mathscr{A}$.

(ii) If $A \in \mathscr{A}$, then so does its complement, A^c, i.e., $A \in \mathscr{A}$ implies $A^c \in \mathscr{A}$.

(iii) If $A_1, A_2, \ldots$ belong to $\mathscr{A}$, then so does their union, $\bigcup_{n=1}^{\infty} A_n$.

Given that $P(A)$ may not be defined for all sets A, we now list some of the technical benefits of defining $P(A)$ for sets A in a σ-field. First, since a σ-field contains $\varnothing$, it makes sense in axiom (i) to talk about $P(\varnothing)$. Second, since the complement of a set in $\mathscr{A}$ is also in $\mathscr{A}$, we have $\Omega = \varnothing^c \in \mathscr{A}$, and so it makes sense in axiom (iv) to talk about $P(\Omega)$. Third, if $A_1, A_2, \ldots$ are in $\mathscr{A}$, then so is their union; hence, it makes sense in axiom (iii) to talk about $P\left(\bigcup_{n=1}^{\infty} A_n\right)$. Fourth, again with regard to $A_n \in \mathscr{A}$, by the indentity
$$\bigcap_{n=1}^{\infty} A_n = \left(\bigcup_{n=1}^{\infty} A_n^c\right)^c,$$
we see that the left-hand side must also belong to $\mathscr{A}$; hence, it makes sense to talk about $P\left(\bigcap_{n=1}^{\infty} A_n\right)$.

Given any set Ω, let 2^Ω denote the collection of all subsets of Ω. We call 2^Ω the **power set** of Ω. This notation is used for both finite and infinite sets. The notation is motivated by the fact that if Ω is a finite set, then there are $2^{|\Omega|}$ different subsets of Ω.[i] Since the power set obviously satisfies the three properties above, the power set is a σ-field.

> Let $\mathscr{C}$ be any collection of subsets of Ω. We do not assume $\mathscr{C}$ is a σ-field. Define $\sigma(\mathscr{C})$ to be the **smallest σ-field that contains** $\mathscr{C}$. By this we mean that if $\mathscr{D}$ is any σ-field with $\mathscr{C} \subset \mathscr{D}$, then $\sigma(\mathscr{C}) \subset \mathscr{D}$.

Example 1.48. Let A be a nonempty subset of Ω, and put $\mathscr{C} = \{A\}$ so that the collection $\mathscr{C}$ consists of a single subset. Find $\sigma(\mathscr{C})$.

Solution. From the three properties of a σ-field, any σ-field that contains A must also contain A^c, $\varnothing$, and Ω. We claim

$$\sigma(\mathscr{C}) = \{\varnothing, A, A^c, \Omega\}.$$

Since $A \cup A^c = \Omega$, it is easy to see that our choice satisfies the three properties of a σ-field. It is also clear that if $\mathscr{D}$ is any σ-field such that $\mathscr{C} \subset \mathscr{D}$, then every subset in our choice for $\sigma(\mathscr{C})$ must belong to $\mathscr{D}$; i.e., $\sigma(\mathscr{C}) \subset \mathscr{D}$.

More generally, if $A_1, \ldots, A_n$ is a partition of Ω, then $\sigma(\{A_1, \ldots, A_n\})$ consists of the empty set along with the $2^n - 1$ subsets constructed by taking all possible unions of the A_i. See Problem 40.

For general collections $\mathscr{C}$ of subsets of Ω, all we can say is that (Problem 45)

$$\sigma(\mathscr{C}) = \bigcap_{\mathscr{A} : \mathscr{C} \subset \mathscr{A}} \mathscr{A},$$

where the intersection is over all σ-fields $\mathscr{A}$ that contain $\mathscr{C}$. Note that there is always at least one σ-field $\mathscr{A}$ that contains $\mathscr{C}$; e.g., the power set.

Note 2. The alert reader will observe that axiom (*i*) is redundant. In axiom (*iii*), take $A_1 = \Omega$ and $A_n = \varnothing$ for $n \geq 2$ so that $\bigcup_{n=1}^\infty A_n = \Omega$ and we can write

$$\mathsf{P}(\Omega) = \mathsf{P}(\Omega) + \sum_{n=2}^\infty \mathsf{P}(\varnothing)$$

$$\geq \sum_{n=2}^\infty \mathsf{P}(\varnothing),$$

By axiom (*ii*), either $\mathsf{P}(\varnothing) = 0$ (which we want to prove) or $\mathsf{P}(\varnothing) > 0$. If $\mathsf{P}(\varnothing) > 0$, then the above right-hand side is infinite, telling us that $\mathsf{P}(\Omega) = \infty$. Since this contradicts axiom (*iv*) that $\mathsf{P}(\Omega) = 1$, we must have $\mathsf{P}(\varnothing) = 0$.

---

[i]Suppose $\Omega = \{\omega_1, \ldots, \omega_n\}$. Each subset of Ω can be associated with an n-bit word. A point ω_i is in the subset if and only if the ith bit in the word is a 1. For example, if $n = 5$, we associate 01011 with the subset $\{\omega_2, \omega_4, \omega_5\}$ since bits 2, 4, and 5 are ones. In particular, 00000 corresponds to the empty set and 11111 corresponds to Ω itself. Since there are 2^n n-bit words, there are 2^n subsets of Ω.

Since axiom (*i*) is redundant, why did we include it? It turns out that axioms (*i*)–(*iii*) characterize what is called a **measure**. If the measure of the whole space is finite, then the foregoing argument can be trivially modified to show that axiom (*i*) is again redundant. However, sometimes we want to have the measure of the whole space be infinite. For example, **Lebesgue measure** on $\mathbb{R}$ takes the measure of an interval to be its length. In this case, the length of $\mathbb{R} = (-\infty, \infty)$ is infinite. Thus, axioms (*i*)–(*iii*) characterize general measures, and a finite measure satisfies these three axioms along with the additional condition that the measure of the whole space is finite.

Note 3. In light of Note 1, we see that to guarantee that $P(\{\omega_n\})$ is defined in Example 1.18, it is necessary to assume that the singleton sets $\{\omega_n\}$ are events, i.e., $\{\omega_n\} \in \mathscr{A}$.

Note 4. Here is a set-theoretic derivation of (1.12). First note that (see Figure 1.13)
$$A = (A \cap B^c) \cup (A \cap B)$$
and
$$B = (A \cap B) \cup (A^c \cap B).$$
Hence,
$$A \cup B = [(A \cap B^c) \cup (A \cap B)] \cup [(A \cap B) \cup (A^c \cap B)].$$
The two copies of $A \cap B$ can be reduced to one using the identity $F \cup F = F$ for any set F. Thus,
$$A \cup B = (A \cap B^c) \cup (A \cap B) \cup (A^c \cap B).$$
A Venn diagram depicting this last decomposition is shown in Figure 1.16. Taking proba-

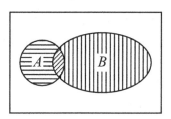

Figure 1.16. Decomposition $A \cup B = (A \cap B^c) \cup (A \cap B) \cup (A^c \cap B)$.

bilities of the preceding equations, which involve disjoint unions, we find that
$$P(A) = P(A \cap B^c) + P(A \cap B),$$
$$P(B) = P(A \cap B) + P(A^c \cap B),$$
$$P(A \cup B) = P(A \cap B^c) + P(A \cap B) + P(A^c \cap B).$$

Using the first two equations, solve for $P(A \cap B^c)$ and $P(A^c \cap B)$, respectively, and then substitute into the first and third terms on the right-hand side of the last equation. This results in
$$P(A \cup B) = [P(A) - P(A \cap B)] + P(A \cap B)$$
$$+ [P(B) - P(A \cap B)]$$
$$= P(A) + P(B) - P(A \cap B).$$

1.5: Conditional probability

Note 5. Here is a choice for Ω and P for Example 1.21. Let

$$\Omega := \{(e,d) : e,d = 0 \text{ or } 1\},$$

where $e = 1$ corresponds to being routed through El Paso, and $d = 1$ corresponds to a dropped packet. We then take

$$E := \{(e,d) : e = 1\} = \{(1,0), (1,1)\},$$

and

$$D := \{(e,d) : d = 1\} = \{(0,1), (1,1)\}.$$

It follows that

$$E^c = \{(0,1), (0,0)\} \quad \text{and} \quad D^c = \{(1,0), (0,0)\}.$$

Hence, $E \cap D = \{(1,1)\}$, $E \cap D^c = \{(1,0)\}$, $E^c \cap D = \{(0,1)\}$, and $E^c \cap D^c = \{(0,0)\}$.

In order to specify a suitable probability measure on Ω, we work backwards. First, *if* a measure P on Ω exists such that (1.24) holds, then

$$P(\{(1,1)\}) = P(E \cap D) = P(D|E)P(E) = 1/4,$$
$$P(\{(0,1)\}) = P(E^c \cap D) = P(D|E^c)P(E^c) = 1/16,$$
$$P(\{(1,0)\}) = P(E \cap D^c) = P(D^c|E)P(E) = 1/2,$$
$$P(\{(0,0)\}) = P(E^c \cap D^c) = P(D^c|E^c)P(E^c) = 3/16.$$

This suggests that we *define* P by

$$P(A) := \sum_{\omega \in A} p(\omega),$$

where $p(\omega) = p(e,d)$ is given by $p(1,1) := 1/4$, $p(0,1) := 1/16$, $p(1,0) := 1/2$, and $p(0,0) := 3/16$. Starting from this definition of P, it is not hard to check that (1.24) holds.

1.6: Independence

Note 6. Here is an example of three events that are pairwise independent, but not mutually independent. Let

$$\Omega := \{1,2,3,4,5,6,7\},$$

and put $P(\{\omega\}) := 1/8$ for $\omega \neq 7$, and $P(\{7\}) := 1/4$. Take $A := \{1,2,7\}$, $B := \{3,4,7\}$, and $C := \{5,6,7\}$. Then $P(A) = P(B) = P(C) = 1/2$. and $P(A \cap B) = P(A \cap C) = P(B \cap C) = P(\{7\}) = 1/4$. Hence, A and B, A and C, and B and C are pairwise independent. However, since $P(A \cap B \cap C) = P(\{7\}) = 1/4$, and since $P(A)P(B)P(C) = 1/8$, A, B, and C are not mutually independent. *Exercise:* Modify this example to use a sample space with only four elements.

Note 7. Here is an example of three events for which $P(A \cap B \cap C) = P(A)P(B)P(C)$ but no pair is independent. Let $\Omega := \{1,2,3,4\}$. Put $P(\{1\}) = P(\{2\}) = P(\{3\}) = p$ and $P(\{4\}) = q$, where $3p+q = 1$ and $0 \le p,q \le 1$. Put $A := \{1,4\}$, $B := \{2,4\}$, and $C := \{3,4\}$. Then the intersection of any pair is $\{4\}$, as is the intersection of all three sets. Also, $P(\{4\}) = q$. Since $P(A) = P(B) = P(C) = p+q$, we require $(p+q)^3 = q$ and $(p+q)^2 \ne q$. Solving $3p+q=1$ and $(p+q)^3 = q$ for q reduces to solving $8q^3 + 12q^2 - 21q + 1 = 0$. Now, $q=1$ is obviously a root, but this results in $p=0$, which implies mutual independence. However, since $q=1$ is a root, it is easy to verify that

$$8q^3 + 12q^2 - 21q + 1 = (q-1)(8q^2 + 20q - 1).$$

By the quadratic formula, the desired root is $q = (-5 + 3\sqrt{3})/4$. It then follows that $p = (3 - \sqrt{3})/4$ and that $p+q = (-1+\sqrt{3})/2$. Now just observe that $(p+q)^2 \ne q$.

Note 8. Here is a choice for Ω and P for Example 1.25. Let

$$\Omega := \{(i,j,k) : i,j,k = 0 \text{ or } 1\},$$

with 1 corresponding to correct reception and 0 to incorrect reception. Now put

$$C_1 := \{(i,j,k) : i = 1\},$$
$$C_2 := \{(i,j,k) : j = 1\},$$
$$C_3 := \{(i,j,k) : k = 1\},$$

and observe that

$$C_1 = \{(1,0,0), (1,0,1), (1,1,0), (1,1,1)\},$$
$$C_2 = \{(0,1,0), (0,1,1), (1,1,0), (1,1,1)\},$$
$$C_3 = \{(0,0,1), (0,1,1), (1,0,1), (1,1,1)\}.$$

Next, let $P(\{(i,j,k)\}) := \lambda^{i+j+k}(1-\lambda)^{3-(i+j+k)}$. Since

$$C_3^c = \{(0,0,0), (1,0,0), (0,1,0), (1,1,0)\},$$

$C_1 \cap C_2 \cap C_3^c = \{(1,1,0)\}$. Similarly, $C_1 \cap C_2^c \cap C_3 = \{(1,0,1)\}$, and $C_1^c \cap C_2 \cap C_3 = \{(0,1,1)\}$. Hence,

$$S_2 = \{(1,1,0), (1,0,1), (0,1,1)\}$$
$$= \{(1,1,0)\} \cup \{(1,0,1)\} \cup \{(0,1,1)\},$$

and thus, $P(S_2) = 3\lambda^2(1-\lambda)$.

Note 9. To show the *existence* of a sample space and probability measure with such independent events is beyond the scope of this book. Such constructions can be found in more advanced texts such as [3, Section 36].

Problems

1.1: Sample spaces, outcomes, and events

1. A computer job scheduler chooses one of six processors to assign programs to. Suggest a sample space to model all possible choices of the job scheduler.

2. A class of 25 students is surveyed to find out how many own an MP3 player. Suggest a sample space to model all possible results of the survey.

3. The ping command is used to measure round-trip times for Internet packets. Suggest a sample space to model all possible round-trip times. What is the event that a round-trip time exceeds 10 ms?

4. A cell-phone tower has a circular coverage area of radius 10 km. We observe the source locations of calls received by the tower.

    (a) Suggest a sample space to model all possible source locations of calls that the tower can receive.

    (b) Using your sample space from part (a), what is the event that the source location of a call is between 2 and 5 km from the tower.

1.2: Review of set notation

5. For real numbers $-\infty < a < b < \infty$, we use the following notation.

$$\begin{aligned}(a,b] &:= \{x : a < x \leq b\} \\ (a,b) &:= \{x : a < x < b\} \\ [a,b) &:= \{x : a \leq x < b\} \\ [a,b] &:= \{x : a \leq x \leq b\}.\end{aligned}$$

We also use

$$\begin{aligned}(-\infty,b] &:= \{x : x \leq b\} \\ (-\infty,b) &:= \{x : x < b\} \\ (a,\infty) &:= \{x : x > a\} \\ [a,\infty) &:= \{x : x \geq a\}.\end{aligned}$$

For example, with this notation, $(0,1]^c = (-\infty, 0] \cup (1, \infty)$ and $(0,2] \cup [1,3) = (0,3)$. Now analyze

   (a) $[2,3]^c$,

   (b) $(1,3) \cup (2,4)$,

   (c) $(1,3) \cap [2,4)$,

   (d) $(3,6] \setminus (5,7)$.

6. Sketch the following subsets of the x-y plane.

(a) $B_z := \{(x,y) : x+y \leq z\}$ for $z = 0, -1, +1$.

(b) $C_z := \{(x,y) : x > 0, y > 0, \text{ and } xy \leq z\}$ for $z = 1$.

(c) $H_z := \{(x,y) : x \leq z\}$ for $z = 3$.

(d) $J_z := \{(x,y) : y \leq z\}$ for $z = 3$.

(e) $H_z \cap J_z$ for $z = 3$.

(f) $H_z \cup J_z$ for $z = 3$.

(g) $M_z := \{(x,y) : \max(x,y) \leq z\}$ for $z = 3$, where $\max(x,y)$ is the larger of x and y. For example, $\max(7,9) = 9$. Of course, $\max(9,7) = 9$ too.

(h) $N_z := \{(x,y) : \min(x,y) \leq z\}$ for $z = 3$, where $\min(x,y)$ is the smaller of x and y. For example, $\min(7,9) = 7 = \min(9,7)$.

(i) $M_2 \cap N_3$.

(j) $M_4 \cap N_3$.

7. Let Ω denote the set of real numbers, $\Omega = (-\infty, \infty)$.

   (a) Use the distributive law to simplify
   $$[1,4] \cap \Big([0,2] \cup [3,5]\Big).$$

   (b) Use De Morgan's law to simplify $\Big([0,1] \cup [2,3]\Big)^c$.

   (c) Simplify $\bigcap_{n=1}^{\infty} (-1/n, 1/n)$.

   (d) Simplify $\bigcap_{n=1}^{\infty} [0, 3 + 1/(2n))$.

   (e) Simplify $\bigcup_{n=1}^{\infty} [5, 7 - (3n)^{-1}]$.

   (f) Simplify $\bigcup_{n=1}^{\infty} [0, n]$.

8. Fix two sets A and C. If $C \subset A$, show that for every set B,
$$(A \cap B) \cup C = A \cap (B \cup C). \quad (1.35)$$
Also show that if (1.35) holds for some set B, then $C \subset A$ (and thus (1.35) holds for all sets B).

*9. Let A and B be subsets of Ω. Put
$$I := \{\omega \in \Omega : \omega \in A \text{ implies } \omega \in B\}.$$
Show that $A \cap I = A \cap B$.

*10. Explain why $f:(-\infty,\infty) \to [0,\infty)$ with $f(x) = x^3$ is not well defined.

*11. Consider the formula $f(x) = \sin(x)$ for $x \in [-\pi/2, \pi/2]$.

   (a) Determine, if possible, a choice of co-domain Y such that $f:[-\pi/2, \pi/2] \to Y$ is invertible. *Hint:* You may find it helpful to sketch the curve.

   (b) Find $\{x : f(x) \leq 1/2\}$.

   (c) Find $\{x : f(x) < 0\}$.

*12. Consider the formula $f(x) = \sin(x)$ for $x \in [0, \pi]$.

   (a) Determine, if possible, a choice of co-domain Y such that $f:[0, \pi] \to Y$ is invertible. *Hint:* You may find it helpful to sketch the curve.

   (b) Find $\{x : f(x) \leq 1/2\}$.

   (c) Find $\{x : f(x) < 0\}$.

*13. Let X be any set, and let $A \subset X$. Define the real-valued function f by

$$f(x) := \begin{cases} 1, & x \in A, \\ 0, & x \notin A. \end{cases}$$

Thus, $f: X \to \mathbb{R}$, where $\mathbb{R} := (-\infty, \infty)$ denotes the real numbers. For arbitrary $B \subset \mathbb{R}$, find $f^{-1}(B)$. *Hint:* There are four cases to consider, depending on whether 0 or 1 belong to B.

*14. Let $f: X \to Y$ be a function such that f takes only n distinct values, say $y_1, \ldots, y_n$. Define

$$A_i := f^{-1}(\{y_i\}) = \{x \in X : f(x) = y_i\}.$$

Let $B \subset Y$. Show that if $f^{-1}(B)$ is not empty, then it can be expressed as a union of the A_i. (It then follows that there are only 2^n possibilities for $f^{-1}(B)$.)

*15. If $f: X \to Y$, show that inverse images preserve the following set operations.

   (a) If $B \subset Y$, show that $f^{-1}(B^c) = f^{-1}(B)^c$.

   (b) If B_n is a sequence of subsets of Y, show that

   $$f^{-1}\left(\bigcup_{n=1}^{\infty} B_n\right) = \bigcup_{n=1}^{\infty} f^{-1}(B_n).$$

   (c) If B_n is a sequence of subsets of Y, show that

   $$f^{-1}\left(\bigcap_{n=1}^{\infty} B_n\right) = \bigcap_{n=1}^{\infty} f^{-1}(B_n).$$

*16. Show that if $B = \bigcup_i \{b_i\}$ and $C = \bigcup_i \{c_i\}$ are countable sets, then so is $A := B \cup C$.

*17. Let $C_1, C_2, \ldots$ be countable sets. Show that

$$B := \bigcup_{i=1}^{\infty} C_i$$

is a countable set.

*18. Show that any subset of a countable set is countable.

*19. Show that if $A \subset B$ and A is uncountable, then so is B.

*20. Show that the union of a countable set and an uncountable set is uncountable.

1.3: Probability models

21. **MATLAB.** At the beginning of Section 1.3, we developed a mathematical model for the toss of a single die. The probability of any one of the six faces landing up is $1/6 \approx 0.167$. If we toss a die 100 times, we expect that each face should land up between 16 and 17 times. Save the following MATLAB script in an M-file and run it to simulate the toss of a fair die. For now, do not worry about how the script works. You will learn more about histograms in Chapter 6.

    ```
    % Simulation of Tossing a Fair Die
    %
    n = 100;                    % Number of tosses.
    X = ceil(6*rand(1,n));
    minX = min(X);              % Save to avoid re-
    maxX = max(X);              % computing min & max.
    e = [minX:maxX+1]-0.5;
    H = histc(X,e);
    nbins = length(e) - 1;
    bin_centers = [minX:maxX];
    bar(bin_centers,H(1:nbins),'hist')
    ```

    Did each face land up between 16 and 17 times? Modify your M-file to try again with $n = 1000$ and $n = 10000$.

22. **MATLAB.** What happens if you toss a pair of dice and add the number of dots on each face — you get a number from 2 to 12. But if you do this 100 times, how many times do you expect each number to appear? In this problem you can investigate using simulation. In the script for the preceding problem, replace the line

    ```
    X = ceil(6*rand(1,n));
    ```

    with the three lines

    ```
    Y = ceil(6*rand(1,n));
    Z = ceil(6*rand(1,n));
    X = Y + Z;
    ```

Run the script with $n = 100$, $n = 1000$, and $n = 10\,000$. Give an intuitive explanation of your results.

23. A letter of the alphabet (a–z) is generated at random. Specify a sample space Ω and a probability measure P. Compute the probability that a vowel (a, e, i, o, u) is generated.

24. A collection of plastic letters, a–z, is mixed in a jar. Two letters are drawn at random, one after the other. What is the probability of drawing a vowel (a, e, i, o, u) and a consonant in either order? Two vowels in any order? Specify your sample space Ω and probability P.

25. MATLAB. Put the following MATLAB script into an M-file, and use it to simulate Example 1.12.

```
% Simulation of Drawing an Ace
%
n = 10000;                  % Number of draws.
X = ceil(52*rand(1,n));
aces = (1 <= X & X <= 4);
naces = sum(aces);
fprintf('There were %g aces in %g draws.\n',naces,n)
```

In Example 1.12, we showed that the probability of drawing an ace is $1/13 \approx 0.0769$. Hence, if we repeat the experiment of drawing a card 10 000 times, we expect to see about 769 aces. What do you get when you run the script? Modify the script to simulate the drawing of a face card. Since the probability of drawing a face card is 0.2308, in 10 000 draws, you should expect about 2308 face cards. What do you get when you run the script?

26. A new baby wakes up exactly once every night. The time at which the baby wakes up occurs at random between 9 pm and 7 am. If the parents go to sleep at 11 pm, what is the probability that the parents are not awakened by the baby before they would normally get up at 7 am? Specify your sample space Ω and probability P.

27. For any real or complex number $z \neq 1$ and any positive integer N, derive the **geometric series** formula

$$\sum_{k=0}^{N-1} z^k = \frac{1-z^N}{1-z}, \quad z \neq 1.$$

Hint: Let $S_N := 1 + z + \cdots + z^{N-1}$, and show that $S_N - zS_N = 1 - z^N$. Then solve for S_N.

Remark. If $|z| < 1$, $|z|^N \to 0$ as $N \to \infty$. Hence,

$$\sum_{k=0}^{\infty} z^k = \frac{1}{1-z}, \quad \text{for } |z| < 1.$$

28. Let $\Omega := \{1,\ldots,6\}$. If $p(\omega) = 2p(\omega - 1)$ for $\omega = 2,\ldots,6$, and if $\sum_{\omega=1}^{6} p(\omega) = 1$, show that $p(\omega) = 2^{\omega-1}/63$. *Hint:* Use Problem 27.

1.4: Axioms and properties of probability

29. Let A and B be events for which $P(A)$, $P(B)$, and $P(A \cup B)$ are known. Express the following in terms of these probabilities:

   (a) $P(A \cap B)$.

   (b) $P(A \cap B^c)$.

   (c) $P(B \cup (A \cap B^c))$.

   (d) $P(A^c \cap B^c)$.

30. Let Ω be a sample space equipped with two probability measures, P_1 and P_2. Given any $0 \leq \lambda \leq 1$, show that if $P(A) := \lambda P_1(A) + (1-\lambda)P_2(A)$, then P satisfies the four axioms of a probability measure.

31. Let Ω be a sample space, and fix any point $\omega_0 \in \Omega$. For any event A, put

$$\mu(A) := \begin{cases} 1, & \omega_0 \in A, \\ 0, & \text{otherwise.} \end{cases}$$

   Show that μ satisfies the axioms of a probability measure.

32. Suppose that instead of axiom (iii) of Section 1.4, we assume only that for any two disjoint events A and B, $P(A \cup B) = P(A) + P(B)$. Use this assumption and induction[j] on N to show that for any *finite* sequence of pairwise disjoint events $A_1, \ldots, A_N$,

$$P\left(\bigcup_{n=1}^{N} A_n\right) = \sum_{n=1}^{N} P(A_n).$$

   Using this result for finite N, it is not possible to derive axiom (iii), which is the assumption needed to derive the limit results of Section 1.4.

*33. The purpose of this problem is to show that any countable union can be written as a union of pairwise disjoint sets. Given any sequence of sets F_n, define a new sequence by $A_1 := F_1$, and

$$A_n := F_n \cap F_{n-1}^c \cap \cdots \cap F_1^c, \quad n \geq 2.$$

Note that the A_n are pairwise disjoint. For finite $N \geq 1$, show that

$$\bigcup_{n=1}^{N} F_n = \bigcup_{n=1}^{N} A_n.$$

Also show that

$$\bigcup_{n=1}^{\infty} F_n = \bigcup_{n=1}^{\infty} A_n.$$

---

[j] In this case, using induction on N means that you first verify the desired result for $N = 2$. Second, you assume the result is true for some arbitrary $N \geq 2$ and then prove the desired result is true for $N+1$.

*34. Use the preceding problem to show that for any sequence of events F_n,

$$\mathsf{P}\left(\bigcup_{n=1}^{\infty} F_n\right) = \lim_{N \to \infty} \mathsf{P}\left(\bigcup_{n=1}^{N} F_n\right).$$

*35. Use the preceding problem to show that for any sequence of events G_n,

$$\mathsf{P}\left(\bigcap_{n=1}^{\infty} G_n\right) = \lim_{N \to \infty} \mathsf{P}\left(\bigcap_{n=1}^{N} G_n\right).$$

36. The finite union bound. Show that for any finite sequence of events $F_1, \ldots, F_N$,

$$\mathsf{P}\left(\bigcup_{n=1}^{N} F_n\right) \leq \sum_{n=1}^{N} \mathsf{P}(F_n).$$

Hint: Use the inclusion–exclusion formula (1.12) and induction on N. See the last footnote for information on induction.

*37. The infinite union bound. Show that for any infinite sequence of events F_n,

$$\mathsf{P}\left(\bigcup_{n=1}^{\infty} F_n\right) \leq \sum_{n=1}^{\infty} \mathsf{P}(F_n).$$

Hint: Combine Problems 34 and 36.

*38. First Borel–Cantelli lemma. Show that if B_n is a sequence of events for which

$$\sum_{n=1}^{\infty} \mathsf{P}(B_n) < \infty, \qquad (1.36)$$

then

$$\mathsf{P}\left(\bigcap_{n=1}^{\infty} \bigcup_{k=n}^{\infty} B_k\right) = 0.$$

Hint: Let $G := \bigcap_{n=1}^{\infty} G_n$, where $G_n := \bigcup_{k=n}^{\infty} B_k$. Now use Problem 35, the union bound of the preceding problem, and the fact that (1.36) implies

$$\lim_{N \to \infty} \sum_{n=N}^{\infty} \mathsf{P}(B_n) = 0.$$

*39. Let $\Omega = [0,1]$, and for $A \subset \Omega$, put $\mathsf{P}(A) := \int_A 1\,d\omega$. In particular, this implies $\mathsf{P}([a,b]) = b - a$. Consider the following sequence of sets. Put $A_0 := \Omega = [0,1]$. Define $A_1 \subset A_0$ by removing the middle third from A_0. In other words,

$$A_1 = [0, 1/3] \cup [2/3, 1].$$

Now define $A_2 \subset A_1$ by removing the middle third of each of the intervals making up A_1. An easy way to do this is to first rewrite

$$A_1 = [0, 3/9] \cup [6/9, 9/9].$$

Then
$$A_2 = \Big([0,1/9] \cup [2/9,3/9]\Big) \cup \Big([6/9,7/9] \cup [8/9,9/9]\Big).$$

Similarly, define A_3 by removing the middle third from each of the above four intervals. Thus,

$$\begin{aligned}A_3 := &[0,1/27] \cup [2/27,3/27]\\ &\cup [6/27,7/27] \cup [8/27,9/27]\\ &\cup [18/27,19/27] \cup [20/27,21/27]\\ &\cup [24/27,25/27] \cup [26/27,27/27].\end{aligned}$$

Continuing in this way, we can define $A_4, A_5, \ldots$.

(a) Compute $P(A_0)$, $P(A_1)$, $P(A_2)$, and $P(A_3)$.

(b) What is the general formula for $P(A_n)$?

(c) The **Cantor set** is defined by $A := \bigcap_{n=0}^{\infty} A_n$. Find $P(A)$. Justify your answer.

*40. **This problem assumes you have read Note 1.** Let $A_1, \ldots, A_n$ be a partition of Ω. If $\mathscr{C} := \{A_1, \ldots, A_n\}$, show that $\sigma(\mathscr{C})$ consists of the empty set along with all unions of the form
$$\bigcup_i A_{k_i}$$
where k_i is a finite subsequence of distinct elements from $\{1, \ldots, n\}$.

*41. **This problem assumes you have read Note 1.** Let $\Omega := [0,1)$, and for $n = 1, 2, \ldots$, let $\mathscr{C}_n$ denote the partition
$$\mathscr{C}_n := \left\{ \left[\frac{k-1}{2^n}, \frac{k}{2^n}\right), k = 1, \ldots, 2^n \right\}.$$
Let $\mathscr{A}_n := \sigma(\mathscr{C}_n)$, and put $\mathscr{A} := \bigcup_{n=1}^{\infty} \mathscr{A}_n$. Determine whether or not $\mathscr{A}$ is a σ-field.

*42. **This problem assumes you have read Note 1.** Let Ω be a sample space, and let $X: \Omega \to \mathbb{R}$, where $\mathbb{R}$ denotes the set of real numbers. Suppose the mapping X takes finitely many distinct values $x_1, \ldots, x_n$. Find the smallest σ-field $\mathscr{A}$ of subsets of Ω such that for all $B \subset \mathbb{R}$, $X^{-1}(B) \in \mathscr{A}$. *Hint:* Problems 14 and 15.

*43. **This problem assumes you have read Note 1.** Let $\Omega := \{1,2,3,4,5\}$, and put $A := \{1,2,3\}$ and $B := \{3,4,5\}$. Put $P(A) := 5/8$ and $P(B) := 7/8$.

(a) Find $\mathscr{F} := \sigma(\{A,B\})$, the smallest σ-field containing the sets A and B.

(b) Compute $P(F)$ for all $F \in \mathscr{F}$.

(c) **Trick question.** What is $P(\{1\})$?

*44. **This problem assumes you have read Note 1.** Show that a σ-field cannot be countably infinite; i.e., show that if a σ-field contains an infinite number of sets, then it contains an uncountable number of sets.

*45. **This problem assumes you have read Note 1.**

   (a) Let $\mathscr{A}_\alpha$ be any indexed collection of σ-fields. Show that $\bigcap_\alpha \mathscr{A}_\alpha$ is also a σ-field.

   (b) Illustrate part (a) as follows. Let $\Omega := \{1,2,3,4\}$,
   $$\mathscr{A}_1 := \sigma(\{1\},\{2\},\{3,4\}) \quad \text{and} \quad \mathscr{A}_2 := \sigma(\{1\},\{3\},\{2,4\}).$$
   Find $\mathscr{A}_1 \cap \mathscr{A}_2$.

   (c) Let $\mathscr{C}$ be any collection of subsets of Ω, and let $\sigma(\mathscr{C})$ denote the smallest σ-field containing $\mathscr{C}$. Show that
   $$\sigma(\mathscr{C}) = \bigcap_{\mathscr{A}:\mathscr{C}\subset\mathscr{A}} \mathscr{A},$$
   where the intersection is over all σ-fields $\mathscr{A}$ that contain $\mathscr{C}$.

*46. **This problem assumes you have read Note 1.** Let Ω be a nonempty set, and let $\mathscr{F}$ and $\mathscr{G}$ be σ-fields. Is $\mathscr{F} \cup \mathscr{G}$ a σ-field? If "yes," prove it. If "no," give a counterexample.

*47. **This problem assumes you have read Note 1.** Let Ω denote the positive integers. Let $\mathscr{A}$ denote the collection of all subsets A such that either A is finite or A^c is finite.

   (a) Let E denote the positive integers that are even. Does E belong to $\mathscr{A}$?

   (b) Show that $\mathscr{A}$ is closed under finite unions. In other words, if $A_1,\ldots,A_n$ are in $\mathscr{A}$, show that $\bigcup_{i=1}^n A_i$ is also in $\mathscr{A}$.

   (c) Determine whether or not $\mathscr{A}$ is a σ-field.

*48. **This problem assumes you have read Note 1.** Let Ω be an uncountable set. Let $\mathscr{A}$ denote the collection of all subsets A such that either A is countable or A^c is countable. Determine whether or not $\mathscr{A}$ is a σ-field.

*49. **The Borel σ-field.** **This problem assumes you have read Note 1.** Let $\mathscr{B}$ denote the smallest σ-field containing all the open subsets of $\mathbb{R} := (-\infty,\infty)$. This collection $\mathscr{B}$ is called the **Borel σ-field**. The sets in $\mathscr{B}$ are called **Borel sets**. Hence, every open set, and every open interval, is a Borel set.

   (a) Show that every interval of the form $(a,b]$ is also a Borel set. *Hint:* Write $(a,b]$ as a countable intersection of open intervals and use the properties of a σ-field.

   (b) Show that every singleton set $\{a\}$ is a Borel set.

   (c) Let $a_1,a_2,\ldots$ be distinct real numbers. Put
   $$A := \bigcup_{k=1}^\infty \{a_k\}.$$
   Determine whether or not A is a Borel set.

(d) **Lebesgue measure** λ on the Borel subsets of $(0,1)$ is a probability measure that is completely characterized by the property that the Lebesgue measure of an open interval $(a,b) \subset (0,1)$ is its length; i.e., $\lambda((a,b)) = b-a$. Show that $\lambda((a,b])$ is also equal to $b-a$. Find $\lambda(\{a\})$ for any singleton set. If the set A in part (c) is a Borel set, compute $\lambda(A)$.

Remark. Note **5** in Chapter 5 contains more details on the construction of probability measures on the Borel subsets of $\mathbb{R}$.

*50. **The Borel σ-field, continued.** This problem assumes you have read Note 1.

Background: Recall that a set $U \subset \mathbb{R}$ is **open** if for every $x \in U$, there is a positive number ε_x, depending on x, such that $(x - \varepsilon_x, x + \varepsilon_x) \subset U$. Hence, an open set U can always be written in the form
$$U = \bigcup_{x \in U} (x - \varepsilon_x, x + \varepsilon_x).$$

Now observe that if $(x - \varepsilon_x, x + \varepsilon_x) \subset U$, we can find a rational number q_x close to x and a rational number $\rho_x < \varepsilon_x$ such that
$$x \in (q_x - \rho_x, q_x + \rho_x) \subset (x - \varepsilon_x, x + \varepsilon_x) \subset U.$$

Thus, every open set can be written in the form
$$U = \bigcup_{x \in U} (q_x - \rho_x, q_x + \rho_x),$$

where each q_x and each ρ_x is a rational number. Since the rational numbers form a countable set, there are only countably many such intervals with rational centers and rational lengths; hence, the union is really a countable one.

Problem: Show that the smallest σ-field containing all the open intervals is equal to the Borel σ-field defined in Problem 49.

1.5: Conditional probability

51. MATLAB. Save the following MATLAB script in an M-file to simulate chips from suppliers S1 and S2. Do not worry about how the script works. Run it, and based on your output, tell which supplier you think has more reliable chips.

```
% Chips from suppliers S1 and S2.
%
NOS1 = 983;        % Number of chips from S1
NOS2 = 871;        % Number of chips from S2
NOWS1 = sum(rand(1,NOS1) >= 0.2); NODS1 = NOS1-NOWS1;
NOWS2 = sum(rand(1,NOS2) >= 0.3); NODS2 = NOS2-NOWS2;
Nmat = [ NOWS1 NOWS2; NODS1 NODS2 ]
NOS  = [ NOS1 NOS2 ]
fprintf('Rel freq working chips from S1 is %4.2f.\n',...
   NOWS1/NOS1)
fprintf('Rel freq working chips from S2 is %4.2f.\n',...
   NOWS2/NOS2)
```

52. If
$$\frac{N(O_{d,S1})}{N(O_{S1})}, \quad N(O_{S1}), \quad \frac{N(O_{d,S2})}{N(O_{S2})}, \quad \text{and} \quad N(O_{S2})$$
are given, compute $N(O_{w,S1})$ and $N(O_{w,S2})$ in terms of them.

53. If $P(C)$ and $P(B \cap C)$ are positive, derive the **chain rule of conditional probability**,
$$P(A \cap B | C) = P(A | B \cap C) P(B | C).$$
Also show that
$$P(A \cap B \cap C) = P(A | B \cap C) P(B | C) P(C).$$

54. The university buys workstations from two different suppliers, Mini Micros (MM) and Highest Technology (HT). On delivery, 10% of MM's workstations are defective, while 20% of HT's workstations are defective. The university buys 140 MM workstations and 60 HT workstations for its computer lab. Suppose you walk into the computer lab and randomly sit down at a workstation.

   (a) What is the probability that your workstation is from MM? From HT?
   (b) What is the probability that your workstation is defective? *Answer:* 0.13.
   (c) Given that your workstation is defective, what is the probability that it came from Mini Micros? *Answer:* 7/13.

55. The probability that a cell in a wireless system is overloaded is $1/3$. Given that it is overloaded, the probability of a blocked call is 0.3. Given that it is not overloaded, the probability of a blocked call is 0.1. Find the conditional probability that the system is overloaded given that your call is blocked. *Answer:* 0.6.

56. The binary channel shown in Figure 1.17 operates as follows. Given that a 0 is transmitted, the conditional probability that a 1 is received is ε. Given that a 1 is transmitted, the conditional probability that a 0 is received is δ. Assume that the probability of transmitting a 0 is the same as the probability of transmitting a 1. Given that a 1 is received, find the conditional probability that a 1 was transmitted. *Hint:* Use the notation
$$T_i := \{i \text{ is transmitted}\}, \quad i = 0, 1,$$
and
$$R_j := \{j \text{ is received}\}, \quad j = 0, 1.$$

Remark. If $\delta = \varepsilon$, this channel is called the **binary symmetric channel**.

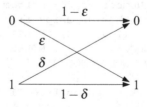

Figure 1.17. Binary channel with crossover probabilities ε and δ. If $\delta = \varepsilon$, this is called a binary symmetric channel.

57. Professor Random has taught probability for many years. She has found that 80% of students who do the homework pass the exam, while 10% of students who don't do the homework pass the exam. If 60% of the students do the homework, what percent of students pass the exam? Of students who pass the exam, what percent did the homework? *Answer:* 12/13.

58. A certain jet aircraft's autopilot has conditional probability 1/3 of failure given that it employs a faulty microprocessor chip. The autopilot has conditional probability 1/10 of failure given that it employs a nonfaulty chip. According to the chip manufacturer, the probability of a customer's receiving a faulty chip is 1/4. Given that an autopilot failure has occurred, find the conditional probability that a faulty chip was used. Use the following notation:

$$A_F = \{\text{autopilot fails}\}$$
$$C_F = \{\text{chip is faulty}\}.$$

Answer: 10/19.

59. Sue, Minnie, and Robin are medical assistants at a local clinic. Sue sees 20% of the patients, while Minnie and Robin each see 40% of the patients. Suppose that 60% of Sue's patients receive flu shots, while 30% of Minnie's patients receive flu shots and 10% of Robin's patients receive flu shots. Given that a patient receives a flu shot, find the conditional probability that Sue gave the shot. *Answer:* 3/7.

*60. You have five computer chips, two of which are known to be defective.

   (a) You test one of the chips; what is the probability that it is defective?

   (b) Your friend tests two chips at random and reports that one is defective and one is not. Given this information, you test one of the three remaining chips at random; what is the conditional probability that the chip you test is defective?

   (c) Consider the following modification of the preceding scenario. Your friend takes away two chips at random for testing; before your friend tells you the results, you test one of the three remaining chips at random; given this (lack of) information, what is the conditional probability that the chip you test is defective? Since you have not yet learned the results of your friend's tests, intuition suggests that your conditional probability should be the same as your answer to part (a). Is your intuition correct?

1.6: Independence

61. (a) If two sets A and B are disjoint, what equation must they satisfy?

    (b) If two events A and B are independent, what equation must they satisfy?

    (c) Suppose two events A and B are disjoint. Give conditions under which they are also independent. Give conditions under which they are not independent.

62. A certain binary communication system has a bit-error rate of 0.1; i.e., in transmitting a single bit, the probability of receiving the bit in error is 0.1. To transmit messages,

a three-bit repetition code is used. In other words, to send the message **1**, 111 is transmitted, and to send the message **0**, 000 is transmitted. At the receiver, if two or more 1s are received, the decoder decides that message **1** was sent; otherwise, i.e., if two or more zeros are received, it decides that message **0** was sent. Assuming bit errors occur independently, find the probability that the decoder puts out the wrong message. *Answer:* 0.028.

63. You and your neighbor attempt to use your cordless phones at the same time. Your phones independently select one of ten channels at random to connect to the base unit. What is the probability that both phones pick the same channel?

64. A new car is equipped with dual airbags. Suppose that they fail independently with probability p. What is the probability that at least one airbag functions properly?

65. A dart is repeatedly thrown at random toward a circular dartboard of radius 10 cm. Assume the thrower never misses the board. Let A_n denote the event that the dart lands within 2 cm of the center on the nth throw. Suppose that the A_n are mutually independent and that $P(A_n) = p$ for some $0 < p < 1$. Find the probability that the dart never lands within 2 cm of the center.

66. Each time you play the lottery, your probability of winning is p. You play the lottery n times, and plays are independent. How large should n be to make the probability of winning at least once more than $1/2$? *Answer:* For $p = 1/10^6$, $n \geq 693\,147$.

67. Anne and Betty go fishing. Find the conditional probability that Anne catches no fish given that at least one of them catches no fish. Assume they catch fish independently and that each has probability $0 < p < 1$ of catching no fish.

68. Suppose that A and B are independent events, and suppose that A and C are independent events. If $C \subset B$, determine whether or not A and $B \setminus C$ are independent.

69. Consider the sample space $\Omega = [0,1)$ equipped with the probability measure

$$P(A) := \int_A 1\,d\omega, \quad A \subset \Omega.$$

For $A = [0,1/2)$, $B = [0,1/4) \cup [1/2,3/4)$, and $C = [0,1/8) \cup [1/4,3/8) \cup [1/2,5/8) \cup [3/4,7/8)$, determine whether or not A, B, and C are mutually independent.

70. Given events A, B, and C, show that

$$P(A \cap C | B) = P(A|B)P(C|B)$$

if and only if

$$P(A|B \cap C) = P(A|B).$$

In this case, A and C are **conditionally independent** given B.

*71. **Second Borel–Cantelli lemma.** Show that if B_n is a sequence of *independent* events for which

$$\sum_{n=1}^{\infty} P(B_n) = \infty,$$

then
$$P\left(\bigcap_{n=1}^{\infty}\bigcup_{k=n}^{\infty} B_k\right) = 1.$$

Hint: The inequality $1 - P(B_k) \leq \exp[-P(B_k)]$ may be helpful.[k]

1.7: Combinatorics and probability

72. An electronics store carries three brands of computers, five brands of flat screens, and seven brands of printers. How many different systems (computer, flat screen, and printer) can the store sell?

73. If we use binary digits, how many n-bit numbers are there?

74. A certain Internet message consists of four header packets followed by 96 data packets. Unfortunately, a faulty router randomly re-orders all of the packets. What is the probability that the first header-type packet to be received is the 10th packet to arrive? *Answer:* 0.02996.

75. Joe has five cats and wants to have pictures taken of him holding one cat in each arm. How many pictures are needed so that every pair of cats appears in one picture? *Answer:* 10.

76. In a pick-4 lottery game, a player selects four digits, each one from $0, \ldots, 9$. If the four digits selected by the player match the random four digits of the lottery drawing in any order, the player wins. If the player has selected four distinct digits, what is the probability of winning? *Answer:* 0.0024.

77. How many 8-bit words are there with three ones (and five zeros)? *Answer:* 56.

78. A faulty computer memory location reads out random 8-bit bytes. What is the probability that a random word has four ones and four zeros? *Answer:* 0.2734.

79. Suppose 41 people enter a contest in which three winners are chosen at random. The first contestant chosen wins $500, the second contestant chosen wins $400, and the third contestant chosen wins $250. How many different outcomes are possible? If all three winners receive $250, how many different outcomes are possible? *Answers:* 63 960 and 10 660.

80. From a well-shuffled deck of 52 playing cards you are dealt 14 cards. What is the probability that two cards are spades, three are hearts, four are diamonds, and five are clubs? *Answer:* 0.0116.

81. From a well-shuffled deck of 52 playing cards you are dealt five cards. What is the probability that all five cards are of the same suit? *Answer:* 0.00198.

82. A finite set D of n elements is to be partitioned into m disjoint subsets, $D_1, \ldots, D_m$ in which $|D_i| = k_i$. How many different partitions are possible?

---

[k] The inequality $1 - x \leq e^{-x}$ for $x \geq 0$ can be derived by showing that the function $f(x) := e^{-x} - (1-x)$ satisfies $f(0) \geq 0$ and is nondecreasing for $x \geq 0$, e.g., its derivative, denoted by f', satisfies $f'(x) \geq 0$ for $x \geq 0$.

83. **m-ary pick-n lottery.** In this game, a player chooses n m-ary digits. In the lottery drawing, n m-ary digits are chosen at random. If the n digits selected by the player match the random n digits of the lottery drawing in any order, the player wins. If the player has selected n digits with k_0 zeros, k_1 ones, ..., and k_{m-1} copies of digit $m-1$, where $k_0 + \cdots + k_{m-1} = n$, what is the probability of winning? In the case of $n = 4$, $m = 10$, and a player's choice of the form xyz, what is the probability of winning; for $xxyy$; for $xxxy$? *Answers:* 0.0012, 0.0006, 0.0004.

84. In Example 1.46, what 7-bit sequence corresponds to two apples and three carrots? What sequence corresponds to two apples and three bananas? What sequence corresponds to five apples?

Exam preparation

You may use the following suggestions to prepare a study sheet, including formulas mentioned that you have trouble remembering. You may also want to ask your instructor for additional suggestions.

1.1. Sample spaces, outcomes, and events. Be able to suggest sample spaces to model simple systems with uncertain measurements. Know the difference between an outcome and an event. Understand the difference between the outcome ω, which is a point in the sample space, and the singleton event $\{\omega\}$, which is a subset of the sample space.

1.2. Review of set notation. Be familiar with set notation, operations, and identities. If required, be familiar with the precise definition of a function and the notions of countable and uncountable sets.

1.3. Probability models. Know how to construct and use probability models for simple problems.

1.4. Axioms and properties of probability. Know the axioms and properties of probability. Important formulas include (1.9) for disjoint unions, and (1.10)–(1.12). If required, understand and know how to use (1.13)–(1.17); in addition, your instructor may also require familiarity with Note **1** and related problems concerning σ-fields.

1.5. Conditional probability. What is important is the **law of total probability** (1.23) and *and being able to use it to solve problems.*

1.6. Independence. Do not confuse independent sets with disjoint sets. If $A_1, A_2, \ldots$ are independent, then so are $\tilde{A}_1, \tilde{A}_2, \ldots$, where each $\tilde{A}_i$ is either A_i or A_i^c.

1.7. Combinatorics and probability. The four kinds of counting problems are:

(i) ordered sampling of k out of n items with replacement: n^k;
(ii) ordered sampling of $k \leq n$ out of n items without replacement: $n!/(n-k)!$;
(iii) unordered sampling of $k \leq n$ out of n items without replacement: $\binom{n}{k}$; and
(iv) unordered sampling of k out of n items with replacement: $\binom{k+n-1}{k}$.

Know also the multinomial coefficient.

Work any review problems assigned by your instructor. If you finish them, re-work your homework assignments.

2
Introduction to discrete random variables

In most scientific and technological applications, measurements and observations are expressed as numerical quantities. Traditionally, numerical measurements or observations that have uncertain variability each time they are repeated are called **random variables**. We typically denote numerical-valued random quantities by uppercase letters such as X and Y. The advantage of working with numerical quantities is that we can perform mathematical operations on them such as

$$X+Y, \quad XY, \quad \max(X,Y), \quad \text{and} \quad \min(X,Y).$$

For example, in a telephone channel the signal X is corrupted by additive noise Y. In a wireless channel, the signal X is corrupted by fading (multiplicative noise). If X and Y are the traffic rates at two different routers of an Internet service provider, it is desirable to have these rates less than the router capacity, say c; i.e., we want $\max(X,Y) \leq c$. If X and Y are sensor voltages, we may want to trigger an alarm if at least one of the sensor voltages falls below a threshold v; e.g., if $\min(X,Y) \leq v$. See Figure 2.1.

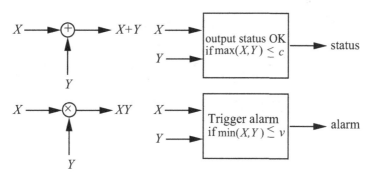

Figure 2.1. Systems represented by operations on random variables.

In order to exploit the axioms and properties of probability that we studied in Chapter 1, we technically define random variables as functions on an underlying sample space Ω. Fortunately, once some basic results are derived, we can think of random variables in the traditional manner, and not worry about, or even mention the underlying sample space.

This chapter introduces the student to random variables. Emphasis is on discrete random variables and basic concepts and tools for working with them, such as probability mass functions, expectation, and moments.

2.1 Probabilities involving random variables

A real-valued function $X(\omega)$ defined for points ω in a sample space Ω is called a **random variable**. Random variables are important because they provide a compact way of referring to events via their numerical attributes. For example, if X models the number of visits to a website, it is much easier to write $\mathsf{P}(X > 1000)$ than to write

$$\mathsf{P}(\text{number of visits} > 1000).$$

We now make this more precise and relate it all back to the properties of P that we developed in Chapter 1.

Example 2.1. Let us construct a model for counting the number of heads in a sequence of three coin tosses. For the underlying sample space, we take

$$\Omega := \{\text{TTT, TTH, THT, HTT, THH, HTH, HHT, HHH}\},$$

which contains the eight possible sequences of tosses. However, since we are only interested in the number of heads in each sequence, we define the random variable (function) X by

$$X(\omega) := \begin{cases} 0, & \omega = \text{TTT}, \\ 1, & \omega \in \{\text{TTH, THT, HTT}\}, \\ 2, & \omega \in \{\text{THH, HTH, HHT}\}, \\ 3, & \omega = \text{HHH}. \end{cases}$$

This is illustrated in Figure 2.2.

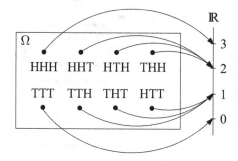

Figure **2.2.** Illustration of a random variable X that counts the number of heads in a sequence of three coin tosses.

With the setup of the previous example, let us assume for specificity that the sequences are equally likely. Now let us find the probability that the number of heads X is less than 2. In other words, we want to find $\mathsf{P}(X < 2)$. But what does this mean? Let us agree that $\mathsf{P}(X < 2)$ is shorthand for

$$\mathsf{P}(\{\omega \in \Omega : X(\omega) < 2\}).$$

Then the first step is to identify the event $\{\omega \in \Omega : X(\omega) < 2\}$. In Figure 2.2, the only lines pointing to numbers less than 2 are the lines pointing to 0 and 1. Tracing these lines backwards from $\mathbb{R}$ into Ω, we see that

$$\{\omega \in \Omega : X(\omega) < 2\} = \{\text{TTT, TTH, THT, HTT}\}.$$

2.1 Probabilities involving random variables

Since the sequences are equally likely,

$$P(\{TTT, TTH, THT, HTT\}) = \frac{|\{TTT, TTH, THT, HTT\}|}{|\Omega|}$$

$$= \frac{4}{8} = \frac{1}{2}.$$

Example 2.2. On the sample space Ω of the preceding example, define a random variable to describe the event that the number of heads in three tosses is even.

Solution. Define the random variable Y by

$$Y(\omega) := \begin{cases} 0, & \omega \in \{TTT, THH, HTH, HHT\}, \\ 1, & \omega \in \{TTH, THT, HTT, HHH\}. \end{cases}$$

Then $Y(\omega) = 0$ if the number of heads is even (0 or 2), and $Y(\omega) = 1$ if the number of heads is odd (1 or 3).

The probability that the number of heads is less than two and odd is $P(X < 2, Y = 1)$, by which we mean the probability of the event

$$\{\omega \in \Omega : X(\omega) < 2 \text{ and } Y(\omega) = 1\}.$$

This is equal to

$$\{\omega \in \Omega : X(\omega) < 2\} \cap \{\omega \in \Omega : Y(\omega) = 1\},$$

or just

$$\{TTT, TTH, THT, HTT\} \cap \{TTH, THT, HTT, HHH\},$$

which is equal to $\{TTH, THT, HTT\}$. The probability of this event, again assuming all sequences are equally likely, is $3/8$.

The shorthand introduced above is standard in probability theory. More generally, if $B \subset \mathbb{R}$, we use the shorthand

$$\{X \in B\} := \{\omega \in \Omega : X(\omega) \in B\}$$

and[1]

$$P(X \in B) := P(\{X \in B\}) = P(\{\omega \in \Omega : X(\omega) \in B\}).$$

If B is an interval such as $B = (a, b]$,

$$\{X \in (a, b]\} := \{a < X \leq b\} := \{\omega \in \Omega : a < X(\omega) \leq b\}$$

and

$$P(a < X \leq b) = P(\{\omega \in \Omega : a < X(\omega) \leq b\}).$$

Analogous notation applies to intervals such as $[a, b]$, $[a, b)$, (a, b), $(-\infty, b)$, $(-\infty, b]$, (a, ∞), and $[a, \infty)$.

***Example* 2.3.** A key step in manufacturing integrated circuits requires baking the chips in a special oven in a certain temperature range. Let T be a random variable modeling the oven temperature. Show that the probability the oven temperature is in the range $a < T \leq b$ can be expressed as
$$\mathsf{P}(a < T \leq b) = \mathsf{P}(T \leq b) - \mathsf{P}(T \leq a).$$

Solution. It is convenient to first rewrite the desired equation as
$$\mathsf{P}(T \leq b) = \mathsf{P}(T \leq a) + \mathsf{P}(a < T \leq b). \tag{2.1}$$

Now observe that
$$\{\omega \in \Omega : T(\omega) \leq b\} = \{\omega \in \Omega : T(\omega) \leq a\} \cup \{\omega \in \Omega : a < T(\omega) \leq b\}.$$

Since we cannot have an ω with $T(\omega) \leq a$ and $T(\omega) > a$ at the same time, the events in the union are disjoint. Taking probabilities of both sides yields (2.1).

If B is a singleton set, say $B = \{x_0\}$, we write $\{X = x_0\}$ instead of $\{X \in \{x_0\}\}$.

***Example* 2.4.** A computer has three disk drives numbered $0, 1, 2$. When the computer is booted, it randomly selects a drive to store temporary files on. If we model the selected drive number with the random variable X, show that the probability drive 0 or drive 1 is selected is given by
$$\mathsf{P}(X = 0 \text{ or } X = 1) = \mathsf{P}(X = 0) + \mathsf{P}(X = 1).$$

Solution. First note that the word "or" means "union." Hence, we are trying to find the probability of $\{X = 0\} \cup \{X = 1\}$. If we expand our shorthand, this union becomes
$$\{\omega \in \Omega : X(\omega) = 0\} \cup \{\omega \in \Omega : X(\omega) = 1\}.$$

Since we cannot have an ω with $X(\omega) = 0$ and $X(\omega) = 1$ at the same time, these events are disjoint. Hence, their probabilities add, and we obtain
$$\mathsf{P}(\{X = 0\} \cup \{X = 1\}) = \mathsf{P}(X = 0) + \mathsf{P}(X = 1). \tag{2.2}$$

2.2 Discrete random variables

We say X is a **discrete random variable** if there exist distinct real numbers x_i such that
$$\sum_i \mathsf{P}(X = x_i) = 1. \tag{2.3}$$

For discrete random variables, it can be shown using the law of total probability that[2]
$$\boxed{\mathsf{P}(X \in B) = \sum_{i : x_i \in B} \mathsf{P}(X = x_i).} \tag{2.4}$$

Integer-valued random variables

An **integer-valued random variable** is a discrete random variable whose distinct values are $x_i = i$. For integer-valued random variables,

$$P(X \in B) = \sum_{i \in B} P(X = i).$$

Here are some simple probability calculations involving integer-valued random variables.

$$P(X \leq 7) = \sum_{i \leq 7} P(X = i) = \sum_{i=-\infty}^{7} P(X = i).$$

Similarly,

$$P(X \geq 7) = \sum_{i \geq 7} P(X = i) = \sum_{i=7}^{\infty} P(X = i).$$

However,

$$P(X > 7) = \sum_{i > 7} P(X = i) = \sum_{i=8}^{\infty} P(X = i),$$

which is equal to $P(X \geq 8)$. Similarly

$$P(X < 7) = \sum_{i < 7} P(X = i) = \sum_{i=-\infty}^{6} P(X = i),$$

which is equal to $P(X \leq 6)$.

Probability mass functions

When X is a discrete random variable taking distinct values x_i, we define its **probability mass function** (pmf) by

$$\boxed{p_X(x_i) := P(X = x_i).}$$

Since $p_X(x_i)$ is a probability, it is a number satisfying

$$\boxed{0 \leq p_X(x_i) \leq 1,}$$

and, on account of (2.3),

$$\boxed{\sum_i p_X(x_i) = 1.}$$

Example 2.5. Let X be the random variable of Example 2.1. Assuming all sequences are equally likely, find the pmf of X.

Solution. From the definition of X or from Figure 2.2, we see that X takes the distinct values $0, 1, 2, 3$. Hence, we must compute $p_X(0)$, $p_X(1)$, $p_X(2)$, and $p_X(3)$. We begin with

$p_X(0) = \mathsf{P}(X = 0)$. The first step is to identify the outcomes ω that constitute the event $\{\omega : X(\omega) = 0\}$. From the definition of X or from Figure 2.2, we see that

$$\{\omega : X(\omega) = 0\} = \{\text{TTT}\}.$$

Hence,

$$p_X(0) = \mathsf{P}(X = 0) = \mathsf{P}(\{\text{TTT}\}) = \frac{|\{\text{TTT}\}|}{|\Omega|} = \frac{1}{8}.$$

Similarly,

$$p_X(1) = \mathsf{P}(X = 1) = \mathsf{P}(\{\text{HTT,THT,TTH}\}) = \frac{3}{8},$$

$$p_X(2) = \mathsf{P}(X = 2) = \mathsf{P}(\{\text{HHT,HTH,HHT}\}) = \frac{3}{8},$$

and

$$p_X(3) = \mathsf{P}(X = 3) = \mathsf{P}(\{\text{HHH}\}) = \frac{1}{8}.$$

This pmf is sketched in Figure 2.3.

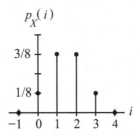

Figure 2.3. Pmf of the random variable X in Example 2.5.

Uniform random variables

When an experiment results in a finite number of "equally likely" or "totally random" outcomes, we model it with a **uniform** random variable. We say that X is uniformly distributed on $1,\ldots,n$ if

$$\mathsf{P}(X = k) = \frac{1}{n}, \quad k = 1,\ldots,n.$$

In other words, its pmf takes only two values:

$$p_X(k) = \begin{cases} 1/n, & k = 1,\ldots,n, \\ 0, & \text{otherwise.} \end{cases}$$

For example, to model the toss of a fair die we would use $p_X(k) = 1/6$ for $k = 1,\ldots,6$ sketched in Figure 2.4. To model the selection of one card from a well-shuffled deck of playing cards we would use $p_X(k) = 1/52$ for $k = 1,\ldots,52$. More generally, we can let k vary over any subset of n integers. A common alternative to $1,\ldots,n$ is $0,\ldots,n-1$. For k not in the range of experimental outcomes, $p_X(k) = 0$.

2.2 Discrete random variables

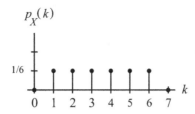

Figure 2.4. Uniform pmf on 1,2,3,4,5,6.

Example 2.6. Ten neighbors each have a cordless phone. The number of people using their cordless phones at the same time is totally random. Find the probability that more than half of the phones are in use at the same time.

Solution. We model the number of phones in use at the same time as a uniformly distributed random variable X taking values $0,\ldots,10$. Zero is included because we allow for the possibility that no phones are in use. We must compute

$$\mathsf{P}(X > 5) = \sum_{i=6}^{10} p_X(i) = \sum_{i=6}^{10} \frac{1}{11} = \frac{5}{11}.$$

If the preceding example had asked for the probability that at least half the phones are in use, then the answer would have been $\mathsf{P}(X \geq 5) = 6/11$.

The Poisson random variable

The **Poisson** random variable is used to model many different physical phenomena ranging from the photoelectric effect and radioactive decay[a] to computer message traffic arriving at a queue for transmission. A random variable X is said to have a Poisson probability mass function with parameter $\lambda > 0$, denoted by $X \sim \text{Poisson}(\lambda)$, if

$$p_X(k) = \frac{\lambda^k e^{-\lambda}}{k!}, \quad k = 0, 1, 2, \ldots.$$

A graph of $p_X(k)$ is shown in Figure 2.5 for $\lambda = 10, 30$, and 50. To see that these probabilities sum to one, recall that for any real or complex number z, the power series for e^z is

$$e^z = \sum_{k=0}^{\infty} \frac{z^k}{k!}.$$

Example 2.7. The number of hits to a popular website during a 1-minute interval is given by a Poisson(λ) random variable. Find the probability that there is at least one hit between 3:00 am and 3:01 am if $\lambda = 2$. Then find the probability that there are at least 2 hits during this time interval.

---

[a]The Poisson probability mass function arises naturally in this case, as shown in Example 3.7.

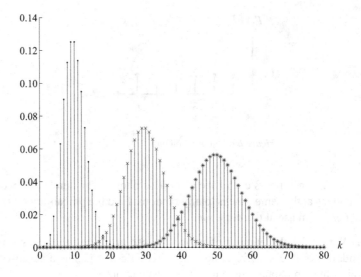

Figure 2.5. The Poisson(λ) pmf $p_X(k) = \lambda^k e^{-\lambda}/k!$ for $\lambda = 10, 30$, and 50 from left to right, respectively.

Solution. Let X denote the number of hits. Then
$$P(X \geq 1) = 1 - P(X = 0) = 1 - e^{-\lambda} = 1 - e^{-2} \approx 0.865.$$
Similarly,
$$\begin{aligned} P(X \geq 2) &= 1 - P(X = 0) - P(X = 1) \\ &= 1 - e^{-\lambda} - \lambda e^{-\lambda} \\ &= 1 - e^{-\lambda}(1 + \lambda) \\ &= 1 - e^{-2}(1 + 2) \approx 0.594. \end{aligned}$$

2.3 Multiple random variables

If X and Y are random variables, we use the shorthand
$$\{X \in B, Y \in C\} := \{\omega \in \Omega : X(\omega) \in B \text{ and } Y(\omega) \in C\},$$
which is equal to
$$\{\omega \in \Omega : X(\omega) \in B\} \cap \{\omega \in \Omega : Y(\omega) \in C\}.$$
Putting all of our shorthand together, we can write
$$\{X \in B, Y \in C\} = \{X \in B\} \cap \{Y \in C\}.$$
We also have
$$\begin{aligned} P(X \in B, Y \in C) &:= P(\{X \in B, Y \in C\}) \\ &= P(\{X \in B\} \cap \{Y \in C\}). \end{aligned}$$

2.3 Multiple random variables

Independence

If the events $\{X \in B\}$ and $\{Y \in C\}$ are independent for all sets B and C, we say that X and Y are **independent random variables**. In light of this definition and the above shorthand, we see that X and Y are independent random variables if and only if

$$P(X \in B, Y \in C) = P(X \in B)P(Y \in C) \qquad (2.5)$$

for all sets[3] B and C.

Example **2.8.** On a certain aircraft, the main control circuit on an autopilot fails with probability p. A redundant backup circuit fails independently with probability q. The aircraft can fly if at least one of the circuits is functioning. Find the probability that the aircraft cannot fly.

Solution. We introduce two random variables, X and Y. We set $X = 1$ if the main circuit fails, and $X = 0$ otherwise. We set $Y = 1$ if the backup circuit fails, and $Y = 0$ otherwise. Then $P(X = 1) = p$ and $P(Y = 1) = q$. We assume X and Y are independent random variables. Then the event that the aircraft cannot fly is modeled by

$$\{X = 1\} \cap \{Y = 1\}.$$

Using the independence of X and Y, $P(X = 1, Y = 1) = P(X = 1)P(Y = 1) = pq$.

The random variables X and Y of the preceding example are said to be **Bernoulli**. To indicate the relevant parameters, we write $X \sim \text{Bernoulli}(p)$ and $Y \sim \text{Bernoulli}(q)$. Bernoulli random variables are good for modeling the result of an experiment having two possible outcomes (numerically represented by 0 and 1), e.g., a coin toss, testing whether a certain block on a computer disk is bad, whether a new radar system detects a stealth aircraft, whether a certain Internet packet is dropped due to congestion at a router, etc. The Bernoulli(p) pmf is sketched in Figure 2.6.

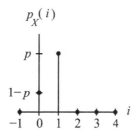

Figure 2.6. Bernoulli(p) probability mass function with $p > 1/2$.

Given any finite number of random variables, say $X_1,\ldots,X_n$, we say they are **independent** if

$$\mathsf{P}\left(\bigcap_{j=1}^n \{X_j \in B_j\}\right) = \prod_{j=1}^n \mathsf{P}(X_j \in B_j), \quad \text{for all choices of the sets } B_1,\ldots,B_n. \quad (2.6)$$

If $X_1,\ldots,X_n$ are independent, then so is any subset of them, e.g., X_1, X_3, and X_n.[4] If $X_1,X_2,\ldots$ is an infinite sequence of random variables, we say that they are independent if (2.6) holds for every finite $n = 1,2,\ldots$.

If for every B, $\mathsf{P}(X_j \in B)$ does not depend on j, then we say the X_j are **identically distributed**. If the X_j are both independent and identically distributed, we say they are **i.i.d.**

Example 2.9. Let X, Y, and Z be the number of hits at a website on three consecutive days. Assuming they are i.i.d. Poisson(λ) random variables, find the probability that on each day the number of hits is at most n.

Solution. The probability that on each day the number of hits is at most n is

$$\mathsf{P}(X \leq n, Y \leq n, Z \leq n).$$

By independence, this is equal to

$$\mathsf{P}(X \leq n)\mathsf{P}(Y \leq n)\mathsf{P}(Z \leq n).$$

Since the random variables are identically distributed, each factor has the same value. Since the random variables are Poisson(λ), each factor is equal to

$$\mathsf{P}(X \leq n) = \sum_{k=0}^n \mathsf{P}(X = k) = \sum_{k=0}^n \frac{\lambda^k}{k!}e^{-\lambda},$$

and so

$$\mathsf{P}(X \leq n, Y \leq n, Z \leq n) = \left(\sum_{k=0}^n \frac{\lambda^k}{k!}e^{-\lambda}\right)^3.$$

Example 2.10. A webpage server can handle r requests per day. Find the probability that the server gets more than r requests at least once in n days. Assume that the number of requests on day i is $X_i \sim$ Poisson(λ) and that $X_1,\ldots,X_n$ are independent.

Solution. We need to compute

$$\mathsf{P}\left(\bigcup_{i=1}^n \{X_i > r\}\right) = 1 - \mathsf{P}\left(\bigcap_{i=1}^n \{X_i \leq r\}\right)$$

$$= 1 - \prod_{i=1}^n \mathsf{P}(X_i \leq r)$$

$$= 1 - \prod_{i=1}^n \left(\sum_{k=0}^r \frac{\lambda^k e^{-\lambda}}{k!}\right)$$

$$= 1 - \left(\sum_{k=0}^r \frac{\lambda^k e^{-\lambda}}{k!}\right)^n.$$

Max and min problems

Calculations similar to those in the preceding example can be used to find probabilities involving the maximum or minimum of several independent random variables.

***Example* 2.11.** For $i = 1,\ldots,n$, let X_i model the yield on the ith production run of an integrated circuit manufacturer. Assume yields on different runs are independent. Find the probability that the highest yield obtained is less than or equal to z, and find the probability that the lowest yield obtained is less than or equal to z.

Solution. We must evaluate

$$P(\max(X_1,\ldots,X_n) \leq z) \quad \text{and} \quad P(\min(X_1,\ldots,X_n) \leq z).$$

Observe that $\max(X_1,\ldots,X_n) \leq z$ if and only if all of the X_k are less than or equal to z; i.e.,

$$\boxed{\{\max(X_1,\ldots,X_n) \leq z\} = \bigcap_{k=1}^{n} \{X_k \leq z\}.}$$

It then follows that

$$P(\max(X_1,\ldots,X_n) \leq z) = P\left(\bigcap_{k=1}^{n} \{X_k \leq z\}\right)$$
$$= \prod_{k=1}^{n} P(X_k \leq z),$$

where the second equation follows by independence.

For the min problem, observe that $\min(X_1,\ldots,X_n) \leq z$ if and only if at least one of the X_i is less than or equal to z; i.e.,

$$\boxed{\{\min(X_1,\ldots,X_n) \leq z\} = \bigcup_{k=1}^{n} \{X_k \leq z\}.}$$

Hence,

$$P(\min(X_1,\ldots,X_n) \leq z) = P\left(\bigcup_{k=1}^{n} \{X_k \leq z\}\right)$$
$$= 1 - P\left(\bigcap_{k=1}^{n} \{X_k > z\}\right)$$
$$= 1 - \prod_{k=1}^{n} P(X_k > z).$$

Geometric random variables

For $0 \le p < 1$, we define two kinds of **geometric** random variables.
We write $X \sim \text{geometric}_1(p)$ if

$$P(X = k) = (1-p)p^{k-1}, \quad k = 1, 2, \ldots.$$

As the example below shows, this kind of random variable arises when we ask how many times an experiment has to be performed until a certain outcome is observed.

We write $X \sim \text{geometric}_0(p)$ if

$$P(X = k) = (1-p)p^k, \quad k = 0, 1, \ldots.$$

This kind of random variable arises in Chapter 12 as the number of packets queued up at an idealized router with an infinite buffer. A plot of the $\text{geometric}_0(p)$ pmf is shown in Figure 2.7.

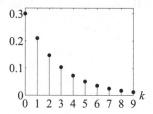

Figure 2.7. The $\text{geometric}_0(p)$ pmf $p_X(k) = (1-p)p^k$ with $p = 0.7$.

By the geometric series formula (Problem 27 in Chapter 1), it is easy to see that the probabilities of both kinds of random variable sum to one (Problem 16).

If we put $q = 1 - p$, then $0 < q \le 1$, and we can write $P(X = k) = q(1-q)^{k-1}$ in the $\text{geometric}_1(p)$ case and $P(X = k) = q(1-q)^k$ in the $\text{geometric}_0(p)$ case.

Example 2.12. When a certain computer accesses memory, the desired data is in the cache with probability p. Find the probability that the first cache miss occurs on the kth memory access. Assume presence in the cache of the requested data is independent for each access.

Solution. Let $T = k$ if the first time a cache miss occurs is on the kth memory access. For $i = 1, 2, \ldots$, let $X_i = 1$ if the ith memory request is in the cache, and let $X_i = 0$ otherwise. Then $P(X_i = 1) = p$ and $P(X_i = 0) = 1 - p$. The key observation is that the *first* cache miss occurs on the kth access if and only if the first $k-1$ accesses result in cache hits and the kth access results in a cache miss. In terms of events,

$$\{T = k\} = \{X_1 = 1\} \cap \cdots \cap \{X_{k-1} = 1\} \cap \{X_k = 0\}.$$

Since the X_i are independent, taking probabilities of both sides yields

$$\begin{aligned} P(T = k) &= P(\{X_1 = 1\} \cap \cdots \cap \{X_{k-1} = 1\} \cap \{X_k = 0\}) \\ &= P(X_1 = 1) \cdots P(X_{k-1} = 1) \cdot P(X_k = 0) \\ &= p^{k-1}(1-p). \end{aligned}$$

2.3 Multiple random variables

Example 2.13. In the preceding example, what is the probability that the first cache miss occurs after the third memory access?

Solution. We need to find

$$P(T > 3) = \sum_{k=4}^{\infty} P(T = k).$$

However, since $P(T = k) = 0$ for $k \leq 0$, a finite series is obtained by writing

$$P(T > 3) = 1 - P(T \leq 3)$$
$$= 1 - \sum_{k=1}^{3} P(T = k)$$
$$= 1 - (1-p)[1 + p + p^2].$$

Joint probability mass functions

The **joint probability mass function** of X and Y is defined by

$$p_{XY}(x_i, y_j) := P(X = x_i, Y = y_j). \quad (2.7)$$

An example for integer-valued random variables is sketched in Figure 2.8.

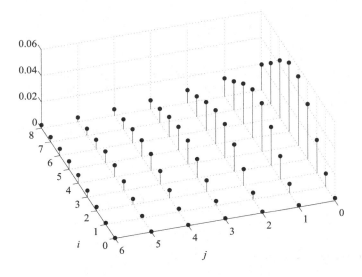

Figure 2.8. Sketch of bivariate probability mass function $p_{XY}(i, j)$.

It turns out that we can extract the **marginal probability mass functions** $p_X(x_i)$ and $p_Y(y_j)$ from the joint pmf $p_{XY}(x_i, y_j)$ using the formulas

$$p_X(x_i) = \sum_j p_{XY}(x_i, y_j) \quad (2.8)$$

and

$$p_Y(y_j) = \sum_i p_{XY}(x_i, y_j), \qquad (2.9)$$

which we derive later in the section.

Another important fact that we derive below is that a pair of discrete random variables is **independent** if and only if their joint pmf factors into the product of their marginal pmfs:

$$p_{XY}(x_i, y_j) = p_X(x_i) p_Y(y_j).$$

When X and Y take finitely many values, say $x_1, \ldots, x_m$ and $y_1, \ldots, y_n$, respectively, we can arrange the probabilities $p_{XY}(x_i, y_j)$ in the $m \times n$ matrix

$$\begin{bmatrix} p_{XY}(x_1,y_1) & p_{XY}(x_1,y_2) & \cdots & p_{XY}(x_1,y_n) \\ p_{XY}(x_2,y_1) & p_{XY}(x_2,y_2) & & p_{XY}(x_2,y_n) \\ \vdots & & \ddots & \vdots \\ p_{XY}(x_m,y_1) & p_{XY}(x_m,y_2) & \cdots & p_{XY}(x_m,y_n) \end{bmatrix}.$$

Notice that the sum of the entries in the top row is

$$\sum_{j=1}^n p_{XY}(x_1, y_j) = p_X(x_1).$$

In general, the sum of the entries in the ith row is $p_X(x_i)$, and the sum of the entries in the jth column is $p_Y(y_j)$. Since the sum of either marginal is one, it follows that the sum of all the entries in the matrix is one as well.

When X or Y takes infinitely many values, a little more thought is required.

Example 2.14. Find the marginal probability mass function $p_X(i)$ if

$$p_{XY}(i,j) := \begin{cases} 2\dfrac{[i/(i+1)]^j}{n(n+1)}, & j \geq 0, i = 0, \ldots, n-1, \\ 0, & \text{otherwise.} \end{cases}$$

Solution. For i in the range $0, \ldots, n-1$, write

$$\begin{aligned} p_X(i) &= \sum_{j=-\infty}^{\infty} p_{XY}(i,j) \\ &= \sum_{j=0}^{\infty} 2\frac{[i/(i+1)]^j}{n(n+1)} \\ &= \frac{2}{n(n+1)} \cdot \frac{1}{1 - i/(i+1)}, \end{aligned}$$

2.3 Multiple random variables

by the geometric series. This further simplifies to $2(i+1)/[n(n+1)]$. Thus,

$$p_X(i) = \begin{cases} 2\dfrac{i+1}{n(n+1)}, & i = 0, \ldots, n-1, \\ 0, & \text{otherwise.} \end{cases}$$

Remark. Since it is easily checked by induction that $\sum_{i=1}^n i = n(n+1)/2$, we can verify that $\sum_{i=0}^{n-1} p_X(i) = 1$.

*Derivation of marginal formulas (2.8) and (2.9)

Since the shorthand in (2.7) can be expanded to

$$p_{XY}(x_i, y_j) = \mathsf{P}(\{X = x_i\} \cap \{Y = y_j\}), \tag{2.10}$$

two applications of the law of total probability as in (1.27) can be used to show that[5]

$$\mathsf{P}(X \in B, Y \in C) = \sum_{i: x_i \in B} \sum_{j: y_j \in C} p_{XY}(x_i, y_j). \tag{2.11}$$

Let us now specialize (2.11) to the case that B is the singleton set $B = \{x_k\}$ and C is the biggest set possible, $C = \mathbb{R}$. Then (2.11) becomes

$$\mathsf{P}(X = x_k, Y \in \mathbb{R}) = \sum_{i: x_i = x_k} \sum_{j: y_j \in \mathbb{R}} p_{XY}(x_i, y_j).$$

To simplify the left-hand side, we use the fact that

$$\{Y \in \mathbb{R}\} := \{\omega \in \Omega : Y(\omega) \in \mathbb{R}\} = \Omega$$

to write

$$\mathsf{P}(X = x_k, Y \in \mathbb{R}) = \mathsf{P}(\{X = x_k\} \cap \Omega) = \mathsf{P}(X = x_k) = p_X(x_k).$$

To simplify the double sum on the right, note that the sum over i contains only one term, the term with $i = k$. Also, the sum over j is unrestricted. Putting this all together yields.

$$p_X(x_k) = \sum_j p_{XY}(x_k, y_j).$$

This is the same as (2.8) if we change k to i. Thus, the pmf of X can be recovered from the joint pmf of X and Y by summing over all values of Y. The derivation of (2.9) is similar.

*Joint PMFs and independence

Recall that X and Y are independent if

$$\mathsf{P}(X \in B, Y \in C) = \mathsf{P}(X \in B)\mathsf{P}(Y \in C) \tag{2.12}$$

for all sets B and C. In particular, taking $B = \{x_i\}$ and $C = \{y_j\}$ shows that

$$P(X = x_i, Y = y_j) = P(X = x_i)P(Y = y_j)$$

or, in terms of pmfs,

$$p_{XY}(x_i, y_j) = p_X(x_i) p_Y(y_j). \qquad (2.13)$$

We now show that the converse is also true; i.e., if (2.13) holds for all i and j, then (2.12) holds for all sets B and C. To see this, write

$$\begin{aligned} P(X \in B, Y \in C) &= \sum_{i:x_i \in B} \sum_{j:y_j \in C} p_{XY}(x_i, y_j), \quad \text{by (2.11)},\\ &= \sum_{i:x_i \in B} \sum_{j:y_j \in C} p_X(x_i) p_Y(y_j), \quad \text{by (2.13)},\\ &= \left[\sum_{i:x_i \in B} p_X(x_i)\right]\left[\sum_{j:y_j \in C} p_Y(y_j)\right]\\ &= P(X \in B)P(Y \in C). \end{aligned}$$

Computing probabilities with MATLAB

Example 2.15. If $X \sim \text{geometric}_0(p)$ with $p = 0.8$, compute the probability that X takes the value of an odd integer between 5 and 13.

Solution. We must compute

$$(1-p)[p^5 + p^7 + p^9 + p^{11} + p^{13}].$$

The straightforward solution is

```
p = 0.8;
s = 0;
for k = 5:2:13    % loop from 5 to 13 by steps of 2
    s = s + p^k;
end
fprintf('The answer is %g\n',(1-p)*s)
```

However, we can avoid using the `for` loop with the commands[b]

```
p = 0.8;
pvec = (1-p)*p.^[5:2:13];
fprintf('The answer is %g\n',sum(pvec))
```

The answer is 0.162. In this script, the expression `[5:2:13]` generates the vector [5 7 9 11 13]. Next, the "dot notation" `p.^[5 7 9 11 13]` means that MATLAB should do exponentiation on each component of the vector. In this case, MATLAB computes $[p^5 p^7 p^9 p^{11} p^{13}]$. Then each component of this vector is multiplied by the scalar $1-p$. This new vector is stored in `pvec`. Finally, the command `sum(pvec)` adds up the components of the vector.

---

[b]Because MATLAB programs are usually not compiled but run through the interpreter, loops require a lot of execution time. By using vectorized commands instead of loops, programs run much faster.

2.3 Multiple random variables

***Example* 2.16.** A light sensor uses a photodetector whose output is modeled as a Poisson(λ) random variable X. The sensor triggers an alarm if $X > 15$. If $\lambda = 10$, compute $P(X > 15)$.

Solution. First note that

$$P(X > 15) = 1 - P(X \leq 15) = 1 - e^{-\lambda}\left(1 + \lambda + \frac{\lambda^2}{2!} + \cdots + \frac{\lambda^{15}}{15!}\right).$$

Next, since $k! = \Gamma(k+1)$, where Γ is the **gamma function**, we can compute the required probability with the commands

```
lambda = 10;
k = [0:15];   % k = [ 0 1 2 ... 15 ]
pvec = exp(-lambda)*lambda.^k./gamma(k+1);
fprintf('The answer is %g\n',1-sum(pvec))
```

The answer is 0.0487. Note the operator ./ which computes the quotients of corresponding vector components; thus,

$$\begin{aligned}\text{pvec} &= e^{-\lambda}[\lambda^0\, \lambda^1\, \lambda^2 \cdots \lambda^{15}]./[0!\, 1!\, 2! \cdots 15!] \\ &= e^{-\lambda}\left[\frac{\lambda^0}{0!}\ \frac{\lambda^1}{1!}\ \frac{\lambda^2}{2!}\ \cdots\ \frac{\lambda^{15}}{15!}\right].\end{aligned}$$

We can use MATLAB for more sophisticated calculations such as $P(g(X) \leq y)$ in many cases in which X is a discrete random variable and $g(x)$ is a function that MATLAB can compute.

***Example* 2.17.** Let X be a uniform random variable on $0, \ldots, 100$. Assuming that $g(x) := \cos(2\pi x/10)$, compute $P(g(X) \leq 1/2)$.

Solution. This can be done with the simple script

```
p = ones(1,101)/101;  % p(i) = P(X=i) = 1/101, i = 0,...,100
k=[0:100];
i = find(cos(2*pi*k/10) <= 1/2);
fprintf('The answer is %g\n',sum(p(i)))
```

The answer is 0.693.

Remark. The MATLAB Statistics Toolbox provides commands for computing several probability mass functions. In particular, we could have used `geopdf(k,1-p)` for the geometric$_0(p)$ pmf and `poisspdf(k,lambda)` for the Poisson(λ) pmf.

We next use MATLAB for calculations involving pairs of random variables.

***Example* 2.18.** The input X and output Y of a system subject to random perturbations are described probabilistically by the joint pmf $p_{XY}(i,j)$, where $i = 1,2,3$ and $j = 1,2,3,4,5$. Let P denote the matrix whose ij entry is $p_{XY}(i,j)$, and suppose that

$$P = \frac{1}{71}\begin{bmatrix} 7 & 2 & 8 & 5 & 4 \\ 4 & 2 & 5 & 5 & 9 \\ 2 & 4 & 8 & 5 & 1 \end{bmatrix}.$$

Find the marginal pmfs $p_X(i)$ and $p_Y(j)$.

Solution. The marginal $p_Y(j)$ is obtained by summing the columns of the matrix. This is exactly what the MATLAB command sum does with a matrix. Thus, if P is already defined, the commands

```
format rat % print numbers as ratios of small integers
pY = sum(P)
```

yield

$$pY =$$
$$13/71 \quad 8/71 \quad 21/71 \quad 15/71 \quad 14/71$$

Similarly, the marginal $p_X(i)$ is obtained by summing the rows of P. Since sum computes column sums, the easy way around this is to use the transpose of P instead of P. The apostrophe ' is used to compute transposes. Hence, the command pX = sum(P')' computes column sums on the transpose of P, which yields a row vector; the second transpose operation converts the row into a column. We find that

$$pX =$$
$$26/71$$
$$25/71$$
$$20/71$$

Example 2.19. Let X and Y be as in the previous example, and let $g(x,y)$ be a given function. Find $P(g(X,Y) < 6)$.

Solution. The first step is to create a 3×5 matrix with entries $g(i,j)$. We then find those pairs (i,j) with $g(i,j) < 6$ and then sum the corresponding entries of P. Here is one way to do this, assuming P and the function g are already defined.

```
for i = 1:3
   for j = 1:5
      gmat(i,j) = g(i,j);
   end
end
prob = sum(P(find(gmat<6)))
```

If $g(x,y) = xy$, the answer is 34/71. A way of computing gmat without loops is given in the problems.

2.4 Expectation

The definition of expectation is motivated by the conventional idea of numerical average. Recall that the numerical average of n numbers, say $a_1, \ldots, a_n$, is

$$\frac{1}{n} \sum_{k=1}^{n} a_k.$$

We use the average to summarize or characterize the entire collection of numbers $a_1, \ldots, a_n$ with a single "typical" value.

2.4 Expectation

***Example* 2.20.** The average of the 10 numbers $5, 2, 3, 2, 5, -2, 3, 2, 5, 2$ is

$$\frac{5+2+3+2+5+(-2)+3+2+5+2}{10} = \frac{27}{10} = 2.7.$$

Notice that in our collection of numbers, -2 occurs once, 2 occurs four times, 3 occurs two times, and 5 occurs three times. In other words, their relative frequencies are

$$-2: 1/10$$
$$2: 4/10$$
$$3: 2/10$$
$$5: 3/10.$$

We can rewrite the average of the ten numbers in terms of their relative frequencies as

$$-2 \cdot \frac{1}{10} + 2 \cdot \frac{4}{10} + 3 \cdot \frac{2}{10} + 5 \cdot \frac{3}{10} = \frac{27}{10} = 2.7.$$

Since probabilities model relative frequencies, if X is a discrete random variable taking distinct values x_i with probabilities $\mathsf{P}(X = x_i)$, we define the **expectation** or **mean** of X by

$$\boxed{\mathsf{E}[X] := \sum_i x_i \mathsf{P}(X = x_i),}$$

or, using the pmf notation $p_X(x_i) = \mathsf{P}(X = x_i)$,

$$\mathsf{E}[X] = \sum_i x_i p_X(x_i).$$

***Example* 2.21.** Find the mean of a Bernoulli(p) random variable X.

Solution. Since X takes only the values $x_0 = 0$ and $x_1 = 1$, we can write

$$\mathsf{E}[X] = \sum_{i=0}^{1} i \mathsf{P}(X = i) = 0 \cdot (1-p) + 1 \cdot p = p.$$

Note that, since X takes only the values 0 and 1, its "typical" value p is never seen (unless $p = 0$ or $p = 1$).

***Example* 2.22.** When light of intensity λ is incident on a photodetector, the number of photoelectrons generated is Poisson with parameter λ. Find the mean number of photoelectrons generated.

Solution. Let X denote the number of photoelectrons generated. We need to calculate $\mathsf{E}[X]$. Since a Poisson random variable takes only nonnegative integer values with positive probability,

$$\mathsf{E}[X] = \sum_{n=0}^{\infty} n \mathsf{P}(X = n).$$

Since the term with $n = 0$ is zero, it can be dropped. Hence,

$$E[X] = \sum_{n=1}^{\infty} n \frac{\lambda^n e^{-\lambda}}{n!}$$

$$= \sum_{n=1}^{\infty} \frac{\lambda^n e^{-\lambda}}{(n-1)!}$$

$$= \lambda e^{-\lambda} \sum_{n=1}^{\infty} \frac{\lambda^{n-1}}{(n-1)!}.$$

Now change the index of summation from n to $k = n - 1$. This results in

$$E[X] = \lambda e^{-\lambda} \sum_{k=0}^{\infty} \frac{\lambda^k}{k!} = \lambda e^{-\lambda} e^{\lambda} = \lambda.$$

Example 2.23. If X is a discrete random variable taking finitely many values, say $x_1, \ldots, x_n$ with corresponding probabilities $p_1, \ldots, p_n$, then it is easy to compute $E[X]$ with MATLAB. If the value x_k is stored in x(k) and its probability p_k is stored in p(k), then $E[X]$ is given by x'*p, assuming both x and p are column vectors. If they are both row vectors, then the appropriate expression is x*p'.

Example 2.24 (infinite expectation). **Zipf** random variables arise in the analysis of website popularity and web caching. Here is an example of a Zipf random variable with infinite expectation. Suppose that $P(X = k) = C^{-1}/k^2$, $k = 1, 2, \ldots$, where[c]

$$C := \sum_{n=1}^{\infty} \frac{1}{n^2}.$$

Then

$$E[X] = \sum_{k=1}^{\infty} k P(X = k) = \sum_{k=1}^{\infty} k \frac{C^{-1}}{k^2} = C^{-1} \sum_{k=1}^{\infty} \frac{1}{k} = \infty$$

as shown in Problem 48.

Some care is necessary when computing expectations of signed random variables that take more than finitely many values. It is the convention in probability theory that $E[X]$ should be evaluated as

$$E[X] = \sum_{i: x_i \geq 0} x_i P(X = x_i) + \sum_{i: x_i < 0} x_i P(X = x_i),$$

assuming that at least one of these sums is finite. If the first sum is $+\infty$ and the second one is $-\infty$, then no value is assigned to $E[X]$, and we say that $E[X]$ is undefined.

---

[c]Note that C is finite by Problem 48. This is important since if $C = \infty$, then $C^{-1} = 0$ and the probabilities would sum to zero instead of one.

2.4 Expectation

Example 2.25 (undefined expectation). With C as in the previous example, suppose that for $k = 1, 2, \ldots$, $\mathsf{P}(X = k) = \mathsf{P}(X = -k) = \tfrac{1}{2}C^{-1}/k^2$. Then

$$\begin{aligned}
\mathsf{E}[X] &= \sum_{k=1}^{\infty} k\mathsf{P}(X=k) + \sum_{k=-\infty}^{-1} k\mathsf{P}(X=k) \\
&= \frac{1}{2C}\sum_{k=1}^{\infty}\frac{1}{k} + \frac{1}{2C}\sum_{k=-\infty}^{-1}\frac{1}{k} \\
&= \frac{\infty}{2C} + \frac{-\infty}{2C} \\
&= \text{undefined}.
\end{aligned}$$

Expectation of a function of a random variable, or the law of the unconscious statistician (LOTUS)

Given a random variable X, we will often have occasion to define a new random variable by $Z := g(X)$, where $g(x)$ is a real-valued function of the real variable x. More precisely, recall that a random variable X is actually a function taking points of the sample space, $\omega \in \Omega$, into real numbers $X(\omega)$. Hence, the notation $Z = g(X)$ is actually shorthand for $Z(\omega) := g(X(\omega))$. If we want to compute $\mathsf{E}[Z]$, it might seem that we first have to find the pmf of Z. Typically, this requires a detailed analysis of g. However, as we show below, we can compute $\mathsf{E}[Z] = \mathsf{E}[g(X)]$ without actually finding the pmf of Z. The precise formula is

$$\mathsf{E}[g(X)] = \sum_i g(x_i)\, p_X(x_i). \tag{2.14}$$

Because it is so much easier to use (2.14) than to first find the pmf of Z, (2.14) is sometimes called the **law of the unconscious statistician** (LOTUS) [23]. As a simple example of its use, we can write, for any constant a,

$$\mathsf{E}[aX] = \sum_i ax_i\, p_X(x_i) = a\sum_i x_i\, p_X(x_i) = a\mathsf{E}[X].$$

In other words, constant factors can be pulled out of the expectation; technically, we say that expectation is a **homogeneous operator**. As we show later, expectation is also additive. An operator that is both homogeneous and additive is said to be linear. Thus, expectation is a linear operator.

*Derivation of LOTUS

To derive (2.14), we proceed as follows. Let X take distinct values x_i. Then Z takes values $g(x_i)$. However, the values $g(x_i)$ may not be distinct. For example, if $g(x) = x^2$, and X takes the four distinct values ± 1 and ± 2, then Z takes only the two distinct values 1 and 4. In any case, let z_k denote the distinct values of Z and observe that

$$\mathsf{P}(Z = z_k) = \sum_{i:g(x_i)=z_k} \mathsf{P}(X = x_i).$$

We can now write

$$E[Z] = \sum_k z_k P(Z = z_k)$$
$$= \sum_k z_k \left(\sum_{i:g(x_i)=z_k} P(X = x_i) \right)$$
$$= \sum_k \left(\sum_{i:g(x_i)=z_k} z_k P(X = x_i) \right)$$
$$= \sum_k \left(\sum_{i:g(x_i)=z_k} g(x_i) P(X = x_i) \right)$$
$$= \sum_i g(x_i) P(X = x_i),$$

since the last double sum is just a special way of summing over all values of i.

Linearity of expectation

The derivation of the law of the unconscious statistician can be generalized to show that if $g(x,y)$ is a real-valued function of two variables x and y, then

$$\boxed{E[g(X,Y)] = \sum_i \sum_j g(x_i, y_j) p_{XY}(x_i, y_j).}$$

In particular, taking $g(x,y) = x+y$, it is a simple exercise to show that $E[X+Y] = E[X] + E[Y]$. Thus, expectation is an **additive operator**. Since we showed earlier that expectation is also homogeneous, it follows that expectation is **linear**; i.e., for constants a and b,

$$E[aX + bY] = E[aX] + E[bY] = aE[X] + bE[Y]. \tag{2.15}$$

Example 2.26. A binary communication link has bit-error probability p. What is the expected number of bit errors in a transmission of n bits?

Solution. For $i = 1, \ldots, n$, let $X_i = 1$ if the ith bit is received incorrectly, and let $X_i = 0$ otherwise. Then $X_i \sim \text{Bernoulli}(p)$, and $Y := X_1 + \cdots + X_n$ is the total number of errors in the transmission of n bits. We know from Example 2.21 that $E[X_i] = p$. Hence,

$$E[Y] = E\left[\sum_{i=1}^n X_i\right] = \sum_{i=1}^n E[X_i] = \sum_{i=1}^n p = np.$$

Moments

The nth **moment**, $n \geq 1$, of a real-valued random variable X is defined to be $E[X^n]$. The first moment of X is its mean, $E[X]$. Letting $m = E[X]$, we define the **variance** of X by

$$\boxed{\text{var}(X) := E[(X-m)^2].} \tag{2.16}$$

2.4 Expectation

The variance is the average squared deviation of X about its mean. The variance characterizes how likely it is to observe values of the random variable far from its mean. For example, consider the two pmfs shown in Figure 2.9. More probability mass is concentrated near zero in the graph at the left than in the graph at the right.

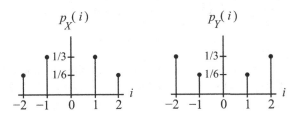

Figure 2.9. Example 2.27 shows that the random variable X with pmf at the left has a smaller variance than the random variable Y with pmf at the right.

Example 2.27. Let X and Y be the random variables with respective pmfs shown in Figure 2.9. Compute $\text{var}(X)$ and $\text{var}(Y)$.

Solution. By symmetry, both X and Y have zero mean, and so $\text{var}(X) = \mathsf{E}[X^2]$ and $\text{var}(Y) = \mathsf{E}[Y^2]$. Write

$$\mathsf{E}[X^2] = (-2)^2 \tfrac{1}{6} + (-1)^2 \tfrac{1}{3} + (1)^2 \tfrac{1}{3} + (2)^2 \tfrac{1}{6} = 2,$$

and

$$\mathsf{E}[Y^2] = (-2)^2 \tfrac{1}{3} + (-1)^2 \tfrac{1}{6} + (1)^2 \tfrac{1}{6} + (2)^2 \tfrac{1}{3} = 3.$$

Thus, X and Y are both zero-mean random variables taking the values ± 1 and ± 2. But Y is more likely to take values far from its mean. This is reflected by the fact that $\text{var}(Y) > \text{var}(X)$.

When a random variable does not have zero mean, it is often convenient to use the **variance formula**,

$$\text{var}(X) = \mathsf{E}[X^2] - (\mathsf{E}[X])^2, \tag{2.17}$$

which says that the variance is equal to the second moment minus the square of the first moment. To derive the variance formula, write

$$\begin{aligned}\text{var}(X) &:= \mathsf{E}[(X-m)^2] \\ &= \mathsf{E}[X^2 - 2mX + m^2] \\ &= \mathsf{E}[X^2] - 2m\mathsf{E}[X] + m^2, \quad \text{by linearity,} \\ &= \mathsf{E}[X^2] - m^2 \\ &= \mathsf{E}[X^2] - (\mathsf{E}[X])^2.\end{aligned}$$

The **standard deviation** of X is defined to be the positive square root of the variance. Since the variance of a random variable is often denoted by the symbol σ^2, the standard deviation is denoted by σ.

Example 2.28. Find the second moment and the variance of X if $X \sim$ Bernoulli(p).

Solution. Since X takes only the values 0 and 1, it has the unusual property that $X^2 = X$. Hence, $\mathsf{E}[X^2] = \mathsf{E}[X] = p$. It now follows that

$$\text{var}(X) = \mathsf{E}[X^2] - (\mathsf{E}[X])^2 = p - p^2 = p(1-p).$$

Example 2.29. An optical communication system employs a photodetector whose output is modeled as a Poisson(λ) random variable X. Find the second moment and the variance of X.

Solution. Observe that $\mathsf{E}[X(X-1)] + \mathsf{E}[X] = \mathsf{E}[X^2]$. Since we know that $\mathsf{E}[X] = \lambda$ from Example 2.22, it suffices to compute

$$\mathsf{E}[X(X-1)] = \sum_{n=0}^{\infty} n(n-1) \frac{\lambda^n e^{-\lambda}}{n!}$$

$$= \sum_{n=2}^{\infty} \frac{\lambda^n e^{-\lambda}}{(n-2)!}$$

$$= \lambda^2 e^{-\lambda} \sum_{n=2}^{\infty} \frac{\lambda^{n-2}}{(n-2)!}.$$

Making the change of summation $k = n - 2$, we have

$$\mathsf{E}[X(X-1)] = \lambda^2 e^{-\lambda} \sum_{k=0}^{\infty} \frac{\lambda^k}{k!}$$

$$= \lambda^2.$$

It follows that $\mathsf{E}[X^2] = \lambda^2 + \lambda$, and

$$\text{var}(X) = \mathsf{E}[X^2] - (\mathsf{E}[X])^2 = (\lambda^2 + \lambda) - \lambda^2 = \lambda.$$

Thus, the Poisson(λ) random variable is unusual in that the values of its mean and variance are the same.

A generalization of the variance is the nth **central moment** of X, which is defined to be $\mathsf{E}[(X-m)^n]$. Hence, the second central moment is the variance. If $\sigma^2 := \text{var}(X)$, then the **skewness** of X is defined to be $\mathsf{E}[(X-m)^3]/\sigma^3$, and the **kurtosis** of X is defined to be $\mathsf{E}[(X-m)^4]/\sigma^4$.

Example 2.30. If X has mean m and variance σ^2, it is sometimes convenient to introduce the normalized random variable

$$Y := \frac{X-m}{\sigma}.$$

2.4 Expectation

It is easy to see that Y has zero mean. Hence,

$$\operatorname{var}(Y) = \mathsf{E}[Y^2] = \mathsf{E}\left[\left(\frac{X-m}{\sigma}\right)^2\right] = \frac{\mathsf{E}[(X-m)^2]}{\sigma^2} = 1.$$

Thus, Y always has zero mean and unit variance. Furthermore, the third moment of Y is

$$\mathsf{E}[Y^3] = \mathsf{E}\left[\left(\frac{X-m}{\sigma}\right)^3\right] = \frac{\mathsf{E}[(X-m)^3]}{\sigma^3},$$

which is the skewness of X, and similarly, the fourth moment of Y is the kurtosis of X.

Indicator functions

Given a set $B \subset \mathbb{R}$, there is a very special function of x, denoted by $I_B(x)$, for which we will be interested in computing $\mathsf{E}[I_B(X)]$ for various random variables X. The **indicator function** of B, denoted by $I_B(x)$, is defined by

$$I_B(x) := \begin{cases} 1, & x \in B, \\ 0, & x \notin B. \end{cases}$$

For example $I_{[a,b)}(x)$ is shown in Figure 2.10(a), and $I_{(a,b]}(x)$ is shown in Figure 2.10(b).

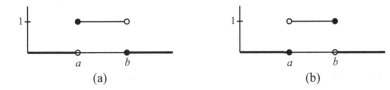

Figure 2.10. (a) Indicator function $I_{[a,b)}(x)$. (b) Indicator function $I_{(a,b]}(x)$.

Readers familiar with the **unit-step function**,

$$u(x) := \begin{cases} 1, & x \geq 0, \\ 0, & x < 0, \end{cases}$$

will note that $u(x) = I_{[0,\infty)}(x)$. However, the indicator notation is often more compact. For example, if $a < b$, it is easier to write $I_{[a,b)}(x)$ than $u(x-a) - u(x-b)$. How would you write $I_{(a,b]}(x)$ in terms of the unit step?

Example 2.31 (every probability is an expectation). If X is any random variable and B is any set, then $I_B(X)$ is a discrete random variable taking the values zero and one. Thus, $I_B(X)$ is a Bernoulli random variable, and

$$\mathsf{E}[I_B(X)] = \mathsf{P}(I_B(X) = 1) = \mathsf{P}(X \in B).$$

We also point out that if X is a discrete random variable taking distinct values x_i, then

$$\sum_i I_B(x_i) \mathsf{P}(X = x_i) = \sum_{i:x_i \in B} \mathsf{P}(X = x_i) = \mathsf{P}(X \in B).$$

The advantage of using the left-hand expression is that the summation is over all i and is not restricted by the set B. This can be useful when we want to interchange the order of summation in more complicated expressions.

The Markov and Chebyshev inequalities

Many practical applications require knowledge of probabilities of the form $\mathsf{P}(X \geq a)$ for some threshold a. For example, an optical communication receiver may decide an incoming bit is a 1 if the number X of photoelectrons detected exceeds a and decide 0 otherwise. Or a could be the buffer size of an Internet router, and packets would be dropped if the packet backlog X exceeds a. Unfortunately, it is often difficult to compute such probabilities when X is the output of a complicated system. The Markov and Chebyshev inequalities provide bounds on probabilities in terms of expectations that are more readily computable.

The **Markov inequality** says that if X is a nonnegative random variable, then for any $a > 0$,

$$\mathsf{P}(X \geq a) \leq \frac{\mathsf{E}[X]}{a}. \qquad (2.18)$$

Since we always have $\mathsf{P}(X \geq a) \leq 1$, the Markov inequality is useful only when the right-hand side is less than one. Following the next two examples, we derive the inequality

$$\mathsf{P}(X \geq a) \leq \frac{\mathsf{E}[X^r]}{a^r}, \quad r > 0. \qquad (2.19)$$

Taking $r = 1$ yields the Markov inequality.[d]

***Example* 2.32.** A cellular company study shows that the expected number of simultaneous calls at a base station is $C_{\text{avg}} = 100$. However, since the actual number of calls is random, the station is designed to handle up to $C_{\text{max}} = 150$ calls. Use the Markov inequality to bound the probability that the station receives more than $C_{\text{max}} = 150$ calls.

Solution. Let X denote the actual number of calls. Then $\mathsf{E}[X] = C_{\text{avg}}$. By the Markov inequality,

$$\mathsf{P}(X > 150) = \mathsf{P}(X \geq 151) \leq \frac{\mathsf{E}[X]}{151} = \frac{C_{\text{avg}}}{151} = 0.662.$$

***Example* 2.33.** In the preceding example, suppose you are given the additional information that the variance of the number of calls is 50. Can you give a better bound on the probability that the base station receives more than $C_{\text{max}} = 150$ calls?

Solution. This time we use the more general result (2.19) with $r = 2$ to write

$$\mathsf{P}(X \geq 151) \leq \frac{\mathsf{E}[X^2]}{151^2} = \frac{\text{var}(X) + (\mathsf{E}[X])^2}{22\,801} = \frac{50 + 100^2}{22\,801} = \frac{10\,050}{22\,801} = 0.441.$$

---

[d]We have derived (2.18) from (2.19). It is also possible to derive (2.19) from (2.18). Since $\{X \geq a\} = \{X^r \geq a^r\}$, write $\mathsf{P}(X \geq a) = \mathsf{P}(X^r \geq a^r) \leq \mathsf{E}[X^r]/a^r$ by (2.18).

*Derivation of (2.19)

We now derive (2.19) using the following two key ideas. First, since every probability can be written as an expectation,

$$\mathsf{P}(X \geq a) = \mathsf{E}[I_{[a,\infty)}(X)]. \tag{2.20}$$

Second, from Figure 2.11, we see that for $x \geq 0$, $I_{[a,\infty)}(x)$ (solid line) is less than or equal to

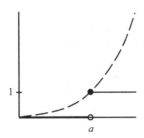

Figure 2.11. Graph showing that $I_{[a,\infty)}(x)$ (solid line) is upper bounded by $(x/a)^r$ (dashed line) for any positive r.

$(x/a)^r$ (dashed line). Since X is a nonnegative random variable,

$$I_{[a,\infty)}(X) \leq (X/a)^r.$$

Now take expectations of both sides to obtain

$$\mathsf{E}[I_{[a,\infty)}(X)] \leq \mathsf{E}[X^r]/a^r.$$

Combining this with (2.20) yields (2.19).

The **Chebyshev inequality** says that for any random variable Y and any $a > 0$,

$$\boxed{\mathsf{P}(|Y| \geq a) \leq \frac{\mathsf{E}[Y^2]}{a^2}.} \tag{2.21}$$

This is an easy consequence of (2.19). As in the case of the Markov inequality, it is useful only when the right-hand side is less than one. To derive the Chebyshev inequality, take $X = |Y|$ and $r = 2$ in (2.19) to get

$$\mathsf{P}(|Y| \geq a) \leq \frac{\mathsf{E}[|Y|^2]}{a^2} = \frac{\mathsf{E}[Y^2]}{a^2}.$$

The following special cases of the Chebyshev inequality are sometimes of interest. If $m := \mathsf{E}[X]$ is finite, then taking $Y = X - m$ yields

$$\boxed{\mathsf{P}(|X - m| \geq a) \leq \frac{\mathrm{var}(X)}{a^2}.} \tag{2.22}$$

If $\sigma^2 := \mathrm{var}(X)$ is also finite, taking $a = k\sigma$ yields

$$\mathsf{P}(|X - m| \geq k\sigma) \leq \frac{1}{k^2}.$$

These two inequalities give bounds on the probability that X is far from its mean value.

We will be using the Chebyshev inequality (2.22) in Section 3.3 to derive the weak law of large numbers.

Example 2.34. A circuit is designed to handle a nominal current of 20 mA plus or minus a deviation of less than 5 mA. If the applied current has mean 20 mA and variance 4 mA^2, use the Chebyshev inequality to bound the probability that the applied current violates the design parameters.

Solution. Let X denote the applied current. Then X is within the design parameters if and only if $|X - 20| < 5$. To bound the probability that this does not happen, write

$$P(|X - 20| \geq 5) \leq \frac{\text{var}(X)}{5^2} = \frac{4}{25} = 0.16.$$

Hence, the probability of violating the design parameters is at most 16%.

Expectations of products of functions of independent random variables

We show that X and Y are independent if and only if

$$E[h(X)k(Y)] = E[h(X)]E[k(Y)] \quad (2.23)$$

for all functions $h(x)$ and $k(y)$. In other words, X and Y are independent if and only if for every pair of functions $h(x)$ and $k(y)$, the expectation of the product $h(X)k(Y)$ is equal to the product of the individual expectations.[6]

There are two claims to be established. We must show that if (2.23) holds for every pair of functions $h(x)$ and $k(y)$, then X and Y are independent, and we must show that if X and Y are independent, then (2.23) holds for every pair of functions $h(x)$ and $k(y)$.

The first claim is easy to show by taking $h(x) = I_B(x)$ and $k(y) = I_C(y)$ for any sets B and C. Then (2.23) becomes

$$E[I_B(X)I_C(Y)] = E[I_B(X)]E[I_C(Y)]$$
$$= P(X \in B)P(Y \in C).$$

Since $I_B(X)I_C(Y) = 1$ if and only if $X \in B$ and $Y \in C$, the left-hand side is simply $P(X \in B, Y \in C)$. It then follows that

$$P(X \in B, Y \in C) = P(X \in B)P(Y \in C),$$

which is the definition of independence.

To derive the second claim, we use the fact that

$$p_{XY}(x_i, y_j) := P(X = x_i, Y = y_j)$$
$$= P(X = x_i)P(Y = y_j), \text{ by independence,}$$
$$= p_X(x_i)p_Y(y_j).$$

2.4 Expectation

Now write

$$\begin{aligned}
\mathsf{E}[h(X)k(Y)] &= \sum_i \sum_j h(x_i) k(y_j) p_{XY}(x_i, y_j) \\
&= \sum_i \sum_j h(x_i) k(y_j) p_X(x_i) p_Y(y_j) \\
&= \sum_i h(x_i) p_X(x_i) \left[\sum_j k(y_j) p_Y(y_j) \right] \\
&= \mathsf{E}[h(X)] \mathsf{E}[k(Y)].
\end{aligned}$$

Example 2.35. Let X and Y be independent random variables with $X \sim \text{Poisson}(\lambda)$ and $Y \sim \text{Poisson}(\mu)$. Find $\mathsf{E}[XY^2]$.

Solution. By independence, $\mathsf{E}[XY^2] = \mathsf{E}[X]\mathsf{E}[Y^2]$. From Example 2.22, $\mathsf{E}[X] = \lambda$, and from Example 2.29, $\mathsf{E}[Y^2] = \mu^2 + \mu$. Hence, $\mathsf{E}[XY^2] = \lambda(\mu^2 + \mu)$.

Correlation and covariance

The **correlation** between two random variables X and Y is defined to be $\mathsf{E}[XY]$. The correlation is important because it determines when two random variables are linearly related; namely, when one is a linear function of the other.

Example 2.36. Let X have zero mean and unit variance, and put $Y := 3X$. Find the correlation between X and Y.

Solution. First note that since X has zero mean, $\mathsf{E}[X^2] = \text{var}(X) = 1$. Then write $\mathsf{E}[XY] = \mathsf{E}[X(3X)] = 3\mathsf{E}[X^2] = 3$. If we had put $Y := -3X$, then $\mathsf{E}[XY] = -3$.

Example 2.37. The input X and output Y of a system subject to random perturbations are described probabilistically by the joint pmf $p_{XY}(i,j)$, where $i = 1, 2, 3$ and $j = 1, 2, 3, 4, 5$. Let P denote the matrix whose ij entry is $p_{XY}(i,j)$, and suppose that

$$P = \frac{1}{71} \begin{bmatrix} 7 & 2 & 8 & 5 & 4 \\ 4 & 2 & 5 & 5 & 9 \\ 2 & 4 & 8 & 5 & 1 \end{bmatrix}.$$

Use MATLAB to compute the correlation $\mathsf{E}[XY]$.

Solution. Assuming P is already defined, we use the script

```
s = 0;
for i = 1:3
   for j = 1:5
      s = s + i*j*P(i,j);
   end
end
[n,d] = rat(s); % to express answer fraction
fprintf('E[XY] = %i/%i = %g\n',n,d,s)
```

and we find that $E[XY] = 428/71 = 6.02817$.

An important property of correlation is the **Cauchy–Schwarz inequality**, which says that
$$|E[XY]| \leq \sqrt{E[X^2]E[Y^2]}, \qquad (2.24)$$
where equality holds if and only if X and Y are linearly related. This result provides an important bound on the correlation between two random variables.

To derive (2.24), let λ be any constant and write
$$\begin{aligned} 0 &\leq E[(X - \lambda Y)^2] \\ &= E[X^2 - 2\lambda XY + \lambda^2 Y^2] \\ &= E[X^2] - 2\lambda E[XY] + \lambda^2 E[Y^2]. \end{aligned} \qquad (2.25)$$

To make further progress, take
$$\lambda = \frac{E[XY]}{E[Y^2]}.$$

Then
$$\begin{aligned} 0 &\leq E[X^2] - 2\frac{E[XY]^2}{E[Y^2]} + \frac{E[XY]^2}{E[Y^2]^2}E[Y^2] \\ &= E[X^2] - \frac{E[XY]^2}{E[Y^2]}. \end{aligned}$$

This can be rearranged to get
$$E[XY]^2 \leq E[X^2]E[Y^2]. \qquad (2.26)$$

Taking square roots yields (2.24). We can also show that if (2.24) holds with equality, then X and Y are linearly related. If (2.24) holds with equality, then so does (2.26). Since the steps leading from (2.25) to (2.26) are reversible, it follows that (2.25) must hold with equality. But $E[(X - \lambda Y)^2] = 0$ implies $X = \lambda Y$.[7]

When X and Y have different means and variances, say $m_X := E[X]$, $m_Y := E[Y]$, $\sigma_X^2 := \text{var}(X)$ and $\sigma_Y^2 := \text{var}(Y)$, we sometimes look at the correlation between the "normalized" random variables
$$\frac{X - m_X}{\sigma_X} \quad \text{and} \quad \frac{Y - m_Y}{\sigma_Y},$$
which each have zero mean and unit variance.

The **correlation coefficient** of random variables X and Y is defined to be the correlation of their normalized versions,
$$\rho_{XY} := E\left[\left(\frac{X - m_X}{\sigma_X}\right)\left(\frac{Y - m_Y}{\sigma_Y}\right)\right].$$

2.4 Expectation

Furthermore, $|\rho_{XY}| \leq 1$, with equality if and only if X and Y are related by a linear function plus a constant. A pair of random variables is said to be **uncorrelated** if their correlation coefficient is zero.

Example 2.38. For the random variables X and Y of Example 2.37, use MATLAB to compute ρ_{XY}.

Solution. First note that the formula for ρ_{XY} can be expanded as

$$\rho_{XY} = \frac{\mathsf{E}[XY] - m_X m_Y}{\sigma_X \sigma_Y}.$$

Next, except for the term $\mathsf{E}[XY]$, the remaining quantities can be computed using marginal pmfs, which can be computed easily with the sum command as done in Example 2.18. Since $\mathsf{E}[XY]$ was computed in Example 2.37 and was called s, the following additional script will compute rhoxy.

```
format rat
pY = sum(P)
y = [ 1 2 3 4 5 ]
mY = y*pY'
varY = ((y-mY).^2)*pY'
pX = sum(P')
x = [ 1 2 3 ]
mX = x*pX'
varX = ((x-mX).^2)*pX'
rhoxy = (s-mX*mY)/sqrt(varX*varY)
```

We find that $m_X = 136/71$, $m_Y = 222/71$, $\text{var}(X) = 412/643$, $\text{var}(Y) = 1337/731$, and $\rho_{XY} = 286/7963 = 0.0359161$.

Example 2.39. If X and Y are zero mean, then $\sigma_X^2 = \mathsf{E}[X^2]$ and $\sigma_Y^2 = \mathsf{E}[Y^2]$. It now follows that

$$\rho_{XY} = \frac{\mathsf{E}[XY]}{\sqrt{\mathsf{E}[X^2]\mathsf{E}[Y^2]}}.$$

Example 2.40. Let U, W_1, and W_2 be independent with zero means. Put

$$X := U + W_1,$$
$$Y := -U + W_2.$$

Find the correlation coefficient between X and Y.

Solution. It is clear that $m_X = m_Y = 0$. Now write

$$\begin{aligned}\mathsf{E}[XY] &= \mathsf{E}[(U+W_1)(-U+W_2)] \\ &= \mathsf{E}[-U^2 + UW_2 - W_1 U + W_1 W_2] \\ &= -\mathsf{E}[U^2],\end{aligned}$$

using independence and the fact that U, W_1, and W_2 are all zero mean. We next calculate
$$\mathsf{E}[X^2] = \mathsf{E}[(U+W_1)^2] = \mathsf{E}[U^2 + 2UW_1 + W_1^2] = \mathsf{E}[U^2] + \mathsf{E}[W_1^2].$$
A similar calculation shows that $\mathsf{E}[Y^2] = \mathsf{E}[U^2] + \mathsf{E}[W_2^2]$. It then follows that
$$\rho_{XY} = \frac{-\mathsf{E}[U^2]}{\sqrt{(\mathsf{E}[U^2] + \mathsf{E}[W_1^2])(\mathsf{E}[U^2] + \mathsf{E}[W_2^2])}}.$$
If W_1 and W_2 have the same variance, say $\mathsf{E}[W_1^2] = \mathsf{E}[W_2^2] = \sigma^2$, then
$$\rho_{XY} = \frac{-\mathsf{E}[U^2]}{\mathsf{E}[U^2] + \sigma^2}. \tag{2.27}$$
If we define the **signal-to-noise ratio** (SNR) by
$$\text{SNR} := \frac{\mathsf{E}[U^2]}{\sigma^2},$$
then
$$\rho_{XY} = \frac{-\text{SNR}}{1 + \text{SNR}}.$$
As the signal-to-noise ratio goes to infinity, say by letting $\sigma^2 \to 0$, we have from (2.27) that $\rho_{XY} \to -1$. If $0 = \sigma^2 = \mathsf{E}[W_1^2] = \mathsf{E}[W_2^2]$, then $W_1 = W_2 \equiv 0$. This means that $X = U$ and $Y = -U$, which implies $Y = -X$; i.e., X and Y are linearly related.

It is frequently more convenient to work with the numerator of the correlation coefficient and to forget about the denominators. This leads to the following definition.

The **covariance** between X and Y is defined by
$$\text{cov}(X,Y) := \mathsf{E}[(X - m_X)(Y - m_Y)].$$
With this definition, we can write
$$\rho_{XY} = \frac{\text{cov}(X,Y)}{\sqrt{\text{var}(X)\text{var}(Y)}}.$$
Hence, X and Y are uncorrelated if and only if their covariance is zero.

Let $X_1, X_2, \ldots$ be a sequence of uncorrelated random variables; more precisely, for $i \neq j$, X_i and X_j are uncorrelated. We show next that
$$\text{var}\left(\sum_{i=1}^{n} X_i\right) = \sum_{i=1}^{n} \text{var}(X_i). \tag{2.28}$$
In other words, *for uncorrelated random variables, the variance of the sum is the sum of the variances.*

2.4 Expectation

Let $m_i := \mathsf{E}[X_i]$ and $m_j := \mathsf{E}[X_j]$. Then uncorrelated means that

$$\mathsf{E}[(X_i - m_i)(X_j - m_j)] = 0 \quad \text{for all } i \neq j.$$

Put

$$X := \sum_{i=1}^{n} X_i.$$

Then

$$\mathsf{E}[X] = \mathsf{E}\left[\sum_{i=1}^{n} X_i\right] = \sum_{i=1}^{n} \mathsf{E}[X_i] = \sum_{i=1}^{n} m_i,$$

and

$$X - \mathsf{E}[X] = \sum_{i=1}^{n} X_i - \sum_{i=1}^{n} m_i = \sum_{i=1}^{n} (X_i - m_i).$$

Now write

$$\begin{aligned}
\text{var}(X) &= \mathsf{E}[(X - \mathsf{E}[X])^2] \\
&= \mathsf{E}\left[\left(\sum_{i=1}^{n}(X_i - m_i)\right)\left(\sum_{j=1}^{n}(X_j - m_j)\right)\right] \\
&= \sum_{i=1}^{n}\left(\sum_{j=1}^{n} \mathsf{E}[(X_i - m_i)(X_j - m_j)]\right).
\end{aligned}$$

For fixed i, consider the sum over j. When $j \neq i$, which is the case for $n-1$ values of j, the expectation is zero because X_j and X_i are uncorrelated. Hence, of all the terms in the inner sum, only the term with $j = i$ survives. Thus,

$$\begin{aligned}
\text{var}(X) &= \sum_{i=1}^{n} \left(\mathsf{E}[(X_i - m_i)(X_i - m_i)]\right) \\
&= \sum_{i=1}^{n} \mathsf{E}[(X_i - m_i)^2] \\
&= \sum_{i=1}^{n} \text{var}(X_i).
\end{aligned}$$

***Example* 2.41.** Show that X and Y are uncorrelated if and only if

$$\mathsf{E}[XY] = \mathsf{E}[X]\mathsf{E}[Y]. \tag{2.29}$$

Solution. The result is obvious if we expand

$$\begin{aligned}
\mathsf{E}[(X - m_X)(Y - m_Y)] &= \mathsf{E}[XY - m_X Y - X m_Y + m_X m_Y] \\
&= \mathsf{E}[XY] - m_X \mathsf{E}[Y] - \mathsf{E}[X] m_Y + m_X m_Y \\
&= \mathsf{E}[XY] - m_X m_Y - m_X m_Y + m_X m_Y \\
&= \mathsf{E}[XY] - m_X m_Y.
\end{aligned}$$

From this we see that $\text{cov}(X, Y) = 0$ if and only if (2.29) holds.

From (2.29), we see that if X and Y are independent, then they are uncorrelated. Intuitively, the property of being uncorrelated is weaker than the property of independence. For independent random variables, recall that

$$\mathsf{E}[h(X)k(Y)] = \mathsf{E}[h(X)]\mathsf{E}[k(Y)]$$

for *all* functions $h(x)$ and $k(y)$, while for uncorrelated random variables, we only require that this hold for $h(x) = x$ and $k(y) = y$. For an example of uncorrelated random variables that are not independent, see Problem 44. For additional examples, see Problems 20 and 51 in Chapter 7.

Notes

2.1: Probabilities involving random variables

Note **1.** According to Note **1** in Chapter 1, $\mathsf{P}(A)$ is only defined for certain subsets $A \in \mathscr{A}$. Hence, in order that the probability

$$\mathsf{P}(\{\omega \in \Omega : X(\omega) \in B\})$$

be defined, it is necessary that

$$\{\omega \in \Omega : X(\omega) \in B\} \in \mathscr{A}. \tag{2.30}$$

To illustrate the problem, consider the sample space $\Omega := \{1,2,3\}$ equipped with the σ-field

$$\mathscr{A} := \{\varnothing, \{1,2\}, \{3\}, \Omega\}. \tag{2.31}$$

Take $\mathsf{P}(\{1,2\}) = 2/3$ and $\mathsf{P}(\{3\}) = 1/3$. Now define two functions $X(\omega) := \omega$ and

$$Y(\omega) := \begin{cases} 2, & \omega = 1,2, \\ 3, & \omega = 3. \end{cases}$$

Observe that

$$\{\omega \in \Omega : X(\omega) = 2\} = \{2\} \notin \mathscr{A},$$

while

$$\{\omega \in \Omega : Y(\omega) = 2\} = \{1,2\} \in \mathscr{A}.$$

Since $\{2\} \notin \mathscr{A}$, $\mathsf{P}(\{\omega \in \Omega : X(\omega) = 2\})$ is not defined. However, since $\{1,2\} \in \mathscr{A}$, $\mathsf{P}(\{\omega \in \Omega : Y(\omega) = 2\}) = 2/3$.

In the general case, to guarantee that (2.30) holds, it is convenient to consider $\mathsf{P}(X \in B)$ only for sets B in some σ-field $\mathscr{B}$ of subsets of $\mathbb{R}$. The technical definition of a random variable is then as follows. A function X from Ω into $\mathbb{R}$ is a **random variable** if and only if (2.30) holds for every $B \in \mathscr{B}$. Usually $\mathscr{B}$ is taken to be the **Borel σ-field**; i.e., $\mathscr{B}$ is the smallest σ-field containing all the open subsets of $\mathbb{R}$. If $B \in \mathscr{B}$, then B is called a **Borel set**. It can be shown [3, pp. 182–183] that a real-valued function X satisfies (2.30) for all Borel sets B if and only if

$$\{\omega \in \Omega : X(\omega) \leq x\} \in \mathscr{A}, \quad \text{for all } x \in \mathbb{R}.$$

Notes

With reference to the functions $X(\omega)$ and $Y(\omega)$ defined above, observe that $\{\omega \in \Omega : X(\omega) \leq 1\} = \{1\} \notin \mathscr{A}$ defined in (2.31), and so X is *not* a random variable. However, it is easy to check that Y does satisfy $\{\omega \in \Omega : Y(\omega) \leq y\} \in \mathscr{A}$ defined in (2.31) for all y; hence, Y *is* a random variable.

For $B \in \mathscr{B}$, if we put

$$\mu(B) := \mathsf{P}(X \in B) = \mathsf{P}(\{\omega \in \Omega : X(\omega) \in B\}), \tag{2.32}$$

then μ satisfies the axioms of a probability measure on $\mathbb{R}$. This follows because μ "inherits" the properties of P through the random variable X (Problem 4). Once we know that μ is a measure, formulas (2.1) and (2.2) become obvious. For example, (2.1) says that

$$\mu((-\infty, b]) = \mu((-\infty, a]) + \mu((a, b]).$$

This is immediate since $(-\infty, b] = (-\infty, a] \cup (a, b]$ is a disjoint union. Similarly, (2.2) says that

$$\mu(\{0\} \cup \{1\}) = \mu(\{0\}) + \mu(\{1\}).$$

Again, since this union is disjoint, the result is immediate.

Since μ depends on X, if more than one random variable is under discussion, we write $\mu_X(B)$ instead. We thus see that different random variables induce different probability measures on $\mathbb{R}$. Another term for measure is **distribution**. Hence, we call μ_X the distribution of X. More generally, the term "distribution" refers to how probability is spread out. As we will see later, for discrete random variables, once we know the probability mass function, we can compute $\mu_X(B) = \mathsf{P}(X \in B)$ for all B of interest. Similarly, for the continuous random variables of Chapter 4, once we know the probability density function, we can compute $\mu_X(B) = \mathsf{P}(X \in B)$ for all B of interest. In this sense, probability mass functions, probability density functions, and distributions are just different ways of describing how to compute $\mathsf{P}(X \in B)$.

2.2: Discrete random variables

Note 2. To derive (2.4), we apply the law of total probability as given in (1.27) with $A = \{X \in B\}$ and $B_i = \{X = x_i\}$. Since the x_i are distinct, the B_i are disjoint, and (1.27) says that

$$\mathsf{P}(X \in B) = \sum_i \mathsf{P}(\{X \in B\} \cap \{X = x_i\}). \tag{2.33}$$

Now observe that if $x_i \in B$, then $X = x_i$ implies $X \in B$, and so

$$\{X = x_i\} \subset \{X \in B\}.$$

This monotonicity tells us that

$$\{X \in B\} \cap \{X = x_i\} = \{X = x_i\}.$$

On the other hand, if $x_i \notin B$, then we cannot have $X = x_i$ and $X \in B$ at the same time; in other words

$$\{X \in B\} \cap \{X = x_i\} = \emptyset.$$

It now follows that

$$P(\{X \in B\} \cap \{X = x_i\}) = \begin{cases} P(X = x_i), & x_i \in B, \\ 0, & x_i \notin B. \end{cases}$$

Substituting this in (2.33) yields (2.4).

2.3: Multiple random variables

Note 3. In light of Note **1** above, we do not require that (2.5) hold for *all* sets B and C, but only for all *Borel* sets B and C.

Note 4. If $X_1, \ldots, X_n$ are independent, we show that any subset of them must also be independent. Since (2.6) must hold for all choices of $B_1, \ldots, B_n$, put $B_j = \mathbb{R}$ for the X_j that we do *not* care about. Then use the fact that $\{X_j \in \mathbb{R}\} = \Omega$ and $P(\Omega) = 1$ to make these variables "disappear." For example, if $B_n = \mathbb{R}$ in (2.6), we get

$$P\left(\left[\bigcap_{j=1}^{n-1}\{X_j \in B_j\}\right] \cap \{X_n \in \mathbb{R}\}\right) = \left[\prod_{j=1}^{n-1} P(X_j \in B_j)\right] P(X_n \in \mathbb{R})$$

or

$$P\left(\left[\bigcap_{j=1}^{n-1}\{X_j \in B_j\}\right] \cap \Omega\right) = \left[\prod_{j=1}^{n-1} P(X_j \in B_j)\right] P(\Omega),$$

which simplifies to

$$P\left(\bigcap_{j=1}^{n-1}\{X_j \in B_j\}\right) = \prod_{j=1}^{n-1} P(X_j \in B_j).$$

This shows that $X_1, \ldots, X_{n-1}$ are independent.

Note 5. We show that

$$P(X \in B, Y \in C) = \sum_{i:x_i \in B} \sum_{j:y_j \in C} p_{XY}(x_i, y_j).$$

Consider the disjoint events $\{Y = y_j\}$. Since $\sum_j P(Y = y_j) = 1$, we can use the law of total probability as in (1.27) with $A = \{X = x_i, Y \in C\}$ to write

$$P(X = x_i, Y \in C) = \sum_j P(X = x_i, Y \in C, Y = y_j).$$

Now observe that

$$\{Y \in C\} \cap \{Y = y_j\} = \begin{cases} \{Y = y_j\}, & y_j \in C, \\ \varnothing, & y_j \notin C. \end{cases}$$

Hence,

$$P(X = x_i, Y \in C) = \sum_{j:y_j \in C} P(X = x_i, Y = y_j) = \sum_{j:y_j \in C} p_{XY}(x_i, y_j).$$

The next step is to use (1.27) again, but this time with the disjoint events $\{X = x_i\}$ and $A = \{X \in B, Y \in C\}$. Then,

$$P(X \in B, Y \in C) = \sum_i P(X \in B, Y \in C, X = x_i).$$

Now observe that

$$\{X \in B\} \cap \{X = x_i\} = \begin{cases} \{X = x_i\}, & x_i \in B, \\ \varnothing, & x_i \notin B. \end{cases}$$

Hence,

$$P(X \in B, Y \in C) = \sum_{i:x_i \in B} P(X = x_i, Y \in C)$$

$$= \sum_{i:x_i \in B} \sum_{j:y_j \in C} p_{XY}(x_i, y_j).$$

2.4: Expectation

Note 6. To be technically correct, in (2.23), we cannot allow $h(x)$ and $k(y)$ to be completely arbitrary. We must restrict them so that

$$E[|h(X)|] < \infty, \quad E[|k(Y)|] < \infty, \quad \text{and} \quad E[|h(X)k(Y)|] < \infty.$$

Note 7. Strictly speaking, we can only conclude that $X = \lambda Y$ with probability one; i.e., $P(X = \lambda Y) = 1$.

Problems

2.1: Probabilities involving random variables

1. On the probability space of Example 1.10, define the random variable $X(\omega) := \omega$.

   (a) Find all the outcomes ω that belong to the event $\{\omega : X(\omega) \leq 3\}$.

   (b) Find all the outcomes ω that belong to the event $\{\omega : X(\omega) > 4\}$.

   (c) Compute $P(X \leq 3)$ and $P(X > 4)$.

2. On the probability space of Example 1.12, define the random variable

   $$X(\omega) := \begin{cases} 2, & \text{if } \omega \text{ corresponds to an ace,} \\ 1, & \text{if } \omega \text{ corresponds to a face card,} \\ 0, & \text{otherwise.} \end{cases}$$

   (a) Find all the outcomes ω that belong to the event $\{\omega : X(\omega) = 2\}$.

   (b) Find all the outcomes ω that belong to the event $\{\omega : X(\omega) = 1\}$.

   (c) Compute $P(X = 1 \text{ or } X = 2)$.

3. On the probability space of Example 1.15, define the random variable $X(\omega) := \omega$. Thus, X is the duration of a cell-phone call.

   (a) Find all the outcomes ω that belong to the event $\{\omega : X(\omega) \leq 1\}$.

   (b) Find all the outcomes ω that belong to the event $\{\omega : X(\omega) \leq 3\}$.

   (c) Compute $P(X \leq 1)$, $P(X \leq 3)$, and $P(1 < X \leq 3)$.

*4. **This problem assumes you have read Note 1.** Show that the distribution μ defined in (2.32) satisfies the axioms of a probability measure on $\mathbb{R}$. *Hints:* Use the fact that $\mu(B) = P(X^{-1}(B))$; use the inverse-image properties of Problem 15 in Chapter 1; the axioms of a probability measure were defined in Section 1.4.

2.2: Discrete random variables

5. Let Y be an integer-valued random variable. Show that

$$P(Y = n) = P(Y > n-1) - P(Y > n).$$

6. Find the pmf of the random variable Y defined in Example 2.2 assuming that all sequences in Ω are equally likely.

7. Find the pmf of the random variable of Problem 1.

8. Find the pmf of the random variable of Problem 2.

9. Consider the sample space $\Omega := \{-2,-1,0,1,2,3,4\}$. For an event $A \subset \Omega$, suppose that $P(A) = |A|/|\Omega|$. Define the random variable $X(\omega) := \omega^2$. Find the probability mass function of X.

10. Let $X \sim \text{Poisson}(\lambda)$. Evaluate $P(X > 1)$; your answer should be in terms of λ. Then compute the numerical value of $P(X > 1)$ when $\lambda = 1$. *Answer:* 0.264.

11. A certain photo-sensor fails to activate if it receives fewer than four photons in a certain time interval. If the number of photons is modeled by a Poisson(2) random variable X, find the probability that the sensor activates. *Answer:* 0.143.

2.3: Multiple random variables

12. A class consists of 15 students. Each student has probability $p = 0.1$ of getting an "A" in the course. Find the probability that exactly one student receives an "A." Assume the students' grades are independent. *Answer:* 0.343.

13. In a certain lottery game, the player chooses three digits. The player wins if at least two out of three digits match the random drawing for that day in both position and value. Find the probability that the player wins. Assume that the digits of the random drawing are independent and equally likely. *Answer:* 0.028.

14. At the Chicago IRS office, there are m independent auditors. The kth auditor processes X_k tax returns per day, where X_k is Poisson distributed with parameter $\lambda > 0$. The office's performance is unsatisfactory if any auditor processes fewer than 2 tax returns per day. Find the probability that the office performance is unsatisfactory.

15. An astronomer has recently discovered n similar galaxies. For $i = 1,\ldots,n$, let X_i denote the number of black holes in the ith galaxy, and assume the X_i are independent Poisson(λ) random variables.

    (a) Find the probability that at least one of the galaxies contains two or more black holes.

    (b) Find the probability that all n galaxies have at least one black hole.

    (c) Find the probability that all n galaxies have exactly one black hole.

    Your answers should be in terms of n and λ.

16. Show that the geometric$_0(p)$ pmf $p_X(k) = (1-p)p^k, k = 0, 1, \ldots$ sums to one. Repeat for the geometric$_1(p)$ pmf $p_X(k) = (1-p)p^{k-1}, k = 1, 2, \ldots$. *Hint:* Use the geometric series formula from Problem 27 in Chapter 1.

17. There are 29 stocks on the Get Rich Quick Stock Exchange. The price of each stock (in whole dollars) is geometric$_0(p)$ (same p for all stocks). Prices of different stocks are independent. If $p = 0.7$, find the probability that at least one stock costs more than 10 dollars. *Answer:* 0.44.

18. Suppose that $X_1, \ldots, X_n$ are independent, geometric$_1(p)$ random variables. Evaluate $P(\min(X_1, \ldots, X_n) > \ell)$ and $P(\max(X_1, \ldots, X_n) \le \ell)$.

19. In a class of 25 students, the number of coins in each student's pocket is uniformly distributed between zero and twenty. Suppose the numbers of coins in different students' pockets are independent.

    (a) Find the probability that no student has fewer than 5 coins in his/her pocket. *Answer:* 1.12×10^{-3}.

    (b) Find the probability that at least one student has at least 19 coins in his/her pocket. *Answer:* 0.918.

    (c) Find the probability that exactly one student has exactly 19 coins in his/her pocket. *Answer:* 0.369.

20. Blocks on a computer disk are good with probability p and faulty with probability $1-p$. Blocks are good or bad independently of each other. Let Y denote the location (starting from 1) of the first bad block. Find the pmf of Y.

21. Let $X \sim$ geometric$_1(p)$.

    (a) Show that $P(X > n) = p^n$.

    (b) Compute $P(\{X > n+k\}|\{X > n\})$. *Hint:* If $A \subset B$, then $A \cap B = A$.

    Remark. Your answer to (b) should not depend on n. For this reason, the geometric random variable is said to have the **memoryless property**. For example, let X model the number of the toss on which the first heads occurs in a sequence of coin tosses. Then given a heads has not occurred up to and including time n, the conditional probability that a heads does not occur in the next k tosses does not depend on n. In other words, given that no heads occurs on tosses $1, \ldots, n$ has no effect on the conditional probability of heads occurring in the future. Future tosses do not remember the past.

*22. From your solution of Problem 21(b), you can see that if $X \sim$ geometric$_1(p)$, then $P(\{X > n+k\}|\{X > n\}) = P(X > k)$. Now prove the converse; i.e., show that if Y is a positive integer-valued random variable such that $P(\{Y > n+k\}|\{Y > n\}) = P(Y > k)$, then $Y \sim$ geometric$_1(p)$, where $p = P(Y > 1)$. *Hint:* First show that $P(Y > n) = P(Y > 1)^n$; then apply Problem 5.

23. Let X and Y be ternary random variables taking values 1, 2, and 3 with joint probabilities $p_{XY}(i,j)$ given by the matrix

$$\begin{bmatrix} 1/8 & 0 & 1/8 \\ 0 & 1/2 & 0 \\ 1/8 & 0 & 1/8 \end{bmatrix}.$$

   (a) Find $p_X(i)$ and $p_Y(j)$ for $i,j = 1,2,3$.
   (b) Compute $\mathsf{P}(X < Y)$.
   (c) Determine whether or not X and Y are independent.

24. Repeat the previous problem if the $p_{XY}(i,j)$ are given by

$$\begin{bmatrix} 1/24 & 1/6 & 1/24 \\ 1/12 & 1/3 & 1/12 \\ 1/24 & 1/6 & 1/24 \end{bmatrix}.$$

25. Let X and Y be jointly discrete, integer-valued random variables with joint pmf

$$p_{XY}(i,j) = \begin{cases} \dfrac{3^{j-1}e^{-3}}{j!}, & i=1, j \geq 0, \\ 4\dfrac{6^{j-1}e^{-6}}{j!}, & i=2, j \geq 0, \\ 0, & \text{otherwise.} \end{cases}$$

Find the marginal pmfs $p_X(i)$ and $p_Y(j)$, and determine whether or not X and Y are independent.

26. Let X and Y have joint pmf

$$p_{XY}(k,n) := \begin{cases} \dfrac{(1-p)p^{k-1}k^n e^{-k}}{n!}, & k \geq 1, n \geq 0, \\ 0, & \text{otherwise.} \end{cases}$$

   (a) Compute $p_X(k)$ for $k \geq 1$.
   (b) Compute $p_Y(0)$.
   (c) Determine whether or not X and Y are independent.

27. **MATLAB.** Write a MATLAB script to compute $\mathsf{P}(g(X) \geq -16)$ if X is a uniform random variable on $0,\ldots,50$ and $g(x) = 5x(x-10)(x-20)(x-30)(x-40)(x-50)/10^6$. *Answer:* 0.6275.

28. **MATLAB.** Let $g(x)$ be as in the preceding problem. Write a MATLAB script to compute $\mathsf{P}(g(X) \geq -16)$ if $X \sim \text{geometric}_0(p)$ with $p = 0.95$. *Answer:* 0.5732.

29. **MATLAB.** Suppose x is a column vector of m numbers and y is a column vector of n numbers and you want to compute $g(x(i),y(j))$, where i ranges from 1 to m, j ranges from 1 to n, and g is a given function. Here is a simple way to do this without any `for` loops. Store the following function in an M-file called `allpairs.m`.

```
function [x1,y1] = allpairs(x,y)
lx = length(x);
ly = length(y);
x1 = kron(ones(ly,1),x);
y1 = kron(y,ones(lx,1));
```

(The MATLAB command kron computes the **Kronecker product**.) Then issue the following commands and print out your results.

```
x = [ 1 2 3 ]'
y = [ 10 20 30 ]'
[x1,y1] = allpairs(x,y);
pairs = [x1 y1]
allsums = x1+y1
```

30. **MATLAB.** Let X and Y have the joint pmf of Example 2.18. Use the following script to compute $P(XY < 6)$.

```
i = [1:3]';
j = [1:5]';
[x,y]=allpairs(i,j);
prob = sum(P(find(x.*y<6)))
```

Do you get the same answer as in Example 2.19?

31. **MATLAB.** Write MATLAB scripts to solve Problems 23 and 24.

2.4: Expectation

32. Compute $E[X]$ if $P(X=2) = 1/3$ and $P(X=5) = 2/3$.

33. If $X \sim \text{geometric}_0(1/2)$, compute $E[I_{(2,6)}(X)]$.

34. If X is Poisson(λ), compute $E[1/(X+1)]$.

35. A random variable X has mean 2 and variance 7. Find $E[X^2]$.

36. If X has mean m and variance σ^2, and if $Y := cX$, find the variance of Y.

37. Compute $E[(X+Y)^3]$ if $X \sim \text{Bernoulli}(p)$ and $Y \sim \text{Bernoulli}(q)$ are independent.

38. Let X be a random variable with mean m and variance σ^2. Find the constant c that best approximates the random variable X in the sense that c minimizes the **mean-squared error** $E[(X-c)^2]$.

39. The general shape of $(x/a)^r$ in Figure 2.11 is correct for $r > 1$. Find out how Figure 2.11 changes for $0 < r \leq 1$ by sketching $(x/a)^r$ and $I_{[a,\infty)}(x)$ for $r = 1/2$ and $r = 1$.

40. Let $X \sim \text{Poisson}(3/4)$. Compute both sides of the Markov inequality,

$$P(X \geq 2) \leq \frac{E[X]}{2}.$$

41. Let $X \sim \text{Poisson}(3/4)$. Compute both sides of the Chebyshev inequality,
$$P(X \geq 2) \leq \frac{E[X^2]}{4}.$$

42. Let X and Y be two random variables with means m_X and m_Y and variances σ_X^2 and σ_Y^2. Let ρ_{XY} denote their correlation coefficient. Show that $\text{cov}(X,Y) = \sigma_X \sigma_Y \rho_{XY}$. Show that $\text{cov}(X,X) = \text{var}(X)$.

43. Let X and Y be two random variables with means m_X and m_Y, variances σ_X^2 and σ_Y^2, and correlation coefficient ρ. Suppose X cannot be observed, but we are able to measure Y. We wish to estimate X by using the quantity aY, where a is a suitable constant. Assuming $m_X = m_Y = 0$, find the constant a that minimizes the mean-squared error $E[(X - aY)^2]$. Your answer should depend on σ_X, σ_Y, and ρ.

44. Show by counterexample that being uncorrelated does not imply independence. *Hint:* Let $P(X = \pm 1) = P(X = \pm 2) = 1/4$, and put $Y := |X|$. Show that $E[XY] = E[X]E[Y]$, but $P(X = 1, Y = 1) \neq P(X = 1)P(Y = 1)$.

45. Suppose that $Y := X_1 + \cdots + X_M$, where the X_k are i.i.d. $\text{geometric}_1(p)$ random variables. Find $E[Y^2]$.

46. **Betting on fair games.** Let $X \sim \text{Bernoulli}(p)$. For example, we could let $X = 1$ model the result of a coin toss being heads. Or we could let $X = 1$ model your winning the lottery. In general, a bettor wagers a stake of s dollars that $X = 1$ with a bookmaker who agrees to pay d dollars to the bettor if $X = 1$ occurs; if $X = 0$, the stake s is kept by the bookmaker. Thus, the net income of the bettor is
$$Y := dX - s(1 - X),$$
since if $X = 1$, the bettor receives $Y = d$ dollars, and if $X = 0$, the bettor receives $Y = -s$ dollars; i.e., loses s dollars. Of course the net income to the bookmaker is $-Y$. If the wager is fair to both the bettor and the bookmaker, then we should have $E[Y] = 0$. In other words, on average, the net income to either party is zero. Show that a fair wager requires that $d/s = (1 - p)/p$.

47. **Odds.** Let $X \sim \text{Bernoulli}(p)$. We say that the (fair) **odds** *against* $X = 1$ are n_2 to n_1 (written $n_2 : n_1$) if n_2 and n_1 are positive integers satisfying $n_2/n_1 = (1 - p)/p$. Typically, n_2 and n_1 are chosen to have no common factors. Conversely, we say that the odds *for* $X = 1$ are n_1 to n_2 if $n_1/n_2 = p/(1 - p)$. Consider a state lottery game in which players wager one dollar that they can correctly guess a randomly selected three-digit number in the range 000–999. The state offers a payoff of \$500 for a correct guess.

    (a) What is the probability of correctly guessing the number?

    (b) What are the (fair) odds against guessing correctly?

    (c) The odds against actually offered by the state are determined by the ratio of the payoff divided by the stake, in this case, 500:1. Is the game fair to the bettor? If not, what should the payoff be to make it fair? (See the preceding problem for the notion of "fair.")

*48. These results are used in Examples 2.24 and 2.25. Show that the sum

$$C_p := \sum_{k=1}^{\infty} \frac{1}{k^p}$$

diverges for $0 < p \leq 1$, but is finite for $p > 1$. *Hint:* For $0 < p \leq 1$, use the inequality

$$\int_k^{k+1} \frac{1}{t^p} dt \leq \int_k^{k+1} \frac{1}{k^p} dt = \frac{1}{k^p},$$

and for $p > 1$, use the inequality

$$\int_k^{k+1} \frac{1}{t^p} dt \geq \int_k^{k+1} \frac{1}{(k+1)^p} dt = \frac{1}{(k+1)^p}.$$

*49. For C_p as defined in Problem 48, if $P(X = k) = C_p^{-1}/k^p$ for some $p > 1$, then X is called a **zeta** or **Zipf** random variable. Show that $E[X^n] < \infty$ for $n < p-1$, and $E[X^n] = \infty$ for $n \geq p-1$.

50. Let X be a discrete random variable taking finitely many distinct values $x_1, \ldots, x_n$. Let $p_i := P(X = x_i)$ be the corresponding probability mass function. Consider the function

$$g(x) := -\log P(X = x).$$

Observe that $g(x_i) = -\log p_i$. The **entropy** of X is defined by

$$H(X) := E[g(X)] = \sum_{i=1}^{n} g(x_i) P(X = x_i) = \sum_{i=1}^{n} p_i \log \frac{1}{p_i}.$$

If all outcomes are equally likely, i.e., $p_i = 1/n$, find $H(X)$. If X is a constant random variable, i.e., $p_j = 1$ for some j and $p_i = 0$ for $i \neq j$, find $H(X)$.

*51. **Jensen's inequality.** Recall that a real-valued function g defined on an interval I is **convex** if for all $x, y \in I$ and all $0 \leq \lambda \leq 1$,

$$g(\lambda x + (1-\lambda)y) \leq \lambda g(x) + (1-\lambda)g(y).$$

Let g be a convex function, and let X be a discrete random variable taking finitely many values, say n values, all in I. Derive **Jensen's inequality**,

$$E[g(X)] \geq g(E[X]).$$

Hint: Use induction on n.

*52. Derive **Lyapunov's inequality**,

$$E[|Z|^\alpha]^{1/\alpha} \leq E[|Z|^\beta]^{1/\beta}, \quad 0 < \alpha \leq \beta < \infty.$$

Hint: Apply Jensen's inequality to the convex function $g(x) = x^{\beta/\alpha}$ and the random variable $X = |Z|^\alpha$.

*53. A discrete random variable is said to be nonnegative, denoted by $X \geq 0$, if $P(X \geq 0) = 1$; i.e., if
$$\sum_i I_{[0,\infty)}(x_i) P(X = x_i) = 1.$$

(a) Show that for a nonnegative random variable, if $x_k < 0$ for some k, then $P(X = x_k) = 0$.

(b) Show that for a nonnegative random variable, $E[X] \geq 0$.

(c) If X and Y are discrete random variables, we write $X \geq Y$ if $X - Y \geq 0$. Show that if $X \geq Y$, then $E[X] \geq E[Y]$; i.e., expectation is **monotone**.

Exam preparation

You may use the following suggestions to prepare a study sheet, including formulas mentioned that you have trouble remembering. You may also want to ask your instructor for additional suggestions.

2.1. Probabilities involving random variables. Know how to do basic probability calculations involving a random variable given as an explicit function on a sample space.

2.2. Discrete random variables. Be able to do simple calculations with probability mass functions, especially the uniform and the Poisson.

2.3. Multiple random variables. Recall that X and Y are independent if $P(X \in A, Y \in B) = P(X \in A) P(Y \in B)$ for all sets A and B. However, for discrete random variables, all we need to check is whether or not $P(X = x_i, Y = y_j) = P(X = x_i) P(Y = y_j)$, or, in terms of pmfs, whether or not $p_{XY}(x_i, y_j) = p_X(x_i) p_Y(y_j)$ for all x_i and y_j. Remember that the marginals p_X and p_Y are computed using (2.8) and (2.9), respectively. Be able to solve problems with intersections and unions of events involving independent random variables. Know how the geometric$_1(p)$ random variable arises.

2.4. Expectation. Important formulas include LOTUS (2.14), linearity of expectation (2.15), the definition of variance (2.16) as well as the variance formula (2.17), and expectation of functions of products of independent random variables (2.23). For sequences of uncorrelated random variables, the variance of the sum is the sum of the variances (2.28). Know the difference between uncorrelated and independent. A list of common pmfs and their means and variances can be found inside the front cover. The Poisson(λ) random variable arises so often that it is worth remembering, even if you are allowed to bring a formula sheet to the exam, that its mean and variance are both λ and that by the variance formula, its second moment is $\lambda + \lambda^2$. Similarly, the mean p and variance $p(1-p)$ of the Bernoulli(p) are also worth remembering. Your instructor may suggest others to memorize. Know the Markov inequality (for nonnegative random variables only) (2.18) and the Chebyshev inequality (for any random variable) (2.21) and also (2.22).

A discrete random variable is completely characterized by its pmf, which is the collection of numbers $p_X(x_i)$. In many problems we do not know the pmf. However, the next best things to know are the mean and variance; they can be used to bound probabilities as in the Markov and Chebyshev inequalities, and they can be used for approximation and estimation as in Problems 38 and 43.

Work any review problems assigned by your instructor. If you finish them, re-work your homework assignments.

3
More about discrete random variables

This chapter develops more tools for working with random variables. The probability generating function is the key tool for working with sums of nonnegative integer-valued random variables that are independent. When random variables are only uncorrelated, we can work with averages (normalized sums) by using the weak law of large numbers. We emphasize that the weak law makes the connection between probability theory and the every-day practice of using averages of observations to estimate probabilities of real-world measurements. The last two sections introduce conditional probability and conditional expectation. The three important tools here are the law of total probability, the law of substitution, and, for independent random variables, "dropping the conditioning."

The foregoing concepts are developed here for discrete random variables, but they will all be extended to more general settings in later chapters.

3.1 Probability generating functions

In many problems we have a sum of independent random variables, and we would like to know the probability mass function of their sum. For example, in an optical communication system, the received signal might be $Y = X + W$, where X is the number of photoelectrons due to incident light on a photodetector, and W is the number of electrons due to dark current noise in the detector. An important tool for solving these kinds of problems is the probability generating function. The name derives from the fact that it can be used to compute the probability mass function. Additionally, the probability generating function can be used to compute the mean and variance in a simple way.

Let X be a discrete random variable taking only nonnegative integer values. The **probability generating function** (pgf) of X is[1]

$$G_X(z) := \mathsf{E}[z^X] = \sum_{n=0}^{\infty} z^n \mathsf{P}(X=n). \tag{3.1}$$

Readers familiar with the *z* **transform** will note that $G(z^{-1})$ is the *z* transform of the probability mass function $p_X(n) := \mathsf{P}(X = n)$.

Example 3.1. Find the probability generating function of X if it is Poisson with parameter λ.

Solution. Write

$$\begin{aligned} G_X(z) &= \mathsf{E}[z^X] \\ &= \sum_{n=0}^{\infty} z^n \mathsf{P}(X=n) \\ &= \sum_{n=0}^{\infty} z^n \frac{\lambda^n e^{-\lambda}}{n!} \end{aligned}$$

3.1 Probability generating functions

$$= e^{-\lambda} \sum_{n=0}^{\infty} \frac{(z\lambda)^n}{n!}$$
$$= e^{-\lambda} e^{z\lambda}$$
$$= e^{\lambda(z-1)}.$$

An important property of probability generating functions is that the pgf of a sum of *independent* random variables is the product of the individual pgfs. To see this, let $Y := X_1 + \cdots + X_n$, where the X_i are independent with corresponding pgfs $G_{X_i}(z)$. Then

$$\begin{aligned} G_Y(z) &:= \mathsf{E}[z^Y] \\ &= \mathsf{E}[z^{X_1 + \cdots + X_n}] \\ &= \mathsf{E}[z^{X_1} \cdots z^{X_n}] \\ &= \mathsf{E}[z^{X_1}] \cdots \mathsf{E}[z^{X_n}], \quad \text{by independence,} \\ &= G_{X_1}(z) \cdots G_{X_n}(z). \end{aligned} \quad (3.2)$$

We call this the **factorization property**. Remember, it works only for *sums* of *independent* random variables.

Example 3.2. Let $Y = X + W$, where X and W are independent Poisson random variables with respective parameters λ and μ. Here X represents the signal and W the dark current in the optical communication systems described at the beginning of the section. Find the pgf of Y.

Solution. Write

$$\begin{aligned} G_Y(z) &= \mathsf{E}[z^Y] \\ &= \mathsf{E}[z^{X+W}] \\ &= \mathsf{E}[z^X z^W] \\ &= G_X(z) G_W(z), \quad \text{by independence,} \\ &= e^{\lambda(z-1)} e^{\mu(z-1)}, \quad \text{by Example 3.1,} \\ &= e^{(\lambda+\mu)(z-1)}, \end{aligned}$$

which is the pgf of a Poisson random variable with parameter $\lambda + \mu$.

The foregoing example shows that the pgf of Y is that of a Poisson random variable. We would like to conclude that Y must have the Poisson($\lambda + \mu$) probability mass function. Is this a justifiable conclusion? For example, if two different probability mass functions can have the same pgf, then we are in trouble. Fortunately, we can show this is not the case. We do this by showing that the probability mass function can be recovered from the pgf as follows.

Let $G_X(z)$ be a probability generating function. Since for $|z| \leq 1$,

$$\left|\sum_{n=0}^{\infty} z^n P(X=n)\right| \leq \sum_{n=0}^{\infty} |z^n P(X=n)|$$

$$= \sum_{n=0}^{\infty} |z|^n P(X=n)$$

$$\leq \sum_{n=0}^{\infty} P(X=n) = 1, \qquad (3.3)$$

the power series for G_X has radius of convergence at least one. Writing

$$G_X(z) = P(X=0) + zP(X=1) + z^2 P(X=2) + \cdots,$$

we immediately see that $G_X(0) = P(X=0)$. If we differentiate the above expression with respect to z, we get

$$G'_X(z) = P(X=1) + 2zP(X=2) + 3z^2 P(X=3) + \cdots,$$

and we see that $G'_X(0) = P(X=1)$. Continuing in this way shows that

$$G_X^{(k)}(z)|_{z=0} = k! P(X=k),$$

or equivalently,

$$\boxed{\frac{G_X^{(k)}(z)|_{z=0}}{k!} = P(X=k).} \qquad (3.4)$$

Example 3.3. If $G_X(z) = \left(\frac{1+z+z^2}{3}\right)^2$, find $P(X=2)$.

Solution. First write

$$G'_X(z) = 2\left(\frac{1+z+z^2}{3}\right)\left(\frac{1+2z}{3}\right),$$

and then

$$G''_X(z) = 2\left(\frac{1+z+z^2}{3}\right)\left(\frac{2}{3}\right) + 2\left(\frac{1+2z}{3}\right)\left(\frac{1+2z}{3}\right).$$

It follows that

$$P(X=2) = \frac{G''_X(0)}{2!} = \frac{1}{2!}\left(\frac{4}{9} + \frac{2}{9}\right) = \frac{1}{3}.$$

The probability generating function can also be used to find moments. Starting from

$$G_X(z) = \sum_{n=0}^{\infty} z^n P(X=n),$$

we compute

$$G'_X(z) = \sum_{n=1}^{\infty} n z^{n-1} P(X=n).$$

Setting $z = 1$ yields

$$G'_X(1) = \sum_{n=1}^{\infty} nP(X=n) = E[X].$$

Similarly, since

$$G''_X(z) = \sum_{n=2}^{\infty} n(n-1)z^{n-2}P(X=n),$$

setting $z = 1$ yields

$$G''_X(1) = \sum_{n=2}^{\infty} n(n-1)P(X=n) = E[X(X-1)] = E[X^2] - E[X].$$

In general, since

$$G^{(k)}_X(z) = \sum_{n=k}^{\infty} n(n-1)\cdots(n-[k-1])z^{n-k}P(X=n),$$

setting $z = 1$ yields[2]

$$G^{(k)}_X(z)|_{z=1} = E[X(X-1)(X-2)\cdots(X-[k-1])]. \quad (3.5)$$

The right-hand side is called the kth **factorial moment** of X.

Example 3.4. The probability generating function of $X \sim \text{Poisson}(\lambda)$ was found in Example 3.1 to be $G_X(z) = e^{\lambda(z-1)}$. Use $G_X(z)$ to find $E[X]$ and $\text{var}(X)$.

Solution. Since $G'_X(z) = e^{\lambda(z-1)}\lambda$, $E[X] = G'_X(1) = \lambda$. Since $G''_X(z) = e^{\lambda(z-1)}\lambda^2$, $E[X^2] - E[X] = \lambda^2$, $E[X^2] = \lambda^2 + \lambda$. For the variance, write

$$\text{var}(X) = E[X^2] - (E[X])^2 = (\lambda^2 + \lambda) - \lambda^2 = \lambda.$$

3.2 The binomial random variable

In many problems, the key quantity of interest can be expressed in the form $Y = X_1 + \cdots + X_n$, where the X_i are i.i.d. Bernoulli(p) random variables.

Example 3.5. A certain communication network consists of n links. Suppose that each link goes down with probability p independently of the other links. Show that the number of links that are down is a sum of independent Bernoulli random variables.

Solution. Let $X_i = 1$ if the ith link is down and $X_i = 0$ otherwise. Then the X_i are independent Bernoulli(p), and $Y := X_1 + \cdots + X_n$ counts the number of links that are down.

Example 3.6. A sample of radioactive material is composed of n molecules. Each molecule has probability p of emitting an alpha particle, and the particles are emitted independently. Show that the number of particles emitted is a sum of independent Bernoulli random variables.

Solution. Let $X_i = 1$ if the ith molecule emits an alpha particle, and $X_i = 0$ otherwise. Then the X_i are independent Bernoulli(p), and $Y := X_1 + \cdots + X_n$ counts the number of alpha particles emitted.

There are several ways to find the probability mass function of Y. The most common method uses a combinatorial argument, which we give in the next paragraph; following that, we give a different derivation using probability generating functions. A third derivation using techniques from Section 3.4 is given in the Notes.[3]

Observe that the only way to have $Y = k$ is to have k of the $X_i = 1$ and the other $n-k$ $X_i = 0$. Let B_k denote the set of all sequences of zeros and ones, say $(b_1,\ldots,b_n)$, in which k of the $b_i = 1$ and the other $n-k$ $b_i = 0$. Then

$$\begin{aligned} \mathsf{P}(Y = k) &= \mathsf{P}((X_1,\ldots,X_n) \in B_k) \\ &= \sum_{(b_1,\ldots,b_n) \in B_k} \mathsf{P}(X_1 = b_1,\ldots,X_n = b_n) \\ &= \sum_{(b_1,\ldots,b_n) \in B_k} \mathsf{P}(X_1 = b_1)\cdots \mathsf{P}(X_n = b_n), \end{aligned}$$

where the last step follows because the X_i are independent. Now each factor in the above product is either p or $1-p$ according to whether each b_i equals zero or one. Since the sum is over $(b_1,\ldots,b_n) \in B_k$, there are k factors equal to p and $n-k$ factors equal to $1-p$. Hence,

$$\begin{aligned} \mathsf{P}(Y = k) &= \sum_{(b_1,\ldots,b_n) \in B_k} p^k(1-p)^{n-k} \\ &= |B_k| p^k(1-p)^{n-k}, \end{aligned}$$

where $|B_k|$ denotes the number of sequences in the set B_k. From the discussion in Section 1.7,

$$|B_k| = \binom{n}{k} = \frac{n!}{k!(n-k)!}.$$

We now see that

$$\mathsf{P}(Y = k) = \binom{n}{k} p^k(1-p)^{n-k}, \quad k = 0,\ldots,n.$$

Another way to derive the formula for $\mathsf{P}(Y = k)$ is to use the theory of probability generating functions as developed in Section 3.1. In this method, we first find $G_Y(z)$ and then use the formula $G_Y^{(k)}(z)|_{z=0}/k! = \mathsf{P}(Y = k)$. To find $G_Y(z)$, we use the factorization property for pgfs of sums of independent random variables. Write

$$\begin{aligned} G_Y(z) &= \mathsf{E}[z^Y] \\ &= \mathsf{E}[z^{X_1+\cdots+X_n}] \\ &= \mathsf{E}[z^{X_1}]\cdots \mathsf{E}[z^{X_n}], \quad \text{by independence,} \\ &= G_{X_1}(z) \cdots G_{X_n}(z). \end{aligned}$$

3.2 The binomial random variable

For the Bernoulli(p) random variables X_i,

$$G_{X_i}(z) := \mathsf{E}[z^{X_i}] = z^0(1-p) + z^1 p = (1-p) + pz.$$

Thus,

$$G_Y(z) = [(1-p) + pz]^n.$$

Next, we need the derivatives of $G_Y(z)$. The first derivative is

$$G'_Y(z) = n[(1-p) + pz]^{n-1} p,$$

and in general, the kth derivative is

$$G_Y^{(k)}(z) = n(n-1) \cdots (n - [k-1])[(1-p) + pz]^{n-k} p^k.$$

It follows that

$$\begin{aligned} \mathsf{P}(Y = k) &= \frac{G_Y^{(k)}(0)}{k!} \\ &= \frac{n(n-1) \cdots (n - [k-1])}{k!} (1-p)^{n-k} p^k \\ &= \frac{n!}{k!(n-k)!} p^k (1-p)^{n-k} \\ &= \binom{n}{k} p^k (1-p)^{n-k}. \end{aligned}$$

Since the formula for $G_Y(z)$ is a polynomial of degree n, $G_Y^{(k)}(z) = 0$ for all $k > n$. Thus, $\mathsf{P}(Y = k) = 0$ for $k > n$.

The preceding random variable Y is called a **binomial**(n, p) random variable. Its probability mass function is usually written using the notation

$$p_Y(k) = \binom{n}{k} p^k (1-p)^{n-k}, \quad k = 0, \ldots, n.$$

In MATLAB, $\binom{n}{k}$ = nchoosek(n,k). A graph of $p_Y(k)$ is shown in Figure 3.1.

The **binomial theorem** says that for any complex numbers a and b,

$$\sum_{k=0}^{n} \binom{n}{k} a^k b^{n-k} = (a+b)^n.$$

A derivation using induction on n along with the easily verified identity

$$\binom{n}{k-1} + \binom{n}{k} = \binom{n+1}{k}, \quad k = 1, \ldots, n, \tag{3.6}$$

can be given. However, for nonnegative a and b with $a + b > 0$, the result is an easy consequence of our knowledge of the binomial random variable (see Problem 10).

More about discrete random variables

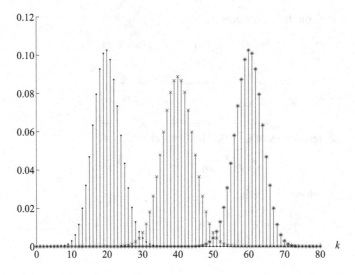

Figure 3.1. The binomial(n,p) pmf $p_Y(k) = \binom{n}{k}p^k(1-p)^{n-k}$ for $n=80$ and $p=0.25, 0.5$, and 0.75 from left to right, respectively.

On account of the binomial theorem, the quantity $\binom{n}{k}$ is sometimes called the **binomial coefficient**. It is convenient to know that the binomial coefficients can be read off from the nth row of **Pascal's triangle** in Figure 3.2. Noting that the top row is row 0, it is immediately seen, for example, that

$$(a+b)^5 = a^5 + 5a^4b + 10a^3b^2 + 10a^2b^3 + 5ab^4 + b^5.$$

To generate the triangle, observe that except for the entries that are ones, each entry is equal to the sum of the two numbers above it to the left and right. Thus, the triangle is a graphical depiction of (3.6).

Figure 3.2. Pascal's triangle.

Poisson approximation of binomial probabilities

If we let $\lambda := np$, then the probability generating function of a binomial(n,p) random variable can be written as

$$[(1-p)+pz]^n = [1+p(z-1)]^n$$
$$= \left[1+\frac{\lambda(z-1)}{n}\right]^n.$$

From calculus, recall the formula

$$\lim_{n\to\infty}\left(1+\frac{x}{n}\right)^n = e^x.$$

So, for large n,

$$\left[1+\frac{\lambda(z-1)}{n}\right]^n \approx \exp[\lambda(z-1)],$$

which is the probability generating function of a Poisson(λ) random variable (Example 3.4). In making this approximation, n should be large compared to $\lambda(z-1)$. Since $\lambda := np$, as n becomes large, so does $\lambda(z-1)$. To keep the size of λ small enough to be useful, we should keep p small. Under this assumption, the binomial(n,p) probability generating function is close to the Poisson(np) probability generating function. This suggests the **Poisson approximation**[a]

$$\binom{n}{k}p^k(1-p)^{n-k} \approx \frac{(np)^k e^{-np}}{k!}, \quad n \text{ large}, p \text{ small}.$$

Example 3.7. As noted in Example 3.6, the number of alpha particles emitted from a radioactive sample is a binomial(n,p) random variable. However, since n is large, say 10^{23}, even if the expected number of particles, np (Problem 8), is in the billions, say 10^9, $p \approx 10^{-14}$ is still very small, and the Poisson approximation is justified.[b]

3.3 The weak law of large numbers

Let $X_1, X_2, \ldots$ be a sequence of random variables with a common mean $\mathsf{E}[X_i] = m$ for all i. In practice, since we do not know m, we use the numerical average, or **sample mean**,

$$M_n := \frac{1}{n}\sum_{i=1}^{n} X_i,$$

in place of the true, but unknown value, m. Can this procedure of using M_n as an estimate of m be justified in some sense?

---

[a]This approximation is justified rigorously in Problems 20 and 21(a) in Chapter 14. It is also derived directly without probability generating functions in Problem 22 in Chapter 14.

[b]If the sample's mass m is measured in grams, then the number of atoms in the sample is $n = mA/w$, where $A = 6.022 \times 10^{23}$ is **Avogadro's number**, and w is the **atomic weight** of the material. For example, the atomic weight of radium is 226.

Example 3.8. You are given a coin which may or may not be fair, and you want to determine the probability of heads, p. If you toss the coin n times and use the fraction of times that heads appears as an estimate of p, how does this fit into the above framework?

Solution. Let $X_i = 1$ if the ith toss results in heads, and let $X_i = 0$ otherwise. Then $P(X_i = 1) = p$ and $m := E[X_i] = p$ as well. Note that $X_1 + \cdots + X_n$ is the number of heads, and M_n is the fraction of heads. Are we justified in using M_n as an estimate of p?

One way to answer these questions is with a **weak law of large numbers (WLLN)**. A weak law of large numbers gives conditions under which

$$\lim_{n \to \infty} P(|M_n - m| \geq \varepsilon) = 0$$

for every $\varepsilon > 0$. This is a complicated formula. However, it can be interpreted as follows. Suppose that based on physical considerations, m is between 30 and 70. Let us agree that if M_n is within $\varepsilon = 1/2$ of m, we are "close enough" to the unknown value m. For example, if $M_n = 45.7$, and if we know that M_n is within $1/2$ of m, then m is between 45.2 and 46.2. Knowing this would be an improvement over the starting point $30 \leq m \leq 70$. So, if $|M_n - m| < \varepsilon$, we are "close enough," while if $|M_n - m| \geq \varepsilon$ we are not "close enough." A weak law says that by making n large (averaging lots of measurements), the probability of not being close enough can be made as small as we like; equivalently, the probability of being close enough can be made as close to one as we like. For example, if $P(|M_n - m| \geq \varepsilon) \leq 0.1$, then

$$P(|M_n - m| < \varepsilon) = 1 - P(|M_n - m| \geq \varepsilon) \geq 0.9,$$

and we would be 90% sure that M_n is "close enough" to the true, but unknown, value of m.

Conditions for the weak law

We now give sufficient conditions for a version of the **weak law of large numbers (WLLN)**. Suppose that the X_i all have the same mean m and the same variance σ^2. Assume also that the X_i are *uncorrelated* random variables. Then for every $\varepsilon > 0$,

$$\lim_{n \to \infty} P(|M_n - m| \geq \varepsilon) = 0.$$

This is an immediate consequence of the following two facts. First, by the Chebyshev inequality (2.22),

$$P(|M_n - m| \geq \varepsilon) \leq \frac{\text{var}(M_n)}{\varepsilon^2}.$$

Second, since the X_i are uncorrelated, a slight extension of (2.28) gives

$$\text{var}(M_n) = \text{var}\left(\frac{1}{n}\sum_{i=1}^{n} X_i\right) = \left(\frac{1}{n}\right)^2 \sum_{i=1}^{n} \text{var}(X_i) = \frac{n\sigma^2}{n^2} = \frac{\sigma^2}{n}. \quad (3.7)$$

Thus,

$$P(|M_n - m| \geq \varepsilon) \leq \frac{\sigma^2}{n\varepsilon^2}, \quad (3.8)$$

which goes to zero as $n \to \infty$. Note that the bound $\sigma^2/n\varepsilon^2$ can be used to select a suitable value of n.

Remark. The weak law was first proved around 1700 for $X_i \sim$ Bernoulli(p) random variables by Jacob (a.k.a. James or Jacques) Bernoulli.

Example 3.9. Given ε and σ^2, determine how large n should be so the probability that M_n is within ε of m is at least 0.9.

Solution. We want to have

$$\mathsf{P}(|M_n - m| < \varepsilon) \geq 0.9.$$

Rewrite this as

$$1 - \mathsf{P}(|M_n - m| \geq \varepsilon) \geq 0.9,$$

or

$$\mathsf{P}(|M_n - m| \geq \varepsilon) \leq 0.1.$$

By (3.8), it suffices to take

$$\frac{\sigma^2}{n\varepsilon^2} \leq 0.1,$$

or $n \geq 10\sigma^2/\varepsilon^2$.

Remark. In using (3.8) as suggested, it would seem that we are smart enough to know σ^2, but not m. In practice, we may replace σ^2 in (3.8) with an upper bound. For example, if $X_i \sim$ Bernoulli(p), then $m = p$ and $\sigma^2 = p(1-p)$. Since, $0 \leq p \leq 1$, it is easy to show that $\sigma^2 \leq 1/4$.

Remark. If $Z_1, Z_2, \ldots$ are arbitrary, *independent*, identically distributed random variables, then for any set $B \subset \mathbb{R}$, taking $X_i := I_B(Z_i)$ gives an independent, and therefore uncorrelated, sequence of Bernoulli(p) random variables, where $p = \mathsf{E}[X_i] = \mathsf{P}(Z_i \in B)$. Hence, the weak law can be used to estimate probabilities as well as expected values. See Problems 18 and 19. This topic is pursued in more detail in Section 6.8.

3.4 Conditional probability

We introduce two main applications of conditional probability for random variables. One application is as an extremely powerful computational tool. In this connection, you will learn how to use

- the law of total probability for random variables,
- the substitution law, and
- independence (if you have it).

The other application of conditional probability is as a tool that uses observational data to estimate data that cannot be directly observed. For example, when data is sent over a noisy channel, we use the received measurements along with knowledge of the channel statistics to estimate the data that was actually transmitted.

For conditional probabilities involving random variables, we use the notation

$$P(X \in B | Y \in C) := P(\{X \in B\} | \{Y \in C\})$$
$$= \frac{P(\{X \in B\} \cap \{Y \in C\})}{P(\{Y \in C\})}$$
$$= \frac{P(X \in B, Y \in C)}{P(Y \in C)}.$$

For discrete random variables, we define the **conditional probability mass functions**,

$$p_{X|Y}(x_i|y_j) := P(X = x_i | Y = y_j)$$
$$= \frac{P(X = x_i, Y = y_j)}{P(Y = y_j)}$$
$$= \frac{p_{XY}(x_i, y_j)}{p_Y(y_j)},$$

and

$$p_{Y|X}(y_j|x_i) := P(Y = y_j | X = x_i)$$
$$= \frac{P(X = x_i, Y = y_j)}{P(X = x_i)}$$
$$= \frac{p_{XY}(x_i, y_j)}{p_X(x_i)}.$$

For future reference, we record these two formulas,

$$\boxed{p_{X|Y}(x_i|y_j) = \frac{p_{XY}(x_i, y_j)}{p_Y(y_j)}} \qquad (3.9)$$

and

$$\boxed{p_{Y|X}(y_j|x_i) = \frac{p_{XY}(x_i, y_j)}{p_X(x_i)}}, \qquad (3.10)$$

noting that they make sense only when the denominators are not zero. We call $p_{X|Y}$ the conditional probability mass function (pmf) of X given Y. Similarly, $p_{Y|X}$ is called the conditional pmf of Y given X. Notice that by multiplying through by the denominators, we obtain

$$\boxed{p_{XY}(x_i, y_j) = p_{X|Y}(x_i|y_j) p_Y(y_j) = p_{Y|X}(y_j|x_i) p_X(x_i).} \qquad (3.11)$$

Note that if either $p_X(x_i) = 0$ or $p_Y(y_j) = 0$, then $p_{XY}(x_i, y_j) = 0$ from the discussion following Example 1.23.

Example 3.10. Find the conditional probability mass function $p_{Y|X}(j|i)$ if

$$p_{XY}(i,j) := \begin{cases} 2\dfrac{[i/(i+1)]^j}{n(n+1)}, & j \geq 0, \, i = 0, \ldots, n-1, \\ 0, & \text{otherwise.} \end{cases}$$

3.4 Conditional probability

Solution. Recall from Example 2.14 that

$$p_X(i) = \begin{cases} 2\dfrac{i+1}{n(n+1)}, & i = 0, \ldots, n-1, \\ 0, & \text{otherwise}. \end{cases}$$

Hence, for $i = 0, \ldots, n-1$,

$$p_{Y|X}(j|i) = \begin{cases} \dfrac{1}{i+1}\left(\dfrac{i}{i+1}\right)^j, & j \geq 0, \\ 0, & j < 0. \end{cases}$$

In other words, given $X = i$ for i in the range $0, \ldots, n-1$, we have that Y is conditionally geometric$_0(i/(i+1))$.

The general formula $p_{Y|X}(y_j|x_i) = p_{XY}(x_i, y_j)/p_X(x_i)$ shows that for fixed x_i, $p_{Y|X}(y_j|x_i)$ as a function of y_j has the same shape as a slice of $p_{XY}(x_i, y_j)$. For the pmfs of Example 3.10 with $n = 5$, this is illustrated in Figure 3.3. Here we see that for fixed i, $p_{XY}(i,j)$ as a function of j has the shape of the geometric$_0(i/(i+1))$ pmf $p_{Y|X}(j|i)$.

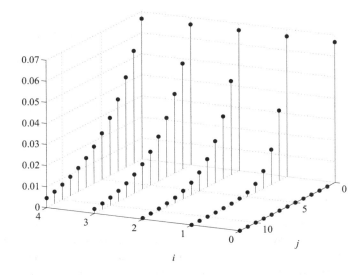

Figure 3.3. Sketch of bivariate probability mass function $p_{XY}(i,j)$ of Example 3.10 with $n = 5$. For fixed i, $p_{XY}(i,j)$ as a function of j is proportional to $p_{Y|X}(j|i)$, which is geometric$_0(i/(i+1))$. The special case $i = 0$ results in $p_{Y|X}(j|0) \sim$ geometric$_0(0)$, which corresponds to a constant random variable that takes the value $j = 0$ with probability one.

Conditional pmfs are important because we can use them to compute conditional probabilities just as we use marginal pmfs to compute ordinary probabilities. For example,

$$\mathsf{P}(Y \in C | X = x_k) = \sum_{j: y_j \in C} p_{Y|X}(y_j | x_k).$$

This formula is derived by taking $B = \{x_k\}$ in (2.11), and then dividing the result by $\mathsf{P}(X = x_k) = p_X(x_k)$.

Example 3.11 (optical channel). To transmit message i using an optical communication system, light of intensity λ_i is directed at a photodetector. When light of intensity λ_i strikes the photodetector, the number of photoelectrons generated is a Poisson(λ_i) random variable. Find the conditional probability that the number of photoelectrons observed at the photodetector is less than 2 given that message i was sent.

Solution. Let X denote the message to be sent, and let Y denote the number of photoelectrons generated by the photodetector. The problem statement is telling us that

$$\mathsf{P}(Y = n | X = i) = \frac{\lambda_i^n e^{-\lambda_i}}{n!}, \quad n = 0, 1, 2, \ldots.$$

The conditional probability to be calculated is

$$\begin{aligned}\mathsf{P}(Y < 2 | X = i) &= \mathsf{P}(Y = 0 \text{ or } Y = 1 | X = i) \\ &= \mathsf{P}(Y = 0 | X = i) + \mathsf{P}(Y = 1 | X = i) \\ &= e^{-\lambda_i} + \lambda_i e^{-\lambda_i}.\end{aligned}$$

Example 3.12. For the random variables X and Y used in the solution of the previous example, write down their joint pmf if $X \sim \text{geometric}_0(p)$.

Solution. The joint pmf is

$$p_{XY}(i, n) = p_X(i) p_{Y|X}(n|i) = (1-p)p^i \frac{\lambda_i^n e^{-\lambda_i}}{n!},$$

for $i, n \geq 0$, and $p_{XY}(i, n) = 0$ otherwise.

The law of total probability

In Chapter 1, we used the law of total probability to compute the probability of an event that can occur in different ways. In this chapter, we adapt the law to handle the case in which the events we condition on are described by discrete random variables. For example, the Internet traffic generated at a university depends on how many students are logged in. Even if we know the number of students logged in, the traffic they generate is random. However, the number of students logged in is a random variable. The law of total probability can help us analyze these situations.

Let $A \subset \Omega$ be any event, and let X be any discrete random variable taking distinct values x_i. Then the events

$$B_i := \{X = x_i\} = \{\omega \in \Omega : X(\omega) = x_i\}.$$

are pairwise disjoint, and $\sum_i \mathsf{P}(B_i) = \sum_i \mathsf{P}(X = x_i) = 1$. The law of total probability as in (1.27) yields

$$\mathsf{P}(A) = \sum_i \mathsf{P}(A \cap B_i) = \sum_i \mathsf{P}(A | X = x_i) \mathsf{P}(X = x_i).$$

3.4 Conditional probability

If Y is an arbitrary random variable, and we take $A = \{Y \in C\}$, where $C \subset \mathbb{R}$, then

$$P(Y \in C) = \sum_i P(Y \in C | X = x_i) P(X = x_i), \qquad (3.12)$$

which we again call the law of total probability. If Y is a discrete random variable taking distinct values y_j, then setting $C = \{y_j\}$ yields

$$P(Y = y_j) = \sum_i P(Y = y_j | X = x_i) P(X = x_i)$$
$$= \sum_i p_{Y|X}(y_j | x_i) p_X(x_i).$$

Example 3.13 (binary channel). If the input to the binary channel shown in Figure 3.4 is a Bernoulli(p) random variable X, and the output is the random variable Y, find $P(Y = j)$ for $j = 0, 1$.

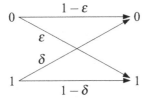

Figure 3.4. Binary channel with crossover probabilities ε and δ. If $\delta = \varepsilon$, this is called the binary symmetric channel.

Solution. The diagram is telling us that $P(Y = 1 | X = 0) = \varepsilon$ and $P(Y = 0 | X = 1) = \delta$. These are called **crossover probabilities**. The diagram also supplies the redundant information that $P(Y = 0 | X = 0) = 1 - \varepsilon$ and $P(Y = 1 | X = 1) = 1 - \delta$. Using the law of total probability, we have

$$P(Y = j) = P(Y = j | X = 0) P(X = 0) + P(Y = j | X = 1) P(X = 1).$$

In particular,

$$P(Y = 0) = P(Y = 0 | X = 0) P(X = 0) + P(Y = 0 | X = 1) P(X = 1)$$
$$= (1 - \varepsilon)(1 - p) + \delta p,$$

and

$$P(Y = 1) = P(Y = 1 | X = 0) P(X = 0) + P(Y = 1 | X = 1) P(X = 1)$$
$$= \varepsilon(1 - p) + (1 - \delta) p.$$

Example 3.14. In the preceding example, suppose $p = 1/2$, $\delta = 1/3$, and $\varepsilon = 1/4$. Compute $P(Y = 0)$ and $P(Y = 1)$.

Solution. We leave it to the reader to verify that $P(Y=0) = 13/24$ and $P(Y=1) = 11/24$. Since the crossover probabilities are small, the effect of the channel on the data is minimal. Since the input bit values are equally likely, we expect the output bit values to be almost equally likely, which they are.

Example 3.15. Radioactive samples give off alpha-particles at a rate based on the size of the sample. For a sample of size k, suppose that the number of particles observed is a Poisson random variable Y with parameter k. If the sample size is a geometric$_1(p)$ random variable X, find $P(Y=0)$ and $P(X=1|Y=0)$.

Solution. The first step is to realize that the problem statement is telling us that as a function of n, $P(Y=n|X=k)$ is the Poisson pmf with parameter k. In other words,

$$P(Y=n|X=k) = \frac{k^n e^{-k}}{n!}, \quad n = 0, 1, \ldots.$$

In particular, note that $P(Y=0|X=k) = e^{-k}$. Now use the law of total probability to write

$$\begin{aligned}
P(Y=0) &= \sum_{k=1}^{\infty} P(Y=0|X=k) \cdot P(X=k) \\
&= \sum_{k=1}^{\infty} e^{-k} \cdot (1-p)p^{k-1} \\
&= \frac{1-p}{p} \sum_{k=1}^{\infty} (p/e)^k \\
&= \frac{1-p}{p} \frac{p/e}{1-p/e} = \frac{1-p}{e-p}.
\end{aligned}$$

Next,

$$\begin{aligned}
P(X=1|Y=0) &= \frac{P(X=1, Y=0)}{P(Y=0)} \\
&= \frac{P(Y=0|X=1)P(X=1)}{P(Y=0)} \\
&= e^{-1} \cdot (1-p) \cdot \frac{e-p}{1-p} \\
&= 1 - p/e.
\end{aligned}$$

Example 3.16. A certain electric eye employs a photodetector whose efficiency occasionally drops in half. When operating properly, the detector outputs photoelectrons according to a Poisson(λ) pmf. When the detector malfunctions, it outputs photoelectrons according to a Poisson$(\lambda/2)$ pmf. Let $p < 1$ denote the probability that the detector is operating properly. Find the pmf of the observed number of photoelectrons. Also find the conditional probability that the circuit is malfunctioning given that n output photoelectrons are observed.

3.4 Conditional probability

Solution. Let Y denote the detector output, and let $X = 1$ indicate that the detector is operating properly. Let $X = 0$ indicate that it is malfunctioning. Then the problem statement is telling us that $P(X=1) = p$ and

$$P(Y=n|X=1) = \frac{\lambda^n e^{-\lambda}}{n!} \quad \text{and} \quad P(Y=n|X=0) = \frac{(\lambda/2)^n e^{-\lambda/2}}{n!}.$$

Now, using the law of total probability,

$$\begin{aligned} P(Y=n) &= P(Y=n|X=1)P(X=1) + P(Y=n|X=0)P(X=0) \\ &= \frac{\lambda^n e^{-\lambda}}{n!} p + \frac{(\lambda/2)^n e^{-\lambda/2}}{n!}(1-p). \end{aligned}$$

This is the pmf of the observed number of photoelectrons.

The above formulas can be used to find $P(X=0|Y=n)$. Write

$$\begin{aligned} P(X=0|Y=n) &= \frac{P(X=0, Y=n)}{P(Y=n)} \\ &= \frac{P(Y=n|X=0)P(X=0)}{P(Y=n)} \\ &= \frac{\dfrac{(\lambda/2)^n e^{-\lambda/2}}{n!}(1-p)}{\dfrac{\lambda^n e^{-\lambda}}{n!}p + \dfrac{(\lambda/2)^n e^{-\lambda/2}}{n!}(1-p)} \\ &= \frac{1}{\dfrac{2^n e^{-\lambda/2} p}{(1-p)} + 1}, \end{aligned}$$

which is clearly a number between zero and one as a probability should be. Notice that as we observe a greater output $Y = n$, the conditional probability that the detector is malfunctioning decreases.

The substitution law

It is often the case that Z is a function of X and some other discrete random variable Y, say $Z = g(X,Y)$, and we are interested in $P(Z=z)$. In this case, the law of total probability becomes

$$\begin{aligned} P(Z=z) &= \sum_i P(Z=z|X=x_i)P(X=x_i). \\ &= \sum_i P(g(X,Y)=z|X=x_i)P(X=x_i). \end{aligned}$$

We claim that

$$P\big(g(X,Y)=z \big| X=x_i\big) = P\big(g(x_i,Y)=z \big| X=x_i\big). \tag{3.13}$$

This property is known as the **substitution law** of conditional probability. To derive it, we need the observation

$$\{g(X,Y)=z\} \cap \{X=x_i\} = \{g(x_i,Y)=z\} \cap \{X=x_i\}.$$

From this we see that

$$\begin{aligned}
\mathsf{P}(g(X,Y)=z|X=x_i) &= \frac{\mathsf{P}(\{g(X,Y)=z\} \cap \{X=x_i\})}{\mathsf{P}(\{X=x_i\})} \\
&= \frac{\mathsf{P}(\{g(x_i,Y)=z\} \cap \{X=x_i\})}{\mathsf{P}(\{X=x_i\})} \\
&= \mathsf{P}(g(x_i,Y)=z|X=x_i).
\end{aligned}$$

We can make further simplifications if X and Y are independent. In this case,

$$\begin{aligned}
\mathsf{P}(g(x_i,Y)=z|X=x_i) &= \frac{\mathsf{P}(g(x_i,Y)=z,X=x_i)}{\mathsf{P}(X=x_i)} \\
&= \frac{\mathsf{P}(g(x_i,Y)=z)\mathsf{P}(X=x_i)}{\mathsf{P}(X=x_i)} \\
&= \mathsf{P}(g(x_i,Y)=z).
\end{aligned}$$

Thus, when X and Y are independent, we can write

$$\mathsf{P}(g(x_i,Y)=z|X=x_i) = \mathsf{P}(g(x_i,Y)=z), \qquad (3.14)$$

and we say that we "drop the conditioning."

Example 3.17 (signal in additive noise). A random, integer-valued signal X is transmitted over a channel subject to independent, additive, integer-valued noise Y. The received signal is $Z = X+Y$ as shown in Figure 3.5. To estimate X based on the received value Z, the system designer wants to use the conditional pmf $p_{X|Z}$. Find the desired conditional pmf.

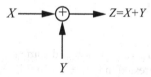

Figure 3.5. Signal X subjected to additive noise Y.

Solution. Let X and Y be independent, discrete, integer-valued random variables with pmfs p_X and p_Y, respectively. Put $Z := X+Y$. We begin by writing out the formula for the desired pmf

$$\begin{aligned}
p_{X|Z}(i|j) &= \mathsf{P}(X=i|Z=j) \\
&= \frac{\mathsf{P}(X=i,Z=j)}{\mathsf{P}(Z=j)}
\end{aligned}$$

3.4 Conditional probability

$$= \frac{\mathsf{P}(Z=j|X=i)\mathsf{P}(X=i)}{\mathsf{P}(Z=j)}$$
$$= \frac{\mathsf{P}(Z=j|X=i)p_X(i)}{\mathsf{P}(Z=j)}. \qquad (3.15)$$

To continue the analysis, we use the substitution law followed by independence to write

$$\begin{aligned}\mathsf{P}(Z=j|X=i) &= \mathsf{P}(X+Y=j|X=i)\\ &= \mathsf{P}(i+Y=j|X=i)\\ &= \mathsf{P}(Y=j-i|X=i)\\ &= \mathsf{P}(Y=j-i)\\ &= p_Y(j-i). \qquad (3.16)\end{aligned}$$

This result can also be combined with the law of total probability to compute the denominator in (3.15). Just write

$$p_Z(j) = \sum_i \mathsf{P}(Z=j|X=i)\mathsf{P}(X=i) = \sum_i p_Y(j-i)p_X(i). \qquad (3.17)$$

In other words, if X and Y are independent, discrete, integer-valued random variables, the pmf of $Z = X + Y$ is the discrete **convolution** of p_X and p_Y.

It now follows that

$$p_{X|Z}(i|j) = \frac{p_Y(j-i)p_X(i)}{\sum_k p_Y(j-k)p_X(k)},$$

where in the denominator we have changed the dummy index of summation to k to avoid confusion with the i in the numerator.

The Poisson(λ) random variable is a good model for the number of photoelectrons generated in a photodetector when the incident light intensity is λ. Now suppose that an additional light source of intensity μ is also directed at the photodetector. Then we expect that the number of photoelectrons generated should be related to the total light intensity $\lambda + \mu$. The next example illustrates the corresponding probabilistic model.

Example 3.18 (Poisson channel). If X and Y are independent Poisson random variables with respective parameters λ and μ, use the results of the preceding example to show that $Z := X + Y$ is Poisson($\lambda + \mu$). Also show that as a function of i, $p_{X|Z}(i|j)$ is a binomial$(j, \lambda/(\lambda + \mu))$ pmf.

Solution. To find $p_Z(j)$, we apply (3.17) as follows. Since $p_X(i) = 0$ for $i < 0$ and since $p_Y(j-i) = 0$ for $j < i$, (3.17) becomes

$$\begin{aligned}p_Z(j) &= \sum_{i=0}^{j} \frac{\lambda^i e^{-\lambda}}{i!} \cdot \frac{\mu^{i-i}e^{-\mu}}{(j-i)!}\\ &= \frac{e^{-(\lambda+\mu)}}{j!} \sum_{i=0}^{j} \frac{j!}{i!(j-i)!}\lambda^i \mu^{j-i}\end{aligned}$$

$$= \frac{e^{-(\lambda+\mu)}}{j!} \sum_{i=0}^{j} \binom{j}{i} \lambda^i \mu^{j-i}$$

$$= \frac{e^{-(\lambda+\mu)}}{j!} (\lambda+\mu)^j, \quad j = 0, 1, \ldots,$$

where the last step follows by the binomial theorem.

Our second task is to compute

$$p_{X|Z}(i|j) = \frac{\mathsf{P}(Z=j|X=i)\mathsf{P}(X=i)}{\mathsf{P}(Z=j)}$$

$$= \frac{\mathsf{P}(Z=j|X=i)p_X(i)}{p_Z(j)}.$$

Since we have already found $p_Z(j)$, all we need is $\mathsf{P}(Z=j|X=i)$, which, using (3.16), is simply $p_Y(j-i)$. Thus,

$$p_{X|Z}(i|j) = \frac{\mu^{j-i}e^{-\mu}}{(j-i)!} \cdot \frac{\lambda^i e^{-\lambda}}{i!} \bigg/ \left[\frac{e^{-(\lambda+\mu)}}{j!}(\lambda+\mu)^j \right]$$

$$= \frac{\lambda^i \mu^{j-i}}{(\lambda+\mu)^j} \binom{j}{i}$$

$$= \binom{j}{i} \left[\frac{\lambda}{(\lambda+\mu)}\right]^i \left[\frac{\mu}{(\lambda+\mu)}\right]^{j-i},$$

for $i = 0, \ldots, j$.

Binary channel receiver design

Consider the problem of a receiver using the binary channel in Figure 3.4. The receiver has access to the channel output Y, and must estimate, or guess, the value of X. What decision rule should be used? It turns out that no decision rule can have a smaller probability of error than the **maximum a posteriori probability** (MAP) rule.[4] Having observed $Y = j$, the MAP rule says to decide $X = 1$ if

$$\mathsf{P}(X=1|Y=j) \geq \mathsf{P}(X=0|Y=j), \tag{3.18}$$

and to decide $X = 0$ otherwise. In other words, the MAP rule decides $X = 1$ if the posterior probability of $X = 1$ given the observation $Y = j$ is greater than the posterior probability of $X = 0$ given the observation $Y = j$.

Observe that since

$$\mathsf{P}(X=i|Y=j) = \frac{\mathsf{P}(X=i,Y=j)}{\mathsf{P}(Y=j)} = \frac{\mathsf{P}(Y=j|X=i)\mathsf{P}(X=i)}{\mathsf{P}(Y=j)},$$

(3.18) can be rewritten as

$$\frac{\mathsf{P}(Y=j|X=1)\mathsf{P}(X=1)}{\mathsf{P}(Y=j)} \geq \frac{\mathsf{P}(Y=j|X=0)\mathsf{P}(X=0)}{\mathsf{P}(Y=j)}.$$

Canceling the common denominator, we have

$$P(Y=j|X=1)P(X=1) \geq P(Y=j|X=0)P(X=0). \quad (3.19)$$

This is an important observation. It says that we do not need to compute the denominator to implement the MAP rule.

Next observe that if the inputs $X = 0$ and $X = 1$ are equally likely, we can cancel these common factors as well and get

$$P(Y=j|X=1) \geq P(Y=j|X=0). \quad (3.20)$$

Sometimes we do not know the prior probabilities $P(X = i)$. In this case, we sometimes use (3.20) anyway. The rule that decides $X = 1$ when (3.20) holds and $X = 0$ otherwise is called the **maximum-likelihood** (ML) rule. In this context, $P(Y = j|X = i)$ is called the **likelihood** of $Y = j$. The maximum-likelihood rule decides $X = i$ if i maximizes the likelihood of the observation $Y = j$.

A final thing to note about the MAP rule is that (3.19) can be rearranged as

$$\frac{P(Y=j|X=1)}{P(Y=j|X=0)} \geq \frac{P(X=0)}{P(X=1)}.$$

Since the left-hand side is the ratio of the likelihoods, this quotient is called the **likelihood ratio**. The right-hand side does not depend on j and is just a constant, sometimes called a threshold. The MAP rule compares the likelihood ratio against this specific threshold. The ML rule compares the likelihood ratio against the threshold one. Both the MAP rule and ML rule are sometimes called **likelihood-ratio tests**. The reason for writing the tests in terms of the likelihood ratio is that the form of the test can be greatly simplified; e.g., as in Problems 35 and 36.

3.5 Conditional expectation

Just as we developed expectation for discrete random variables in Section 2.4, including the law of the unconscious statistician, we can develop conditional expectation in the same way. This leads to the formula

$$\boxed{E[g(Y)|X=x_i] = \sum_j g(y_j) p_{Y|X}(y_j|x_i).} \quad (3.21)$$

Example **3.19.** The random number Y of alpha particles emitted by a radioactive sample is conditionally Poisson(k) given that the sample size $X = k$. Find $E[Y|X = k]$.

Solution. We must compute

$$E[Y|X=k] = \sum_n n P(Y=n|X=k),$$

where (cf. Example 3.15)

$$P(Y=n|X=k) = \frac{k^n e^{-k}}{n!}, \quad n=0,1,\ldots.$$

Hence,
$$E[Y|X=k] = \sum_{n=0}^{\infty} n \frac{k^n e^{-k}}{n!}.$$
Now observe that the right-hand side is exactly ordinary expectation of a Poisson random variable with parameter k (cf. the calculation in Example 2.22). Therefore, $E[Y|X=k] = k$.

***Example* 3.20.** Let Z be the output of the Poisson channel of Example 3.18, and let X be the transmitted signal. Compute $E[X|Z=j]$ using the conditional pmf $p_{X|Z}(i|j)$ found in Example 3.18.

Solution. We must compute
$$E[X|Z=j] = \sum_{i=0}^{j} i P(X=i|Z=j),$$
where, letting $p := \lambda/(\lambda + \mu)$,
$$P(X=i|Z=j) = \binom{j}{i} p^i (1-p)^{j-i}.$$
Hence,
$$E[X|Z=j] = \sum_{i=0}^{j} i \binom{j}{i} p^i (1-p)^{j-i}.$$
Now observe that the right-hand side is exactly the ordinary expectation of a binomial(j, p) random variable. It is shown in Problem 8 that the mean of such a random variable is jp. Therefore, $E[X|Z=j] = jp = j\lambda/(\lambda + \mu)$.

Substitution law for conditional expectation

For functions of two variables, we have the following conditional law of the unconscious statistician,
$$E[g(X,Y)|X=x_i] = \sum_k \sum_j g(x_k, y_j) p_{XY|X}(x_k, y_j|x_i).$$
However,
$$p_{XY|X}(x_k, y_j|x_i) = P(X=x_k, Y=y_j|X=x_i)$$
$$= \frac{P(X=x_k, Y=y_j, X=x_i)}{P(X=x_i)}.$$
Now, when $k \neq i$, the intersection
$$\{X=x_k\} \cap \{Y=y_j\} \cap \{X=x_i\}$$
is empty, and has zero probability. Hence, the numerator above is zero for $k \neq i$. When $k = i$, the above intersections reduce to $\{X=x_i\} \cap \{Y=y_j\}$, and so
$$p_{XY|X}(x_k, y_j|x_i) = p_{Y|X}(y_j|x_i), \quad \text{for } k=i.$$

3.5 Conditional expectation

It now follows that

$$\mathsf{E}[g(X,Y)|X=x_i] = \sum_j g(x_i,y_j) p_{Y|X}(y_j|x_i)$$
$$= \mathsf{E}[g(x_i,Y)|X=x_i].$$

We call

$$\mathsf{E}[g(X,Y)|X=x_i] = \mathsf{E}[g(x_i,Y)|X=x_i] \qquad (3.22)$$

the **substitution law for conditional expectation**. Note that if g in (3.22) is a function of Y only, then (3.22) reduces to (3.21). Also, if g is of product form, say $g(x,y) = h(x)k(y)$, then

$$\mathsf{E}[h(X)k(Y)|X=x_i] = h(x_i)\mathsf{E}[k(Y)|X=x_i].$$

Law of total probability for expectation

In Section 3.4 we discussed the law of total probability, which shows how to compute probabilities in terms of conditional probabilities. We now derive the analogous formula for expectation. Write

$$\sum_i \mathsf{E}[g(X,Y)|X=x_i] p_X(x_i) = \sum_i \left[\sum_j g(x_i,y_j) p_{Y|X}(y_j|x_i) \right] p_X(x_i)$$
$$= \sum_i \sum_j g(x_i,y_j) p_{XY}(x_i,y_j)$$
$$= \mathsf{E}[g(X,Y)].$$

Hence, the **law of total probability for expectation** is

$$\mathsf{E}[g(X,Y)] = \sum_i \mathsf{E}[g(X,Y)|X=x_i] p_X(x_i). \qquad (3.23)$$

In particular, if g is a function of Y only, then

$$\mathsf{E}[g(Y)] = \sum_i \mathsf{E}[g(Y)|X=x_i] p_X(x_i).$$

Example 3.21. Light of intensity λ is directed at a photomultiplier that generates $X \sim$ Poisson(λ) primaries. The photomultiplier then generates Y secondaries, where given $X = n$, Y is conditionally geometric$_1((n+2)^{-1})$. Find the expected number of secondaries and the correlation between the primaries and the secondaries.

Solution. The law of total probability for expectations says that

$$\mathsf{E}[Y] = \sum_{n=0}^{\infty} \mathsf{E}[Y|X=n] p_X(n),$$

where the range of summation follows because X is Poisson(λ). The next step is to compute the conditional expectation. The conditional pmf of Y is geometric$_1(p)$, where, in this case,

$p = (n+2)^{-1}$, and the mean of such a pmf is, by Problem 4, $1/(1-p)$. Hence,

$$\mathsf{E}[Y] = \sum_{n=0}^{\infty} \left[1 + \frac{1}{n+1}\right] p_X(n) = \mathsf{E}\left[1 + \frac{1}{X+1}\right].$$

An easy calculation (Problem 34 in Chapter 2) shows that for $X \sim \text{Poisson}(\lambda)$,

$$\mathsf{E}\left[\frac{1}{X+1}\right] = [1 - e^{-\lambda}]/\lambda,$$

and so $\mathsf{E}[Y] = 1 + [1 - e^{-\lambda}]/\lambda$.

The correlation between X and Y is

$$\mathsf{E}[XY] = \sum_{n=0}^{\infty} \mathsf{E}[XY|X=n] p_X(n)$$

$$= \sum_{n=0}^{\infty} n \mathsf{E}[Y|X=n] p_X(n)$$

$$= \sum_{n=0}^{\infty} n \left[1 + \frac{1}{n+1}\right] p_X(n)$$

$$= \mathsf{E}\left[X\left(1 + \frac{1}{X+1}\right)\right].$$

Now observe that

$$X\left(1 + \frac{1}{X+1}\right) = X + 1 - \frac{1}{X+1}.$$

It follows that

$$\mathsf{E}[XY] = \lambda + 1 - [1 - e^{-\lambda}]/\lambda.$$

Notes

3.1: Probability generating functions

Note 1. When z is complex,

$$\mathsf{E}[z^X] := \mathsf{E}[\text{Re}(z^X)] + j\mathsf{E}[\text{Im}(z^X)].$$

By writing

$$z^n = r^n e^{jn\theta} = r^n[\cos(n\theta) + j\sin(n\theta)],$$

it is easy to check that for $|z| \leq 1$, the above expectations are finite (cf. (3.3)) and that

$$\mathsf{E}[z^X] = \sum_{n=0}^{\infty} z^n \mathsf{P}(X=n).$$

Note 2. Although $G_X(z)$ is well defined for $|z| \leq 1$, the existence of its derivatives is only guaranteed for $|z| < 1$. Hence, $G_X^{(k)}(1)$ may have to be understood as $\lim_{z \uparrow 1} G_X^{(k)}(z)$. By **Abel's theorem** [32, pp. 64–65], this limit is equal to the kth factorial moment on the right-hand side of (3.5), even if it is infinite.

3.4: Conditional probability

Note 3. Here is an alternative derivation of the fact that the sum of independent Bernoulli random variables is a binomial random variable. Let $X_1, X_2, \ldots$ be independent Bernoulli(p) random variables. Put

$$Y_n := \sum_{i=1}^{n} X_i.$$

We need to show that $Y_n \sim \text{binomial}(n, p)$. The case $n = 1$ is trivial. Suppose the result is true for some $n \geq 1$. We show that it must be true for $n + 1$. Use the law of total probability to write

$$P(Y_{n+1} = k) = \sum_{i=0}^{n} P(Y_{n+1} = k | Y_n = i) P(Y_n = i). \qquad (3.24)$$

To compute the conditional probability, we first observe that $Y_{n+1} = Y_n + X_{n+1}$. Also, since the X_i are independent, and since Y_n depends only on $X_1, \ldots, X_n$, we see that Y_n and X_{n+1} are independent. Keeping this in mind, we apply the substitution law and write

$$\begin{aligned} P(Y_{n+1} = k | Y_n = i) &= P(Y_n + X_{n+1} = k | Y_n = i) \\ &= P(i + X_{n+1} = k | Y_n = i) \\ &= P(X_{n+1} = k - i | Y_n = i) \\ &= P(X_{n+1} = k - i). \end{aligned}$$

Since X_{n+1} takes only the values zero and one, this last probability is zero unless $i = k$ or $i = k - 1$. Returning to (3.24), we can write[c]

$$P(Y_{n+1} = k) = \sum_{i=k-1}^{k} P(X_{n+1} = k - i) P(Y_n = i).$$

Assuming that $Y_n \sim \text{binomial}(n, p)$, this becomes

$$P(Y_{n+1} = k) = p \binom{n}{k-1} p^{k-1}(1-p)^{n-(k-1)} + (1-p)\binom{n}{k} p^k (1-p)^{n-k}.$$

Using the easily verified identity,

$$\binom{n}{k-1} + \binom{n}{k} = \binom{n+1}{k},$$

we see that $Y_{n+1} \sim \text{binomial}(n+1, p)$.

Note 4. We show that the MAP rule is optimal for minimizing the probability of a decision error. Consider a communication system whose input X takes values $1, \ldots, M$ with given probabilities $p_X(i) = P(X = i)$. The channel output is an integer-valued random variable Y. Assume that the conditional probability mass function $p_{Y|X}(j|i) = P(Y = j | X = i)$ is also known. The receiver decision rule is $\psi(Y) = i$ if $Y \in D_i$, where $D_1, \ldots, D_M$ is

---

[c] When $k = 0$ or $k = n+1$, this sum actually has only one term, since $P(Y_n = -1) = P(Y_n = n+1) = 0$.

a partition of IR. The problem is to characterize the choice for the partition sets D_i that minimizes the probability of a decision error, or, equivalently, maximizes the probability of a correct decision. Use the laws of total probability and substitution to write the probability of a correct decision as

$$P(\psi(Y) = X) = \sum_{i=1}^{M} P(\psi(Y) = X | X = i) P(X = i)$$

$$= \sum_{i=1}^{M} P(\psi(Y) = i | X = i) p_X(i)$$

$$= \sum_{i=1}^{M} P(Y \in D_i | X = i) p_X(i)$$

$$= \sum_{i=1}^{M} \left[\sum_j I_{D_i}(j) p_{Y|X}(j|i) \right] p_X(i)$$

$$= \sum_j \left[\sum_{i=1}^{M} I_{D_i}(j) p_{Y|X}(j|i) p_X(i) \right].$$

For fixed j, consider the inner sum. Since the D_i form a partition, the only term that is not zero is the one for which $j \in D_i$. To maximize this value, we should put $j \in D_i$ if and only if the weight $p_{Y|X}(j|i)p_X(i)$ is greater than or equal to $p_{Y|X}(j|i')p_X(i')$ for all $i' \neq i$. This is exactly the MAP rule (cf. (3.19)).

Problems

3.1: Probability generating functions

1. Find $\text{var}(X)$ if X has probability generating function

$$G_X(z) = \tfrac{1}{6} + \tfrac{1}{6}z + \tfrac{2}{3}z^2.$$

2. If $G_X(z)$ is as in the preceding problem, find the probability mass function of X.

3. Find $\text{var}(X)$ if X has probability generating function

$$G_X(z) = \left(\frac{2+z}{3}\right)^5.$$

4. Evaluate $G_X(z)$ for the cases $X \sim \text{geometric}_0(p)$ and $X \sim \text{geometric}_1(p)$. Use your results to find the mean and variance of X in each case.

5. For $i = 1, \ldots, n$, let $X_i \sim \text{Poisson}(\lambda_i)$. Put

$$Y := \sum_{i=1}^{n} X_i.$$

Find $P(Y = 2)$ if the X_i are independent.

6. Let $a_0, \ldots, a_n$ be nonnegative and not all zero. Let m be any positive integer. Find a constant D such that

$$G_X(z) := (a_0 + a_1 z + a_2 z^2 + \cdots + a_n z^n)^m / D$$

is a valid probability generating function.

7. Let $X_1, X_2, \ldots, X_n$ be i.i.d. geometric$_1(p)$ random variables, and put $Y := X_1 + \cdots + X_n$. Find $\mathsf{E}[Y]$, $\text{var}(Y)$, and $\mathsf{E}[Y^2]$. Also find the probability generating function of Y.

   Remark. We say that Y is a **negative binomial** or **Pascal** random variable with parameters n and p.

3.2: The binomial random variable

8. Use the probability generating function of $Y \sim \text{binomial}(n,p)$ to find the mean and variance of Y.

9. Show that the binomial(n,p) probabilities sum to one. *Hint:* Use the fact that for any nonnegative integer-valued random variable, $G_Y(z)|_{z=1} = 1$.

10. The **binomial theorem** says that

$$\sum_{k=0}^{n} \binom{n}{k} a^k b^{n-k} = (a+b)^n.$$

Derive this result for nonnegative a and b with $a+b > 0$ by using the fact that the binomial(n,p) probabilities sum to one. *Hint:* Take $p = a/(a+b)$.

11. A certain digital communication link has bit-error probability p. In a transmission of n bits, find the probability that k bits are received incorrectly, assuming bit errors occur independently.

12. A new school has M classrooms. For $i = 1, \ldots, M$, let n_i denote the number of seats in the ith classroom. Suppose that the number of students in the ith classroom is binomial(n_i, p) and independent. Let Y denote the total number of students in the school. Find $\mathsf{P}(Y = k)$.

13. Let $X_1, \ldots, X_n$ be i.i.d. with $\mathsf{P}(X_i = 1) = 1 - p$ and $\mathsf{P}(X_i = 2) = p$. If $Y := X_1 + \cdots + X_n$, find $\mathsf{P}(Y = k)$ for all k.

14. Ten-bit codewords are transmitted over a noisy channel. Bits are flipped independently with probability p. If no more than two bits of a codeword are flipped, the codeword can be correctly decoded. Find the probability that a codeword cannot be correctly decoded.

15. Make a table comparing both sides of the Poisson approximation of binomial probabilities,

$$\binom{n}{k} p^k (1-p)^{n-k} \approx \frac{(np)^k e^{-np}}{k!}, \quad n \text{ large, } p \text{ small,}$$

for $k = 0, 1, 2, 3, 4, 5$ if $n = 150$ and $p = 1/100$. *Hint:* If MATLAB is available, the binomial probability can be written

$$\text{nchoosek(n,k)} * \text{p\^{}k} * (1-\text{p})\text{\^{}(n-k)}$$

and the Poisson probability can be written

$$(\text{n}*\text{p})\text{\^{}k} * \exp(-\text{n}*\text{p})/\text{factorial(k)}.$$

3.3: The weak law of large numbers

16. Show that $\mathsf{E}[M_n] = m$. Also show that for any constant c, $\text{var}(cX) = c^2 \text{var}(X)$.

17. Student heights range from 120 to 220 cm. To estimate the average height, determine how many students' heights should be measured to make the sample mean within 0.25 cm of the true mean height with probability at least 0.9. Assume measurements are uncorrelated and have variance $\sigma^2 = 1$. What if you only want to be within 1 cm of the true mean height with probability at least 0.9?

18. Let $Z_1, Z_2, \ldots$ be i.i.d. random variables, and for any set $B \subset \mathbb{R}$, put $X_i := I_B(Z_i)$.

    (a) Find $\mathsf{E}[X_i]$ and $\text{var}(X_i)$.

    (b) Show that the X_i are uncorrelated.

    Observe that

    $$M_n = \frac{1}{n} \sum_{i=1}^{n} X_i = \frac{1}{n} \sum_{i=1}^{n} I_B(Z_i)$$

    counts the fraction of times Z_i lies in B. By the weak law of large numbers, for large n this fraction should be close to $\mathsf{P}(Z_i \in B)$.

19. With regard to the preceding problem, put $p := \mathsf{P}(Z_i \in B)$. If p is very small, and n is not large enough, it is likely that $M_n = 0$, which is useless as an estimate of p. If $p = 1/1000$, and $n = 100$, find $\mathsf{P}(M_{100} = 0)$.

20. Let X_i be a sequence of random variables, and put $M_n := (1/n) \sum_{i=1}^{n} X_i$. Assume that each X_i has mean m. Show that it is not always true that for every $\varepsilon > 0$,

    $$\lim_{n \to \infty} \mathsf{P}(|M_n - m| \geq \varepsilon) = 0.$$

    Hint: Let Z be a nonconstant random variable and take $X_i := Z$ for $i = 1, 2, \ldots$. To be specific, try $Z \sim \text{Bernoulli}(1/2)$ and $\varepsilon = 1/4$.

*21. Let $X_1, X_2, \ldots$ be uncorrelated random variables with common mean m and common variance σ^2. Let ε_n be a sequence of positive numbers with $\varepsilon_n \to 0$. With $M_n := (1/n) \sum_{i=1}^{n} X_i$, give sufficient conditions on ε_n such that

    $$\mathsf{P}(|M_n - m| \geq \varepsilon_n) \to 0.$$

3.4: Conditional probability

22. If $Z = X + Y$ as in the Poisson channel Example 3.18, find $E[X|Z = j]$.

23. Let X and Y be integer-valued random variables. Suppose that conditioned on $X = i$, $Y \sim \text{binomial}(n, p_i)$, where $0 < p_i < 1$. Evaluate $P(Y < 2|X = i)$.

24. Let X and Y be integer-valued random variables. Suppose that conditioned on $Y = j$, $X \sim \text{Poisson}(\lambda_j)$. Evaluate $P(X > 2|Y = j)$.

25. Let X and Y be independent random variables. Show that $p_{X|Y}(x_i|y_j) = p_X(x_i)$ and $p_{Y|X}(y_j|x_i) = p_Y(y_j)$.

26. Let X and Y be independent with $X \sim \text{geometric}_0(p)$ and $Y \sim \text{geometric}_0(q)$. Put $T := X - Y$, and find $P(T = n)$ for all n.

27. When a binary optical communication system transmits a 1, the receiver output is a Poisson(μ) random variable. When a 2 is transmitted, the receiver output is a Poisson(ν) random variable. Given that the receiver output is equal to 2, find the conditional probability that a 1 was sent. Assume messages are equally likely.

28. In a binary communication system, when a 0 is sent, the receiver outputs a random variable Y that is geometric$_0(p)$. When a 1 is sent, the receiver output $Y \sim$ geometric$_0(q)$, where $q \neq p$. Given that the receiver outputs $Y = k$, find the conditional probability that the message sent was a 1. Assume messages are equally likely.

29. Apple crates are supposed to contain only red apples, but occasionally a few green apples are found. Assume that the number of red apples and the number of green apples are independent Poisson random variables with parameters ρ and γ, respectively. Given that a crate contains a total of k apples, find the conditional probability that none of the apples is green.

30. Let $X \sim \text{Poisson}(\lambda)$, and suppose that given $X = n$, $Y \sim \text{Bernoulli}(1/(n+1))$. Find $P(X = n|Y = 1)$.

31. Let $X \sim \text{Poisson}(\lambda)$, and suppose that given $X = n$, $Y \sim \text{binomial}(n, p)$. Find $P(X = n|Y = k)$ for $n \geq k$.

32. Let X and Y be independent binomial(n, p) random variables. Find the conditional probability of $X > k$ given that $\max(X, Y) > k$ if $n = 100$, $p = 0.01$, and $k = 1$. *Answer:* 0.576.

33. Let $X \sim \text{geometric}_0(p)$ and $Y \sim \text{geometric}_0(q)$, and assume X and Y are independent.

    (a) Find $P(XY = 4)$.

    (b) Put $Z := X + Y$ and find $p_Z(j)$ for all j using the discrete convolution formula (3.17). Treat the cases $p \neq q$ and $p = q$ separately.

34. Let X and Y be independent random variables, each taking the values $0, 1, 2, 3$ with equal probability. Put $Z := X + Y$ and find $p_Z(j)$ for all j. *Hint:* Use the discrete convolution formula (3.17) and pay careful attention to the limits of summation.

35. Let $X \sim$ Bernoulli(p), and suppose that given $X = i$, Y is conditionally Poisson(λ_i), where $\lambda_1 > \lambda_0$. Express the likelihood-ratio test

$$\frac{\mathsf{P}(Y = j|X = 1)}{\mathsf{P}(Y = j|X = 0)} \geq \frac{\mathsf{P}(X = 0)}{\mathsf{P}(X = 1)}$$

in as simple a form as possible.

36. Let $X \sim$ Bernoulli(p), and suppose that given $X = i$, Y is conditionally geometric$_0(q_i)$, where $q_1 < q_0$. Express the likelihood-ratio test

$$\frac{\mathsf{P}(Y = j|X = 1)}{\mathsf{P}(Y = j|X = 0)} \geq \frac{\mathsf{P}(X = 0)}{\mathsf{P}(X = 1)}$$

in as simple a form as possible.

*37. Show that if $\mathsf{P}(X = x_i|Y = y_j) = h(x_i)$ for all j and some function h, then X and Y are independent.

3.5: Conditional expectation

38. Let X and Y be jointly discrete, integer-valued random variables with joint pmf

$$p_{XY}(i,j) = \begin{cases} \dfrac{3^{j-1}e^{-3}}{j!}, & i = 1, j \geq 0, \\ 4\dfrac{6^{j-1}e^{-6}}{j!}, & i = 2, j \geq 0, \\ 0, & \text{otherwise.} \end{cases}$$

Compute $\mathsf{E}[Y|X = i]$, $\mathsf{E}[Y]$, and $\mathsf{E}[X|Y = j]$.

39. Let X and Y be as in Example 3.15. Find $\mathsf{E}[Y]$, $\mathsf{E}[XY]$, $\mathsf{E}[Y^2]$, and var(Y).

40. Let X and Y be as in Example 3.16. Find $\mathsf{E}[Y|X = 1]$, $\mathsf{E}[Y|X = 0]$, $\mathsf{E}[Y]$, $\mathsf{E}[Y^2]$, and var(Y).

41. Let $X \sim$ Bernoulli$(2/3)$, and suppose that given $X = i$, $Y \sim$ Poisson$(3(i+1))$. Find $\mathsf{E}[(X+1)Y^2]$.

42. Let $X \sim$ Poisson(λ), and suppose that given $X = n$, $Y \sim$ Bernoulli$(1/(n+1))$. Find $\mathsf{E}[XY]$.

43. Let $X \sim$ geometric$_1(p)$, and suppose that given $X = n$, $Y \sim$ Pascal(n,q). Find $\mathsf{E}[XY]$.

44. Let X and Y be integer-valued random variables, with Y being positive. Suppose that given $Y = k$, X is conditionally Poisson with parameter k. If Y has mean m and variance r, find $\mathsf{E}[X^2]$.

45. Let X and Y be independent random variables, with $X \sim$ binomial(n,p), and let $Y \sim$ binomial(m,p). Put $V := X + Y$. Find the pmf of V. Find $\mathsf{P}(V = 10|X = 4)$ (assume $n \geq 4$ and $m \geq 6$).

46. Let X and Y be as in Example 3.15. Find $G_Y(z)$.

Exam preparation

You may use the following suggestions to prepare a study sheet, including formulas mentioned that you have trouble remembering. You may also want to ask your instructor for additional suggestions.

3.1. Probability generating functions. Important formulas include the definition (3.1), the factorization property for pgfs of sums of independent random variables (3.2), and the probability formula (3.4). The factorial moment formula (3.5) is most useful in its special cases

$$G'_X(z)|_{z=1} = \mathsf{E}[X] \quad \text{and} \quad G''_X(z)|_{z=1} = \mathsf{E}[X^2] - \mathsf{E}[X].$$

The pgfs of common discrete random variables can be found inside the front cover.

3.2. The binomial random variable. The binomial(n,p) random variable arises as the sum of n i.i.d. Bernoulli(p) random variables. The binomial(n,p) pmf, mean, variance, and pgf can be found inside the front cover. It is sometimes convenient to remember how to generate and use Pascal's triangle for computing the binomial coefficient $\binom{n}{k} = n!/[k!(n-k)!]$.

3.3. The weak law of large numbers. Understand what it means if $\mathsf{P}(|M_n - m| \geq \varepsilon)$ is small.

3.4. Conditional probability. I often tell my students that the three most important things in probability are:

(i) the law of total probability (3.12);
(ii) the substitution law (3.13); and
(iii) independence for "dropping the conditioning" as in (3.14).

3.5. Conditional expectation. I again tell my students that the three most important things in probability are:

(i) the law of total probability (for expectations) (3.23);
(ii) the substitution law (3.22); and
(iii) independence for "dropping the conditioning."

If the conditional pmf of Y given X is listed in the table inside the front cover (this table includes moments), then $\mathsf{E}[Y|X=i]$ or $\mathsf{E}[Y^2|X=i]$ can often be found by inspection. This is a *very* useful skill.

Work any review problems assigned by your instructor. If you finish them, re-work your homework assignments.

4
Continuous random variables

In Chapters 2 and 3, the only random variables we considered specifically were discrete ones such as the Bernoulli, binomial, Poisson, and geometric. In this chapter we consider a class of random variables allowed to take a continuum of values. These random variables are called continuous random variables and are introduced in Section 4.1. Continuous random variables are important models for integrator output voltages in communication receivers, file download times on the Internet, velocity and position of an airliner on radar, etc. Expectation and moments of continuous random variables are computed in Section 4.2. Section 4.3 develops the concepts of moment generating function (Laplace transform) and characteristic function (Fourier transform). In Section 4.4 expectation of multiple random variables is considered. Applications of characteristic functions to sums of independent random variables are illustrated. In Section 4.5 the Markov inequality, the Chebyshev inequality, and the Chernoff bound illustrate simple techniques for bounding probabilities in terms of expectations.

4.1 Densities and probabilities

Introduction

Suppose that a random voltage in the range $[0,1)$ is applied to a voltmeter with a one-digit display. Then the display output can be modeled by a discrete random variable Y taking values $.0, .1, .2, \ldots, .9$ with $\mathsf{P}(Y = k/10) = 1/10$ for $k = 0, \ldots, 9$. If this same random voltage is applied to a voltmeter with a two-digit display, then we can model its display output by a discrete random variable Z taking values $.00, .01, \ldots, .99$ with $\mathsf{P}(Z = k/100) = 1/100$ for $k = 0, \ldots, 99$. But how can we model the voltage itself? The voltage itself, call it X, can be any number in range $[0,1)$. For example, if $0.15 \leq X < 0.25$, the one-digit voltmeter would round to the tens place and show $Y = 0.2$. In other words, we want to be able to write

$$\mathsf{P}\left(\frac{k}{10} - 0.05 \leq X < \frac{k}{10} + 0.05\right) = \mathsf{P}\left(Y = \frac{k}{10}\right) = \frac{1}{10}.$$

Notice that $1/10$ is the length of the interval

$$\left[\frac{k}{10} - 0.05, \frac{k}{10} + 0.05\right).$$

This suggests that probabilities involving X can be computed via

$$\mathsf{P}(a \leq X < b) = \int_a^b 1\, dx = b - a,$$

which is the length of the interval $[a,b)$. This observation motivates the concept of a continuous random variable.

4.1 Densities and probabilities

Definition

We say that X is a **continuous random variable** if $\mathsf{P}(X \in B)$ has the form

$$\mathsf{P}(X \in B) = \int_B f(t)\,dt := \int_{-\infty}^{\infty} I_B(t) f(t)\,dt \qquad (4.1)$$

for some integrable function f.[a] Since $\mathsf{P}(X \in \mathbb{R}) = 1$, the function f must integrate to one; i.e., $\int_{-\infty}^{\infty} f(t)\,dt = 1$. Further, since $\mathsf{P}(X \in B) \geq 0$ for all B, it can be shown that f must be nonnegative.[1] A nonnegative function that integrates to one is called a **probability density function** (pdf).

Usually, the set B is an interval such as $B = [a,b]$. In this case,

$$\mathsf{P}(a \leq X \leq b) = \int_a^b f(t)\,dt.$$

See Figure 4.1(a). Computing such probabilities is analogous to determining the mass of a piece of wire stretching from a to b by integrating its mass density per unit length from a to b. Since most probability densities we work with are continuous, for a small interval, say $[x, x + \Delta x]$, we have

$$\mathsf{P}(x \leq X \leq x + \Delta x) = \int_x^{x + \Delta x} f(t)\,dt \approx f(x)\,\Delta x.$$

See Figure 4.1(b).

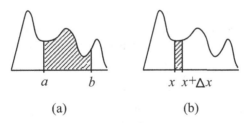

Figure 4.1. (a) $\mathsf{P}(a \leq X \leq b) = \int_a^b f(t)\,dt$ is the area of the shaded region under the density $f(t)$. (b) $\mathsf{P}(x \leq X \leq x + \Delta x) = \int_x^{x+\Delta x} f(t)\,dt$ is the area of the shaded vertical strip.

Note that for random variables with a density,

$$\mathsf{P}(a \leq X \leq b) = \mathsf{P}(a < X \leq b) = \mathsf{P}(a \leq X < b) = \mathsf{P}(a < X < b)$$

since the corresponding integrals over an interval are not affected by whether or not the endpoints are included or excluded.

Some common densities

Here are some examples of continuous random variables. A summary of the more common ones can be found on the inside of the back cover.

---

[a] Later, when more than one random variable is involved, we write $f_X(x)$ instead of $f(x)$.

Continuous random variables

Uniform. The simplest continuous random variable is the uniform. It is used to model experiments in which the outcome is constrained to lie in a known interval, say $[a,b]$, and all outcomes are equally likely. We write $f \sim \text{uniform}[a,b]$ if $a < b$ and

$$f(x) = \begin{cases} \dfrac{1}{b-a}, & a \leq x \leq b, \\ 0, & \text{otherwise.} \end{cases}$$

This density is shown in Figure 4.2. To verify that f integrates to one, first note that since

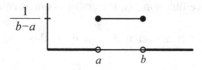

Figure 4.2. The uniform density on $[a,b]$.

$f(x) = 0$ for $x < a$ and $x > b$, we can write

$$\int_{-\infty}^{\infty} f(x)\,dx = \int_{a}^{b} f(x)\,dx.$$

Next, for $a \leq x \leq b$, $f(x) = 1/(b-a)$, and so

$$\int_{a}^{b} f(x)\,dx = \int_{a}^{b} \frac{1}{b-a}\,dx = 1.$$

This calculation illustrates an important technique that is often incorrectly carried out by novice students: *First* modify the limits of integration, *then* substitute the appropriate formula for $f(x)$. For example, it is quite common to see the *incorrect* calculation,

$$\int_{-\infty}^{\infty} f(x)\,dx = \int_{-\infty}^{\infty} \frac{1}{b-a}\,dx = \infty.$$

***Example* 4.1.** In coherent radio communications, the phase difference between the transmitter and the receiver, denoted by Θ, is modeled as having a density $f \sim \text{uniform}[-\pi, \pi]$. Find $\mathsf{P}(\Theta \leq 0)$ and $\mathsf{P}(\Theta \leq \pi/2)$.

Solution. To begin, write

$$\mathsf{P}(\Theta \leq 0) = \int_{-\infty}^{0} f(\theta)\,d\theta = \int_{-\pi}^{0} f(\theta)\,d\theta,$$

where the second equality follows because $f(\theta) = 0$ for $\theta < -\pi$. Now that we have restricted the limits of integration to be inside the region where the density is positive, we can write

$$\mathsf{P}(\Theta \leq 0) = \int_{-\pi}^{0} \frac{1}{2\pi}\,d\theta = 1/2.$$

The second probability is treated in the same way. First write

$$\mathsf{P}(\Theta \leq \pi/2) = \int_{-\infty}^{\pi/2} f(\theta)\,d\theta = \int_{-\pi}^{\pi/2} f(\theta)\,d\theta.$$

It then follows that

$$P(\Theta \leq \pi/2) = \int_{-\pi}^{\pi/2} \frac{1}{2\pi} d\theta = 3/4.$$

Example 4.2. Use the results of the preceding example to compute $P(\Theta > \pi/2 | \Theta > 0)$.

Solution. To calculate

$$P(\Theta > \pi/2 | \Theta > 0) = \frac{P(\{\Theta > \pi/2\} \cap \{\Theta > 0\})}{P(\Theta > 0)},$$

first observe that the denominator is simply $P(\Theta > 0) = 1 - P(\Theta \leq 0) = 1 - 1/2 = 1/2$. As for the numerator, note that

$$\{\Theta > \pi/2\} \subset \{\Theta > 0\}.$$

Then use the fact that $A \subset B$ implies $A \cap B = A$ to write

$$P(\{\Theta > \pi/2\} \cap \{\Theta > 0\}) = P(\Theta > \pi/2) = 1 - P(\Theta \leq \pi/2) = 1/4.$$

Thus, $P(\Theta > \pi/2 | \Theta > 0) = (1/4)/(1/2) = 1/2$.

Exponential. Another simple continuous random variable is the exponential with parameter $\lambda > 0$. We write $f \sim \exp(\lambda)$ if

$$f(x) = \begin{cases} \lambda e^{-\lambda x}, & x \geq 0, \\ 0, & x < 0. \end{cases}$$

This density is shown in Figure 4.3. As λ increases, the height increases and the width decreases. It is easy to check that f integrates to one. The exponential random variable is often used to model lifetimes, such as how long a cell-phone call lasts or how long it takes a computer network to transmit a message from one node to another. The exponential random variable also arises as a function of other random variables. For example, in Problem 4.3 you will show that if $U \sim \text{uniform}(0,1)$, then $X = \ln(1/U)$ is $\exp(1)$. We also point out that if U and V are independent Gaussian random variables, which are defined later in this section, then $U^2 + V^2$ is exponential and $\sqrt{U^2 + V^2}$ is Rayleigh (defined in Problem 30).[2]

Example 4.3. Given that a cell-phone call has lasted more than t seconds so far, suppose the conditional probability that the call ends by $t + \Delta t$ is approximately $\lambda \Delta t$ when Δt is small. Show that the call duration is an $\exp(\lambda)$ random variable.

Solution. Let T denote the call duration. We treat the problem assumption as saying that

$$P(T \leq t + \Delta t | T > t) \approx \lambda \Delta t.$$

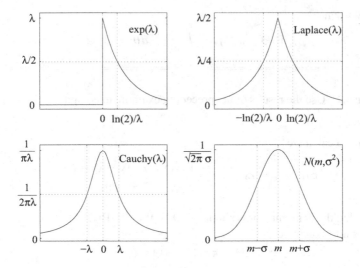

Figure 4.3. Several common density functions.

To find the density of T, we proceed as follows. Let $t \geq 0$ and write

$$\mathsf{P}(T \leq t+\Delta t | T > t) = \frac{\mathsf{P}(\{T \leq t+\Delta t\} \cap \{T > t\})}{\mathsf{P}(T > t)}$$

$$= \frac{\mathsf{P}(t < T \leq t+\Delta t)}{\mathsf{P}(T > t)}$$

$$= \frac{\int_t^{t+\Delta t} f_T(\theta)\, d\theta}{\mathsf{P}(T > t)}.$$

For small Δt, the left-hand side is approximately $\lambda \Delta t$, and the right-hand side is approximately $f_T(t)\Delta t / \mathsf{P}(T > t)$; i.e.,

$$\lambda \Delta t = \frac{f_T(t)\Delta t}{\mathsf{P}(T > t)}.$$

Now cancel Δt on both sides and multiply both sides by $\mathsf{P}(T > t)$ to get $\lambda \mathsf{P}(T > t) = f_T(t)$. In this equation, write $\mathsf{P}(T > t)$ as an integral to obtain

$$\lambda \int_t^\infty f_T(\theta)\, d\theta = f_T(t).$$

Differentiating both sides with respect to t shows that

$$-\lambda f_T(t) = f_T'(t), \quad t \geq 0.$$

The solution of this differential equation is easily seen to be $f_T(t) = ce^{-\lambda t}$ for some constant c. However, since $f_T(t)$ is a density and since its integral from zero to infinity must be one, it follows that $c = \lambda$.

4.1 Densities and probabilities

Remark. In the preceding example, T was the duration of a cell-phone call. However, if T were the lifetime or time-to-failure of a device or system, then in reliability theory, the quantity

$$\lim_{\Delta t \to 0} \frac{\mathsf{P}(T \le t + \Delta t | T > t)}{\Delta t}$$

is called the failure rate. If this limit does not depend on t, then the calculation of the preceding example shows that the density of T must be exponential. Time-varying failure rates are considered in Section 5.7.

Laplace / double-sided exponential. Related to the exponential is the Laplace, sometimes called the double-sided exponential. For $\lambda > 0$, we write $f \sim \text{Laplace}(\lambda)$ if

$$f(x) = \tfrac{\lambda}{2} e^{-\lambda |x|}.$$

This density is shown in Figure 4.3. As λ increases, the height increases and the width decreases. You will show in Problem 54 that the difference of two independent $\exp(\lambda)$ random variables is a $\text{Laplace}(\lambda)$ random variable.

Example 4.4. An Internet router can send packets via route 1 or route 2. The packet delays on each route are independent $\exp(\lambda)$ random variables, and so the difference in delay between route 1 and route 2, denoted by X, has a $\text{Laplace}(\lambda)$ density. Find

$$\mathsf{P}(-3 \le X \le -2 \text{ or } 0 \le X \le 3).$$

Solution. The desired probability can be written as

$$\mathsf{P}(\{-3 \le X \le -2\} \cup \{0 \le X \le 3\}).$$

Since these are disjoint events, the probability of the union is the sum of the individual probabilities. We therefore need to compute

$$\mathsf{P}(-3 \le X \le -2) \quad \text{and} \quad \mathsf{P}(0 \le X \le 3).$$

Since X has a $\text{Laplace}(\lambda)$ density, these probabilities are equal to the areas of the corresponding shaded regions in Figure 4.4. We first compute

$$\mathsf{P}(-3 \le X \le -2) = \int_{-3}^{-2} \tfrac{\lambda}{2} e^{-\lambda |x|} dx = \tfrac{\lambda}{2} \int_{-3}^{-2} e^{\lambda x} dx,$$

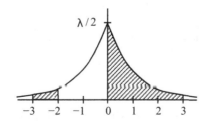

Figure 4.4. $\text{Laplace}(\lambda)$ density for Example 4.4.

where we have used the fact that since x is negative in the range of integration, $|x| = -x$. This last integral is equal to $(e^{-2\lambda} - e^{-3\lambda})/2$. It remains to compute

$$P(0 \leq X \leq 3) = \int_0^3 \tfrac{\lambda}{2} e^{-\lambda |x|} \, dx = \tfrac{\lambda}{2} \int_0^3 e^{-\lambda x} \, dx,$$

which is equal to $(1 - e^{-3\lambda})/2$. The desired probability is then

$$\frac{1 - 2e^{-3\lambda} + e^{-2\lambda}}{2}.$$

Cauchy. The Cauchy random variable with parameter $\lambda > 0$ is also easy to work with. We write $f \sim \text{Cauchy}(\lambda)$ if

$$f(x) = \frac{\lambda/\pi}{\lambda^2 + x^2}.$$

This density is shown in Figure 4.3. As λ increases, the height decreases and the width increases. Since $(1/\pi)(d/dx) \tan^{-1}(x/\lambda) = f(x)$, and since $\tan^{-1}(\infty) = \pi/2$, it is easy to check that f integrates to one. The Cauchy random variable arises as the tangent of a uniform random variable (Example 5.10) and also as the quotient of independent Gaussian random variables (Problem 33 in Chapter 7).

Example 4.5. In the λ-lottery you choose a number λ with $1 \leq \lambda \leq 10$. Then a random variable X is chosen according to the Cauchy density with parameter λ. If $|X| \geq 1$, then you win the lottery. Which value of λ should you choose to maximize your probability of winning?

Solution. Your probability of winning is

$$P(|X| \geq 1) = P(X \geq 1 \text{ or } X \leq -1)$$
$$= \int_1^\infty f(x) \, dx + \int_{-\infty}^{-1} f(x) \, dx,$$

where $f(x) = (\lambda/\pi)/(\lambda^2 + x^2)$ is the Cauchy density. Since the Cauchy density is an even function,

$$P(|X| \geq 1) = 2 \int_1^\infty \frac{\lambda/\pi}{\lambda^2 + x^2} \, dx.$$

Now make the change of variable $y = x/\lambda$, $dy = dx/\lambda$, to get

$$P(|X| \geq 1) = 2 \int_{1/\lambda}^\infty \frac{1/\pi}{1 + y^2} \, dy.$$

Since the integrand is nonnegative, the integral is maximized by minimizing $1/\lambda$ or by maximizing λ. Hence, choosing $\lambda = 10$ maximizes your probability of winning.

4.1 Densities and probabilities

Gaussian / normal. The most important density is the Gaussian or normal. For $\sigma^2 > 0$, we write $f \sim N(m, \sigma^2)$ if

$$f(x) = \frac{1}{\sqrt{2\pi}\,\sigma} \exp\left[-\frac{1}{2}\left(\frac{x-m}{\sigma}\right)^2\right], \qquad (4.2)$$

where σ is the positive square root of σ^2. A graph of the $N(m, \sigma^2)$ density is sketched in Figure 4.3. It is shown in Problems 9 and 10 that the density is concave for $x \in [m-\sigma, m+\sigma]$ and convex for x outside this interval. As σ increases, the height of the density decreases and it becomes wider as illustrated in Figure 4.5. If $m = 0$ and $\sigma^2 = 1$, we say that f is a **standard normal density**.

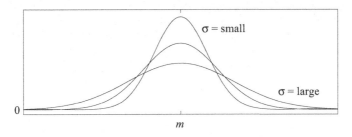

Figure 4.5. $N(m, \sigma^2)$ densities with different values of σ.

As a consequence of the central limit theorem, whose discussion is taken up in Chapter 5, the Gaussian density is a good approximation for computing probabilities involving a sum of many independent random variables; this is true whether the random variables are continuous or discrete! For example, let

$$X := X_1 + \cdots + X_n,$$

where the X_i are i.i.d. with common mean m and common variance σ^2. For large n, it is shown in Chapter 5 that if the X_i are continuous random variables, then

$$f_X(x) \approx \frac{1}{\sqrt{2\pi}\,\sigma\sqrt{n}} \exp\left[-\frac{1}{2}\left(\frac{x-nm}{\sigma\sqrt{n}}\right)^2\right],$$

while if the X_i are integer-valued,

$$p_X(k) \approx \frac{1}{\sqrt{2\pi}\,\sigma\sqrt{n}} \exp\left[-\frac{1}{2}\left(\frac{k-nm}{\sigma\sqrt{n}}\right)^2\right].$$

In particular, since the macroscopic noise current measured in a circuit results from the sum of forces of many independent collisions on an atomic scale, noise current is well-described by the Gaussian density. For this reason, Gaussian random variables are the noise model of choice in electronic communication and control systems.

To verify that an arbitrary normal density integrates to one, we proceed as follows. (For an alternative derivation, see Problem 17.) First, making the change of variable $t = (x-m)/\sigma$ shows that

$$\int_{-\infty}^{\infty} f(x)\,dx = \frac{1}{\sqrt{2\pi}} \int_{-\infty}^{\infty} e^{-t^2/2}\,dt.$$

So, without loss of generality, we may assume f is a standard normal density with $m = 0$ and $\sigma = 1$. We then need to show that $I := \int_{-\infty}^{\infty} e^{-x^2/2}\,dx = \sqrt{2\pi}$. The trick is to show instead that $I^2 = 2\pi$. First write

$$I^2 = \left(\int_{-\infty}^{\infty} e^{-x^2/2}\,dx\right)\left(\int_{-\infty}^{\infty} e^{-y^2/2}\,dy\right).$$

Now write the product of integrals as the iterated integral

$$I^2 = \int_{-\infty}^{\infty}\int_{-\infty}^{\infty} e^{-(x^2+y^2)/2}\,dx\,dy.$$

Next, we interpret this as a double integral over the whole plane and change from Cartesian coordinates x and y to polar coordinates r and θ. To integrate over the whole plane in polar coordinates, the radius r ranges from 0 to ∞, and the angle θ ranges from 0 to 2π. The substitution is $x = r\cos\theta$ and $y = r\sin\theta$. We also change $dx\,dy$ to $r\,dr\,d\theta$. This yields

$$I^2 = \int_0^{2\pi}\int_0^{\infty} e^{-r^2/2} r\,dr\,d\theta$$

$$= \int_0^{2\pi} \left(-e^{-r^2/2}\bigg|_0^{\infty}\right) d\theta$$

$$= \int_0^{2\pi} 1\,d\theta$$

$$= 2\pi.$$

Example 4.6. The noise voltage in a certain amplifier has the standard normal density. Show that the noise is as likely to be positive as it is to be negative.

Solution. In terms of the density, which we denote by f, we must show that

$$\int_{-\infty}^0 f(x)\,dx = \int_0^{\infty} f(x)\,dx.$$

Since $f(x) = e^{-x^2/2}/\sqrt{2\pi}$ is an even function of x, the two integrals are equal. Furthermore, we point out that since the sum of the two integrals is $\int_{-\infty}^{\infty} f(x)\,dx = 1$, each individual integral must be $1/2$.

Location and scale parameters and the gamma densities

Since a probability density function can be any nonnegative function that integrates to one, it is easy to create a whole family of density functions starting with just one density function. Let f be any nonnegative function that integrates to one. For any real number c and any positive number λ, consider the nonnegative function

$$\lambda f(\lambda(x-c)).$$

Here c is called a **location parameter** and λ is called a **scale parameter**. To show that this new function is a probability density, all we have to do is show that it integrates to one. In the integral

$$\int_{-\infty}^{\infty} \lambda f(\lambda(x-c))\,dx$$

4.1 Densities and probabilities

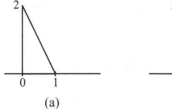
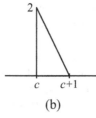
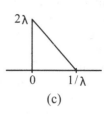

(a)   (b)   (c)

Figure 4.6. (a) Triangular density $f(x)$. (b) Shifted density $f(x-c)$. (c) Scaled density $\lambda f(\lambda x)$ shown for $0 < \lambda < 1$.

make the change of variable $t = \lambda(x-c)$, $dt = \lambda\, dx$ to get

$$\int_{-\infty}^{\infty} \lambda f\bigl(\lambda(x-c)\bigr)\, dx \;=\; \int_{-\infty}^{\infty} f(t)\, dt \;=\; 1.$$

Let us first focus on the case $\lambda = 1$. Then our new density reduces to $f(x-c)$. For example, if f is the triangular density shown in Figure 4.6(a), then $f(x-c)$ is the density shown in Figure 4.6(b). If c is positive, then $f(x-c)$ is $f(x)$ shifted to the right, and if c is negative, then $f(x-c)$ is $f(x)$ shifted to the left.

Next consider the case $c = 0$ and $\lambda > 0$. In this case, the main effect of λ is to shrink (if $\lambda > 1$) or to expand (if $\lambda < 1$) the density. The second effect of λ is to increase or decrease the height of the density. For example, if f is again the triangular density of Figure 4.6(a), then $\lambda f(\lambda x)$ is shown in Figure 4.6(c) for $0 < \lambda < 1$.

To see what happens both $c \neq 0$ and $\lambda > 0$, first put $h(x) := \lambda f(\lambda x)$. Then observe that $h(x-c) = \lambda f(\lambda(x-c))$. In other words, first find the picture for $\lambda f(\lambda x)$, and then shift this picture by c.

In the exponential and Laplace densities, λ is a scale parameter, while in the Cauchy density, $1/\lambda$ is a scale parameter. In the Gaussian, if we write $f(x) = e^{-x^2/2}/\sqrt{2\pi}$, then

$$\lambda f\bigl(\lambda(x-c)\bigr) \;=\; \lambda \frac{e^{-(\lambda(x-c))^2/2}}{\sqrt{2\pi}}.$$

Comparing this with (4.2) shows that $c = m$ and $\lambda = 1/\sigma$. In other words for an $N(m,\sigma^2)$ random variable, m is a location parameter and $1/\sigma$ is a scale parameter. Note in particular that as σ increases, the density becomes shorter and wider, while as σ decreases, the density becomes taller and narrower (recall Figure 4.5).

An important application of the scale parameter arises with the basic **gamma density** with parameter $p > 0$. This density is given by

$$g_p(x) \;:=\; \begin{cases} \dfrac{x^{p-1} e^{-x}}{\Gamma(p)}, & x > 0, \\ 0, & \text{otherwise}, \end{cases}$$

where

$$\Gamma(p) \;:=\; \int_0^{\infty} x^{p-1} e^{-x}\, dx, \quad p > 0,$$

is the **gamma function**. In other words, the gamma function is defined to make the gamma density integrate to one.[3] (Properties of the gamma function are derived in Problem 14.) Graphs of $g_p(x)$ for $p = 1/2$, $p = 1$, $p = 3/2$, $p = 2$, and $p = 3$ are shown in Figure 4.7. To explain the shapes of these curves, observe that for x near zero, $e^{-x} \approx 1$, and so the behavior is determined by the factor x^{p-1}. For the values of p in the figure, this factor is $x^{-1/2}$, x^0, $x^{1/2}$, x, and x^2. In the first case, $x^{-1/2}$ blows up as x approaches the origin, and decreases as x moves to the right. Of course, $x^0 = 1$ is a constant, and in the remaining cases, x^{p-1} is zero for $x = 0$ and then increases. In all cases, as x moves to the right, eventually, the decaying nature of the factor e^{-x} dominates, and the curve decreases to zero as $x \to \infty$.

Setting $g_{p,\lambda}(x) := \lambda g_p(\lambda x)$ defines the general gamma density, and the following special cases are of great importance. When $p = m$ is a positive integer, $g_{m,\lambda}$ is called an **Erlang**(m, λ) density (see Problem 15). As shown in Problem 55(c), the sum of m i.i.d. $\exp(\lambda)$ random variables is an Erlang(m, λ) random variable. For example, if m customers are waiting in a queue, and the service time for each one is $\exp(\lambda)$, then the time to serve all m is Erlang(m, λ). The Erlang densities for $m = 1, 2, 3$ and $\lambda = 1$ are $g_1(x)$, $g_2(x)$, and $g_3(x)$ shown in Figure 4.7. When $p = k/2$ and $\lambda = 1/2$, $g_{p,\lambda}$ is called a **chi-squared** density with k degrees of freedom. As you will see in Problem 46, the chi-squared random variable arises as the square of a normal random variable. In communication systems employing noncoherent receivers, the incoming signal is squared before further processing. Since the thermal noise in these receivers is Gaussian, chi-squared random variables naturally appear. Since chi-squared densities are scaled versions of $g_{k/2}$, the chi-squared densities for $k = 1, 2, 3, 4$, and 6 are *scaled versions* of $g_{1/2}$, g_1, $g_{3/2}$, g_2, and g_3 shown in Figure 4.7.

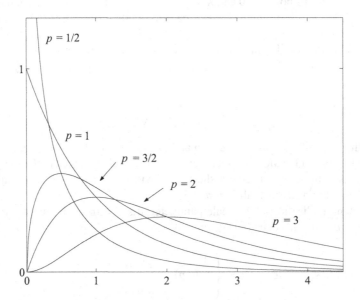

Figure 4.7. The gamma densities $g_p(x)$ for $p = 1/2$, $p = 1$, $p = 3/2$, $p = 2$, and $p = 3$.

The paradox of continuous random variables

Let X be a continuous random variable. For any given x_0, write

$$1 = \int_{-\infty}^{\infty} f(t)\,dt = \int_{-\infty}^{x_0} f(t)\,dt + \int_{x_0}^{\infty} f(t)\,dt$$
$$= P(X \le x_0) + P(X \ge x_0)$$
$$= P(X \le x_0) + P(X = x_0) + P(X > x_0).$$

Since $P(X \le x_0) + P(X > x_0) = P(X \in \mathbb{R}) = 1$, it follows that $P(X = x_0) = 0$. We are thus confronted with the fact that continuous random variables take no fixed value with positive probability! The way to understand this apparent paradox is to realize that continuous random variables are an *idealized model* of what we normally think of as continuous-valued measurements. For example, a voltmeter only shows a certain number of digits after the decimal point, say 5.127 volts because physical devices have limited precision. Hence, the measurement $X = 5.127$ should be understood as saying that

$$5.1265 \le X < 5.1275,$$

since all numbers in this range round to 5.127. Now there is no paradox since $P(5.1265 \le X < 5.1275)$ has positive probability.

You may still ask, "Why not just use a discrete random variable taking the distinct values $k/1000$, where k is any integer?" After all, this would model the voltmeter in question. One answer is that if you get a better voltmeter, you need to redefine the random variable, while with the idealized, continuous-random-variable model, even if the voltmeter changes, the random variable does not. Also, the continuous-random-variable model is often mathematically simpler to work with.

Remark. If B is any set with finitely many points, or even countably many points, then $P(X \in B) = 0$ when X is a continuous random variable. To see this, suppose $B = \{x_1, x_2, \ldots\}$ where the x_i are distinct real numbers. Then

$$P(X \in B) = P\left(\bigcup_{i=1}^{\infty} \{x_i\}\right) = \sum_{i=1}^{\infty} P(X = x_i) = 0,$$

since, as argued above, each term is zero.

4.2 Expectation of a single random variable

For a discrete random variable X with probability mass function p, we computed expectations using the **law of the unconscious statistician** (LOTUS)

$$E[g(X)] = \sum_i g(x_i) p(x_i).$$

Analogously, for a continuous random variable X with density f, we have

$$E[g(X)] = \int_{-\infty}^{\infty} g(x) f(x)\,dx. \tag{4.3}$$

In particular, taking $g(x) = x$ yields

$$E[X] = \int_{-\infty}^{\infty} x f(x)\, dx.$$

We derive these formulas later in this section. For now, we illustrate LOTUS with several examples.

Example 4.7. If X is a uniform$[a,b]$ random variable, find $E[X]$, $E[X^2]$, and $\text{var}(X)$.

Solution. To find $E[X]$, write

$$E[X] = \int_{-\infty}^{\infty} x f(x)\, dx = \int_{a}^{b} x \frac{1}{b-a}\, dx = \left. \frac{x^2}{2(b-a)} \right|_{a}^{b},$$

which simplifies to

$$\frac{b^2 - a^2}{2(b-a)} = \frac{(b+a)(b-a)}{2(b-a)} = \frac{a+b}{2},$$

which is simply the numerical average of a and b.

To compute the second moment, write

$$E[X^2] = \int_{-\infty}^{\infty} x^2 f(x)\, dx = \int_{a}^{b} x^2 \frac{1}{b-a}\, dx = \left. \frac{x^3}{3(b-a)} \right|_{a}^{b},$$

which simplifies to

$$\frac{b^3 - a^3}{3(b-a)} = \frac{(b-a)(b^2 + ba + a^2)}{3(b-a)} = \frac{b^2 + ba + a^2}{3}.$$

Since $\text{var}(X) = E[X^2] - (E[X])^2$, we have

$$\text{var}(X) = \frac{b^2 + ba + a^2}{3} - \frac{a^2 + 2ab + b^2}{4}$$

$$= \frac{b^2 - 2ba + a^2}{12}$$

$$= \frac{(b-a)^2}{12}.$$

Example 4.8 (quantizer noise). An analog-to-digital converter or **quantizer** with resolution or step size Δ volts rounds its input to the nearest multiple of Δ volts as shown in Figure 4.8. If the input is a random voltage V_{in} and the output is denoted by V_{out}, then the performance of the device is characterized by its mean squared error, $E[|V_{\text{in}} - V_{\text{out}}|^2]$. In general it is difficult to compute this quantity. However, since the converter is just rounding to the nearest multiple of Δ volts, the error always lies between $\pm \Delta/2$. Hence, in many cases it is assumed that the error $V_{\text{in}} - V_{\text{out}}$ is approximated by a uniform$[-\Delta/2, \Delta/2]$ random variable [18]. In this case, evaluate the converter's performance.

4.2 Expectation of a single random variable

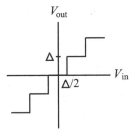

Figure 4.8. Input–output relationship of an analog-to-digital converter or quantizer with resolution Δ.

Solution. If $X \sim$ uniform$[-\Delta/2, \Delta/2]$, then the uniform approximation allows us to write
$$\begin{aligned}
\mathsf{E}[|V_{\text{in}} - V_{\text{out}}|^2] &\approx \mathsf{E}[X^2] \\
&= \text{var}(X), \qquad \text{since } X \text{ has zero mean,} \\
&= \frac{(\Delta/2 - (-\Delta/2))^2}{12}, \qquad \text{by the preceding example,} \\
&= \frac{\Delta^2}{12}.
\end{aligned}$$

Example 4.9. If X is an exponential random variable with parameter $\lambda = 1$, find all moments of X.

Solution. We need to compute
$$\mathsf{E}[X^n] = \int_0^\infty x^n e^{-x} \, dx.$$

Use integration by parts (see Note **4** for a refresher) with $u = x^n$ and $dv = e^{-x} dx$. Then $du = nx^{n-1} dx$, $v = -e^{-x}$, and
$$\mathsf{E}[X^n] = -x^n e^{-x} \Big|_0^\infty + n \int_0^\infty x^{n-1} e^{-x} \, dx.$$

Using the fact that $x^n e^{-x} = 0$ both for $x = 0$ and for $x \to \infty$, we have
$$\mathsf{E}[X^n] = n \int_0^\infty x^{n-1} e^{-x} \, dx = n\mathsf{E}[X^{n-1}], \tag{4.4}$$

Taking $n = 1$ yields $\mathsf{E}[X] = \mathsf{E}[X^0] = \mathsf{E}[1] = 1$. Taking $n = 2$ yields $\mathsf{E}[X^2] = 2 \cdot 1$, and $n = 3$ yields $\mathsf{E}[X^3] = 3 \cdot 2 \cdot 1$. The general result is that $\mathsf{E}[X^n] = n!$.

Observe that
$$\int_0^\infty x^n e^{-x} \, dx = \int_0^\infty x^{(n+1)-1} e^{-x} \, dx = \Gamma(n+1).$$

Hence, the preceding example shows that $\Gamma(n+1) = n!$. In Problem 14(a) you will generalize the calculations leading to (4.4) to show that $\Gamma(p+1) = p \cdot \Gamma(p)$ for $p > 0$.

Example 4.10. Find the mean and variance of an exp(λ) random variable.

Solution. Since var$(X) = \mathsf{E}[X^2] - (\mathsf{E}[X])^2$, we need the first two moments of X. The nth moment is
$$\mathsf{E}[X^n] = \int_0^\infty x^n \cdot \lambda e^{-\lambda x}\, dx.$$
Making the change of variable $y = \lambda x$, $dy = \lambda\, dx$, we have
$$\mathsf{E}[X^n] = \int_0^\infty (y/\lambda)^n e^{-y}\, dy = \frac{1}{\lambda^n}\int_0^\infty y^n e^{-y}\, dy.$$
Since this last integral is the nth moment of the exp(1) random variable, which is $n!$ by the last example, it follows that
$$\mathsf{E}[X^n] = \frac{n!}{\lambda^n}.$$
Hence,
$$\text{var}(X) = \mathsf{E}[X^2] - (\mathsf{E}[X])^2 = \frac{2}{\lambda^2} - \left(\frac{1}{\lambda}\right)^2 = \frac{1}{\lambda^2}.$$

Example 4.11. Let X be a continuous random variable with standard Gaussian density $f \sim N(0,1)$. Compute $\mathsf{E}[X^n]$ for all $n \geq 1$.

Solution. Write
$$\mathsf{E}[X^n] = \int_{-\infty}^\infty x^n f(x)\, dx,$$
where $f(x) = \exp(-x^2/2)/\sqrt{2\pi}$. Since f is an even function of x, the above integrand is odd for n odd. Hence, all the odd moments are zero. For $n \geq 2$, write
$$\mathsf{E}[X^n] = \int_{-\infty}^\infty x^n \frac{e^{-x^2/2}}{\sqrt{2\pi}}\, dx$$
$$= \frac{1}{\sqrt{2\pi}}\int_{-\infty}^\infty x^{n-1}\cdot xe^{-x^2/2}\, dx.$$
Integration by parts shows that this last integral is equal to
$$-x^{n-1}e^{-x^2/2}\Big|_{-\infty}^\infty + (n-1)\int_{-\infty}^\infty x^{n-2}e^{-x^2/2}\, dx.$$
Since $e^{-x^2/2}$ decays faster than any power of x, the first term is zero. Thus,
$$\mathsf{E}[X^n] = (n-1)\int_{-\infty}^\infty x^{n-2}\frac{e^{-x^2/2}}{\sqrt{2\pi}}\, dx$$
$$= (n-1)\mathsf{E}[X^{n-2}],$$
where, from the integral with $n = 2$, we see that $\mathsf{E}[X^0] = 1$. When $n = 2$ this yields $\mathsf{E}[X^2] = 1$, and when $n = 4$ this yields $\mathsf{E}[X^4] = 3$. The general result is
$$\mathsf{E}[X^n] = \begin{cases} 1\cdot 3\cdots (n-3)(n-1), & n \text{ even}, \\ 0, & n \text{ odd}. \end{cases}$$

4.2 Expectation of a single random variable

At this point, it is convenient to introduce the **double factorial** notation,

$$n!! := \begin{cases} 1 \cdot 3 \cdots (n-2) \cdot n, & n > 0 \text{ and odd,} \\ 2 \cdot 4 \cdots (n-2) \cdot n, & n > 0 \text{ and even.} \end{cases}$$

In particular, with odd $n = 2m - 1$, we see that

$$\begin{aligned}(2m-1)!! &= 1 \cdot 3 \cdots (2m-3)(2m-1) \\ &= \frac{1 \cdot 2 \cdot 3 \cdot 4 \cdots (2m-1)(2m)}{2 \cdot 4 \cdots (2m)} \\ &= \frac{(2m)!}{(2 \cdot 1) \cdot (2 \cdot 2) \cdot (2 \cdot 3) \cdots (2m)} \\ &= \frac{(2m)!}{2^m \, m!}.\end{aligned}$$

Hence, if $X \sim N(0,1)$,

$$\mathsf{E}[X^{2m}] = 1 \cdot 3 \cdots (2m-3)(2m-1) = (2m-1)!! = \frac{(2m)!}{2^m \, m!}. \qquad (4.5)$$

Example 4.12. If X has density $f \sim N(m, \sigma^2)$, show that $\mathsf{E}[X] = m$ and $\text{var}(X) = \sigma^2$.

Solution. Instead of showing $\mathsf{E}[X] = m$, it is easier to show that $\mathsf{E}[X - m] = 0$. Write

$$\begin{aligned}\mathsf{E}[X - m] &= \int_{-\infty}^{\infty} (x - m) f(x) \, dx \\ &= \int_{-\infty}^{\infty} (x - m) \frac{\exp\left[-\frac{1}{2}\left(\frac{x-m}{\sigma}\right)^2\right]}{\sqrt{2\pi} \, \sigma} \, dx \\ &= \int_{-\infty}^{\infty} \frac{x - m}{\sigma} \cdot \frac{\exp\left[-\frac{1}{2}\left(\frac{x-m}{\sigma}\right)^2\right]}{\sqrt{2\pi}} \, dx.\end{aligned}$$

Make the change of variable $y = (x - m)/\sigma$, noting that $dy = dx/\sigma$. Then

$$\mathsf{E}[X - m] = \sigma \int_{-\infty}^{\infty} y \frac{e^{-y^2/2}}{\sqrt{2\pi}} \, dy,$$

which is seen to be zero once we recognize the integral as having the form of the mean of an $N(0,1)$ random variable Y.

To compute $\text{var}(X)$, write

$$\begin{aligned}\mathsf{E}[(X-m)^2] &= \int_{-\infty}^{\infty} (x-m)^2 f(x) \, dx \\ &= \int_{-\infty}^{\infty} (x-m)^2 \frac{\exp\left[-\frac{1}{2}\left(\frac{x-m}{\sigma}\right)^2\right]}{\sqrt{2\pi} \, \sigma} \, dx \\ &= \sigma \int_{-\infty}^{\infty} \left(\frac{x-m}{\sigma}\right)^2 \cdot \frac{\exp\left[-\frac{1}{2}\left(\frac{x-m}{\sigma}\right)^2\right]}{\sqrt{2\pi}} \, dx.\end{aligned}$$

Making the same change of variable as before, we obtain

$$E[(X-m)^2] = \sigma^2 \int_{-\infty}^{\infty} y^2 \frac{e^{-y^2/2}}{\sqrt{2\pi}} \, dy.$$

Now recognize this integral as having the form of the second moment of an $N(0,1)$ random variable Y. By the previous example, $E[Y^2] = 1$. Hence, $E[(X-m)^2] = \sigma^2$.

Example **4.13** (infinite expectation). **Pareto** densities[b] have been used to model packet delay, files sizes, and other Internet characteristics. Let X have the Pareto density $f(x) = 1/x^2, x \geq 1$. Find $E[X]$.

Solution. Write

$$E[X] = \int_1^{\infty} x \cdot \frac{1}{x^2} \, dx = \int_1^{\infty} \frac{1}{x} \, dx = \ln x \bigg|_1^{\infty} = \infty.$$

Example **4.14.** Determine $E[X]$ if X has a Cauchy density with parameter $\lambda = 1$.

Solution. This is a trick question. Recall that as noted following Example 2.24, for signed discrete random variables,

$$E[X] = \sum_{i:x_i \geq 0} x_i P(X = x_i) + \sum_{i:x_i < 0} x_i P(X = x_i),$$

if at least one of the sums is finite. The analogous formula for continuous random variables is

$$E[X] = \int_0^{\infty} x f(x) \, dx + \int_{-\infty}^0 x f(x) \, dx,$$

assuming at least one of the integrals is finite. Otherwise we say that $E[X]$ is undefined. Since f is the Cauchy(1) density,

$$x f(x) = x \cdot \frac{1/\pi}{1+x^2}.$$

Since this integrand has anti-derivative

$$\frac{1}{2\pi} \ln(1+x^2),$$

we have

$$E[X] = \frac{1}{2\pi} \ln(1+x^2) \bigg|_0^{\infty} + \frac{1}{2\pi} \ln(1+x^2) \bigg|_{-\infty}^0$$
$$= (\infty - 0) + (0 - \infty)$$
$$= \infty - \infty$$
$$= \text{undefined}.$$

---

[b]Additional Pareto densities are considered in Problems 2, 23, and 26 and in Problem 59 in Chapter 5.

4.2 Expectation of a single random variable

Derivation of LOTUS

Let X be a continuous random variable with density f. We first show that if g is a real-valued function taking finitely many distinct values $y_j \in \mathbb{R}$, then

$$E[g(X)] = \int_{-\infty}^{\infty} g(x) f(x)\, dx. \tag{4.6}$$

To begin, observe that

$$P(Y = y_j) = P(g(X) = y_j) = \int_{\{x: g(x) = y_j\}} f(x)\, dx.$$

Then write

$$\begin{aligned}
E[Y] &= \sum_j y_j P(Y = y_j) \\
&= \sum_j y_j \int_{\{x: g(x) = y_j\}} f(x)\, dx \\
&= \sum_j \int_{\{x: g(x) = y_j\}} y_j f(x)\, dx \\
&= \sum_j \int_{\{x: g(x) = y_j\}} g(x) f(x)\, dx \\
&= \int_{-\infty}^{\infty} g(x) f(x)\, dx,
\end{aligned}$$

since the last sum of integrals is just a special way of integrating over all values of x.

We would like to apply (4.6) for more arbitrary functions g. However, if $g(X)$ is not a discrete random variable, its expectation has not yet been defined! This raises the question of how to define the expectation of an arbitrary random variable. The approach is to approximate X by a sequence of discrete random variables (for which expectation was defined in Chapter 2) and then define $E[X]$ to be the limit of the expectations of the approximations. To be more precise about this, consider the sequence of functions q_n sketched in Figure 4.9.

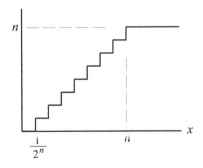

Figure 4.9. Finite-step quantizer $q_n(x)$ for approximating arbitrary random variables by discrete random variables. The number of steps is $n2^n$. In the figure, $n = 2$ and so there are eight steps.

Since each q_n takes only finitely many distinct values, $q_n(X)$ is a discrete random variable for which $\mathsf{E}[q_n(X)]$ is defined. Since $q_n(X) \to X$, we then *define*[5] $\mathsf{E}[X] := \lim_{n \to \infty} \mathsf{E}[q_n(X)]$.

Now suppose X is a continuous random variable. Since $q_n(X)$ is a discrete random variable, (4.6) applies, and we can write

$$\mathsf{E}[X] := \lim_{n \to \infty} \mathsf{E}[q_n(X)] = \lim_{n \to \infty} \int_{-\infty}^{\infty} q_n(x) f(x) \, dx.$$

Now bring the limit inside the integral,[6] and then use the fact that $q_n(x) \to x$. This yields

$$\mathsf{E}[X] = \int_{-\infty}^{\infty} \lim_{n \to \infty} q_n(x) f(x) \, dx = \int_{-\infty}^{\infty} x f(x) \, dx.$$

The same technique can be used to show that (4.6) holds even if g takes more than finitely many values. Write

$$\mathsf{E}[g(X)] := \lim_{n \to \infty} \mathsf{E}[q_n(g(X))]$$
$$= \lim_{n \to \infty} \int_{-\infty}^{\infty} q_n(g(x)) f(x) \, dx$$
$$= \int_{-\infty}^{\infty} \lim_{n \to \infty} q_n(g(x)) f(x) \, dx$$
$$= \int_{-\infty}^{\infty} g(x) f(x) \, dx.$$

4.3 Transform methods

In Chapter 3, we made extensive use of probability generating functions to compute moments and to analyze sums of independent random variables. However, probability generating functions apply only to nonnegative, integer-valued random variables. To handle other kinds of random variables, we now introduce moment generating functions and characteristic functions.

Moment generating functions

The **moment generating function** (mgf) of a real-valued random variable X is defined by

$$\boxed{M_X(s) := \mathsf{E}[e^{sX}].} \qquad (4.7)$$

This generalizes the concept of probability generating function because if X is discrete taking only nonnegative integer values, then

$$M_X(s) = \mathsf{E}[e^{sX}] = \mathsf{E}[(e^s)^X] = G_X(e^s).$$

To see why M_X is called the moment generating function, we differentiate (4.7) with respect to s and obtain

$$M_X'(s) = \frac{d}{ds} \mathsf{E}[e^{sX}]$$
$$= \mathsf{E}\left[\frac{d}{ds} e^{sX}\right]$$
$$= \mathsf{E}[X e^{sX}].$$

4.3 Transform methods

Taking $s = 0$, we have
$$M'_X(s)|_{s=0} = \mathsf{E}[X].$$

Differentiating k times and then setting $s = 0$ yields
$$M_X^{(k)}(s)|_{s=0} = \mathsf{E}[X^k], \qquad (4.8)$$

assuming $M_X(s)$ is finite in a neighborhood of $s = 0$ [3, p. 278].

Example 4.15. If $X \sim N(0,1)$, find its moment generating function, and use it to find the first four moments of X.

Solution. To begin, write
$$M_X(s) = \frac{1}{\sqrt{2\pi}} \int_{-\infty}^{\infty} e^{sx} e^{-x^2/2} \, dx.$$

Combining the exponents in the above integral and completing the square we find
$$M_X(s) = e^{s^2/2} \int_{-\infty}^{\infty} \frac{e^{-(x-s)^2/2}}{\sqrt{2\pi}} \, dx.$$

The above integrand is a normal density with mean s and unit variance. Since densities integrate to one, $M_X(s) = e^{s^2/2}$.

For the first derivative of $M_X(s)$, we have $M'_X(x) = e^{s^2/2} \cdot s$, and so $\mathsf{E}[X] = M'_X(0) = 0$. Before computing the remaining derivatives, it is convenient to notice that $M'_X(s) = M_X(s)s$. Using the product rule,
$$M''_X(s) = M_X(s) + M'_X(s)s = M_X(s) + M_X(s)s^2.$$

Since we always have $M_X(0) = 1$, we see that $\mathsf{E}[X^2] = M''_X(0) = 1$. Rewriting $M''_X(s) = M_X(s)(1+s^2)$, we get for the third derivative,
$$M_X^{(3)}(s) = M_X(s)(2s) + M'_X(s)(1+s^2) = M_X(s)(3s+s^3).$$

Hence, $\mathsf{E}[X^3] = M_X^{(3)}(0) = 0$. For the fourth derivative,
$$M_X^{(4)}(s) = M_X(s)(3+3s^2) + M'_X(s)(3s+s^3).$$

Since we are stopping with the fourth moment, no further simplification is necessary since we are going to set $s = 0$. We find that $\mathsf{E}[X^4] = M_X^{(4)}(0) = 3$.

It is now convenient to allow s to be complex. This means that we need to define the expectation of complex-valued functions of X. If $g(x) = u(x) + jv(x)$, where u and v are real-valued functions of x, then
$$\mathsf{E}[g(X)] := \mathsf{E}[u(X)] + j\mathsf{E}[v(X)].$$

If X has density f, then

$$\begin{aligned}
\mathsf{E}[g(X)] &:= \mathsf{E}[u(X)] + j\mathsf{E}[v(X)] \\
&= \int_{-\infty}^{\infty} u(x)f(x)\,dx + j\int_{-\infty}^{\infty} v(x)f(x)\,dx \\
&= \int_{-\infty}^{\infty} [u(x) + jv(x)]f(x)\,dx \\
&= \int_{-\infty}^{\infty} g(x)f(x)\,dx.
\end{aligned}$$

It now follows that if X is a continuous random variable with density f, then

$$M_X(s) = \mathsf{E}[e^{sX}] = \int_{-\infty}^{\infty} e^{sx} f(x)\,dx,$$

which is just the **Laplace transform**[c] of f.

Example 4.16. If X is an exponential random variable with parameter $\lambda > 0$, find its moment generating function.

Solution. Write

$$\begin{aligned}
M_X(s) &= \int_0^{\infty} e^{sx} \lambda e^{-\lambda x}\,dx \\
&= \lambda \int_0^{\infty} e^{x(s-\lambda)}\,dx \quad (4.9) \\
&= \frac{\lambda}{\lambda - s}.
\end{aligned}$$

For real s, the integral in (4.9) is finite if and only if $s < \lambda$. For complex s, the analogous condition is $\operatorname{Re} s < \lambda$. Hence, the moment generating function of an $\exp(\lambda)$ random variable is defined only for $\operatorname{Re} s < \lambda$.

If $M_X(s)$ is finite for all real s in a neighborhood of the origin, say for $-r < s < r$ for some $0 < r \le \infty$, then X has finite moments of all orders, and the following calculation using the power series $e^\xi = \sum_{n=0}^{\infty} \xi^n/n!$ is valid for complex s with $|s| < r$ [3, p. 278]:

$$\begin{aligned}
\mathsf{E}[e^{sX}] &= \mathsf{E}\left[\sum_{n=0}^{\infty} \frac{(sX)^n}{n!}\right] \\
&= \sum_{n=0}^{\infty} \frac{s^n}{n!} \mathsf{E}[X^n], \quad |s| < r. \quad (4.10)
\end{aligned}$$

Example 4.17. For the exponential random variable of the previous example, we can obtain the power series as follows. Recalling the geometric series formula (Problem 27 in Chapter 1), write

$$\frac{\lambda}{\lambda - s} = \frac{1}{1 - s/\lambda} = \sum_{n=0}^{\infty} (s/\lambda)^n,$$

---

[c] Signals and systems textbooks define the Laplace transform of f by $\int_{-\infty}^{\infty} e^{-sx} f(x)\,dx$. Hence, to be precise, we should say that $M_X(s)$ is the Laplace transform of f evaluated at $-s$.

4.3 Transform methods

which is finite for all complex s with $|s| < \lambda$. Comparing the above sum with (4.10) and equating the coefficients of the powers of s, we see by inspection that $\mathsf{E}[X^n] = n!/\lambda^n$. In particular, we have $\mathsf{E}[X] = 1/\lambda$ and $\mathsf{E}[X^2] = 2/\lambda^2$. Since $\text{var}(X) = \mathsf{E}[X^2] - (\mathsf{E}[X])^2$, it follows that $\text{var}(X) = 1/\lambda^2$.

In the preceding example, we computed $M_X(s)$ directly, and since we knew its power series expansion, we could pick off the moments by inspection. As the next example shows, the reverse procedure is sometimes possible; i.e., if we know all the moments, we can write down the power series, and if we are lucky, we can find a closed-form expression for the sum.

Example 4.18. If $X \sim N(0,1)$, sum the power series expansion for $M_X(s)$.

Solution. Since we know from Example 4.15 that $M_X(s)$ is finite for all real s, we can use the power series expansion (4.10) to find $M_X(s)$ for all complex s. The moments of X were determined in Example 4.11. Recalling that the odd moments are zero,

$$\begin{aligned}
M_X(s) &= \sum_{m=0}^{\infty} \frac{s^{2m}}{(2m)!} \mathsf{E}[X^{2m}] \\
&= \sum_{m=0}^{\infty} \frac{s^{2m}}{(2m)!} \cdot \frac{(2m)!}{2^m m!}, \quad \text{by (4.5),} \\
&= \sum_{m=0}^{\infty} \frac{(s^2/2)^m}{m!} \\
&= e^{s^2/2}.
\end{aligned}$$

Remark. Lest the reader think that Example 4.18 is redundant in light of Example 4.15, we point out that the solution of Example 4.15 works only for *real* s because we treated s as the mean of a real-valued random variable.[7]

Characteristic functions

In Example 4.16, the moment generating function was guaranteed finite only for $\text{Re}\,s < \lambda$. It is possible to have random variables for which $M_X(s)$ is defined only for $\text{Re}\,s = 0$; i.e., $M_X(s)$ is only defined for imaginary s. For example, if X is a Cauchy random variable, then it is easy to see that $M_X(s) = \infty$ for all real $s \neq 0$. In order to develop transform methods that always work for *any* random variable X, we introduce the **characteristic function** of X, defined by

$$\varphi_X(\nu) := \mathsf{E}[e^{j\nu X}]. \qquad (4.11)$$

Note that $\varphi_X(\nu) = M_X(j\nu)$. Also, since $|e^{j\nu X}| = 1$, $|\varphi_X(\nu)| \leq \mathsf{E}[|e^{j\nu X}|] = 1$. Hence, the characteristic function always exists and is bounded in magnitude by one.

If X is a continuous random variable with density f, then

$$\varphi_X(v) = \int_{-\infty}^{\infty} e^{jvx} f(x) \, dx,$$

which is just the **Fourier transform**[d] of f. Using the Fourier **inversion formula**,

$$f(x) = \frac{1}{2\pi} \int_{-\infty}^{\infty} e^{-jvx} \varphi_X(v) \, dv. \tag{4.12}$$

Example **4.19.** If X is an $N(0,1)$ random variable, then by Example 4.18, $M_X(s) = e^{s^2/2}$. Thus, $\varphi_X(v) = M_X(jv) = e^{(jv)^2/2} = e^{-v^2/2}$. In terms of Fourier transforms,

$$\frac{1}{\sqrt{2\pi}} \int_{-\infty}^{\infty} e^{jvx} e^{-x^2/2} \, dx = e^{-v^2/2}.$$

In signal processing terms, the Fourier transform of a **Gaussian pulse** is a Gaussian pulse. An alternative derivation of the $N(0,1)$ characteristic function is given in Problem 50.

Example **4.20.** If X has the gamma density $g_p(x) = x^{p-1} e^{-x} / \Gamma(p)$, $x > 0$, find the characteristic function of X.

Solution. It is shown in Problem 44 that $M_X(s) = 1/(1-s)^p$. Taking $s = jv$ shows that $\varphi_X(v) = 1/(1-jv)^p$ is the characteristic function.[8] An alternative derivation is given in Problem 51.

Example **4.21.** As noted above, the characteristic function of an $N(0,1)$ random variable is $e^{-v^2/2}$. Show that

$$\varphi_X(v) = e^{jvm - \sigma^2 v^2/2}, \quad \text{if } X \sim N(m, \sigma^2).$$

Solution. Let f_0 denote the $N(0,1)$ density. If $X \sim N(m, \sigma^2)$, then $f_X(x) = f_0((x-m)/\sigma)/\sigma$. Now write

$$\varphi_X(v) = \mathsf{E}[e^{jvX}]$$
$$= \int_{-\infty}^{\infty} e^{jvx} f_X(x) \, dx$$
$$= \int_{-\infty}^{\infty} e^{jvx} \cdot \frac{1}{\sigma} f_0\left(\frac{x-m}{\sigma}\right) dx.$$

---

[d]Signals and systems textbooks define the Fourier transform of f by $\int_{-\infty}^{\infty} e^{-j\omega x} f(x) \, dx$. Hence, to be precise, we should say that $\varphi_X(v)$ is the Fourier transform of f evaluated at $-v$.

4.3 Transform methods

Then apply the change of variable $y = (x-m)/\sigma$, $dy = dx/\sigma$ and obtain

$$\begin{aligned}\varphi_X(v) &= \int_{-\infty}^{\infty} e^{jv(\sigma y+m)} f_0(y)\,dy \\ &= e^{jvm}\int_{-\infty}^{\infty} e^{j(v\sigma)y} f_0(y)\,dy \\ &= e^{jvm} e^{-(v\sigma)^2/2} \\ &= e^{jvm-\sigma^2 v^2/2}.\end{aligned}$$

If X is an *integer-valued* random variable, then

$$\varphi_X(v) = \mathsf{E}[e^{jvX}] = \sum_n e^{jvn} \mathsf{P}(X=n)$$

is a 2π-periodic **Fourier series**. Given φ_X, the coefficients can be recovered by the formula for Fourier series coefficients,

$$\mathsf{P}(X=n) = \frac{1}{2\pi}\int_{-\pi}^{\pi} e^{-jvn} \varphi_X(v)\,dv. \qquad (4.13)$$

When the moment generating function is not finite in a neighborhood of the origin, the moments of X cannot be obtained from (4.8). However, the moments can sometimes be obtained from the characteristic function. For example, if we differentiate (4.11) with respect to v, we obtain

$$\varphi_X'(v) = \frac{d}{dv}\mathsf{E}[e^{jvX}] = \mathsf{E}[jXe^{jvX}].$$

Taking $v = 0$ yields $\varphi_X'(0) = j\mathsf{E}[X]$. The general result is

$$\boxed{\varphi_X^{(k)}(v)|_{v=0} = j^k \mathsf{E}[X^k],} \qquad (4.14)$$

assuming $\mathsf{E}[|X|^k] < \infty$ [3, pp. 344–345].

Why so many transforms?

We have now discussed probability generating functions, moment generating functions, and characteristic functions. Why do we need them all? After all, the characteristic function exists for all random variables, and we can use it to recover probability mass functions and densities and to find expectations.

In the case of nonnegative, integer-valued random variables, there are two reasons for using the probability generating function. One is economy of notation. Since $\varphi_X(v) = G_X(e^{jv})$, the formula for the probability generating function is simpler to derive and to remember. The second reason is that it easier to compute the nth derivative $\mathsf{P}(X=n) = G_X^{(n)}(0)/n!$ than the integral (4.13).

There are three reasons for using the moment generating function when it exists. First, we again have economy of notation due to the fact that $\varphi_X(v) = M_X(jv)$. Second, for computing moments, the formula $\mathsf{E}[X^n] = M_X^{(n)}(0)$ is easier to use than $\mathsf{E}[X^n] = j^n \varphi_X^{(n)}(0)$, and is much easier to use than the factorial moment formulas, e.g., $\mathsf{E}[X(X-1)(X-2)] = G_X^{(3)}(1)$. Third, the Chernoff bound, discussed later in Section 4.5, and importance sampling (Section 6.8) require the use of $M_X(s)$ for positive values of s; imaginary values are not useful.

To summarize, for some random variables, such as the Cauchy, the moment generating function does not exist, and we have to use the characteristic function. Otherwise we should exploit the benefits of the probability and moment generating functions.

4.4 Expectation of multiple random variables

In Chapter 2 we showed that for discrete random variables, expectation is linear and monotone. We also showed that the expectation of a product of independent discrete random variables is the product of the individual expectations. These properties continue to hold for general random variables. Before deriving these results, we illustrate them with some examples.

Example 4.22. Let $Z := X + Y$, were X and Y are independent, with $X \sim N(0,1)$ and $Y \sim \text{Laplace}(1)$ random variables. Find $\text{cov}(X,Z)$.

Solution. Recall from Section 2.4 that

$$\text{cov}(X,Z) := \mathsf{E}[(X-m_X)(Z-m_Z)] = \mathsf{E}[XZ] - m_X m_Z.$$

Since $m_X = 0$ in this example,

$$\text{cov}(X,Z) = \mathsf{E}[XZ] = \mathsf{E}[X(X+Y)] = \mathsf{E}[X^2] + \mathsf{E}[XY].$$

Since $\mathsf{E}[X^2] = 1$ and since X and Y are independent,

$$\text{cov}(X,Z) = 1 + \mathsf{E}[X]\mathsf{E}[Y] = 1.$$

Example 4.23. Let $Z := X + Y$, where X and Y are independent random variables. Show that the characteristic function of Z is the product of the characteristic functions of X and Y.

Solution. The characteristic function of Z is

$$\varphi_Z(v) := \mathsf{E}[e^{jvZ}] = \mathsf{E}[e^{jv(X+Y)}] = \mathsf{E}[e^{jvX} e^{jvY}].$$

Now use independence to write

$$\varphi_Z(v) = \mathsf{E}[e^{jvX}]\mathsf{E}[e^{jvY}] = \varphi_X(v)\varphi_Y(v). \tag{4.15}$$

4.4 Expectation of multiple random variables

An immediate consequence of the preceding example is that if X and Y are independent continuous random variables, then the density of their sum $Z = X + Y$ is the **convolution** of their densities,

$$f_Z(z) = \int_{-\infty}^{\infty} f_X(z-y) f_Y(y) \, dy. \tag{4.16}$$

This follows by inverse Fourier transforming (4.15).[9]

Example 4.24. In the preceding example, suppose that X and Y are Cauchy with parameters λ and μ, respectively. Find the density of $Z := X + Y$.

Solution. The characteristic functions of X and Y are, by Problem 49, $\varphi_X(v) = e^{-\lambda|v|}$ and $\varphi_Y(v) = e^{-\mu|v|}$. Hence,

$$\varphi_Z(v) = \varphi_X(v)\varphi_Y(v) = e^{-\lambda|v|}e^{-\mu|v|} = e^{-(\lambda+\mu)|v|},$$

which is the characteristic function of a Cauchy random variable with parameter $\lambda + \mu$.

*Derivations

Recall that for an arbitrary random variable X, $\mathsf{E}[X] := \lim_{n\to\infty} \mathsf{E}[q_n(X)]$, where $q_n(x)$ is sketched in Figure 4.9, $q_n(x) \to x$, and for each n, $q_n(X)$ is a discrete random variable taking finitely many values.

To establish linearity, write

$$\begin{aligned}
a\mathsf{E}[X] + b\mathsf{E}[Y] &:= a \lim_{n\to\infty} \mathsf{E}[q_n(X)] + b \lim_{n\to\infty} \mathsf{E}[q_n(Y)] \\
&= \lim_{n\to\infty} a\mathsf{E}[q_n(X)] + b\mathsf{E}[q_n(Y)] \\
&= \lim_{n\to\infty} \mathsf{E}[aq_n(X) + bq_n(Y)] \\
&= \mathsf{E}[\lim_{n\to\infty} aq_n(X) + bq_n(Y)] \\
&= \mathsf{E}[aX + bY],
\end{aligned}$$

where the third equality is justified because expectation is linear for discrete random variables.

From $\mathsf{E}[X] := \lim_{n\to\infty} \mathsf{E}[q_n(X)]$ and the definition of q_n (Figure 4.9), it is clear that if $X \geq 0$, then so is $\mathsf{E}[X]$. Combining this with linearity shows that monotonicity holds for general random variables; i.e., $X \geq Y$ implies $\mathsf{E}[X] \geq \mathsf{E}[Y]$.

Example 4.25. Show that

$$\boxed{|\mathsf{E}[X]| \leq \mathsf{E}[|X|].}$$

Solution. We use the fact that for $p > 0$, the condition $|t| \leq p$ is equivalent to the condition $-p \leq t \leq p$. Since $X \leq |X|$,

$$\mathsf{E}[X] \leq \mathsf{E}[|X|]. \tag{4.17}$$

Since $-X \leq |X|$,
$$\mathsf{E}[-X] \leq \mathsf{E}[|X|].$$
Multiplying through by a minus sign yields
$$-\mathsf{E}[|X|] \leq \mathsf{E}[X],$$
which combined with (4.17) gives the desired result.

Suppose X and Y are independent random variables. For any functions $h(x)$ and $k(y)$ we show that $\mathsf{E}[h(X)k(Y)] = \mathsf{E}[h(X)]\mathsf{E}[k(Y)]$. Write

$$\begin{aligned}
\mathsf{E}[h(X)]\mathsf{E}[k(Y)] &:= \lim_{n\to\infty} \mathsf{E}[q_n(h(X))] \lim_{n\to\infty} \mathsf{E}[q_n(k(Y))] \\
&= \lim_{n\to\infty} \mathsf{E}[q_n(h(X))] \mathsf{E}[q_n(k(Y))] \\
&= \lim_{n\to\infty} \mathsf{E}[q_n(h(X))q_n(k(Y))] \\
&= \mathsf{E}[\lim_{n\to\infty} q_n(h(X))q_n(k(Y))] \\
&= \mathsf{E}[h(X)k(Y)],
\end{aligned}$$

where the third equality is justified because $q_n(h(X))$ and $q_n(k(X))$ are independent *discrete* random variables.

4.5 *Probability bounds

In many applications, it is difficult to compute the probability of an event exactly. However, bounds on the probability can often be obtained in terms of various expectations. For example, the Markov and Chebyshev inequalities were derived in Chapter 2. Below we derive a much stronger result known as the **Chernoff bound**.[e]

***Example* 4.26** (using the Markov inequality). Let X be a Poisson random variable with parameter $\lambda = 1/2$. Use the Markov inequality to bound $\mathsf{P}(X > 2)$. Compare your bound with the exact result.

Solution. First note that since X takes only integer values, $\mathsf{P}(X > 2) = \mathsf{P}(X \geq 3)$. Hence, by the Markov inequality and the fact that $\mathsf{E}[X] = \lambda = 1/2$ from Example 2.22,

$$\mathsf{P}(X \geq 3) \leq \frac{\mathsf{E}[X]}{3} = \frac{1/2}{3} = 0.167.$$

The exact answer can be obtained by noting that $\mathsf{P}(X \geq 3) = 1 - \mathsf{P}(X < 3) = 1 - \mathsf{P}(X = 0) - \mathsf{P}(X = 1) - \mathsf{P}(X = 2)$. For a Poisson($\lambda$) random variable with $\lambda = 1/2$, $\mathsf{P}(X \geq 3) = 0.0144$. So the Markov inequality gives quite a loose bound.

---

[e]This bound, often attributed to Chernoff (1952) [6], was used earlier by Cramér (1938) [11].

4.5 Probability bounds

Example 4.27 (using the Chebyshev inequality). Let X be a Poisson random variable with parameter $\lambda = 1/2$. Use the Chebyshev inequality to bound $P(X > 2)$. Compare your bound with the result of using the Markov inequality in Example 4.26.

Solution. Since X is nonnegative, we don't have to worry about the absolute value signs. Using the Chebyshev inequality and the fact that $E[X^2] = \lambda^2 + \lambda = 0.75$ from Example 2.29,

$$P(X \geq 3) \leq \frac{E[X^2]}{3^2} = \frac{3/4}{9} \approx 0.0833.$$

From Example 4.26, the exact probability is 0.0144 and the Markov bound is 0.167.

We now derive the Chernoff bound. As in the derivation of the Markov inequality, there are two key ideas. First, since every probability can be written as an expectation,

$$P(X \geq a) = E[I_{[a,\infty)}(X)]. \tag{4.18}$$

Second, from Figure 4.10, we see that for all x, $I_{[a,\infty)}(x)$ (solid line) is less than or equal to $e^{s(x-a)}$ (dashed line) for any $s \geq 0$. Taking expectations of

$$I_{[a,\infty)}(X) \leq e^{s(X-a)}$$

yields

$$E[I_{[a,\infty)}(X)] \leq E[e^{s(X-a)}] = e^{-sa}E[e^{sX}] = e^{-sa}M_X(s).$$

Combining this with (4.18) yields[10]

$$P(X \geq a) \leq e^{-sa}M_X(s). \tag{4.19}$$

Now observe that this inequality holds for all $s \geq 0$, and the left-hand side does not depend on s. Hence, we can minimize the right-hand side to get as tight a bound as possible. The **Chernoff bound** is given by

$$\boxed{P(X \geq a) \leq \min_{s \geq 0}\left[e^{-sa}M_X(s)\right],} \tag{4.20}$$

where the minimum is over all $s \geq 0$ for which $M_X(s)$ is finite.

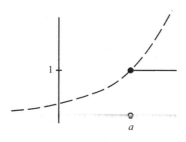

Figure 4.10. Graph showing that $I_{[a,\infty)}(x)$ (solid line) is upper bounded by $e^{s(x-a)}$ (dashed line) for any positive s. Note that the inequality $I_{[a,\infty)}(x) \leq e^{s(x-a)}$ holds even if $s = 0$.

Example 4.28. Let X be a Poisson random variable with parameter $\lambda = 1/2$. Bound $P(X > 2)$ using the Chernoff bound. Compare your result with the exact probability and with the bound obtained via the Chebyshev inequality in Example 4.27 and with the bound obtained via the Markov inequality in Example 4.26.

Solution. First recall that $M_X(s) = G_X(e^s)$, where $G_X(z) = \exp[\lambda(z-1)]$ was derived in Example 3.4. Hence,

$$e^{-sa} M_X(s) = e^{-sa} \exp[\lambda(e^s - 1)] = \exp[\lambda(e^s - 1) - as].$$

The desired Chernoff bound when $a = 3$ is

$$P(X \geq 3) \leq \min_{s \geq 0} \exp[\lambda(e^s - 1) - 3s].$$

We must now minimize the exponential. Since exp is an increasing function, it suffices to minimize its argument. Taking the derivative of the argument and setting it equal to zero requires us to solve $\lambda e^s - 3 = 0$. The solution is $s = \ln(3/\lambda)$. Substituting this value of s into $\exp[\lambda(e^s - 1) - 3s]$ and simplifying the exponent yields

$$P(X \geq 3) \leq \exp[3 - \lambda - 3\ln(3/\lambda)].$$

Since $\lambda = 1/2$,

$$P(X \geq 3) \leq \exp[2.5 - 3\ln 6] = 0.0564.$$

Recall that from Example 4.26, the exact probability is 0.0144 and the Markov inequality yielded the bound 0.167. From Example 4.27, the Chebyshev inequality yielded the bound 0.0833.

Example 4.29. Let X be a continuous random variable having exponential density with parameter $\lambda = 1$. Compute $P(X \geq 7)$ and the corresponding Markov, Chebyshev, and Chernoff bounds.

Solution. The exact probability is $P(X \geq 7) = \int_7^\infty e^{-x} dx = e^{-7} = 0.00091$. For the Markov and Chebyshev inequalities, recall that from Example 4.17, $E[X] = 1/\lambda$ and $E[X^2] = 2/\lambda^2$. For the Chernoff bound, we need $M_X(s) = \lambda/(\lambda - s)$ for $s < \lambda$, which was derived in Example 4.16. Armed with these formulas, we find that the Markov inequality yields $P(X \geq 7) \leq E[X]/7 = 1/7 = 0.143$ and the Chebyshev inequality yields $P(X \geq 7) \leq E[X^2]/7^2 = 2/49 = 0.041$. For the Chernoff bound, write

$$P(X \geq 7) \leq \min_s e^{-7s}/(1-s),$$

where the minimization is over $0 \leq s < 1$. The derivative of $e^{-7s}/(1-s)$ with respect to s is

$$\frac{e^{-7s}(7s - 6)}{(1-s)^2}.$$

Setting this equal to zero requires that $s = 6/7$. Hence, the Chernoff bound is

$$P(X \geq 7) \leq \left.\frac{e^{-7s}}{(1-s)}\right|_{s=6/7} = 7e^{-6} = 0.017.$$

Notes

4.1: Densities and probabilities

Note 1. Strictly speaking, it can only be shown that f in (4.1) is nonnegative almost everywhere; that is

$$\int_{\{t\in\mathbb{R}: f(t)<0\}} 1\, dt = 0.$$

For example, f could be negative at a finite or countably infinite set of points.

Note 2. If U and V are $N(0,1)$, then by Problem 46, U^2 and V^2 are each chi-squared with one degree of freedom (defined in Problem 15). By the Remark following Problem 55(c), $U^2 + V^2$ is chi-squared with two degrees of freedom, which is the same as $\exp(1/2)$.

Note 3. In the formula $g_p(x) = x^{p-1}e^{-x}/\Gamma(p)$, it is important that $\Gamma(p)$ be finite; otherwise $g_p(x) = 0$ for all x and would not be a valid density. We claim that the gamma function integral is finite for $p > 0$, but infinite for $p \leq 0$. To begin, write

$$\Gamma(p) = \int_0^\infty x^{p-1} e^{-x}\, dx = \int_0^1 x^{p-1} e^{-x}\, dx + \int_1^\infty x^{p-1} e^{-x}\, dx.$$

The integral from zero to one is finite for $p \geq 1$ since in this case the integrand $x^{p-1}e^{-x}$ is bounded. However, for $p < 1$, the factor x^{p-1} blows up as x approaches zero. Observe that

$$\int_0^1 x^{p-1} e^{-x}\, dx \leq \int_0^1 x^{p-1}\, dx$$

and

$$\int_0^1 x^{p-1} e^{-x}\, dx \geq e^{-1} \int_0^1 x^{p-1}\, dx.$$

Now note that

$$\int_0^1 x^{p-1}\, dx = \tfrac{1}{p} x^p \Big|_0^1$$

is finite for $0 < p < 1$ and infinite for $p < 0$. For $p = 0$, the anti-derivative is $\ln x$, and the integral is again infinite.

It remains to consider the integral from 1 to ∞. For $p \leq 1$, this integral is finite because it is upper bounded by $\int_1^\infty e^{-x}\, dx < \infty$. For $p > 1$, we use the fact that

$$e^x = \sum_{k=0}^\infty \frac{x^k}{k!} \geq \frac{x^n}{n!}, \quad x \geq 0, n \geq 0.$$

This implies $e^{-x} \leq n!/x^n$. Then
$$x^{p-1}e^{-x} \leq x^{p-1}n!/x^n = n!x^{(p-1)-n}.$$
Now, for $x > 1$, if we take $n > (p-1)+2$, we can write $x^{p-1}e^{-x} \leq n!/x^2$. Hence,
$$\int_1^\infty x^{p-1}e^{-x}\,dx \leq \int_1^\infty n!/x^2\,dx < \infty.$$

We now see that of the two integrals making up $\Gamma(p)$, the second integral is always finite, but the first one is finite only for $p > 0$. Hence, the sum of the two integrals is finite if and only if $p > 0$.

4.2: Expectation of a single random variable

Note 4. Integration by parts. The formula for integration by parts is
$$\int_a^b u\,dv = uv\Big|_a^b - \int_a^b v\,du.$$
This is shorthand for
$$\int_a^b u(t)v'(t)\,dt = u(t)v(t)\Big|_a^b - \int_a^b v(t)u'(t)\,dt.$$
It is obtained by integrating the derivative of the product $u(t)v(t)$ and rearranging the result. If we integrate the formula
$$\frac{d}{dt}u(t)v(t) = u'(t)v(t) + u(t)v'(t),$$
we get
$$u(t)v(t)\Big|_a^b = \int_a^b \frac{d}{dt}[u(t)v(t)]\,dt = \int_a^b u'(t)v(t)\,dt + \int_a^b u(t)v'(t)\,dt.$$
Rearranging yields the integration-by-parts formula.

To apply this formula, you need to break the integrand into two factors, where you know the anti-derivative of one of them. The other factor you can almost always differentiate. For example, to integrate $t^n e^{-t}$, take $u(t) = t^n$ and $v'(t) = e^{-t}$, since you know that the anti-derivative of e^{-t} is $v(t) = -e^{-t}$.

Another useful example in this book is $t^n e^{-t^2/2}$, where $n \geq 1$. Although there is no closed-form anti-derivative of $e^{-t^2/2}$, observe that
$$\frac{d}{dt}e^{-t^2/2} = -te^{-t^2/2}.$$

In other words, the anti-derivative of $te^{-t^2/2}$ is $-e^{-t^2/2}$. This means that to integrate $t^n e^{-t^2/2}$, first write it as $t^{n-1} \cdot te^{-t^2/2}$. Then take $u(t) = t^{n-1}$ and $v'(t) = te^{-t^2/2}$ and use $v(t) = -e^{-t^2/2}$.

Note 5. Since q_n in Figure 4.9 is defined only for $x \geq 0$, the definition of expectation in the text applies only to arbitrary nonnegative random variables. However, for signed random variables,

$$X = \frac{|X|+X}{2} - \frac{|X|-X}{2},$$

and we define

$$\mathsf{E}[X] := \mathsf{E}\left[\frac{|X|+X}{2}\right] - \mathsf{E}\left[\frac{|X|-X}{2}\right],$$

assuming the difference is not of the form $\infty - \infty$. Otherwise, we say $\mathsf{E}[X]$ is undefined.

We also point out that for $x \geq 0$, $q_n(x) \to x$. To see this, fix any $x \geq 0$, and let $n > x$. Then x will lie under one of the steps in Figure 4.9. If x lies under the kth step, then

$$\frac{k-1}{2^n} \leq x < \frac{k}{2^n},$$

For x in this range, the value of $q_n(x)$ is $(k-1)/2^n$. Hence, $0 \leq x - q_n(x) < 1/2^n$.

Another important fact to note is that for each $x \geq 0$, $q_n(x) \leq q_{n+1}(x)$. Hence, $q_n(X) \leq q_{n+1}(X)$, and so $\mathsf{E}[q_n(X)] \leq \mathsf{E}[q_{n+1}(X)]$ as well. In other words, the sequence of real numbers $\mathsf{E}[q_n(X)]$ is nondecreasing. This implies that $\lim_{n \to \infty} \mathsf{E}[q_n(X)]$ exists either as a finite real number or the extended real number ∞ [51, p. 55].

Note 6. In light of the preceding note, we are using Lebesgue's monotone convergence theorem [3, p. 208], which applies to nonnegative functions.

4.3: Transform methods

Note 7. If s were complex, we could interpret $\int_{-\infty}^{\infty} e^{-(x-s)^2/2} dx$ as a contour integral in the complex plane. By appealing to the Cauchy–Goursat theorem [9, pp. 115–121], one could then show that this integral is equal to $\int_{-\infty}^{\infty} e^{-t^2/2} dt = \sqrt{2\pi}$. Alternatively, one can use a **permanence of form argument** [9, pp. 286–287]. In this approach, one shows that $M_X(s)$ is analytic in some region, in this case the whole complex plane. One then obtains a formula for $M_X(s)$ on a contour in this region, in this case, the contour is the entire real axis. The permanence of form theorem then states that the formula is valid in the entire region.

Note 8. Problem 44(a) only shows that $M_X(s) = 1/(1-s)^p$ for real s with $s < 1$. However, since $M_X(s)$ is analytic for complex s with $\operatorname{Re} s < 1$, the permanence of form argument mentioned in Note 7 shows that $M_X(s) = 1/(1-s)^p$ holds all such s. In particular, the formula holds for $s = jv$, since in this case, $\operatorname{Re} s = 0 < 1$.

4.4: Expectation of multiple random variables

Note 9. We show that if X and Y are independent with densities f_X and f_Y, then the density of $Z := X + Y$ is given by the convolution of f_X and f_Y. We have from (4.13) that the characteristic functions of X, Y, and Z satisfy $\varphi_Z(v) = \varphi_X(v)\varphi_Y(v)$. Now write

$$f_Z(z) = \frac{1}{2\pi} \int_{-\infty}^{\infty} e^{-jvz} \varphi_Z(v) dv$$

$$= \frac{1}{2\pi} \int_{-\infty}^{\infty} e^{-jvz} \varphi_X(v) \varphi_Y(v) \, dv$$

$$= \frac{1}{2\pi} \int_{-\infty}^{\infty} e^{-jvz} \varphi_X(v) \left[\int_{-\infty}^{\infty} e^{jvy} f_Y(y) \, dy \right] dv$$

$$= \int_{-\infty}^{\infty} \left[\frac{1}{2\pi} \int_{-\infty}^{\infty} e^{-jv(z-y)} \varphi_X(v) \, dv \right] f_Y(y) \, dy$$

$$= \int_{-\infty}^{\infty} f_X(z-y) f_Y(y) \, dy.$$

4.5: *Probability bounds

Note 10. We note that (4.19) follows directly from the Markov inequality. Observe that for $s > 0$,

$$\{X \geq a\} = \{sX \geq sa\} = \{e^{sX} \geq e^{sa}\}.$$

Hence,

$$\mathsf{P}(X \geq a) = \mathsf{P}(e^{sX} \geq e^{sa}) \leq \frac{\mathsf{E}[e^{sX}]}{e^{sa}} = e^{-sa} M_X(s),$$

which is exactly (4.19). The reason for using the derivation in the text is to emphasize the idea of bounding $I_{[a,\infty)}(x)$ by different functions of x. For (2.19), we used $(x/a)^r$ for $x \geq 0$. For (4.19), we used $e^{s(x-a)}$ for all x.

Problems

4.1: Densities and probabilities

1. A certain burglar alarm goes off if its input voltage exceeds 5 V at three consecutive sampling times. If the voltage samples are independent and uniformly distributed on $[0,7]$, find the probability that the alarm sounds.

2. Let X have the **Pareto** density

$$f(x) = \begin{cases} 2/x^3, & x > 1, \\ 0, & \text{otherwise,} \end{cases}$$

The **median** of X is the number t satisfying $\mathsf{P}(X > t) = 1/2$. Find the median of X.

3. Let X have density

$$f(x) = \begin{cases} cx^{-1/2}, & 0 < x \leq 1, \\ 0, & \text{otherwise,} \end{cases}$$

shown in Figure 4.11. Find the constant c and the median of X.

4. Let X have an $\exp(\lambda)$ density.

   (a) Show that $\mathsf{P}(X > t) = e^{-\lambda t}$ for $t \geq 0$.

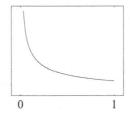

Figure **4.11.** Density of Problem 3. Even though the density blows up as it approaches the origin, the area under the curve between 0 and 1 is unity.

   (b) Compute $P(X > t + \Delta t | X > t)$ for $t \geq 0$ and $\Delta t > 0$. *Hint:* If $A \subset B$, then $A \cap B = A$.

   Remark. Observe that X has a memoryless property similar to that of the geometric$_1(p)$ random variable. See the remark following Problem 21 in Chapter 2.

5. A company produces independent voltage regulators whose outputs are $\exp(\lambda)$ random variables. In a batch of 10 voltage regulators, find the probability that exactly three of them produce outputs greater than v volts.

6. Let $X_1, \ldots, X_n$ be i.i.d. $\exp(\lambda)$ random variables.

   (a) Find the probability that $\min(X_1, \ldots, X_n) > 2$.
   (b) Find the probability that $\max(X_1, \ldots, X_n) > 2$.

   Hint: Example 2.11 may be helpful.

7. A certain computer is equipped with a hard drive whose lifetime, measured in months, is $X \sim \exp(\lambda)$. The lifetime of the monitor (also measured in months) is $Y \sim \exp(\mu)$. Assume the lifetimes are independent.

   (a) Find the probability that the monitor fails during the first 2 months.
   (b) Find the probability that both the hard drive and the monitor fail during the first year.
   (c) Find the probability that either the hard drive or the monitor fails during the first year.

8. A random variable X has the **Weibull** density with parameters $p > 0$ and $\lambda > 0$, denoted by $X \sim \text{Weibull}(p, \lambda)$, if its density is given by $f(x) := \lambda p x^{p-1} e^{-\lambda x^p}$ for $x > 0$, and $f(x) := 0$ for $x \leq 0$.

   (a) Show that this density integrates to one.
   (b) If $X \sim \text{Weibull}(p, \lambda)$, evaluate $P(X > t)$ for $t > 0$.
   (c) Let $X_1, \ldots, X_n$ be i.i.d. $\text{Weibull}(p, \lambda)$ random variables. Find the probability that none of them exceeds 3. Find the probability that at least one of them exceeds 3.

   Remark. The Weibull density arises in the study of reliability in Chapter 5. Note that $\text{Weibull}(1, \lambda)$ is the same as $\exp(\lambda)$.

*9. The standard normal density $f \sim N(0,1)$ is given by $f(x) := e^{-x^2/2}/\sqrt{2\pi}$. The following steps provide a mathematical proof that the normal density is indeed "bell-shaped" as shown in Figure 4.3.

   (a) Use the derivative of f to show that f is decreasing for $x > 0$ and increasing for $x < 0$. (It then follows that f has a global maximum at $x = 0$.)

   (b) Show that f is concave for $|x| < 1$ and convex for $|x| > 1$. *Hint:* Show that the second derivative of f is negative for $|x| < 1$ and positive for $|x| > 1$.

   (c) Since $e^z = \sum_{n=0}^{\infty} z^n/n!$, for positive z, $e^z \geq z$. Hence, $e^{+x^2/2} \geq x^2/2$. Use this fact to show that $e^{-x^2/2} \to 0$ as $|x| \to \infty$.

*10. Use the results of parts (a) and (b) of the preceding problem to obtain the corresponding results for the general normal density $f \sim N(m, \sigma^2)$. *Hint:* Let $\varphi(t) := e^{-t^2/2}/\sqrt{2\pi}$ denote the $N(0,1)$ density, and observe that $f(x) = \varphi((x-m)/\sigma)/\sigma$.

*11. As in the preceding problem, let $f \sim N(m, \sigma^2)$. Keeping in mind that $f(x)$ depends on $\sigma > 0$, show that $\lim_{\sigma \to \infty} f(x) = 0$. Using the result of part (c) of Problem 9, show that for $x \neq m$, $\lim_{\sigma \to 0} f(x) = 0$, whereas for $x = m$, $\lim_{\sigma \to 0} f(x) = \infty$.

12. For $n = 1, 2, \ldots$, let $f_n(x)$ be a probability density function, and let p_n be a sequence of nonnegative numbers summing to one; i.e., a probability mass function.

   (a) Show that
   $$f(x) := \sum_n p_n f_n(x)$$
   is a probability density function. When f has this form, it is called a **mixture density**.

   Remark. When the $f_n(x)$ are chi-squared densities and the p_n are appropriate Poisson probabilities, the resulting mixture f is called a noncentral chi-squared density. See Problem 65 for details.

   (b) If $f_1 \sim \text{uniform}[0,1]$ and $f_2 \sim \text{uniform}[2,3]$, sketch the mixture density
   $$f(x) = 0.25 f_1(x) + 0.75 f_2(x).$$

   (c) If $f_1 \sim \text{uniform}[0,2]$ and $f_2 \sim \text{uniform}[1,3]$, sketch the mixture density
   $$f(x) = 0.5 f_1(x) + 0.5 f_2(x).$$

13. If g and h are probability densities, show that their convolution,
   $$(g * h)(x) := \int_{-\infty}^{\infty} g(y) h(x-y) \, dy,$$
   is also a probability density; i.e., show that $(g * h)(x)$ is nonnegative and when integrated with respect to x yields one.

14. The **gamma** density with parameter $p > 0$ is given by

$$g_p(x) := \begin{cases} \dfrac{x^{p-1} e^{-x}}{\Gamma(p)}, & x > 0, \\ 0, & x \leq 0, \end{cases}$$

where $\Gamma(p)$ is the **gamma function**,

$$\Gamma(p) := \int_0^\infty x^{p-1} e^{-x} \, dx, \quad p > 0.$$

In other words, the gamma *function* is defined exactly so that the gamma *density* integrates to one. Note that the gamma density is a generalization of the exponential since g_1 is the exp(1) density. Sketches of g_p for several values of p were shown in Figure 4.7.

Remark. In MATLAB, $\Gamma(p) = \text{gamma}(p)$.

(a) Use integration by parts as in Example 4.9 to show that

$$\Gamma(p) = (p-1) \cdot \Gamma(p-1), \quad p > 1.$$

Since $\Gamma(1)$ can be directly evaluated and is equal to one, it follows that

$$\Gamma(n) = (n-1)!, \quad n = 1, 2, \ldots.$$

Thus Γ is sometimes called the **factorial function**.

(b) Show that $\Gamma(1/2) = \sqrt{\pi}$ as follows. In the defining integral, use the change of variable $y = \sqrt{2x}$. Write the result in terms of the standard normal density, which integrates to one, in order to obtain the answer by inspection.

(c) Show that

$$\Gamma\left(\frac{2n+1}{2}\right) = \frac{(2n-1) \cdots 5 \cdot 3 \cdot 1}{2^n} \sqrt{\pi} = \frac{(2n-1)!!}{2^n} \sqrt{\pi}, \quad n \geq 1.$$

(d) Show that $(g_p * g_q)(x) = g_{p+q}(x)$. *Hints:* First show that for $x > 0$,

$$(g_p * g_q)(x) = \int_0^x g_p(y) g_q(x-y) \, dy$$

$$= \frac{x^{p+q-1} e^{-x}}{\Gamma(p) \Gamma(q)} \int_0^1 \theta^{p-1} (1-\theta)^{q-1} d\theta. \quad (4.21)$$

Now integrate this equation with respect to x; use the definition of the gamma function and the result of Problem 13.

Remark. The integral definition of $\Gamma(p)$ makes sense only for $p > 0$. However, the recursion $\Gamma(p) = (p-1)\Gamma(p-1)$ suggests a simple way to define Γ for negative, noninteger arguments. For $0 < \varepsilon < 1$, the right-hand side of $\Gamma(\varepsilon) = (\varepsilon-1)\Gamma(\varepsilon-1)$ is undefined. However, we rearrange this equation and make the *definition*,

$$\Gamma(\varepsilon - 1) := -\frac{\Gamma(\varepsilon)}{1 - \varepsilon}.$$

Similarly writing $\Gamma(\varepsilon-1) = (\varepsilon-2)\Gamma(\varepsilon-2)$, and so on, leads to

$$\Gamma(\varepsilon-n) = \frac{(-1)^n \Gamma(\varepsilon)}{(n-\varepsilon)\cdots(2-\varepsilon)(1-\varepsilon)}.$$

Note also that

$$\Gamma(n+1-\varepsilon) = (n-\varepsilon)\Gamma(n-\varepsilon)$$
$$= (n-\varepsilon)\cdots(1-\varepsilon)\Gamma(1-\varepsilon).$$

Hence,

$$\Gamma(\varepsilon-n) = \frac{(-1)^n \Gamma(\varepsilon)\Gamma(1-\varepsilon)}{\Gamma(n+1-\varepsilon)}.$$

15. Important generalizations of the gamma density g_p of the preceding problem arise if we include a **scale parameter**. For $\lambda > 0$, put

$$g_{p,\lambda}(x) := \lambda g_p(\lambda x) = \lambda \frac{(\lambda x)^{p-1} e^{-\lambda x}}{\Gamma(p)}, \quad x > 0.$$

We write $X \sim \text{gamma}(p, \lambda)$ if X has density $g_{p,\lambda}$, which is called the gamma density with parameters p and λ.

(a) Let f be any probability density. For $\lambda > 0$, show that

$$f_\lambda(x) := \lambda f(\lambda x)$$

is also a probability density.

(b) For $p = m$ a positive integer, $g_{m,\lambda}$ is called the **Erlang** density with parameters m and λ. We write $X \sim \text{Erlang}(m, \lambda)$ if X has density

$$g_{m,\lambda}(x) = \lambda \frac{(\lambda x)^{m-1} e^{-\lambda x}}{(m-1)!}, \quad x > 0.$$

What kind of density is $g_{1,\lambda}(x)$?

(c) If $X \sim \text{Erlang}(m, \lambda)$, show that

$$P(X > t) = \sum_{k=0}^{m-1} \frac{(\lambda t)^k}{k!} e^{-\lambda t}, \quad t \geq 0.$$

In other words, if $Y \sim \text{Poisson}(\lambda t)$, then $P(X > t) = P(Y < m)$. *Hint:* Use repeated integration by parts.

(d) For $p = k/2$ and $\lambda = 1/2$, $g_{k/2, 1/2}$ is called the **chi-squared** density with k degrees of freedom. It is not required that k be an integer. Of course, the chi-squared density with an even number of degrees of freedom, say $k = 2m$, is the same as the Erlang$(m, 1/2)$ density. Using Problem 14(b), it is also clear that for $k = 1$,

$$g_{1/2, 1/2}(x) = \frac{e^{-x/2}}{\sqrt{2\pi x}}, \quad x > 0.$$

For an odd number of degrees of freedom, say $k = 2m+1$, where $m \geq 1$, show that
$$g_{\frac{2m+1}{2}, \frac{1}{2}}(x) = \frac{x^{m-1/2} e^{-x/2}}{(2m-1)\cdots 5 \cdot 3 \cdot 1 \sqrt{2\pi}}$$
for $x > 0$. *Hint:* Use Problem 14(c).

16. The **beta** density with parameters $p > 0$ and $q > 0$ is defined by
$$b_{p,q}(x) := \frac{\Gamma(p+q)}{\Gamma(p)\Gamma(q)} x^{p-1}(1-x)^{q-1}, \quad 0 < x < 1,$$
where Γ is the gamma function defined in Problem 14. We note that if $X \sim \text{gamma}(p, \lambda)$ and $Y \sim \text{gamma}(q, \lambda)$ are independent random variables, then $X/(X+Y)$ has the beta density with parameters p and q (Problem 42 in Chapter 7).

   (a) Find simplified formulas and sketch the beta density for the following sets of parameter values: (*i*) $p = 1, q = 1$. (*ii*) $p = 2, q = 2$. (*iii*) $p = 1/2, q = 1$.

   (b) Use the result of Problem 14(d), including equation (4.21), to show that the beta density integrates to one.

Remark. The fact that the beta density integrates to one can be rewritten as
$$\Gamma(p)\Gamma(q) = \Gamma(p+q) \int_0^1 u^{p-1}(1-u)^{q-1}\, du. \tag{4.22}$$
This integral, which is a function of p and q, is usually called the **beta function**, and is denoted by $B(p,q)$. Thus,
$$B(p,q) = \frac{\Gamma(p)\Gamma(q)}{\Gamma(p+q)}, \tag{4.23}$$
and
$$b_{p,q}(x) = \frac{x^{p-1}(1-x)^{q-1}}{B(p,q)}, \quad 0 < x < 1.$$

*17. Use equation (4.22) in the preceding problem to show that $\Gamma(1/2) = \sqrt{\pi}$. *Hint:* Make the change of variable $u = \sin^2\theta$. Then take $p = q = 1/2$.

Remark. In Problem 14(b), you used the fact that the normal density integrates to one to show that $\Gamma(1/2) = \sqrt{\pi}$. Since your derivation there is reversible, it follows that the normal density integrates to one if and only if $\Gamma(1/2) = \sqrt{\pi}$. In this problem, you used the fact that the beta density integrates to one to show that $\Gamma(1/2) = \sqrt{\pi}$. Thus, you have an alternative derivation of the fact that the normal density integrates to one.

*18. Show that
$$\int_0^{\pi/2} \sin^n \theta\, d\theta = \frac{\Gamma\left(\frac{n+1}{2}\right)\sqrt{\pi}}{2\Gamma\left(\frac{n+2}{2}\right)}$$
Hint: Use equation (4.22) in Problem 16 with $p = (n+1)/2$ and $q = 1/2$, and make the substitution $u = \sin^2\theta$.

*19. The beta function $B(p,q)$ is defined as the integral in (4.22) in Problem 16. Show that

$$B(p,q) = \int_0^\infty (1-e^{-\theta})^{p-1} e^{-q\theta} \, d\theta.$$

*20. **Student's t density** with ν degrees of freedom is given by

$$f_\nu(x) := \frac{\left(1+\frac{x^2}{\nu}\right)^{-(\nu+1)/2}}{\sqrt{\nu}B(\frac{1}{2},\frac{\nu}{2})}, \quad -\infty < x < \infty,$$

where B is the beta function. Show that f_ν integrates to one. *Hint:* The change of variable $e^\theta = 1+x^2/\nu$ may be useful. Also, the result of the preceding problem may be useful.

Remark. (*i*) Note that $f_1 \sim$ Cauchy(1).

(*ii*) It is shown in Problem 44 in Chapter 7 that if X and Y are independent with $X \sim N(0,1)$ and Y chi-squared with k degrees of freedom, then $X/\sqrt{Y/k}$ has Student's t density with k degrees of freedom, a result of crucial importance in the study of confidence intervals.

(*iii*) This density was reported by William Sealey Gosset in the journal paper,

Student, "The probable error of a mean," *Biometrika*, vol. VI, no. 1, pp. 1–25, Mar. 1908.

Gosset obtained his results from statistical studies at the Guinness brewery in Dublin. He used a pseudonym because Guinness did not allow employees to publish.

*21. As illustrated in Figure 4.12, Student's t density $f_\nu(x)$ defined in Problem 20 converges to the standard normal density as $\nu \to \infty$. In this problem you will demonstrate this mathematically.

(a) **Stirling's formula** says that $\Gamma(x) \approx \sqrt{2\pi} x^{x-1/2} e^{-x}$. Use Stirling's formula to show that

$$\frac{\Gamma\left(\frac{1+\nu}{2}\right)}{\sqrt{\nu}\,\Gamma\left(\frac{\nu}{2}\right)} \approx \frac{1}{\sqrt{2}}.$$

(b) Use the fact that $(1+\xi/n)^n \to e^\xi$ to show that

$$\left(1+\frac{x^2}{\nu}\right)^{(\nu+1)/2} \to e^{x^2/2}.$$

Then combine this with part (a) to show that $f_\nu(x) \to e^{-x^2/2}/\sqrt{2\pi}$.

*22. For p and q positive, let $B(p,q)$ denote the beta function defined by the integral in (4.22) in Problem 16. Show that

$$f_Z(z) := \frac{1}{B(p,q)} \cdot \frac{z^{p-1}}{(1+z)^{p+q}}, \quad z > 0,$$

is a valid density (i.e., integrates to one) on $(0,\infty)$. *Hint:* Make the change of variable $t = 1/(1+z)$.

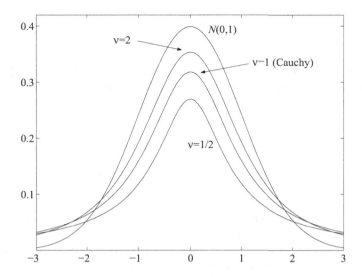

Figure 4.12. Comparision of standard normal density and Student's t density for $v = 1/2$, 1, and 2.

4.2: Expectation of a single random variable

23. Let X have the **Pareto** density $f(x) = 2/x^3$ for $x \geq 1$ and $f(x) = 0$ otherwise. Compute $\mathsf{E}[X]$.

24. The quantizer input–output relation shown in Figure 4.8 in Example 4.8 has five levels, but in applications, the number of levels n is a power of 2, say $n = 2^b$. If V_{in} lies between $\pm V_{\text{max}}$, find the smallest number of bits b required to achieve a performance of $\mathsf{E}[|V_{\text{in}} - V_{\text{out}}|^2] < \varepsilon$.

25. Let X be a continuous random variable with density f, and suppose that $\mathsf{E}[X] = 0$. If Z is another random variable with density $f_Z(z) := f(z - m)$, find $\mathsf{E}[Z]$.

*26. Let X have the **Pareto** density $f(x) = 2/x^3$ for $x \geq 1$ and $f(x) = 0$ otherwise. Find $\mathsf{E}[X^2]$.

*27. Let X have Student's t density with v degrees of freedom, as defined in Problem 20. Show that $\mathsf{E}[|X|^k]$ is finite if and only if $k < v$.

28. Let $Z \sim N(0,1)$, and put $Y = Z + n$ for some constant n. Show that $\mathsf{E}[Y^4] = n^4 + 6n^2 + 3$.

29. Let $X \sim \text{gamma}(p, 1)$ as in Problem 15. Show that

$$\mathsf{E}[X^n] = \frac{\Gamma(n+p)}{\Gamma(p)} = p(p+1)(p+2)\cdots(p+[n-1])$$

30. Let X have the standard **Rayleigh** density, $f(x) := xe^{-x^2/2}$ for $x \geq 0$ and $f(x) := 0$ for $x < 0$.

(a) Show that $E[X] = \sqrt{\pi/2}$.

(b) For $n \geq 2$, show that $E[X^n] = 2^{n/2}\Gamma(1+n/2)$.

31. Consider an Internet router with n input links. Assume that the flows in the links are independent standard Rayleigh random variables as defined in the preceding problem. Suppose that the router's buffer overflows if more than two links have flows greater than β. Find the probability of buffer overflow.

32. Let $X \sim \text{Weibull}(p, \lambda)$ as in Problem 8. Show that $E[X^n] = \Gamma(1+n/p)/\lambda^{n/p}$.

33. A certain nonlinear circuit has random input $X \sim \exp(1)$, and output $Y = X^{1/4}$. Find the second moment of the output.

34. High-Mileage Cars has just begun producing its new Lambda Series, which averages μ miles per gallon. Al's Excellent Autos has a limited supply of n cars on its lot. Actual mileage of the ith car is given by an exponential random variable X_i with $E[X_i] = \mu$. Assume actual mileages of different cars are independent. Find the probability that at least one car on Al's lot gets less than $\mu/2$ miles per gallon.

35. A small airline makes five flights a day from Chicago to Denver. The number of passengers on each flight is approximated by an exponential random variable with mean 20. A flight makes money if it has more than 25 passengers. Find the probability that at least one flight a day makes money. Assume that the numbers of passengers on different flights are independent.

36. The **differential entropy** of a continuous random variable X with density f is

$$h(X) := E[-\log f(X)] = \int_{-\infty}^{\infty} f(x) \log \frac{1}{f(x)} dx.$$

If $X \sim \text{uniform}[0,2]$, find $h(X)$. Repeat for $X \sim \text{uniform}[0, \frac{1}{2}]$ and for $X \sim N(m, \sigma^2)$.

*37. Let X have Student's t density with v degrees of freedom, as defined in Problem 20. For n a positive integer less than $v/2$, show that

$$E[X^{2n}] = v^n \frac{\Gamma(\frac{2n+1}{2})\Gamma(\frac{v-2n}{2})}{\Gamma(\frac{1}{2})\Gamma(\frac{v}{2})}.$$

4.3: Transform methods

38. Let X have moment generating function $M_X(s) = e^{\sigma^2 s^2/2}$. Use formula (4.8) to find $E[X^2]$.

*39. Recall that the moment generating function of an $N(0,1)$ random variable $e^{s^2/2}$. Use this fact to find the moment generating function of an $N(m, \sigma^2)$ random variable.

40. If $X \sim \text{uniform}(0,1)$, show that $Y = \ln(1/X) \sim \exp(1)$ by finding its moment generating function for $s < 1$.

41. Find a closed-form expression for $M_X(s)$ if $X \sim \text{Laplace}(\lambda)$. Use your result to find $\text{var}(X)$.

*42. Let X have the **Pareto** density $f(x) = 2/x^3$ for $x \geq 1$ and $f(x) = 0$ otherwise. For what real values of s is $M_X(s)$ finite? *Hint:* It is not necessary to evaluate $M_X(s)$ to answer the question.

43. Let $M_p(s)$ denote the moment generating function of the gamma density g_p defined in Problem 14. Show that

$$M_p(s) = \frac{1}{1-s} M_{p-1}(s), \quad p > 1.$$

Remark. Since $g_1(x)$ is the $\exp(1)$ density, and $M_1(s) = 1/(1-s)$ by direct calculation, it now follows that the moment generating function of an $\text{Erlang}(m,1)$ random variable is $1/(1-s)^m$.

*44. Let X have the gamma density g_p given in Problem 14.

(a) For real $s < 1$, show that $M_X(s) = 1/(1-s)^p$.

(b) The moments of X are given in Problem 29. Hence, from (4.10), we have for complex s,

$$M_X(s) = \sum_{n=0}^{\infty} \frac{s^n}{n!} \cdot \frac{\Gamma(n+p)}{\Gamma(p)}, \quad |s| < 1.$$

For complex s with $|s| < 1$, derive the Taylor series for $1/(1-s)^p$ and show that it is equal to the above series. Thus, $M_X(s) = 1/(1-s)^p$ for all complex s with $|s| < 1$. (This formula actually holds for all complex s with $\text{Re}\, s < 1$; see Note **8**.)

45. As shown in the preceding problem, the basic gamma density with parameter p, $g_p(x)$, has moment generating function $1/(1-s)^p$. The more general gamma density defined by $g_{p,\lambda}(x) := \lambda g_p(\lambda x)$ is given in Problem 15.

(a) Find the moment generating function and then the characteristic function of $g_{p,\lambda}(x)$.

(b) Use the answer to (a) to find the moment generating function and the characteristic function of the Erlang density with parameters m and λ, $g_{m,\lambda}(x)$.

(c) Use the answer to (a) to find the moment generating function and the characteristic function of the chi-squared density with k degrees of freedom, $g_{k/2,1/2}(x)$.

46. Let $X \sim N(0,1)$, and put $Y = X^2$. For real values of $s < 1/2$, show that

$$M_Y(s) = \left(\frac{1}{1-2s}\right)^{1/2}.$$

By Problem 45(c), it follows that Y is chi-squared with one degree of freedom.

47. Let $X \sim N(m,1)$, and put $Y = X^2$. For real values of $s < 1/2$, show that

$$M_Y(s) = \frac{e^{sm^2/(1-2s)}}{\sqrt{1-2s}}.$$

Remark. For $m \neq 0$, Y is said to be **noncentral chi-squared** with one degree of freedom and **noncentrality parameter** m^2. For $m = 0$, this reduces to the result of the previous problem.

48. Let X have characteristic function $\varphi_X(v)$. If $Y := aX + b$ for constants a and b, express the characteristic function of Y in terms of a, b, and φ_X.

49. Apply the Fourier inversion formula to $\varphi_X(v) = e^{-\lambda|v|}$ to verify that this is the characteristic function of a Cauchy(λ) random variable.

*50. Use the following approach to find the characteristic function of the $N(0,1)$ density [62, pp. 138–139]. Let $f(x) := e^{-x^2/2}/\sqrt{2\pi}$.

   (a) Show that $f'(x) = -xf(x)$.
   (b) Starting with $\varphi_X(v) = \int_{-\infty}^{\infty} e^{jvx} f(x)\,dx$, compute $\varphi_X'(v)$. Then use part (a) to show that $\varphi_X'(v) = -j\int_{-\infty}^{\infty} e^{jvx} f'(x)\,dx$.
   (c) Using integration by parts, show that this last integral is $-jv\varphi_X(v)$.
   (d) Show that $\varphi_X'(v) = -v\varphi_X(v)$.
   (e) Show that $K(v) := \varphi_X(v) e^{v^2/2}$ satisfies $K'(v) = 0$.
   (f) Show that $K(v) = 1$ for all v. (It then follows that $\varphi_X(v) = e^{-v^2/2}$.)

*51. Use the method of Problem 50 to find the characteristic function of the gamma density $g_p(x) = x^{p-1}e^{-x}/\Gamma(p)$, $x > 0$. *Hints:* Show that $(d/dx)xg_p(x) = (p-x)g_p(x)$. Use integration by parts to show that $\varphi_X'(v) = -(p/v)\varphi_X(v) + (1/jv)\varphi_X'(v)$. Show that $K(v) := \varphi_X(v)(1-jv)^p$ satisfies $K'(v) = 0$.

4.4: Expectation of multiple random variables

52. Let $Z := X + Y$, where X and Y are independent with $X \sim \exp(1)$ and $Y \sim \text{Laplace}(1)$. Find $\text{cov}(X,Z)$ and $\text{var}(Z)$.

53. Find $\text{var}(Z)$ for the random variable Z of Example 4.22.

54. Let X and Y be independent random variables with moment generating functions $M_X(s)$ and $M_Y(s)$. If $Z := X - Y$, show that $M_Z(s) = M_X(s)M_Y(-s)$. Show that if both X and Y are $\exp(\lambda)$, then $Z \sim \text{Laplace}(\lambda)$.

55. Let $X_1, \ldots, X_n$ be independent, and put $Y_n := X_1 + \cdots + X_n$.

   (a) If $X_i \sim N(m_i, \sigma_i^2)$, show that $Y_n \sim N(m, \sigma^2)$, and identify m and σ^2. In other words, "The sum of independent Gaussian random variables is Gaussian."
   (b) If $X_i \sim \text{Cauchy}(\lambda_i)$, show that $Y_n \sim \text{Cauchy}(\lambda)$, and identify λ. In other words, "The sum of independent Cauchy random variables is Cauchy."

(c) If X_i is a gamma random variable with parameters p_i and λ (same λ for all i), show that Y_n is gamma with parameters p and λ, and identify p. In other words, "The sum of independent gamma random variables (with the same scale factor) is gamma (with the same scale factor)."

Remark. Note the following special cases of this result. If all the $p_i = 1$, then the X_i are exponential with parameter λ, and Y_n is Erlang with parameters n and λ. If all the $p_i = 1/2$ and $\lambda = 1/2$, then the X_i are chi-squared with one degree of freedom, and Y_n is chi-squared with n degrees of freedom.

56. Let $X_1, \ldots, X_r$ be i.i.d. gamma random variables with parameters p and λ. Let $Y = X_1 + \cdots + X_r$. Find $\mathsf{E}[Y^n]$.

57. Packet transmission times on a certain network link are i.i.d. with an exponential density of parameter λ. Suppose n packets are transmitted. Find the density of the time to transmit n packets.

58. The random number generator on a computer produces i.i.d. uniform$(0,1)$ random variables $X_1, \ldots, X_n$. Find the probability density of

$$Y = \ln\left(\prod_{i=1}^{n} \frac{1}{X_i}\right).$$

59. Let $X_1, \ldots, X_n$ be i.i.d. Cauchy(λ). Find the density of $Y := \beta_1 X_1 + \cdots + \beta_n X_n$, where the β_i are given positive constants.

60. Two particles arrive at a detector at random, independent positions X and Y lying on a straight line. The particles are resolvable if the absolute difference in their positions is greater than two. Find the probability that the two particles are *not* resolvable if X and Y are both Cauchy(1) random variables. Give a numerical answer.

61. Three independent pressure sensors produce output voltages U, V, and W, each $\exp(\lambda)$ random variables. The three voltages are summed and fed into an alarm that sounds if the sum is greater than x volts. Find the probability that the alarm sounds.

62. A certain electric power substation has n power lines. The line loads are independent Cauchy(λ) random variables. The substation automatically shuts down if the total load is greater than ℓ. Find the probability of automatic shutdown.

63. The new outpost on Mars extracts water from the surrounding soil. There are 13 extractors. Each extractor produces water with a random efficiency that is uniformly distributed on $[0,1]$. The outpost operates normally if fewer than three extractors produce water with efficiency less than 0.25. If the efficiencies are independent, find the probability that the outpost operates normally.

64. The time to send an Internet packet is a chi-squared random variable T with one degree of freedom. The time to receive the acknowledgment A is also chi-squared with one degree of freedom. If T and A are independent, find the probability that the round trip time $R := T + A$ is more than r.

*65. In this problem we generalize the **noncentral chi-squared** density of Problem 47. To distinguish these new densities from the original chi-squared densities defined in Problem 15, we refer to the original ones as **central** chi-squared densities. The noncentral chi-squared density with k degrees of freedom and **noncentrality parameter** λ^2 is defined by[f]

$$c_{k,\lambda^2}(x) := \sum_{n=0}^{\infty} \frac{(\lambda^2/2)^n e^{-\lambda^2/2}}{n!} c_{2n+k}(x), \quad x > 0,$$

where c_{2n+k} denotes the central chi-squared density with $2n+k$ degrees of freedom. Hence, $c_{k,\lambda^2}(x)$ is a **mixture density** (Problem 12) with $p_n = (\lambda^2/2)^n e^{-\lambda^2/2}/n!$ being a Poisson($\lambda^2/2$) pmf.

(a) Show that $\int_0^\infty c_{k,\lambda^2}(x)\,dx = 1$.

(b) If X is a noncentral chi-squared random variable with k degrees of freedom and noncentrality parameter λ^2, show that X has moment generating function

$$M_{k,\lambda^2}(s) = \frac{\exp[s\lambda^2/(1-2s)]}{(1-2s)^{k/2}}.$$

Hint: Problem 45 may be helpful.

Remark. When $k = 1$, this agrees with Problem 47.

(c) Use part (b) to show that if $X \sim c_{k,\lambda^2}$, then $\mathsf{E}[X] = k + \lambda^2$.

(d) Let $X_1, \ldots, X_n$ be independent random variables with $X_i \sim c_{k_i, \lambda_i^2}$. Show that $Y := X_1 + \cdots + X_n$ has the c_{k,λ^2} density, and identify k and λ^2.

Remark. By part (b), if each $k_i = 1$, we could assume that each X_i is the *square* of an $N(\lambda_i, 1)$ random variable.

(e) Show that

$$\frac{e^{-(x+\lambda^2)/2}}{\sqrt{2\pi x}} \cdot \frac{e^{\lambda\sqrt{x}} + e^{-\lambda\sqrt{x}}}{2} = c_{1,\lambda^2}(x).$$

(Note that if $\lambda = 0$, the left-hand side reduces to the central chi-squared density with one degree of freedom.) *Hint:* Use the power series $e^\xi = \sum_{n=0}^\infty \xi^n/n!$ for the two exponentials involving $\sqrt{x}$.

4.5: *Probability bounds

66. Let X have the **Pareto** density $f(x) = 2/x^3$ for $x \geq 1$ and $f(x) = 0$ otherwise. For $a \geq 1$, compare $\mathsf{P}(X \geq a)$ and the bound obtained via Markov inequality.

*67. Let X be an exponential random variable with parameter $\lambda = 1$. Compute the Markov inequality, the Chebyshev inequality, and the Chernoff bound to obtain bounds on $\mathsf{P}(X \geq a)$ as a function of a. Also compute $\mathsf{P}(X \geq a)$.

---

[f] A closed-form expression is derived in Problem 25 of Chapter 5.

(a) For what values of a is the Markov inequality smaller than the Chebyshev inequality?

(b) **MATLAB.** Plot the Markov bound, the Chebyshev bound, the Chernoff bound, and $P(X \geq a)$ for $0 \leq a \leq 6$ on the same graph. For what range of a is the Markov bound the smallest? the Chebyshev? Now use the MATLAB command `semilogy` to draw the same four curves for $6 \leq a \leq 20$. Which bound is the smallest?

Exam preparation

You may use the following suggestions to prepare a study sheet, including formulas mentioned that you have trouble remembering. You may also want to ask your instructor for additional suggestions.

4.1. Densities and probabilities. Know how to compute probabilities involving a random variable with a density (4.1). A list of the more common densities can be found inside the back cover. Remember, density functions can never be negative and must integrate to one.

4.2. Expectation. LOTUS (4.3), especially for computing moments. The table inside the back cover contains moments of many of the more common densities.

4.3. Transform methods. Moment generating function definition (4.7) and moment formula (4.8). For continuous random variables, the mgf is essentially the Laplace transform of the density. Characteristic function definition (4.11) and moment formula (4.14). For continuous random variables, the density can be recovered with the inverse Fourier transform (4.12). For integer-valued random variables, the pmf can be recovered with the formula for Fourier series coefficients (4.13). The table inside the back cover contains the mgf (or characteristic function) of many of the more common densities. Remember that $\varphi_X(v) = M_X(s)|_{s=jv}$.

4.4. Expectation of multiple random variables. If X and Y are independent, then we have $E[h(X)k(Y)] = E[h(X)]E[k(Y)]$ for any functions $h(x)$ and $k(y)$. If $X_1,\ldots,X_n$ are independent random variables, then the moment generating function of the sum is the product of the moment generating functions, e.g., Example 4.23. If the X_i are continuous random variables, then the density of their sum is the convolution of their densities, e.g., (4.16).

4.5. *Probability bounds. The Markov inequality (2.18) and the Chebyshev inequality (2.21) were derived in Section 2.4. The Chernoff bound (4.20).

Work any review problems assigned by your instructor. If you finish them, re-work your homework assignments.

5
Cumulative distribution functions and their applications

In this chapter we introduce the **cumulative distribution function** (cdf) of a random variable X. The cdf is defined by[a]

$$F_X(x) := \mathsf{P}(X \leq x).$$

As we shall see, knowing the cdf is equivalent to knowing the density or pmf of a random variable. By this we mean that if you know the cdf, then you can find the density or pmf, and if you know the density or pmf, then you can find the cdf. This is the same sense in which knowing the characteristic function is equivalent to knowing the density or pmf. Similarly, just as some problems are more easily solved using characteristic functions instead of densities, there are some problems that are more easily solved using cdfs instead of densities.

This chapter emphasizes three applications in which cdfs figure prominently: (i) Finding the probability density of $Y = g(X)$ when the function g and the density of X are given; (ii) The central limit theorem; and (iii) Reliability.

The first application concerns what happens when the input of a system g is modeled as a random variable. The system output $Y = g(X)$ is another random variable, and we would like to compute probabilities involving Y. For example, g could be an amplifier, and we might need to find the probability that the output exceeds some danger level. If we knew the probability mass function or the density of Y, we would know what to do next. It turns out that we can easily find the probability mass function or density of Y if we know its cdf, $F_Y(y) = \mathsf{P}(Y \leq y)$, for all y. Section 5.1 focuses on the problem $Y = g(X)$ when X has a density and g is a fairly simple function to analyze. We note that Example 5.9 motivates a discussion of the maximum a posteriori probability (MAP) and maximum likelihood (ML) rules for detecting discrete signals in continuous noise. We also show how to simulate a continuous random variable by applying the inverse cdf to a uniform random variable. Section 5.2 introduces cdfs of discrete random variables. It is also shown how to simulate a discrete random variable as a function of a uniform random variable. Section 5.3 introduces cdfs of mixed random variables. Mixed random variables frequently appear in the form $Y = g(X)$ when X is continuous, but g has "flat spots." For example, most amplifiers have a linear region, say $-v \leq x \leq v$, wherein $g(x) = \alpha x$. However, if $x > v$, then $g(x) = \alpha v$, and if $x < -v$, then $g(x) = -\alpha v$. If a continuous random variable is applied to such a device, the output will be a mixed random variable, which can be thought of as a random variable whose "generalized density" contains Dirac impulses. The problem of finding the cdf and generalized density of $Y = g(X)$ is studied in Section 5.4. At this point, having seen several

---

[a]As we have defined it, the cdf is a right-continuous function of x (see Section 5.5). However, we alert the reader that some texts put $F_X(x) = \mathsf{P}(X < x)$, which is left-continuous in x.

5.1 Continuous random variables

generalized densities and their corresponding cdfs, Section 5.5 summarizes and derives the general properties that characterize arbitrary cdfs.

Section 5.6 contains our second application of cdfs, the **central limit theorem**. (This section can be covered immediately after Section 5.1 if desired.) Although we have seen many examples for which we can explicitly write down probabilities involving a sum of i.i.d. random variables, in general, the problem is quite hard. The central limit theorem provides an approximation of probabilities involving the sum of i.i.d. random variables — even when the density of the individual random variables is unknown! This is crucial in parameter-estimation problems where we need to compute confidence intervals as in Chapter 6.

Section 5.7 contains our third application of cdfs. This section, which is a brief diversion into reliability theory, can be covered immediately after Section 5.1 if desired. With the exception of the formula

$$\mathsf{E}[T] = \int_0^\infty \mathsf{P}(T > t)\,dt$$

for nonnegative random variables, which is derived at the beginning of Section 5.7, the remaining material on reliability is not used in the rest of the book.

5.1 Continuous random variables

If X is a continuous random variable with density f, then[b]

$$F(x) = \mathsf{P}(X \le x) = \int_{-\infty}^x f(t)\,dt.$$

Pictorially, $F(x)$ is the area under the density $f(t)$ from $-\infty < t \le x$. This is the area of the shaded region in Figure 5.1. Since the total area under a density is one, the area of the unshaded region, which is $\int_x^\infty f(t)\,dt$, must be $1 - F(x)$. Thus,

$$1 - F(x) = \int_x^\infty f(t)\,dt = \mathsf{P}(X \ge x).$$

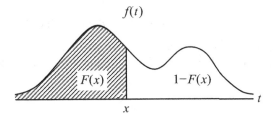

Figure 5.1. The area under the density $f(t)$ from $-\infty < t \le x$ is $\int_{-\infty}^x f(t)\,dt = \mathsf{P}(X \le x) = F(x)$. Since the total area under the density is one, the area of the unshaded region is $1 - F(x)$.

---

[b]When only one random variable is under discussion, we simplify the notation by writing $F(x)$ instead of $F_X(x)$.

For $a < b$, we can use the cdf to compute probabilities of the form

$$P(a \leq X \leq b) = \int_a^b f(t)\,dt$$
$$= \int_{-\infty}^b f(t)\,dt - \int_{-\infty}^a f(t)\,dt$$
$$= F(b) - F(a).$$

Thus, $F(b) - F(a)$ is the area of the shaded region in Figure 5.2.

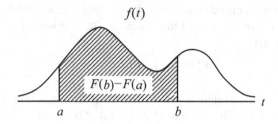

Figure 5.2. The area of the shaded region is $\int_a^b f(t)\,dt = F(b) - F(a)$.

Example 5.1. Find the cdf of a Cauchy random variable X with parameter $\lambda = 1$.

Solution. Write

$$F(x) = \int_{-\infty}^x \frac{1/\pi}{1+t^2}\,dt$$
$$= \frac{1}{\pi}\tan^{-1}(t)\Big|_{-\infty}^x$$
$$= \frac{1}{\pi}\left(\tan^{-1}(x) - \frac{-\pi}{2}\right)$$
$$= \frac{1}{\pi}\tan^{-1}(x) + \frac{1}{2}.$$

A graph of F is shown in Figure 5.3.

Example 5.2. Find the cdf of a uniform$[a,b]$ random variable X.

Solution. Since $f(t) = 0$ for $t < a$ and $t > b$, we see that $F(x) = \int_{-\infty}^x f(t)\,dt$ is equal to 0 for $x < a$, and is equal to $\int_{-\infty}^\infty f(t)\,dt = 1$ for $x > b$. For $a \leq x \leq b$, we have

$$F(x) = \int_{-\infty}^x f(t)\,dt = \int_a^x \frac{1}{b-a}\,dt = \frac{x-a}{b-a}.$$

Hence, for $a \leq x \leq b$, $F(x)$ is an **affine**[c] function of x. A graph of F when $X \sim \text{uniform}[0,1]$ is shown in Figure 5.3.

---

[c]A function is affine if it is equal to a linear function plus a constant.

5.1 Continuous random variables

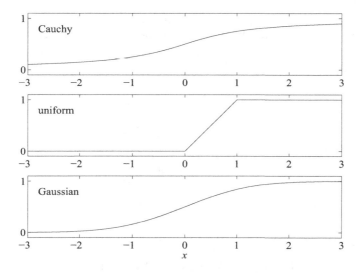

Figure 5.3. Cumulative distribution functions of Cauchy(1), uniform[0, 1], and standard normal random variables.

We now consider the cdf of a Gaussian random variable. If $X \sim N(m, \sigma^2)$, then

$$F(x) = \int_{-\infty}^{x} \frac{1}{\sqrt{2\pi}\,\sigma} \exp\left[-\frac{1}{2}\left(\frac{t-m}{\sigma}\right)^2\right] dt. \tag{5.1}$$

Unfortunately, there is no closed-form expression for this integral. However, it can be computed numerically, and there are many subroutines available for doing it. For example, in MATLAB, the above integral can be computed with `normcdf(x,m,sigma)`.

We next show that the $N(m, \sigma^2)$ cdf can always be expressed using the standard normal cdf,

$$\boxed{\Phi(y) := \frac{1}{\sqrt{2\pi}} \int_{-\infty}^{y} e^{-\theta^2/2}\, d\theta,}$$

which is graphed in Figure 5.3. In (5.1), make change of variable $\theta = (t-m)/\sigma$ to get

$$F(x) = \frac{1}{\sqrt{2\pi}} \int_{-\infty}^{(x-m)/\sigma} e^{-\theta^2/2}\, d\theta$$
$$= \Phi\left(\frac{x-m}{\sigma}\right).$$

It is also convenient to define the **complementary cumulative distribution function** (ccdf),

$$\boxed{Q(y) := 1 - \Phi(y) = \frac{1}{\sqrt{2\pi}} \int_{y}^{\infty} e^{-\theta^2/2}\, d\theta.}$$

Example 5.3 (bit-error probability). At the receiver of a digital communication system, thermal noise in the amplifier sometimes causes an incorrect decision to be made. For example, if antipodal signals of energy $\mathscr{E}$ are used, then the bit-error probability can be shown to be $\mathsf{P}(X > \sqrt{\mathcal{E}})$, where $X \sim N(0, \sigma^2)$ represents the noise, and σ^2 is the noise power. Express the bit-error probability in terms of the standard normal cdf Φ and in terms of Q.

Solution. Write

$$\mathsf{P}(X > \sqrt{\mathcal{E}}) = 1 - F_X(\sqrt{\mathcal{E}})$$

$$= 1 - \Phi\left(\frac{\sqrt{\mathcal{E}}}{\sigma}\right)$$

$$= 1 - \Phi\left(\sqrt{\frac{\mathcal{E}}{\sigma^2}}\right)$$

$$= Q\left(\sqrt{\frac{\mathcal{E}}{\sigma^2}}\right).$$

This calculation shows that the bit-error probability is completely determined by $\mathcal{E}/\sigma^2$, which is called the **signal-to-noise ratio** (SNR). As the SNR increases, so does Φ, while Q decreases, and the error probability as well. In other words, *increasing the SNR decreases the error probability*. Hence, the only ways to improve performance are to use higher-energy signals or lower-noise amplifiers.

Because every Gaussian cdf can be expressed in terms of the standard normal cdf Φ, we can compute any Gaussian probability if we have a table of values of $\Phi(x)$ or a program to compute $\Phi(x)$. For example, a small table of values of $\Phi(x)$ and $Q(x) = 1 - \Phi(x)$ is shown in Table 5.1. Fortunately, since Φ can be expressed in terms of the **error function**,[1] which is available in most numerical subroutine libraries, tables are rarely needed.

Example 5.4. Compute the bit-error probability in the preceding example if the signal-to-noise ratio is 6 dB.

Solution. As shown in the preceding example, the bit-error probability is $Q(\sqrt{\mathcal{E}/\sigma^2})$. The problem statement is telling us that

$$10 \log_{10} \frac{\mathcal{E}}{\sigma^2} = 6,$$

or $\mathcal{E}/\sigma^2 = 10^{6/10} \approx 3.98$ and $\sqrt{\mathcal{E}/\sigma^2} \approx 2.0$. Hence, from Table 5.1, the error probability is $Q(2) = 0.0228$.

For continuous random variables, the density can be recovered from the cdf by differentiation. Since

$$F(x) = \int_{-\infty}^{x} f(t)\,dt,$$

differentiation yields

$$F'(x) = f(x).$$

5.1 Continuous random variables

x	$\Phi(x)$	$Q(x)$	x	$\Phi(x)$	$Q(x)$
0.0	0.5000	0.5000	2.0	0.9772	0.0228
0.1	0.5398	0.4602	2.1	0.9821	0.0179
0.2	0.5793	0.4207	2.2	0.9861	0.0139
0.3	0.6179	0.3821	2.3	0.9893	0.0107
0.4	0.6554	0.3446	2.4	0.9918	0.0082
0.5	0.6915	0.3085	2.5	0.9938	0.0062
0.6	0.7257	0.2743	2.6	0.9953	0.0047
0.7	0.7580	0.2420	2.7	0.9965	0.0035
0.8	0.7881	0.2119	2.8	0.9974	0.0026
0.9	0.8159	0.1841	2.9	0.9981	0.0019
1.0	0.8413	0.1587	3.0	0.9987	0.0013
1.1	0.8643	0.1357	3.1	0.9990	0.0010
1.2	0.8849	0.1151	3.2	0.9993	0.0007
1.3	0.9032	0.0968	3.3	0.9995	0.0005
1.4	0.9192	0.0808	3.4	0.9997	0.0003
1.5	0.9332	0.0668	3.5	0.9998	0.0002
1.6	0.9452	0.0548	3.6	0.9998	0.0002
1.7	0.9554	0.0446	3.7	0.9999	0.0001
1.8	0.9641	0.0359	3.8	0.9999	0.0001
1.9	0.9713	0.0287	3.9	1.0000	0.0000

Table 5.1. Values of the standard normal cumulative distribution function $\Phi(x)$ and complementary cumulative distribution function $Q(x) := 1 - \Phi(x)$. To evaluate Φ and Q for negative arguments, use the fact that since the standard normal density is even, $\Phi(-x) = Q(x)$.

Example 5.5. Let the random variable X have cdf

$$F(x) := \begin{cases} \sqrt{x}, & 0 < x < 1, \\ 1, & x \geq 1, \\ 0, & x \leq 0. \end{cases}$$

Find the density and sketch both the cdf and pdf.

Solution. For $0 < x < 1$, $f(x) = F'(x) = \frac{1}{2}x^{-1/2}$, while for other values of x, $F(x)$ is piecewise constant with value zero or one; for these values of x, $F'(x) = 0$. Hence,[2]

$$f(x) := \begin{cases} \dfrac{1}{2\sqrt{x}}, & 0 < x < 1, \\ 0, & \text{otherwise.} \end{cases}$$

The cdf and pdf are sketched in Figure 5.4.

The observation that the density of a continuous random variable can be recovered from its cdf is of tremendous importance, as the following examples illustrate.

Example 5.6. Consider an electrical circuit whose random input voltage X is first amplified by a gain $\mu > 0$ and then added to a constant offset voltage β. Then the output

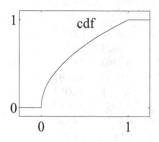 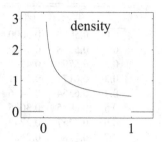

Figure 5.4. Cumulative distribution function $F(x)$ (left) and density $f(x)$ (right) of Example 5.5.

voltage is $Y = \mu X + \beta$. If the input voltage is a continuous random variable X, find the density of the output voltage Y.

Solution. Although the question asks for the density of Y, it is more advantageous to find the cdf first and then differentiate to obtain the density. Write

$$\begin{aligned} F_Y(y) &= \mathsf{P}(Y \leq y) \\ &= \mathsf{P}(\mu X + \beta \leq y) \\ &= \mathsf{P}(X \leq (y-\beta)/\mu), \quad \text{since } \mu > 0, \\ &= F_X((y-\beta)/\mu). \end{aligned}$$

If X has density f_X, then[d]

$$f_Y(y) = \frac{d}{dy}F_X\left(\frac{y-\beta}{\mu}\right) = \frac{1}{\mu}F_X'\left(\frac{y-\beta}{\mu}\right) = \frac{1}{\mu}f_X\left(\frac{y-\beta}{\mu}\right).$$

Example 5.7. In wireless communications systems, fading is sometimes modeled by **lognormal** random variables. We say that a positive random variable Y is lognormal if $\ln Y$ is a normal random variable. Find the density of Y if $\ln Y \sim N(m, \sigma^2)$.

Solution. Put $X := \ln Y$ so that $Y = e^X$, where $X \sim N(m, \sigma^2)$. Although the question asks for the density of Y, it is more advantageous to find the cdf first and then differentiate to obtain the density. To begin, note that since $Y = e^X$ is positive, if $y \leq 0$, $F_Y(y) = \mathsf{P}(Y \leq y) = 0$. For $y > 0$, write

$$F_Y(y) = \mathsf{P}(Y \leq y) = \mathsf{P}(e^X \leq y) = \mathsf{P}(X \leq \ln y) = F_X(\ln y).$$

By the chain rule,

$$f_Y(y) = f_X(\ln y)\tfrac{1}{y}.$$

---

[d]Recall the **chain rule**,

$$\frac{d}{dy}F(G(y)) = F'(G(y))G'(y).$$

In the present case, $G(y) = (y-\beta)/\mu$ and $G'(y) = 1/\mu$.

Using the fact that $X \sim N(m, \sigma^2)$,

$$f_Y(y) = \frac{e^{-[(\ln y - m)/\sigma]^2/2}}{\sqrt{2\pi}\,\sigma y}, \quad y > 0.$$

The functions $g(x) = \mu x + \beta$ and $g(x) = e^x$ of the preceding examples are continuous, strictly increasing functions of x. In general, if $g(x)$ is continuous and strictly increasing (or strictly decreasing), it can be shown[3] that if $Y = g(X)$, then

$$f_Y(y) = f_X(h(y))|h'(y)|, \tag{5.2}$$

where $h(y) := g^{-1}(y)$. Since we have from calculus that $h'(y) = 1/g'(g^{-1}(y))$, (5.2) is sometimes written as

$$f_Y(y) = \frac{f_X(g^{-1}(y))}{|g'(g^{-1}(y))|}.$$

Although (5.2) is a nice formula, it is of limited use because it only applies to continuous, strictly-increasing or strictly-decreasing functions. Even simple functions like $g(x) = x^2$ do not qualify (note that x^2 is decreasing for $x < 0$ and increasing for $x > 0$). These kinds of functions can be handled as follows.

***Example* 5.8.** Amplitude modulation in certain communication systems can be accomplished using various nonlinear devices such as a semiconductor diode. Suppose we model the nonlinear device by the function $Y = X^2$. If the input X is a continuous random variable, find the density of the output $Y = X^2$.

Solution. Although the question asks for the density of Y, it is more advantageous to find the cdf first and then differentiate to obtain the density. To begin, note that since $Y = X^2$ is nonnegative, for $y < 0$, $F_Y(y) = P(Y \leq y) = 0$. For nonnegative y, write

$$\begin{aligned} F_Y(y) &= P(Y \leq y) \\ &= P(X^2 \leq y) \\ &= P(-\sqrt{y} \leq X \leq \sqrt{y}) \\ &= \int_{-\sqrt{y}}^{\sqrt{y}} f_X(t)\,dt. \end{aligned}$$

The density is[e]

$$\begin{aligned} f_Y(y) &= \frac{d}{dy} \int_{-\sqrt{y}}^{\sqrt{y}} f_X(t)\,dt \\ &= \frac{1}{2\sqrt{y}}[f_X(\sqrt{y}) + f_X(-\sqrt{y})], \quad y > 0. \end{aligned}$$

Since $P(Y \leq y) = 0$ for $y < 0$, $f_Y(y) = 0$ for $y < 0$.

---

[e]Recall **Leibniz' rule**,
$$\frac{d}{dy} \int_{a(y)}^{b(y)} f(t)\,dt = f(b(y))b'(y) - f(a(y))a'(y).$$
The general form is derived in Note 7 in Chapter 7.

When the diode input voltage X of the preceding example is $N(0,1)$, it turns out that Y is chi-squared with one degree of freedom (Problem 11). If X is $N(m,1)$ with $m \neq 0$, then Y is noncentral chi-squared with one degree of freedom (Problem 12). These results are frequently used in the analysis of digital communication systems.

The two preceding examples illustrate the problem of finding the density of $Y = g(X)$ when X is a continuous random variable. The next example illustrates the problem of finding the density of $Z = g(X,Y)$ when X is discrete and Y is continuous.

Example **5.9** (signal in additive noise). Let X and Y be independent random variables, with X being discrete with pmf p_X and Y being continuous with density f_Y. Put $Z := X + Y$ and find the density of Z.

Solution. Although the question asks for the density of Z, it is more advantageous to find the cdf first and then differentiate to obtain the density. This time we use the law of total probability, substitution, and independence. Write

$$\begin{aligned}
F_Z(z) &= \mathsf{P}(Z \leq z) \\
&= \sum_i \mathsf{P}(Z \leq z | X = x_i) \mathsf{P}(X = x_i) \\
&= \sum_i \mathsf{P}(X + Y \leq z | X = x_i) \mathsf{P}(X = x_i) \\
&= \sum_i \mathsf{P}(x_i + Y \leq z | X = x_i) \mathsf{P}(X = x_i) \\
&= \sum_i \mathsf{P}(Y \leq z - x_i | X = x_i) \mathsf{P}(X = x_i) \\
&= \sum_i \mathsf{P}(Y \leq z - x_i) \mathsf{P}(X = x_i) \\
&= \sum_i F_Y(z - x_i) p_X(x_i).
\end{aligned}$$

Differentiating this expression yields

$$f_Z(z) = \sum_i f_Y(z - x_i) p_X(x_i).$$

We should also note that $F_{Z|X}(z|x_i) := \mathsf{P}(Z \leq z | X = x_i)$ is called the **conditional cdf** of Z given X. When $F_{Z|X}(z|x_i)$ is differentiable with respect to z, we call this derivative the **conditional density** of Z given X, and we denote it by $f_{Z|X}(z|x_i)$. In the case of the preceding example, $f_{Z|X}(z|x_i) = f_Y(z - x_i)$. In analogy with the discussion at the end of Section 3.4, $f_{Z|X}(z|x_i)$ is sometimes called the **likelihood** of $Z = z$.

Receiver design for discrete signals in continuous noise

Considering the situation in the preceding example, how should a receiver estimate or guess the transmitted message $X = x_i$ based only on observing a value $Z = z$? The design goal is to minimize the probability of a decision error. If we proceed as in the case of discrete random variables,[f] we are led to the continuous analog of the **Maximum A posteriori**

---

[f]See Note 4 in Chapter 3. The derivation there carries over to the present case if the sum using the conditional pmf is replaced by an integral using the conditional density.

5.1 Continuous random variables

Probability (MAP) rule in (3.19); that is, we should decide $X = x_i$ if

$$f_{Z|X}(z|x_i) \, \mathsf{P}(X = x_i) \geq f_{Z|X}(z|x_j) \, \mathsf{P}(X = x_j) \tag{5.3}$$

for all $j \neq i$. If X takes only M values, and if they are equally likely, we can cancel the common factors $\mathsf{P}(X = x_i) = 1/M = \mathsf{P}(X = x_j)$ and obtain the **maximum likelihood** (ML) rule, which says to decide $X = x_i$ if

$$f_{Z|X}(z|x_i) \geq f_{Z|X}(z|x_j)$$

for all $j \neq i$. If X takes only two values, say 0 and 1, the MAP rule (5.3) says to decide $X = 1$ if and only if

$$\frac{f_{Z|X}(z|1)}{f_{Z|X}(z|0)} \geq \frac{\mathsf{P}(X=0)}{\mathsf{P}(X=1)}.$$

The corresponding ML rule takes the ratio on the right to be one. As in the discrete case, the ratio on the left is again called the **likelihood ratio**. Both the MAP and ML rules are sometimes called **likelihood-ratio tests**. The reason for writing the tests in terms of the likelihood ratio is that the form of the test can be greatly simplified; for example, as in Problem 17.

Simulation

Virtually all computers have routines for generating uniformly distributed random numbers on $(0,1)$, and most computers have routines for generating random numbers from the more common densities and probability mass functions. What if you need random numbers from a density or mass function for which no routine is available on your computer? There is a vast literature of methods for generating random numbers, such as [15], [45], [47]. If you cannot find anything in the literature, you can use the methods discussed in this section and later in the text. We caution, however, that while the methods we present always work in theory, they may not always be the most computationally efficient.

If $X \sim \text{uniform}(0,1)$, we can always perform a transformation $Y = g(X)$ so that Y is any kind of random variable we want. Below we show how to do this when Y is to have a continuous, strictly increasing cdf $F(y)$. In Section 5.2, we show how to do this when Y is to be a discrete random variable. The general case is more complicated, and is covered in Problems 37–39 in Chapter 11.

If $F(y)$ is a continuous, strictly increasing cdf, it has an inverse F^{-1} such that for all $0 < x < 1$, $F(y) = x$ can be solved for y with $y = F^{-1}(x)$. If $X \sim \text{uniform}(0,1)$, and we put $Y = F^{-1}(X)$, then

$$F_Y(y) = \mathsf{P}(Y \leq y) = \mathsf{P}\big(F^{-1}(X) \leq y\big).$$

Since

$$\{F^{-1}(X) \leq y\} = \{X \leq F(y)\},$$

we can further write

$$F_Y(y) = \mathsf{P}\big(X \leq F(y)\big) = \int_0^{F(y)} 1 \, dx = F(y)$$

as required.

Example 5.10. Find a transformation to convert $X \sim \text{uniform}(0,1)$ into a Cauchy(1) random variable.

Solution. We have to solve $F(y) = x$ when $F(y)$ is the Cauchy(1) cdf of Example 5.1. From

$$\frac{1}{\pi}\tan^{-1}(y) + \frac{1}{2} = x,$$

we find that

$$y = \tan[\pi(x-1/2)].$$

Thus, the desired transformation is $Y = \tan[\pi(X-1/2)]$.

In MATLAB, we can generate a vector of k Cauchy(1) random variables with the commands

```
X = rand(1,k);
Y = tan(pi*(X-1/2));
```

where `rand(1,k)` returns a $1 \times k$ matrix of uniform(0,1) random numbers.

Other cdfs that can be easily inverted include the exponential, the Rayleigh, and the Weibull.[g] If the cdf is not invertible in closed form, the inverse can be computed numerically by applying a root-finding algorithm to $F(y) - x = 0$. The Gaussian cdf, which cannot be expressed in closed form, much less inverted in closed form, is difficult to simulate with this approach. Fortunately, there is a simple alternative transformation of uniform(0,1) random variables that yields $N(0,1)$ random variables; this transformation is given in Problem 24 of Chapter 8. In MATLAB, even this is not necessary since `randn(1,k)` returns a $1 \times k$ matrix of $N(0,1)$ random numbers.

5.2 Discrete random variables

For continuous random variables, the cdf and density are related by

$$F(x) = \int_{-\infty}^{x} f(t)\,dt \quad \text{and} \quad f(x) = F'(x).$$

In this section we show that for a discrete random variable taking distinct values x_i with probabilities $p(x_i) := \mathsf{P}(X = x_i)$, the analogous formulas are

$$F(x) = \mathsf{P}(X \leq x) = \sum_{i: x_i \leq x} p(x_i),$$

and for two adjacent values $x_{j-1} < x_j$,

$$p(x_j) = F(x_j) - F(x_{j-1}).$$

---

[g]In these cases, the result can be further simplified by taking advantage of the fact that if $X \sim \text{uniform}(0,1)$, then $1 - X$ is also uniform(0,1) (cf. Problems 6, 7, and 8).

5.2 Discrete random variables

For the cdf, the analogy between the continuous and discrete cases is clear: The density becomes a pmf, and the integral becomes a sum. The analogy between the density and pmf formulas becomes clear if we write the derivative as a derivative from the left:

$$f(x) = F'(x) = \lim_{y \uparrow x} \frac{F(x) - F(y)}{x - y}.$$

The formulas for the cdf and pmf of discrete random variables are illustrated in the following examples.

Example 5.11. Find the cdf of a Bernoulli(p) random variable.

Solution. Since the Bernoulli random variable takes only the values zero and one, there are three ranges of x that we need to worry about: $x < 0$, $0 \leq x < 1$, and $x \geq 1$. Consider an x with $0 \leq x < 1$. The only way we can have $X \leq x$ for such x is if $X = 0$. Hence, for such x, $F(x) = \mathsf{P}(X \leq x) = \mathsf{P}(X = 0) = 1 - p$. Next consider an $x < 0$. Since we never have $X < 0$, we cannot have $X \leq x$. Therefore, $F(x) = \mathsf{P}(X \leq x) = \mathsf{P}(\emptyset) = 0$. Finally, since we always have $X \leq 1$, if $x \geq 1$, we always have $X \leq x$. Thus, $F(x) = \mathsf{P}(X \leq x) = \mathsf{P}(\Omega) = 1$. We now have

$$F(x) = \begin{cases} 0, & x < 0, \\ 1-p, & 0 \leq x < 1, \\ 1, & x \geq 1, \end{cases}$$

which is sketched in Figure 5.5. Notice that $F(1) - F(0) = p = \mathsf{P}(X = 1)$.

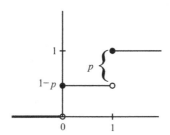

Figure 5.5. Cumulative distribution function of a Bernoulli(p) random variable.

Example 5.12. Find the cdf of a discrete random variable taking the values 0, 1, and 2 with probabilities p_0, p_1, and p_2, where the p_i are nonnegative and sum to one.

Solution. Since X takes three values, there are four ranges to worry about: $x < 0$, $0 \leq x < 1$, $1 \leq x < 2$, and $x \geq 2$. As in the previous example, for x less than the minimum possible value of X, $\mathsf{P}(X \leq x) = \mathsf{P}(\emptyset) = 0$. Similarly, for x greater than or equal to the maximum value of X, we have $\mathsf{P}(X \leq x) = \mathsf{P}(\Omega) = 1$. For $0 \leq x < 1$, the only way we can have $X \leq x$ is to have $X = 0$. Thus, $F(x) = \mathsf{P}(X \leq x) = \mathsf{P}(X = 0) = p_0$. For $1 \leq x < 2$, the only way we can have $X \leq x$ is to have $X = 0$ or $X = 1$. Thus,

$$F(x) = \mathsf{P}(X \leq x) = \mathsf{P}(\{X = 0\} \cup \{X = 1\}) = p_0 + p_1.$$

In summary,

$$F(x) = \begin{cases} 0, & x < 0, \\ p_0, & 0 \le x < 1, \\ p_0 + p_1, & 1 \le x < 2, \\ 1, & x \ge 2, \end{cases}$$

which is sketched in Figure 5.6. Notice that

$$F(1) - F(0) = p_1,$$

and

$$F(2) - F(1) = 1 - (p_0 + p_1) = p_2 = \mathsf{P}(X = 2).$$

Thus, each of the probability masses can be recovered from the cdf.

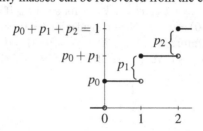

Figure 5.6. Cumulative distribution function of the discrete random variable in Example 5.12.

Simulation

Suppose we need to simulate a discrete random variable taking distinct values y_i with probabilities p_i. If $X \sim \text{uniform}(0,1)$, observe that

$$\mathsf{P}(X \le p_1) = \int_0^{p_1} 1\, dx = p_1.$$

Similarly,

$$\mathsf{P}(p_1 < X \le p_1 + p_2) = \int_{p_1}^{p_1+p_2} 1\, dx = p_2,$$

and $\mathsf{P}(p_1 + p_2 < X \le p_1 + p_2 + p_3) = p_3$, and so on. For example, to simulate a Bernoulli(p) random variable Y, we would take $y_1 = 0$, $p_1 = 1-p$, $y_2 = 1$, and $p_2 = p$. This suggests the following MATLAB script for generating a vector of n Bernoulli(p) random variables. Try typing it in yourself!

```
p = 0.3
n = 5
X = rand(1,n)
Y = zeros(1,n)
i = find(X>1-p)
Y(i) = ones(size(i))
```

5.3 Mixed random variables

In this script, `rand(1,n)` returns a $1 \times n$ matrix of uniform$(0,1)$ random numbers; `zeros(1,n)` returns a $1 \times n$ matrix of zeros; `find(X>1-p)` returns the positions in X that have values greater than `1-p`; the command `ones(size(i))` puts a 1 at the positions in Y that correspond to the positions in X that have values greater than `1-p`. By adding the command `Z = sum(Y)`, you can create a single binomial(n,p) random variable Z. (The command `sum(Y)` returns the sum of the elements of the vector Y.)

Now suppose we wanted to generate a vector of m binomial(n,p) random numbers. An easy way to do this is to first generate an $m \times n$ matrix of independent Bernoulli(p) random numbers, and then sum the rows. The sum of each row will be a binomial(n,p) random number. To take advantage of MATLAB'S vector and matrix operations, we first create an M-file containing a function that returns an $m \times n$ matrix of Bernoulli(p) random numbers.

```
% M-file with function to generate an
% m-by-n matrix of Bernoulli(p) random numbers.
%
function Y = bernrnd(p,m,n)
X = rand(m,n);
Y = zeros(m,n);
i = find(X>1-p);
Y(i) = ones(size(i));
```

Once you have created the above M-file, you can try the following commands.

```
bernmat = bernrnd(.5,10,4)
X = sum(bernmat')
```

Since the default operation of `sum` on a matrix is to compute *column sums*, we included the apostrophe (') to transpose `bernmat`. Be sure to include the semicolons (;) so that large vectors and matrices will not be printed out.

5.3 Mixed random variables

We begin with an example. Consider the function

$$g(x) = \begin{cases} x, & x \geq 0, \\ 0, & x < 0, \end{cases} \quad (5.4)$$

which is sketched in Figure 5.7. The function g operates like a half-wave rectifier in that if a positive voltage x is applied, the output is $y = x$, while if a negative voltage x is applied, the output is $y = 0$. Suppose $Y = g(X)$, where $X \sim \text{uniform}[-1,1]$. We now find the cdf of

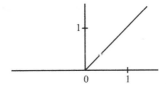

Figure 5.7. Half-wave-rectifier transformation $g(x)$ defined in (5.4).

Y, $F_Y(y) := \mathsf{P}(Y \leq y)$. The first step is to identify the event $\{Y \leq y\}$ for all values of y. As X ranges over $[-1,1]$, $Y = g(X)$ ranges over $[0,1]$. It is important to note that Y is never less than zero and never greater than one. Hence, we easily have

$$\{Y \leq y\} = \begin{cases} \emptyset, & y < 0, \\ \Omega, & y \geq 1. \end{cases}$$

This immediately gives us

$$F_Y(y) = \mathsf{P}(Y \leq y) = \begin{cases} \mathsf{P}(\emptyset) = 0, & y < 0, \\ \mathsf{P}(\Omega) = 1, & y \geq 1. \end{cases}$$

It remains to compute $F_Y(y)$ for $0 \leq y < 1$. For such y, Figure 5.7 tells us that $g(x)$ is less than or equal to some level y if and only if $x \leq y$. Hence,

$$F_Y(y) = \mathsf{P}(Y \leq y) = \mathsf{P}(X \leq y).$$

Now use the fact that since $X \sim \text{uniform}[-1,1]$, X is never less than -1; i.e., $\mathsf{P}(X < -1) = 0$. Hence,

$$\begin{aligned} F_Y(y) &= \mathsf{P}(X \leq y) \\ &= \mathsf{P}(X < -1) + \mathsf{P}(-1 \leq X \leq y) \\ &= \mathsf{P}(-1 \leq X \leq y) \\ &= \frac{y - (-1)}{2}. \end{aligned}$$

In summary,

$$F_Y(y) = \begin{cases} 0, & y < 0, \\ (y+1)/2, & 0 \leq y < 1, \\ 1, & y \geq 1, \end{cases}$$

which is sketched in Figure 5.8(a). The derivative, $f_Y(y)$, is shown in Figure 5.8(b). Its formula is

$$f_Y(y) = \tilde{f}_Y(y) + \frac{1}{2}\delta(y),$$

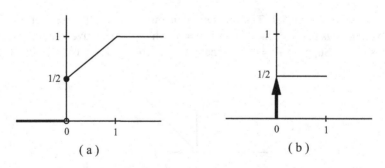

(a)   (b)

Figure 5.8. (a) Cumulative distribution function of a mixed random variable. (b) The corresponding impulsive density.

5.3 Mixed random variables

where

$$\tilde{f}_Y(y) := \begin{cases} 1/2, & 0 < y < 1, \\ 0, & \text{otherwise.} \end{cases}$$

Notice that we need an **impulse function**[h] in $f_Y(y)$ at $y = 0$ since $F_Y(y)$ has a jump discontinuity there. The strength of the impulse is the size of the jump discontinuity.

A random variable whose density contains impulse terms as well as an "ordinary" part is called a **mixed random variable**, and the density, $f_Y(y)$, is said to be **impulsive**. Sometimes we say that a mixed random variable has a **generalized density**. The typical form of a generalized density is

$$f_Y(y) = \tilde{f}_Y(y) + \sum_i \mathsf{P}(Y = y_i)\delta(y - y_i), \tag{5.5}$$

where the y_i are the distinct points at which $F_Y(y)$ has jump discontinuities, and $\tilde{f}_Y(y)$ is an ordinary, nonnegative function without impulses. The ordinary part $\tilde{f}_Y(y)$ is obtained by differentiating $F_Y(y)$ at y-values where there are no jump discontinuities. Expectations $\mathsf{E}[k(Y)]$ when Y has the above generalized density can be computed with the formula

$$\mathsf{E}[k(Y)] = \int_{-\infty}^{\infty} k(y)f_Y(y)\,dy = \int_{-\infty}^{\infty} k(y)\tilde{f}_Y(y)\,dy + \sum_i k(y_i)\mathsf{P}(Y = y_i).$$

Example 5.13. Consider the generalized density

$$f_Y(y) = \frac{1}{4}e^{-|y|} + \frac{1}{3}\delta(y) + \frac{1}{6}\delta(y - 7).$$

Compute $\mathsf{P}(0 < Y \le 7)$, $\mathsf{P}(Y = 0)$, and $\mathsf{E}[Y^2]$.

Solution. In computing

$$\mathsf{P}(0 < Y \le 7) = \int_{0+}^{7} f_Y(y)\,dy,$$

the impulse at the origin makes no contribution, but the impulse at 7 does. Thus,

$$\begin{aligned}\mathsf{P}(0 < Y \le 7) &= \int_{0+}^{7} \left[\frac{1}{4}e^{-|y|} + \frac{1}{3}\delta(y) + \frac{1}{6}\delta(y - 7)\right]dy \\ &= \frac{1}{4}\int_{0}^{7} e^{-y}\,dy + \frac{1}{6} \\ &= \frac{1 - e^{-7}}{4} + \frac{1}{6} = \frac{5}{12} - \frac{e^{-7}}{4}.\end{aligned}$$

---

[h]The **unit impulse** or **Dirac delta function**, denoted by δ, is defined by the two properties

$$\delta(t) = 0 \text{ for } t \ne 0 \quad \text{and} \quad \int_{-\infty}^{\infty} \delta(t)\,dt = 1.$$

Using these properties, it can be shown that for any function $h(t)$ and any t_0,

$$\int_{-\infty}^{\infty} h(t)\delta(t - t_0)\,dt = h(t_0).$$

Similarly, in computing $P(Y = 0) = P(Y \in \{0\})$, only the impulse at zero makes a contribution. Thus,
$$P(Y = 0) = \int_{\{0\}} f_Y(y)\,dy = \int_{\{0\}} \frac{1}{3}\delta(y)\,dy = \frac{1}{3}.$$

To conclude, write
$$\begin{aligned}
\mathsf{E}[Y^2] &= \int_{-\infty}^{\infty} y^2 f_Y(y)\,dy \\
&= \int_{-\infty}^{\infty} y^2 \left(\frac{1}{4}e^{-|y|} + \frac{1}{3}\delta(y) + \frac{1}{6}\delta(y-7)\right) dy \\
&= \int_{-\infty}^{\infty} \frac{y^2}{4}e^{-|y|} + \frac{y^2}{3}\delta(y) + \frac{y^2}{6}\delta(y-7)\,dy \\
&= \frac{1}{4}\int_{-\infty}^{\infty} y^2 e^{-|y|}\,dy + \frac{0^2}{3} + \frac{7^2}{6} \\
&= \frac{1}{2}\int_{0}^{\infty} y^2 e^{-y}\,dy + \frac{49}{6}.
\end{aligned}$$

Since this last integral is the second moment of an exp(1) random variable, which is 2 by Example 4.17, we find that
$$\mathsf{E}[Y^2] = \frac{2}{2} + \frac{49}{6} = \frac{55}{6}.$$

5.4 Functions of random variables and their cdfs

Most modern systems today are composed of many subsystems in which the output of one system serves as the input to another. When the input to a system or a subsystem is random, so is the output. To evaluate system performance, it is necessary to take into account this randomness. The first step in this process is to find the cdf of the system output if we know the pmf or density of the random input. In many cases, the output will be a mixed random variable with a generalized impulsive density.

We consider systems modeled by real-valued functions $g(x)$. The system input is a random variable X, and the system output is the random variable $Y = g(X)$. To find $F_Y(y)$, observe that
$$F_Y(y) := P(Y \leq y) = P(g(X) \leq y) = P(X \in B_y),$$
where
$$B_y := \{x \in \mathbb{R} : g(x) \leq y\}.$$

If X has density f_X, then
$$F_Y(y) = P(X \in B_y) = \int_{B_y} f_X(x)\,dx.$$

The difficulty is to identify the set B_y. However, if we first sketch the function $g(x)$, the problem becomes manageable.

5.4 Functions of random variables and their cdfs

Example 5.14. Find the cdf and density of $Y = g(X)$ if $X \sim$ uniform$[0,4]$, and

$$g(x) := \begin{cases} x, & 0 \leq x < 1, \\ 1, & 1 \leq x < 2, \\ 3-x, & 2 \leq x < 3, \\ 0, & \text{otherwise.} \end{cases}$$

Solution. We begin by sketching g as shown in Figure 5.9(a). Since $0 \leq g(x) \leq 1$, we can never have $Y = g(X) < 0$, and we always have $Y = g(X) \leq 1$. Hence, we immediately have

$$F_Y(y) = P(Y \leq y) = \begin{cases} P(\emptyset) = 0, & y < 0, \\ P(\Omega) = 1, & y \geq 1. \end{cases}$$

To deal with $0 \leq y < 1$, draw a horizontal line at level y as shown in Figure 5.9(b). Also drop vertical lines where the level crosses the curve $g(x)$. In Figure 5.9(b) the vertical lines intersect the x-axis at u and v. Observe also that $g(x) \leq y$ if and only if $x \leq u$ or $x \geq v$. Hence, for $0 \leq y < 1$,

$$F_Y(y) = P(Y \leq y) = P(g(X) \leq y) = P(\{X \leq u\} \cup \{X \geq v\}).$$

Since $X \sim$ uniform$[0,4]$,

$$P(\{X \leq u\} \cup \{X \geq v\}) = \frac{u-0}{4} + \frac{4-v}{4}.$$

It remains to find u and v. From Figure 5.9(b), we see that $g(u) = y$, and since $0 \leq u < 1$, the formula for $g(u)$ is $g(u) = u$. Hence, $g(u) = y$ implies $u = y$. Similarly, since $g(v) = y$ and $2 \leq v < 3$, the formula for $g(v)$ is $g(v) = 3 - v$. Solving $3 - v = y$ yields $v = 3 - y$. We can now simplify

$$F_Y(y) = \frac{y + (4 - [3-y])}{4} = \frac{2y+1}{4} = \frac{y}{2} + \frac{1}{4}, \quad 0 \leq y < 1.$$

The complete formula for $F_Y(y)$ is

$$F_Y(y) := \begin{cases} 0, & y < 0, \\ \frac{y}{2} + \frac{1}{4}, & 0 \leq y < 1, \\ 1, & y \geq 1. \end{cases}$$

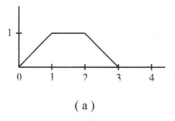
(a)

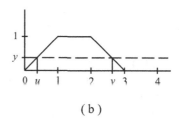
(b)

Figure 5.9. (a) The function g of Example 5.14. (b) Drawing a horizontal line at level y.

Examination of this formula shows that there are jump discontinuities at $y = 0$ and $y = 1$. Both jumps are of height $1/4$. See Figure 5.10(a). Jumps in the cdf mean there are corresponding impulses in the density. The complete density formula is

$$f_Y(y) = \tilde{f}_Y(y) + \frac{1}{4}\delta(y) + \frac{1}{4}\delta(y-1),$$

where

$$\tilde{f}_Y(y) := \begin{cases} 1/2, & 0 < y < 1, \\ 0, & \text{otherwise}. \end{cases}$$

Figure 5.10(b) shows $f_Y(y)$.

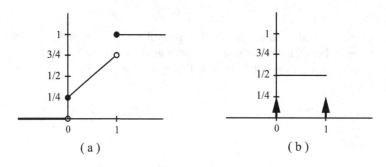

Figure 5.10. (a) Cumulative distribution of Y in Example 5.14. (b) Corresponding impulsive density.

Example 5.15. Suppose g is given by

$$g(x) := \begin{cases} 1, & -2 \le x < -1, \\ x^2, & -1 \le x < 0, \\ x, & 0 \le x < 2, \\ 2, & 2 \le x < 3, \\ 0, & \text{otherwise}. \end{cases}$$

If $Y = g(X)$ and $X \sim \text{uniform}[-4, 4]$, find the cdf and density of Y.

Solution. To begin, we sketch g in Figure 5.11. Since $0 \le g(x) \le 2$, we can never have $Y < 0$, and we always have $Y \le 2$. Hence, we immediately have

$$F_Y(y) = P(Y \le y) = \begin{cases} P(\emptyset) = 0, & y < 0, \\ P(\Omega) = 1, & y \ge 2. \end{cases}$$

To deal with $0 \le y < 2$, we see from Figure 5.11 that there are two interesting places to draw a horizontal level y: $1 \le y < 2$ and $0 \le y < 1$.

Fix any y with $1 \le y < 2$. On the graph of g, draw a horizontal line at level y. At the intersection of the horizontal line and the curve g, drop a vertical line to the x-axis. This vertical line hits the x-axis at the point marked × in Figure 5.12. Observe that for all x to the left of this point, and for all $x \ge 3$, $g(x) \le y$. To find the x-coordinate of ×, we solve

5.4 Functions of random variables and their cdfs

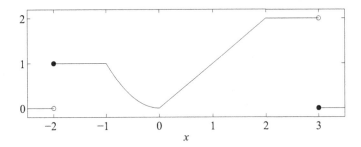

Figure 5.11. The function $g(x)$ from Example 5.15.

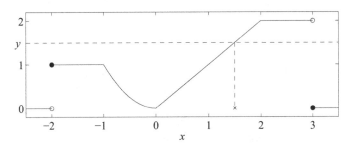

Figure 5.12. Drawing a horizontal line at level y, $1 \le y < 2$.

$g(x) = y$ for x. For the y-value in question, the formula for $g(x)$ is $g(x) = x$. Hence, the x-coordinate of × is simply y. Thus, $g(x) \le y \Leftrightarrow x \le y$ or $x \ge 3$, and so,

$$\begin{aligned} F_Y(y) &= \mathsf{P}(g(X) \le y) \\ &= \mathsf{P}(\{X \le y\} \cup \{X \ge 3\}) \\ &= \frac{y-(-4)}{8} + \frac{4-3}{8} \\ &= \frac{y+5}{8}. \end{aligned}$$

Now fix any y with $0 \le y < 1$, and draw a horizontal line at level y as shown in Figure 5.13. This time the horizontal line intersects the curve g in two places, and there are two points marked × on the x-axis. Call the x-coordinate of the left one x_1 and that of the right one x_2. We must solve $g(x_1) = y$, where x_1 is negative and $g(x_1) = x_1^2$. We must also solve $g(x_2) = y$, where $g(x_2) = x_2$. We conclude that $g(x) \le y \Leftrightarrow x < -2$ or $-\sqrt{y} \le x \le y$ or $x \ge 3$. Thus,

$$\begin{aligned} F_Y(y) &= \mathsf{P}(g(X) \le y) \\ &= \mathsf{P}(\{X < -2\} \cup \{-\sqrt{y} < X < y\} \cup \{X \ge 3\}) \\ &= \frac{(-2)-(-4)}{8} + \frac{y-(-\sqrt{y})}{8} + \frac{4-3}{8} \\ &= \frac{y+\sqrt{y}+3}{8}. \end{aligned}$$

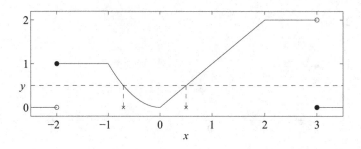

Figure 5.13. Drawing a horizontal line at level y, $0 \leq y < 1$.

Putting all this together,

$$F_Y(y) = \begin{cases} 0, & y < 0, \\ (y+\sqrt{y}+3)/8, & 0 \leq y < 1, \\ (y+5)/8, & 1 \leq y < 2, \\ 1, & y \geq 2. \end{cases}$$

In sketching $F_Y(y)$, we note from the formula that it is 0 for $y < 0$ and 1 for $y \geq 2$. Also from the formula, there is a jump discontinuity of $3/8$ at $y = 0$ and a jump of $1/8$ at $y = 1$ and at $y = 2$. See Figure 5.14.

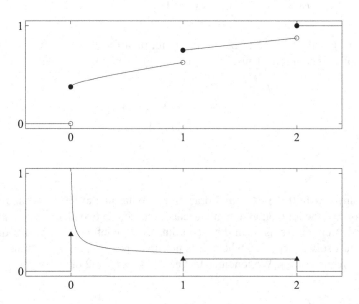

Figure 5.14. Cumulative distribution function $F_Y(y)$ (top) and impulsive density $f_Y(y)$ (bottom) of Example 5.15. The strength of the impulse at zero is $3/8$; the other two impulses are both $1/8$.

From the observations used in graphing F_Y, we can easily obtain the generalized density,

$$f_Y(y) = \frac{3}{8}\delta(y) + \frac{1}{8}\delta(y-1) + \frac{1}{8}\delta(y-2) + \tilde{f}_Y(y),$$

where

$$\tilde{f}_Y(y) = \begin{cases} [1+1/(2\sqrt{y})]/8, & 0 < y < 1, \\ 1/8, & 1 < y < 2, \\ 0, & \text{otherwise,} \end{cases}$$

is obtained by differentiating $F_Y(y)$ at non-jump points y. A sketch of f_Y is shown in Figure 5.14.

5.5 Properties of cdfs

Given an arbitrary real-valued random variable X, its cumulative distribution function is defined by

$$F(x) := \mathsf{P}(X \leq x), \quad -\infty < x < \infty.$$

We show below that F satisfies eight properties. For help in visualizing these properties, the reader should consult Figures 5.3, 5.4(top), 5.8(a), 5.10(a), and 5.14(top).

(i) $0 \leq F(x) \leq 1$.
(ii) For $a < b$, $\mathsf{P}(a < X \leq b) = F(b) - F(a)$.
(iii) F is nondecreasing, i.e., $a \leq b$ implies $F(a) \leq F(b)$.
(iv) $\lim_{x \uparrow \infty} F(x) = 1$.

Since this is a statement about limits, it does not require that $F(x) = 1$ for any finite value of x. For example, the Gaussian and Cauchy cdfs never take the value one for finite values of x. However, all of the other cdfs in the figures mentioned above do take the value one for finite values of x. In particular, by properties (i) and (iii), if $F(x) = 1$ for some finite x, then for all $y \geq x$, $F(y) = 1$.

(v) $\lim_{x \downarrow -\infty} F(x) = 0$.

Again, since this is a statement about limits, it does not require that $F(x) = 0$ for any finite value of x. The Gaussian and Cauchy cdfs never take the value zero for finite values of x, while all the other cdfs in the figures do. Moreover, if $F(x) = 0$ for some finite x, then for all $y \leq x$, $F(y) = 0$.

(vi) $F(x_0+) := \lim_{x \downarrow x_0} F(x) = \mathsf{P}(X \leq x_0) = F(x_0)$.

This says that F is right-continuous.

(vii) $F(x_0-) := \lim_{x \uparrow x_0} F(x) = \mathsf{P}(X < x_0)$.

(viii) $\mathsf{P}(X = x_0) = F(x_0) - F(x_0-)$.

This says that X can take the value x_0 with positive probability if and only if the cdf has a jump discontinuity at x_0. The height of the jump is the value of $\mathsf{P}(X = x_0)$.

We also point out that

$$\mathsf{P}(X > x_0) = 1 - \mathsf{P}(X \leq x_0) = 1 - F(x_0),$$

and

$$\mathsf{P}(X \geq x_0) = 1 - \mathsf{P}(X < x_0) = 1 - F(x_0-).$$

If $F(x)$ is continuous at $x = x_0$, i.e., $F(x_0-) = F(x_0)$, then this last equation becomes

$$\mathsf{P}(X \geq x_0) = 1 - F(x_0).$$

Another consequence of the continuity of $F(x)$ at $x = x_0$ is that $P(X = x_0) = 0$. Hence, if a random variable has a nonimpulsive density, then its cumulative distribution is continuous everywhere.

We now derive the eight properties of cumulative distribution functions.
(i) The properties of P imply that $F(x) = P(X \leq x)$ satisfies $0 \leq F(x) \leq 1$.
(ii) First consider the disjoint union $(-\infty, b] = (-\infty, a] \cup (a, b]$. It then follows that

$$\{X \leq b\} = \{X \leq a\} \cup \{a < X \leq b\}$$

is a disjoint union of events in Ω. Now write

$$\begin{aligned} F(b) &= P(X \leq b) \\ &= P(\{X \leq a\} \cup \{a < X \leq b\}) \\ &= P(X \leq a) + P(a < X \leq b) \\ &= F(a) + P(a < X \leq b). \end{aligned}$$

Now subtract $F(a)$ from both sides.
(iii) This follows from (ii) since $P(a < X \leq b) \geq 0$.
(iv) We prove the simpler result $\lim_{N \to \infty} F(N) = 1$. Starting with

$$\mathbb{R} = (-\infty, \infty) = \bigcup_{n=1}^{\infty} (-\infty, n],$$

we can write

$$\begin{aligned} 1 &= P(X \in \mathbb{R}) \\ &= P\left(\bigcup_{n=1}^{\infty} \{X \leq n\}\right) \\ &= \lim_{N \to \infty} P(X \leq N), \quad \text{by limit property (1.15),} \\ &= \lim_{N \to \infty} F(N). \end{aligned}$$

(v) We prove the simpler result, $\lim_{N \to \infty} F(-N) = 0$. Starting with

$$\emptyset = \bigcap_{n=1}^{\infty} (-\infty, -n],$$

we can write

$$\begin{aligned} 0 &= P(X \in \emptyset) \\ &= P\left(\bigcap_{n=1}^{\infty} \{X \leq -n\}\right) \\ &= \lim_{N \to \infty} P(X \leq -N), \quad \text{by limit property (1.16),} \\ &= \lim_{N \to \infty} F(-N). \end{aligned}$$

(vi) We prove the simpler result, $P(X \leq x_0) = \lim_{N \to \infty} F(x_0 + \frac{1}{N})$. Starting with

$$(-\infty, x_0] = \bigcap_{n=1}^{\infty} (-\infty, x_0 + \tfrac{1}{n}],$$

we can write

$$P(X \leq x_0) = P\left(\bigcap_{n=1}^{\infty} \{X \leq x_0 + \tfrac{1}{n}\}\right)$$
$$= \lim_{N \to \infty} P(X \leq x_0 + \tfrac{1}{N}), \quad \text{by (1.16),}$$
$$= \lim_{N \to \infty} F(x_0 + \tfrac{1}{N}).$$

(vii) We prove the simpler result, $P(X < x_0) = \lim_{N \to \infty} F(x_0 - \frac{1}{N})$. Starting with

$$(-\infty, x_0) = \bigcup_{n=1}^{\infty} (-\infty, x_0 - \tfrac{1}{n}],$$

we can write

$$P(X < x_0) = P\left(\bigcup_{n=1}^{\infty} \{X \leq x_0 - \tfrac{1}{n}\}\right)$$
$$= \lim_{N \to \infty} P(X \leq x_0 - \tfrac{1}{N}), \quad \text{by (1.15),}$$
$$= \lim_{N \to \infty} F(x_0 - \tfrac{1}{N}).$$

(viii) First consider the disjoint union $(-\infty, x_0] = (-\infty, x_0) \cup \{x_0\}$. It then follows that

$$\{X \leq x_0\} = \{X < x_0\} \cup \{X = x_0\}$$

is a disjoint union of events in Ω. Using Property (vii), it follows that

$$F(x_0) = F(x_0-) + P(X = x_0).$$

Some additional technical information on cdfs can be found in the Notes.[4,5]

5.6 The central limit theorem

Let $X_1, X_2, \ldots$ be i.i.d. with common mean m and common variance σ^2. There are many cases for which we know the probability mass function or density of

$$\sum_{i=1}^{n} X_i.$$

For example, if the X_i are Bernoulli, binomial, Poisson, gamma, or Gaussian, we know the cdf of the sum (see Section 3.2, Problems 5 and 12 in Chapter 3, and Problem 55 in Chapter 4). Note that the exponential and chi-squared are special cases of the gamma (see

Problem 15 in Chapter 4). In general, however, finding the cdf of a sum of i.i.d. random variables is not computationally feasible. Furthermore, in parameter-estimation problems, we do not even know the common probability mass function or density of the X_i. In this case, finding the cdf of the sum is impossible, and the central limit theorem stated below is a rather amazing result.

Before stating the central limit theorem, we make a few observations. First note that

$$\mathsf{E}\left[\sum_{i=1}^n X_i\right] = \sum_{i=1}^n \mathsf{E}[X_i] = nm.$$

As $n \to \infty$, nm does not converge if $m \neq 0$. Hence, if we are to get any kind of limit result, it might be better to consider

$$\sum_{i=1}^n (X_i - m),$$

which has zero mean for all n. The second thing to note is that since the above terms are independent, the variance of the sum is the sum of the variances (Eq. (2.28)). Hence, the variance of the above sum is $n\sigma^2$. As $n \to \infty$, $n\sigma^2 \to \infty$. This suggests that we focus our analysis on

$$Y_n := \frac{1}{\sqrt{n}} \sum_{i=1}^n \left(\frac{X_i - m}{\sigma}\right), \quad (5.6)$$

which has zero mean and unit variance for all n (Problem 51).

Central limit theorem (CLT). Let $X_1, X_2, \ldots$ be independent, identically distributed random variables with finite mean m and finite variance σ^2. If Y_n is defined by (5.6), then

$$\lim_{n \to \infty} F_{Y_n}(y) = \Phi(y),$$

where $\Phi(y) := \int_{-\infty}^y e^{-t^2/2}/\sqrt{2\pi}\, dt$ is the standard normal cdf.

Remark. When the X_i are Bernoulli$(1/2)$, the CLT was derived by Abraham de Moivre around 1733. The case of Bernoulli(p) for $0 < p < 1$ was considered by Pierre-Simon Laplace. The CLT as stated above is known as the **Lindeberg–Lévy theorem**.

To get some idea of how large n should be, we compare $F_{Y_n}(y)$ and $\Phi(y)$ in cases where F_{Y_n} is known. To do this, we need the following result.

Example 5.16. Show that if G_n is the cdf of $\sum_{i=1}^n X_i$, then

$$F_{Y_n}(y) = G_n(y\sigma\sqrt{n} + nm). \quad (5.7)$$

Solution. Write

$$F_{Y_n}(y) = \mathsf{P}\left(\frac{1}{\sqrt{n}} \sum_{i=1}^n \left(\frac{X_i - m}{\sigma}\right) \leq y\right)$$

$$= \mathsf{P}\left(\sum_{i=1}^n (X_i - m) \leq y\sigma\sqrt{n}\right)$$

5.6 The central limit theorem

$$= \mathsf{P}\left(\sum_{i=1}^{n} X_i \leq y\sigma\sqrt{n} + nm\right)$$
$$= G_n(y\sigma\sqrt{n} + nm).$$

When the X_i are exp(1), G_n is the Erlang$(n, 1)$ cdf given in Problem 15(c) in Chapter 4. With $n = 30$, we plot in Figure 5.15 $F_{Y_{30}}(y)$ (dashed line) and the $N(0,1)$ cdf $\Phi(y)$ (solid line).

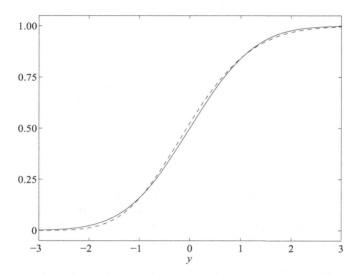

Figure 5.15. Illustration of the central limit theorem when the X_i are exponential with parameter 1. The dashed line is $F_{Y_{30}}(y)$, and the solid line is the standard normal cumulative distribution, $\Phi(y)$.

A typical calculation using the central limit theorem is as follows. To approximate

$$\mathsf{P}\left(\sum_{i=1}^{n} X_i > t\right),$$

write

$$\mathsf{P}\left(\sum_{i=1}^{n} X_i > t\right) = \mathsf{P}\left(\sum_{i=1}^{n} (X_i - m) > t - nm\right)$$
$$= \mathsf{P}\left(\sum_{i=1}^{n} \left(\frac{X_i - m}{\sigma}\right) > \frac{t - nm}{\sigma}\right)$$
$$= \mathsf{P}\left(Y_n > \frac{t - nm}{\sigma\sqrt{n}}\right)$$
$$\approx 1 - \Phi\left(\frac{t - nm}{\sigma\sqrt{n}}\right). \quad (5.8)$$

For example, if $X_i \sim$ exp(1), then the probability that $\sum_{i=1}^{30} X_i$ (whose expected value is 30) is greater than $t = 35$ is 0.177, while the central limit approximation is $1 - \Phi(0.91287)$

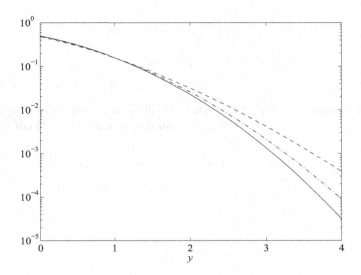

Figure 5.16. Plots of $\log_{10}(1-F_{Y_{30}}(y))$ (dashed line), $\log_{10}(1-F_{Y_{300}}(y))$ (dash-dotted line), and $\log_{10}(1-\Phi(y))$ (solid line).

$= 0.181$. This is not surprising since Figure 5.15 shows good agreement between $F_{Y_{30}}(y)$ and $\Phi(y)$ for $|y| \leq 3$. Unfortunately, this agreement deteriorates rapidly as $|y|$ gets large. This is most easily seen if we plot $\log_{10}(1-F_{Y_n}(y))$ and $\log_{10}(1-\Phi(y))$ as shown in Figure 5.16. Notice that for $y = 4$, the $n = 30$ curve differs from the limit by more than an order of magnitude.

These observations do not mean the central limit theorem is wrong, only that we need to interpret it properly. The theorem says that for any given y, $F_{Y_n}(y) \to \Phi(y)$ as $n \to \infty$. However, in practice, n is fixed, and we use the approximation for different values of y. For values of y near the origin, the approximation is better than for values of y away from the origin. We must be careful not to use the central limit approximation when y is too far away from the origin for the value of n we may be stuck with.

Example 5.17. A certain digital communication link has bit-error probability p. Use the central limit theorem to approximate the probability that in transmitting a word of n bits, more than k bits are received incorrectly.

Solution. Let $X_i = 1$ if bit i is received in error, and $X_i = 0$ otherwise. We assume the X_i are independent Bernoulli(p) random variables. Hence, $m = p$ and $\sigma^2 = p(1-p)$. The number of errors in n bits is $\sum_{i=1}^{n} X_i$. We must compute (5.8) with $t = k$. However, since the X_i are integer valued, the left-hand side of (5.8) is the same for all $t \in [k, k+1)$. It turns out we get a better approximation using $t = k + 1/2$. Taking $t = k + 1/2$, $m = p$, and $\sigma^2 = p(1-p)$ in (5.8), we have

$$\mathsf{P}\left(\sum_{i=1}^{n} X_i > k + \frac{1}{2}\right) \approx 1 - \Phi\left(\frac{k+1/2-np}{\sqrt{np(1-p)}}\right). \tag{5.9}$$

5.6 The central limit theorem

Let us consider the preceding example with $n = 30$ and $p = 1/30$. On average, we expect that one out of 30 bits will be received incorrectly. What is the probability that more than 2 bits will be received incorrectly? With $k = 2$, the exact probability is 0.077, and the approximation is 0.064. What is the probability that more than 6 bits will be received incorrectly? With $k = 6$, the exact probability is 5×10^{-5}, and the approximation is 6×10^{-9}. Clearly, the central limit approximation is not useful for estimating very small probabilities.

Approximation of densities and pmfs using the CLT

Above we have used the central limit theorem to compute probabilities. However, we can also gain insight into the density or pmf of $X_1 + \cdots + X_n$. In addition, by considering special cases (Example 5.18 and Problem 54), we get Stirling's formula for free.

Suppose $F_{Y_n}(y) \approx \Phi(y)$. Fix a small Δy, and suppose $F_{Y_n}(y + \Delta y) \approx \Phi(y + \Delta y)$ as well. Then

$$F_{Y_n}(y + \Delta y) - F_{Y_n}(y) \approx \Phi(y + \Delta y) - \Phi(y)$$
$$= \int_y^{y+\Delta y} \frac{e^{-t^2/2}}{\sqrt{2\pi}} dt$$
$$\approx \frac{e^{-y^2/2}}{\sqrt{2\pi}} \Delta y, \qquad (5.10)$$

since the Gaussian density is continuous.

If F_{Y_n} has density f_{Y_n}, the above left-hand side can be replaced by $\int_y^{y+\Delta y} f_{Y_n}(t)\,dt$. If the density f_{Y_n} is continuous, this integral is approximately $f_{Y_n}(y)\Delta y$. We are thus led to the approximation

$$f_{Y_n}(y)\Delta y \approx \frac{e^{-y^2/2}}{\sqrt{2\pi}} \Delta y,$$

and then

$$f_{Y_n}(y) \approx \frac{e^{-y^2/2}}{\sqrt{2\pi}}. \qquad (5.11)$$

This is illustrated in Figure 5.17 when the X_i are i.i.d. exp(1). Figure 5.17 shows $f_{Y_n}(y)$ for $n = 1, 2, 5, 30$ along with the $N(0,1)$ density.

In practice, it is not Y_n that we are usually interested in, but the cdf of $X_1 + \cdots + X_n$, which we denote by G_n. Using (5.7) we find that

$$\boxed{G_n(x) = F_{Y_n}\left(\frac{x - nm}{\sigma\sqrt{n}}\right) \approx \Phi\left(\frac{x - nm}{\sigma\sqrt{n}}\right),}$$

Thus, G_n is approximated by the cdf of a Gaussian random variable with mean nm and variance $n\sigma^2$. Differentiating $G_n(x)$ and denoting the corresponding density by $g_n(x)$, we have

$$\boxed{g_n(x) \approx \frac{1}{\sqrt{2\pi}\,\sigma\sqrt{n}} \exp\left[-\frac{1}{2}\left(\frac{x - nm}{\sigma\sqrt{n}}\right)^2\right],} \qquad (5.12)$$

which is the $N(nm, n\sigma^2)$ density. Just as the approximation $F_{Y_n}(y) \approx \Phi(y)$ is best for y near zero, the approximation of $G_n(x)$ and $g_n(x)$ is best for x near nm.

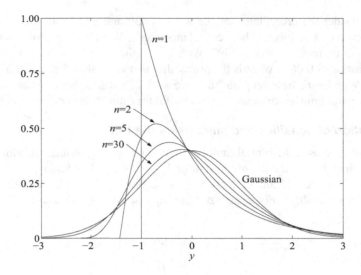

Figure 5.17. For X_i i.i.d. exp(1), sketch of $f_{Y_n}(y)$ for $n = 1,2,5,30$ and the $N(0,1)$ density.

Example 5.18. Let $X_1,\ldots,X_n$ be i.i.d. exp(1) so that in (5.12) g_n is the Erlang$(n,1)$ density (Problem 55(c) in Chapter 4). Since $m = \sigma^2 = 1$, (5.12) becomes

$$\frac{x^{n-1}e^{-x}}{(n-1)!} \approx \frac{1}{\sqrt{2\pi n}} \exp\left[-\frac{1}{2}\left(\frac{x-n}{\sqrt{n}}\right)^2\right]. \tag{5.13}$$

Since the approximation is best for x close to n, let us take $x = n$ to get

$$\frac{n^{n-1}e^{-n}}{(n-1)!} \approx \frac{1}{\sqrt{2\pi n}},$$

which we can rewrite as

$$\sqrt{2\pi}\, n^{n-1/2} e^{-n} \approx (n-1)!.$$

Multiplying through by n yields **Stirling's formula**,

$$n! \approx \sqrt{2\pi}\, n^{n+1/2} e^{-n}.$$

Remark. A more precise version of Stirling's formula is [16, pp. 50–53]

$$\sqrt{2\pi}\, n^{n+1/2} e^{-n+1/(12n+1)} < n! < \sqrt{2\pi}\, n^{n+1/2} e^{-n+1/(12n)}.$$

Remark. Since in (5.13) we have the exact formula on the left-hand side, we can see why the central limit theorem provides a bad approximation for large x when n is fixed. The left-hand side is dominated by e^{-x}, while the right-hand side is dominated by $e^{-x^2/2n}$. As x increases, $e^{-x^2/2n}$ decays much faster than e^{-x}. Although in this case the central limit

5.6 The central limit theorem

theorem density decays more quickly than the true density g_n, there are other examples in which the central limit theorem density decays more slowly than g_n. See Problem 55.

If the X_i are discrete, then

$$T_n := X_1 + \cdots + X_n$$

is also discrete, and its cdf G_n has no density. However, if the X_i are integer valued, then so is T_n, and we can write

$$\begin{aligned} P(T_n = k) &= P(k-1/2 < T_n \leq k+1/2) \\ &= G_n(k+1/2) - G_n(k-1/2) \\ &= F_{Y_n}\left(\frac{k+\frac{1}{2}-nm}{\sigma\sqrt{n}}\right) - F_{Y_n}\left(\frac{k-\frac{1}{2}-nm}{\sigma\sqrt{n}}\right). \end{aligned}$$

Proceeding as in the derivation of (5.10), we have

$$\begin{aligned} F_{Y_n}(y+\delta/2) - F_{Y_n}(y-\delta/2) &\approx \Phi(y+\delta/2) - \Phi(y-\delta/2) \\ &= \int_{y-\delta/2}^{y+\delta/2} \frac{e^{-t^2/2}}{\sqrt{2\pi}} \, dt \\ &\approx \frac{e^{-y^2/2}}{\sqrt{2\pi}} \delta. \end{aligned}$$

Taking $y = (k - nm)/\sigma\sqrt{n}$ and $\delta = 1/\sigma\sqrt{n}$ shows that

$$P(T_n = k) \approx \frac{1}{\sqrt{2\pi}} \exp\left[-\frac{1}{2}\left(\frac{k-nm}{\sigma\sqrt{n}}\right)^2\right] \frac{1}{\sigma\sqrt{n}}. \quad (5.14)$$

Just as the approximation $F_{Y_n}(y) \approx \Phi(y)$ is best for y near zero, the above approximation of $P(T_n = k)$ is best for k near nm.

Example 5.19 (normal approximation of the binomial). Let X_i be i.i.d. Bernoulli(p) random variables. Since $m = p$ and $\sigma^2 = p(1-p)$ (Example 2.28), (5.14) gives us the approximation

$$P(T_n = k) \approx \frac{1}{\sqrt{2\pi}} \exp\left[-\frac{1}{2}\left(\frac{k-np}{\sqrt{np(1-p)}}\right)^2\right] \frac{1}{\sqrt{np(1-p)}}.$$

We also know that T_n is binomial(n,p) (Section 3.2). Hence, $P(T_n = k) = \binom{n}{k} p^k (1-p)^{n-k}$, and it follows that

$$\binom{n}{k} p^k (1-p)^{n-k} \approx \frac{1}{\sqrt{2\pi}} \exp\left[-\frac{1}{2}\left(\frac{k-np}{\sqrt{np(1-p)}}\right)^2\right] \frac{1}{\sqrt{np(1-p)}},$$

as claimed in Chapter 1. The approximation is best for k near $nm = np$. The approximation can be bad for large k. In fact, notice that $T_n = X_1 + \cdots + X_n \leq n$ since the X_i are either zero or one. Hence, $P(T_n = k) = 0$ for $k > n$ while the above right-hand side is positive.

Derivation of the central limit theorem

It is instructive to consider first the following special case, which illustrates the key steps of the general derivation. Suppose that the X_i are i.i.d. Laplace with parameter $\lambda = \sqrt{2}$. Then $m = 0$, $\sigma^2 = 1$, and (5.6) becomes

$$Y_n = \frac{1}{\sqrt{n}} \sum_{i=1}^n X_i.$$

The characteristic function of Y_n is

$$\varphi_{Y_n}(v) = \mathsf{E}[e^{jvY_n}] = \mathsf{E}\left[e^{j(v/\sqrt{n})\sum_{i=1}^n X_i}\right] = \prod_{i=1}^n \mathsf{E}\left[e^{j(v/\sqrt{n})X_i}\right].$$

Of course, $\mathsf{E}\left[e^{j(v/\sqrt{n})X_i}\right] = \varphi_{X_i}(v/\sqrt{n})$, where, for the Laplace$(\sqrt{2})$ random variable X_i,

$$\varphi_{X_i}(v) = \frac{2}{2+v^2} = \frac{1}{1+v^2/2}.$$

Thus,

$$\mathsf{E}\left[e^{j(v/\sqrt{n})X_i}\right] = \varphi_{X_i}\left(\frac{v}{\sqrt{n}}\right) = \frac{1}{1+\frac{v^2/2}{n}},$$

and

$$\varphi_{Y_n}(v) = \left(\frac{1}{1+\frac{v^2/2}{n}}\right)^n = \frac{1}{\left(1+\frac{v^2/2}{n}\right)^n}.$$

We now use the fact that for any number ξ,

$$\left(1+\frac{\xi}{n}\right)^n \to e^\xi.$$

It follows that

$$\varphi_{Y_n}(v) = \frac{1}{\left(1+\frac{v^2/2}{n}\right)^n} \to \frac{1}{e^{v^2/2}} = e^{-v^2/2},$$

which is the characteristic function of an $N(0,1)$ random variable.

We now turn to the derivation in the general case. Letting $Z_i := (X_i - m)/\sigma$, (5.6) becomes

$$Y_n = \frac{1}{\sqrt{n}} \sum_{i=1}^n Z_i,$$

where the Z_i are i.i.d. zero mean and unit variance. Let $\varphi_Z(v) := \mathsf{E}[e^{jvZ_i}]$ denote their common characteristic function. We can write the characteristic function of Y_n as

$$\varphi_{Y_n}(v) := \mathsf{E}[e^{jvY_n}]$$
$$= \mathsf{E}\left[\exp\left(j\frac{v}{\sqrt{n}}\sum_{i=1}^n Z_i\right)\right]$$

$$= \mathsf{E}\left[\prod_{i=1}^{n} \exp\left(j\frac{v}{\sqrt{n}}Z_i\right)\right]$$

$$= \prod_{i=1}^{n} \mathsf{E}\left[\exp\left(j\frac{v}{\sqrt{n}}Z_i\right)\right]$$

$$= \prod_{i=1}^{n} \varphi_Z\left(\frac{v}{\sqrt{n}}\right)$$

$$= \varphi_Z\left(\frac{v}{\sqrt{n}}\right)^n.$$

Now recall that for any complex ξ,

$$e^\xi = 1 + \xi + \frac{1}{2}\xi^2 + R(\xi).$$

Thus,

$$\varphi_Z\left(\frac{v}{\sqrt{n}}\right) = \mathsf{E}[e^{j(v/\sqrt{n})Z_i}]$$

$$= \mathsf{E}\left[1 + j\frac{v}{\sqrt{n}}Z_i + \frac{1}{2}\left(j\frac{v}{\sqrt{n}}Z_i\right)^2 + R\left(j\frac{v}{\sqrt{n}}Z_i\right)\right].$$

Since Z_i is zero mean and unit variance,

$$\varphi_Z\left(\frac{v}{\sqrt{n}}\right) = 1 - \frac{1}{2}\cdot\frac{v^2}{n} + \mathsf{E}\left[R\left(j\frac{v}{\sqrt{n}}Z_i\right)\right].$$

It can be shown that the last term on the right is asymptotically negligible [3, pp. 357–358], and so

$$\varphi_Z\left(\frac{v}{\sqrt{n}}\right) \approx 1 - \frac{v^2/2}{n}.$$

We now have

$$\varphi_{Y_n}(v) = \varphi_Z\left(\frac{v}{\sqrt{n}}\right)^n \approx \left(1 - \frac{v^2/2}{n}\right)^n \to e^{-v^2/2},$$

which is the $N(0,1)$ characteristic function. Since the characteristic function of Y_n converges to the $N(0,1)$ characteristic function, it follows that $F_{Y_n}(y) \to \Phi(y)$ [3, p. 349, Theorem 26.3].

5.7 Reliability

Let T be the lifetime of a device or system. The **reliability function** of the device or system is defined by

$$R(t) := \mathsf{P}(T > t) = 1 - F_T(t). \quad (5.15)$$

The reliability at time t is the probability that the lifetime is greater than t.

The **mean time to failure** (MTTF) is defined to be the expected lifetime, $\mathsf{E}[T]$. Since lifetimes are nonnegative random variables, we claim that

$$\boxed{\mathsf{E}[T] = \int_0^\infty \mathsf{P}(T>t)\,dt.} \qquad (5.16)$$

It then follows that

$$\mathsf{E}[T] = \int_0^\infty R(t)\,dt;$$

namely, the MTTF is the integral of the reliability. To derive (5.16), first recall that every probability can be written as an expectation (Example 2.31). Hence,

$$\int_0^\infty \mathsf{P}(T>t)\,dt = \int_0^\infty \mathsf{E}[I_{(t,\infty)}(T)]\,dt = \mathsf{E}\left[\int_0^\infty I_{(t,\infty)}(T)\,dt\right].$$

Next, observe that as a function of t, $I_{(t,\infty)}(T) = I_{(-\infty,T)}(t)$; just check the cases $t < T$ and $t \geq T$. It follows that

$$\int_0^\infty \mathsf{P}(T>t)\,dt = \mathsf{E}\left[\int_0^\infty I_{(-\infty,T)}(t)\,dt\right].$$

To evaluate this last integral, observe that since T is nonnegative, the intersection of $[0,\infty)$ and $(-\infty,T)$ is $[0,T)$. Hence,

$$\int_0^\infty \mathsf{P}(T>t)\,dt = \mathsf{E}\left[\int_0^T dt\right] = \mathsf{E}[T].$$

The **failure rate** of a device or system with lifetime T is

$$r(t) := \lim_{\Delta t \downarrow 0} \frac{\mathsf{P}(T \leq t+\Delta t \mid T > t)}{\Delta t}.$$

This can be rewritten as

$$\mathsf{P}(T \leq t+\Delta t \mid T > t) \approx r(t)\Delta t.$$

In other words, given that the device or system has operated for more than t units of time, the conditional probability of failure before time $t+\Delta t$ is approximately $r(t)\Delta t$. Intuitively, the form of a failure rate function should be as shown in Figure 5.18. For small values of t, $r(t)$ is relatively large when pre-existing defects are likely to appear. Then for intermediate values of t, $r(t)$ is flat indicating a constant failure rate. For large t, as the device gets older, $r(t)$ increases indicating that failure is more likely.

To say more about the failure rate, write

$$\begin{aligned}
\mathsf{P}(T \leq t+\Delta t \mid T > t) &= \frac{\mathsf{P}(\{T \leq t+\Delta t\} \cap \{T > t\})}{\mathsf{P}(T > t)} \\
&= \frac{\mathsf{P}(t < T \leq t+\Delta t)}{\mathsf{P}(T > t)} \\
&= \frac{F_T(t+\Delta t) - F_T(t)}{R(t)}.
\end{aligned}$$

5.7 Reliability

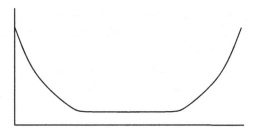

Figure 5.18. Typical form of a failure rate function $r(t)$.

Since $F_T(t) = 1 - R(t)$, we can rewrite this as

$$P(T \leq t + \Delta t | T > t) = -\frac{R(t + \Delta t) - R(t)}{R(t)}.$$

Dividing both sides by Δt and letting $\Delta t \downarrow 0$ yields the differential equation

$$r(t) = -\frac{R'(t)}{R(t)}.$$

Now suppose T is a continuous random variable with density f_T. Then

$$R(t) = P(T > t) = \int_t^\infty f_T(\theta) d\theta,$$

and

$$R'(t) = -f_T(t).$$

We can now write

$$\boxed{r(t) = -\frac{R'(t)}{R(t)} = \frac{f_T(t)}{\int_t^\infty f_T(\theta) d\theta}.} \quad (5.17)$$

In this case, the failure rate $r(t)$ is completely determined by the density $f_T(t)$. The converse is also true; namely, given the failure rate $r(t)$, we can recover the density $f_T(t)$. To see this, rewrite the above differential equation as

$$-r(t) = \frac{R'(t)}{R(t)} = \frac{d}{dt} \ln R(t).$$

Integrating the left and right-hand formulas from zero to t yields

$$\int_0^t r(\tau) d\tau = \ln R(t) - \ln R(0).$$

Then

$$e^{-\int_0^t r(\tau) d\tau} = \frac{R(t)}{R(0)} = R(t), \quad (5.18)$$

where we have used the fact that for a nonnegative, continuous random variable, $R(0) = P(T > 0) = P(T \geq 0) = 1$. If we differentiate the left and right-hand sides of (5.18) and use the fact that $R'(t) = -f_T(t)$, we find that

$$f_T(t) = r(t)e^{-\int_0^t r(\tau)\,d\tau}. \tag{5.19}$$

***Example* 5.20.** In some problems, you are given the failure rate and have to find the density using (5.19). If the failure rate is constant, say $r(t) = \lambda$, then

$$\int_0^t r(\tau)\,d\tau = \int_0^t \lambda\,d\tau = \lambda t.$$

It follows that

$$f_T(t) = \lambda e^{-\lambda t},$$

and we see that T has an exponential density with parameter λ.

A more complicated failure rate is $r(t) = t/\lambda^2$. In this case,

$$\int_0^t r(\tau)\,d\tau = \int_0^t \frac{\tau}{\lambda^2}\,d\tau = \frac{t^2}{2\lambda^2}.$$

It then follows that

$$f_T(t) = \frac{t}{\lambda^2} e^{-(t/\lambda)^2/2},$$

which we recognize as the Rayleigh(λ) density.

***Example* 5.21.** In other problems you are given the density of T and have to find the failure rate using (5.17). For example, if $T \sim \exp(\lambda)$, the denominator in (5.17) is

$$\int_t^\infty f_T(\theta)\,d\theta = \int_t^\infty \lambda e^{-\lambda\theta}\,d\theta = -e^{-\lambda\theta}\Big|_t^\infty = e^{-\lambda t}.$$

It follows that

$$r(t) = \frac{f_T(t)}{\int_t^\infty f_T(\theta)\,d\theta} = \frac{\lambda e^{-\lambda t}}{e^{-\lambda t}} = \lambda.$$

If $T \sim$ Rayleigh(λ), the denominator in (5.17) is

$$\int_t^\infty \frac{\theta}{\lambda^2} e^{-(\theta/\lambda)^2/2}\,d\theta = -e^{-(\theta/\lambda)^2/2}\Big|_t^\infty = e^{-(t/\lambda)^2/2}.$$

It follows that

$$r(t) = \frac{f_T(t)}{\int_t^\infty f_T(\theta)\,d\theta} = \frac{\frac{t}{\lambda^2} e^{-(t/\lambda)^2/2}}{e^{-(t/\lambda)^2/2}} = t/\lambda^2.$$

Notes

5.1: Continuous random variables

Note 1. The normal cdf and the error function. We begin by writing

$$Q(y) := 1 - \Phi(y) = \frac{1}{\sqrt{2\pi}} \int_y^\infty e^{-\theta^2/2}\, d\theta.$$

Then make the change of variable $\xi = \theta/\sqrt{2}$ to get

$$Q(y) = \frac{1}{\sqrt{\pi}} \int_{y/\sqrt{2}}^\infty e^{-\xi^2}\, d\xi.$$

Since the **complementary error function** is given by

$$\mathrm{erfc}(z) := \frac{2}{\sqrt{\pi}} \int_z^\infty e^{-\xi^2}\, d\xi,$$

we can write

$$\boxed{Q(y) = \tfrac{1}{2} \mathrm{erfc}(y/\sqrt{2}).}$$

The MATLAB command for erfc(z) is `erfc(z)`.

We next use the fact that since the Gaussian density is even,

$$Q(-y) = \frac{1}{\sqrt{2\pi}} \int_{-y}^\infty e^{-\theta^2/2}\, d\theta = \frac{1}{\sqrt{2\pi}} \int_{-\infty}^y e^{-t^2/2}\, dt = \Phi(y).$$

Hence,

$$\boxed{\Phi(y) = \tfrac{1}{2} \mathrm{erfc}(-y/\sqrt{2}).}$$

The **error function** is defined by

$$\mathrm{erf}(z) := \frac{2}{\sqrt{\pi}} \int_0^z e^{-\xi^2}\, d\xi.$$

It is easy to check that $\mathrm{erf}(z) + \mathrm{erfc}(z) = 1$. Since erf is odd, $\Phi(y) = \tfrac{1}{2}[1 + \mathrm{erf}(y/\sqrt{2})]$. However, this formula is not recommended because $\mathrm{erf}(z)$ is negative for $z < 0$, and this could result in a loss of significant digits in numerical computation.

Note 2. In Example 5.5, the cdf has "corners" at $x = 0$ and at $x = 1$. In other words, left and right derivatives are not equal at these points. Hence, strictly speaking, $F'(x)$ does not exist at $x = 0$ or at $x = 1$.

Note 3. Derivation of (5.2). Let $g(x)$ be a continuous, strictly-increasing function. By strictly increasing, we mean that for $x_1 < x_2$, $g(x_2) < g(x_2)$. Such a function always has an inverse $g^{-1}(y)$, which is also strictly increasing. So, if $Y = g(X)$, we can find the cdf of Y by writing

$$\begin{aligned} F_Y(y) &= \mathsf{P}(Y \leq y) \\ &= \mathsf{P}(g(X) \leq y) \\ &= \mathsf{P}(X \leq g^{-1}(y)) \\ &= F_X(g^{-1}(y)), \end{aligned}$$

where the third equation follows because

$$\{g(X) \leq y\} = \{X \leq g^{-1}(y)\}. \tag{5.20}$$

It is now convenient to put $h(y) := g^{-1}(y)$ so that we can write

$$F_Y(y) = F_X(h(y)).$$

Differentiating both sides, we have

$$f_Y(y) = f_X(h(y))h'(y). \tag{5.21}$$

We next consider continuous, strictly-decreasing functions. By strictly decreasing, we mean that for $x_1 < x_2$, $g(x_1) > g(x_2)$. For such functions, the inverse is also strictly decreasing. In this case, instead of (5.20), we have $\{g(X) \leq y\} = \{X \geq g^{-1}(y)\}$. This leads to

$$\begin{aligned} F_Y(y) &= \mathsf{P}(Y \leq y) \\ &= \mathsf{P}(g(X) \leq y) \\ &= \mathsf{P}(X \geq g^{-1}(y)) \\ &= 1 - F_X(g^{-1}(y)). \end{aligned}$$

Again using the notation $h(y) := g^{-1}(y)$, we can write

$$F_Y(y) = 1 - F_X(h(y)).$$

Differentiating yields

$$f_Y(y) = -f_X(h(y))h'(y). \tag{5.22}$$

We further note that if h is increasing, then[i]

$$h'(x) := \lim_{\Delta x \to 0} \frac{h(x + \Delta x) - h(x)}{\Delta x} \geq 0, \tag{5.23}$$

while if h is decreasing, $h'(x) \leq 0$. Since densities are nonnegative, this explains the minus sign in (5.22). We can combine (5.21) and (5.22) into the single expression

$$f_Y(y) = f_X(h(y))|h'(y)|.$$

---

[i]If h is increasing, then the numerator in (5.23) is nonnegative for $\Delta x > 0$ and nonpositive for $\Delta x < 0$. In either case, the quotient is nonnegative, and so the limit is too.

5.5: Properties of cdfs

Note 4. So far we have discussed random variables that are discrete, continuous, or mixed, noting that the discrete and continuous are special cases of the mixed. By allowing density functions to contain impulses, any of the these cdfs can be expressed in the form

$$F(x) = \int_{-\infty}^{x} f(t)\,dt. \tag{5.24}$$

With this representation, if f is continuous at a point x_0, then $f(x_0) = F'(x_0)$, while if f has an impulse at x_0, then F has a jump at x_0, and the size of the jump, $P(X = x_0) = F(x_0) - F(x_0-)$, is magnitude of the impulse. This suffices for most applications. However, we mention that it is possible to have a random variable whose cdf is continuous, strictly increasing, but whose derivative is the zero function [3]. Such a cdf cannot be written in the above form: Since F is continuous, f cannot have impulses; since F' is the zero function, $F'(x) = f(x)$ is zero too; but then (5.24) would say $F(x) = 0$ for all x, contradicting $F(x) \to 1$ as $x \to \infty$. A random variable whose cdf is continuous but whose derivative is the zero function is said to be **singular**. Since both singular random variables and continuous random variables have continuous cdfs, in advanced texts, continuous random variables are sometimes called **absolutely continuous**.

Note 5. If X is a random variable defined on Ω, then $\mu(B) := P(\{\omega \in \Omega : X(\omega) \in B\})$ satisfies the axioms of a probability measure on the Borel subsets of $\mathbb{R}$ (Problem 4 in Chapter 2). (Also recall Note 1 and Problems 49 and 50 in Chapter 1 and Note 1 in Chapter 2.) Taking $B = (-\infty, x]$ shows that the cdf of X is $F(x) = \mu((-\infty, x])$. Thus, μ determines F. The converse is also true in the sense that if F is a right-continuous, nondecreasing function satisfying $F(x) \to 1$ as $x \to \infty$ and $F(x) \to 0$ as $x \to -\infty$, then there is a unique probability measure μ on the Borel sets of $\mathbb{R}$ such that $\mu((-\infty, x]) = F(x)$ for all $x \in \mathbb{R}$. A complete proof of this fact is beyond the scope of this book, but here is a sketch of the main ideas. Given such a function F, for $a < b$, put $\mu((a,b]) := F(b) - F(a)$. For more general Borel sets B, we proceed as follows. Suppose we have a collection of intervals $(a_i, b_i]$ such that[j]

$$B \subset \bigcup_{i=1}^{\infty} (a_i, b_i].$$

Such a collection is called a **covering of intervals**. Note that we always have the covering $B \subset (-\infty, \infty)$. We then define

$$\mu(B) := \inf_{B \subset \bigcup_i (a_i, b_i]} \sum_{i=1}^{\infty} F(b_i) - F(a_i),$$

where the infimum is over all coverings of intervals. Uniqueness is a consequence of the fact that if two probability measures agree on intervals, then they agree on all the Borel sets. This fact follows from the π–λ theorem [3].

---
[j] If $b_i = \infty$, it is understood that $(a_i, b_i]$ means (a_i, ∞).

Problems

5.1: Continuous random variables

1. Find the cumulative distribution function $F(x)$ of an exponential random variable X with parameter λ.

2. The **Rayleigh** density with parameter λ is defined by

$$f(x) := \begin{cases} \dfrac{x}{\lambda^2} e^{-(x/\lambda)^2/2}, & x \geq 0, \\ 0, & x < 0. \end{cases}$$

   Find the cumulative distribution function.

3. Find the cdf of the Weibull(p, λ) density defined in Problem 8 of Chapter 4.

*4. The **Maxwell** density with parameter λ is defined by

$$f(x) := \begin{cases} \sqrt{\dfrac{2}{\pi}} \dfrac{x^2}{\lambda^3} e^{-(x/\lambda)^2/2}, & x \geq 0, \\ 0, & x < 0. \end{cases}$$

   Show that the cdf $F(x)$ can be expressed in terms of the standard normal cdf

$$\Phi(y) := \frac{1}{\sqrt{2\pi}} \int_{-\infty}^{y} e^{-\theta^2/2} \, d\theta.$$

5. If Z has density $f_Z(z)$ and $Y = e^Z$, find $f_Y(y)$.

6. If $Y = 1 - X$, and X has density f_X, show that $f_Y(y) = f_X(1-y)$. In particular, show that if $X \sim \text{uniform}(0,1)$, then $Y \sim \text{uniform}(0,1)$.

7. If $X \sim \text{uniform}(0,1)$, find the density of $Y = \ln(1/X)$.

8. Let $X \sim \text{Weibull}(p, \lambda)$. Find the density of $Y = \lambda X^p$.

9. If X is exponential with parameter $\lambda = 1$, show that $Y = \sqrt{X}$ is Rayleigh$(1/\sqrt{2})$.

10. If $X \sim N(m, \sigma^2)$, then $Y := e^X$ is said to be a **lognormal** random variable. Find the moments $\mathsf{E}[Y^n]$.

11. The input to a squaring circuit is a Gaussian random variable X with mean zero and variance one. Use the methods of this chapter to show that the output $Y = X^2$ has the chi-squared density with one degree of freedom,

$$f_Y(y) = \frac{e^{-y/2}}{\sqrt{2\pi y}}, \quad y > 0.$$

12. If the input to the squaring circuit of Problem 11 includes a fixed bias, say m, then the output is given by $Y = (X+m)^2$, where again $X \sim N(0,1)$. Use the methods of this chapter to show that Y has the noncentral chi-squared density with one degree of freedom and noncentrality parameter m^2,

$$f_Y(y) = \frac{e^{-(y+m^2)/2}}{\sqrt{2\pi y}} \cdot \frac{e^{m\sqrt{y}} + e^{-m\sqrt{y}}}{2}, \quad y > 0.$$

Note that if $m = 0$, we recover the result of Problem 11.

13. Let $X_1, \ldots, X_n$ be independent with common cumulative distribution function $F(x)$. Let us define $X_{\max} := \max(X_1, \ldots, X_n)$ and $X_{\min} := \min(X_1, \ldots, X_n)$. Express the cumulative distributions of $X_{\max}$ and $X_{\min}$ in terms of $F(x)$. *Hint:* Example 2.11 may be helpful.

14. If X and Y are independent $\exp(\lambda)$ random variables, find $\mathsf{E}[\max(X,Y)]$.

15. Let $X \sim \text{Poisson}(\mu)$, and suppose that given $X = m$, $Y \sim \text{Erlang}(m, \lambda)$. Find the correlation $\mathsf{E}[XY]$.

16. Let $X \sim \text{Poisson}(\lambda)$, and suppose that given $X = n$, Y is conditionally an exponential random variable with parameter n. Find $\mathsf{P}(Y > y)$ for $y \geq 0$.

17. **Digital communication system.** The received voltage in a digital communication system is $Z = X + Y$, where $X \sim \text{Bernoulli}(p)$ is a random message, and $Y \sim N(0,1)$ is a Gaussian noise voltage. Assume X and Y are independent.

    (a) Find the conditional cdf $F_{Z|X}(z|i)$ for $i = 0, 1$, the cdf $F_Z(z)$, and the density $f_Z(z)$.

    (b) Find $f_{Z|X}(z|1)$, $f_{Z|X}(z|0)$, and then express the likelihood-ratio test

    $$\frac{f_{Z|X}(z|1)}{f_{Z|X}(z|0)} \geq \frac{\mathsf{P}(X=0)}{\mathsf{P}(X=1)}$$

    in as simple a form as possible.

18. **Fading channel.** Let X and Y be as in the preceding problem, but now suppose $Z = X/A + Y$, where A, X, and Y are independent, and A takes the values 1 and 2 with equal probability. Find the conditional cdf $F_{Z|A,X}(z|a,i)$ for $a = 1, 2$ and $i = 0, 1$.

19. **Generalized Rayleigh densities.** Let Y_n be chi-squared with $n > 0$ degrees of freedom as defined in Problem 15 of Chapter 4. Put $Z_n := \sqrt{Y_n}$.

    (a) Express the cdf of Z_n in terms of the cdf of Y_n.
    (b) Find the density of Z_1.
    (c) Show that Z_2 has a Rayleigh density, as defined in Problem 2, with $\lambda = 1$.
    (d) Show that Z_3 has a Maxwell density, as defined in Problem 4, with $\lambda = 1$.

(e) Show that Z_{2m} has a **Nakagami-m** density

$$f(z) := \begin{cases} \dfrac{2}{2^m \Gamma(m)} \dfrac{z^{2m-1}}{\lambda^{2m}} e^{-(z/\lambda)^2/2}, & z \geq 0, \\ 0, & z < 0, \end{cases}$$

with $\lambda = 1$.

Remark. For the general chi-squared random variable Y_n, it is not necessary that n be an integer. However, if n is a positive integer, and if $X_1,\ldots,X_n$ are i.i.d. $N(0,1)$, then the X_i^2 are chi-squared with one degree of freedom by Problem 11, and $Y_n := X_1^2 + \cdots + X_n^2$ is chi-squared with n degrees of freedom by Problem 55(c) in Chapter 4. Hence, the above densities usually arise from taking the square root of the sum of squares of standard normal random variables. For example, (X_1,X_2) can be regarded as a random point in the plane whose horizontal and vertical coordinates are independent $N(0,1)$. The distance of this point from the origin is $\sqrt{X_1^2 + X_2^2} = Z_2$, which is a Rayleigh random variable. As another example, consider an ideal gas. The velocity of a given particle is obtained by adding up the results of many collisions with other particles. By the central limit theorem (Section 5.6), each component of the given particle's velocity vector, say (X_1,X_2,X_3) should be i.i.d. $N(0,1)$. The speed of the particle is $\sqrt{X_1^2 + X_2^2 + X_3^2} = Z_3$, which has the Maxwell density. When the Nakagami-m density is used as a model for fading in wireless communication channels, m is often not an integer.

20. Let $X_1,\ldots,X_n$ be i.i.d. $N(0,1)$ random variables. Find the density of

$$Y := (X_1 + \cdots + X_n)^2.$$

21. **Generalized gamma densities.**

    (a) For positive p and q, let $X \sim \text{gamma}(p,1)$, and put $Y := X^{1/q}$. Show that

    $$f_Y(y) = \frac{q y^{pq-1} e^{-y^q}}{\Gamma(p)}, \quad y > 0.$$

    (b) If in part (a) we replace p with p/q, we find that

    $$f_Y(y) = \frac{q y^{p-1} e^{-y^q}}{\Gamma(p/q)}, \quad y > 0.$$

    Evaluate $\lim_{y \to 0} f_Y(y)$ for the three cases $0 < p < 1$, $p = 1$, and $p > 1$.

    (c) If we introduce a scale parameter $\lambda > 0$, we have the **generalized gamma** density [60]. More precisely, we say that $Y \sim \text{g-gamma}(p,\lambda,q)$ if Y has density

    $$f_Y(y) = \frac{\lambda q (\lambda y)^{p-1} e^{-(\lambda y)^q}}{\Gamma(p/q)}, \quad y > 0.$$

    Clearly, g-gamma$(p,\lambda,1) = \text{gamma}(p,\lambda)$, which includes the exponential, Erlang, and the chi-squared as special cases. Show that

(i) g-gamma$(p, \lambda^{1/p}, p)$ = Weibull(p, λ).
(ii) g-gamma$(2, 1/(\sqrt{2}\lambda), 2)$ is the Rayleigh density defined in Problem 2.
(iii) g-gamma$(3, 1/(\sqrt{2}\lambda), 2)$ is the Maxwell density defined in Problem 4.

(d) If $Y \sim$ g-gamma(p, λ, q), show that

$$E[Y^n] = \frac{\Gamma((n+p)/q)}{\Gamma(p/q)\lambda^n},$$

and conclude that

$$M_Y(s) = \sum_{n=0}^{\infty} \frac{s^n}{n!} \cdot \frac{\Gamma((n+p)/q)}{\Gamma(p/q)\lambda^n}.$$

(e) Show that the g-gamma(p, λ, q) cdf is given by $F_Y(y) = G_{p/q}((\lambda y)^q)$, where G_p is the cdf of the gamma$(p, 1)$ random variable,[k]

$$G_p(x) := \frac{1}{\Gamma(p)} \int_0^x t^{p-1} e^{-t} \, dt. \qquad (5.25)$$

Remark. In MATLAB, $G_p(x)$ = gamcdf(x,p). Hence, you can easily compute the cdf of any gamma random variable such as the Erlang or chi-squared or any g-gamma random variable such as the Rayleigh, Maxwell, or Weibull. Note, however, that the Rayleigh and Weibull cdfs have closed forms (Problems 2 and 3). Note also that MATLAB provides the command chi2cdf(x,k) to compute the cdf of a chi-squared random variable with k degrees of freedom.

*22. In the analysis of communication systems, one is often interested in $P(X > x) = 1 - F(x)$ for some voltage threshold x. We call $F^c(x) := 1 - F(x)$ the **complementary cumulative distribution function** (ccdf) of X. Of particular interest is the ccdf of the standard normal, which is often denoted by

$$Q(x) := 1 - \Phi(x) = \frac{1}{\sqrt{2\pi}} \int_x^{\infty} e^{-t^2/2} \, dt.$$

Using the hints below, show that for $x > 0$,

$$\frac{e^{-x^2/2}}{\sqrt{2\pi}} \left(\frac{1}{x} - \frac{1}{x^3} \right) < Q(x) < \frac{e^{-x^2/2}}{x\sqrt{2\pi}}.$$

Hints: To derive the upper bound, apply integration by parts to

$$\int_x^{\infty} \frac{1}{t} \cdot t e^{-t^2/2} \, dt,$$

and then drop the new integral term (which is positive),

$$\int_x^{\infty} \frac{1}{t^2} e^{-t^2/2} \, dt.$$

---

[k]The integral in (5.25), as well as $\int_x^{\infty} t^{p-1} e^{-t} \, dt$, are sometimes referred to as **incomplete gamma functions**. MATLAB actually uses (5.25) as the definition of the incomplete gamma function. Hence, in MATLAB, $G_p(x)$ = gammainc(x,p).

If you do *not* drop the above term, you can derive the lower bound by applying integration by parts one more time (after dividing and multiplying by t again) and then dropping the final integral.

*23. In wireless communications, it is often necessary to compute $E[Q(Z)]$, where Q is the complementary cdf (ccdf) of the standard normal defined in the previous problem, and Z is some random variable that arises in the detector. The formulas in parts (c)–(f) can be inferred from [65, eqs. (3.60), (3.65), (3.61), (3.64)], respectively. Additional formulas can be found in [65, Section 3.3].

(a) If X and Z are continuous random variables, show that

$$E[F_X^c(Z)] = E[F_Z(X)].$$

(b) If $Z \sim \text{Erlang}(m, \lambda)$, show that

$$E[F_Z(X)] = P(X \geq 0) - \sum_{k=0}^{m-1} \frac{\lambda^k}{k!} E[X^k e^{-\lambda X} I_{[0,\infty)}(X)].$$

Hint: Problem 15(c) in Chapter 4.

(c) If $Z \sim \exp(\lambda)$, show that

$$E[Q(Z)] = \frac{1}{2} - e^{\lambda^2/2} Q(\lambda).$$

Hint: This is a special case of part (b).

(d) If Y is chi-squared with $2m$ degrees of freedom, show that

$$E[Q(\sigma\sqrt{Y})] = \frac{1}{2} - \frac{1}{2\sqrt{1+\sigma^{-2}}} \sum_{k=0}^{m-1} \frac{1 \cdot 3 \cdot 5 \cdots (2k-1)}{k! 2^k (1+\sigma^2)^k}.$$

(e) If $Z \sim \text{Rayleigh}(\sigma)$, show that

$$E[Q(Z)] = \frac{1}{2}\left(1 - \frac{1}{\sqrt{1+\sigma^{-2}}}\right).$$

Hint: This is a special case of part (d).

(f) Let $V_1, \ldots, V_m$ be independent, $\exp(\lambda_i)$ random variables for distinct, positive values of λ_i. Show that

$$E[Q(\sqrt{V_1 + \cdots + V_m})] = \sum_{i=1}^{m} \frac{c_i}{2}[1 - (1 + 2\lambda_i)^{-1/2}],$$

where $c_i := \prod_{k \neq i} \lambda_k / (\lambda_k - \lambda_i)$. *Hint:* The first step is to put $Y := V_1 + \cdots + V_m$, and find f_Y by first expanding its moment generating function using partial fractions.

Remark. In applications, V_i arises as $V_i = U_i^2 + W_i^2$, where U_i and W_i are independent $N(0, \sigma_i^2/2)$, and represent the real and imaginary parts of a complex Gaussian random variable $U_i + jW_i$ (Section 9.6). In this case, $\lambda_i = 1/\sigma_i^2$.

*24. Let $C_k(x)$ denote the chi-squared cdf with k degrees of freedom. Show that the noncentral chi-squared cdf with k degrees of freedom and noncentrality parameter λ^2 is given by (recall Problem 65 in Chapter 4)

$$C_{k,\lambda^2}(x) = \sum_{n=0}^{\infty} \frac{(\lambda^2/2)^n e^{-\lambda^2/2}}{n!} C_{2n+k}(x).$$

Remark. In MATLAB we have that $C_k(x) = \mathtt{chi2cdf(x,k)}$ and that $C_{k,\lambda^2}(x) = \mathtt{ncx2cdf(x,k,lambda\hat{\ }2)}$.

*25. **Generalized Rice or noncentral Rayleigh densities.** Let Y_n be noncentral chi-squared with $n > 0$ degrees of freedom and noncentrality parameter m^2 as defined in Problem 65 in Chapter 4. (In general, n need not be an integer, but if it is, and if $X_1, \ldots, X_n$ are i.i.d. normal random variables with $X_i \sim N(m_i, 1)$, then by Problem 12, X_i^2 is noncentral chi-squared with one degree of freedom and noncentrality parameter m_i^2, and by Problem 65 in Chapter 4, $X_1^2 + \cdots + X_n^2$ is noncentral chi-squared with n degrees of freedom and noncentrality parameter $m^2 = m_1^2 + \cdots + m_n^2$.)

(a) Show that $Z_n := \sqrt{Y_n}$ has the **generalized Rice** density,

$$f_{Z_n}(z) = \frac{z^{n/2}}{m^{n/2-1}} e^{-(m^2+z^2)/2} I_{n/2-1}(mz), \quad z > 0,$$

where I_ν is the modified **Bessel function** of the first kind, order ν,

$$I_\nu(x) := \sum_{\ell=0}^{\infty} \frac{(x/2)^{2\ell+\nu}}{\ell! \, \Gamma(\ell+\nu+1)}.$$

Graphs of $f_{Z_n}(z)$ for different values of n and m are shown in Figures 5.19–5.21. Graphs of I_ν for different values of ν are shown in Figure 5.22. In MATLAB, $I_\nu(x) = \mathtt{besseli(nu,x)}$.

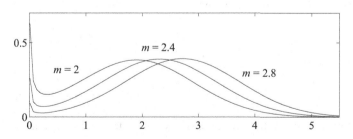

Figure 5.19. Rice density $f_{Z_{1/2}}(z)$ for different values of m.

(b) Show that Z_2 has the original Rice density,

$$f_{Z_2}(z) = z e^{-(m^2+z^2)/2} I_0(mz), \quad z > 0.$$

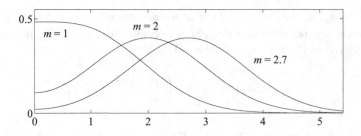

Figure 5.20. Rice density $f_{Z_1}(z)$ for different values of m.

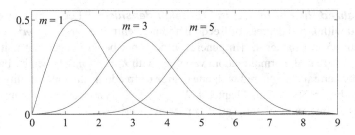

Figure 5.21. Rice density $f_{Z_2}(z)$ for different values of m.

(c) Show that

$$f_{Y_n}(y) = \frac{1}{2}\left(\frac{\sqrt{y}}{m}\right)^{n/2-1} e^{-(m^2+y)/2} I_{n/2-1}(m\sqrt{y}), \quad y > 0,$$

giving a closed-form expression for the noncentral chi-squared density. Recall that you already have a closed-form expression for the moment generating function and characteristic function of a noncentral chi-squared random variable (see Problem 65(b) in Chapter 4).

Remark. In MATLAB, the cdf of Y_n is given by $F_{Y_n}(y) = \text{ncx2cdf}(y,n,m\texttt{\^{}}2)$.

(d) Denote the complementary cumulative distribution of Z_n by

$$F_{Z_n}^c(z) := \mathsf{P}(Z_n > z) = \int_z^\infty f_{Z_n}(t)\,dt.$$

Show that

$$F_{Z_n}^c(z) = \left(\frac{z}{m}\right)^{n/2-1} e^{-(m^2+z^2)/2} I_{n/2-1}(mz) + F_{Z_{n-2}}^c(z).$$

Hint: Use integration by parts; you will need the easily-verified fact that

$$\frac{d}{dx}\left(x^\nu I_\nu(x)\right) = x^\nu I_{\nu-1}(x).$$

(e) The complementary cdf of Z_2, $F_{Z_2}^c(z)$, is known as the **Marcum Q function**,

$$Q(m,z) := \int_z^\infty t e^{-(m^2+t^2)/2} I_0(mt)\,dt.$$

Show that if $n \geq 4$ is an even integer, then

$$F_{Z_n}^c(z) = Q(m,z) + e^{-(m^2+z^2)/2} \sum_{k=1}^{n/2-1} \left(\frac{z}{m}\right)^k I_k(mz).$$

(f) Show that $Q(m,z) = \widetilde{Q}(m,z)$, where

$$\widetilde{Q}(m,z) := e^{-(m^2+z^2)/2} \sum_{k=0}^{\infty} (m/z)^k I_k(mz).$$

Hint: [27, p. 450] Show that $Q(0,z) = \widetilde{Q}(0,z) = e^{-z^2/2}$. It then suffices to prove that

$$\frac{\partial}{\partial m} Q(m,z) = \frac{\partial}{\partial m} \widetilde{Q}(m,z).$$

To this end, use the derivative formula in the hint of part (d) to show that

$$\frac{\partial}{\partial m} \widetilde{Q}(m,z) = z e^{-(m^2+z^2)/2} I_1(mz);$$

you will also need the fact (derived in the next problem) that $I_{-1}(x) = I_1(x)$. Now take the same partial derivative of $Q(m,z)$ as defined in part (e), and then use integration by parts on the term involving I_1.

*26. **Properties of modified Bessel functions.*** In this problem you will derive some basic properties of the modified Bessel functions of the first kind,

$$I_\nu(x) := \sum_{\ell=0}^{\infty} \frac{(x/2)^{2\ell+\nu}}{\ell!\,\Gamma(\ell+\nu+1)},$$

several of which are sketched in Figure 5.22.

(a) Show that

$$\lim_{x \downarrow 0} \frac{I_\nu(x)}{(x/2)^\nu} = \frac{1}{\Gamma(\nu+1)},$$

and use result to evaluate

$$\lim_{z \downarrow 0} f_{Z_n}(z),$$

where f_{Z_n} is the Rice density of the previous problem. *Hint:* Remembering that $n > 0$ need not be an integer, the three cases to consider are $0 < n < 1$, $n = 1$, and $n > 1$.

(b) Show that

$$I'_\nu(x) = \tfrac{1}{2}[I_{\nu-1}(x) + I_{\nu+1}(x)]$$

and that

$$I_{\nu-1}(x) - I_{\nu+1}(x) = 2(\nu/x) I_\nu(x).$$

Note that the second identity implies the recursion,

$$I_{\nu+1}(x) = I_{\nu-1}(x) - 2(\nu/x) I_\nu(x).$$

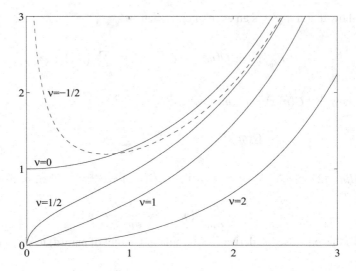

Figure 5.22. Graphs if $I_\nu(x)$ for different values of ν.

Hence, once $I_0(x)$ and $I_1(x)$ are known, $I_n(x)$ can be computed for $n = 2, 3, \ldots$. We also mention that the second identity with $\nu = 0$ implies $I_{-1}(x) = I_1(x)$. Using this in the first identity shows that $I_0'(x) = I_1(x)$.

(c) Parts (c) and (d) of this problem are devoted to showing that for integers $n \geq 0$,

$$I_n(x) = \frac{1}{2\pi} \int_{-\pi}^{\pi} e^{x \cos \theta} \cos(n\theta) \, d\theta.$$

To this end, denote the above integral by $\tilde{I}_n(x)$. Use integration by parts and then a trigonometric identity to show that

$$\tilde{I}_n(x) = \frac{x}{2n} [\tilde{I}_{n-1}(x) - \tilde{I}_{n+1}(x)].$$

Hence, in Part (d) it will be enough to show that $\tilde{I}_0(x) = I_0(x)$ and $\tilde{I}_1(x) = I_1(x)$.

(d) As noted in part (b), $I_0'(x) = I_1(x)$. From the integral definition of $\tilde{I}_n(x)$, it is clear that $\tilde{I}_0'(x) = \tilde{I}_1(x)$ as well. Hence, it is enough to show that $\tilde{I}_0(x) = I_0(x)$. Since the integrand defining $\tilde{I}_0(x)$ is even,

$$\tilde{I}_0(x) = \frac{1}{\pi} \int_0^{\pi} e^{x \cos \theta} \, d\theta.$$

Show that

$$\tilde{I}_0(x) = \frac{1}{\pi} \int_{-\pi/2}^{\pi/2} e^{-x \sin t} \, dt.$$

Then use the power series $e^\xi = \sum_{k=0}^{\infty} \xi^k / k!$ in the above integrand and integrate term by term. Then use the results of Problems 18 and 14 of Chapter 4 to show that $\tilde{I}_0(x) = I_0(x)$.

(e) Use the integral formula for $I_n(x)$ to show that

$$I'_n(x) = \tfrac{1}{2}[I_{n-1}(x) + I_{n+1}(x)].$$

5.2: Discrete random variables

27. MATLAB. Computing the binomial probability

$$\binom{n}{k} p^k (1-p)^{n-k} = \frac{n!}{k!(n-k)!} p^k (1-p)^{n-k}$$

numerically can cause overflow problems if the factorials in the numerator and denominator are computed separately. However, since the log of the right-hand side is

$$\ln(n!) - \ln(k!) - \ln[(n-k)!] + k \ln p + (n-k) \ln(1-p),$$

this suggests an alternative way to calculate the probability. We can do even more to reduce overflow problems. Since $n! = \Gamma(n+1)$, we use the built-in MATLAB function gammaln, which computes the log of the gamma function. Enter the following MATLAB M-file containing the function binpmf(n,p) for computing the required probability.

```
% M-file with function for computing the
% binomial(n,p) pmf.
%
function y = binpmf(k,n,p)
nk = n-k;
p1 = 1-p;
w = gammaln(n+1) - gammaln(nk+1) - gammaln(k+1) + ...
    log(p)*k + log(p1)*nk;
y = exp(w);
```

Now type in the commands

```
n = 4
p = 0.75
k = [0:n]
prob = binpmf(k,n,p)
stem(k,prob,'filled')
```

to generate a stem plot of the binomial$(4, 3/4)$ pmf.

28. Let $X \sim \text{binomial}(n,p)$ with $n = 4$ and $p = 3/4$. Sketch the graph of the cumulative distribution function of X, $F(x)$.

5.3: Mixed random variables

29. A random variable X has generalized density
$$f(t) = \tfrac{1}{3}e^{-t}u(t) + \tfrac{1}{2}\delta(t) + \tfrac{1}{6}\delta(t-1),$$
where u is the unit step function defined in Section 2.1, and δ is the Dirac delta function defined in Section 5.3.

   (a) Sketch $f(t)$.
   (b) Compute $P(X=0)$ and $P(X=1)$.
   (c) Compute $P(0 < X < 1)$ and $P(X > 1)$.
   (d) Use your above results to compute $P(0 \leq X \leq 1)$ and $P(X \geq 1)$.
   (e) Compute $E[X]$.

30. If X has generalized density $f(t) = \tfrac{1}{2}[\delta(t) + I_{(0,1]}(t)]$, evaluate $E[e^X]$ and $P(X=0|X \leq 1/2)$.

31. Show that $E[X] = 7/12$ if X has cdf
$$F_X(x) = \begin{cases} 0, & x < 0, \\ x^2, & 0 \leq x < 1/2, \\ x, & 1/2 \leq x < 1, \\ 1, & x \geq 1. \end{cases}$$

32. Show that $E[\sqrt{X}] = 49/30$ if X has cdf
$$F_X(x) := \begin{cases} 0, & x < 0, \\ \sqrt{x}/4, & 0 \leq x < 4, \\ (x+11)/20, & 4 \leq x < 9, \\ 1, & x \geq 9. \end{cases}$$

33. Find and sketch the cumulative distribution function of Example 5.13.

*34. A certain computer monitor contains a loose connection. The connection is loose with probability $1/2$. When the connection is loose, the monitor displays a blank screen (brightness $= 0$). When the connection is not loose, the brightness is uniformly distributed on $(0,1]$. Let X denote the observed brightness. Find formulas and plot the cdf and generalized density of X.

5.4: Functions of random variables and their cdfs

35. Let $\Theta \sim \text{uniform}[-\pi, \pi]$, and put $X := \cos\Theta$ and $Y := \sin\Theta$.

   (a) Show that $F_X(x) = 1 - \tfrac{1}{\pi}\cos^{-1}x$ for $-1 \leq x \leq 1$.
   (b) Show that $F_Y(y) = \tfrac{1}{2} + \tfrac{1}{\pi}\sin^{-1}y$ for $-1 \leq y \leq 1$.

(c) Show that $f_X(x) = (1/\pi)/\sqrt{1-x^2}$ and $f_Y(y) = (1/\pi)/\sqrt{1-y^2}$. Since X and Y have the same density, they have the same cumulative distribution function. Hence, both X and Y are called **arcsine random variables**.

(d) Show that $Z = (Y+1)/2$ has the beta density of Problem 16 in Chapter 4.

36. Find the cdf and density of $Y = X(X+2)$ if X is uniformly distributed on $[-3,1]$.

37. Let $X \sim \text{uniform}[-3,3]$, and suppose $Y = g(X)$, where

$$g(x) = \begin{cases} 2, & -1 \le x \le 1, \\ 2/x^2, & 1 < |x| \le 2, \\ 0, & \text{otherwise}. \end{cases}$$

Find $f_Y(y)$ for $-\infty < y < \infty$.

38. Consider the series RLC circuit shown in Figure 5.23. The voltage transfer function

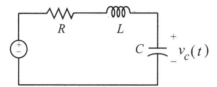

Figure 5.23. Series RLC circuit. The output is the capacitor voltage $v_c(t)$.

between the source and the capacitor is

$$H(\omega) = \frac{1}{(1-\omega^2 LC) + j\omega RC}.$$

A plot of

$$|H(\omega)|^2 = \frac{1}{(1-\omega^2 LC)^2 + (\omega RC)^2}$$

is shown in Figure 5.24. The resonant frequency of the circuit, ω_0, is the value of ω

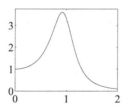

Figure 5.24. Plot of $|H(\omega)|^2$ for $L=1$, $C=1$, and $R=\sqrt{0.3}$.

that maximizes $|H(\omega)|^2$. It is not hard to show that

$$\omega_0 = \sqrt{\frac{1}{LC} - \frac{1}{2}\left(\frac{R}{L}\right)^2}.$$

If these circuits are mass produced, then the actual values of R, L, and C in a particular device vary somewhat from their design specifications, and hence, so does the resonant frequency. Assuming $L = 1$, $C = 1$, and $R \sim$ uniform $[0, \sqrt{2}]$, find the probability density function of the resonant frequency $Y = \sqrt{1 - R^2/2}$.

39. With the setup of the previous problem, find the probability density of the resonant peak
$$Z = |H(\omega_0)|^2 = \frac{1}{R^2(1 - R^2/4)},$$
where $R \sim$ uniform $[0, \sqrt{2}]$, and we again take $L = 1$ and $C = 1$.

40. Suppose that a unit step voltage is applied to the circuit in Figure 5.23. When the system is underdamped, the capacitor voltage has the form in Figure 5.25. When

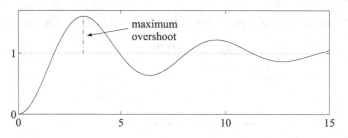

Figure 5.25. Capacitor voltage $v_c(t)$ when a unit step source voltage is applied to the circuit in Figure 5.23 and the circuit is underdamped. The horizontal dashed line is the limiting capacitor voltage $v_c(t)$ as $t \to \infty$.

$L = 1$ and $C = 1$, the *time* at which the maximum overshoot occurs is
$$T = \frac{\pi}{\sqrt{1 - R^2/4}}.$$

If $R \sim$ uniform $[0, \sqrt{2}]$, find the probability density of T. Also find the probability density of the maximum overshoot,
$$M = e^{-\pi(R/2)/\sqrt{1 - R^2/4}}.$$

41. Let g be as in Example 5.15. Find the cdf and density of $Y = g(X)$ if
    (a) $X \sim$ uniform$[-1, 1]$;
    (b) $X \sim$ uniform$[-1, 2]$;
    (c) $X \sim$ uniform$[-2, 3]$;
    (d) $X \sim \exp(\lambda)$.

42. Let
$$g(x) := \begin{cases} 0, & |x| < 1, \\ |x| - 1, & 1 \le |x| \le 2, \\ 1, & |x| > 2. \end{cases}$$
Find the cdf and density of $Y = g(X)$ if

(a) $X \sim \text{uniform}[-1,1]$;
(b) $X \sim \text{uniform}[-2,2]$;
(c) $X \sim \text{uniform}[-3,3]$;
(d) $X \sim \text{Laplace}(\lambda)$.

43. Let
$$g(x) := \begin{cases} -x-2, & x < -1, \\ -x^2, & -1 \leq x < 0, \\ x^3, & 0 \leq x < 1, \\ 1, & x \geq 1. \end{cases}$$

Find the cdf and density of $Y = g(X)$ if

(a) $X \sim \text{uniform}[-3,2]$;
(b) $X \sim \text{uniform}[-3,1]$;
(c) $X \sim \text{uniform}[-1,1]$.

44. Consider the function g given by
$$g(x) = \begin{cases} x^2 - 1, & x < 0, \\ x - 1, & 0 \leq x < 2, \\ 1, & x \geq 2. \end{cases}$$

If X is uniform$[-3,3]$, find the cdf and density of $Y = g(X)$.

45. Let X be a uniformly distributed random variable on the interval $[-3,1]$. Let $Y = g(X)$, where
$$g(x) = \begin{cases} 0, & x < -2, \\ x+2, & -2 \leq x < -1, \\ x^2, & -1 \leq x < 0, \\ \sqrt{x}, & x \geq 0. \end{cases}$$

Find the cdf and density of Y.

46. Let $X \sim \text{uniform}[-6,0]$, and suppose that $Y = g(X)$, where
$$g(x) = \begin{cases} |x| - 1, & 1 \leq |x| < 2, \\ 1 - \sqrt{|x|-2}, & |x| \geq 2, \\ 0, & \text{otherwise}. \end{cases}$$

Find the cdf and density of Y.

47. Let $X \sim \text{uniform}[-2,1]$, and suppose that $Y = g(X)$, where
$$g(x) = \begin{cases} x+2, & -2 \leq x < -1, \\ \dfrac{2x^2}{1+x^2}, & -1 \leq x < 0, \\ 0, & \text{otherwise}. \end{cases}$$

Find the cdf and density of Y.

*48. For $x \geq 0$, let $g(x)$ denote the fractional part of x. For example, $g(5.649) = 0.649$, and $g(0.123) = 0.123$. Find the cdf and density of $Y = g(X)$ if

(a) $X \sim \exp(1)$;

(b) $X \sim \text{uniform}[0, 1)$;

(c) $X \sim \text{uniform}[v, v+1)$, where $v = m + \delta$ for some integer $m \geq 0$ and some $0 < \delta < 1$.

5.5: Properties of cdfs

*49. Show that $G(x) := P(X < x)$ is a left-continuous function of x. Also show that $P(X = x_0) = G(x_0+) - G(x_0)$.

*50. From your solution of Problem 4(b) in Chapter 4, you can see that if $X \sim \exp(\lambda)$, then $P(X > t + \Delta t | X > t) = P(X > \Delta t)$. Now prove the converse; i.e., show that if Y is a nonnegative random variable such that $P(Y > t + \Delta t | Y > t) = P(Y > \Delta t)$, then $Y \sim \exp(\lambda)$, where $\lambda = -\ln[1 - F_Y(1)]$, assuming that $P(Y > t) > 0$ for all $t \geq 0$. *Hints:* Put $h(t) := \ln P(Y > t)$, which is a right-continuous function of t (Why?). Show that $h(t + \Delta t) = h(t) + h(\Delta t)$ for all $t, \Delta t \geq 0$.

5.6: The central limit theorem

51. Let $X_1, \ldots, X_n$ be i.i.d. with mean m and variance σ^2. Show that

$$Y_n := \frac{1}{\sqrt{n}} \sum_{i=1}^{n} \left(\frac{X_i - m}{\sigma} \right)$$

has zero mean and unit variance.

52. Packet transmission times on a certain Internet link are i.i.d. with mean m and variance σ^2. Suppose n packets are transmitted. Then the total expected transmission time for n packets is nm. Use the central limit theorem to approximate the probability that the total transmission time for the n packets exceeds twice the expected transmission time.

53. To combat noise in a digital communication channel with bit-error probability p, the use of an error-correcting code is proposed. Suppose that the code allows correct decoding of a received binary codeword if the fraction of bits in error is less than or equal to t. Use the central limit theorem to approximate the probability that a received word cannot be reliably decoded.

54. If $X_1, \ldots, X_n$ are i.i.d. Poisson(1), evaluate both sides of (5.14). Then rearrange your result to obtain Stirling's formula, $n! \approx \sqrt{2\pi}\, n^{n+1/2} e^{-n}$.

55. Following Example 5.18, we remarked that when the X_i are i.i.d. $\exp(1)$, the central limit theorem density decays faster than $g_n(x)$ as $x \to \infty$. Here is an example in which the central limit theorem density decays more slowly than $g_n(x)$. If the X_i are i.i.d. uniform$[-1, 1]$, find $x_{\max}$ such that for $x > x_{\max}$, $g_n(x) = 0$, while the central limit density is always positive.

56. Let $X_i = \pm 1$ with equal probability. Then the X_i are zero mean and have unit variance. Put
$$Y_n = \sum_{i=1}^{n} \frac{X_i}{\sqrt{n}}.$$
Derive the central limit theorem for this case; i.e., show that $\varphi_{Y_n}(v) \to e^{-v^2/2}$. *Hint:* Use the Taylor series approximation $\cos(\xi) \approx 1 - \xi^2/2$.

5.7: Reliability

57. The lifetime T of a Model n Internet router has an Erlang$(n,1)$ density, $f_T(t) = t^{n-1} e^{-t}/(n-1)!$.

   (a) What is the router's mean time to failure?

   (b) Show that the reliability of the router after t time units of operation is
$$R(t) = \sum_{k=0}^{n-1} \frac{t^k}{k!} e^{-t}.$$

   (c) Find the failure rate (known as the **Erlang** failure rate). Sketch the failure rate for $n = 2$.

58. A certain device has the **Weibull** failure rate
$$r(t) = \lambda p t^{p-1}, \quad t > 0.$$

   (a) Sketch the failure rate for $\lambda = 1$ and the cases $p = 1/2, p = 1, p = 3/2, p = 2$, and $p = 3$.

   (b) Find the reliability $R(t)$.

   (c) Find the mean time to failure.

   (d) Find the density $f_T(t)$.

59. A certain device has the **Pareto** failure rate
$$r(t) = \begin{cases} p/t, & t \geq t_0, \\ 0, & t < t_0. \end{cases}$$

   (a) Find the reliability $R(t)$ for $t \geq 0$.

   (b) Sketch $R(t)$ if $t_0 = 1$ and $p = 2$.

   (c) Find the mean time to failure if $p > 1$.

   (d) Find the Pareto density $f_T(t)$.

60. A certain device has failure rate $r(t) = t^2 - 2t + 2$ for $t \geq 0$.

   (a) Sketch $r(t)$ for $t \geq 0$.

   (b) Find the corresponding density $f_T(t)$ in closed form (no integrals).

61. Suppose that the lifetime T of a device is uniformly distributed on the interval $[1,2]$.

    (a) Find and sketch the reliability $R(t)$ for $t \geq 0$.

    (b) Find the failure rate $r(t)$ for $1 < t < 2$.

    (c) Find the mean time to failure.

62. Consider a system composed of two devices with respective lifetimes T_1 and T_2. Let T denote the lifetime of the composite system. Suppose that the system operates properly if and only if *both* devices are functioning. In other words, $T > t$ if and only if $T_1 > t$ and $T_2 > t$. Express the reliability of the overall system $R(t)$ in terms of $R_1(t)$ and $R_2(t)$, where $R_1(t)$ and $R_2(t)$ are the reliabilities of the individual devices. Assume T_1 and T_2 are independent.

63. Consider a system composed of two devices with respective lifetimes T_1 and T_2. Let T denote the lifetime of the composite system. Suppose that the system operates properly if and only if *at least one of* the devices is functioning. In other words, $T > t$ if and only if $T_1 > t$ or $T_2 > t$. Express the reliability of the overall system $R(t)$ in terms of $R_1(t)$ and $R_2(t)$, where $R_1(t)$ and $R_2(t)$ are the reliabilities of the individual devices. Assume T_1 and T_2 are independent.

64. Let Y be a nonnegative random variable. Show that

$$\mathsf{E}[Y^n] = \int_0^\infty ny^{n-1}\mathsf{P}(Y>y)\,dy.$$

Hint: Put $T = Y^n$ in (5.16).

Exam preparation

You may use the following suggestions to prepare a study sheet, including formulas mentioned that you have trouble remembering. You may also want to ask your instructor for additional suggestions.

5.1. Continuous random variables. Know that a continuous random variable is completely characterized by its cdf because the density is given by the derivative of the cdf. Be able to find the cdf of $Y = g(X)$ in terms of F_X when g is a simple function such as $g(x) = x^2$ or $g(x) = \sqrt{x}$. Then use the formula $f_Y(y) = (d/dy)F_Y(y)$. You should be aware of the MAP and ML rules. You should also be aware of how to use the inverse cdf to simulate a random variable starting with a uniform$(0,1)$ random variable.

5.2. Discrete random variables. Know that a discrete random variable is completely characterized by its cdf since $\mathsf{P}(X = x_j) = F_X(x_j) - F_X(x_{j-1})$. Be aware of how to simulate a discrete random variable starting with a uniform$(0,1)$ random variable.

5.3. Mixed random variables. Know that a mixed random variable is completely characterized by its cdf since the generalized density is given by (5.5), where $\tilde{f}_Y(y) = F_Y'(y)$ for $y \neq y_i$, and $\mathsf{P}(Y = y_i)$ is the size of the jump discontinuity in the cdf at y_i.

5.4. *Functions of random variables and their cdfs.* When $Y = g(X)$, be able to use graphical methods to find the cdf $F_Y(y)$. Then differentiate to find $f_Y(y)$, **but be careful to account for jumps in the cdf**. Jumps in the cdf correspond to impulses in the density.

5.5. *Properties of cdfs.* Be familiar with the eight properties.

5.6. *The central limit theorem.* When the X_i are i.i.d. with finite first and second moments, the key formulas are (5.8) for continuous random variables and (5.9) for integer-valued random variables.

5.7. *Reliability.* Key formulas are the reliability function $R(t)$ (5.15), the mean time to failure formula for $\mathsf{E}[T]$ (5.16), the differential equation for the failure rate function $r(t)$ and its representation in terms of the density $f_T(t)$ (5.17), and the density $f_T(t)$ in terms of the failure rate function (5.19).

Work any review problems assigned by your instructor. If you finish them, re-work your homework assignments.

6
Statistics†

As we have seen, most problems in probability textbooks start out with random variables having a given probability mass function or density. However, in the real world, problems start out with a finite amount of data, $X_1, X_2, \ldots, X_n$, about which very little is known based on the physical situation. We are still interested in computing probabilities, but we first have to find the pmf or density with which to do the calculations. Sometimes the physical situation determines the form of the pmf or density up to a few unknown parameters. For example, the number of alpha particles given off by a radioactive sample is Poisson(λ), but we need to estimate λ from measured data. In other situations, we may have no information about the pmf or density. In this case, we collect data and look at histograms to suggest possibilities. In this chapter, we not only look at parameter estimators and histograms, we also try to quantify how confident we are that our estimate or density choice is a good one.

Section 6.1 introduces the sample mean and sample variance as unbiased estimators of the true mean and variance. The concept of strong consistency is introduced and used to show that estimators based on the sample mean and sample variance inherit strong consistency. Section 6.2 introduces histograms and the chi-squared statistic for testing the goodness-of-fit of a hypothesized pmf or density to a histogram. Sections 6.3 and 6.4 focus on how good a sample mean estimator is; namely, how confident we are that it is close to the true mean. This is made precise through the notion of a confidence interval. Section 6.5 considers estimation of the mean and variance for Gaussian data. While the results of Sections 6.3 and 6.4 use approximations based on the central limit theorem, for Gaussian data, no such approximation is required. Section 6.6 uses our knowledge of confidence intervals to develop one-tailed and two-tailed hypothesis tests for the mean. Section 6.7 gives a quick introduction to curve fitting under the name of regression. Although formulas are developed using both variational arguments (derivatives) and the orthogonality principle, the emphasis is on using MATLAB to do the calculations. Section 6.8 provides a brief introduction to the estimation of probabilities using Monte Carlo simulation. Confidence intervals are used to assess the estimates. Particular attention is paid to the difficulties of estimating very small probabilities, say 10^{-4} and smaller. The use of importance sampling is suggested for estimating very small probabilities.

6.1 Parameter estimators and their properties

A sequence of observations or data measurements, say $X_1, \ldots, X_n$, is called a **sample**. A **statistic** is any function of the data. The **sample mean**,

$$M_n := \frac{1}{n} \sum_{i=1}^{n} X_i, \qquad (6.1)$$

---

†The material in this chapter is not used elsewhere in the book, with the exception of Problem 16 in Chapter 11, Problem 6 in Chapter 15, and Section 8.5. The present chapter can be covered at any time after Chapter 5.

6.1 Parameter estimators and their properties

is a statistic. Another useful statistic is the **sample variance**,

$$S_n^2 := \frac{1}{n-1} \sum_{i=1}^{n} (X_i - M_n)^2. \tag{6.2}$$

The **sample standard deviation** is $S_n := \sqrt{S_n^2}$. In MATLAB, if X is a vector of data, try the following commands to compute the sample mean, the sample standard deviation, and the sample variance.

```
X = [ 5 2 7 3 8 ]
Mn = mean(X)
Sn = std(X)
Sn2 = var(X)
```

Now suppose that the X_i all have the same mean m and the same variance σ^2. To distinguish between the sample mean M_n and the parameter m, m is called the **population mean** or the **ensemble mean**. Similarly, σ^2 is called the **population variance** or the **ensemble variance**. Is there a relationship between the *random variable* M_n and the *constant* m? What about the *random variable* S_n^2 and the *constant* σ^2? With regard to M_n and m, it is easy to see that

$$\mathsf{E}[M_n] = \mathsf{E}\left[\frac{1}{n}\sum_{i=1}^{n} X_i\right] = \frac{1}{n}\sum_{i=1}^{n}\mathsf{E}[X_i] = \frac{1}{n}\sum_{i=1}^{n} m = m.$$

In other words, the expected value of the sample mean is the population mean. For this reason, we say that the sample mean M_n is an **unbiased estimator** of the population mean m. With regard to S_n^2 and σ^2, if we make the additional assumption that the X_i are uncorrelated, then the formula[a]

$$S_n^2 = \frac{1}{n-1}\left[\left(\sum_{i=1}^{n} X_i^2\right) - nM_n^2\right], \tag{6.3}$$

can be used to show that $\mathsf{E}[S_n^2] = \sigma^2$ (Problem 1). In other words, the sample variance S_n^2 is an unbiased estimator of the ensemble variance σ^2. To derive (6.3), write

$$S_n^2 := \frac{1}{n-1} \sum_{i=1}^{n}(X_i - M_n)^2$$

$$= \frac{1}{n-1} \sum_{i=1}^{n}(X_i^2 - 2X_i M_n + M_n^2)$$

$$= \frac{1}{n-1}\left[\left(\sum_{i=1}^{n} X_i^2\right) - 2\left(\sum_{i=1}^{n} X_i\right)M_n + nM_n^2\right]$$

$$= \frac{1}{n-1}\left[\left(\sum_{i=1}^{n} X_i^2\right) - 2(nM_n)M_n + nM_n^2\right]$$

$$= \frac{1}{n-1}\left[\left(\sum_{i=1}^{n} X_i^2\right) - nM_n^2\right].$$

Up to this point we have assumed only that the X_i are uncorrelated. However, to establish the main theoretical results to follow, we need a stronger assumption. **Therefore, in the**

---
[a] For Bernoulli random variables, since $X_i^2 = X_i$, (6.3) simplifies to $S_n^2 = M_n(1-M_n)n/(n-1)$.

rest of this chapter, we make the assumption that the X_i are independent, identically distributed (i.i.d.) with common mean m and common variance σ^2. Then the strong law of large numbers implies[1]

$$\lim_{n \to \infty} M_n = m \quad \text{and} \quad \lim_{n \to \infty} S_n^2 = \sigma^2. \tag{6.4}$$

In other words, for large n, the *random variables* M_n and S_n^2 are close to the *constants* m and σ^2, respectively. When an estimator converges to the desired parameter, the estimator is said to be **strongly consistent**.

Example 6.1. Let $X_1, \ldots, X_n$ be i.i.d. with known mean m, but unknown variance σ^2. Determine whether or not

$$\frac{1}{n} \sum_{i=1}^{n} (X_i - m)^2.$$

is an unbiased estimator of σ^2. Is it strongly consistent?

Solution. To see if the estimator is unbiased, write

$$\mathsf{E}\left[\frac{1}{n} \sum_{i=1}^{n} (X_i - m)^2\right] = \frac{1}{n} \sum_{i=1}^{n} \mathsf{E}[(X_i - m)^2] = \frac{1}{n} \sum_{i=1}^{n} \sigma^2 = \sigma^2.$$

Hence, the proposed formula is an unbiased estimator of σ^2. To assess consistency, write

$$\frac{1}{n} \sum_{i=1}^{n} (X_i - m)^2 = \sum_{i=1}^{n} (X_i^2 - 2X_i m + m^2)$$

$$= \left(\frac{1}{n} \sum_{i=1}^{n} X_i^2\right) - 2m\left(\frac{1}{n} \sum_{i=1}^{n} X_i\right) + \frac{1}{n} \sum_{i=1}^{n} m^2$$

$$= \left(\frac{1}{n} \sum_{i=1}^{n} X_i^2\right) - 2m M_n + m^2.$$

We already know that by the strong law of large numbers,

$$M_n = \frac{1}{n} \sum_{i=1}^{n} X_i \to m.$$

Similarly,

$$\frac{1}{n} \sum_{i=1}^{n} X_i^2 \to \mathsf{E}[X_i^2] = \sigma^2 + m^2,$$

since the X_i^2 are i.i.d. on account of the fact that the X_i are i.i.d. Hence,

$$\lim_{n \to \infty} \frac{1}{n} \sum_{i=1}^{n} (X_i - m)^2 = (\sigma^2 + m^2) - 2m^2 + m^2 = \sigma^2.$$

Thus, the proposed estimator is strongly consistent.

6.1 Parameter estimators and their properties

Example 6.2. Let $X_1,\ldots,X_n$ be i.i.d. with unknown mean m and unknown variance σ^2. Since we do not know m, we cannot use the estimator of the previous example. However, in that estimator, let us replace m by the estimator M_n; i.e., we propose the estimator

$$\widetilde{S}_n^2 := \frac{1}{n}\sum_{i=1}^{n}(X_i - M_n)^2$$

Determine whether or not $\widetilde{S}_n^2$ is an unbiased estimator of σ^2. Is it strongly consistent?

Solution. It is helpful to observe that from (6.2)

$$\widetilde{S}_n^2 = \frac{n-1}{n}S_n^2.$$

Then

$$\mathsf{E}[\widetilde{S}_n^2] = \frac{n-1}{n}\mathsf{E}[S_n^2] = \frac{n-1}{n}\sigma^2 \neq \sigma^2.$$

Thus, $\widetilde{S}_n^2$ is *not* an unbiased estimator of σ^2 (we say that $\widetilde{S}_n^2$ is a **biased estimator** of σ^2). However, since $\mathsf{E}[\widetilde{S}_n^2] \to \sigma^2$, we say that $\widetilde{S}_n^2$ is an **asymptotically unbiased estimator** of σ^2. We also point out that since $S_n^2 \to \sigma^2$,

$$\lim_{n\to\infty}\widetilde{S}_n^2 = \lim_{n\to\infty}\frac{n-1}{n}\lim_{n\to\infty}S_n^2 = 1\cdot\sigma^2 = \sigma^2.$$

Thus, both $\widetilde{S}_n^2$ and S_n^2 are strongly consistent estimators of σ^2, with S_n^2 being unbiased, and $\widetilde{S}_n^2$ being only asymptotically unbiased.

Example 6.3. Let $X_1,\ldots,X_n$ be i.i.d. binomial(N,p) where N is known and p is not known. Find an unbiased, strongly consistent estimator of p.

Solution. Since $\mathsf{E}[X_i] = Np$ (Problem 8 in Chapter 3),

$$p = \frac{\mathsf{E}[X_i]}{N}.$$

This suggests the estimator

$$p_n = \frac{M_n}{N}.$$

The estimator is unbiased because its expectation is $Np/N = p$. The estimator is strongly consistent because $M_n \to Np$ implies $M_n/N \to Np/N = p$.

Example 6.4. Let $X_1,\ldots,X_n$ be i.i.d. $\exp(\lambda)$ where λ is not known. Find a strongly consistent estimator of λ.

Solution. Recall that $\mathsf{E}[X_i] = 1/\lambda$. Rewrite this as

$$\lambda = \frac{1}{\mathsf{E}[X_i]}.$$

This suggests the estimator

$$\lambda_n = \frac{1}{M_n}.$$

Since $M_n \to 1/\lambda$, $\lambda_n \to \lambda$, and we see that the estimator is strongly consistent as required.

6.2 Histograms

In the preceding section, we showed how to estimate various parameters from a collection of i.i.d. data $X_1, \ldots, X_n$. In this section, we show how to estimate the entire probability mass function or density of the X_i.

Given data $X_1, \ldots, X_n$, we create a **histogram** as follows. We first select m intervals called **bins**, denoted by $[e_j, e_{j+1})$, where[b]

$$e_1 < \cdots < e_{m+1},$$

and e_1 and e_{m+1} satisfy

$$e_1 \leq \min_i X_i \quad \text{and} \quad \max_i X_i \leq e_{m+1}.$$

When $\max_i X_i = e_{m+1}$, we use the interval $[e_m, e_{m+1}]$ instead of $[e_m, e_{m+1})$ so that no data is lost. The sequence e is called the **edge sequence**. Notice that the number of edges is equal to one plus the number of bins. The **histogram count** for bin j is

$$H_j := \sum_{i=1}^n I_{[e_j, e_{j+1})}(X_i).$$

In other words, H_j is the number of data samples X_i that lie in bin j; i.e., the number of data samples X_i that satisfy $e_j \leq X_i < e_{j+1}$.

For each j, the term $I_{[e_j, e_{j+1})}(X_i)$ takes only the values zero and one. It is therefore a Bernoulli random variable with parameter equal to its expectation,

$$\mathsf{E}[I_{[e_j, e_{j+1})}(X_i)] = \mathsf{P}(e_j \leq X_i < e_{j+1}). \qquad (6.5)$$

We assume that the X_i are i.i.d. to guarantee that for each j, the $I_{[e_j, e_{j+1})}(X_i)$ are also i.i.d. Since H_j/n is just the sample mean of the $I_{[e_j, e_{j+1})}(X_i)$, we have from the discussion in Section 6.1 that H_j/n converges to (6.5); i.e., for large n,

$$\frac{H_j}{n} \approx \mathsf{P}(e_j \leq X_i < e_{j+1}). \qquad (6.6)$$

If the X_i are integer-valued random variables, it is convenient to use bins centered on the integers, e.g., $e_j = j - 1/2$ and $e_{j+1} = j + 1/2$. Then

$$\frac{H_j}{n} \approx \mathsf{P}(X_i = j).$$

Such an edge sequence and histogram counts are easily constructed in MATLAB with the commands

---

[b] We have used intervals of the form $[a, b)$ because this is the form used by the MATLAB function `histc`.

6.2 Histograms

```
e = [min(X):max(X)+1]-0.5;
H = histc(X,e);
```

where X is a previously defined vector of integer-valued data. To plot the histogram H normalized by the sample size n, use the commands

```
n = length(X);
nbins = length(e) - 1;
bin_centers = [min(X):max(X)];
bar(bin_centers,H(1:nbins)/n,'hist')
```

Remark. The reason we have to write H(1:nbins) instead of just H is that the vector returned by histc has the length of e, which is one plus the number of bins.

After we plot the normalized histogram, it is convenient to overlay it with the corresponding probability mass function. For example, if we know the data is i.i.d. binomial(10, 0.3), we can use the function binpmf defined in Problem 27 of Chapter 5 to compute the binomial pmf.

```
hold on                     % prevent erasure of last plot
k = [0:10];                 % range for plotting pmf
prob = binpmf(k,10,0.3);    % compute binomial(10,0.3) pmf
stem(k,prob,'filled')       % make stem plot of pmf
```

In a real situation, even if we have good reasons for believing the data is, say, binomial $(10, p)$, we do not know p. In this case, we can use the estimator developed in Example 6.3. The MATLAB commands for doing this are

```
Mn = mean(X);
pn = Mn/10;
```

Example 6.5. Let us apply the above procedure to a sequence of $n = 1000$ binomial$(10, 0.3)$ random variables. We use the function bernrnd defined in Section 5.2 to generate an array of Bernoulli random variables. The sum of each row gives one binomial random variable.

```
n = 1000;                         % sample size
Bernmat = bernrnd(0.3,n,10);      % generate n binomial
X = sum(Bernmat');                % random numbers in X
minX = min(X);                    % save to avoid re-
maxX = max(X);                    % computing min & max
e = [minX:maxX+1]-0.5;
H = histc(X,e);
nbins = length(e) - 1;
bin_centers = [minX:maxX];
bar(bin_centers,H(1:nbins)/n,'hist')
hold on
k = [0:10];                       % range of pmf
Mn = mean(X);
pn = Mn/10;                       % estimate p
prob = binpmf(k,10,pn);           % pmf w/ estimated p
stem(k,prob,'filled')
```

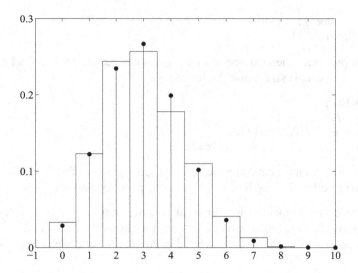

Figure 6.1. Normalized histogram of 1000 i.i.d. binomial(10,0.3) random numbers. Stem plot shows pmf using $p_n = 0.2989$ estimated from the data.

```
fprintf('Mn = %g   pn = %g\n',Mn,pn)
hold off
```

The command `hold off` allows the next run to erase the current figure. The plot is shown in Figure 6.1. Notice that our particular realization of $X_1, \ldots, X_n$ did not have any occurrences of $X_i = 9$ or $X_i = 10$. This is not surprising since with $N = 10$ and $p = 0.3$, the probability that $X_i = 10$ is $0.3^{10} = 6 \times 10^{-6}$, while we used only 1000 samples.

For continuous random variables X_i, the edge sequence, histogram counts, and bin centers are computed in MATLAB as follows, assuming that X and nbins have already been defined.

```
minX = min(X);
maxX = max(X);
e = linspace(minX,maxX,nbins+1);
H = histc(X,e);
H(nbins) = H(nbins)+H(nbins+1);   % explained below
H = H(1:nbins);                    % resize H
bw = (maxX-minX)/nbins;            % bin width
a = e(1:nbins);                    % left  edge sequence
b = e(2:nbins+1);                  % right edge sequence
bin_centers = (a+b)/2;             % bin centers
```

Since we have set e(nbins+1) = max(X), histc will not count the value of max(X) as belonging to the interval [e(nbins),e(nbins+1)). What histc does is use H(nbins+1) is to count how many elements of X are exactly equal to e(nbins+1). Hence, we add H(nbins+1) to H(nbins) to get the correct count for [e(nbins),

6.2 Histograms

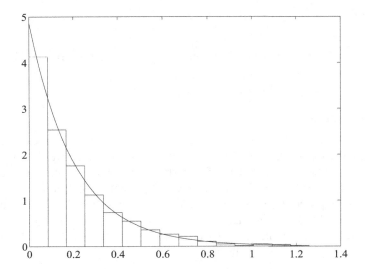

Figure 6.2. Normalized histogram of 1000 i.i.d. exponential random numbers and the exp(λ) density with the value of λ estimated from the data.

e(nbins+1)]. For later use, note that a(j) is the left edge of bin j and b(j) is the right edge of bin j.

We normalize the histogram as follows. If c_j is the center of bin j, and Δx_j is the bin width, then we want

$$\frac{H_j}{n} \approx \mathsf{P}(e_j \leq X_i < e_{j+1}) = \int_{e_j}^{e_{j+1}} f_X(x)\,dx \approx f_X(c_j)\Delta x_j.$$

So, if we are planning to draw a density function over the histogram, we should plot the normalized values $H_j/(n\Delta x_j)$. The appropriate MATLAB commands are

```
n = length(X);
bar(bin_centers,H/(bw*n),'hist')
```

Example 6.6. You are given $n = 1000$ i.i.d. measurements $X_1,\ldots,X_n$ stored in a MATLAB vector X. You plot the normalized histogram shown in Figure 6.2, and you believe the X_i to be exponential random variables with unknown parameter λ. Use MATLAB to compute the estimator in Example 6.4 to estimate λ. Then plot the corresponding density over the histogram.

Solution. The estimator in Example 6.4 was $\lambda_n = 1/M_n$. The following MATLAB code performs the desired task assuming X is already given and the normalized histogram already plotted.

```
hold on
Mn = mean(X);
lambdan = 1/Mn;
t = linspace(minX,maxX,150); % range to plot pdf
plot(t,lambdan*exp(-lambdan*t))
```

```
fprintf('Mn = %g   lambdan = %g\n',Mn,lambdan)
hold off
```

The histogram and density are shown in Figure 6.2. The value of Mn was 0.206 and the value of lambdan was 4.86.

*The chi-squared test

So far we have been plotting histograms, and by subjective observation selected a pmf or density that would overlay the histogram nicely, in our opinion. Here we develop objective criteria, which are known as **goodness-of-fit tests**. The first step is to select a candidate pmf or density that we subjectively think might be a good fit to the data. We refer to this candidate as the **hypothesis**. Then we compute a statistic, call it Z, and compare it to a threshold, call it z_α. If $Z \leq z_\alpha$, we agree that our hypothesis is a reasonable fit to the data. If $Z > z_\alpha$, we reject our hypothesis and try another one. The threshold z_α is chosen so that if the data actually is i.i.d. p_X or f_X, then

$$P(Z > z_\alpha) = \alpha.$$

Thus α is the probability of rejecting the hypothesis when it is actually correct. Hence, α is usually taken to be a small number such as 0.01 or 0.05. We call α the **significance level**, and specify it as a percentage, e.g., 1% or 5%.

The **chi-squared test**[c] is based on the histogram. The first step is to use the hypothesized pmf or density to compute

$$p_j := P(e_j \leq X_i < e_{j+1}).$$

If our candidate pmf or density is a good one, then (6.6) tells us that

$$\sum_{j=1}^{m} |H_j - np_j|^2$$

should be small. However, as we shall see, it is advantageous to use the normalized statistic

$$Z := \sum_{j=1}^{m} \frac{|H_j - np_j|^2}{np_j}.$$

This normalization is motivated by the fact that (see Problem 16)

$$\frac{H_j - np_j}{\sqrt{np_j}}$$

has zero mean and variance $1 - p_j$ for all n and m, which implies $E[Z] = m - 1$ for all n.

Example 6.7. In Example 6.5, we plotted a histogram and overlayed a binomial(10, 0.2989) pmf, where 0.2989 was the estimated from the data. Compute the statistic Z.

Solution. This is easily done in MATLAB.

---

[c]Other popular tests include the **Kolmogorov–Smirnov test** and the **Anderson–Darling test**. However, they apply only to continuous cdfs.

```
p = binpmf(bin_centers,10,0.2989);
Z = sum((H(1:nbins)-n*p).^2./(n*p))
```

We found that $Z = 7.01$.

Example 6.8. In Example 6.6, we plotted a histogram and fitted an $\exp(\lambda)$ density. The estimated value of λ was 4.86. Compute the statistic Z under this hypothesis.

Solution. Recall that for continuous random variables,

$$p_j = \mathsf{P}(e_j \leq X_i < e_{j+1}) = \int_{e_j}^{e_{j+1}} f_X(x)\,dx = F_X(e_{j+1}) - F_X(e_j).$$

Recalling the left and right edge sequences defined earlier, we can also write

$$p_j = \mathsf{P}(e_j \leq X_i < e_{j+1}) = \int_{a_j}^{b_j} f_X(x)\,dx = F_X(b_j) - F_X(a_j).$$

Since the cdf of $X \sim \exp(\lambda)$ is $F_X(x) = 1 - e^{-\lambda x}$, we first create an M-file containing the MATLAB function F to compute $F_X(x)$.

```
function y = F(x)
y = 1 - exp(-4.86*x);
```

The following MATLAB commands compute Z.

```
p = F(b) - F(a);
Z = sum((H-n*p).^2./(n*p))
```

We found that $Z = 8.33$.

The only remaining problem is to choose α and to find the threshold z_α, called the **critical value** of the test. It turns out that for large n, the cdf of Z is approximately that of a chi-squared random variable (Problem 15(d) in Chapter 4) with $m - 1$ degrees of freedom [3, p. 386, Problem 29.8]. However, if you use r estimated parameters, then Z has only $m - 1 - r$ degrees of freedom [12], [47, pp. 205–206].[d] Hence, to solve

$$\mathsf{P}(Z > z_\alpha) = \alpha$$

for z_α we must solve $1 - F_Z(z_\alpha) = \alpha$ or $F_Z(z_\alpha) = 1 - \alpha$. This can be done by applying a root-finding algorithm to the equation $F_Z(z_\alpha) - 1 + \alpha = 0$ or in MATLAB with the command `chi2inv(1-alpha,k)`, where k is the number of degrees of freedom of Z. Some solutions are also shown in Table 6.1.

Example 6.9. Find z_α for $\alpha = 0.05$ in Examples 6.5 and 6.6.

Solution. In the first case, the number of degrees of freedom is $m - 1 - 1$, where m is the number of bins, which from Figure 6.1, is 9, and the extra 1 is subtracted because

---

[d]The basic Kolmogorov–Smirnov test does not account for using estimated parameters. If estimated parameters are used, the critical value z_α must be determined by simulation [47, p. 208].

k	$z_\alpha(\alpha = 5\%)$	$z_\alpha(\alpha = 1\%)$
1	3.841	6.635
2	5.991	9.210
3	7.815	11.345
4	9.488	13.277
5	11.070	15.086
6	12.592	16.812
7	14.067	18.475
8	15.507	20.090
9	16.919	21.666
10	18.307	23.209
11	19.675	24.725
12	21.026	26.217
13	22.362	27.688
14	23.685	29.141
15	24.996	30.578
16	26.296	32.000
17	27.587	33.409
18	28.869	34.805
19	30.144	36.191
20	31.410	37.566

Table 6.1. Thresholds z_α for the chi-squared test with k degrees of freedom and significance levels $\alpha = 5\%$ and $\alpha = 1\%$.

we used one estimated parameter. Hence, $k = 7$, and from Table 6.1, $z_\alpha = 14.067$. From Example 6.7, the statistic Z was 7.01, which is well below the threshold. We conclude that the binomial$(10, 0.2989)$ is a good fit to the data. In the second case, the number of degrees of freedom is $m - 1 - 1$, where m is the number of bins, and the extra 1 is subtracted because we estimated one parameter. From Figure 6.2, $m = 15$. Hence, $k = 13$, and from Table 6.1, $z_\alpha = 22.362$. From Example 6.8, Z was 8.33. Since Z is much smaller than the threshold, we conclude that the exp(4.86) density is a good fit to the data.

6.3 Confidence intervals for the mean – known variance

As noted in Section 6.1, the sample mean M_n converges to the population mean m as $n \to \infty$. However, in practice, n is finite, and we would like to say something about how close the random variable M_n is to the unknown constant m. One way to do this is with a confidence interval. For theory purposes, we write

$$P(m \in [M_n - \delta, M_n + \delta]) = 1 - \alpha, \tag{6.7}$$

where $[M_n - \delta, M_n + \delta]$ is the **confidence interval**, and $1 - \alpha$ is the **confidence level**. Thus, a confidence interval is a random set, and the confidence level $1 - \alpha$ is the probability that the random set contains the unknown parameter m. Commonly used values of $1 - \alpha$ range

6.3 Confidence intervals for the mean – known variance

from 0.90 to 0.99. In applications, we usually write (6.7) in the form

$$m = M_n \pm \delta \quad \text{with } 100(1-\alpha)\% \text{ probability.}$$

The next problem we consider is how to choose δ so that equation (6.7) holds.[e] From (6.7), the left-hand side depends on $M_n = (X_1 + \cdots + X_n)/n$. Hence, the first step would be to find the pmf or density of the sum of i.i.d. random variables. This can only be done in special cases (e.g., Problem 55 in Chapter 4). We need a more general approach. To proceed, we first rewrite the condition $m \in [M_n - \delta, M_n + \delta]$ as $|M_n - m| \leq \delta$. To see that these conditions are equivalent, observe that $m \in [M_n - \delta, M_n + \delta]$ if and only if

$$M_n - \delta \leq m \leq M_n + \delta.$$

Multiplying through by -1 yields

$$-M_n + \delta \geq -m \geq -M_n - \delta,$$

from which we get

$$\delta \geq M_n - m \geq -\delta.$$

This is more compactly rewritten as[f]

$$|M_n - m| \leq \delta.$$

It follows that the left-hand side of (6.7) is equal to

$$\mathsf{P}(|M_n - m| \leq \delta). \tag{6.8}$$

Now take

$$\delta = \sigma y/\sqrt{n}$$

so that (6.8) becomes

$$\mathsf{P}\!\left(|M_n - m| \leq \frac{\sigma y}{\sqrt{n}}\right),$$

or

$$\mathsf{P}\!\left(\left|\frac{M_n - m}{\sigma/\sqrt{n}}\right| \leq y\right). \tag{6.9}$$

Setting

$$Y_n := \frac{M_n - m}{\sigma/\sqrt{n}},$$

we have[2]

$$\mathsf{P}\!\left(\left|\frac{M_n - m}{\sigma/\sqrt{n}}\right| \leq y\right) = \mathsf{P}(|Y_n| \leq y) = \mathsf{P}(-y \leq Y_n \leq y) = F_{Y_n}(y) - F_{Y_n}(-y).$$

---

[e] Alternatively, one could specify δ and then compute $1 - \alpha$.
[f] To see that $|t| \leq \delta$ is equivalent to $-\delta \leq t \leq \delta$, consider separately the two cases $t \geq 0$ and $t < 0$, and note that for $t < 0$, $|t| = -t$.

By the **central limit theorem** (Section 5.6),[g] $F_{Y_n}(y) \to \Phi(y)$, where Φ is the standard normal cdf,

$$\Phi(y) = \frac{1}{\sqrt{2\pi}} \int_{-\infty}^{y} e^{-t^2/2}\, dt.$$

Thus, for large n,

$$\mathsf{P}\left(\left|\frac{M_n - m}{\sigma/\sqrt{n}}\right| \le y\right) \approx \Phi(y) - \Phi(-y) = 2\Phi(y) - 1, \qquad (6.10)$$

where the last step uses the fact that the standard normal density is even (Problem 17). The importance of this formula is that if we want the left-hand side to be $1 - \alpha$, all we have to do is solve for y in the equation

$$1 - \alpha = 2\Phi(y) - 1,$$

or

$$\Phi(y) = 1 - \alpha/2.$$

Notice that this equation does not depend on n or on the pmf or density of the X_i! We denote the solution of this equation by $y_{\alpha/2}$. It can be found from tables, e.g., Table 6.2, or numerically by finding the unique root of the equation $\Phi(y) + \alpha/2 - 1 = 0$, or in MATLAB by $y_{\alpha/2} = \mathtt{norminv(1 - alpha/2)}$.[h]

We now summarize the procedure. Fix a confidence level $1 - \alpha$. Find the corresponding $y_{\alpha/2}$ from Table 6.2. Then write

$$m = M_n \pm \frac{\sigma y_{\alpha/2}}{\sqrt{n}} \quad \text{with } 100(1-\alpha)\% \text{ probability,} \qquad (6.11)$$

and the corresponding confidence interval is

$$\left[M_n - \frac{\sigma y_{\alpha/2}}{\sqrt{n}},\, M_n + \frac{\sigma y_{\alpha/2}}{\sqrt{n}}\right]. \qquad (6.12)$$

Example 6.10. Let $X_1, X_2, \ldots$ be i.i.d. random variables with variance $\sigma^2 = 2$. If $M_{100} = 7.129$, find the 93 and 97% confidence intervals for the population mean.

Solution. In Table 6.2 we scan the $1 - \alpha$ column until we find 0.93. The corresponding value of $y_{\alpha/2}$ is 1.812. Since $y_{\alpha/2}\sigma/\sqrt{n} = 1.812\sqrt{2}/\sqrt{100} = 0.256$, we write

$$m = 7.129 \pm 0.256 \text{ with 93\% probability,}$$

and the corresponding confidence interval is $[6.873, 7.385]$.

---

[g]The reader should verify that Y_n as defined here is equal to Y_n as defined in (5.6).

[h]Since Φ can be related to the error function erf (see Note **1** in Chapter 5), $y_{\alpha/2}$ can also be found using the inverse of the error function. Hence, $y_{\alpha/2} = \sqrt{2}\,\mathrm{erf}^{-1}(1-\alpha)$. The MATLAB command for $\mathrm{erf}^{-1}(z)$ is $\mathtt{erfinv(z)}$.

6.4 Confidence intervals for the mean – unknown variance

$1-\alpha$	$y_{\alpha/2}$
0.90	1.645
0.91	1.695
0.92	1.751
0.93	1.812
0.94	1.881
0.95	1.960
0.96	2.054
0.97	2.170
0.98	2.326
0.99	2.576

Table 6.2. Confidence levels $1-\alpha$ and corresponding $y_{\alpha/2}$ such that $2\Phi(y_{\alpha/2}) - 1 = 1 - \alpha$.

For the 97% confidence interval, we use $y_{\alpha/2} = 2.170$. Then $y_{\alpha/2}\sigma/\sqrt{n}$ is equal to $2.170\sqrt{2}/\sqrt{100} = 0.307$, and we write

$$m = 7.129 \pm 0.307 \text{ with 97\% probability,}$$

and the corresponding confidence interval is $[6.822, 7.436]$.

This example illustrates the general result that if we want more confidence, we have to use a wider interval. Equivalently, if we use a smaller interval, we will be less confident that it contains the unknown parameter. Mathematically, the width of the confidence interval is

$$2 \cdot \frac{\sigma y_{\alpha/2}}{\sqrt{n}}.$$

From Table 6.2, we see that as $1-\alpha$ increases, so does $y_{\alpha/2}$, and hence the width of the confidence interval. It should also be noted that for fixed $1-\alpha$, we can reduce the width of the confidence interval by increasing n; i.e., taking more measurements.

6.4 Confidence intervals for the mean – unknown variance

In practice, we usually do not know σ. Hence, we replace it by S_n. To justify this, first observe that the argument showing the left-hand side of (6.7) is equal to (6.8) can be carried out with $\delta = S_n y_{\alpha/2}/\sqrt{n}$. Hence,

$$\mathsf{P}\left(m \in \left[M_n - \frac{S_n y_{\alpha/2}}{\sqrt{n}}, M_n + \frac{S_n y_{\alpha/2}}{\sqrt{n}}\right]\right)$$

is equal to

$$\mathsf{P}\left(|M_n - m| < \frac{S_n y_{\alpha/2}}{\sqrt{n}}\right),$$

Now observe that this is equal to

$$\mathsf{P}\left(\left|\frac{M_n - m}{\sigma/\sqrt{n}}\right| \le \frac{S_n y_{\alpha/2}}{\sigma}\right). \tag{6.13}$$

Recalling our definition $Y_n := (M_n - m)/(\sigma/\sqrt{n})$, the above probability is

$$P\left(|Y_n| \leq \frac{S_n y_{\alpha/2}}{\sigma}\right).$$

Since $S_n \to \sigma$, the ratio $S_n/\sigma \to 1$. This suggests[3] that for large n

$$P\left(|Y_n| \leq \frac{S_n y_{\alpha/2}}{\sigma}\right) \approx P(|Y_n| \leq y_{\alpha/2}) \qquad (6.14)$$
$$\approx 2\Phi(y_{\alpha/2}) - 1, \quad \text{by the central limit theorem,}$$
$$= 1 - \alpha, \quad \text{by the definition of } y_{\alpha/2}. \qquad (6.15)$$

We now summarize the new procedure. Fix a confidence level $1 - \alpha$. Find the corresponding $y_{\alpha/2}$ from Table 6.2. Then write

$$m = M_n \pm \frac{S_n y_{\alpha/2}}{\sqrt{n}} \quad \text{with } 100(1-\alpha)\% \text{ probability}, \qquad (6.16)$$

and the corresponding confidence interval is

$$\left[M_n - \frac{S_n y_{\alpha/2}}{\sqrt{n}}, M_n + \frac{S_n y_{\alpha/2}}{\sqrt{n}}\right]. \qquad (6.17)$$

Example 6.11. Let $X_1, X_2, \ldots$ be i.i.d. Bernoulli(p) random variables. Find the 95% confidence interval for p if $M_{100} = 0.28$ and $S_{100} = 0.451$.

Solution. Observe that since $m := \mathsf{E}[X_i] = p$, we can use M_n to estimate p. From Table 6.2, $y_{\alpha/2} = 1.960$, $S_{100} y_{\alpha/2}/\sqrt{100} = 0.088$, and

$$p = 0.28 \pm 0.088 \text{ with 95\% probability.}$$

The corresponding confidence interval is $[0.192, 0.368]$.

Applications

Estimating the number of defective products in a lot. Consider a production run of N cellular phones, of which, say d are defective. The only way to determine d exactly and for certain is to test every phone. This is not practical if N is large. So we consider the following procedure to estimate the fraction of defectives, $p := d/N$, based on testing only n phones, where n is large, but smaller than N.

```
FOR i = 1 TO n
  Select a phone at random from the lot of N phones;
  IF the ith phone selected is defective
    LET X_i = 1;
  ELSE
    LET X_i = 0;
```

6.4 Confidence intervals for the mean – unknown variance

END IF
Return the phone to the lot;
END FOR

Because phones are returned to the lot (sampling with replacement), it is possible to test the same phone more than once. However, because the phones are always chosen from the same set of N phones, the X_i are i.i.d. with $P(X_i = 1) = d/N = p$. Hence, the central limit theorem applies, and we can use the method of Example 6.11 to estimate p and $d = Np$. For example, if $N = 1000$ and we use the numbers from Example 6.11, we would estimate that

$$d = 280 \pm 88 \text{ with 95\% probability.}$$

In other words, we are 95% sure that the number of defectives is between 192 and 368 for this particular lot of 1000 phones.

If the phones were not returned to the lot after testing (sampling without replacement), the X_i would not be i.i.d. as required by the central limit theorem. However, in sampling with replacement when n is much smaller than N, the chances of testing the same phone twice are negligible. Hence, we can actually sample without replacement and proceed as above.

Predicting the outcome of an election. In order to predict the outcome of a presidential election, 4000 registered voters are surveyed at random. In total, 2104 (more than half) say they will vote for candidate A, and the rest say they will vote for candidate B. To predict the outcome of the election, let p be the fraction of votes actually received by candidate A out of the total number of voters N (millions). Our poll samples $n = 4000$, and $M_{4000} = 2104/4000 = 0.526$. Suppose that $S_{4000} = 0.499$. For a 95% confidence interval for p, $y_{\alpha/2} = 1.960$, $S_{4000} y_{\alpha/2}/\sqrt{4000} = 0.015$, and

$$p = 0.526 \pm 0.015 \text{ with 95\% probability.}$$

Rounding off, we would predict that candidate A will receive 53% of the vote, with a margin of error of 2%. Thus, we are 95% sure that candidate A will win the election.

Sampling with and without replacement

Consider sampling n items from a batch of N items, d of which are defective. If we sample with replacement, then the theory above worked out rather simply. We also argued briefly that if n is much smaller than N, then sampling without replacement would give essentially the same results. We now make this statement more precise.

To begin, recall that the central limit theorem says that for large n, $F_{Y_n}(y) \approx \Phi(y)$, where $Y_n = (M_n - m)/(\sigma/\sqrt{n})$, and $M_n = (1/n)\sum_{i=1}^n X_i$. If we sample with replacement and set $X_i = 1$ if the ith item is defective, then the X_i are i.i.d. Bernoulli(p) with $p = d/N$. When $X_1, X_2, \ldots$ are i.i.d. Bernoulli(p), we know from Section 3.2 that $\sum_{i=1}^n X_i$ is binomial(n,p). Putting this all together, we obtain the **de Moivre–Laplace theorem**, which says that if $V \sim$ binomial(n,p) and n is large, then the cdf of $(V/n - p)/\sqrt{p(1-p)/n}$ is approximately standard normal.

Now suppose we sample n items without replacement. Let U denote the number of defectives out of the n samples. It was shown in Example 1.41 that U is a **hypergeometric**(N,

$d, n)$ random variable with pmf[i]

$$P(U = k) = \frac{\binom{d}{k}\binom{N-d}{n-k}}{\binom{N}{n}}, \quad k = 0, \ldots, n.$$

In the next paragraph we show that if n is much smaller than d, $N-d$, and N, then $P(U = k) \approx P(V = k)$. It then follows that the cdf of $(U/n - p)/\sqrt{p(1-p)/n}$ is close to the cdf of $(V/n - p)/\sqrt{p(1-p)/n}$, which is close to the standard normal cdf if n is large. (Thus, to make it all work we need n large, but still much smaller than d, $N-d$, and N.)

To show that $P(U = k) \approx P(V = k)$, write out $P(U = k)$ as

$$\frac{d!}{k!(d-k)!} \cdot \frac{(N-d)!}{(n-k)![(N-d)-(n-k)]!} \cdot \frac{n!(N-n)!}{N!}.$$

We can easily identify the factor $\binom{n}{k}$. Next, since $0 \le k \le n \ll d$,

$$\frac{d!}{(d-k)!} = d(d-1)\cdots(d-k+1) \approx d^k.$$

Similarly, since $0 \le k \le n \ll (N-d)$,

$$\frac{(N-d)!}{[(N-d)-(n-k)]!} = (N-d)\cdots[(N-d)-(n-k)+1] \approx (N-d)^{n-k}.$$

Finally, since $n \ll N$,

$$\frac{(N-n)!}{N!} = \frac{1}{N(N-1)\cdots(N-n+1)} \approx \frac{1}{N^n}.$$

Writing $p = d/N$, we have

$$P(U = k) \approx \binom{n}{k} p^k (1-p)^{n-k}.$$

6.5 Confidence intervals for Gaussian data

In this section we assume that $X_1, X_2, \ldots$ are i.i.d. $N(m, \sigma^2)$.

Estimating the mean

If the X_i are i.i.d. $N(m, \sigma^2)$, then by Problem 30, $Y_n = (M_n - m)/(\sigma/\sqrt{n})$ is $N(0, 1)$. Hence, the analysis in Sections 6.3 shows that

$$P\left(m \in \left[M_n - \frac{\sigma y}{\sqrt{n}}, M_n + \frac{\sigma y}{\sqrt{n}}\right]\right) = P(|Y_n| \le y)$$
$$= 2\Phi(y) - 1.$$

The point is that for normal data there is no central limit theorem approximation. Hence, we can determine confidence intervals as in Section 6.3 even if n is not large.

---

[i]See the Notes[4] for an alternative derivation using the law of total probability.

6.5 Confidence intervals for Gaussian data

Example 6.12. Let $X_1, X_2, \ldots$ be i.i.d. $N(m,2)$. If $M_{10} = 5.287$, find the 90% confidence interval for m.

Solution. From Table 6.2 for $1 - \alpha = 0.90$, $y_{\alpha/2}$ is 1.645. We then have $y_{\alpha/2}\sigma/\sqrt{10} = 1.645\sqrt{2}/\sqrt{10} = 0.736$,

$$m = 5.287 \pm 0.736 \text{ with 90\% probability,}$$

and the corresponding confidence interval is $[4.551, 6.023]$.

Unfortunately, formula (6.14) in Section 6.4 still involves the approximation $S_n/\sigma \approx 1$ even if the X_i are normal. However, let us rewrite (6.13) as

$$\mathsf{P}\left(\left|\frac{M_n - m}{S_n/\sqrt{n}}\right| \leq y_{\alpha/2}\right),$$

and put

$$T := \frac{M_n - m}{S_n/\sqrt{n}}.$$

As shown later, if the X_i are i.i.d. $N(m, \sigma^2)$, then T has Student's t density with $\nu = n - 1$ degrees of freedom (defined in Problem 20 in Chapter 4). To compute $100(1 - \alpha)\%$ confidence intervals, we must solve

$$\mathsf{P}(|T| \leq y) = 1 - \alpha$$

or

$$F_T(y) - F_T(-y) = 1 - \alpha.$$

Since the density f_T is even, $F_T(-y) = 1 - F_T(y)$, and we must solve

$$2F_T(y) - 1 = 1 - \alpha,$$

or $F_T(y) = 1 - \alpha/2$. This can be solved using tables, e.g., Table 6.3, or numerically by finding the unique root of the equation $F_T(y) + \alpha/2 - 1 = 0$, or in MATLAB by $y_{\alpha/2} = \texttt{tinv}(1-\texttt{alpha}/2,\texttt{n}-1)$.

Example 6.13. Let $X_1, X_2, \ldots$ be i.i.d. $N(m, \sigma^2)$ random variables, and suppose $M_{10} = 5.287$. Further suppose that $S_{10} = 1.564$. Find the 90% confidence interval for m.

Solution. In Table 6.3 with $n = 10$, we see that for $1 - \alpha = 0.90$, $y_{\alpha/2}$ is 1.833. Since $S_{10} y_{\alpha/2}/\sqrt{10} = 0.907$,

$$m = 5.287 \pm 0.907 \text{ with 90\% probability.}$$

The corresponding confidence interval is $[4.380, 6.194]$.

$1-\alpha$	$y_{\alpha/2}(n=10)$		$1-\alpha$	$y_{\alpha/2}(n=100)$
0.90	1.833		0.90	1.660
0.91	1.899		0.91	1.712
0.92	1.973		0.92	1.769
0.93	2.055		0.93	1.832
0.94	2.150		0.94	1.903
0.95	2.262		0.95	1.984
0.96	2.398		0.96	2.081
0.97	2.574		0.97	2.202
0.98	2.821		0.98	2.365
0.99	3.250		0.99	2.626

Table 6.3. Confidence levels $1-\alpha$ and corresponding $y_{\alpha/2}$ such that $P(|T| \le y_{\alpha/2}) = 1-\alpha$. The left-hand table is for $n=10$ observations with T having $n-1=9$ degrees of freedom, and the right-hand table is for $n=100$ observations with T having $n-1=99$ degrees of freedom.

Limiting t distribution

If we compare the $n=100$ table in Table 6.3 with Table 6.2, we see they are almost the same. This is a consequence of the fact that as n increases, the t cdf converges to the standard normal cdf. We can see this by writing

$$\begin{aligned} P(T \le t) &= P\left(\frac{M_n - m}{S_n/\sqrt{n}} \le t\right) \\ &= P\left(\frac{M_n - m}{\sigma/\sqrt{n}} \le \frac{S_n}{\sigma}t\right) \\ &= P\left(Y_n \le \frac{S_n}{\sigma}t\right) \\ &\approx P(Y_n \le t), \end{aligned}$$

since[3] S_n converges to σ. Finally, since the X_i are independent and normal, $F_{Y_n}(t) = \Phi(t)$.

We also recall from Problem 21 in Chapter 4 and Figure 4.12 there that the t density converges to the standard normal density.

Estimating the variance – known mean

Suppose that $X_1, X_2, \ldots$ are i.i.d. $N(m, \sigma^2)$ with m known but σ^2 unknown. We use

$$V_n^2 := \frac{1}{n}\sum_{i=1}^{n}(X_i - m)^2 \tag{6.18}$$

as our estimator of the variance σ^2. It is easy to see that V_n^2 is a strongly consistent estimator of σ^2.

For determining confidence intervals, it is easier to work with

$$\frac{n}{\sigma^2}V_n^2 = \sum_{i=1}^{n}\left(\frac{X_i - m}{\sigma}\right)^2.$$

6.5 Confidence intervals for Gaussian data

$1-\alpha$	ℓ	u
0.90	77.929	124.342
0.91	77.326	125.170
0.92	76.671	126.079
0.93	75.949	127.092
0.94	75.142	128.237
0.95	74.222	129.561
0.96	73.142	131.142
0.97	71.818	133.120
0.98	70.065	135.807
0.99	67.328	140.169

Table 6.4. Confidence levels $1-\alpha$ and corresponding values of ℓ and u such that $\mathsf{P}(\ell \le nV_n^2/\sigma^2 \le u) = 1-\alpha$ and such that $\mathsf{P}(nV_n^2/\sigma^2 \le \ell) = \mathsf{P}(nV_n^2/\sigma^2 \ge u) = \alpha/2$ for $n = 100$ observations.

Since $(X_i - m)/\sigma$ is $N(0,1)$, its square is chi-squared with one degree of freedom (Problem 46 in Chapter 4 or Problem 11 in Chapter 5). It then follows that nV_n^2/σ^2 is chi-squared with n degrees of freedom (see Problem 55(c) and its Remark in Chapter 4).

Choose $0 < \ell < u$, and consider the equation

$$\mathsf{P}\left(\ell \le \frac{n}{\sigma^2}V_n^2 \le u\right) = 1-\alpha.$$

We can rewrite this as

$$\mathsf{P}\left(\frac{nV_n^2}{\ell} \ge \sigma^2 \ge \frac{nV_n^2}{u}\right) = 1-\alpha.$$

This suggests the confidence interval

$$\left[\frac{nV_n^2}{u}, \frac{nV_n^2}{\ell}\right]. \tag{6.19}$$

Then the probability that σ^2 lies in this interval is

$$F(u) - F(\ell) = 1-\alpha,$$

where F is the chi-squared cdf with n degrees of freedom. We usually choose ℓ and u to solve

$$F(\ell) = \alpha/2 \quad \text{and} \quad F(u) = 1-\alpha/2.$$

These equations can be solved using tables, e.g., Table 6.4, or numerically by root finding, or in MATLAB with the commands

$$\ell = \texttt{chi2inv(alpha/2,n)} \quad \text{and} \quad u = \texttt{chi2inv(1-alpha/2,n)}.$$

Example 6.14. Let $X_1, X_2, \ldots$ be i.i.d. $N(5, \sigma^2)$ random variables. Suppose that $V_{100}^2 = 1.645$. Find the 90% confidence interval for σ^2.

Solution. From Table 6.4 we see that for $1-\alpha = 0.90$, $\ell = 77.929$ and $u = 124.342$. The 90% confidence interval is

$$\left[\frac{100(1.645)}{124.342}, \frac{100(1.645)}{77.929}\right] = [1.323, 2.111].$$

Estimating the variance – unknown mean

Let $X_1, X_2, \ldots$ be i.i.d. $N(m, \sigma^2)$, where both the mean and the variance are unknown, but we are interested only in estimating the variance. Since we do not know m, we cannot use the estimator V_n^2 above. Instead we use S_n^2. However, for determining confidence intervals, it is easier to work with $((n-1)/\sigma^2)S_n^2$. As argued below, $(n-1)S_n^2/\sigma^2$ is a chi-squared random variable with $n-1$ degrees of freedom.

Choose $0 < \ell < u$, and consider the equation

$$\mathsf{P}\left(\ell \leq \frac{n-1}{\sigma^2}S_n^2 \leq u\right) = 1 - \alpha.$$

We can rewrite this as

$$\mathsf{P}\left(\frac{(n-1)S_n^2}{\ell} \geq \sigma^2 \geq \frac{(n-1)S_n^2}{u}\right) = 1 - \alpha.$$

This suggests the confidence interval

$$\left[\frac{(n-1)S_n^2}{u}, \frac{(n-1)S_n^2}{\ell}\right]. \tag{6.20}$$

Then the probability that σ^2 lies in this interval is

$$F(u) - F(\ell) = 1 - \alpha,$$

where now F is the chi-squared cdf with $n-1$ degrees of freedom. We usually choose ℓ and u to solve

$$F(\ell) = \alpha/2 \quad \text{and} \quad F(u) = 1 - \alpha/2.$$

These equations can be solved using tables, e.g., Table 6.5, or numerically by root finding, or in MATLAB with the commands ℓ = chi2inv(alpha/2, n-1) and u = chi2inv(1-alpha/2, n-1).

Example 6.15. Let $X_1, X_2, \ldots$ be i.i.d. $N(m, \sigma^2)$ random variables. If $S_{100}^2 = 1.608$, find the 90% confidence interval for σ^2.

Solution. From Table 6.5 we see that for $1-\alpha = 0.90$, $\ell = 77.046$ and $u = 123.225$. The 90% confidence interval is

$$\left[\frac{99(1.608)}{123.225}, \frac{99(1.608)}{77.046}\right] = [1.292, 2.067].$$

6.5 Confidence intervals for Gaussian data

$1-\alpha$	ℓ	u
0.90	77.046	123.225
0.91	76.447	124.049
0.92	75.795	124.955
0.93	75.077	125.963
0.94	74.275	127.103
0.95	73.361	128.422
0.96	72.288	129.996
0.97	70.972	131.966
0.98	69.230	134.642
0.99	66.510	138.987

Table 6.5. Confidence levels $1-\alpha$ and corresponding values of ℓ and u such that $\mathsf{P}(\ell \le (n-1)S_n^2/\sigma^2 \le u) = 1-\alpha$ and such that $\mathsf{P}((n-1)S_n^2/\sigma^2 \le \ell) = \mathsf{P}((n-1)S_n^2/\sigma^2 \ge u) = \alpha/2$ for $n = 100$ observations ($n-1 = 99$ degrees of freedom).

*Derivations

The remainder of this section is devoted to deriving the distributions of S_n^2 and

$$T := \frac{M_n - m}{S_n/\sqrt{n}} = \frac{(M_n - m)/(\sigma/\sqrt{n})}{\sqrt{\frac{n-1}{\sigma^2} S_n^2 / (n-1)}}$$

under the assumption that the X_i are i.i.d. $N(m, \sigma^2)$.

We begin with the numerator in T. By Problem 30, $Y_n := (M_n - m)/(\sigma/\sqrt{n})$ is $\sim N(0, 1)$.

For the denominator, we show that the density of $(n-1)S_n^2/\sigma^2$ is chi-squared with $n-1$ degrees of freedom. We begin by recalling the derivation of (6.3). If we replace the first line of the derivation with

$$S_n^2 = \frac{1}{n-1} \sum_{i=1}^{n} ([X_i - m] - [M_n - m])^2,$$

then we end up with

$$S_n^2 = \frac{1}{n-1} \left[\left(\sum_{i=1}^{n} [X_i - m]^2 \right) - n[M_n - m]^2 \right].$$

Using the notation $Z_i := (X_i - m)/\sigma$, we have

$$\frac{n-1}{\sigma^2} S_n^2 = \sum_{i=1}^{n} Z_i^2 - n \left(\frac{M_n - m}{\sigma} \right)^2,$$

or

$$\frac{n-1}{\sigma^2} S_n^2 + \left(\frac{M_n - m}{\sigma/\sqrt{n}} \right)^2 = \sum_{i=1}^{n} Z_i^2.$$

As we argue below, the two terms on the left are independent. It then follows that the density of $\sum_{i=1}^{n} Z_i^2$ is equal to the convolution of the densities of the other two terms. To find the density of $(n-1)S_n^2/\sigma^2$, we use moment generating functions. Now, the second term on the left is the square of an $N(0,1)$ random variable. It is therefore chi-squared with one degree of freedom and has moment generating function is $1/(1-2s)^{1/2}$ (Problem 46 in Chapter 4). The same holds for each Z_i^2. Since the Z_i are independent, $\sum_{i=1}^{n} Z_i^2$ is chi-squared with n degrees of freedom and has moment generating function is $1/(1-2s)^{n/2}$ (Problem 55(c) in Chapter 4). It now follows that the moment generating function of $(n-1)S_n^2/\sigma^2$ is the quotient

$$\frac{1/(1-2s)^{n/2}}{1/(1-2s)^{1/2}} = \frac{1}{(1-2s)^{(n-1)/2}},$$

which is the moment generating function of a chi-squared random variable with $n-1$ degrees of freedom.

It remains to show that S_n^2 and M_n are independent. Observe that S_n^2 is a function of the vector

$$W := [(X_1 - M_n), \ldots, (X_n - M_n)]'.$$

In fact, $S_n^2 = W'W/(n-1)$. By Example 9.6, the vector W and the sample mean M_n are independent. It then follows that any function of W and any function of M_n are independent.

We can now find the density of

$$T = \frac{(M_n - m)/(\sigma/\sqrt{n})}{\sqrt{\frac{n-1}{\sigma^2} S_n^2 / (n-1)}}.$$

If the X_i are i.i.d. $N(m,\sigma^2)$, then the numerator and the denominator are independent; the numerator is $N(0,1)$, and in the denominator $(n-1)S_n^2/\sigma^2$ is chi-squared with $n-1$ degrees of freedom. By Problem 44 in Chapter 7, T has Student's t density with $v = n-1$ degrees of freedom.

6.6 Hypothesis tests for the mean

Let $X_1, X_2, \ldots$ be i.i.d. with mean m and variance σ^2. Consider the problem of deciding between two possibilities such as

$$m \leq m_0 \quad \text{or} \quad m > m_0,$$

where m_0 is a threshold. Other pairs of possibilities include

$$m = m_0 \quad \text{or} \quad m \neq m_0$$

and

$$m = m_0 \quad \text{or} \quad m > m_0.$$

It is not required that each possibility be the negation of the other, although we usually do so here.

6.6 Hypothesis tests for the mean

Example 6.16. A telecommunications satellite maker claims that its new satellite has a bit-error probability of no more that p_0. To put this information into the above framework, let $X_i = 1$ if the ith bit transmitted is received in error, and let $X_i = 0$ otherwise. Then $p := \mathsf{P}(X_i = 1) = \mathsf{E}[X_i]$, and the claim is that $p \leq p_0$. For the other possibility, we allow $p > p_0$.

The problem of deciding which of several possibilities is correct is called **hypothesis testing**. In this section we restrict attention to the case of two possibilities. In assessing two competing claims, it is usually natural give the benefit of the doubt to one and the burden of proof to the other. The possibility that is given the benefit of the doubt is called the **null hypothesis**, and the possibility that is given the burden of proof is called the **alternative hypothesis**.

Example 6.17. In the preceding example, the satellite maker claims that $p \leq p_0$. If the government is considering buying such a satellite, it will put the burden of proof on the manufacturer. Hence, the government will take $p > p_0$ as the null hypothesis, and it will be up to the data to give compelling evidence that the alternative hypothesis $p \leq p_0$ is true.

Decision rules

To assess claims about m, it is natural to use the sample mean M_n, since we know that M_n converges to m as $n \to \infty$. However, for finite n, M_n is usually not exactly equal to m. In fact, M_n may be quite far from m. To account for this, when we test the null hypothesis $m \leq m_0$ against the alternative hypothesis $m > m_0$, we select $\delta > 0$ and agree that if

$$M_n \leq m_0 + \delta, \tag{6.21}$$

then we declare that the null hypothesis $m \leq m_0$ is true, while if $M_n > m_0 + \delta$, we declare that the alternative hypothesis $m > m_0$ is true. Clearly, the burden of proof is on the alternative hypothesis $m > m_0$, since we do not believe it unless M_n is substantially greater than m_0. The null hypothesis $m \leq m_0$ gets the benefit of the doubt, since we accept it even if $m_0 < M_n \leq m_0 + \delta$.

Proceeding similarly to test the null hypothesis $m > m_0$ against the alternative hypothesis $m \leq m_0$, we agree that if

$$M_n > m_0 - \delta, \tag{6.22}$$

then we declare $m > m_0$ to be true, while if $M_n \leq m_0 - \delta$, we declare $m \leq m_0$. Again, the burden of proof is on the alternative hypothesis, and the benefit of the doubt is on the null hypothesis.

To test the null hypothesis $m = m_0$ against the alternative hypothesis $m \neq m_0$, we declare $m = m_0$ to be true if

$$|M_n - m_0| \leq \delta, \tag{6.23}$$

and $m \neq m_0$ otherwise.

Acceptance and rejection regions

As in the case of confidence intervals, it is sensible to have δ depend on the variance and on the number of observations n. For this reason, if the variance is known, we take δ to be of the form $\delta = y\sigma/\sqrt{n}$, while if the variance is unknown, we take $\delta = yS_n/\sqrt{n}$. This amounts to working with the statistic

$$\frac{M_n - m_0}{\sigma/\sqrt{n}} \qquad (6.24)$$

if σ known, or the statistic

$$\frac{M_n - m_0}{S_n/\sqrt{n}} \qquad (6.25)$$

otherwise. If the appropriate statistic is denoted by Z_n, then (6.21)–(6.23) become

$$Z_n \leq y, \quad Z_n > -y, \quad \text{and} \quad |Z_n| \leq y.$$

The value with which Z_n or $|Z_n|$ is compared is called the **critical value**. The corresponding **acceptance regions** are the intervals

$$(-\infty, y], \quad (-y, \infty), \quad \text{and} \quad [-y, y].$$

In other words, if Z_n lies in the acceptance region, we accept the null hypothesis as true. The complement of the acceptance region is called the **rejection region** or the **critical region**. If Z_n lies in the rejection region, we reject the null hypothesis and declare the alternative hypothesis to be true.

Types of errors

In deciding between the null hypothesis and the alternative hypothesis, there are two kinds of erroneous decisions we can make. We say that a **Type I error** occurs if we declare the alternative hypothesis to be true when the null hypothesis is true. We say that a **Type II error** occurs if we declare the null hypothesis to be true when the alternative hypothesis is true.

Since the burden of proof is on the alternative hypothesis, we want to bound the probability of mistakenly declaring it to be true. In other words, we want to bound the Type I error by some value, which we denote by α. The number α is called the **significance level** of the test.

Finding the critical value

We first treat the case of testing the null hypothesis $m = m_0$ against the alternative hypothesis $m \neq m_0$. In this problem, we declare the alternative hypothesis to be true if $|Z_n| > y$. So, we need to choose the critical value y so that $\mathsf{P}(|Z_n| > y) \leq \alpha$. If $m = m_0$, then as argued in Sections 6.3 and 6.4, the cdf of Z_n in either (6.24) or (6.25) can be approximated by the standard normal cdf Φ as long as the number of observations n is large enough. Hence,

$$\mathsf{P}(|Z_n| > y) \approx 1 - [\Phi(y) - \Phi(-y)] = 2[1 - \Phi(y)], \qquad (6.26)$$

since the $N(0,1)$ density is even. The right-hand side is equal to α if and only if $\Phi(y) = 1 - \alpha/2$. The solution, which we denote by $y_{\alpha/2}$, can be obtained in MATLAB with the

6.6 Hypothesis tests for the mean

α	$y_{\alpha/2}$	y_α
0.01	2.576	2.326
0.02	2.326	2.054
0.03	2.170	1.881
0.04	2.054	1.751
0.05	1.960	1.645
0.06	1.881	1.555
0.07	1.812	1.476
0.08	1.751	1.405
0.09	1.695	1.341
0.10	1.645	1.282

Table 6.6. Significance levels α and corresponding critical values $y_{\alpha/2}$ such that $\Phi(y_{\alpha/2}) = 1 - \alpha/2$ for a two-tailed test and y_α such that $\Phi(y_\alpha) = 1 - \alpha$ for the one-tailed test of the hypothesis $m \leq m_0$. For the one-tailed test $m > m_0$, use $-y_\alpha$.

command `norminv(1-alpha/2)` or from Table 6.6. The formula $P(|Z_n| > y_{\alpha/2}) = \alpha$ is illustrated in Figure 6.3; notice that the rejection region lies under the tails of the density. For this reason, testing $m = m_0$ against $m \neq m_0$ is called a **two-tailed** or **two-sided** test.

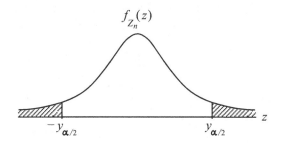

Figure 6.3. Illustration of the condition $P(|Z_n| > y_{\alpha/2}) = \alpha$. Each of the two shaded regions has area $\alpha/2$. Their union has total area α. The acceptance region is the interval $[-y_{\alpha/2}, y_{\alpha/2}]$, and the rejection region is its complement.

We next find the critical value for testing the null hypothesis $m \leq m_0$ against the alternative hypothesis $m > m_0$. Since we accept the null hypothesis if $Z_n \leq y$ and reject it if $Z_n > y$, the Type I error probability is $P(Z_n > y)$. To analyze this, it is helpful to expand Z_n into two terms. We treat only the unknown-variance case with Z_n given by (6.25); the known-variance case is similar. When Z_n is given by (6.25), we have

$$Z_n = \frac{M_n - m}{S_n/\sqrt{n}} + \frac{m - m_0}{S_n/\sqrt{n}}.$$

If the X_i have mean m, then we know from the discussion in Section 6.4 that the cdf of the first term on the right can be approximated by the standard normal cdf Φ if n is large. Hence,

$$P(Z_n > y) = P\left(\frac{M_n - m}{S_n/\sqrt{n}} > y - \frac{m - m_0}{S_n/\sqrt{n}}\right)$$

$$\leq \mathsf{P}\!\left(\frac{M_n - m}{S_n/\sqrt{n}} > y\right), \quad \text{since } m \leq m_0,$$
$$\approx 1 - \Phi(y). \tag{6.27}$$

The value of y that achieves $1 - \Phi(y) = \alpha$ is denoted by y_α and can be obtained in MATLAB with the command y = norminv(1-alpha) or found in Table 6.6. The formula $\mathsf{P}(Z_n > y_\alpha) = \alpha$ is illustrated in Figure 6.4. Since the rejection region is the interval under the upper tail, testing $m \leq m_0$ against $m > m_0$ is called a **one-tailed** or **one-sided** test.

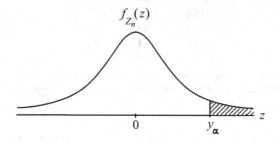

Figure 6.4. The shaded region has area α. The rejection region is (y_α, ∞) under the upper tail of the density.

To find the critical value for testing the null hypothesis $m > m_0$ against the alternative hypothesis $m \leq m_0$, write

$$\mathsf{P}(Z_n \leq -y) = \mathsf{P}\!\left(\frac{M_n - m}{S_n/\sqrt{n}} \leq -y - \frac{m - m_0}{S_n/\sqrt{n}}\right)$$
$$\leq \mathsf{P}\!\left(\frac{M_n - m}{S_n/\sqrt{n}} \leq -y\right), \quad \text{since } m > m_0,$$
$$\approx \Phi(-y), \tag{6.28}$$

Since the $N(0,1)$ density is even, the value of y that solves $\Phi(-y) = \alpha$ is the same as the one that solves $1 - \Phi(y) = \alpha$ (Problem 35). This is the value y_α defined above. It can be obtained in MATLAB with the command y = norminv(1-alpha) or found in Table 6.6.

Example 6.18 (Zener diodes). An electronics manufacturer sells Zener diodes to maintain a nominal voltage no greater than m_0 when reverse biased. You receive a shipment of n diodes that maintain voltages $X_1, \ldots, X_n$, which are assumed i.i.d. with mean m. You want to assess the manufacturer's claim that $m \leq m_0$. (a) Would you take Z_n to be the statistic in (6.24) or in (6.25)? (b) If the burden of proof is on the manufacturer, what should you choose for the alternative hypothesis? For the null hypothesis? (c) For the test in (b), what critical value is needed for a significance level of $\alpha = 0.05$? What is the critical region?

Solution. (a) Since the variance of the X_i is not given, we take Z_n as in (6.25).
(b) The manufacturer claims $m \leq m_0$. To put the burden of proof on the manufacturer, we make $m \leq m_0$ the alternative hypothesis and $m > m_0$ the null hypothesis.
(c) The acceptance region for such a null hypothesis is $(-y, \infty)$. To achieve a significance level of $\alpha = 0.05$, we should take $y = y_\alpha$ from Table 6.6. In this case, $y_\alpha = 1.645$. The

Small samples

The approximations in (6.26)–(6.28) are based on the central limit theorem, which is valid only asymptotically as $n \to \infty$. However, if the X_i are Gaussian with known variance and Z_n is given by (6.24), then Z_n has the $N(0,1)$ cdf Φ for all values of n. In this case, there is no approximation in (6.26)–(6.28), and the foregoing results hold exactly for all values of n. If the X_i are Gaussian with unknown variance and Z_n is given by (6.25), then Z_n has Student's t density with $n-1$ degrees of freedom. In this case, if Φ is replaced by Student's t cdf with $n-1$ degrees of freedom in (6.26)–(6.28), then there is no approximation and the foregoing results are exact for all values of n. This can be accomplished in MATLAB if y_α is computed with the command `tinv(1-alpha,n-1)` and if $y_{\alpha/2}$ is computed with the command `tinv(1-alpha/2,n-1)`.

6.7 Regression and curve fitting

In analyzing a physical system, we often formulate an idealized model of the form $y = g(x)$ that relates the output y to the input x. If we apply a particular input x to the system, we do not expect the output we measure to be exactly equal to $g(x)$ for two reasons. First, the formula $g(x)$ is only a mathematical approximation of the physical system. Second, there is measurement error. To account for this, we assume measurements are corrupted by additive noise. For example, if we apply inputs $x_1, \ldots, x_n$ and measure corresponding outputs $Y_1, \ldots, Y_n$, we assume that

$$Y_k = g(x_k) + W_k, \quad k = 1, \ldots, n, \quad (6.29)$$

where the W_k are noise random variables with zero mean.

When we have a model of the form $y = g(x)$, the structural form of $g(x)$ is often known, but there are unknown parameters that we need to estimate based on physical measurements. In this situation, the function $g(x)$ is called the **regression curve** of Y on x. The procedure of finding the best parameters to use in the function $g(x)$ is called **regression**. It is also called **curve fitting**.

Example 6.19. A resistor can be viewed as a system whose input is the applied current, i, and whose output is the resulting voltage drop v. In this case, the output is related to the input by the formula $v = iR$. Suppose we apply a sequence of currents, $i_1, \ldots, i_n$, and measure corresponding voltages, $V_1, \ldots, V_n$, as shown in Figure 6.5. If we draw the best straight line through the data points, the slope of that line would be our estimate of R.

Example 6.20. The current–voltage relationship for an ideal diode is of the form

$$i = i_s(e^{av} - 1),$$

where i_s and a are constants to be estimated. In this example, we can avoid working with the exponential function if we restrict attention to large v. For large v, we use the approximation

$$i \approx i_s e^{av}.$$

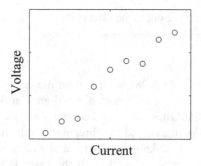

Figure 6.5. Scatter plot of voltage versus current in a resistor.

Taking logarithms and letting $y := \ln i$ and $b := \ln i_s$ suggests the linear model

$$y = av + b.$$

When we have measurements (x_k, Y_k) modeled by (6.29), and we have specified the structural form of g up to some unknown constants, our goal is to choose those constants so as to minimize the **sum of squared errors**,

$$e(g) := \sum_{k=1}^{n} |Y_k - g(x_k)|^2. \tag{6.30}$$

Example 6.21 (linear regression). To fit a straight line through the data means that $g(x)$ has the form $g(x) = ax + b$. Then the sum of squared errors becomes

$$e(a,b) = \sum_{k=1}^{n} |Y_k - [ax_k + b])|^2.$$

To find the minimizing values of a and b we could compute the partial derivatives of $e(a,b)$ with respect to a and b and set them equal to zero. Solving the system of equations would yield

$$a = \frac{S_{xY}}{S_{xx}} \quad \text{and} \quad b = \overline{Y} - a\overline{x},$$

where

$$\overline{x} := \frac{1}{n}\sum_{k=1}^{n} x_k, \quad \text{and} \quad \overline{Y} := \frac{1}{n}\sum_{k=1}^{n} Y_k,$$

and

$$S_{xx} := \sum_{k=1}^{n}(x_k - \overline{x})^2, \quad \text{and} \quad S_{xY} := \sum_{k=1}^{n}(x_k - \overline{x})(Y_k - \overline{Y}).$$

Fortunately, there is an easier and more systematic way of deriving these equations, which we discuss shortly.

6.7 Regression and curve fitting

In many cases, such as in the preceding example, the set of all functions $g(x)$ with a given structure forms a subspace. In other words, the set is closed under linear combinations.

Example 6.22. Let $\mathscr{G}_p$ denote the set of all functions $g(x)$ that are polynomials of degree p or less. Show that $\mathscr{G}_p$ is a subspace.

Solution. Let $g_1(x) = a_0 + a_1 x + \cdots + a_p x^p$, and let $g_2(x) = b_0 + b_1 x + \cdots + b_p x^p$. We must show that any linear combination $\lambda g_1(x) + \mu g_2(x)$ is a polynomial of degree p or less. Write

$$\begin{aligned}\lambda g_1(x) + \mu g_2(x) &= \lambda(a_0 + a_1 x + \cdots + a_p x^p) + \mu(b_0 + b_1 x + \cdots + b_p x^p) \\ &= (\lambda a_0 + \mu b_0) + (\lambda a_1 + \mu b_1)x + \cdots + (\lambda a_p + \mu b_p)x^p,\end{aligned}$$

which is a polynomial of degree p or less.

When we want to minimize $e(g)$ in (6.30) as g ranges over a subspace of functions $\mathscr{G}$, we can show that the minimizing g, denoted by $\widehat{g}$, is characterized by the property

$$\sum_{k=1}^{n} [Y_k - \widehat{g}(x_k)] g(x_k) = 0, \quad \text{for all } g \in \mathscr{G}. \tag{6.31}$$

This result is known as the **orthogonality principle** because (6.31) says that the n-dimensional vectors

$$[Y_1 - \widehat{g}(x_1), \ldots, Y_n - \widehat{g}(x_n)] \quad \text{and} \quad [g(x_1), \ldots, g(x_n)]$$

are orthogonal.

To show that (6.31) implies $e(\widehat{g}) \leq e(g)$ for all $g \in \mathscr{G}$, let $g \in \mathscr{G}$ and write

$$\begin{aligned}e(g) &= \sum_{k=1}^{n} |Y_k - g(x_k)|^2 \\ &= \sum_{k=1}^{n} |[Y_k - \widehat{g}(x_k)] + [\widehat{g}(x_k) - g(x_k)]|^2 \\ &= \sum_{k=1}^{n} |Y_k - \widehat{g}(x_k)|^2 + 2[Y_k - \widehat{g}(x_k)][\widehat{g}(x_k) - g(x_k)] + |\widehat{g}(x_k) - g(x_k)|^2.\end{aligned}$$

Since $\mathscr{G}$ is a subspace, the function $\widehat{g} - g \in \mathscr{G}$. Hence, in (6.31), we can replace the factor $g(x_k)$ by $\widehat{g}(x_k) - g(x_k)$. This tells us that

$$\sum_{k=1}^{n} [Y_k - \widehat{g}(x_k)][\widehat{g}(x_k) - g(x_k)] = 0.$$

We then have

$$\begin{aligned}e(g) &= \sum_{k=1}^{n} |Y_k - \widehat{g}(x_k)|^2 + \sum_{k=1}^{n} |\widehat{g}(x_k) - g(x_k)|^2 \\ &\geq \sum_{k=1}^{n} |Y_k - \widehat{g}(x_k)|^2 \\ &= e(\widehat{g}).\end{aligned}$$

We have thus shown that if $\widehat{g}$ satisfies (6.31), then $e(\widehat{g}) \leq e(g)$ for all $g \in \mathscr{G}$.

Example 6.23 (linear regression again). We can use (6.31) to derive the formulas in Example 6.21 as follows. In the linear case, (6.31) says that

$$\sum_{k=1}^{n}[Y_k - (\widehat{a}x_k + \widehat{b})](ax_k + b) = 0 \qquad (6.32)$$

has to hold for all values of a and b. In particular, taking $a = 0$ and $b = 1$ implies

$$\sum_{k=1}^{n}[Y_k - (\widehat{a}x_k + \widehat{b})] = 0.$$

Using the notation of Example 6.21, this says that

$$n\overline{Y} - \widehat{a}n\overline{x} - n\widehat{b} = 0,$$

or $\widehat{b} = \overline{Y} - \widehat{a}\overline{x}$. Now substitute this formula for $\widehat{b}$ into (6.32) and take $a = 1$ and $b = -\overline{x}$. We then find that

$$\sum_{k=1}^{n}\left[(Y_k - \overline{Y}) - \widehat{a}(x_k - \overline{x})\right](x_k - \overline{x}) = 0.$$

Using the notation of Example 6.21, this says that

$$S_{xY} - \widehat{a}S_{xx} = 0,$$

or $\widehat{a} = S_{xY}/S_{xx}$. It is shown in Problem 39 that $e(\widehat{g}) = S_{YY} - S_{xY}^2/S_{xx}$, where

$$S_{YY} := \sum_{k=1}^{n}(Y_k - \overline{Y})^2.$$

In general, to find the polynomial of degree p that minimizes the sum of squared errors, we can take a similar approach as in the preceding example and derive $p + 1$ equations in the $p + 1$ unknown coefficients of the desired polynomial. Fortunately, there are MATLAB routines that do all the work for us automatically.

Suppose x and Y are MATLAB vectors containing the data points x_k and Y_k, respectively. If

$$\widehat{g}(x) = a_1 x^p + a_2 x^{p-1} + \cdots a_p x + a_{p+1}$$

denotes the best-fit polynomial of degree p, then the vector $a = [a_1, \ldots, a_{p+1}]$ can be obtained with the command

$$a = \texttt{polyfit(x,Y,p)}.$$

To compute $\widehat{g}(t)$ at a point t or a vector of points $t = [t_1, \ldots, t_m]$, use the command,

$$\texttt{polyval(a,t)}.$$

For example, these commands can be used to plot the best-fit straight line through the points in Figure 6.5. The result is shown in Figure 6.6. As another example, at the left in Figure 6.7, a scatter plot of some data (x_k, Y_k) is shown. At the right is the best-fit cubic polynomial.

6.8 Monte Carlo estimation

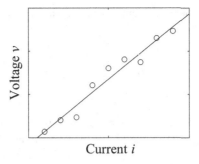

Figure 6.6. Best-fit line through points in Figure 6.5.

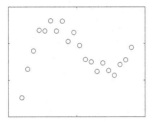 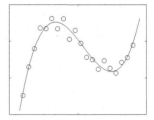

Figure 6.7. Scatter plot (left) and best-fit cubic (right).

6.8 *Monte Carlo estimation

Suppose we would like to know the value of $P(Z > t)$ for some random variable Z and some threshold t. For example, t could be the size of a buffer in an Internet router, and if the number of packets received, Z, exceeds t, it will be necessary to drop packets. Or Z could be a signal voltage in a communications receiver, and $\{Z > t\}$ could correspond to a decoding error. In complicated systems, there is no hope of finding the cdf or density of Z. However, we can repeatedly simulate the operation of the system to obtain i.i.d. simulated values $Z_1, Z_2, \ldots$. The fraction of times that $Z_i > t$ can be used as an estimate of $P(Z > t)$. More precisely, put

$$X_i := I_{(t,\infty)}(Z_i).$$

Then the X_i are i.i.d. Bernoulli(p) with $p := P(Z > t)$, and

$$M_n := \frac{1}{n}\sum_{i=1}^{n} X_i = \frac{1}{n}\sum_{i=1}^{n} I_{(t,\infty)}(Z_i)$$

is the fraction of times that $Z_i > t$. Also,[j]

$$E[M_n] = E[X_i] = E[I_{(t,\infty)}(Z_i)] = P(Z_i > t).$$

---

[j] More generally, we might consider $X_i = h(Z_i)$ for some function h. Then $E[M_n] = E[h(Z_i)]$, and M_n would be an estimate of $E[h(Z)]$.

Hence, we can even use the theory of confidence intervals to assess the quality of the probability estimate M_n, e.g.,

$$P(Z > t) = M_n \pm \frac{S_n y_{\alpha/2}}{\sqrt{n}} \quad \text{with } 100(1-\alpha)\% \text{ probability,}$$

where $y_{\alpha/2}$ is chosen from Table 6.2.

Example 6.24. Suppose the Z_i are i.i.d. exponential with parameter $\lambda = 1$, and we want to estimate $P(Z > 2)$. If $M_{100} = 0.15$ and $S_{100} = 0.359$, find the 95% confidence interval for $P(Z > 2)$.

Solution. The estimate is

$$P(Z > 2) = 0.15 \pm \frac{0.359(1.96)}{\sqrt{100}} = 0.15 \pm 0.07,$$

which corresponds to the interval $[0.08, 0.22]$. This interval happens to contain the true value $e^{-2} = 0.135$.

Caution. The foregoing does not work well if $P(Z > t)$ is very small, unless n is correspondingly large. The reason is that if $P(Z > t)$ is small, it is likely that we will have all $X_1, \ldots, X_n$ equal to zero. This forces both M_n and S_n to be zero too, which is not a useful estimate. The probability that all X_i are zero is $[1 - P(Z > t)]^n$. For example, with $Z \sim \exp(1)$, $P(Z > 7) = e^{-7} = 0.000912$, and so $[1 - P(Z > 7)]^{100} = 0.9$. In other words, 90% of simulations provide no information about $P(Z > 7)$. In fact, using the values of Z_i of the preceding example to estimate $P(Z > 7)$ did result in $M_n = 0$ and $S_n = 0$.

To estimate small probabilities without requiring n to be unreasonably large requires more sophisticated strategies such as **importance sampling** [26], [47], [58]. The idea of importance sampling is to redefine

$$X_i := I_{(t,\infty)}(\widetilde{Z}_i) \frac{f_Z(\widetilde{Z}_i)}{f_{\widetilde{Z}}(\widetilde{Z}_i)},$$

where the $\widetilde{Z}_i$ have a different density $f_{\widetilde{Z}}$ such that $P(\widetilde{Z} > t)$ is much bigger than $P(Z > t)$. If $P(\widetilde{Z} > t)$ is large, then very likely, many of the X_i will be nonzero. Observe that we still have

$$\begin{aligned} E[X_i] &= \int_{-\infty}^{\infty} \left[I_{(t,\infty)}(z) \frac{f_Z(z)}{f_{\widetilde{Z}}(z)} \right] f_{\widetilde{Z}}(z) \, dz \\ &= \int_{-\infty}^{\infty} I_{(t,\infty)}(z) f_Z(z) \, dz \\ &= P(Z > t). \end{aligned}$$

Thus, the X_i are no longer Bernoulli, but they still have the desired expected value. Our choice for $f_{\widetilde{Z}}$ is

$$f_{\widetilde{Z}}(z) := e^{sz} f_Z(z) / M_Z(s), \tag{6.33}$$

where the real parameter s is to be chosen later. This choice for $f_{\widetilde{Z}}$ is called a **tilted** or a **twisted** density. Since integrating $e^{sz}f_Z(z)$ is just computing the moment generating function of Z, we need $M_Z(s)$ in the denominator above to make $f_{\widetilde{Z}}$ integrate to one. Our goal is to adjust s so that a greater amount of probability is located near t. For $Z \sim \exp(1)$,

$$f_{\widetilde{Z}}(z) = \frac{e^{sz}f_Z(z)}{M_Z(s)} = \frac{e^{sz}e^{-z}}{1/(1-s)} = (1-s)e^{-z(1-s)}.$$

In other words, $\widetilde{Z} \sim \exp(1-s)$. Hence, we can easily adjust s so that $E[\widetilde{Z}] = 7$ by taking $s = 1 - 1/7 = 0.8571$. We simulated $n = 100$ values of $\widetilde{Z}_i$ and found $M_{100} = 0.0007$, $S_{100} = 0.002$, and

$$P(Z > 7) = 0.0007 \pm \frac{0.002(1.96)}{\sqrt{100}} = 0.0007 \pm 0.0004.$$

This corresponds to the interval $[0.0003, 0.0011]$. Thus, still using only 100 simulations, we obtained a nonzero estimate and an informative confidence interval.

We also point out that even in the search for $P(Z > 2)$, importance sampling can result in a smaller confidence interval. In Example 6.24, the width of the confidence interval is 0.14. With importance sampling (still with $n = 100$), we found the width was only about 0.08.

Notes

6.1: Parameter estimators and their properties

Note **1.** The limits in (6.4) are in the almost-sure sense of Section 14.3. By the first-moment **strong law of large numbers** (stated following Example 14.15), M_n converges almost surely to m. Similarly $(1/n)\sum_{i=1}^{n} X_i^2$ converges almost surely to $\sigma^2 + m^2$. Using (6.3),

$$S_n^2 = \frac{n}{n-1}\left[\left(\frac{1}{n}\sum_{i=1}^{n}X_i^2\right) - M_n^2\right].$$

Then

$$\lim_{n \to \infty} S_n^2 = \lim_{n \to \infty} \frac{n}{n-1}\left[\lim_{n \to \infty}\left(\frac{1}{n}\sum_{i=1}^{n}X_i^2\right) - \lim_{n \to \infty} M_n^2\right]$$
$$= 1 \cdot [(\sigma^2 + m^2) - m^2]$$
$$= \sigma^2.$$

6.3: Confidence intervals for the mean – known variance

Note **2.** The derivation of (6.10) used the formula

$$P(-y \leq Y_n \leq y) = F_{Y_n}(y) - F_{Y_n}(-y),$$

which is valid when F_{Y_n} is a continuous cdf. In the general case, it suffices to write

$$P(-y \leq Y_n \leq y) = P(Y_n = -y) + P(-y < Y_n \leq y) = P(Y_n = -y) + F_{Y_n}(y) - F_{Y_n}(-y)$$

and then show that $P(Y_n = -y) \to 0$. To do this fix any $\varepsilon > 0$ and write

$$\begin{aligned} P(Y_n = -y) &\leq P(-y - \varepsilon < Y_n \leq -y + \varepsilon) \\ &= F_{Y_n}(-y + \varepsilon) - F_{Y_n}(-y - \varepsilon) \\ &\to \Phi(-y + \varepsilon) - \Phi(-y - \varepsilon), \quad \text{by the central limit theorem.} \end{aligned}$$

To conclude, write

$$\Phi(-y + \varepsilon) - \Phi(-y - \varepsilon) = [\Phi(-y + \varepsilon) - \Phi(-y)] + [\Phi(-y) - \Phi(-y - \varepsilon)],$$

which goes to zero as $\varepsilon \to 0$ on account of the continuity of Φ. Hence, $P(-y \leq Y_n \leq y) \to 0$.

6.4: Confidence intervals for the mean – unknown variance

Note 3. Since the X_i have finite mean and variance, they have finite second moment. Thus, the X_i^2 have finite first moment $\sigma^2 + m^2$. By the first-moment weak law of large numbers stated following Example 14.15, $\sum_{i=1}^{n} X_i^2$ converges in probability to $\sigma^2 + m^2$. Using (6.3), Example 14.2, and Problem 2 in Chapter 14, it follows that S_n converges in probability to σ. Now appeal to the fact that if the cdf of Y_n, say F_n, converges to a continuous cdf F, and if U_n converges in probability to 1, then

$$P(Y_n \leq y U_n) \to F(y).$$

This result, which is proved in Example 14.11, is a version of **Slutsky's theorem**.

Note 4. The hypergeometric random variable arises in the following situation. We have a collection of N items, d of which are defective. Rather than test all N items, we select at random a small number of items, say $n < N$. Let Y_n denote the number of defectives out the n items tested. We show that

$$P(Y_n = k) = \frac{\binom{d}{k}\binom{N-d}{n-k}}{\binom{N}{n}}, \quad k = 0, \ldots, n.$$

We denote this by $Y_n \sim \text{hypergeometric}(N, d, n)$.

Remark. In the typical case, $d \geq n$ and $N - d \geq n$; however, if these conditions do not hold in the above formula, it is understood that $\binom{d}{k} = 0$ if $d < k \leq n$, and $\binom{N-d}{n-k} = 0$ if $n - k > N - d$, i.e., if $0 \leq k < n - (N - d)$.

For $i = 1, \ldots, n$, draw at random an item from the collection and test it. If the ith item is defective, let $X_i = 1$, and put $X_i = 0$ otherwise. In either case, do *not* put the tested item back into the collection (sampling without replacement). Then the total number of defectives among the first n items tested is

$$Y_n := \sum_{i=1}^{n} X_i.$$

We show that $Y_n \sim \text{hypergeometric}(N, d, n)$.

Consider the case $n=1$. Then $Y_1 = X_1$, and the chance of drawing a defective item at random is simply the ratio of the number of defectives to the total number of items in the collection; i.e., $\mathsf{P}(Y_1 = 1) = \mathsf{P}(X_1 = 1) = d/N$. Now in general, suppose the result is true for some $n \geq 1$. We show it is true for $n+1$. Use the law of total probability to write

$$\mathsf{P}(Y_{n+1} = k) = \sum_{i=0}^{n} \mathsf{P}(Y_{n+1} = k | Y_n = i) \mathsf{P}(Y_n = i). \tag{6.34}$$

Since $Y_{n+1} = Y_n + X_{n+1}$, we can use the substitution law to write

$$\begin{aligned}
\mathsf{P}(Y_{n+1} = k | Y_n = i) &= \mathsf{P}(Y_n + X_{n+1} = k | Y_n = i) \\
&= \mathsf{P}(i + X_{n+1} = k | Y_n = i) \\
&= \mathsf{P}(X_{n+1} = k - i | Y_n = i).
\end{aligned}$$

Since X_{n+1} takes only the values zero and one, this last expression is zero unless $i = k$ or $i = k-1$. Returning to (6.34), we can write

$$\mathsf{P}(Y_{n+1} = k) = \sum_{i=k-1}^{k} \mathsf{P}(X_{n+1} = k - i | Y_n = i) \mathsf{P}(Y_n = i). \tag{6.35}$$

When $i = k-1$, the above conditional probability is

$$\mathsf{P}(X_{n+1} = 1 | Y_n = k-1) = \frac{d - (k-1)}{N - n},$$

since given $Y_n = k-1$, there are $N-n$ items left in the collection, and of those, the number of defectives remaining is $d - (k-1)$. When $i = k$, the needed conditional probability is

$$\mathsf{P}(X_{n+1} = 0 | Y_n = k) = \frac{(N-d) - (n-k)}{N - n},$$

since given $Y_n = k$, there are $N-n$ items left in the collection, and of those, the number of *non*defectives remaining is $(N-d) - (n-k)$. If we now assume that $Y_n \sim$ hypergeometric (N, d, n), we can expand (6.35) to get

$$\begin{aligned}
\mathsf{P}(Y_{n+1} = k) &= \frac{d - (k-1)}{N - n} \cdot \frac{\binom{d}{k-1}\binom{N-d}{n-(k-1)}}{\binom{N}{n}} \\
&\quad + \frac{(N-d) - (n-k)}{N - n} \cdot \frac{\binom{d}{k}\binom{N-d}{n-k}}{\binom{N}{n}}.
\end{aligned}$$

It is a simple calculation to see that the first term on the right is equal to

$$\left(1 - \frac{k}{n+1}\right) \cdot \frac{\binom{d}{k}\binom{N-d}{[n+1]-k}}{\binom{N}{n+1}},$$

and the second term is equal to

$$\frac{k}{n+1} \cdot \frac{\binom{d}{k}\binom{N-d}{[n+1]-k}}{\binom{N}{n+1}}.$$

Thus, $Y_{n+1} \sim$ hypergeometric$(N,d,n+1)$.

Problems

6.1: Parameter estimators and their properties

1. Use formula (6.3) to show that S_n^2 is unbiased, assuming the X_i are uncorrelated.

2. (a) If $X_1, X_2, \ldots$ are i.i.d. Rayleigh(λ), find an unbiased, strongly consistent estimator of λ.

   (b) **MATLAB.** Modify the MATLAB code below to generate $n = 1000$ Rayleigh(λ) random variables with $\lambda = 3$, and use your answer in part (a) to estimate λ from the data.

   ```
   n = 1000;
   U = rand(1,n);
   X = sqrt(-2*log(U));   % X is Rayleigh(1)
   X = 3*X;               % Make X Rayleigh(3)
   ```

   (c) **MATLAB.** Since $E[X_i] = \lambda\sqrt{\pi/2}$, and since we know the value of λ used to generate the data, we can regard π as the unknown parameter. Modify your code in part (b) to estimate π from simulation data with $n = 100\,000$. What is your estimate of π?

3. (a) If $X_1, X_2, \ldots$ are i.i.d. gamma(p, λ), where λ is known, find an unbiased, strongly consistent estimator of p.

   (b) **MATLAB.** Modify the MATLAB code below to generate $n = 1000$ chi-squared random variables with $k = 5$ degrees of freedom, and use your answer in part (a) to estimate k from the data. Remember, chi-squared with k degrees of freedom is the same as gamma$(k/2, 1/2)$. Recall also Problems 46 and 55 in Chapter 4.

   ```
   n = 1000;
   U = randn(5,n);         % U is N(0,1)
   U2 = U.^2;              % U2 is chi-squared
   % with one degree of freedom
   X = sum(U2);            % column sums are
   % chi-squared with 5 degrees of freedom
   ```

   Here the expression U.^2 squares each element of the matrix U.

4. (a) If $X_1, X_2, \ldots$ are i.i.d. noncentral chi-squared with k degrees of freedom and noncentrality parameter λ^2, where k is known, find an unbiased, strongly consistent estimator of λ^2. *Hint:* Use Problem 65(c) in Chapter 4.

(b) **MATLAB.** Modify the following MATLAB code to generate $n = 1000$ noncentral chi-squared random variables with $k = 5$ degrees of freedom, and noncentrality parameter $\lambda^2 = 4$, and use your answer in part (a) to estimate λ^2 from the data.

```
n = 1000;
U = randn(5,n);        % U is N(0,1)
U = U + 2/sqrt(5);     % U is N(m,1)
% with m = 2/sqrt(5)
U2 = U.^2;             % U2 is noncentral
% chi-squared with one degree of freedom
% and noncentrality parameter 4/5
X = sum(U2);           % column sums are
% noncentral chi-squared with 5 degrees
% of freedom and noncentrality parameter 4
```

Here the expression U.^2 squares each element of the matrix U.

5. (a) If $X_1, X_2, \ldots$ are i.i.d. gamma(p, λ), where p is known, find a strongly consistent estimator of λ.

   (b) **MATLAB.** Modify the following MATLAB code to generate $n = 1000$ gamma random variables with $p = 3$ and $\lambda = 1/5$, and use your answer in part (a) to estimate λ from the data. Recall Problem 55 in Chapter 4.

```
n = 1000;
U = rand(3,n);      % U is uniform(0,1)
V = -log(U);        % V is exp(1)
V = 5*V;            % V is exp(1/5)
X = sum(V);         % column sums are Erlang(3,1/5)
```

Remark. This suggests a faster method to simulate chi-squared random variables than the one used in Problem 3. If k is even, then the chi-squared is Erlang$(k/2, 1/2)$. If k is odd, then the chi-squared is equal to the sum of an Erlang$(k/2 - 1/2, 1/2)$ and the square of a single $N(0,1)$.

6. (a) If $X_1, X_2, \ldots$ are i.i.d. Laplace(λ), find a strongly consistent estimator of λ.

   (b) **MATLAB.** Modify the MATLAB code below to generate $n = 1000$ Laplace(λ) random variables with $\lambda = 2$, and use your answer in part (a) to estimate λ from the data. Recall Problem 54 in Chapter 4.

```
n = 1000;
U1 = rand(1,n);      % U1 is uniform(0,1)
V1 = -log(U1)/2;     % V1 is exp(2)
U2 = rand(1,n);      % U2 is uniform(0,1)
V2 = -log(U2)/2;     % V2 is exp(2)
X = V1-V2;           % X is Laplace(2)
```

7. (a) If $X_1, X_2, \ldots$ are i.i.d. gamma(p, λ), find strongly consistent estimators of p and λ. *Hint:* Consider both $\mathsf{E}[X_i]$ and $\mathsf{E}[X_i^2]$.

   (b) **MATLAB.** Modify the MATLAB code in Problem 5 to generate $n = 1000$ gamma random variables with $p = 3$ and $\lambda = 1/5$, and use your answer in part (a) to estimate p and λ from the data.

8. If $X_1, X_2, \ldots$ are i.i.d. generalized gamma with parameters p, λ, and q, where p and q are known, find a strongly consistent estimator of λ. (The generalized gamma was defined in Problem 21 in Chapter 5.) *Hint:* Consider $E[X^q]$.

*9. In the preceding problem, assume that only q is known and that both p and λ are unknown. Find strongly consistent estimators of p and λ.

6.2: Histograms

10. **MATLAB.** Use the following MATLAB code to generate $n = 1000$ $N(0,1)$ random variables, plot a histogram and the true density over it.

```
n = 1000;
X = randn(1,n);         % X is N(0,1)
nbins = 15;
minX = min(X);
maxX = max(X);
e = linspace(minX,maxX,nbins+1);
H = histc(X,e);
H(nbins) = H(nbins)+H(nbins+1);
H = H(1:nbins);                     % resize H
bw = (maxX-minX)/nbins;             % bin width
a = e(1:nbins);                     % left  edge sequence
b = e(2:nbins+1);                   % right edge sequence
bin_centers = (a+b)/2;              % bin centers
bar(bin_centers,H/(bw*n),'hist')
hold on
t = linspace(min(X),max(X),150);
y = exp(-t.^2/2)/sqrt(2*pi);
plot(t,y)
hold off
```

11. **MATLAB.** Modify the code in Problem 2 to plot a histogram of X, and using the estimated parameter value, draw the density on top of the histogram. If you studied the subsection on the chi-squared test, print out the chi-squared statistic Z, the critical value z_α for $\alpha = 0.05$, and whether or not the test accepts the density as a good fit to the data.

12. **MATLAB.** Modify the code in Problem 3 to plot a histogram of X, and using the estimated parameter value, draw the density on top of the histogram. If you studied the subsection on the chi-squared test, print out the chi-squared statistic Z, the critical value z_α for $\alpha = 0.05$, and whether or not the test accepts the density as a good fit to the data.

*13. **MATLAB.** Modify the code in Problem 4 to plot a histogram of X, and using the estimated parameter value, draw the density on top of the histogram. Use the noncentral chi-squared density formula given in Problem 25(c) in Chapter 5. If you studied the subsection on the chi-squared test, print out the chi-squared statistic Z, the critical value z_α for $\alpha = 0.05$, and whether or not the test accepts the density as a good fit to the data.

14. **MATLAB.** Modify the code in Problem 5 to plot a histogram of X, and using the estimated parameter value, draw the density on top of the histogram. If you studied the subsection on the chi-squared test, print out the chi-squared statistic Z, the critical value z_α for $\alpha = 0.05$, and whether or not the test accepts the density as a good fit to the data.

15. **MATLAB.** Modify the code in Problem 6 to plot a histogram of X, and using the estimated parameter value, draw the density on top of the histogram. If you studied the subsection on the chi-squared test, print out the chi-squared statistic Z, the critical value z_α for $\alpha = 0.05$, and whether or not the test accepts the density as a good fit to the data.

16. Show that
$$\frac{H_j - np_j}{\sqrt{np_j}}$$
has zero mean and variance $1 - p_j$.

6.3: Confidence intervals for the mean – known variance

17. Let F be the cdf any *even* density function f. Show that $F(-x) = 1 - F(x)$. In particular, note that the standard normal density is even.

18. If $\sigma^2 = 4$ and $n = 100$, how wide is the 99% confidence interval? How large would n have to be to have a 99% confidence interval of width less than or equal to $1/4$?

19. Let $W_1, W_2, \ldots$ be i.i.d. with zero mean and variance 4. Let $X_i = m + W_i$, where m is an unknown constant. If $M_{100} = 14.846$, find the 95% confidence interval.

20. Let $X_i = m + W_i$, where m is an unknown constant, and the W_i are i.i.d. Cauchy with parameter 1. Find $\delta > 0$ such that the probability is $2/3$ that the confidence interval $[M_n - \delta, M_n + \delta]$ contains m; i.e., find $\delta > 0$ such that
$$P(|M_n - m| \leq \delta) = 2/3.$$
Hints: Since $\mathsf{E}[W_i^2] = \infty$, the central limit theorem does not apply. However, you can solve for δ exactly if you can find the cdf of $M_n - m$. The cdf of W_i is $F(w) = \frac{1}{\pi}\tan^{-1}(w) + 1/2$, and the characteristic function of W_i is $\mathsf{E}[e^{jvW_i}] = e^{-|v|}$.

21. **MATLAB.** Use the following script to generate a vector of $n = 100$ Gaussian random numbers with mean $m = 3$ and variance one. Then compute the 95% confidence interval for the mean.

```
n = 100
X = randn(1,n);    % N(0,1) random numbers
X = X + 3;         % Change mean to 3
Mn = mean(X)
sigma = 1
delta = 1.96*sigma/sqrt(n)
fprintf('The 95%% confidence interval is [%g,%g]\n', ...
    Mn-delta,Mn+delta)
```

6.4: Confidence intervals for the mean – unknown variance

22. Let $X_1, X_2, \ldots$ be i.i.d. random variables with unknown, finite mean m and variance σ^2. If $M_{100} = 10.083$ and $S_{100} = 0.568$, find the 95% confidence interval for the population mean.

23. Suppose that 100 engineering freshmen are selected at random and $X_1, \ldots, X_{100}$ are their times (in years) to graduation. If $M_{100} = 4.422$ and $S_{100} = 0.957$, find the 93% confidence interval for their expected time to graduate.

24. From a batch of $N = 10\,000$ computers, $n = 100$ are sampled, and 10 are found defective. Estimate the number of defective computers in the total batch of 10 000, and give the margin of error for 90% probability if $S_{100} = 0.302$.

25. You conduct a presidential preference poll by surveying 3000 voters. You find that 1559 (more than half) say they plan to vote for candidate A, and the others say they plan to vote for candidate B. If $S_{3000} = 0.500$, are you 90% sure that candidate A will win the election? Are you 99% sure?

26. From a batch of 100 000 airbags, 500 are sampled, and 48 are found defective. Estimate the number of defective airbags in the total batch of 100 000, and give the margin of error for 94% probability if $S_{100} = 0.295$.

27. A new vaccine has just been developed at your company. You need to be 97% sure that side effects do not occur more than 10% of the time.

    (a) In order to estimate the probability p of side effects, the vaccine is tested on 100 volunteers. Side effects are experienced by 6 of the volunteers. Using the value $S_{100} = 0.239$, find the 97% confidence interval for p if $S_{100} = 0.239$. Are you 97% sure that $p \leq 0.1$?

    (b) Another study is performed, this time with 1000 volunteers. Side effects occur in 71 volunteers. Find the 97% confidence interval for the probability p of side effects if $S_{1000} = 0.257$. Are you 97% sure that $p \leq 0.1$?

28. Packet transmission times on a certain Internet link are independent and identically distributed. Assume that the times have an exponential density with mean μ.

    (a) Find the probability that in transmitting n packets, at least one of them takes more than t seconds to transmit.

    (b) Let T denote the total time to transmit n packets. Find a closed-form expression for the density of T.

    (c) Your answers to parts (a) and (b) depend on μ, which in practice is unknown and must be estimated. To estimate the expected transmission time, $n = 100$ packets are sent, and the transmission times $T_1, \ldots, T_n$ recorded. It is found that the sample mean $M_{100} = 1.994$, and sample standard deviation $S_{100} = 1.798$, where

    $$M_n := \frac{1}{n}\sum_{i=1}^{n} T_i \quad \text{and} \quad S_n^2 := \frac{1}{n-1}\sum_{i=1}^{n}(T_i - M_n)^2.$$

    Find the 95% confidence interval for the expected transmission time.

29. **MATLAB.** Use the following script to generate $n = 100$ Gaussian random variables with mean $m = 5$ and variance $\sigma^2 = 9$. Compute the 95% confidence interval for the mean.

```
n = 100
X = randn(1,n);    % N(0,1) random numbers
m = 5
sigma = 3
X = sigma*X + m;   % Change to N(m,sigma^2)
Mn = mean(X)
Sn = std(X)
delta = 1.96*Sn/sqrt(n)
fprintf('The 95%% confidence interval is [%g,%g]\n', ...
   Mn-delta,Mn+delta)
```

6.5: Confidence intervals for Gaussian data

30. If $X_1, \ldots, X_n$ are i.i.d. $N(m, \sigma^2)$, show that $Y_n = (M_n - m)/(\sigma/\sqrt{n})$ is $N(0,1)$. *Hint:* Recall Problem 55(a) in Chapter 4.

31. Let $W_1, W_2, \ldots$ be i.i.d. $N(0, \sigma^2)$ with σ^2 unknown. Let $X_i = m + W_i$, where m is an unknown constant. Suppose $M_{10} = 14.832$ and $S_{10} = 1.904$. Find the 95% confidence interval for m.

32. Let $X_1, X_2, \ldots$ be i.i.d. $N(0, \sigma^2)$ with σ^2 unknown. Find the 95% confidence interval for σ^2 if $V_{100}^2 = 4.413$.

33. Let $W_1, W_2, \ldots$ be i.i.d. $N(0, \sigma^2)$ with σ^2 unknown. Let $X_i = m + W_i$, where m is an unknown constant. Find the 95% confidence interval for σ^2 if $S_{100}^2 = 4.736$.

6.6: Hypothesis tests for the mean

34. In a two-sided test of the null hypothesis $m = m_0$ against the alternative hypothesis $m \neq m_0$, the statistic $Z_n = -1.80$ is observed. Is the null hypothesis accepted at the 0.05 significance level? If we are doing a one-sided test of the null hypothesis $m > m_0$ against the alternative hypothesis $m \leq m_0$ and $Z_n = -1.80$ is observed, do we accept the null hypothesis at the 0.05 significance level?

35. Show that if $\Phi(-y) = \alpha$, then $\Phi(y) = 1 - \alpha$.

36. An Internet service provider claims that a certain link has a packet loss probability of at most p_0. To test the claim, you send n packets and let $X_i = 1$ if the ith packet is lost and $X_i = 0$ otherwise. Mathematically, the claim is that $P(X_i = 1) = E[X_i] \leq p_0$. You compute the statistic Z_n in (6.25) and find $Z_n = 1.50$. (a) The Internet service provider takes $E[X_i] \leq p_0$ as the null hypothesis. On the basis of $Z_n = 1.50$, is the claim $E[X_i] \leq p_0$ accepted at the 0.06 significance level? (b) Being skeptical, you take $E[X_i] > p_0$ as the null hypothesis. On the basis of the same data $Z_n = 1.50$ and significance level 0.06, do you accept the Internet service provider's claim?

37. A computer vendor claims that the average waiting time on its technical support hotline is at most m_0 minutes. However, a consumer group claims otherwise based on the following analysis. The consumer group made n calls, letting X_i denote the waiting time on the ith call. It computed the statistic Z_n in (6.25) and found that $Z_n = 1.30$. Assuming that the group used a significance level from Table 6.6, what critical value did they use?

38. A drug company claims that its new medicine relieves pain for more than m_0 hours on average. To justify this claim, the company tested its medicine in n people. The ith person reported pain relief for X_i hours. The company computed the statistic Z_n in (6.25) and found that $Z_n = -1.60$. Can the company justify its claim if a 0.05 significance level is used? Explain your answer.

6.7: Regression and curve fitting

39. For the linear regression problem in Example 6.23, show that the minimum sum of squared errors, $e(\hat{g})$, is equal to $S_{YY} - S_{xY}^2/S_{xx}$, where this notation is defined in Examples 6.21 and 6.23.

40. **Regression and conditional expectation.** Let X and W be independent random variables with W having zero mean. If $Y := g(X) + W$, show that $\mathsf{E}[Y|X=x] = g(x)$.

41. **MATLAB.** Use the script below to plot the best-fit polynomial of degree $p = 2$ to the data. Note that the last two lines compute the sum of squared errors.

```
x = [ 1 2 3 4 5 6 7 8 9 ];
Y = [ 0.2631 0.2318 0.1330 0.6751 1.3649 1.5559, ...
      2.3184 3.7019 5.2953];
p = 2;
a = polyfit(x,Y,p)
subplot(2,2,1);         % Put multiple plots in same fig.
plot(x,Y,'o')           % Plot pnts only; do not connect.
axis([0 10 -1 7]);      % Force plot to use this scale.
subplot(2,2,2)
t=linspace(0,10,50);    % For plotting g from 0 to 10
gt = polyval(a,t);      % at 50 points.
plot(x,Y,'o',t,gt)
axis([0 10 -1 7]);      % Use same scale as prev. plot.
gx = polyval(a,x);      % Compute g(x_k) for each k.
sse = sum((Y-gx).^2)    % Compute sum of squared errors.
```

Do you get a smaller sum of squared errors with $p = 3$? What about $p = 7$ and $p = 8$? Is it a good idea to continue increasing p?

42. **MATLAB.** You can use the methods of this section to find polynomial approximations to nonpolynomial functions. Use the following script to plot the best-fit polynomial of degree $p = 4$ to $\sin(x)$ on $[0, 2\pi]$ based on five equal-spaced samples.

```
T = 2*pi;
x = linspace(0,T,5);
```

```
Y = sin(x);
p = 4;
a = polyfit(x,Y,p);
t = linspace(0,T,50);
st = sin(t);
gt = polyval(a,t);
subplot(2,1,1)
plot(t,st,t,gt)
subplot(2,1,2)
plot(t,st-gt)          % Plot error curve sin(t)-g(t).
```

Since the values of $\sin(x)$ for $\pi/2 \leq x \leq 2\pi$ can be computed using values of $\sin(x)$ for $0 \leq x \leq \pi/2$, modify the above code by setting $T = \pi/2$. Do you get a better approximation now that you have restricted attention to $[0, \pi/2]$?

43. **MATLAB.** The data shown at the left in Figure 6.8 appears to follow a power law of the form c/t^q, where c and q are to be estimated. Instead of fitting a polynomial to the data, consider taking logarithms to get

$$\ln(c/t^q) = \ln c - q \ln t.$$

Let us denote the points at the left in Figure 6.8 by (t_k, Z_k), and put $Y_k := \ln Z_k$ and $x_k := \ln t_k$. A plot of (x_k, Y_k) and the best-fit straight line through it are shown at the right in Figure 6.8. How would you estimate c and q from the best-fit straight line? Use your answer to fill in the two blanks in the code below. Then run the code to see a comparison of the best cubic fit to the data and a plot of $\hat{c}/t^{\hat{q}}$. If you change from a cubic to higher-order polynomials, can you get a better plot than with the log–log method?

```
t = [ 1 1.4444 1.8889 2.3333 2.7778 3.2222 3.6667, ...
      4.1111 4.5556 5 ];
Z = [ 1.0310 0.6395 0.3404 0.2873 0.2090 0.1147, ...
      0.2016 0.1192 0.1297 0.0536 ];
x = log(t);
Y = log(Z);
subplot(2,2,1)
a = polyfit(t,Z,3);      % Fit cubic to data (t_k,Z_k).
```

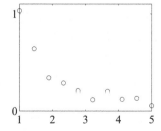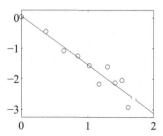

Figure 6.8. Data (t_k, Z_k) for Problem 43 (left). Log of data $(\ln t_k, \ln Z_k)$ and best-fit straight line (right).

```
u = linspace(1,5,50);
v = polyval(a,u);
plot(t,Z,'o',u,v)      % Plot (t_k,Z_k) & cubic.
axis([1 5 0 1.1])
title('Best-fit cubic to data')
subplot(2,2,2)
a = polyfit(x,Y,1);    % Fit st. line to (x_k,Y_k).
u = linspace(0,2,2);
v = polyval(a,u);
plot(x,Y,'o',u,v)      % Plot (x_k,Y_k) & st. line.
title('Best-fit straight line to (ln(t_k),ln(Z_k))')
subplot(2,2,3)
qhat = _____
chat = _____
u = linspace(1,5,50);
v = chat./u.^qhat;
plot(t,Z,'o',u,v)
axis([1 5 0 1.1]) % Plot (t_k,Z_k) & c/t^q using estimates.
title('(estimate of c)/t\^(estimate of q)')
```

6.8: *Monte Carlo estimation

44. For the tilted density $f_{\tilde{Z}}(z) = e^{sz} f_Z(z)/M_Z(s)$, show that

$$\frac{f_Z(z)}{f_{\tilde{Z}}(z)} = e^{-sz} M_Z(s).$$

45. If $Z \sim N(0,1)$, find the tilted density $f_{\tilde{Z}}(z) = e^{sz} f_Z(z)/M_Z(s)$. How would you choose s to make $\mathsf{E}[\tilde{Z}] = t$?

46. If $Z \sim \text{gamma}(p,\lambda)$, find the tilted density $f_{\tilde{Z}}(z) = e^{sz} f_Z(z)/M_Z(s)$. How would you choose s to make $\mathsf{E}[\tilde{Z}] = t$? Note that the gamma includes the Erlang and chi-squared as special cases.

47. **MATLAB.** If $Z \sim N(0,1)$, use the following script to estimate $\mathsf{P}(Z > t)$ for $t = 5$ with 95% confidence.

```
t = 5;
s = t;
n = 100;
Z = randn(1,n); % N(0,1) random numbers
Zt = Z+s; % change mean to s
X = zeros(1,n);
i = find(Zt>t);
X(i) = exp(-s*Zt(i))*exp(s^2/2);
Mn = mean(X);
Sn = std(X);
delta = Sn*1.96/sqrt(n);
fprintf('M(%7i)  = %g +/- %g, Sn = %g\n',...
```

```
        n,Mn,delta,Sn)
fprintf('The 95%% confidence interval is [%g,%g]\n', ...
    Mn-delta,Mn+delta)
```

48. It is also possible to tilt probability mass functions. The formula for tilting the probability mass function of a discrete random variable taking values z_i is

$$p_{\widetilde{Z}}(z_i) := e^{sz_i} p_Z(z_i)/M_Z(s).$$

If $Z \sim \text{Bernoulli}(p)$, find the tilted pmf $p_{\widetilde{Z}}(i)$ for $i = 0, 1$.

Exam preparation

You may use the following suggestions to prepare a study sheet, including formulas mentioned that you have trouble remembering. You may also want to ask your instructor for additional suggestions.

6.1. Parameter estimators and their properties. Know the sample mean (6.1) and sample variance (6.2). Know the meaning of unbiased and the fact that the sample mean and sample variance are unbiased estimators of the population (or ensemble) mean and variance. Know how to derive estimators of parameters that are related to moments.

6.2. Histograms. Understand how the pieces of code in the text can be collected to solve problems in MATLAB. Be able to explain how the chi-squared test works.

6.3. Confidence intervals for the mean – known variance. Know formulas (6.11) and (6.12) and how to find $y_{\alpha/2}$ from Table 6.2.

6.4. Confidence intervals for the mean – unknown variance. Know formulas (6.16) and (6.17) and how to find $y_{\alpha/2}$ from Table 6.2. Know how to apply these results to estimating the number of defective items in a lot.

6.5. Confidence intervals for Gaussian data. For estimating the mean with unknown variance, use formulas (6.16) and (6.17), except that $y_{\alpha/2}$ is chosen from Table 6.3. To estimate the variance when the mean is known, use (6.19) with ℓ and u chosen from Table 6.4. To estimate the variance when the mean is unknown, use (6.20) with ℓ and u chosen from Table 6.4.

6.6. Hypothesis tests for the mean. Know when to use the appropriate statistic (6.24) or (6.25). For testing $m = m_0$, use the critical value $y_{\alpha/2}$ in Table 6.6; accept the hypothesis if $|Z_n| \leq y_{\alpha/2}$. For testing $m \leq m_0$, use the critical value y_α in Table 6.6; accept the hypothesis if $Z_n \leq y_\alpha$. For testing $m > m_0$, use the critical value $-y_\alpha$, where y_α is from Table 6.6; accept the hypothesis if $Z_n > -y_\alpha$.

6.7. Regression and curve fitting. Regression is another name for curve fitting. In the model (6.29), the W_k can account for either measurement noise or inaccuracies in $g(x)$. To give an example of the latter case, consider approximating $\sin(x)$ by a polynomial $g(x)$. If we put $W_k := \sin(x_k) - g(x_k)$ and $Y_k := \sin(x_k)$, then (6.29) holds.

6.8. *Monte Carlo estimation. If you are not using importance sampling or some sophisticated technique, and you want to estimate a very small probability, you will need a correspondingly large number of simulations. Know the formula for the tilted density (6.33).

Work any review problems assigned by your instructor. If you finish them, re-work your homework assignments.

7
Bivariate random variables

The main focus of this chapter is the study of pairs of continuous random variables that are not independent. In particular, conditional probability and conditional expectation along with corresponding laws of total probability and substitution are studied. These tools are used to compute probabilities involving the output of systems with two (and sometimes three or more) random inputs.

7.1 Joint and marginal probabilities

Consider the following functions of two random variables X and Y,

$$X+Y, \quad XY, \quad \max(X,Y), \quad \text{and} \quad \min(X,Y).$$

For example, in a telephone channel the signal X is corrupted by additive noise Y. In a wireless channel, the signal X is corrupted by fading (multiplicative noise). If X and Y are the traffic rates at two different routers of an Internet service provider, it is desirable to have these rates less than the router capacity, say u; i.e., we want $\max(X,Y) \leq u$. If X and Y are sensor voltages, we may want to trigger an alarm if at least one of the sensor voltages falls below a threshold v; e.g., if $\min(X,Y) \leq v$. We now show that the cdfs of these four functions of X and Y can be expressed in the form $P((X,Y) \in A)$ for various sets[1] $A \subset \mathbb{R}^2$. We then argue that such probabilities can be computed in terms of the joint cumulative distribution function to be defined later in the section.

Before proceeding, you should re-work Problem 6 in Chapter 1.

Example 7.1 (signal in additive noise). A random signal X is transmitted over a channel subject to additive noise Y. The received signal is $Z = X+Y$. Express the cdf of Z in the form $P((X,Y) \in A_z)$ for some set A_z.

Solution. Write

$$F_Z(z) = P(Z \leq z) = P(X+Y \leq z) = P((X,Y) \in A_z),$$

where

$$A_z := \{(x,y) : x+y \leq z\}.$$

Since $x+y \leq z$ if and only if $y \leq -x+z$, it is easy to see that A_z is the shaded region in Figure 7.1.

Example 7.2 (signal in multiplicative noise). A random signal X is transmitted over a channel subject to multiplicative noise Y. The received signal is $Z = XY$. Express the cdf of Z in the form $P((X,Y) \in A_z)$ for some set A_z.

Bivariate random variables

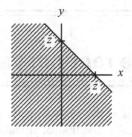

Figure 7.1. The shaded region is $A_z = \{(x,y) : x+y \leq z\}$. The equation of the diagonal line is $y = -x + z$.

Solution. Write

$$F_Z(z) = P(Z \leq z) = P(XY \leq z) = P((X,Y) \in A_z),$$

where now $A_z := \{(x,y) : xy \leq z\}$. To see how to sketch this set, we are tempted to write $A_z = \{(x,y) : y \leq z/x\}$, **but this would be wrong** because if $x < 0$ we need to reverse the inequality. To get around this problem, it is convenient to partition A_z into two disjoint regions, $A_z = A_z^+ \cup A_z^-$, where

$$A_z^+ := A_z \cap \{(x,y) : x > 0\} \quad \text{and} \quad A_z^- := A_z \cap \{(x,y) : x < 0\}.$$

Thus, A_z^+ and A_z^- are similar to A_z, but now we know the sign of x in each set. Hence, it is correct to write

$$A_z^+ := \{(x,y) : y \leq z/x \text{ and } x > 0\}$$

and

$$A_z^- := \{(x,y) : y \geq z/x \text{ and } x < 0\}.$$

These regions are sketched in Figure 7.2.

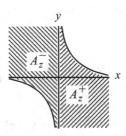

Figure 7.2. The curve is $y = z/x$. The shaded region to the left of the vertical axis is $A_z^- = \{(x,y) : y \geq z/x, x < 0\}$, and the shaded region to the right of the vertical axis is $A_z^+ = \{(x,y) : y \leq z/x, x > 0\}$. The sketch is for the case $z > 0$. How would the sketch need to change if $z = 0$ or if $z < 0$?

Example 7.3. Express the cdf of $U := \max(X,Y)$ in the form $P((X,Y) \in A_u)$ for some set A_u.

7.1 Joint and marginal probabilities

Solution. To find the cdf of U, begin with

$$F_U(u) = \mathsf{P}(U \le u) = \mathsf{P}(\max(X,Y) \le u).$$

Since the larger of X and Y is less than or equal to u if and only if $X \le u$ and $Y \le u$,

$$\mathsf{P}(\max(X,Y) \le u) = \mathsf{P}(X \le u, Y \le u) = \mathsf{P}((X,Y) \in A_u),$$

where $A_u := \{(x,y) : x \le u \text{ and } y \le u\}$ is the shaded "southwest" region shown in Figure 7.3(a).

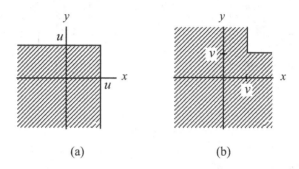

(a) (b)

Figure 7.3. (a) Southwest region $\{(x,y) : x \le u \text{ and } y \le u\}$. (b) The region $\{(x,y) : x \le v \text{ or } y \le v\}$.

Example 7.4. Express the cdf of $V := \min(X,Y)$ in the form $\mathsf{P}((X,Y) \in A_v)$ for some set A_v.

Solution. To find the cdf of V, begin with

$$F_V(v) = \mathsf{P}(V \le v) = \mathsf{P}(\min(X,Y) \le v).$$

Since the smaller of X and Y is less than or equal to v if and only either $X \le v$ or $Y \le v$,

$$\mathsf{P}(\min(X,Y) \le v) = \mathsf{P}(X \le v \text{ or } Y \le v) = \mathsf{P}((X,Y) \in A_v),$$

where $A_v := \{(x,y) : x \le v \text{ or } y \le v\}$ is the shaded region shown in Figure 7.3(b).

Product sets and marginal probabilities

The **Cartesian product** of two univariate sets B and C is defined by

$$B \times C := \{(x,y) : x \in B \text{ and } y \in C\}.$$

In other words,

$$(x,y) \in B \times C \iff x \in B \text{ and } y \in C.$$

For example, if $B = [1,3]$ and $C = [0.5, 3.5]$, then $B \times C$ is the rectangle

$$[1,3] \times [0.5, 3.5] = \{(x,y) : 1 \leq x \leq 3 \text{ and } 0.5 \leq y \leq 3.5\},$$

which is illustrated in Figure 7.4(a). In general, if B and C are intervals, then $B \times C$ is a rectangle or square. If one of the sets is an interval and the other is a singleton, then the product set degenerates to a line segment in the plane. A more complicated example is shown in Figure 7.4(b), which illustrates the product $\big([1,2] \cup [3,4]\big) \times [1,4]$. Figure 7.4(b) also illustrates the general result that $\times$ distributes over $\cup$; i.e., $(B_1 \cup B_2) \times C = (B_1 \times C) \cup (B_2 \times C)$.

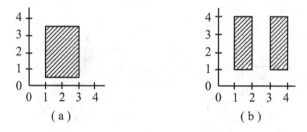

Figure 7.4. The Cartesian products (a) $[1,3] \times [0.5, 3.5]$ and (b) $\big([1,2] \cup [3,4]\big) \times [1,4]$.

Using the notion of product set,

$$\{X \in B, Y \in C\} = \{\omega \in \Omega : X(\omega) \in B \text{ and } Y(\omega) \in C\}$$
$$= \{\omega \in \Omega : (X(\omega), Y(\omega)) \in B \times C\},$$

for which we use the shorthand

$$\{(X,Y) \in B \times C\}.$$

We can therefore write

$$\mathsf{P}(X \in B, Y \in C) = \mathsf{P}((X,Y) \in B \times C).$$

The preceding expression allows us to obtain the **marginal probability** $\mathsf{P}(X \in B)$ as follows. First, for any event E, we have $E \subset \Omega$, and therefore, $E = E \cap \Omega$. Second, Y is assumed to be a real-valued random variable, i.e., $Y(\omega) \in \mathbb{R}$ for all ω. Thus, $\{Y \in \mathbb{R}\} = \Omega$. Now write

$$\mathsf{P}(X \in B) = \mathsf{P}(\{X \in B\} \cap \Omega)$$
$$= \mathsf{P}(\{X \in B\} \cap \{Y \in \mathbb{R}\})$$
$$= \mathsf{P}(X \in B, Y \in \mathbb{R})$$
$$= \mathsf{P}((X,Y) \in B \times \mathbb{R}).$$

Similarly,

$$\mathsf{P}(Y \in C) = \mathsf{P}((X,Y) \in \mathbb{R} \times C). \qquad (7.1)$$

7.1 Joint and marginal probabilities

Joint cumulative distribution functions

The **joint cumulative distribution function** of X and Y is defined by

$$F_{XY}(x,y) := \mathsf{P}(X \leq x, Y \leq y).$$

We can also write this using a Cartesian product set as

$$F_{XY}(x,y) = \mathsf{P}((X,Y) \in (-\infty,x] \times (-\infty,y]).$$

In other words, $F_{XY}(x,y)$ is the probability that (X,Y) lies in the southwest region shown in Figure 7.5(a).

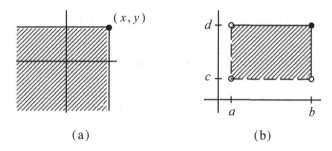

Figure 7.5. (a) Southwest region $(-\infty,x] \times (-\infty,y]$. (b) Rectangle $(a,b] \times (c,d]$.

The joint cdf is important because it can be used to compute $\mathsf{P}((X,Y) \in A)$ for any set A. For example, you will show in Problems 3 and 4 that

$$\mathsf{P}(a < X \leq b, c < Y \leq d),$$

which is the probability that (X,Y) belongs to the rectangle $(a,b] \times (c,d]$ shown in Figure 7.5(b), is given by the **rectangle formula**[2]

$$F_{XY}(b,d) - F_{XY}(a,d) - F_{XY}(b,c) + F_{XY}(a,c). \tag{7.2}$$

Example 7.5. If X and Y have joint cdf F_{XY}, find the joint cdf of $U := \max(X,Y)$ and $V := \min(X,Y)$.

Solution. Begin with
$$F_{UV}(u,v) = \mathsf{P}(U \leq u, V \leq v).$$

From Example 7.3, we know that $U = \max(X,Y) \leq u$ if and only if (X,Y) lies in the southwest region shown in Figure 7.3(a). Similarly, from Example 7.4, we know that $V = \min(X,Y) \leq v$ if and only if (X,Y) lies in the region shown in Figure 7.3(b). Hence, $U \leq u$ and $V \leq v$ if and only if (X,Y) lies in the intersection of these two regions. The form of this intersection depends on whether $u > v$ or $u \leq v$. If $u \leq v$, then the southwest region

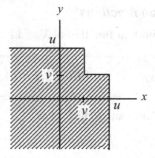

Figure 7.6. The intersection of the shaded regions of Figures 7.3(a) and 7.3(b) when $v < u$.

in Figure 7.3(a) is a subset of the region in Figure 7.3(b). Their intersection is the smaller set, and so

$$P(U \leq u, V \leq v) = P(U \leq u) = F_U(u) = F_{XY}(u,u), \quad u \leq v.$$

If $u > v$, the intersection is shown in Figure 7.6. Since this region can be obtained by removing the rectangle $(v, u] \times (v, u]$ from the southwest region $(-\infty, u] \times (-\infty, u]$,

$$P(U \leq u, V \leq v) = F_{XY}(u,u) - P(v < X \leq u, v < Y \leq u).$$

This last probability is given by the rectangle formula (7.2),

$$F_{XY}(u,u) - F_{XY}(v,u) - F_{XY}(u,v) + F_{XY}(v,v).$$

Hence,

$$F_{UV}(u,v) = F_{XY}(v,u) + F_{XY}(u,v) - F_{XY}(v,v), \quad u > v.$$

The complete joint cdf formula is

$$F_{UV}(u,v) = \begin{cases} F_{XY}(u,u), & u \leq v, \\ F_{XY}(v,u) + F_{XY}(u,v) - F_{XY}(v,v), & u > v. \end{cases}$$

Marginal cumulative distribution functions

It is possible to obtain the **marginal cumulative distributions** F_X and F_Y directly from F_{XY} by setting the unwanted variable to ∞. More precisely, it can be shown that[3]

$$F_X(x) = \lim_{y \to \infty} F_{XY}(x,y) =: F_{XY}(x, \infty), \qquad (7.3)$$

and

$$F_Y(y) = \lim_{x \to \infty} F_{XY}(x,y) =: F_{XY}(\infty, y). \qquad (7.4)$$

7.1 Joint and marginal probabilities

Example 7.6. Use the joint cdf F_{UV} derived in Example 7.5 to compute the marginal cdfs F_U and F_V.

Solution. To compute
$$F_U(u) = \lim_{v \to \infty} F_{UV}(u,v),$$
observe that as v becomes large, eventually it will be greater than u. For $v \geq u$, $F_{UV}(u,v) = F_{XY}(u,u)$. In other words, for $v \geq u$, $F_{UV}(u,v)$ is constant and no longer depends on v. Hence, the limiting value is also $F_{XY}(u,u)$.

To compute
$$F_V(v) = \lim_{u \to \infty} F_{UV}(u,v),$$
observe that as u becomes large, eventually it will be greater than v. For $u > v$,
$$F_{UV}(u,v) = F_{XY}(v,u) + F_{XY}(u,v) - F_{XY}(v,v)$$
$$\to F_X(v) + F_Y(v) - F_{XY}(v,v)$$
as $u \to \infty$.

To check the preceding result, we compute F_U and F_V directly. From Example 7.3,
$$F_U(u) = \mathsf{P}(\max(X,Y) \leq u) = \mathsf{P}(X \leq u, Y \leq u) = F_{XY}(u,u).$$
From Example 7.4,
$$F_V(v) = \mathsf{P}(X \leq v \text{ or } Y \leq v).$$
By the inclusion–exclusion formula (1.12),
$$F_V(v) = \mathsf{P}(X \leq v) + \mathsf{P}(Y \leq v) - \mathsf{P}(X \leq v, Y \leq v)$$
$$= F_X(v) + F_Y(v) - F_{XY}(v,v).$$

The foregoing shows how to compute the cdfs of $\max(X,Y)$ and $\min(X,Y)$ in terms of the joint cdf F_{XY}. Computation of the cdfs of $X+Y$ and XY in terms of F_{XY} can only be done in a limiting sense by chopping up the regions A_z of Figures 7.1 and 7.2 into small rectangles, applying the rectangle formula (7.2) to each rectangle, and adding up the results.

To conclude this subsection, we give another application of (7.3) and (7.4).

Example 7.7. If
$$F_{XY}(x,y) = \begin{cases} \dfrac{y + e^{-x(y+1)}}{y+1} - e^{-x}, & x,y > 0, \\ 0, & \text{otherwise,} \end{cases}$$
find both of the marginal cumulative distribution functions, $F_X(x)$ and $F_Y(y)$.

Solution. For $x,y > 0$,
$$F_{XY}(x,y) = \frac{y}{y+1} + \frac{1}{y+1} \cdot e^{-x(y+1)} - e^{-x}.$$

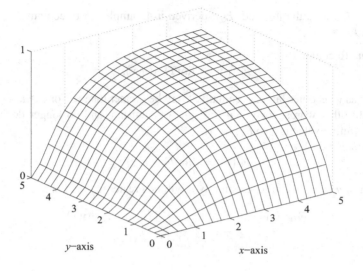

Figure 7.7. Joint cumulative distribution function $F_{XY}(x,y)$ of Example 7.7.

(This surface is shown in Figure 7.7.) Hence, for $x > 0$,
$$\lim_{y \to \infty} F_{XY}(x,y) = 1 + 0 \cdot 0 - e^{-x} = 1 - e^{-x}.$$

For $x \leq 0$, $F_{XY}(x,y) = 0$ for all y. So, for $x \leq 0$, $\lim_{y \to \infty} F_{XY}(x,y) = 0$. The complete formula for the marginal cdf of X is

$$F_X(x) = \begin{cases} 1 - e^{-x}, & x > 0, \\ 0, & x \leq 0, \end{cases} \quad (7.5)$$

which implies $X \sim \exp(1)$. Next, for $y > 0$,
$$\lim_{x \to \infty} F_{XY}(x,y) = \frac{y}{y+1} + \frac{1}{y+1} \cdot 0 - 0 = \frac{y}{y+1}.$$

We then see that the marginal cdf of Y is

$$F_Y(y) = \begin{cases} y/(y+1), & y > 0, \\ 0, & y \leq 0. \end{cases} \quad (7.6)$$

Independent random variables

Recall that X and Y are independent if and only if $\mathsf{P}(X \in B, Y \in C) = \mathsf{P}(X \in B)\mathsf{P}(Y \in C)$ for all sets B and C. In terms of product sets, this says that

$$\mathsf{P}((X,Y) \in B \times C) = \mathsf{P}(X \in B)\mathsf{P}(Y \in C). \quad (7.7)$$

In other words, the probability that (X,Y) belongs to a Cartesian-product set is the product of the individual probabilities. In particular, if X and Y are independent, the joint cdf factors into

$$F_{XY}(x,y) = \mathsf{P}(X \leq x, Y \leq y) = F_X(x)F_Y(y).$$

Example 7.8. Show that X and Y of Example 7.7 are not independent.

Solution. Using the results of Example 7.7, for any $x,y > 0$,

$$F_X(x)F_Y(y) = (1-e^{-x})\frac{y}{y+1} \neq \frac{y+e^{-x(y+1)}}{y+1} - e^{-x}.$$

As noted above, if X and Y are independent, then their joint cdf factors. The converse is also true; i.e., if $F_{XY}(x,y) = F_X(x)F_Y(y)$ for all x,y, then X and Y are independent in the sense that (7.7) holds for all sets B and C. We prove this only for the case of $B = (a,b]$ and $C = (c,d]$. Since $B \times C = (a,b] \times (c,d]$ is a rectangle, the left-hand side of (7.7) is given by the rectangle formula (7.2). Since we are assuming the joint cdf factors, (7.2) becomes

$$F_X(b)F_Y(d) - F_X(a)F_Y(d) - F_X(b)F_Y(c) + F_X(a)F_Y(c),$$

which factors into

$$[F_X(b) - F_X(a)]F_Y(d) - [F_X(b) - F_X(a)]F_Y(c)$$

or

$$[F_X(b) - F_X(a)][F_Y(d) - F_Y(c)],$$

which is the product $\mathsf{P}(X \in (a,b])\mathsf{P}(Y \in (c,d])$ required for the right-hand side of (7.7).

We thus record here that X and Y are **independent** if and only if their joint cdf factors into the product of the marginal cdfs,

$$F_{XY}(x,y) = F_X(x)F_Y(y).$$

7.2 Jointly continuous random variables

In analogy with the univariate case, we say that two random variables X and Y are **jointly continuous**[4] with **joint density** $f_{XY}(x,y)$ if

$$\mathsf{P}((X,Y) \in A) = \iint_A f_{XY}(x,y)\,dx\,dy$$

for some nonnegative function f_{XY} that integrates to one; i.e.,

$$\int_{-\infty}^{\infty}\int_{-\infty}^{\infty} f_{XY}(x,y)\,dx\,dy = 1.$$

Sketches of several joint densities are shown in Figure 7.8 below and in Figures 7.9–7.11 in Section 7.4.

Caution. It is possible to have two continuous random variables X and Y that are not jointly continuous. In other words, X has a density $f_X(x)$ and Y has a density $f_Y(y)$, but there is no joint density $f_{XY}(x,y)$. An example is given at the end of the section.

Example 7.9. Show that

$$f_{XY}(x,y) = \frac{1}{2\pi}e^{-(2x^2-2xy+y^2)/2}$$

is a valid joint probability density.

Solution. Since $f_{XY}(x,y)$ is nonnegative, all we have to do is show that it integrates to one. By completing the square in the exponent, we obtain

$$f_{XY}(x,y) = \frac{e^{-(y-x)^2/2}}{\sqrt{2\pi}} \cdot \frac{e^{-x^2/2}}{\sqrt{2\pi}}.$$

This factorization allows us to write the double integral

$$\int_{-\infty}^{\infty}\int_{-\infty}^{\infty} f_{XY}(x,y)\,dx\,dy = \int_{-\infty}^{\infty}\int_{-\infty}^{\infty} \frac{e^{-(2x^2-2xy+y^2)/2}}{2\pi}\,dx\,dy$$

as the iterated integral

$$\int_{-\infty}^{\infty} \frac{e^{-x^2/2}}{\sqrt{2\pi}} \left(\int_{-\infty}^{\infty} \frac{e^{-(y-x)^2/2}}{\sqrt{2\pi}}\,dy \right) dx.$$

The inner integral, as a function of y, is a normal density with mean x and variance one. Hence, the inner integral is one. But this leaves only the outer integral, whose integrand is an $N(0,1)$ density, which also integrates to one.

Example 7.10 (signal in additive noise, continued). Suppose that a random, continuous-valued signal X is transmitted over a channel subject to additive, continuous-valued noise Y. The received signal is $Z = X + Y$. Find the cdf and density of Z if X and Y are jointly continuous random variables with joint density f_{XY}.

Solution. As in Example 7.1, write

$$F_Z(z) = \mathsf{P}(Z \le z) = \mathsf{P}(X+Y \le z) = \mathsf{P}((X,Y) \in A_z),$$

where $A_z := \{(x,y) : x+y \le z\}$ was sketched in Figure 7.1. With this figure in mind, the double integral for $\mathsf{P}((X,Y) \in A_z)$ can be computed using the iterated integral

$$F_Z(z) = \int_{-\infty}^{\infty} \left[\int_{-\infty}^{z-x} f_{XY}(x,y)\,dy \right] dx.$$

Now carefully differentiate with respect to z. Write[a]

$$f_Z(z) = \frac{\partial}{\partial z} \int_{-\infty}^{\infty} \left[\int_{-\infty}^{z-x} f_{XY}(x,y)\,dy \right] dx$$

$$= \int_{-\infty}^{\infty} \frac{\partial}{\partial z} \left[\int_{-\infty}^{z-x} f_{XY}(x,y)\,dy \right] dx$$

$$= \int_{-\infty}^{\infty} f_{XY}(x, z-x)\,dx.$$

---

[a]Recall that

$$\frac{\partial}{\partial z} \int_{-\infty}^{g(z)} h(y)\,dy = h(g(z))g'(z).$$

If $g(z) = z - x$, then $g'(z) = 1$. See Note **7** for the general case.

7.2 Jointly continuous random variables

***Example* 7.11** (signal in multiplicative noise, continued). A random, continuous-valued signal X is transmitted over a channel subject to multiplicative, continuous-valued noise Y. The received signal is $Z = XY$. Find the cdf and density of Z if X and Y are jointly continuous random variables with joint density f_{XY}.

Solution. As in Example 7.2, write

$$F_Z(z) = \mathsf{P}(Z \leq z) = \mathsf{P}(XY \leq z) = \mathsf{P}((X,Y) \in A_z),$$

where $A_z := \{(x,y) : xy \leq z\}$ is partitioned into two disjoint regions, $A_z = A_z^+ \cup A_z^-$, as sketched in Figure 7.2. Next, since

$$F_Z(z) = \mathsf{P}((X,Y) \in A_z^-) + \mathsf{P}((X,Y) \in A_z^+),$$

we proceed to compute these two terms. Write

$$\mathsf{P}((X,Y) \in A_z^+) = \int_0^\infty \left[\int_{-\infty}^{z/x} f_{XY}(x,y) \, dy \right] dx$$

and

$$\mathsf{P}((X,Y) \in A_z^-) = \int_{-\infty}^0 \left[\int_{z/x}^\infty f_{XY}(x,y) \, dy \right] dx.$$

It follows that[b]

$$f_Z(z) = \int_0^\infty f_{XY}(x, \tfrac{z}{x}) \tfrac{1}{x} \, dx - \int_{-\infty}^0 f_{XY}(x, \tfrac{z}{x}) \tfrac{1}{x} \, dx.$$

In the first integral on the right, the range of integration implies x is positive, and so we can replace $1/x$ with $1/|x|$. In the second integral on the right, the range of integration implies x is negative, and so we can replace $1/(-x)$ with $1/|x|$. Hence,

$$f_Z(z) = \int_0^\infty f_{XY}(x, \tfrac{z}{x}) \tfrac{1}{|x|} \, dx + \int_{-\infty}^0 f_{XY}(x, \tfrac{z}{x}) \tfrac{1}{|x|} \, dx.$$

Now that the integrands are the same, the two integrals can be combined to get

$$f_Z(z) = \int_{-\infty}^\infty f_{XY}(x, \tfrac{z}{x}) \tfrac{1}{|x|} \, dx.$$

Joint and marginal densities

In this section we first show how to obtain the joint density $f_{XY}(x,y)$ from the joint cdf $F_{XY}(x,y)$. Then we show how to obtain the **marginal densities** $f_X(x)$ and $f_Y(y)$ from the joint density $f_{XY}(x,y)$.

---

[b]Recall that

$$\frac{\partial}{\partial z} \int_{g(z)}^\infty h(y) \, dy = -h(g(z)) g'(z).$$

If $g(z) = z/x$, then $g'(z) = 1/x$. See Note **7** for the general case.

To begin, write

$$P(X \in B, Y \in C) = P((X,Y) \in B \times C)$$
$$= \iint_{B \times C} f_{XY}(x,y)\,dx\,dy$$
$$= \int_B \left(\int_C f_{XY}(x,y)\,dy \right) dx \qquad (7.8)$$
$$= \int_C \left(\int_B f_{XY}(x,y)\,dx \right) dy.$$

At this point we would like to substitute $B = (-\infty, x]$ and $C = (-\infty, y]$ in order to obtain expressions for $F_{XY}(x,y)$. However, the preceding integrals already use x and y for the variables of integration. To avoid confusion, we must first replace the variables of integration. We change x to t and y to τ. We then find that

$$F_{XY}(x,y) = \int_{-\infty}^{x} \left(\int_{-\infty}^{y} f_{XY}(t,\tau)\,d\tau \right) dt,$$

or, equivalently,

$$F_{XY}(x,y) = \int_{-\infty}^{y} \left(\int_{-\infty}^{x} f_{XY}(t,\tau)\,dt \right) d\tau.$$

It then follows that

$$\boxed{\frac{\partial^2}{\partial y \partial x} F_{XY}(x,y) = f_{XY}(x,y) \quad \text{and} \quad \frac{\partial^2}{\partial x \partial y} F_{XY}(x,y) = f_{XY}(x,y).} \qquad (7.9)$$

Example 7.12. Let

$$F_{XY}(x,y) = \begin{cases} \dfrac{y + e^{-x(y+1)}}{y+1} - e^{-x}, & x,y > 0, \\ 0, & \text{otherwise,} \end{cases}$$

as in Example 7.7. Find the joint density f_{XY}.

Solution. For $x, y > 0$,

$$\frac{\partial}{\partial x} F_{XY}(x,y) = e^{-x} - e^{-x(y+1)},$$

and

$$\frac{\partial^2}{\partial y \partial x} F_{XY}(x,y) = x e^{-x(y+1)}.$$

Thus,

$$f_{XY}(x,y) = \begin{cases} x e^{-x(y+1)}, & x,y > 0, \\ 0, & \text{otherwise.} \end{cases}$$

This surface is shown in Figure 7.8.

7.2 Jointly continuous random variables

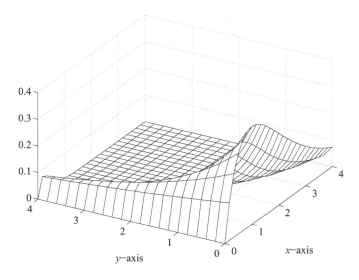

Figure 7.8. The joint density $f_{XY}(x,y) = xe^{-x(y+1)}$ of Example 7.12.

We now show that if X and Y are jointly continuous, then X and Y are individually continuous with marginal densities obtained as follows. Taking $C = \mathbb{R}$ in (7.8), we obtain

$$P(X \in B) = P((X,Y) \in B \times \mathbb{R}) = \int_B \left(\int_{-\infty}^{\infty} f_{XY}(x,y)\, dy \right) dx,$$

which implies that the inner integral is the **marginal density** of X, i.e.,

$$\boxed{f_X(x) = \int_{-\infty}^{\infty} f_{XY}(x,y)\, dy.} \quad (7.10)$$

Similarly,

$$P(Y \in C) = P((X,Y) \in \mathbb{R} \times C) = \int_C \left(\int_{-\infty}^{\infty} f_{XY}(x,y)\, dx \right) dy,$$

and

$$\boxed{f_Y(y) = \int_{-\infty}^{\infty} f_{XY}(x,y)\, dx.}$$

Thus, to obtain the marginal densities, integrate out the unwanted variable.

Example 7.13. Using the joint density f_{XY} obtained in Example 7.12, find the marginal densities f_X and f_Y by integrating out the unneeded variable. To check your answer, also compute the marginal densities by differentiating the marginal cdfs obtained in Example 7.7.

Solution. We first compute $f_X(x)$. To begin, observe that for $x \leq 0$, $f_{XY}(x,y) = 0$. Hence, for $x \leq 0$, the integral in (7.10) is zero. Now suppose $x > 0$. Since $f_{XY}(x,y) = 0$

whenever $y \leq 0$, the lower limit of integration in (7.10) can be changed to zero. For $x > 0$, it remains to compute

$$\int_0^\infty f_{XY}(x,y)\,dy = xe^{-x}\int_0^\infty e^{-xy}\,dy$$
$$= e^{-x}.$$

Hence,
$$f_X(x) = \begin{cases} e^{-x}, & x > 0, \\ 0, & x \leq 0, \end{cases}$$

and we see that X is exponentially distributed with parameter $\lambda = 1$. Note that the same answer can be obtained by differentiating the formula for $F_X(x)$ in (7.5).

We now turn to the calculation of $f_Y(y)$. Arguing as above, we have $f_Y(y) = 0$ for $y \leq 0$, and $f_Y(y) = \int_0^\infty f_{XY}(x,y)\,dx$ for $y > 0$. Write this integral as

$$\int_0^\infty f_{XY}(x,y)\,dx = \frac{1}{y+1}\int_0^\infty x \cdot (y+1)e^{-(y+1)x}\,dx. \qquad (7.11)$$

If we put $\lambda = y+1$, then the integral on the right has the form

$$\int_0^\infty x \cdot \lambda e^{-\lambda x}\,dx,$$

which is the mean of an exponential random variable with parameter λ. This integral is equal to $1/\lambda = 1/(y+1)$, and so the right-hand side of (7.11) is equal to $1/(y+1)^2$. We conclude that

$$f_Y(y) = \begin{cases} 1/(y+1)^2, & y > 0, \\ 0, & y \leq 0. \end{cases}$$

Note that the same answer can be obtained by differentiating the formula for $F_Y(y)$ in (7.6).

Independence

We now consider the joint density of jointly continuous *independent* random variables. As noted in Section 7.1, if X and Y are independent, then $F_{XY}(x,y) = F_X(x)F_Y(y)$ for all x and y. If X and Y are also jointly continuous, then by taking second-order mixed partial derivatives, we find

$$\frac{\partial^2}{\partial y \partial x} F_X(x) F_Y(y) = f_X(x) f_Y(y).$$

In other words, if X and Y are jointly continuous and independent, then the joint density is the product of the marginal densities. Using (7.8), it is easy to see that the converse is also true. If $f_{XY}(x,y) = f_X(x)f_Y(y)$, (7.8) implies

$$P(X \in B, Y \in C) = \int_B \left(\int_C f_{XY}(x,y)\,dy\right) dx$$
$$= \int_B \left(\int_C f_X(x) f_Y(y)\,dy\right) dx$$

7.2 Jointly continuous random variables

$$= \int_B f_X(x) \left(\int_C f_Y(y)\, dy \right) dx$$
$$= \left(\int_B f_X(x)\, dx \right) \mathsf{P}(Y \in C)$$
$$= \mathsf{P}(X \in B)\mathsf{P}(Y \in C).$$

We record here that jointly continuous random variables X and Y are **independent** if and only if their joint density factors into the product of their marginal densities: $f_{XY}(x,y) = f_X(x)f_Y(y)$.

Expectation

If X and Y are jointly continuous with joint density f_{XY}, then the methods of Section 4.2 can easily be used to show that

$$\boxed{\mathsf{E}[g(X,Y)] = \int_{-\infty}^{\infty}\int_{-\infty}^{\infty} g(x,y) f_{XY}(x,y)\, dx\, dy.}$$

For arbitrary random variables X and Y, their **bivariate characteristic function** is defined by

$$\boxed{\varphi_{XY}(v_1, v_2) := \mathsf{E}[e^{j(v_1 X + v_2 Y)}].}$$

If X and Y have joint density f_{XY}, then

$$\varphi_{XY}(v_1, v_2) = \int_{-\infty}^{\infty}\int_{-\infty}^{\infty} f_{XY}(x,y) e^{j(v_1 x + v_2 y)}\, dx\, dy,$$

which is simply the **bivariate Fourier transform** of f_{XY}. By the inversion formula,

$$f_{XY}(x,y) = \frac{1}{(2\pi)^2} \int_{-\infty}^{\infty}\int_{-\infty}^{\infty} \varphi_{XY}(v_1, v_2) e^{-j(v_1 x + v_2 y)}\, dv_1\, dv_2.$$

Now suppose that X and Y are independent. Then

$$\varphi_{XY}(v_1, v_2) = \mathsf{E}[e^{j(v_1 X + v_2 Y)}] = \mathsf{E}[e^{jv_1 X}]\mathsf{E}[e^{jv_2 Y}] = \varphi_X(v_1)\varphi_Y(v_2).$$

In other words, if X and Y are independent, then their joint characteristic function factors. The converse is also true; i.e., if the joint characteristic function factors, then X and Y are independent. The general proof is complicated, but if X and Y are jointly continuous, it suffices to show that the joint density has product form. This is easily done with the inversion formula. Write

$$f_{XY}(x,y) = \frac{1}{(2\pi)^2} \int_{-\infty}^{\infty}\int_{-\infty}^{\infty} \varphi_{XY}(v_1, v_2) e^{-j(v_1 x + v_2 y)}\, dv_1\, dv_2$$

$$= \frac{1}{(2\pi)^2} \int_{-\infty}^{\infty} \int_{-\infty}^{\infty} \varphi_X(v_1) \varphi_Y(v_2) e^{-j(v_1 x + v_2 y)} dv_1 dv_2$$

$$= \left(\frac{1}{2\pi} \int_{-\infty}^{\infty} \varphi_X(v_1) e^{-jv_1 x} dx \right) \left(\frac{1}{2\pi} \int_{-\infty}^{\infty} \varphi_Y(v_2) e^{-jv_2 y} dy \right)$$

$$= f_X(x) f_Y(y).$$

We summarize here that X and Y are **independent** if and only if their joint characteristic function is a product of their marginal characteristic functions; i.e.,

$$\varphi_{XY}(v_1, v_2) = \varphi_X(v_1) \varphi_Y(v_2).$$

*Continuous random variables that are not jointly continuous

Let $\Theta \sim$ uniform$[-\pi, \pi]$, and put $X := \cos \Theta$ and $Y := \sin \Theta$. As shown in Problem 35 in Chapter 5, X and Y are both arcsine random variables, each having density $(1/\pi)/\sqrt{1-x^2}$ for $-1 < x < 1$.

Next, since $X^2 + Y^2 = 1$, the pair (X, Y) takes values only on the unit circle

$$C := \{(x, y) : x^2 + y^2 = 1\}.$$

Thus, $\mathsf{P}((X, Y) \in C) = 1$. On the other hand, if X and Y have a joint density f_{XY}, then

$$\mathsf{P}((X, Y) \in C) = \iint_C f_{XY}(x, y) \, dx \, dy = 0$$

because a double integral over a set of zero area must be zero. So, if X and Y had a joint density, this would imply that $1 = 0$. Since this is not true, there can be no joint density.

Remark. Problem 44 of Chapter 2 provided an example of uncorrelated discrete random variables that are not independent. The foregoing $X = \cos \Theta$ and $Y = \sin \Theta$ provide an example of continuous random variables that are uncorrelated but not independent (Problem 20).

7.3 Conditional probability and expectation

If X is a continuous random variable, then its cdf $F_X(x) := \mathsf{P}(X \leq x) = \int_{-\infty}^{x} f_X(t) \, dt$ is a continuous function of x.[5] It follows from the properties of cdfs in Section 5.5 that $\mathsf{P}(X = x) = 0$ for all x. **Hence, we cannot define $\mathsf{P}(Y \in C | X = x)$ by $\mathsf{P}(X = x, Y \in C)/\mathsf{P}(X = x)$ since this requires division by zero!** Similar problems arise with conditional expectation. How should we define conditional probability and expectation in this case?

Conditional probability

As a first step, let us compute

$$\lim_{\Delta x \to 0} \mathsf{P}(Y \in C | x < X \leq x + \Delta x).$$

7.3 Conditional probability and expectation

For positive Δx, this conditional probability is given by

$$\frac{\mathsf{P}(x < X \leq x+\Delta x, Y \in C)}{\mathsf{P}(x < X \leq x+\Delta x)}.$$

If we write the numerator as $\mathsf{P}((X,Y) \in (x,x+\Delta x] \times C)$, and if we assume X and Y are jointly continuous, the desired conditional probability can be written as

$$\frac{\int_x^{x+\Delta x} \left(\int_C f_{XY}(t,y)\,dy \right) dt}{\int_x^{x+\Delta x} f_X(\tau)\,d\tau}.$$

Now divide the numerator and denominator by Δx to get

$$\frac{\frac{1}{\Delta x}\int_x^{x+\Delta x} \left(\int_C f_{XY}(t,y)\,dy \right) dt}{\frac{1}{\Delta x}\int_x^{x+\Delta x} f_X(\tau)\,d\tau}.$$

Letting $\Delta x \to 0$, we obtain the limit

$$\frac{\int_C f_{XY}(x,y)\,dy}{f_X(x)} = \int_C \frac{f_{XY}(x,y)}{f_X(x)}\,dy.$$

We therefore define the **conditional density** of Y given X by

$$\boxed{f_{Y|X}(y|x) := \frac{f_{XY}(x,y)}{f_X(x)}, \quad \text{for } x \text{ with } f_X(x) > 0,} \tag{7.12}$$

and we define the **conditional probability**

$$\boxed{\mathsf{P}(Y \in C | X = x) := \int_C f_{Y|X}(y|x)\,dy.}$$

The **conditional cdf** is

$$F_{Y|X}(y|x) := \mathsf{P}(Y \leq y|X = x) = \int_{-\infty}^y f_{Y|X}(t|x)\,dt.$$

Note also that if X and Y are independent, the joint density factors, and so

$$f_{Y|X}(y|x) = \frac{f_{XY}(x,y)}{f_X(x)} = \frac{f_X(x)f_Y(y)}{f_X(x)} = f_Y(y).$$

It then follows that $\mathsf{P}(Y \in C|X = x) = \mathsf{P}(Y \in C)$; similarly $F_{Y|X}(y|x) = F_Y(y)$. In other words, we can "drop the conditioning."

Recall that for discrete random variables, conditional pmfs are proportional to slices of the joint pmf (cf. Example 3.10 and Figure 3.3). Similarly, (7.12) shows that conditional densities are proportional to slices of the joint density. For example, the joint density $f_{XY}(x,y) = xe^{-x(y+1)}$ was sketched in Figure 7.8. For fixed x, slices have the shape of an exponential density, while for fixed y, slices have the shape of a gamma density with $p = 2$ shown in Figure 4.7.

We now show that our definition of conditional probability satisfies the following **law of total probability**,

$$P(Y \in C) = \int_{-\infty}^{\infty} P(Y \in C | X = x) f_X(x) \, dx. \quad (7.13)$$

Remark. Notice that although (7.12) only makes sense for those x with $f_X(x) > 0$, these are the only values of x used to evaluate the integral in (7.13).

To derive (7.13), first write

$$\int_{-\infty}^{\infty} P(Y \in C | X = x) f_X(x) \, dx = \int_{-\infty}^{\infty} \left(\int_C f_{Y|X}(y|x) \, dy \right) f_X(x) \, dx.$$

Then from (7.12), observe that

$$f_{Y|X}(y|x) f_X(x) = f_{XY}(x,y).$$

Hence, the above double integral becomes

$$\iint_{\mathbb{R} \times C} f_{XY}(x,y) \, dx \, dy = P((X,Y) \in \mathbb{R} \times C) = P(Y \in C),$$

where the last step uses (7.1).

If we repeat the limit derivation above for $P((X,Y) \in A | x < X \leq x + \Delta x)$, then we are led to define (Problem 24)

$$P((X,Y) \in A | X = x) := \int_{-\infty}^{\infty} I_A(x,y) f_{Y|X}(y|x) \, dy.$$

It is similarly easy to show that the **law of total probability**

$$P((X,Y) \in A) = \int_{-\infty}^{\infty} P((X,Y) \in A | X = x) f_X(x) \, dx \quad (7.14)$$

holds. We also have the **substitution law**,

$$P((X,Y) \in A | X = x) = P((x,Y) \in A | X = x). \quad (7.15)$$

Rather than derive these laws of total probability and substitution here, we point out that they follow immediately from the corresponding results for conditional expectation that we discuss later in this section.[6]

7.3 Conditional probability and expectation

***Example* 7.14** (signal in additive noise). Suppose that a random, continuous-valued signal X is transmitted over a channel subject to additive, continuous-valued noise Y. The received signal is $Z = X + Y$. Find the cdf and density of Z if X and Y are jointly continuous random variables with joint density f_{XY}.

Solution. Since we are not assuming that X and Y are independent, the characteristic-function method of Example 4.23 does not work here. Instead, we use the laws of total probability and substitution. Write

$$\begin{aligned}
F_Z(z) &= \mathsf{P}(Z \le z) \\
&= \int_{-\infty}^{\infty} \mathsf{P}(Z \le z | Y = y) f_Y(y)\, dy \\
&= \int_{-\infty}^{\infty} \mathsf{P}(X + Y \le z | Y = y) f_Y(y)\, dy \\
&= \int_{-\infty}^{\infty} \mathsf{P}(X + y \le z | Y = y) f_Y(y)\, dy \\
&= \int_{-\infty}^{\infty} \mathsf{P}(X \le z - y | Y = y) f_Y(y)\, dy \\
&= \int_{-\infty}^{\infty} F_{X|Y}(z - y | y) f_Y(y)\, dy.
\end{aligned}$$

By differentiating with respect to z,

$$f_Z(z) = \int_{-\infty}^{\infty} f_{X|Y}(z - y | y) f_Y(y)\, dy = \int_{-\infty}^{\infty} f_{XY}(z - y, y)\, dy.$$

This is essentially the formula obtained in Example 7.10; to see the connection, make the change of variable $x = z - y$. We also point out that if X and Y are independent, we can drop the conditioning and obtain the convolution

$$f_Z(z) = \int_{-\infty}^{\infty} f_X(z - y) f_Y(y)\, dy. \tag{7.16}$$

This formula was derived using characteristic functions following Example 4.23.

***Example* 7.15** (signal in multiplicative noise). A random, continuous-valued signal X is transmitted over a channel subject to multiplicative, continuous-valued noise Y. The received signal is $Z = XY$. Find the cdf and density of Z if X and Y are jointly continuous random variables with joint density f_{XY}.

Solution. We proceed as in the previous example. Write

$$\begin{aligned}
F_Z(z) &= \mathsf{P}(Z \le z) \\
&= \int_{-\infty}^{\infty} \mathsf{P}(Z \le z | Y = y) f_Y(y)\, dy \\
&= \int_{-\infty}^{\infty} \mathsf{P}(XY \le z | Y = y) f_Y(y)\, dy \\
&= \int_{-\infty}^{\infty} \mathsf{P}(Xy \le z | Y = y) f_Y(y)\, dy.
\end{aligned}$$

At this point we have a problem when we attempt to divide through by y. If y is negative, we have to reverse the inequality sign. Otherwise, we do not have to reverse the inequality. The solution to this difficulty is to break up the range of integration. Write

$$F_Z(z) = \int_{-\infty}^{0} \mathsf{P}(Xy \leq z | Y = y) f_Y(y) \, dy + \int_{0}^{\infty} \mathsf{P}(Xy \leq z | Y = y) f_Y(y) \, dy.$$

Now we can divide by y separately in each integral. Thus,

$$F_Z(z) = \int_{-\infty}^{0} \mathsf{P}(X \geq z/y | Y = y) f_Y(y) \, dy + \int_{0}^{\infty} \mathsf{P}(X \leq z/y | Y = y) f_Y(y) \, dy$$

$$= \int_{-\infty}^{0} \left[1 - F_{X|Y}\left(\tfrac{z}{y} \big| y\right) \right] f_Y(y) \, dy + \int_{0}^{\infty} F_{X|Y}\left(\tfrac{z}{y} \big| y\right) f_Y(y) \, dy.$$

Differentiating with respect to z yields

$$f_Z(z) = -\int_{-\infty}^{0} f_{X|Y}\left(\tfrac{z}{y} \big| y\right) \tfrac{1}{y} f_Y(y) \, dy + \int_{0}^{\infty} f_{X|Y}\left(\tfrac{z}{y} \big| y\right) \tfrac{1}{y} f_Y(y) \, dy. \tag{7.17}$$

Now observe that in the first integral, the range of integration implies that y is always negative. For such y, $-y = |y|$. In the second integral, y is always positive, and so $y = |y|$. Thus,

$$f_Z(z) = \int_{-\infty}^{0} f_{X|Y}\left(\tfrac{z}{y} \big| y\right) \tfrac{1}{|y|} f_Y(y) \, dy + \int_{0}^{\infty} f_{X|Y}\left(\tfrac{z}{y} \big| y\right) \tfrac{1}{|y|} f_Y(y) \, dy$$

$$= \int_{-\infty}^{\infty} f_{X|Y}\left(\tfrac{z}{y} \big| y\right) \tfrac{1}{|y|} f_Y(y) \, dy$$

$$= \int_{-\infty}^{\infty} f_{XY}\left(\tfrac{z}{y}, y\right) \tfrac{1}{|y|} \, dy.$$

This is essentially the formula obtained in Example 7.11; to see the connection, make the change of variable $x = z/y$ in (7.17) and proceed as before.

Example 7.16. If X and Y are jointly continuous, find the density of $Z := X^2 + Y^2$.

Solution. As always, we first find the cdf.

$$F_Z(z) = \mathsf{P}(Z \leq z)$$

$$= \int_{-\infty}^{\infty} \mathsf{P}(Z \leq z | Y = y) f_Y(y) \, dy$$

$$= \int_{-\infty}^{\infty} \mathsf{P}(X^2 + Y^2 \leq z | Y = y) f_Y(y) \, dy$$

$$= \int_{-\infty}^{\infty} \mathsf{P}(X^2 \leq z - y^2 | Y = y) f_Y(y) \, dy.$$

At this point, we observe that for $y^2 > z$, $\mathsf{P}(X^2 \leq z - y^2 | Y = y) = 0$ since X^2 cannot be negative. We therefore write

$$F_Z(z) = \int_{-\sqrt{z}}^{\sqrt{z}} \mathsf{P}(X^2 \leq z - y^2 | Y = y) f_Y(y) \, dy$$

7.3 Conditional probability and expectation

$$= \int_{-\sqrt{z}}^{\sqrt{z}} P(-\sqrt{z-y^2} \leq X \leq \sqrt{z-y^2} | Y = y) f_Y(y) \, dy$$

$$= \int_{-\sqrt{z}}^{\sqrt{z}} [F_{X|Y}(\sqrt{z-y^2}|y) - F_{X|Y}(-\sqrt{z-y^2}|y)] f_Y(y) \, dy.$$

Using **Leibniz' rule,**[7]

$$\frac{d}{dz} \int_{a(z)}^{b(z)} h(z,y) \, dy = -h(z, a(z)) a'(z) + h(z, b(z)) b'(z) + \int_{a(z)}^{b(z)} \frac{\partial}{\partial z} h(z,y) \, dy,$$

we find that

$$f_Z(z) = \int_{-\sqrt{z}}^{\sqrt{z}} \frac{f_{X|Y}(\sqrt{z-y^2}|y) + f_{X|Y}(-\sqrt{z-y^2}|y)}{2\sqrt{z-y^2}} f_Y(y) \, dy.$$

Example 7.17. Let X and Y be jointly continuous, positive random variables. Find the cdf and density of $Z := \min(X,Y)/\max(X,Y)$.

Solution. First note that since $0 < Z \leq 1$, we only worry about $F_Z(z)$ for $0 < z < 1$. (Why?) Second, note that if $Y \leq X$, $Z = Y/X$, while if $X < Y$, $Z = X/Y$. Our analytical approach is to write

$$P(Z \leq z) = P(Z \leq z, Y \leq X) + P(Z \leq z, X < Y)$$
$$= P(Y/X \leq z, Y \leq X) + P(X/Y \leq z, X < Y),$$

and evaluate each term using the law of total probability. We begin with

$$P(Y/X \leq z, Y \leq X) = \int_0^\infty P(Y/X \leq z, Y \leq X | X = x) f_X(x) \, dx$$
$$= \int_0^\infty P(Y \leq zx, Y \leq x | X = x) f_X(x) \, dx.$$

Since $0 < z < 1$, $\{Y \leq zx\} \subset \{Y \leq x\}$, and so

$$\{Y \leq zx\} \cap \{Y \leq x\} = \{Y \leq zx\}.$$

Hence,

$$P(Y/X \leq z, Y \leq X) = \int_0^\infty P(Y \leq zx | X = x) f_X(x) \, dx$$
$$= \int_0^\infty F_{Y|X}(zx|x) f_X(x) \, dx.$$

Similarly,

$$P(X/Y \leq z, X < Y) = \int_0^\infty P(X \leq zy, X < y | Y = y) f_Y(y) \, dy$$
$$= \int_0^\infty P(X \leq zy | Y = y) f_Y(y) \, dy$$
$$= \int_0^\infty F_{X|Y}(zy|y) f_Y(y) \, dy.$$

It now follows that

$$f_Z(z) = \int_0^\infty x f_{Y|X}(zx|x) f_X(x)\, dx + \int_0^\infty y f_{X|Y}(zy|y) f_Y(y)\, dy.$$

Conditional expectation

Since $P(Y \in C | X = x)$ is computed by integrating the conditional density $f_{Y|X}(y|x)$ over the set C, it is only natural to define[8]

$$\boxed{E[g(Y)|X=x] := \int_{-\infty}^\infty g(y) f_{Y|X}(y|x)\, dy.} \qquad (7.18)$$

To see how $E[g(X,Y)|X=x]$ should be defined so that suitable laws of total probability and substitution can be obtained, write

$$\begin{aligned}
E[g(X,Y)] &= \int_{-\infty}^\infty \int_{-\infty}^\infty g(x,y) f_{XY}(x,y)\, dx\, dy \\
&= \int_{-\infty}^\infty \left(\int_{-\infty}^\infty g(x,y) f_{XY}(x,y)\, dy \right) dx \\
&= \int_{-\infty}^\infty \left(\int_{-\infty}^\infty g(x,y) \frac{f_{XY}(x,y)}{f_X(x)}\, dy \right) f_X(x)\, dx \\
&= \int_{-\infty}^\infty \left(\int_{-\infty}^\infty g(x,y) f_{Y|X}(y|x)\, dy \right) f_X(x)\, dx.
\end{aligned}$$

Thus, defining

$$E[g(X,Y)|X=x] := \int_{-\infty}^\infty g(x,y) f_{Y|X}(y|x)\, dy \qquad (7.19)$$

gives us the **law of total probability**

$$\boxed{E[g(X,Y)] = \int_{-\infty}^\infty E[g(X,Y)|X=x] f_X(x)\, dx.} \qquad (7.20)$$

Furthermore, if we replace $g(y)$ in (7.18) by $g_x(y) := g(x,y)$ and compare the result with (7.19), we obtain the **substitution law**,

$$\boxed{E[g(X,Y)|X=x] = E[g(x,Y)|X=x].} \qquad (7.21)$$

Another important point to note is that if X and Y are independent, then $f_{Y|X}(y|x) = f_Y(y)$. In this case, (7.18) becomes $E[g(Y)|X=x] = E[g(Y)]$. In other words, we can "drop the conditioning."

Example 7.18. Let $X \sim \exp(1)$, and suppose that given $X = x$, Y is conditionally normal with $f_{Y|X}(\cdot|x) \sim N(0, x^2)$. Evaluate $E[Y^2]$ and $E[Y^2 X^3]$.

Solution. We use the law of total probability for expectation. We begin with

$$\mathsf{E}[Y^2] = \int_{-\infty}^{\infty} \mathsf{E}[Y^2|X=x] f_X(x)\,dx.$$

Since $f_{Y|X}(y|x) = e^{-(y/x)^2/2}/(\sqrt{2\pi}x)$ is an $N(0,x^2)$ density in the variable y, $\mathsf{E}[Y^2|X=x] = x^2$. Substituting this into the above integral yields

$$\mathsf{E}[Y^2] = \int_{-\infty}^{\infty} x^2 f_X(x)\,dx = \mathsf{E}[X^2].$$

Since $X \sim \exp(1)$, $\mathsf{E}[X^2] = 2$ by Example 4.17.

To compute $\mathsf{E}[Y^2 X^3]$, we proceed similarly. Write

$$\begin{aligned}
\mathsf{E}[Y^2 X^3] &= \int_{-\infty}^{\infty} \mathsf{E}[Y^2 X^3 | X=x] f_X(x)\,dx \\
&= \int_{-\infty}^{\infty} \mathsf{E}[Y^2 x^3 | X=x] f_X(x)\,dx \\
&= \int_{-\infty}^{\infty} x^3 \mathsf{E}[Y^2 | X=x] f_X(x)\,dx \\
&= \int_{-\infty}^{\infty} x^3 x^2 f_X(x)\,dx \\
&= \mathsf{E}[X^5] \\
&= 5!, \quad \text{by Example 4.17.}
\end{aligned}$$

7.4 The bivariate normal

The **bivariate Gaussian** or **bivariate normal** density is a generalization of the univariate $N(m, \sigma^2)$ density. (The multivariate case is treated in Chapter 9.) Recall that the standard $N(0,1)$ density is given by $\psi(x) := \exp(-x^2/2)/\sqrt{2\pi}$. The general $N(m,\sigma^2)$ density can be written in terms of ψ as

$$\frac{1}{\sqrt{2\pi}\,\sigma} \exp\left[-\frac{1}{2}\left(\frac{x-m}{\sigma}\right)^2\right] = \frac{1}{\sigma}\cdot\psi\left(\frac{x-m}{\sigma}\right).$$

In order to define the general bivariate Gaussian density, it is convenient to define a standard bivariate density first. So, for $|\rho| < 1$, put

$$\psi_\rho(u,v) := \frac{\exp\left(\frac{-1}{2(1-\rho^2)}[u^2 - 2\rho uv + v^2]\right)}{2\pi\sqrt{1-\rho^2}}. \qquad (7.22)$$

For fixed ρ, this function of the two variables u and v defines a surface. The surface corresponding to $\rho = 0$ is shown in Figure 7.9. From the figure and from the formula (7.22), we see that ψ_0 is circularly symmetric; i.e., for all (u,v) on a circle of radius r, in other words, for $u^2 + v^2 = r^2$, $\psi_0(u,v) = e^{-r^2/2}/2\pi$ does not depend on the particular values of u and v,

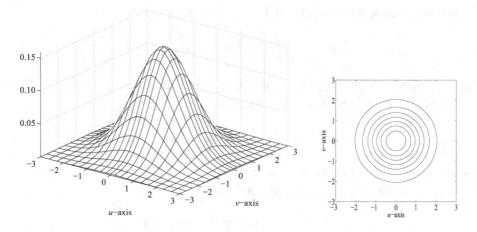

Figure 7.9. The Gaussian surface $\psi_\rho(u,v)$ of (7.22) with $\rho = 0$ (left). The corresponding level curves (right).

but only on the radius of the circle on which they lie. Some of these circles (**level curves**) are shown in Figure 7.9. We also point out that for $\rho = 0$, the formula (7.22) factors into the product of two univariate $N(0,1)$ densities, i.e., $\psi_0(u,v) = \psi(u)\psi(v)$. For $\rho \neq 0$, ψ_ρ does not factor. In other words, if U and V have joint density ψ_ρ, then U and V are independent if and only if $\rho = 0$. A plot of ψ_ρ for $\rho = -0.85$ is shown in Figure 7.10. It turns out that now ψ_ρ is constant on ellipses instead of circles. The axes of the ellipses are not parallel to the coordinate axes, as shown in Figure 7.10. Notice how the major axis of these ellipses and the density are concentrated along the line $v = -u$. As $\rho \to -1$, this concentration becomes more extreme. As $\rho \to +1$, the density concentrates around the line $v = u$. We now show

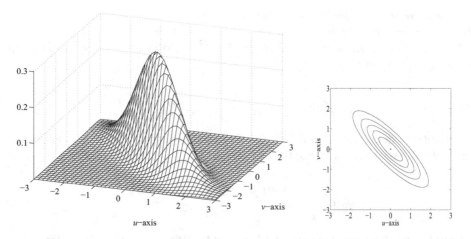

Figure 7.10. The Gaussian surface $\psi_\rho(u,v)$ of (7.22) with $\rho = -0.85$ (left). The corresponding level curves (right).

7.4 The bivariate normal

that the density ψ_ρ integrates to one. To do this, first observe that for all $|\rho| < 1$,

$$u^2 - 2\rho uv + v^2 = u^2(1-\rho^2) + (v-\rho u)^2.$$

It follows that

$$\psi_\rho(u,v) = \frac{e^{-u^2/2}}{\sqrt{2\pi}} \cdot \frac{\exp\left(\frac{-1}{2(1-\rho^2)}[v-\rho u]^2\right)}{\sqrt{2\pi}\sqrt{1-\rho^2}}$$

$$= \psi(u) \cdot \frac{1}{\sqrt{1-\rho^2}} \psi\left(\frac{v-\rho u}{\sqrt{1-\rho^2}}\right). \tag{7.23}$$

Observe that the right-hand factor as a function of v has the form of a univariate normal density with mean ρu and variance $1-\rho^2$. With ψ_ρ factored as in (7.23), we can write $\int_{-\infty}^{\infty}\int_{-\infty}^{\infty} \psi_\rho(u,v)\,du\,dv$ as the iterated integral

$$\int_{-\infty}^{\infty} \psi(u) \left[\int_{-\infty}^{\infty} \frac{1}{\sqrt{1-\rho^2}} \psi\left(\frac{v-\rho u}{\sqrt{1-\rho^2}}\right) dv\right] du.$$

As noted above, the inner integrand, as a function of v, is simply an $N(\rho u, 1-\rho^2)$ density, and therefore integrates to one. Hence, the above iterated integral becomes $\int_{-\infty}^{\infty} \psi(u)\,du = 1$.

We can now easily define the general bivariate Gaussian density with parameters m_X, m_Y, σ_X^2, σ_Y^2, and ρ by

$$f_{XY}(x,y) := \frac{1}{\sigma_X \sigma_Y} \psi_\rho\left(\frac{x-m_X}{\sigma_X}, \frac{y-m_Y}{\sigma_Y}\right). \tag{7.24}$$

More explicitly, this density is

$$\frac{\exp\left(\frac{-1}{2(1-\rho^2)}\left[\left(\frac{x-m_X}{\sigma_X}\right)^2 - 2\rho\left(\frac{x-m_X}{\sigma_X}\right)\left(\frac{y-m_Y}{\sigma_Y}\right) + \left(\frac{y-m_Y}{\sigma_Y}\right)^2\right]\right)}{2\pi\sigma_X\sigma_Y\sqrt{1-\rho^2}}. \tag{7.25}$$

It can be shown that the marginals are $f_X \sim N(m_X, \sigma_X^2)$ and $f_Y \sim N(m_Y, \sigma_Y^2)$ and that

$$\mathsf{E}\left[\left(\frac{X-m_X}{\sigma_X}\right)\left(\frac{Y-m_Y}{\sigma_Y}\right)\right] = \rho$$

(see Problems 47 and 50). Hence, ρ is the **correlation coefficient** between X and Y. From (7.25), we observe that X and Y are independent if and only if $\rho = 0$. A plot of f_{XY} with $m_X = m_Y = 0$, $\sigma_X = 1.5$, $\sigma_Y = 0.6$, and $\rho = 0$ is shown in Figure 7.11. The corresponding elliptical level curves are shown in Figure 7.11. Notice how the level curves and density are concentrated around the x-axis. Also, f_{XY} is constant on ellipses of the form

$$\left(\frac{x}{\sigma_X}\right)^2 + \left(\frac{y}{\sigma_Y}\right)^2 = r^2.$$

To show that $\int_{-\infty}^{\infty}\int_{-\infty}^{\infty} f_{XY}(x,y)\,dx\,dy = 1$ as well, use formula (7.23) for ψ_ρ and proceed as above, integrating with respect to y first and then x. For the inner integral, make the change of variable $v = (y-m_Y)/\sigma_Y$, and in the remaining outer integral make the change of variable $u = (x-m_X)/\sigma_X$.

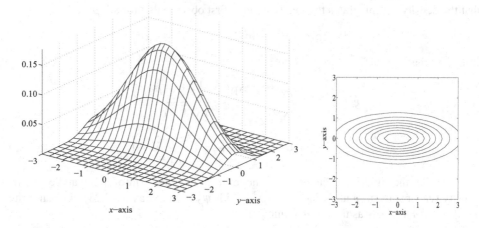

Figure 7.11. The bivariate normal density $f_{XY}(x,y)$ of (7.25) with $m_X = m_Y = 0$, $\sigma_X = 1.5$, $\sigma_Y = 0.6$, and $\rho = 0$ (left). The corresponding level curves (right).

Example 7.19. Let random variables U and V have the standard bivariate normal density ψ_ρ in (7.22). Show that $\mathsf{E}[UV] = \rho$.

Solution. Using the factored form of ψ_ρ in (7.23), write

$$\mathsf{E}[UV] = \int_{-\infty}^{\infty}\int_{-\infty}^{\infty} uv\, \psi_\rho(u,v)\, du\, dv$$

$$= \int_{-\infty}^{\infty} u\, \psi(u) \left[\int_{-\infty}^{\infty} \frac{v}{\sqrt{1-\rho^2}} \psi\left(\frac{v-\rho u}{\sqrt{1-\rho^2}}\right) dv \right] du.$$

The quantity in brackets has the form $\mathsf{E}[\hat{V}]$, where $\hat{V}$ is a univariate normal random variable with mean ρu and variance $1 - \rho^2$. Thus,

$$\mathsf{E}[UV] = \int_{-\infty}^{\infty} u\, \psi(u)[\rho u]\, du$$

$$= \rho \int_{-\infty}^{\infty} u^2\, \psi(u)\, du$$

$$= \rho,$$

since ψ is the $N(0,1)$ density.

Example 7.20. Let U and V have the standard bivariate normal density $f_{UV}(u,v) = \psi_\rho(u,v)$ given in (7.22). Find the conditional densities $f_{V|U}$ and $f_{U|V}$.

Solution. It is shown in Problem 47 that f_U and f_V are both $N(0,1)$. Hence,

$$f_{V|U}(v|u) = \frac{f_{UV}(u,v)}{f_U(u)} = \frac{\psi_\rho(u,v)}{\psi(u)},$$

7.4 The bivariate normal

where ψ is the $N(0,1)$ density. If we now substitute the factored form of $\psi_\rho(u,v)$ given in (7.23), we obtain

$$f_{V|U}(v|u) = \frac{1}{\sqrt{1-\rho^2}}\psi\left(\frac{v-\rho u}{\sqrt{1-\rho^2}}\right);$$

i.e., $f_{V|U}(\cdot|u) \sim N(\rho u, 1-\rho^2)$.

To compute $f_{U|V}$ we need the following alternative factorization of ψ_ρ,

$$\psi_\rho(u,v) = \frac{1}{\sqrt{1-\rho^2}}\psi\left(\frac{u-\rho v}{\sqrt{1-\rho^2}}\right)\cdot\psi(v). \quad (7.26)$$

It then follows that

$$f_{U|V}(u|v) = \frac{1}{\sqrt{1-\rho^2}}\psi\left(\frac{u-\rho v}{\sqrt{1-\rho^2}}\right);$$

i.e., $f_{U|V}(\cdot|v) \sim N(\rho v, 1-\rho^2)$. To see the shape of this density with $\rho = -0.85$, look at slices of Figure 7.10 for fixed values of v. Two slices from Figure 7.10 are shown in Figure 7.12. Notice how the mean value of the different slices depends on v and ρ.

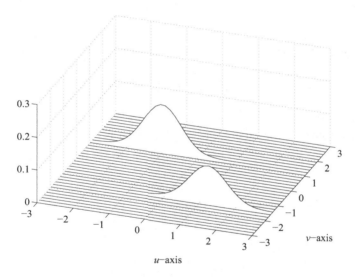

Figure 7.12. Two slices from Figure 7.10.

Example 7.21. If U and V have standard joint normal density $\psi_\rho(u,v)$, find $\mathsf{E}[V|U=u]$.

Solution. Recall that from Example 7.20, $f_{V|U}(\cdot|u) \sim N(\rho u, 1-\rho^2)$. Hence,

$$\mathsf{E}[V|U=u] = \int_{-\infty}^{\infty} v f_{V|U}(v|u)\,dv = \rho u.$$

It is important to note here that $\mathsf{E}[V|U=u] = \rho u$ is a linear function of u. For arbitrary random variables U and V, $\mathsf{E}[V|U=u]$ is usually a much more complicated function of

u. However, for the general bivariate normal, the conditional expectation is either a linear function or a linear function plus a constant, as shown in Problem 48.

7.5 Extension to three or more random variables

The ideas we have developed for pairs of random variables readily extend to any finite number of random variables. However, for ease of notation, we illustrate the case of three random variables. We also point out that the use of vector notation can simplify many of these formulas as shown in Chapters 8 and 9.

Given a joint density $f_{XYZ}(x,y,z)$, if we need to find $f_{XY}(x,y)$, $f_{XZ}(x,z)$, or $f_{YZ}(y,z)$, we just integrate out the unwanted variable; e.g.,

$$f_{YZ}(y,z) = \int_{-\infty}^{\infty} f_{XYZ}(x,y,z)\,dx.$$

If we then need only $f_Z(z)$, we integrate out y:

$$f_Z(z) = \int_{-\infty}^{\infty} f_{YZ}(y,z)\,dy.$$

These two steps can be combined into the double integral

$$f_Z(z) = \int_{-\infty}^{\infty}\int_{-\infty}^{\infty} f_{XYZ}(x,y,z)\,dx\,dy.$$

With more variables, there are more possibilities for conditional densities. In addition to conditional densities of one variable given another such as

$$f_{Y|Z}(y|z) := \frac{f_{YZ}(y,z)}{f_Z(z)}, \qquad (7.27)$$

we also have conditional densities of the form

$$f_{X|YZ}(x|y,z) := \frac{f_{XYZ}(x,y,z)}{f_{YZ}(y,z)} \qquad (7.28)$$

and

$$f_{XY|Z}(x,y|z) := \frac{f_{XYZ}(x,y,z)}{f_Z(z)}.$$

We also point out that (7.27) and (7.28) imply

$$f_{XYZ}(x,y,z) = f_{X|YZ}(x|y,z)\,f_{Y|Z}(y|z)\,f_Z(z).$$

Example 7.22. Let

$$f_{XYZ}(x,y,z) = \frac{3z^2}{7\sqrt{2\pi}} e^{-zy} \exp\left[-\tfrac{1}{2}\left(\tfrac{x-y}{z}\right)^2\right],$$

for $y \geq 0$ and $1 \leq z \leq 2$, and $f_{XYZ}(x,y,z) = 0$ otherwise. Find $f_{YZ}(y,z)$ and $f_{X|YZ}(x|y,z)$. Then find $f_Z(z)$, $f_{Y|Z}(y|z)$, and $f_{XY|Z}(x,y|z)$.

7.5 Extension to three or more random variables

Solution. Observe that the joint density can be written as

$$f_{XYZ}(x,y,z) = \frac{\exp\left[-\frac{1}{2}\left(\frac{x-y}{z}\right)^2\right]}{\sqrt{2\pi}\,z} \cdot ze^{-zy} \cdot \tfrac{3}{7}z^2.$$

The first factor as a function of x is an $N(y,z^2)$ density. Hence,

$$f_{YZ}(y,z) = \int_{-\infty}^{\infty} f_{XYZ}(x,y,z)\,dx = ze^{-zy} \cdot \tfrac{3}{7}z^2,$$

and

$$f_{X|YZ}(x|y,z) = \frac{f_{XYZ}(x,y,z)}{f_{YZ}(y,z)} = \frac{\exp\left[-\frac{1}{2}\left(\frac{x-y}{z}\right)^2\right]}{\sqrt{2\pi}\,z}.$$

Thus, $f_{X|YZ}(\cdot|y,z) \sim N(y,z^2)$. Next, in the above formula for $f_{YZ}(y,z)$, observe that ze^{-zy} as a function of y is an exponential density with parameter z. Thus,

$$f_Z(z) = \int_0^{\infty} f_{YZ}(y,z)\,dy = \tfrac{3}{7}z^2, \quad 1 \le z \le 2.$$

It follows that $f_{Y|Z}(y|z) = f_{YZ}(y,z)/f_Z(z) = ze^{-zy}$; i.e., $f_{Y|Z}(\cdot|z) \sim \exp(z)$. Finally,

$$f_{XY|Z}(x,y|z) = \frac{f_{XYZ}(x,y,z)}{f_Z(z)} = \frac{\exp\left[-\frac{1}{2}\left(\frac{x-y}{z}\right)^2\right]}{\sqrt{2\pi}\,z} \cdot ze^{-zy}.$$

The law of total probability

For expectations, we have

$$\mathsf{E}[g(X,Y,Z)] = \int_{-\infty}^{\infty}\int_{-\infty}^{\infty}\int_{-\infty}^{\infty} g(x,y,z)\,f_{XYZ}(x,y,z)\,dx\,dy\,dz.$$

A little calculation using conditional probabilities shows that with

$$\mathsf{E}[g(X,Y,Z)|Y=y,Z=z] := \int_{-\infty}^{\infty} g(x,y,z)\,f_{X|YZ}(x|y,z)\,dx,$$

we have the **law of total probability**,

$$\mathsf{E}[g(X,Y,Z)] = \int_{-\infty}^{\infty}\int_{-\infty}^{\infty} \mathsf{E}[g(X,Y,Z)|Y=y,Z=z]\,f_{YZ}(y,z)\,dy\,dz. \tag{7.29}$$

In addition, we have the **substitution law**,

$$\mathsf{E}[g(X,Y,Z)|Y=y,Z=z] = \mathsf{E}[g(X,y,z)|Y=y,Z=z]. \tag{7.30}$$

Example 7.23. Let X, Y, and Z be as in Example 7.22. Find $E[X]$ and $E[XZ]$.

Solution. Rather than use the marginal density of X to compute $E[X]$, we use the law of total probability. Write

$$E[X] = \int_{-\infty}^{\infty} \int_{-\infty}^{\infty} E[X|Y=y, Z=z] f_{YZ}(y,z) \, dy \, dz.$$

From Example 7.22, $f_{X|YZ}(\cdot|y,z) \sim N(y, z^2)$, and so $E[X|Y=y, Z=z] = y$. Thus,

$$E[X] = \int_{-\infty}^{\infty} \int_{-\infty}^{\infty} y f_{YZ}(y,z) \, dy \, dz = E[Y],$$

which we compute by again using the law of total probability. Write

$$E[Y] = \int_{-\infty}^{\infty} E[Y|Z=z] f_Z(z) \, dz.$$

From Example 7.22, $f_{Y|Z}(\cdot|z) \sim \exp(z)$; hence, $E[Y|Z=z] = 1/z$. Since $f_Z(z) = 3z^2/7$,

$$E[Y] = \int_1^2 \tfrac{3}{7} z \, dz = \tfrac{9}{14}.$$

Thus, $E[X] = E[Y] = 9/14$.

To find $E[XZ]$, write

$$E[XZ] = \int_{-\infty}^{\infty} \int_{-\infty}^{\infty} E[Xz|Y=y, Z=z] f_{YZ}(y,z) \, dy \, dz.$$

We then note that $E[Xz|Y=y, Z=z] = E[X|Y=y, Z=z]z = yz$. Thus,

$$E[XZ] = \int_{-\infty}^{\infty} \int_{-\infty}^{\infty} yz f_{YZ}(y,z) \, dy \, dz = E[YZ].$$

In Problem 56 the reader is asked to show that $E[YZ] = 1$. Thus, $E[XZ] = 1$ as well.

Example 7.24. Let N be a positive, integer-valued random variable, and let $X_1, X_2, \ldots$ be i.i.d. Further assume that N is independent of $X_1, \ldots, X_n$ for every n. Consider the **random sum**,

$$\sum_{i=1}^{N} X_i.$$

Note that the number of terms in the sum is a random variable. Find the mean value of the random sum.

Solution. Use the law of total probability to write

$$E\left[\sum_{i=1}^{N} X_i\right] = \sum_{n=1}^{\infty} E\left[\sum_{i=1}^{n} X_i \middle| N=n\right] P(N=n).$$

By independence of N and the X_i sequence,

$$\mathsf{E}\!\left[\sum_{i=1}^{n} X_i \Big| N = n\right] \;=\; \mathsf{E}\!\left[\sum_{i=1}^{n} X_i\right] \;=\; \sum_{i=1}^{n} \mathsf{E}[X_i].$$

Since the X_i are i.i.d., they all have the same mean. In particular, for all i, $\mathsf{E}[X_i] = \mathsf{E}[X_1]$. Thus,

$$\mathsf{E}\!\left[\sum_{i=1}^{n} X_i \Big| N = n\right] \;=\; n\,\mathsf{E}[X_1].$$

Now we can write

$$\mathsf{E}\!\left[\sum_{i=1}^{N} X_i\right] \;=\; \sum_{n=1}^{\infty} n\,\mathsf{E}[X_1]\,\mathsf{P}(N=n)$$
$$=\; \mathsf{E}[N]\,\mathsf{E}[X_1].$$

Notes

7.1: Joint and marginal probabilities

Note 1. Comments analogous to Note 1 in Chapter 2 apply here. Specifically, the set A must be restricted to a suitable σ-field $\mathscr{B}$ of subsets of $\mathbb{R}^2$. Typically, $\mathscr{B}$ is taken to be the collection of Borel sets of $\mathbb{R}^2$; i.e., $\mathscr{B}$ is the smallest σ-field containing all the open sets of $\mathbb{R}^2$.

Note 2. While it is easily seen that every joint cdf $F_{XY}(x,y)$ satisfies

(i) $0 \le F_{XY}(x,y) \le 1$,
(ii) For fixed y, $F_{XY}(x,y)$ is nondecreasing in x,
(iii) For fixed x, $F_{XY}(x,y)$ is nondecreasing in y,

it is the rectangle formula

$$\mathsf{P}(a < X \le b, c < Y \le d) \;=\; F_{XY}(b,d) - F_{XY}(a,d) - F_{XY}(b,c) + F_{XY}(a,c)$$

that implies the above right-hand side is nonnegative.

Given a function $F(x,y)$ that satisfies the above three properties, the function may or may not satisfy

$$F(b,d) - F(a,d) - F(b,c) + F(a,c) \;\ge\; 0.$$

In fact, the function

$$F(x,y) \;:=\; \begin{cases} 1, & (x,y) \in \text{quadrants I, II, or IV}, \\ 0, & (x,y) \in \text{quadrant III}, \end{cases}$$

satisfies the three properties, but for

$$(a,b] \times (c,d] \;=\; (-1/2, 1/2] \times (-1/2, 1/2],$$

it is easy to check that

$$F(b,d) - F(a,d) - F(b,c) + F(a,c) \;=\; -1 \;<\; 0.$$

Note 3. We now derive the limit formula for $F_X(x)$ in (7.3); the formula for $F_Y(y)$ can be derived similarly. To begin, write

$$F_X(x) := P(X \leq x) = P((X,Y) \in (-\infty,x] \times \mathbb{R}).$$

Next, observe that $\mathbb{R} = \bigcup_{n=1}^{\infty}(-\infty,n]$, and write

$$(-\infty,x] \times \mathbb{R} = (-\infty,x] \times \bigcup_{n=1}^{\infty}(-\infty,n]$$

$$= \bigcup_{n=1}^{\infty}(-\infty,x] \times (-\infty,n].$$

Since the union is increasing, we can use the limit property (1.15) to show that

$$F_X(x) = P\left((X,Y) \in \bigcup_{n=1}^{\infty}(-\infty,x] \times (-\infty,n]\right)$$

$$= \lim_{N \to \infty} P((X,Y) \in (-\infty,x] \times (-\infty,N])$$

$$= \lim_{N \to \infty} F_{XY}(x,N).$$

7.2: Jointly continuous random variables

Note 4. As illustrated at the end of Section 7.2, it is possible to have X and Y each be continuous random variables but not jointly continuous. When a joint density exists, advanced texts say the pair is **absolutely continuous**. See also Note 4 in Chapter 5.

7.3: Conditional probability and expectation

Note 5. If the density f_X is bounded, say by K, it is easy to see that the cdf $F_X(x) = \int_{-\infty}^{x} f_X(t)\,dt$ is continuous. Just write

$$|F_X(x+\Delta x) - F_X(x)| = \left|\int_{x}^{x+\Delta x} f_X(t)\,dt\right| \leq K|\Delta x|.$$

For the general case, see Problem 6 in Chapter 13.

Note 6. To show that the law of substitution holds for conditional probability, write

$$P(g(X,Y) \in C) = E[I_C(g(X,Y))] = \int_{-\infty}^{\infty} E[I_C(g(X,Y))|X=x] f_X(x)\,dx$$

and reduce the problem to one involving conditional expectation, for which the law of substitution is easily established.

Note 7. Here is a derivation of Leibniz' rule for computing

$$\frac{d}{dz}\int_{a(z)}^{b(z)} h(z,y)\,dy. \qquad (7.31)$$

Recall that by the **chain rule** from calculus, for functions $H(u,v,w)$, $a(z)$, $b(z)$, and $c(z)$,

$$\frac{d}{dz}H(a(z),b(z),c(z)) = \frac{\partial H}{\partial u}a'(z) + \frac{\partial H}{\partial v}b'(z) + \frac{\partial H}{\partial w}c'(z),$$

where occurrences of u, v, and w in the formulas for the partial derivatives are replaced by $u = a(z)$, $v = b(z)$, and $w = c(z)$. Consider the function

$$H(u,v,w) := \int_u^v h(w,y)\,dy,$$

and note that

$$\frac{\partial H}{\partial u} = -h(w,u), \quad \frac{\partial H}{\partial v} = h(w,v), \quad \text{and} \quad \frac{\partial H}{\partial w} = \int_u^v \frac{\partial}{\partial w}h(w,y)\,dy.$$

Now observe that (7.31) is the derivative of $H(a(z),b(z),z)$ with respect to z. It follows that (7.31) is equal to

$$-h(z,a(z))a'(z) + h(z,b(z))b'(z) + \int_{a(z)}^{b(z)} \frac{\partial}{\partial z}h(z,y)\,dy.$$

Note 8. When g takes only finitely many distinct values, (7.18) and (7.19) can be derived by conditioning on $x < X \leq x + \Delta x$ and letting $\Delta x \to 0$. Then the case for general g can be derived in the same way as the law of the unconscious statistician was derived for continuous random variables at the end of Section 4.2.

Problems

7.1: Joint and marginal distributions

1. Express the cdf of $Z := Y - X$ in the form $P((X,Y) \in A_z)$ for some set A_z. Sketch your set A_z.

2. Express the cdf of $Z := Y/X$ in the form $P((X,Y) \in A_z)$ for some set A_z. Sketch your set A_z.

3. For $a < b$ and $c < d$, sketch the following sets.

    (a) $R := (a,b] \times (c,d]$.
    (b) $A := (-\infty,a] \times (-\infty,d]$.
    (c) $B := (-\infty,b] \times (-\infty,c]$.
    (d) $C := (a,b] \times (-\infty,c]$.
    (e) $D := (-\infty,a] \times (c,d]$.
    (f) $A \cap B$.

4. Show that $P(a < X \le b, c < Y \le d)$ is given by

$$F_{XY}(b,d) - F_{XY}(a,d) - F_{XY}(b,c) + F_{XY}(a,c).$$

Hint: Using the notation of the preceding problem, observe that

$$(-\infty, b] \times (-\infty, d] = R \cup (A \cup B),$$

and solve for $P((X,Y) \in R)$.

5. For each of the following two-dimensional sets, determine whether or not it is a Cartesian product. If it is, find the two one-dimensional sets of which it is a product.

   (a) $\{(x,y) : |x| \le y \le 1\}$.
   (b) $\{(x,y) : 2 < x \le 4, 1 \le y < 2\}$.
   (c) $\{(x,y) : 2 < x \le 4, y = 1\}$.
   (d) $\{(x,y) : 2 < x \le 4\}$.
   (e) $\{(x,y) : y = 1\}$.
   (f) $\{(1,1), (2,1), (3,1)\}$.
   (g) The union of $\{(1,3), (2,3), (3,3)\}$ and the set in (f).
   (h) $\{(1,0), (2,0), (3,0), (0,1), (1,1), (2,1), (3,1)\}$.

6. If[c]

$$F_{XY}(x,y) = \begin{cases} x - 1 - \dfrac{e^{-y} - e^{-xy}}{y}, & 1 \le x \le 2, y \ge 0, \\ 1 - \dfrac{e^{-y} - e^{-2y}}{y}, & x > 2, y \ge 0, \\ 0, & \text{otherwise,} \end{cases}$$

find the marginals $F_X(x)$ and $F_Y(y)$ and determine whether or not X and Y are independent.

7. If

$$F_{XY}(x,y) = \begin{cases} \frac{2}{7}(1 - e^{-2y}), & 2 \le x < 3, y \ge 0, \\ \dfrac{(7 - 2e^{-2y} - 5e^{-3y})}{7}, & x \ge 3, y \ge 0, \\ 0, & \text{otherwise.} \end{cases}$$

find the marginals $F_X(x)$ and $F_Y(y)$ and determine whether or not X and Y are independent.

---

[c]The quotients involving division by y are understood as taking their limiting values when $y = 0$.

7.2: Jointly continuous random variables

8. The joint density in Example 7.12 was obtained by differentiating $F_{XY}(x,y)$ first with respect to x and then with respect to y. In this problem, find the joint density by differentiating first with respect to y and then with respect to x.

9. Find the marginal density $f_X(x)$ if
$$f_{XY}(x,y) = \frac{\exp[-|y-x|-x^2/2]}{2\sqrt{2\pi}}.$$

10. Find the marginal density $f_Y(y)$ if
$$f_{XY}(x,y) = \frac{4e^{-(x-y)^2/2}}{y^5\sqrt{2\pi}}, \quad y \geq 1.$$

11. Let X and Y have joint density $f_{XY}(x,y)$. Find the marginal cdf and density of $\max(X,Y)$ and of $\min(X,Y)$. How do your results simplify if X and Y are independent? What if you further assume that the densities of X and Y are the same?

12. Let $X \sim \text{gamma}(p,1)$ and $Y \sim \text{gamma}(q,1)$ be independent random variables. Find the density of $Z := X + Y$. Then compute $P(Z > 1)$ if $p = q = 1/2$.

13. Find the density of $Z := X + Y$, where X and Y are independent Cauchy random variables with parameters λ and μ, respectively. Then compute $P(Z \leq 1)$ if $\lambda = \mu = 1/2$.

14. In Example 7.10, the double integral for $P((X,Y) \in A_z)$, where A_z is sketched in Figure 7.1, was evaluated as an iterated integral with the inner integral with respect to y and the outer integral with respect to x. Re-work the example if the inner integral is with respect to x and the outer integral is with respect to y.

15. Re-work Example 7.11 if instead of the partition $A_z = A_z^+ \cup A_z^-$ shown in Figure 7.2, you use the partition $A_z = B_z^+ \cup B_z^-$, where
$$B_z^+ := \{(x,y) : x \leq z/y, y > 0\} \quad \text{and} \quad B_z^- := \{(x,y) : x \geq z/y, y < 0\}.$$

16. If X and Y have joint density f_{XY}, find the cdf and density of $Z = Y - X$.

17. If X and Y have joint density f_{XY}, find the cdf and density of $Z = Y/X$.

18. Let
$$f_{XY}(x,y) := \begin{cases} Kx^n y^m, & (x,y) \in D, \\ 0, & \text{otherwise,} \end{cases}$$
where n and m are nonnegative integers, K is a constant, and $D := \{(x,y) : |x| \leq y \leq 1\}$.

   (a) Sketch the region D.

   (b) Are there any restrictions on n and m that you need to make in order that f_{XY} be a valid joint density? If n and m are allowable, find K so that f_{XY} is a valid joint density.

(c) For $0 < z < 1$, sketch the region
$$A_z := \{(x,y) : xy > z\}.$$

(d) Sketch the region $A_z \cap D$.

(e) Compute $P((X,Y) \in A_z)$.

*19. A rectangle is drawn with random width being uniform on $[0,w]$ and random height being uniform on $[0,h]$. For fraction $0 < \lambda < 1$, find the probability that the area of the rectangle exceeds λ times the maximum possible area. Assume that the width and the height are independent.

20. Let $X := \cos \Theta$ and $Y := \sin \Theta$, where $\Theta \sim$ uniform$[-\pi, \pi]$. Show that $E[XY] = 0$. Show that $E[X] = E[Y] = 0$. Argue that X and Y cannot be independent. This gives an example of continuous random variables that are uncorrelated, but not independent. *Hint:* Use the results of Problem 35 in Chapter 5.

*21. Suppose that X and Y are random variables with the property that for all bounded continuous functions $h(x)$ and $k(y)$,
$$E[h(X)k(Y)] = E[h(X)]E[k(Y)].$$
Show that X and Y are independent random variables.

*22. If $X \sim N(0,1)$, then the complementary cumulative distribution function (ccdf) of X is
$$Q(x_0) := P(X > x_0) = \int_{x_0}^{\infty} \frac{e^{-x^2/2}}{\sqrt{2\pi}} dx.$$

(a) Show that
$$Q(x_0) = \frac{1}{\pi} \int_0^{\pi/2} \exp\left(\frac{-x_0^2}{2\cos^2 \theta}\right) d\theta, \quad x_0 \geq 0.$$
Hint: For any random variables X and Y, we can always write
$$P(X > x_0) = P(X > x_0, Y \in \mathbb{R}) = P((X,Y) \in D),$$
where D is the half plane $D := \{(x,y) : x > x_0\}$. Now specialize to the case where X and Y are independent and both $N(0,1)$. Then the probability on the right is a double integral that can be evaluated using polar coordinates.

Remark. The procedure outlined in the hint is a generalization of that used in Section 4.1 to show that the standard normal density integrates to one. To see this, note that if $x_0 = -\infty$, then $D = \mathbb{R}^2$.

(b) Use the result of (a) to derive **Craig's formula** [10, p. 572, Eq. (9)],
$$Q(x_0) = \frac{1}{\pi} \int_0^{\pi/2} \exp\left(\frac{-x_0^2}{2\sin^2 t}\right) dt, \quad x_0 \geq 0.$$

Remark. Simon and Alouini [54] have derived a similar result for the Marcum Q function (defined in Problem 25 in Chapter 5) and its higher-order generalizations. See also [56, pp. 1865–1867].

7.3: Conditional probability and expectation

23. Using the definition (7.12) of conditional density, show that

$$\int_{-\infty}^{\infty} f_{Y|X}(y|x)\,dy = 1.$$

*24. If X and Y are jointly continuous, show that

$$\lim_{\Delta x \to 0} \mathsf{P}((X,Y) \in A | x < X \leq x+\Delta x) = \int_{-\infty}^{\infty} I_A(x,y) f_{Y|X}(y|x)\,dy.$$

25. Let $f_{XY}(x,y)$ be as derived in Example 7.12, and note that $f_X(x)$ and $f_Y(y)$ were found in Example 7.13. Find $f_{Y|X}(y|x)$ and $f_{X|Y}(x|y)$ for $x,y > 0$. How do these conditional densities compare with the marginals $f_Y(y)$ and $f_X(x)$; is $f_{Y|X}(y|x)$ similar to $f_Y(y)$ and is $f_{X|Y}(x|y)$ similar to $f_X(x)$?

26. Let $f_{XY}(x,y)$ be as derived in Example 7.12, and note that $f_X(x)$ and $f_Y(y)$ were found in Example 7.13. Compute $\mathsf{E}[Y|X = x]$ for $x > 0$ and $\mathsf{E}[X|Y = y]$ for $y > 0$.

27. Let X and Y be jointly continuous. Show that if

$$\mathsf{P}(X \in B | Y = y) := \int_B f_{X|Y}(x|y)\,dx,$$

then

$$\mathsf{P}(X \in B) = \int_{-\infty}^{\infty} \mathsf{P}(X \in B | Y = y) f_Y(y)\,dy.$$

28. Use the formula of Example 7.16 to compute $f_Z(z)$ if X and Y are independent $N(0, \sigma^2)$.

29. Use the formula of Example 7.17 to compute $f_Z(z)$ if X and Y are independent $\exp(\lambda)$ random variables.

30. Find $\mathsf{P}(X \leq Y)$ if X and Y are independent with $X \sim \exp(\lambda)$ and $Y \sim \exp(\mu)$.

31. Let X and Y be independent random variables with Y being exponential with parameter 1 and X being uniform on $[1,2]$. Find $\mathsf{P}(Y/\ln(1+X^2) > 1)$.

32. Let X and Y be jointly continuous random variables with joint density f_{XY}. Find $f_Z(z)$ if

   (a) $Z = e^X Y$.
   (b) $Z = |X+Y|$.

33. Let X and Y be independent continuous random variables with respective densities f_X and f_Y. Put $Z = Y/X$.

   (a) Find the density of Z. *Hint:* Review Example 7.15.
   (b) If X and Y are both $N(0, \sigma^2)$, show that Z has a Cauchy(1) density that does not depend on σ^2.

(c) If X and Y are both Laplace(λ), find a closed-form expression for $f_Z(z)$ that does not depend on λ.

(d) Find a closed-form expression for the density of Z if Y is uniform on $[-1,1]$ and $X \sim N(0,1)$.

(e) If X and Y are both Rayleigh random variables with parameter λ, find a closed-form expression for the density of Z. Your answer should not depend on λ.

34. Let X and Y be independent with densities $f_X(x)$ and $f_Y(y)$. If X is a positive random variable, and if $Z = Y/\ln(X)$, find the density of Z.

35. Let X, Z, and U be independent random variables with X and Z being independent exp(1) random variables and $U \sim$ uniform$[-1/2, 1/2]$. Compute $\mathsf{E}[e^{(X+Z)U}]$.

36. Let $Y \sim$ uniform$[1,2]$, and given $Y = y$, suppose that $X \sim$ Laplace(y). Find $\mathsf{E}[X^2 Y]$.

37. Let $Y \sim \exp(\lambda)$, and suppose that given $Y = y$, $X \sim$ gamma(p,y). Assuming $r > n$, evaluate $\mathsf{E}[X^n Y^r]$.

38. Let V and U be independent random variables with V being Erlang with parameters $m = 2$ and $\lambda = 1$ and $U \sim$ uniform$[-1/2, 1/2]$. Put $Y := e^{VU}$.

   (a) Find the density $f_Y(y)$ for all y.

   (b) Use your answer to part (a) to compute $\mathsf{E}[Y]$.

   (c) Compute $\mathsf{E}[Y]$ directly by using the laws of total probability and substitution.

   Remark. Your answers to parts (b) and (c) should be the same as your answer to Problem 35. Can you explain why?

39. Use the law of total probability to solve the following problems.

   (a) Evaluate $\mathsf{E}[\cos(X+Y)]$ if given $X = x$, Y is conditionally uniform on $[x - \pi, x + \pi]$.

   (b) Evaluate $\mathsf{P}(Y > y)$ if $X \sim$ uniform$[1,2]$, and given $X = x$, Y is exponential with parameter x.

   (c) Evaluate $\mathsf{E}[Xe^Y]$ if $X \sim$ uniform$[3,7]$, and given $X = x$, $Y \sim N(0, x^2)$.

   (d) Let $X \sim$ uniform$[1,2]$, and suppose that given $X = x$, $Y \sim N(0, 1/x)$. Evaluate $\mathsf{E}[\cos(XY)]$.

40. The Gaussian signal $X \sim N(0, \sigma^2)$ is subjected to independent Rayleigh fading so that the received signal is $Y = ZX$, where $Z \sim$ Rayleigh(1) and X are independent. Use the law of total probability to find the moment generating function of Y. What is the density of Y?

41. Find $\mathsf{E}[X^n Y^m]$ if $Y \sim \exp(\beta)$, and given $Y = y$, $X \sim$ Rayleigh(y).

*42. Let $X \sim$ gamma(p, λ) and $Y \sim$ gamma(q, λ) be independent.

(a) If $Z := X/Y$, show that the density of Z is

$$f_Z(z) = \frac{1}{B(p,q)} \cdot \frac{z^{p-1}}{(1+z)^{p+q}}, \quad z > 0.$$

Observe that $f_Z(z)$ depends on p and q, but not on λ. It was shown in Problem 22 in Chapter 4 that $f_Z(z)$ integrates to one. *Hint:* You will need the fact that $B(p,q) = \Gamma(p)\Gamma(q)/\Gamma(p+q)$, which was shown in Problem 16 in Chapter 4.

(b) Show that

$$V := \frac{X}{X+Y}$$

has a beta density with parameters p and q. In particular, if $p = q = 1$ so that X and Y are $\exp(\lambda)$, then $V \sim \text{uniform}(0,1)$. *Hint:* Observe that $V = Z/(1+Z)$, where $Z = X/Y$ as above.

Remark. If $W := (X/p)/(Y/q)$, then $f_W(w) = (p/q) f_Z(w(p/q))$. If further $p = k_1/2$ and $q = k_2/2$, then W is said to be an **F random variable** with k_1 and k_2 degrees of freedom. If further $\lambda = 1/2$, then X and Y are chi-squared with k_1 and k_2 degrees of freedom, respectively.

*43. Let $X_1, \ldots, X_n$ be i.i.d. gamma(p, λ) random variables, and put

$$Y_i := \frac{X_i}{X_1 + \cdots + X_n}.$$

Use the result of Problem 42(b) to show that Y_i has a beta density with parameters p and $(n-1)p$.

Remark. Note that although the Y_i are not independent, they are identically distributed. Also, $Y_1 + \cdots + Y_n = 1$. Here are two applications. First, the numbers $Y_1, \ldots, Y_n$ can be thought of as a randomly chosen probability mass function on the integers 1 to n. Second, if we let Z be the vector of length n whose ith component is $\sqrt{Y_i}$, then Z has length

$$\sqrt{Z_1^2 + \cdots + Z_n^2} = \sqrt{Y_1 + \cdots + Y_n} = 1.$$

In other words, Z is a randomly chosen vector that always lies on the surface of the unit sphere in n-dimensional space.

*44. Let X and Y be independent with $X \sim N(0,1)$ and Y being chi-squared with k degrees of freedom. Show that the density of $Z := X/\sqrt{Y/k}$ has Student's t density with k degrees of freedom. *Hint:* For this problem, it may be helpful to review the results of Problems 14–16 and 20 in Chapter 4.

*45. The generalized gamma density was introduced in Problem 21 in Chapter 3. Recall that $X \sim \text{g-gamma}(p, \lambda, r)$ if

$$f_X(x) = \frac{\lambda r (\lambda x)^{p-1} e^{-(\lambda x)^r}}{\Gamma(p/r)}, \quad x > 0.$$

If $X \sim$ g-gamma(p,λ,r) and $Y \sim$ g-gamma(q,λ,r) are independent and $Z := X/Y$, show that the density of Z is

$$f_Z(z) = \frac{r}{B(p/r,q/r)} \cdot \frac{z^{p-1}}{(1+z^r)^{(p+q)/r}}, \quad z > 0.$$

Since $r = 1$ is the ordinary gamma density, we can recover the result of Problem 42(a). Since $p = r = 2$ is the Rayleigh density, we can recover the result of Problem 33(e).

*46. Let X and Y be independent, both with the density of Problem 3 in Chapter 4. Put $Z := X + Y$, and use the convolution formula (7.16) to show that

$$f_Z(z) = \begin{cases} \frac{\pi}{4}, & 0 < z \leq 1, \\ \frac{1}{2}\left[\sin^{-1}(1/\sqrt{z}) - \sin^{-1}(\sqrt{1-1/z}\,)\right], & 1 < z \leq 2, \\ 0, & \text{otherwise.} \end{cases}$$

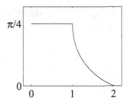

Figure 7.13. Density $f_Z(z)$ of Problem 46.

7.4: The bivariate normal

47. Let U and V have the joint Gaussian density in (7.22). Show that for all ρ with $-1 < \rho < 1$, U and V both have standard univariate $N(0,1)$ marginal densities that do not involve ρ. *Hint:* Use (7.23) and (7.26).

48. Let X and Y be jointly Gaussian with density $f_{XY}(x,y)$ given by (7.25). Find $f_X(x)$, $f_Y(y)$, $f_{X|Y}(x|y)$, and $f_{Y|X}(y|x)$. *Hint:* Apply (7.23) and (7.26) to (7.24).

49. Let X and Y be jointly Gaussian with density $f_{XY}(x,y)$ given by (7.25). Find $\mathsf{E}[Y|X = x]$ and $\mathsf{E}[X|Y = y]$. *Hint:* Use the conditional densities found in Problem 48.

50. If X and Y are jointly normal with parameters, m_X, m_Y, σ_X^2, σ_Y^2, and ρ, compute $\mathsf{E}[X]$, $\mathsf{E}[X^2]$, and

$$\frac{\text{cov}(X,Y)}{\sigma_X \sigma_Y} = \mathsf{E}\left[\left(\frac{X-m_X}{\sigma_X}\right)\left(\frac{Y-m_Y}{\sigma_Y}\right)\right].$$

You may use the results of Problem 48.

*51. Let ψ_ρ be the standard bivariate normal density defined in (7.22). Put

$$f_{UV}(u,v) := \frac{1}{2}[\psi_{\rho_1}(u,v) + \psi_{\rho_2}(u,v)],$$

where $-1 < \rho_1 \neq \rho_2 < 1$.

(a) Show that the marginals f_U and f_V are both $N(0,1)$. (You may use the results of Problem 47.)

(b) Show that $\bar{\rho} := \mathsf{E}[UV] = (\rho_1 + \rho_2)/2$. (You may use the result of Example 7.19.)

(c) Show that U and V cannot be jointly normal. *Hints:* (*i*) To obtain a contradiction, suppose that f_{UV} is a jointly normal density with parameters given by parts (a) and (b). (*ii*) Consider $f_{UV}(u,u)$. (*iii*) Use the following fact: If $\beta_1, \ldots, \beta_n$ are distinct real numbers, and if

$$\sum_{k=1}^{n} \alpha_k e^{\beta_k t} = 0, \quad \text{for all } t \geq 0,$$

then $\alpha_1 = \cdots = \alpha_n = 0$.

(d) By construction, U and V are jointly continuous. If $\rho_1 = -\rho_2$, then part (b) shows that U and V are uncorrelated. However, they are not independent. Show this by arguing as follows. First compute $\mathsf{E}[V^2|U=u]$ and show that even if $\rho_1 = -\rho_2$, this conditional expectation is a function of u unless $\rho_1 = \rho_2 = 0$. Then note that if U and V were independent, $\mathsf{E}[V^2|U=u] = \mathsf{E}[V^2] = 1$ and does not depend on u. *Hint:* Example 7.20 will be helpful.

*52. Let U and V be jointly normal with joint density $\psi_\rho(u,v)$ defined in (7.22). Put

$$Q_\rho(u_0, v_0) := \mathsf{P}(U > u_0, V > v_0).$$

Show that for $u_0, v_0 \geq 0$,

$$Q_\rho(u_0, v_0) = \int_0^{\tan^{-1}(v_0/u_0)} h_\rho(v_0^2, \theta) \, d\theta + \int_0^{\pi/2 - \tan^{-1}(v_0/u_0)} h_\rho(u_0^2, \theta) \, d\theta,$$

where

$$h_\rho(z, \theta) := \frac{\sqrt{1-\rho^2}}{2\pi(1-\rho\sin 2\theta)} \exp\left[\frac{-z(1-\rho\sin 2\theta)}{2(1-\rho^2)\sin^2\theta}\right].$$

This formula for $Q_\rho(u_0, v_0)$ is Simon's [55, eq. (78b)], [56, pp. 1864–1865] bivariate generalization of Craig's univariate formula given in Problem 22. *Hint:* Write $\mathsf{P}(U > u_0, V > v_0)$ as a double integral and convert to polar coordinates. It may be helpful to review your solution of Problem 22 first.

*53. Use Simon's formula in Problem 52 to show that

$$Q(x_0)^2 = \frac{1}{\pi} \int_0^{\pi/4} \exp\left(\frac{-x_0^2}{2\sin^2\theta}\right) d\theta, \quad x_0 \geq 0.$$

In other words, to compute $Q(x_0)^2$, we integrate Craig's integrand (Problem 22) only half as far [55, p. 210], [56, p. 1865]!

7.5: Extension to three or more random variables

54. If
$$f_{XYZ}(x,y,z) = \frac{2\exp[-|x-y|-(y-z)^2/2]}{z^5\sqrt{2\pi}}, \quad z \geq 1,$$
and $f_{XYZ}(x,y,z) = 0$ otherwise, find $f_{YZ}(y,z)$, $f_{X|YZ}(x|y,z)$, $f_Z(z)$, and $f_{Y|Z}(y|z)$.

55. Let
$$f_{XYZ}(x,y,z) = \frac{e^{-(x-y)^2/2}e^{-(y-z)^2/2}e^{-z^2/2}}{(2\pi)^{3/2}}.$$
Find $f_{XY}(x,y)$. Then find the means and variances of X and Y. Also find the correlation, $\mathsf{E}[XY]$.

56. Let X, Y, and Z be as in Example 7.22. Evaluate $\mathsf{E}[XY]$ and $\mathsf{E}[YZ]$.

57. Let X, Y, and Z be as in Problem 54. Evaluate $\mathsf{E}[XYZ]$.

58. Let X, Y, and Z be jointly continuous. Assume that $X \sim$ uniform$[1,2]$; that given $X = x$, $Y \sim \exp(1/x)$; and that given $X = x$ and $Y = y$, Z is $N(x,1)$. Find $\mathsf{E}[XYZ]$.

59. Let N denote the number of primaries in a photomultiplier, and let X_i be the number of secondaries due to the ith primary. Then the total number of secondaries is
$$Y = \sum_{i=1}^N X_i.$$
Express the characteristic function of Y in terms of the probability generating function of N, $G_N(z)$, and the characteristic function of the X_i, assuming that the X_i are i.i.d. with common characteristic function $\varphi_X(\nu)$. Assume that N is independent of the X_i sequence. Find the density of Y if $N \sim$ geometric$_1(p)$ and $X_i \sim \exp(\lambda)$.

Exam preparation

You may use the following suggestions to prepare a study sheet, including formulas mentioned that you have trouble remembering. You may also want to ask your instructor for additional suggestions.

7.1. Joint and marginal cdfs. Know the rectangle formula (7.2). Know how to obtain marginal cdfs from the joint cdf; i.e., (7.3) and (7.4). Know that X and Y are independent if and only if the joint cdf is equal to the product of the marginal cdfs.

7.2. Jointly continuous random variables. Know the mixed partial formula (7.9) for obtaining the joint density from the joint cdf. Know how to integrate out unneeded variables from the joint density to obtain the marginal density (7.10). Know that jointly continuous random variables are independent if and only if their joint density factors as $f_{XY}(x,y) = f_X(x)f_Y(y)$.

7.3. Conditional probability and expectation. Know the formula for conditional densities (7.12). Again, I tell my students that the three most important things in probability are:

(*i*) the laws of total probability (7.13), (7.14), and (7.20);
(*ii*) the substitution laws (7.15) and (7.21); and
(*iii*) independence.

If the conditional density of Y given X is listed in the table inside the back cover (this table includes moments), then $\mathsf{E}[Y|X=x]$ or $\mathsf{E}[Y^2|X=x]$ can often be found by inspection. This is a *very* useful skill.

7.4. *The bivariate normal.* For me, the easiest way to remember the bivariate normal density is in two stages. First, I remember (7.22), and then I use (7.24). Remember that if X and Y are jointly normal, the conditional density of one given the other is also normal, and $\mathsf{E}[X|Y=y]$ has the form $my+b$ for some slope m and some y-intercept b. See Problems 48 and 49.

7.5. *Extension to three or more random variables.* Note the more general forms of the law of total probability (7.29) and the substitution law (7.30).

Work any review problems assigned by your instructor. If you finish them, re-work your homework assignments.

8
Introduction to random vectors[†]

In the previous chapter, we worked mostly with two or three random variables at a time. When we need to work with a larger number of random variables, it is convenient to collect them into a column vector. The notation of vectors and matrices allows us to express powerful formulas in straightforward, compact notation.

8.1 Review of matrix operations

Transpose of a matrix. Recall that if A is a matrix with entries A_{ij}, then its **transpose**, denoted by A', is defined by $(A')_{ij} := A_{ji}$. For example,

$$\begin{bmatrix} 1 & 3 & 5 \\ 2 & 4 & 6 \end{bmatrix}' = \begin{bmatrix} 1 & 2 \\ 3 & 4 \\ 5 & 6 \end{bmatrix}.$$

The transpose operation converts every row into a column, or equivalently, it converts every column into a row. The example

$$\begin{bmatrix} 1 & 3 & 5 \end{bmatrix}' = \begin{bmatrix} 1 \\ 3 \\ 5 \end{bmatrix}$$

shows that an easy way to specify column vectors is to take the transpose of a row vector, a practice we use frequently.

Sum of matrices. If two matrices have the same dimensions, then their sum is computed by adding the corresponding entries. For example,

$$\begin{bmatrix} 1 & 2 & 3 \\ 4 & 5 & 6 \end{bmatrix} + \begin{bmatrix} 10 & 20 & 30 \\ 40 & 50 & 60 \end{bmatrix} = \begin{bmatrix} 11 & 22 & 33 \\ 44 & 55 & 66 \end{bmatrix}.$$

Product of matrices. If A is an $r \times n$ matrix and B is an $n \times p$ matrix, then their product is the $r \times p$ matrix whose entries are given by

$$(AB)_{ij} := \sum_{k=1}^{n} A_{ik} B_{kj},$$

where $i = 1, \ldots, r$ and $j = 1, \ldots, p$. For example, using a piece of scratch paper, you can check that

$$\begin{bmatrix} 7 & 8 & 9 \\ 4 & 5 & 6 \end{bmatrix} \begin{bmatrix} 10 & 40 \\ 20 & 50 \\ 30 & 60 \end{bmatrix} = \begin{bmatrix} 500 & 1220 \\ 320 & 770 \end{bmatrix}. \tag{8.1}$$

You can also check it with the MATLAB commands

[†]This chapter and the next are not required for the study of random processes in Chapter 10. See the Chapter Dependencies graph in the preface.

8.1 Review of matrix operations

```
A = [ 7 8 9 ; 4 5 6 ]
B = [ 10 40 ; 20 50 ; 30 60 ]
A*B
```

Notice how rows are separated with the semicolon ";".

Trace of a matrix. If C is a square $r \times r$ matrix, then the **trace** of C is defined to be the sum of its diagonal elements,

$$\text{tr}(C) := \sum_{k=1}^{r} C_{kk}.$$

For example, the trace of the matrix on the right-hand side of (8.1) is 1270. If A is an $r \times n$ matrix and B is an $n \times r$ matrix, it is shown in Problem 4 that

$$\text{tr}(AB) = \text{tr}(BA),$$

where the left-hand side is the trace of an $r \times r$ matrix, and the right-hand side is the trace of an $n \times n$ matrix. In particular, if $n = 1$, BA is a scalar; in this case, $\text{tr}(AB) = BA$. The MATLAB command for tr is `trace`. For example, if A and B are defined by the above MATLAB commands, you can easily check that `trace(A*B)` gives the same result as `trace(B*A)`.

Norm of a vector. If $x = [x_1, \ldots, x_n]'$, then we define the **norm** of x by

$$\|x\| := (x'x)^{1/2}.$$

Notice that since $\|x\|^2 = x'x$ is a scalar,

$$\|x\|^2 = x'x = \text{tr}(x'x) = \text{tr}(xx'), \tag{8.2}$$

a formula we use later.

Inner product of vectors. If $x = [x_1, \ldots, x_n]'$ and $y = [y_1, \ldots, y_n]'$ are two column vectors, their **inner product** or **dot product** is defined by

$$\langle x, y \rangle := y'x.$$

Taking $y = x$ yields $\langle x, x \rangle = \|x\|^2$. An important property of the inner product is that

$$|\langle x, y \rangle| \leq \|x\| \|y\|, \tag{8.3}$$

with equality if and only if one of them is a scalar multiple of the other. This result is known as the **Cauchy–Schwarz inequality** for column vectors and is derived in Problem 6.

Remark. While $y'x$ is called the inner product, xy' is sometimes called the **outer product**. Since $y'x = \text{tr}(y'x) = \text{tr}(xy')$, the inner product is equal to the trace of the outer product. While the formula $\text{tr}(xy')$ is useful for theoretical analysis, it is computationally inefficient.

Block matrices. Sometimes it is convenient to partition a large matrix so that it can be written in terms of smaller submatrices or blocks. For example, we can partition

$$\begin{bmatrix} 11 & 12 & 13 & 14 \\ 21 & 22 & 23 & 24 \\ 31 & 32 & 33 & 34 \\ 41 & 42 & 43 & 44 \end{bmatrix}$$

in several different ways such as

$$\left[\begin{array}{cc|cc} 11 & 12 & 13 & 14 \\ 21 & 22 & 23 & 24 \\ \hline 31 & 32 & 33 & 34 \\ 41 & 42 & 43 & 44 \end{array}\right], \quad \left[\begin{array}{c|c|cc} 11 & 12 & 13 & 14 \\ \hline 21 & 22 & 23 & 24 \\ \hline 31 & 32 & 33 & 34 \\ 41 & 42 & 43 & 44 \end{array}\right], \quad \text{or} \quad \left[\begin{array}{ccc|c} 11 & 12 & 13 & 14 \\ 21 & 22 & 23 & 24 \\ 31 & 32 & 33 & 34 \\ \hline 41 & 42 & 43 & 44 \end{array}\right].$$

The middle partition has the form

$$\begin{bmatrix} A & B & C \\ D & E & F \\ G & H & K \end{bmatrix}.$$

The first and last partitions both have the form

$$\begin{bmatrix} A & B \\ C & D \end{bmatrix},$$

but the the corresponding blocks have different sizes. For example, in the partition on the left, A would be 2×2, but in the partition on the right, A would be 3×3.

A partitioned matrix can be transposed block by block. For example,

$$\begin{bmatrix} A & B & C \\ D & E & F \end{bmatrix}' = \begin{bmatrix} A' & D' \\ B' & E' \\ C' & F' \end{bmatrix}.$$

A pair of partitioned matrices can be added block by block *if the corresponding blocks have the same dimensions*. For example,

$$\begin{bmatrix} A & B \\ C & D \end{bmatrix} + \begin{bmatrix} \alpha & \beta \\ \gamma & \delta \end{bmatrix} = \begin{bmatrix} A+\alpha & B+\beta \\ C+\gamma & D+\delta \end{bmatrix},$$

provided the dimensions of A and α are the same, the dimensions of B and β are the same, the dimensions of C and γ are the same, and the dimensions of D and δ are the same.

A pair of partitioned matrices can be multiplied blockwise if the blocks being multiplied have the "right" dimensions. For example,

$$\begin{bmatrix} A & B \\ C & D \end{bmatrix} \begin{bmatrix} \alpha & \beta \\ \gamma & \delta \end{bmatrix} = \begin{bmatrix} A\alpha+B\gamma & A\beta+B\delta \\ C\alpha+D\gamma & C\beta+D\delta \end{bmatrix},$$

provided the sizes of the blocks are such that the matrix multiplications

$$A\alpha, B\gamma, A\beta, B\delta, C\alpha, D\gamma, C\beta, \text{ and } D\delta$$

are all defined.

8.2 Random vectors and random matrices

A vector whose entries are random variables is called a **random vector**, and a matrix whose entries are random variables is called a **random matrix**.

Expectation

The expectation of a random vector $X = [X_1,\ldots,X_n]'$ is defined to be the vector of expectations of its entries; i.e.,

$$\mathsf{E}[X] := \begin{bmatrix} \mathsf{E}[X_1] \\ \vdots \\ \mathsf{E}[X_n] \end{bmatrix}.$$

In other words, the **mean vector** $m := \mathsf{E}[X]$ has entries $m_i = \mathsf{E}[X_i]$.

More generally, if X is the $n \times p$ random matrix

$$X = \begin{bmatrix} X_{11} & \cdots & X_{1p} \\ \vdots & & \vdots \\ X_{n1} & \cdots & X_{np} \end{bmatrix},$$

then its **mean matrix** is

$$\mathsf{E}[X] := \begin{bmatrix} \mathsf{E}[X_{11}] & \cdots & \mathsf{E}[X_{1p}] \\ \vdots & & \vdots \\ \mathsf{E}[X_{n1}] & \cdots & \mathsf{E}[X_{np}] \end{bmatrix}.$$

An easy consequence of this definition is that if A is an $r \times n$ matrix with nonrandom entries, then AX is an $r \times p$ random matrix, and $\mathsf{E}[AX] = A\mathsf{E}[X]$. To see this, write

$$\begin{aligned}
\mathsf{E}[(AX)_{ij}] &= \mathsf{E}\left[\sum_{k=1}^n A_{ik}X_{kj}\right] \\
&= \sum_{k=1}^n \mathsf{E}[A_{ik}X_{kj}] \\
&= \sum_{k=1}^n A_{ik}\mathsf{E}[X_{kj}] \\
&= \sum_{k=1}^n A_{ik}(\mathsf{E}[X])_{kj} \\
&= (A\mathsf{E}[X])_{ij}.
\end{aligned}$$

It is similarly easy to show that if B is a $p \times q$ matrix with nonrandom entries, then $\mathsf{E}[XB] = \mathsf{E}[X]B$. Hence, $\mathsf{E}[AXB] = A\mathsf{E}[X]B$. If G is $r \times q$ with nonrandom entries, then

$$\mathsf{E}[AXB+G] = A\mathsf{E}[X]B+G.$$

Correlation

If $X = [X_1, \ldots, X_n]'$ is a random vector with mean vector $m := \mathsf{E}[X]$, then we define the **correlation matrix** of X by

$$R := \mathsf{E}[XX'].$$

We now point out that since XX' is equal to

$$\begin{bmatrix} X_1^2 & \cdots & X_1 X_n \\ \vdots & & \vdots \\ X_n X_1 & \cdots & X_n^2 \end{bmatrix},$$

the ij entry of R is just $\mathsf{E}[X_i X_j]$. Since $R_{ij} = R_{ji}$, R is **symmetric**. In other words, $R' = R$.

Example 8.1. Write out the correlation matrix of the three-dimensional random vector $W := [X, Y, Z]'$.

Solution. The correlation matrix of W is

$$\mathsf{E}[WW'] = \begin{bmatrix} \mathsf{E}[X^2] & \mathsf{E}[XY] & \mathsf{E}[XZ] \\ \mathsf{E}[YX] & \mathsf{E}[Y^2] & \mathsf{E}[YZ] \\ \mathsf{E}[ZX] & \mathsf{E}[ZY] & \mathsf{E}[Z^2] \end{bmatrix}.$$

There is a **Cauchy–Schwarz inequality** for random variables. For scalar random variables U and V,

$$|\mathsf{E}[UV]| \leq \sqrt{\mathsf{E}[U^2]\,\mathsf{E}[V^2]}. \tag{8.4}$$

This is formula (2.24), which was derived in Chapter 2.

Example 8.2. Show that if R is the correlation matrix of X, then

$$|R_{ij}| \leq \sqrt{R_{ii} R_{jj}}.$$

Solution. By the Cauchy–Schwarz inequality,

$$|\mathsf{E}[X_i X_j]| \leq \sqrt{\mathsf{E}[X_i^2]\,\mathsf{E}[X_j^2]}.$$

These expectations are, respectively, R_{ij}, R_{ii}, and R_{jj}.

8.2 Random vectors and random matrices

Covariance

If $X = [X_1, \ldots, X_n]'$ is a random vector with mean vector $m := \mathsf{E}[X]$, then we define the **covariance matrix** of X by

$$\mathrm{cov}(X) := \mathsf{E}[(X-m)(X-m)'].$$

Since $\mathsf{E}[X'] = (\mathsf{E}[X])'$, we see that

$$\begin{aligned}\mathrm{cov}(X) &= \mathsf{E}[XX' - Xm' - mX' + mm']\\ &= \mathsf{E}[XX'] - \mathsf{E}[X]m' - m\mathsf{E}[X'] + mm'\\ &= \mathsf{E}[XX'] - mm',\end{aligned}$$

which generalizes the variance formula (2.17). We often denote the covariance matrix of X by C_X, or just C if X is understood. Since $\mathsf{E}[XX']$ is the correlation matrix of X, we see that the covariance and correlation matrices are equal if and only if the mean vector is zero. We now point out that since $(X-m)(X-m)'$ is equal to

$$\begin{bmatrix} (X_1 - m_1)(X_1 - m_1) & \cdots & (X_1 - m_1)(X_n - m_n) \\ \vdots & & \vdots \\ (X_n - m_n)(X_1 - m_1) & \cdots & (X_n - m_n)(X_n - m_n) \end{bmatrix},$$

the ij entry of $C = \mathrm{cov}(X)$ is just

$$\mathsf{E}[(X_i - m_i)(X_j - m_j)] = \mathrm{cov}(X_i, X_j),$$

the covariance between entries X_i and X_j. Note the distinction between the covariance of a pair of random variables, which is a scalar, and the covariance of a column vector, which is a matrix. We also point out the following facts.

- $C_{ii} = \mathrm{cov}(X_i, X_i) = \mathrm{var}(X_i)$.
- Since $C_{ij} = C_{ji}$, the matrix C is **symmetric**.
- For $i \neq j$, $C_{ij} = 0$ if and only if X_i and X_j are uncorrelated. Thus, C is a diagonal matrix if and only if X_i and X_j are uncorrelated for all $i \neq j$.

Example 8.3. If a random vector X has covariance matrix C, show that $Y := AX$ has covariance matrix ACA'.

Solution. Put $m := \mathsf{E}[X]$ so that $Y - \mathsf{E}[Y] = AX - \mathsf{E}[AX] = A(X - m)$. Then

$$\begin{aligned}\mathrm{cov}(Y) &= \mathsf{E}\big[(Y - \mathsf{E}[Y])(Y - \mathsf{E}[Y])'\big]\\ &= \mathsf{E}[A(X-m)(X-m)'A']\\ &= A\mathsf{E}[(X-m)(X-m)']A'\\ &= ACA'.\end{aligned}$$

Example 8.4. A simple application of Example 8.3 is to the case in which $A = a'$, where a is a column vector. In this case, $Y = a'X$ is a scalar, and $\text{var}(Y) = \text{cov}(Y) = a'Ca$. In particular,
$$a'Ca = \text{var}(Y) \geq 0, \quad \text{for all } a.$$

A symmetric matrix C with the property $a'Ca \geq 0$ for all vectors a is said to be **positive semidefinite**. By Example 8.4, every covariance matrix is positive semidefinite. If $a'Ca > 0$ for all nonzero a, then C is called **positive definite**.

Example 8.5. Let X be a zero-mean random vector whose covariance matrix $C = \mathsf{E}[XX']$ is singular. Show that for some i and some coefficients b_j,
$$X_i = \sum_{j \neq i} b_j X_j.$$
In other words, one of the X_i is a deterministic, linear function of the remaining components.

Solution. Recall that C is singular means that there is a nonzero vector a such that $Ca = 0$. For such a, we have by Example 8.4 that the *scalar* random variable $Y := a'X$ satisfies $\mathsf{E}[Y^2] = a'Ca = 0$. Hence, Y is the zero random variable. In other words,
$$0 = Y = a'X = \sum_{j=1}^n a_j X_j.$$
Since a is not the zero vector, some component, say $a_i \neq 0$, and it follows that
$$X_i = -\frac{1}{a_i} \sum_{j \neq i} a_j X_j.$$
Taking $b_j = -a_j/a_i$ solves the problem.

Remark. The solution of Example 8.5 shows that if X is zero mean with singular covariance matrix C, then there is a nonzero vector a such that $a'X$ is the zero random variable. If X has mean vector m, then the same argument shows that $a'(X - m)$ is the zero random variable, or equivalently, $a'X$ is the constant random variable with value $a'm$.

Cross-covariance

If $X = [X_1, \ldots, X_n]'$ and $Y = [Y_1, \ldots, Y_p]'$ are both random vectors with respective means m_X and m_Y, then their **cross-covariance matrix** is the $n \times p$ matrix
$$\text{cov}(X, Y) := \mathsf{E}[(X - m_X)(Y - m_Y)'],$$

8.2 Random vectors and random matrices

which we denote by C_{XY}. Note that $(C_{XY})_{ij} = \text{cov}(X_i, Y_j)$ is just the covariance between X_i and Y_j. Also,

$$C_{YX} = \mathsf{E}[(Y - m_Y)(X - m_X)'] = (C_{XY})',$$

which is $p \times n$. The **cross-correlation matrix** of X and Y is $R_{XY} := \mathsf{E}[XY']$, and $(R_{XY})_{ij} = \mathsf{E}[X_i Y_j]$.

If we stack X and Y into the $(n+p)$-dimensional composite vector

$$Z := \begin{bmatrix} X \\ Y \end{bmatrix},$$

then the covariance matrix of Z is given by

$$C_Z = \begin{bmatrix} C_X & C_{XY} \\ C_{YX} & C_Y \end{bmatrix},$$

where C_Z is $(n+p) \times (n+p)$, C_X is $n \times n$, C_Y is $p \times p$, C_{XY} is $n \times p$, and C_{YX} is $p \times n$.

Just as two random variables U and V are said to be uncorrelated if $\text{cov}(U,V) = 0$, we say that two random vectors $X = [X_1, \ldots, X_n]'$ and $Y = [Y_1, \ldots, Y_p]'$ are uncorrelated if $\text{cov}(X_i, Y_j) = 0$ for all $i = 1, \ldots, n$ and all $j = 1, \ldots, p$. This is equivalent to the condition that $C_{XY} = \text{cov}(X,Y) = 0$ be the $n \times p$ zero matrix. If this is the case, then the matrix C_Z above is block diagonal; i.e.,

$$C_Z = \begin{bmatrix} C_X & 0 \\ 0 & C_Y \end{bmatrix}.$$

*Characteristic functions

The **joint characteristic function** of $X = [X_1, \ldots, X_n]'$ is defined by

$$\varphi_X(v) := \mathsf{E}[e^{jv'X}] = \mathsf{E}[e^{j(v_1 X_1 + \cdots + v_n X_n)}],$$

where $v = [v_1, \ldots, v_n]'$.

When X has a joint density, $\varphi_X(v) = \mathsf{E}[e^{jv'X}]$ is just the n-dimensional Fourier transform,

$$\varphi_X(v) = \int_{\mathbb{R}^n} e^{jv'x} f_X(x) \, dx, \qquad (8.5)$$

and the joint density can be recovered using the multivariate inverse Fourier transform:

$$f_X(x) = \frac{1}{(2\pi)^n} \int_{\mathbb{R}^n} e^{-jv'x} \varphi_X(v) \, dv.$$

Whether X has a joint density or not, the joint characteristic function can be used to obtain its various moments.

Example 8.6. The components of the mean vector and covariance matrix can be obtained from the characteristic function as follows. Write

$$\frac{\partial}{\partial v_k} \mathsf{E}[e^{jv'X}] = \mathsf{E}[e^{jv'X} j X_k],$$

and
$$\frac{\partial^2}{\partial v_\ell \partial v_k} \mathsf{E}[e^{jv'X}] = \mathsf{E}[e^{jv'X}(jX_\ell)(jX_k)].$$
Then
$$\frac{\partial}{\partial v_k} \mathsf{E}[e^{jv'X}]\bigg|_{v=0} = j\mathsf{E}[X_k],$$
and
$$\frac{\partial^2}{\partial v_\ell \partial v_k} \mathsf{E}[e^{jv'X}]\bigg|_{v=0} = -\mathsf{E}[X_\ell X_k].$$

Higher-order moments can be obtained in a similar fashion.

If the components of $X = [X_1, \ldots, X_n]'$ are independent, then

$$\begin{aligned}
\varphi_X(v) &= \mathsf{E}[e^{jv'X}] \\
&= \mathsf{E}[e^{j(v_1 X_1 + \cdots + v_n X_n)}] \\
&= \mathsf{E}\left[\prod_{k=1}^n e^{jv_k X_k}\right] \\
&= \prod_{k=1}^n \mathsf{E}[e^{jv_k X_k}] \\
&= \prod_{k=1}^n \varphi_{X_k}(v_k).
\end{aligned}$$

We have just shown that if the components of X are **independent**, then the joint characteristic function is the product of the marginal characteristic functions. The converse is also true; i.e., if the joint characteristic function is the product of the marginal characteristic functions, then the random variables are independent [3]. A derivation in the case of two jointly continuous random variables was given in Section 7.2.

*Decorrelation and the Karhunen–Loève expansion

Let X be an n-dimensional random vector with zero mean and covariance matrix C. We show that X has the representation $X = PY$, where the components of Y are uncorrelated and P is an $n \times n$ matrix satisfying $P'P = PP' = I$. (Hence, $P' = P^{-1}$.) This representation is called the **Karhunen–Loève expansion**.

Step 1. Recall that since a covariance matrix is symmetric, it can be diagonalized [30]; i.e., there is a square matrix P such that $P'P = PP' = I$ and such that $P'CP = \Lambda$ is a diagonal matrix, say $\Lambda = \text{diag}(\lambda_1, \ldots, \lambda_n)$.

Step 2. Define a new random variable $Y := P'X$. By Example 8.3, $\text{cov}(Y) = P'CP = \Lambda$. Since $\text{cov}(Y) = \Lambda$ is diagonal, the components of Y are uncorrelated. For this reason, we call P' a **decorrelating transformation**.

Step 3. X and Y are equivalent in that each is a function of the other. By definition, $Y = P'X$. To recover X from Y, write $PY = PP'X = X$.

8.2 Random vectors and random matrices

Step 4. If C is singular, we can actually throw away some components of Y without any loss of information! Writing $C = P\Lambda P'$, we have

$$\det C = \det P \det \Lambda \det P' = \det P' \det P \det \Lambda = \det(P'P)\det \Lambda = \det \Lambda = \lambda_1 \cdots \lambda_n.$$

Thus, C is singular if and only if some of the λ_i are zero. Since $\lambda_i = \mathsf{E}[Y_i^2]$, we see that $\lambda_i = 0$ if and only if Y_i is the zero random variable. Hence, we only need to keep around the Y_i for which $\lambda_i > 0$ — we know that the other Y_i are zero.

***Example* 8.7** (data reduction). Suppose that X is a zero-mean vector of dimension $n = 5$, and suppose that $\lambda_2 = \lambda_3 = 0$. Then we can extract the nonzero Y_i from $Y = P'X$ by writing

$$\begin{bmatrix} Y_1 \\ Y_4 \\ Y_5 \end{bmatrix} = \begin{bmatrix} 1 & 0 & 0 & 0 & 0 \\ 0 & 0 & 0 & 1 & 0 \\ 0 & 0 & 0 & 0 & 1 \end{bmatrix} \begin{bmatrix} Y_1 \\ Y_2 \\ Y_3 \\ Y_4 \\ Y_5 \end{bmatrix}.$$

For this reason, we call the above 3×5 matrix of zeros and ones an "extractor matrix." Since $X = PY$, and since we know Y_2 and Y_3 are zero, we can reconstruct X from Y_1, Y_4, and Y_5 by applying P to

$$\begin{bmatrix} Y_1 \\ 0 \\ 0 \\ Y_4 \\ Y_5 \end{bmatrix} = \begin{bmatrix} 1 & 0 & 0 \\ 0 & 0 & 0 \\ 0 & 0 & 0 \\ 0 & 1 & 0 \\ 0 & 0 & 1 \end{bmatrix} \begin{bmatrix} Y_1 \\ Y_4 \\ Y_5 \end{bmatrix}.$$

We call this 5×3 matrix of zeros and ones a "reconstructor matrix." Notice that it is the transpose of the "extractor matrix."

The foregoing example illustrates the general result. If we let E denote the "extractor matrix" that creates the subvector of nonzero components of Y, then $X = PE'EP'X$, where E' is the "reconstructor matrix" that rebuilds Y from the subvector of its nonzero components. See Problem 19 for more details.

***Example* 8.8** (noiseless detection). Suppose that the random variable X in the previous example is the noise in a channel over which we must send either the signal $m = 0$ or a signal $m \neq 0$. The received vector is $Z = m + X$. Design a signal $m \neq 0$ and a receiver that can distinguish $m \neq 0$ from $m = 0$ without error.

Solution. We consider a receiver that applies the transformation P' to the received vector Z to get $P'Z = P'm + P'X$. Letting $W := P'Z$, $\mu := P'm$, and $Y := P'X$, we can write

$$\begin{bmatrix} W_1 \\ W_2 \\ W_3 \\ W_4 \\ W_5 \end{bmatrix} = \begin{bmatrix} \mu_1 \\ \mu_2 \\ \mu_3 \\ \mu_4 \\ \mu_5 \end{bmatrix} + \begin{bmatrix} Y_1 \\ 0 \\ 0 \\ Y_4 \\ Y_5 \end{bmatrix}.$$

In particular, we observe that $W_2 = \mu_2$ and $W_3 = \mu_3$. Thus, as long as the nonzero m is chosen so that the second and third components of $\mu = P'm$ are not both zero, we can noiselessly distinguish between $m = 0$ and $m \neq 0$. For example, if the nonzero m is any vector of the form $m = P\beta$, where β_2 and β_3 are not both zero, then $\mu = P'm = P'P\beta = \beta$ satisfies the desired condition that μ_2 and μ_3 are not both zero.

Remark. For future reference, we write $X = PY$ in component form as

$$X_i = \sum_{k=1}^{n} P_{ik} Y_k. \tag{8.6}$$

By writing the component form, it will be easier to see the similarity with the Karhunen–Loève expansion of continuous-time random processes derived in Chapter 13.

Remark. Given a covariance matrix C, the matrices P and Λ can be obtained with the MATLAB command

$$[\text{P}, \text{Lambda}] = \text{eig}(\text{C}).$$

To extract the diagonal elements of Lambda as a vector, use the command

$$\text{lambda} = \text{diag}(\text{Lambda}).$$

8.3 Transformations of random vectors

If $G(x)$ is a vector-valued function of $x \in \mathbb{R}^n$, and X is an $\mathbb{R}^n$-valued random vector, we can define a new random vector by $Y = G(X)$. If X has joint density f_X, and G is a suitable invertible mapping, then we can find a relatively explicit formula for the joint density of Y. Suppose that the entries of the vector equation $y = G(x)$ are given by

$$\begin{bmatrix} y_1 \\ \vdots \\ y_n \end{bmatrix} = \begin{bmatrix} g_1(x_1, \ldots, x_n) \\ \vdots \\ g_n(x_1, \ldots, x_n) \end{bmatrix}.$$

If G is invertible, we can apply G^{-1} to both sides of $y = G(x)$ to obtain $G^{-1}(y) = x$. Using the notation $H(y) := G^{-1}(y)$, we can write the entries of the vector equation $x = H(y)$ as

$$\begin{bmatrix} x_1 \\ \vdots \\ x_n \end{bmatrix} = \begin{bmatrix} h_1(y_1, \ldots, y_n) \\ \vdots \\ h_n(y_1, \ldots, y_n) \end{bmatrix}.$$

Assuming that H is continuous and has continuous partial derivatives, let

$$dH(y) := \begin{bmatrix} \frac{\partial h_1}{\partial y_1} & \cdots & \frac{\partial h_1}{\partial y_n} \\ \vdots & & \vdots \\ \frac{\partial h_i}{\partial y_1} & \cdots & \frac{\partial h_i}{\partial y_n} \\ \vdots & & \vdots \\ \frac{\partial h_n}{\partial y_1} & \cdots & \frac{\partial h_n}{\partial y_n} \end{bmatrix}. \tag{8.7}$$

8.3 Transformations of random vectors

To compute $P(Y \in C) = P(G(X) \in C)$, it is convenient to put $B := \{x : G(x) \in C\}$ so that

$$P(Y \in C) = P(G(X) \in C)$$
$$= P(X \in B)$$
$$= \int_{\mathbb{R}^n} I_B(x) f_X(x) \, dx.$$

Now apply the multivariate change of variable $x = H(y)$. Keeping in mind that $dx = |\det dH(y)| \, dy$,

$$P(Y \in C) = \int_{\mathbb{R}^n} I_B(H(y)) f_X(H(y)) |\det dH(y)| \, dy.$$

Observe that $I_B(H(y)) = 1$ if and only if $H(y) \in B$, which happens if and only if $G(H(y)) \in C$. However, since $H = G^{-1}$, $G(H(y)) = y$, and we see that $I_B(H(y)) = I_C(y)$. Thus,

$$P(Y \in C) = \int_C f_X(H(y)) |\det dH(y)| \, dy.$$

Since the set C is arbitrary, the integrand must be the density of Y. Thus,

$$\boxed{f_Y(y) = f_X(H(y)) |\det dH(y)|.}$$

Since $\det dH(y)$ is called the **Jacobian** of H, the preceding equations are sometimes called **Jacobian formulas**. They provide the multivariate generalization of (5.2).

Example 8.9. Let $Y = AX + b$, where A is a square, invertible matrix, b is a column vector, and X has joint density f_X. Find f_Y.

Solution. Since A is invertible, we can solve $Y = AX + b$ for $X = A^{-1}(Y - b)$. In other words, $H(y) = A^{-1}(y - b)$. It is easy to check that $dH(y) = A^{-1}$. Hence,

$$f_Y(y) = f_X(A^{-1}(y-b)) |\det A^{-1}| = \frac{f_X(A^{-1}(y-b))}{|\det A|}.$$

The formula in the preceding example is useful when solving problems for arbitrary A. However, when A is small and given explicitly, it is more convenient to proceed as follows.

Example 8.10. Let X and Y be independent univariate $N(0,1)$ random variables. If $U := 2X - 5Y$ and $V := X - 4Y$, find the joint density of U and V. Are U and V independent?

Solution. The transformation $[u, v]' = G(x, y)$ is given by

$$u = 2x - 5y$$
$$v = x - 4y.$$

By solving these equations for x and y in terms of u and v, we obtain the inverse transformation $[x,y]' = H(u,v)$ given by

$$x = \tfrac{4}{3}u - \tfrac{5}{3}v$$
$$y = \tfrac{1}{3}u - \tfrac{2}{3}v.$$

The matrix $dH(u,v)$ is given by

$$dH(u,v) = \begin{bmatrix} \frac{\partial x}{\partial u} & \frac{\partial x}{\partial v} \\ \frac{\partial y}{\partial u} & \frac{\partial y}{\partial v} \end{bmatrix} = \begin{bmatrix} 4/3 & -5/3 \\ 1/3 & -2/3 \end{bmatrix},$$

and we see that $\det dH(u,v) = -1/3$. Since $f_{XY}(x,y) = e^{-(x^2+y^2)/2}/2\pi$, we can write

$$f_{UV}(u,v) = f_{XY}(x,y)\Big|_{\substack{x=4u/3-5v/3 \\ y=u/3-2v/3}} \cdot |\det dH(u,v)|$$
$$= (2\pi)^{-1} \exp\left[-\tfrac{1}{2}\left(\tfrac{17}{9}u^2 - \tfrac{44}{9}uv + \tfrac{29}{9}v^2\right)\right]\tfrac{1}{3}.$$

Recalling the formula for the bivariate normal density (7.25), we see that U and V have nonzero correlation coefficient; hence, they are not independent.

Example 8.11. Let X and Y be independent random variables where $Y \sim N(0,1)$ and X has the standard Rayleigh density, $f_X(x) = xe^{-x^2/2}$, $x \geq 0$. Find the joint density of $U := \sqrt{X^2+Y^2}$ and $V := \lambda Y/X$, where λ is a positive real number. Are U and V independent?

Solution. The transformation $[u,v]' = G(x,y)$ is given by

$$u = \sqrt{x^2+y^2}$$
$$v = \lambda y/x.$$

By solving these equations for x and y in terms of u and v, we obtain the inverse transformation $[x,y]' = H(u,v)$. To do this, we first write $u^2 = x^2 + y^2$ and $v^2 = \lambda^2 y^2/x^2$. From this second equation, we have $y^2 = v^2 x^2/\lambda^2$. We then write

$$u^2 = x^2 + y^2 = x^2 + v^2 x^2/\lambda^2 = x^2(1+v^2/\lambda^2).$$

It then follows that

$$x^2 = \frac{u^2}{1+v^2/\lambda^2}.$$

Since X is a nonnegative random variable, we take the positive square root to get

$$x = \frac{u}{\sqrt{1+v^2/\lambda^2}}.$$

Next, since $v = \lambda y/x$,

$$y = vx/\lambda = \frac{uv}{\lambda\sqrt{1+v^2/\lambda^2}}.$$

8.3 Transformations of random vectors

A little calculation shows that

$$dH(u,v) = \begin{bmatrix} \frac{\partial x}{\partial u} & \frac{\partial x}{\partial v} \\ \frac{\partial y}{\partial u} & \frac{\partial y}{\partial v} \end{bmatrix} = \begin{bmatrix} \frac{1}{\sqrt{1+v^2/\lambda^2}} & \frac{-uv}{\lambda^2(1+v^2/\lambda^2)^{3/2}} \\ \frac{v}{\lambda\sqrt{1+v^2/\lambda^2}} & \frac{u}{\lambda(1+v^2/\lambda^2)^{3/2}} \end{bmatrix}.$$

It then follows that $\det dH(u,v) = \lambda u/(\lambda^2 + v^2)$. The next step is to write

$$f_{UV}(u,v) = f_{XY}(x,y) \cdot |\det dH(u,v)|$$

and to substitute x and y using the above formulas. Now

$$f_{XY}(x,y) = xe^{-x^2/2}e^{-y^2/2}/\sqrt{2\pi} = xe^{-(x^2+y^2)/2}/\sqrt{2\pi}.$$

From the original definition of U, we know that $u^2 = x^2 + y^2$, and we already solved for x. Hence,

$$\begin{aligned} f_{UV}(u,v) &= \frac{u}{\sqrt{1+v^2/\lambda^2}} \cdot \frac{e^{-u^2/2}}{\sqrt{2\pi}} \cdot \frac{\lambda u}{\lambda^2 + v^2} \\ &= \sqrt{\frac{2}{\pi}} u^2 e^{-u^2/2} \cdot \frac{1}{2\lambda}\left(1 + \frac{v^2}{\lambda^2}\right)^{-3/2} \\ &= f_U(u)f_V(v), \end{aligned}$$

where f_U is the standard Maxwell density defined in Problem 4 in Chapter 5 and f_V is a scaled Student's t density with two degrees of freedom (defined in Problem 20 in Chapter 4). In particular, U and V are independent.[1]

Example 8.12. Let X and Y be independent univariate $N(0,1)$ random variables. Let R denote the length of the vector $[X,Y]'$, and let Θ denote the angle the vector makes with the x-axis. In other words, if X and Y are the Cartesian coordinates of a random point in the plane, then $R \geq 0$ and $-\pi < \Theta \leq \pi$ are the corresponding polar coordinates. Find the joint density of R and Θ.

Solution. The transformation $[r,\theta]' = G(x,y)$ is given by[2]

$$\begin{aligned} r &= \sqrt{x^2 + y^2}, \\ \theta &= \text{angle}(x,y). \end{aligned}$$

The inverse transformation $[x,y]' = H(r,\theta)$ is the mapping that takes polar coordinates into Cartesian coordinates. Hence, $H(r,\theta)$ is given by

$$\begin{aligned} x &= r\cos\theta, \\ y &= r\sin\theta. \end{aligned}$$

The matrix $dH(r,\theta)$ is given by

$$dH(r,\theta) = \begin{bmatrix} \frac{\partial x}{\partial r} & \frac{\partial x}{\partial \theta} \\ \frac{\partial y}{\partial r} & \frac{\partial y}{\partial \theta} \end{bmatrix} = \begin{bmatrix} \cos\theta & -r\sin\theta \\ \sin\theta & r\cos\theta \end{bmatrix},$$

and $\det dH(r,\theta) = r\cos^2\theta + r\sin^2\theta = r$. Then

$$\begin{aligned} f_{R,\Theta}(r,\theta) &= f_{XY}(x,y)\Big|_{\substack{x=r\cos\theta \\ y=r\sin\theta}} \cdot |\det dH(r,\theta)| \\ &= f_{XY}(r\cos\theta, r\sin\theta)\, r. \end{aligned}$$

Now, since X and Y are independent $N(0,1)$, $f_{XY}(x,y) = f_X(x)f_Y(y) = e^{-(x^2+y^2)/2}/(2\pi)$, and

$$f_{R,\Theta}(r,\theta) = re^{-r^2/2}\cdot\frac{1}{2\pi}, \quad r\geq 0, -\pi < \theta \leq \pi.$$

Thus, R and Θ are independent, with R having a Rayleigh density and Θ having a uniform $(-\pi, \pi]$ density.

8.4 *Linear estimation of random vectors (Wiener filters)

Consider a pair of random vectors X and Y, where X is not observed, but Y is observed. For example, X could be the input to a noisy channel, and Y could be the channel output. In this situation, the receiver knows Y and needs to estimate X. By an **estimator** of X based on Y, we mean a function $g(y)$ such that $\widehat{X} := g(Y)$ is our **estimate** or "guess" of the value of X. What is the best function g to use? What do we mean by best? In this section, we define g to be best if it minimizes the **mean-squared error** (MSE) $\mathsf{E}[\|X-g(Y)\|^2]$ for all functions g in some class of functions. Here we restrict attention to the class of functions of the form $g(y) = Ay + b$, where A is a matrix and b is a column vector; we drop this restriction in Section 8.6. A function of the form $Ay+b$ is said to be **affine**. If $b=0$, then g is linear. It is common to say g is linear even if $b\neq 0$ since this only is a slight abuse of terminology, and the meaning is understood. We shall follow this convention. The optimal such function g is called the **linear minimum mean-squared-error estimator**, or more simply, **linear MMSE estimator**. Linear MMSE estimators are sometimes called **Wiener filters**.

To find the best linear estimator is to find the matrix A and the column vector b that minimize the MSE, which for linear estimators has the form

$$\mathsf{E}\big[\|X - (AY+b)\|^2\big].$$

Letting $m_X := \mathsf{E}[X]$ and $m_Y := \mathsf{E}[Y]$, the MSE is equal to

$$\mathsf{E}\big[\|\{(X-m_X) - A(Y-m_Y)\} + \{m_X - Am_Y - b\}\|^2\big].$$

Since the left-hand quantity in braces is zero mean, and since the right-hand quantity in braces is a constant (nonrandom), the MSE simplifies to

$$\mathsf{E}\big[\|(X-m_X) - A(Y-m_Y)\|^2\big] + \|m_X - Am_Y - b\|^2.$$

No matter what matrix A is used, the optimal choice of b is

$$b = m_X - Am_Y,$$

8.4 Linear estimation of random vectors (Wiener filters)

and the estimate is

$$g(Y) = AY + b = A(Y - m_Y) + m_X. \tag{8.8}$$

The estimate is truly linear in Y if and only if $Am_Y = m_X$.

We show later in this section that the optimal choice of A is any solution of

$$AC_Y = C_{XY}. \tag{8.9}$$

When C_Y is invertible, $A = C_{XY}C_Y^{-1}$. This is best computed in MATLAB with the command A = CXY/CY. Even if C_Y is not invertible, there is always a solution of (8.9), as shown in Problem 38.

Remark. If X and Y are uncorrelated, by which we mean $C_{XY} = 0$, then taking $A = 0$ solves (8.9). In this case, the estimate of X reduces to $g(Y) = m_X$. In other words, the value we guess for X based on observing Y does not involve Y! Hence, if X and Y are uncorrelated, then linear signal processing of Y cannot extract any information about X that can minimize the MSE below $\mathsf{E}[\|X - m_X\|^2] = \text{tr}(C_X)$ (cf. Problem 9).

Example 8.13 (signal in additive noise). Let X denote a random signal of zero mean and known covariance matrix C_X. Suppose that in order to estimate X, all we have available is the noisy measurement

$$Y = X + W,$$

where W is a noise vector with zero mean and known, positive-definite covariance matrix C_W. Further assume that the covariance between the signal and noise, C_{XW}, is zero. Find the linear MMSE estimate of X based on Y.

Solution. Since X and W are zero mean, $m_Y = \mathsf{E}[Y] = \mathsf{E}[X + W] = 0$. Next,

$$\begin{aligned} C_{XY} &= \mathsf{E}[(X - m_X)(Y - m_Y)'] \\ &= \mathsf{E}[X(X+W)'] \\ &= C_X, \quad \text{since } C_{XW} = 0. \end{aligned}$$

Similarly,

$$\begin{aligned} C_Y &= \mathsf{E}[(Y - m_Y)(Y - m_Y)'] \\ &= \mathsf{E}[(X+W)(X+W)'] \\ &= C_X + C_W. \end{aligned}$$

It follows that

$$C_{XY}C_Y^{-1}Y = C_X(C_X + C_W)^{-1}Y$$

is the linear MMSE estimate of X based on Y.

Example 8.14 (MATLAB). Use MATLAB to compute $A = C_{XY}C_Y^{-1}$ of the preceding example if

$$C_X = \begin{bmatrix} 10 & 14 \\ 14 & 20 \end{bmatrix} \quad \text{and} \quad C_W = \begin{bmatrix} 1 & 0 \\ 0 & 1 \end{bmatrix}.$$

Solution. We use the commands

```
CX = [ 10 14 ; 14 20 ]
CXY = CX;
CW = eye(2)   % 2 by 2 identity matrix
CY = CX + CW
format rat    % print numbers as ratios of small integers
A = CXY/CY
```

to find that

$$A = \begin{bmatrix} 2/5 & 2/5 \\ 2/5 & 24/35 \end{bmatrix}.$$

Example 8.15 (MATLAB). Let $\widehat{X}$ denote the linear MMSE estimate of X based on Y. It is shown in Problems 34 and 35 that the MSE is given by

$$\mathsf{E}[\|X - \widehat{X}\|^2] = \operatorname{tr}(C_X - AC'_{XY}).$$

Using the data for the previous two examples, compute the MSE and compare it with $\mathsf{E}[\|X - m_X\|^2] = \operatorname{tr}(C_X)$.

Solution. The command `trace(CX-A*(CXY'))` shows that the MSE achieved using $\widehat{X}$ is 1.08571. The MSE achieved using m_X (which makes no use of the observation Y) is $\mathsf{E}[\|X - m_X\|^2] = \operatorname{tr}(C_X) = 30$. Hence, using even a linear function of the data has reduced the error by a factor of about 30.

Example 8.16. Find the linear MMSE estimate of X based on Y if $Y \sim \exp(\lambda)$, and given $Y = y$, X is conditionally Rayleigh(y).

Solution. Using the table of densities inside the back cover, we find that $m_Y = 1/\lambda$, $C_Y = \operatorname{var}(Y) = 1/\lambda^2$, and $\mathsf{E}[X|Y=y] = y\sqrt{\pi/2}$. Using the law of total probability, it is an easy calculation to show that $m_X = \sqrt{\pi/2}/\lambda$ and that $\mathsf{E}[XY] = 2\sqrt{\pi/2}/\lambda^2$. It follows that $C_{XY} = \mathsf{E}[XY] - m_X m_Y = \sqrt{\pi/2}/\lambda^2$. Then

$$A = C_{XY}C_Y^{-1} = \frac{\sqrt{\pi/2}/\lambda^2}{1/\lambda^2} = \sqrt{\pi/2},$$

and the linear MMSE estimate of X based on Y is

$$C_{XY}C_Y^{-1}(Y - m_Y) + m_X = \sqrt{\pi/2}\, Y.$$

8.4 Linear estimation of random vectors (Wiener filters)

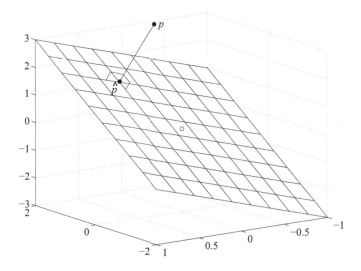

Figure 8.1. The point on the plane that is closest to p is called the projection of p, and is denoted by $\hat{p}$. The orthogonality principle says that $\hat{p}$ is characterized by the property that the line joining $\hat{p}$ to p is orthogonal to the plane. The symbol ○ denotes the origin.

Derivation of the linear MMSE estimator

We now turn to the problem of minimizing

$$E\big[\|(X - m_X) - A(Y - m_Y)\|^2\big].$$

The matrix A is optimal if and only if for all matrices B,

$$E\big[\|(X - m_X) - A(Y - m_Y)\|^2\big] \leq E\big[\|(X - m_X) - B(Y - m_Y)\|^2\big]. \tag{8.10}$$

The following condition is equivalent and is easier to use. This equivalence is known as the **orthogonality principle**. It says that (8.10) holds for all B if and only if

$$E\big[\{B(Y - m_Y)\}'\{(X - m_X) - A(Y - m_Y)\}\big] = 0, \quad \text{for all } B. \tag{8.11}$$

Below we prove that (8.11) implies (8.10). The converse is also true, but we shall not use it in this book.

We first explain the terminology and show geometrically why it is true. A two-dimensional subspace (a plane passing through the origin) is shown in Figure 8.1. A point p not in the subspace is also shown. The point on the plane that is closest to p is the point $\hat{p}$. This closest point is called the **projection** of p. The orthogonality principle says that the projection $\hat{p}$ is characterized by the property that the line joining $\hat{p}$ to p, which is the error vector $p - \hat{p}$, is orthogonal to the subspace. In our situation, the role of p is played by the random variable $X - m_X$, the role of $\hat{p}$ is played by the random variable $A(Y - m_Y)$, and the role of the subspace is played by the set of all random variables of the form $B(Y - m_Y)$ as B runs over all matrices of the right dimensions. Since the inner product between two random vectors U and V can be defined as $E[V'U]$, (8.11) says that

$$(X - m_X) - A(Y - m_Y)$$

is orthogonal to all $B(Y - m_Y)$.

To use (8.11), first note that since it is a scalar equation, the left-hand side is equal to its trace. Bringing the trace inside the expectation and using the fact that $\text{tr}(\alpha\beta) = \text{tr}(\beta\alpha)$, we see that the left-hand side of (8.11) is equal to

$$\mathsf{E}\big[\text{tr}\big(\{(X-m_X) - A(Y-m_Y)\}(Y-m_Y)'B'\big)\big].$$

Taking the trace back out of the expectation shows that (8.11) is equivalent to

$$\text{tr}([C_{XY} - AC_Y]B') = 0, \quad \text{for all } B. \tag{8.12}$$

By Problem 5, it follows that (8.12) holds if and only if $C_{XY} - AC_Y$ is the zero matrix, or equivalently, if and only if A solves the equation

$$AC_Y = C_{XY}.$$

If C_Y is invertible, the unique solution of this equation is

$$A = C_{XY}C_Y^{-1}.$$

In this case, the estimate of X is

$$C_{XY}C_Y^{-1}(Y - m_Y) + m_X.$$

We now show that (8.11) implies (8.10). To simplify the notation, we assume zero means. Write

$$\begin{aligned}
\mathsf{E}[\|X - BY\|^2] &= \mathsf{E}[\|(X - AY) + (AY - BY)\|^2] \\
&= \mathsf{E}[\|(X - AY) + (A - B)Y\|^2] \\
&= \mathsf{E}[\|X - AY\|^2] + \mathsf{E}[\|(A - B)Y\|^2],
\end{aligned}$$

where the cross terms $2\mathsf{E}[\{(A - B)Y\}'(X - AY)]$ vanish by (8.11). If we drop the right-hand term in the above display, we obtain

$$\mathsf{E}[\|X - BY\|^2] \geq \mathsf{E}[\|X - AY\|^2].$$

8.5 *Estimation of covariance matrices†

As we saw in the previous section, covariance matrices are a critical component in the design of linear estimators of random vectors. In real-world problems, however, we may not know these matrices. Instead, we have to estimate them from the data.

Remark. We use the term "estimation" in two ways. In the previous section we estimated *random* vectors. In this section we estimate *nonrandom* parameters, namely, the elements of the covariance matrix.

Recall from Chapter 6 that if $X_1, \ldots, X_n$ are i.i.d. with common mean m and common variance σ^2, then

$$M_n := \frac{1}{n} \sum_{k=1}^{n} X_k$$

---

†Section 8.5 is not used elsewhere in the text and can be skipped without loss of continuity.

8.5 Estimation of covariance matrices

is an unbiased, strongly consistent estimator of the mean m. Similarly,

$$S_n^2 := \frac{1}{n-1}\sum_{k=1}^n (X_k - M_n)^2$$

is an unbiased, strongly consistent estimator of σ^2. If we know a priori that $m = 0$, then

$$\frac{1}{n}\sum_{k=1}^n X_k^2$$

an unbiased and strongly consistent estimator of σ^2.

Suppose we have i.i.d. random *vectors* $X_1,\ldots,X_n$ with zero mean and common covariance matrix C. Our estimator of C is

$$\widehat{C}_n = \frac{1}{n}\sum_{k=1}^n X_k X_k'.$$

Note that since X_k is a column vector, $X_k X_k'$ is a matrix, which makes sense since $\widehat{C}_n$ is an estimate of the covariance matrix C. We can do the above computation efficiently in MATLAB if we arrange the X_k as the columns of a matrix. Observe that if $X := [X_1,\ldots,X_n]$, then

$$XX' = X_1 X_1' + \cdots + X_n X_n'.$$

Example **8.17** (MATLAB). Here is a way to generate simulation examples using i.i.d. zero-mean random vectors of dimension d and covariance matrix $C = GG'$, where G is any $d \times d$ matrix. We will use $d = 5$ and

$$G = \begin{bmatrix} 1 & -2 & -2 & 1 & 0 \\ 0 & 1 & -1 & 3 & -1 \\ 1 & 3 & 3 & -3 & 4 \\ -1 & 1 & 2 & 1 & -4 \\ 0 & 2 & 0 & -4 & 3 \end{bmatrix}.$$

When we ran the script

```
G = [ 1 -2 -2 1 0 ; 0 1 -1  3 -1 ; 1 3 3 -3 4 ; ...
     -1 1 2 1 -4 ; 0 2 0 -4 3 ];
C = G*G'
d = length(G);
n = 1000;
Z = randn(d,n);  % Create d by n array of i.i.d. N(0,1) RVs
X = G*Z;         % Multiply each column by G
Chat = X*X'/n
```

we got

$$C = \begin{bmatrix} 10 & 3 & -14 & -6 & -8 \\ 3 & 12 & -13 & 6 & -13 \\ -14 & -13 & 44 & -11 & 30 \\ -6 & 6 & -11 & 23 & -14 \\ -8 & -13 & 30 & -14 & 29 \end{bmatrix}$$

and

$$\widehat{C} = \begin{bmatrix} 10.4318 & 3.1342 & -15.3540 & -5.6080 & -8.6550 \\ 3.1342 & 11.1334 & -12.9239 & 5.5280 & -12.5036 \\ -15.3540 & -12.9239 & 46.2087 & -10.0815 & 30.5472 \\ -5.6080 & 5.5280 & -10.0815 & 21.2744 & -13.1262 \\ -8.6550 & -12.5036 & 30.5472 & -13.1262 & 29.0639 \end{bmatrix}.$$

When $X_1,\ldots,X_n$ are i.i.d. but have nonzero mean vector, we put

$$M_n := \frac{1}{n}\sum_{k=1}^{n} X_k,$$

and we use

$$\widehat{C}_n := \frac{1}{n-1}\sum_{k=1}^{n}(X_k - M_n)(X_k - M_n)'$$

to estimate C. Note that in MATLAB, if X has $X_1,\ldots,X_n$ for its columns, then the column vector M_n can be computed with the command mean(X,2).

8.6 *Nonlinear estimation of random vectors

In Section 8.4, we had two random vectors X and Y, but we could only observe Y. Based on Y, we used functions of the form $g(Y) = AY + b$ as estimates of X. We characterized the choice of A and b that would minimize $\mathsf{E}[\|X - g(Y)\|^2]$.

In this section, we consider three other estimators of X. The first is the **minimum mean-squared error** (MMSE) estimator. In this method, we no longer restrict $g(y)$ to be of the form $Ay + b$, and we try to further minimize the MSE $\mathsf{E}[\|X - g(Y)\|^2]$. As we show later, the function g that minimizes the MSE is

$$\boxed{g_{\text{MMSE}}(y) \;=\; \mathsf{E}[X|Y=y].}$$

The second estimator uses the conditional density $f_{Y|X}(y|x)$. The **maximum-likelihood** (ML) estimator of X is the function

$$\boxed{g_{\text{ML}}(y) \;:=\; \operatorname*{argmax}_{x} f_{Y|X}(y|x).}$$

In other words, $g_{\text{ML}}(y)$ is the value of x that maximizes $f_{Y|X}(y|x)$. See Problem 43. When the density of X is not positive for all x, we only consider values of x for which $f_X(x) > 0$. See Problem 44. The third estimator uses the conditional density $f_{X|Y}(x|y)$. The **maximum a posteriori probability** (MAP) estimator of X is the function

$$g_{\text{MAP}}(y) \;:=\; \operatorname*{argmax}_{x} f_{X|Y}(x|y).$$

8.6 Nonlinear estimation of random vectors

Notice that since $f_{X|Y}(x|y) = f_{Y|X}(y|x)f_X(x)/f_Y(y)$, the maximizing value of x does not depend on the value of $f_Y(y)$. Hence,

$$g_{\text{MAP}}(y) = \underset{x}{\text{argmax}}\, f_{Y|X}(y|x)f_X(x). \qquad (8.13)$$

When X is a uniform random variable, the constant value of $f_X(x)$ does not affect the maximizing value of x. In this case, $g_{\text{MAP}}(y) = g_{\text{ML}}(y)$.

***Example* 8.18** (signal in additive noise). A signal X with density $f_X(x)$ is transmitted over a noisy channel so that the received vector is $Y = X + W$, where the noise W and the signal X are independent, and the noise has density $f_W(w)$. Find the MMSE estimator of X based on Y.

Solution. To compute $\mathsf{E}[X|Y=y]$, we first need to find $f_{X|Y}(x|y)$. Since[a]

$$\begin{aligned} \mathsf{P}(Y \le y | X = x) &= \mathsf{P}(X + W \le y | X = x) \\ &= \mathsf{P}(W \le y - x | X = x), \quad \text{by substitution,} \\ &= \mathsf{P}(W \le y - x), \quad \text{by independence,} \end{aligned}$$

we see that $f_{Y|X}(y|x) = f_W(y-x)$. Hence,

$$f_{X|Y}(x|y) = \frac{f_{Y|X}(y|x)f_X(x)}{f_Y(y)} = \frac{f_W(y-x)f_X(x)}{f_Y(y)}.$$

It then follows that

$$g_{\text{MMSE}}(y) = \mathsf{E}[X|Y = y] = \int x \frac{f_W(y-x)f_X(x)}{f_Y(y)}\, dx,$$

where, since the density of the sum of independent random variables is the convolution of their densities,

$$f_Y(y) = \int f_X(y-w)f_W(w)\, dw.$$

Even in this context where X and Y are simply related, it is difficult in general to compute $\mathsf{E}[X|Y=y]$. This is actually one of the motivations for developing linear estimation as in Section 8.4.

In the above example, it was relatively easy to find $f_{Y|X}(y|x)$. This is one explanation for the popularity of the ML estimator $g_{\text{ML}}(y)$. We again mention that although the definition of the MAP estimator uses $f_{X|Y}(x|y)$ which requires knowledge of $f_Y(y)$, in fact, by (8.13), $f_Y(y)$ is not really needed; only $f_{Y|X}(y|x)f_X(x)$ is needed.

---

[a]When $Y = [Y_1, \ldots, Y_n]'$ is a random vector and $y = [y_1, \ldots, y_n]'$, the joint cdf is

$$\mathsf{P}(Y \le y) := \mathsf{P}(Y_1 \le y_1, \ldots, Y_n \le y_n)$$

The corresponding density is obtained by computing

$$\frac{\partial^n}{\partial y_1 \cdots \partial y_n} \mathsf{P}(Y_1 \le y_1, \ldots, Y_n \le y_n).$$

Analogous shorthand is used for conditional cdfs.

Example 8.19. If $W \sim N(0, \sigma^2)$ in the previous example, find the ML estimator of X.

Solution. From the solution of Example 8.18,

$$f_{Y|X}(y|x) = f_W(y-x) = \frac{1}{\sqrt{2\pi}\,\sigma} e^{-[(y-x)/\sigma]^2/2}.$$

If we observe $Y = y$, then $g_{\text{ML}}(y) = y$ since taking $x = y$ maximizes the conditional density.

One of the advantages of the ML estimator is that we can compute it even if we do not know the density of X. However, if we do know the density of X, the ML estimator does not make use of that information, while the MAP estimator does.

Example 8.20. If $X \sim N(0,1)$ and $W \sim N(0,\sigma^2)$ in Example 8.18, find the MAP estimator of X.

Solution. This time, given $Y = y$, we need to maximize

$$f_{Y|X}(y|x) f_X(x) = \frac{1}{\sqrt{2\pi}\,\sigma} e^{-[(y-x)/\sigma]^2/2} \cdot \frac{1}{\sqrt{2\pi}} e^{-x^2/2}. \tag{8.14}$$

The coefficients do not affect the maximization, and we can combine the exponents. Also, since e^{-t} is decreasing in t, it suffices to minimize

$$x^2 + (y-x)^2/\sigma^2$$

with respect to x. The minimizing value of x is easily obtained by differentiation and is found to be $y/(1+\sigma^2)$. Hence,

$$g_{\text{MAP}}(y) = \frac{y}{1+\sigma^2}.$$

Example 8.21. If $X \sim N(0,1)$ and $W \sim N(0,\sigma^2)$ in Example 8.18, find the MMSE estimator of X.

Solution. We need to find $f_{X|Y}(x|y)$. Since

$$f_{X|Y}(x|y) = \frac{f_{Y|X}(y|x) f_X(x)}{f_Y(y)},$$

we observe that the numerator was already found in (8.14) above. In general, to find $f_Y(y)$, we would integrate (8.14) with respect to x. However, we can avoid integration by arguing as follows. Since $Y = X + Z$ and since X and Y are independent and Gaussian, Y is also Gaussian by Problem 55(a) in Chapter 4. Furthermore, $E[Y] = 0$ and $\text{var}(Y) = \text{var}(X) + \text{var}(Z) = 1 + \sigma^2$. Thus, $Y \sim N(0, 1+\sigma^2)$, and so

$$f_{X|Y}(x|y) = \frac{\dfrac{1}{\sqrt{2\pi}\,\sigma} e^{-[(y-x)/\sigma]^2/2} \cdot \dfrac{1}{\sqrt{2\pi}} e^{-x^2/2}}{\dfrac{e^{-y^2/[2(1+\sigma^2)]}}{\sqrt{2\pi(1+\sigma^2)}}}$$

$$= \frac{\exp\left[-\frac{1+\sigma^2}{2\sigma^2}\left(x - \frac{y}{1+\sigma^2}\right)^2\right]}{\sqrt{2\pi\sigma^2/(1+\sigma^2)}}.$$

8.6 Nonlinear estimation of random vectors

In other words
$$f_{X|Y}(\cdot|y) \sim N\left(\frac{y}{1+\sigma^2}, \frac{\sigma^2}{1+\sigma^2}\right).$$

It then follows that
$$\mathsf{E}[X|Y=y] = \frac{y}{1+\sigma^2}.$$

The two preceding examples show that it is possible to have $g_{\text{MAP}}(y) = g_{\text{MMSE}}(y)$. However, this is not always the case, as shown in Problem 46.

Derivation of the MMSE estimator

We first establish an orthogonality principle that says if
$$\mathsf{E}[h(Y)'\{X-g(Y)\}] = 0, \quad \text{for all functions } h, \tag{8.15}$$

then
$$\mathsf{E}[\|X-g(Y)\|^2] \leq \mathsf{E}[\|X-h(Y)\|^2], \quad \text{for all functions } h. \tag{8.16}$$

We then show that $g(y) = \mathsf{E}[X|Y=y]$ satisfies (8.15). In fact there is at most one function g that can satisfy (8.15), as shown in Problem 47.

To begin, write
$$\begin{aligned}\mathsf{E}[\|X-h(Y)\|^2] &= \mathsf{E}[\|X-g(Y)+g(Y)-h(Y)\|^2] \\ &= \mathsf{E}[\|X-g(Y)\|^2] - 2\mathsf{E}[\{g(Y)-h(Y)\}'\{X-g(Y)\}] \\ &\quad + \mathsf{E}[\|g(Y)-h(Y)\|^2].\end{aligned}$$

If we put $\tilde{h}(y) := g(y) - h(y)$, we see that the cross term
$$\mathsf{E}[\{g(Y)-h(Y)\}'\{X-g(Y)\}] = \mathsf{E}[\tilde{h}(Y)'\{X-g(Y)\}]$$

is equal to zero if (8.15) holds. We continue with
$$\begin{aligned}\mathsf{E}[\|X-h(Y)\|^2] &= \mathsf{E}[\|X-g(Y)\|^2] + \mathsf{E}[\|g(Y)-h(Y)\|^2] \\ &\geq \mathsf{E}[\|X-g(Y)\|^2].\end{aligned}$$

The last thing to show is that $g(y) = \mathsf{E}[X|Y=y]$ satisfies (8.15). We do this in the case X is a scalar and and Y has a density. Using the law of total probability and the law of substitution,
$$\begin{aligned}\mathsf{E}[h(Y)\{X-g(Y)\}] &= \int \mathsf{E}[h(Y)\{X-g(Y)\}|Y=y]f_Y(y)\,dy \\ &= \int \mathsf{E}[h(y)\{X-g(y)\}|Y=y]f_Y(y)\,dy \\ &= \int h(y)\{\mathsf{E}[X|Y=y]-g(y)\}f_Y(y)\,dy.\end{aligned}$$

Hence, the choice $g(y) = \mathsf{E}[X|Y=y]$ makes (8.15) hold.

Notes

8.3: Transformations of random vectors

Note 1. The most interesting part of Example 8.11 is that U and V are independent. From our work in earlier chapters we could have determined their marginal densities as follows. First, since X is standard Rayleigh, we can infer from Problem 19(c) in Chapter 5 that X^2 is chi-squared with two degrees of freedom. Second, since $Y \sim N(0,1)$, we know from Problem 46 in Chapter 4 or from Problem 11 in Chapter 5 that Y^2 is chi-squared with one degree of freedom. Third, since X and Y are independent, so are X^2 and Y. Fourth, by Problem 55(c) (and the remark following it), sums of independent chi-squared random variables are chi-squared with the degrees of freedom added; hence, $X^2 + Y^2$ is chi-squared with three degrees of freedom. By Problem 19(d) in Chapter 5, $U = \sqrt{X^2 + Y^2}$ has the standard Maxwell density. As for the density of V, Problem 44 in Chapter 7 shows that if we divide an $N(0,1)$ random variable by the square root of a chi-squared, we get a Student's t density (if the numerator and denominator are independent); however, the square root of a chi-squared with two degrees of freedom is the standard Rayleigh by Problem 19(c) in Chapter 5.

Note 2. Let (x,y) be a point in the plane, and consider the line from the origin to (x,y). The angle θ that this line makes with the positive horizontal axis is called the **principal angle** if θ lies in the range $-\pi < \theta \leq \pi$. Recall that the **principal inverse tangent** function takes values in $(-\pi/2, \pi/2)$. Hence, if $x > 0$ so that (x,y) lies in the first or fourth quadrants, angle$(x,y) = \tan^{-1}(y/x)$. If $x < 0$ and $y > 0$ so that (x,y) lies in the second quadrant, then angle$(x,y) = \tan^{-1}(y/x) + \pi$. If $x < 0$ and $y \leq 0$ so that (x,y) lies in the third quadrant, then angle$(x,y) = \tan^{-1}(y/x) - \pi$. Since tan has period π, in all cases we can write $\tan(\text{angle}(x,y)) = y/x$.

Problems

8.1: Review of matrix operations

1. Compute by hand

$$\begin{bmatrix} 10 & 40 \\ 20 & 50 \\ 30 & 60 \end{bmatrix} \begin{bmatrix} 7 & 8 & 9 \\ 4 & 5 & 6 \end{bmatrix}.$$

Then compute the trace of your answer; compare with the trace of the right-hand side of (8.1).

2. **MATLAB.** Check your answers to the previous problem using MATLAB.

3. **MATLAB.** Use the MATLAB commands

   ```
   A = [ 7 8 9 ; 4 5 6 ]
   A'
   ```

   to compute the transpose of

   $$\begin{bmatrix} 7 & 8 & 9 \\ 4 & 5 & 6 \end{bmatrix}.$$

4. Let A be an $r \times n$ matrix, and let B be an $n \times r$ matrix. Derive the formula $\operatorname{tr}(AB) = \operatorname{tr}(BA)$.

5. For column vectors x and y, we defined their inner product by $\langle x,y \rangle := y'x = \operatorname{tr}(y'x) = \operatorname{tr}(xy')$. This suggests that for $r \times n$ matrices A and B, we define their inner product by $\langle A,B \rangle := \operatorname{tr}(AB')$.

   (a) Show that
   $$\operatorname{tr}(AB') = \sum_{i=1}^{r}\sum_{k=1}^{n} A_{ik}B_{ik}.$$

   (b) Show that if A is fixed and $\operatorname{tr}(AB') = 0$ for all matrices B, then $A = 0$.

6. Show that column vectors x and y satisfy the Cauchy–Schwarz inequality,
$$|\langle x,y \rangle| \leq \|x\|\,\|y\|,$$
with equality if and only if one of them is a scalar multiple of the other. *Hint:* The derivation is similar to that of the Cauchy–Schwarz inequality for random variables (2.24) given in Chapter 2: Instead of (2.25), start with
$$0 \leq \|x - \lambda y\|^2$$
$$= \langle x - \lambda y, x - \lambda y \rangle$$
$$\vdots$$

8.2: Random vectors and random matrices

7. Let X be a random $n \times p$ matrix, and let B be a $p \times q$ matrix with nonrandom entries. Show that $\mathsf{E}[XB] = \mathsf{E}[X]B$.

8. If X is an random $n \times n$ matrix, show that $\operatorname{tr}(\mathsf{E}[X]) = \mathsf{E}[\operatorname{tr}(X)]$.

9. Show that if X is an n-dimensional random vector with covariance matrix C, then
$$\mathsf{E}[\|X - \mathsf{E}[X]\|^2] = \operatorname{tr}(C) = \sum_{i=1}^{n} \operatorname{var}(X_i).$$

10. The input U to a certain amplifier is $N(0,1)$, and the output is $X = ZU + Y$, where the amplifier's random gain Z has density
$$f_Z(z) = \tfrac{3}{7}z^2, \quad 1 \leq z \leq 2;$$
and given $Z = z$, the amplifier's random bias Y is conditionally exponential with parameter z. Assuming that the input U is independent of the amplifier parameters Z and Y, find the mean vector and the covariance matrix of $[X,Y,Z]'$.

11. Find the mean vector and covariance matrix of $[X,Y,Z]'$ if

$$f_{XYZ}(x,y,z) = \frac{2\exp[-|x-y|-(y-z)^2/2]}{z^5\sqrt{2\pi}}, \quad z \geq 1,$$

and $f_{XYZ}(x,y,z) = 0$ otherwise.

12. Let X, Y, and Z be jointly continuous. Assume that $X \sim$ uniform$[1,2]$; that given $X = x$, $Y \sim \exp(1/x)$; and that given $X = x$ and $Y = y$, Z is $N(x,1)$. Find the mean vector and covariance matrix of $[X,Y,Z]'$.

13. Find the mean vector and covariance matrix of $[X,Y,Z]'$ if

$$f_{XYZ}(x,y,z) = \frac{e^{-(x-y)^2/2}e^{-(y-z)^2/2}e^{-z^2/2}}{(2\pi)^{3/2}}.$$

14. Find the joint characteristic function of $[X,Y,Z]'$ of the preceding problem.

15. If X has correlation matrix R_X and $Y = AX$, show that $R_Y = ARA'$.

16. If X has correlation matrix R, show that R is positive semidefinite.

17. Show that

$$|(C_{XY})_{ij}| \leq \sqrt{(C_X)_{ii}(C_Y)_{jj}}.$$

*18. Let $[X,Y]'$ be a two-dimensional, zero-mean random vector with $\sigma_X^2 := \text{var}(X)$ and $\sigma_Y^2 := \text{var}(Y)$. Find the decorrelating transformation P'. *Hint:* Determine θ so that the rotation matrix

$$P = \begin{bmatrix} \cos\theta & -\sin\theta \\ \sin\theta & \cos\theta \end{bmatrix}$$

yields

$$\begin{bmatrix} U \\ V \end{bmatrix} := P' \begin{bmatrix} X \\ Y \end{bmatrix}$$

with $E[UV] = 0$. *Answer:* $\theta = \frac{1}{2}\tan^{-1}\left(\frac{2E[XY]}{\sigma_X^2 - \sigma_Y^2}\right)$. In particular, note that if $\sigma_X^2 = \sigma_Y^2$, then $\theta = \pi/4$.

*19. Let e_i denote the ith standard unit vector in $\mathbb{R}^n$.

(a) Show that $e_i e_i' = \text{diag}(0,\ldots,0,1,0,\ldots,0)$, where the 1 is in the ith position.

(b) Show that if

$$E = \begin{bmatrix} e_1' \\ e_4' \\ e_5' \end{bmatrix},$$

then $E'E$ is a diagonal matrix with ones at positions 1, 4, and 5 along the diagonal and zeros elsewhere. Hence, $E'Ex$ is obtained by setting $x_j = 0$ for $j \neq 1,4,5$ and leaving x_1, x_4, and x_5 unchanged. We also remark that $E'Ex$ is the orthogonal projection of x onto the three-dimensional subspace spanned by e_1, e_4, and e_5.

20. Let U be an n-dimensional random vector with zero mean and covariance matrix C_U. Let Q' be a decorrelating transformation for U. In other words, $Q'C_U Q = M = \text{diag}(\mu_1, \ldots, \mu_n)$, where $Q'Q = QQ' = I$. Now put $X := U + V$, where U and V are uncorrelated with V having zero mean and covariance matrix $C_V = I$.

   (a) Find a decorrelating transformation P' for X.

   (b) If P' is the decorrelating transformation from part (a), and $Y := P'X$, find the covariance matrix C_Y.

8.3: Transformations of random vectors

21. Let X and Y have joint density $f_{XY}(x,y)$. Let $U := X + Y$ and $V := X - Y$. Find $f_{UV}(u,v)$.

22. Let X and Y be positive random variables with joint density $f_{XY}(x,y)$. If $U := XY$ and $V := Y/X$, find the joint density of U and V. Also find the marginal densities $f_U(u)$ and $f_V(v)$. Your marginal density $f_U(u)$ should be a special case of the result in Example 7.15, and your marginal density $f_V(v)$ should be a special case of your answer to Problem 33(a) in Chapter 7.

23. Let X and Y be independent Laplace(λ) random variables. Put $U := X$ and $V := Y/X$. Find $f_{UV}(u,v)$ and $f_V(v)$. Compare with Problem 33(c) in Chapter 7.

24. Let X and Y be independent uniform(0,1] random variables. Show that if $U := \sqrt{-2\ln X}\cos(2\pi Y)$ and $V := \sqrt{-2\ln X}\sin(2\pi Y)$, then U and V are independent $N(0,1)$ random variables.

25. Let X and Y have joint density $f_{XY}(x,y)$. Let $U := X+Y$ and $V := X/(X+Y)$. Find $f_{UV}(u,v)$. Apply your result to the case where X and Y are independent gamma random variables $X \sim \text{gamma}(p,\lambda)$ and $Y \sim \text{gamma}(q,\lambda)$. Show that U and V are independent with $U \sim \text{gamma}(p+q,\lambda)$ and $V \sim \text{beta}(p,q)$. Compare with Problem 55 in Chapter 4 and Problem 42(b) in Chapter 7.

26. Let X and Y have joint density $f_{XY}(x,y)$. Let $U := X+Y$ and $V := X/Y$. Find $f_{UV}(u,v)$. Apply your result to the case where X and Y are independent gamma random variables $X \sim \text{gamma}(p,\lambda)$ and $Y \sim \text{gamma}(q,\lambda)$. Show that U and V are independent with $U \sim \text{gamma}(p+q,\lambda)$ and V having the density of Problem 42(a) in Chapter 7.

27. Let X and Y be $N(0,1)$ with $\mathsf{E}[XY] = \rho$. If $R = \sqrt{X^2 + Y^2}$ and $\Theta = \text{angle}(X,Y)$, find $f_{R,\Theta}(r,\theta)$ and $f_\Theta(\theta)$.

8.4: *Linear estimation of random vectors (Wiener filters)

28. Let X and W be independent $N(0,1)$ random variables, and put $Y := X^3 + W$. Find A and b that minimize $\mathsf{E}[|X - \widehat{X}|^2]$, where $\widehat{X} := AY + b$.

29. Let $X \sim N(0,1)$ and $W \sim \text{Laplace}(\lambda)$ be independent, and put $Y := X + W$. Find the linear MMSE estimator of X based on Y.

30. Let X denote a random signal of known mean m_X and known covariance matrix C_X. Suppose that in order to estimate X, all we have available is the noisy measurement

$$Y = GX + W,$$

where G is a known gain matrix, and W is a noise vector with zero mean and known covariance matrix C_W. Further assume that the covariance between the signal and noise, C_{XW}, is zero. Find the linear MMSE estimate of X based on Y assuming that C_Y is invertible.

Remark. It is easy to see that C_Y is invertible if C_W is positive definite or if GC_XG' is positive definite. If C_X is positive definite and if G' is nonsingular, then GC_XG' is positive definite.

31. Let X and Y be as in Problem 30, and let $AC_Y = C_{XY}$. Assuming C_X is invertible, show that $A = (C_X^{-1} + G'C_W^{-1}G)^{-1}G'C_W^{-1}$. *Hint:* Use the **matrix inverse formula**

$$(\alpha + \beta\gamma\delta)^{-1} = \alpha^{-1} - \alpha^{-1}\beta(\gamma^{-1} + \delta\alpha^{-1}\beta)^{-1}\delta\alpha^{-1}.$$

32. Let $\widehat{X}$ denote the linear MMSE estimate of the vector X based on the observation vector Y. Now suppose that $Z := BX$. Let $\widehat{Z}$ denote the linear MMSE estimate of Z based on Y. Show that $\widehat{Z} = B\widehat{X}$.

33. Let X and Y be random vectors with known means and covariance matrices. Do not assume zero means. Find the best *purely linear* estimate of X based on Y; i.e., find the matrix A that minimizes $\mathsf{E}[\|X - AY\|^2]$. Similarly, find the best constant estimate of X; i.e., find the vector b that minimizes $\mathsf{E}[\|X - b\|^2]$.

34. Let X and Y be random vectors with m_X, m_Y, C_X, C_Y, and C_{XY} given. Do *not* assume C_Y is invertible. Let $\widehat{X} = A(Y - m_Y) + m_X$ be the linear MMSE estimate of X based on Y. Show that the error covariance, defined to be $\mathsf{E}[(X - \widehat{X})(X - \widehat{X})']$ has the following representations:

$$C_X - AC_{YX} - C_{XY}A' + AC_YA'$$
$$C_X - C_{XY}A'$$
$$C_X - AC_{YX}$$
$$C_X - AC_YA'.$$

35. Use the result of the preceding problem to show that MSE is

$$\mathsf{E}[\|X - \widehat{X}\|^2] = \mathrm{tr}(C_X - AC_{YX}).$$

36. **MATLAB.** In Problem 30 suppose that $C_X = C_W$ are 4×4 identity matrices and that

$$G = \begin{bmatrix} -2 & 1 & -5 & 11 \\ 9 & -4 & -3 & 11 \\ -10 & -10 & -25 & -13 \\ -3 & -1 & 5 & 0 \end{bmatrix}.$$

Compute $A = C_{XY}C_Y^{-1}$ and the MSE using MATLAB.

37. Let $X = [X_1, \ldots, X_n]'$ be a random vector with zero mean and covariance matrix C_X. Put $Y := [X_1, \ldots, X_m]'$, where $m < n$. Find the linear MMSE estimate of X based on Y. Also find the error covariance matrix and the MSE. Your answers should be in terms of the block components of C_X,

$$C_X = \begin{bmatrix} C_1 & C_2 \\ C_2' & C_3 \end{bmatrix},$$

where C_1 is $m \times m$ and invertible.

*38. In this problem you will show that $AC_Y = C_{XY}$ has a solution even if C_Y is singular. Let P' be the decorrelating transformation of Y. Put $Z := P'Y$ and solve $\tilde{A}C_Z = C_{XZ}$ for $\tilde{A}$. Use the fact that C_Z is diagonal. You also need to use the Cauchy–Schwarz inequality for random variables (8.4) to show that $(C_Z)_{jj} = 0$ implies $(C_{XZ})_{ij} = 0$. To conclude, show that if $\tilde{A}C_Z = C_{XZ}$, then $A = \tilde{A}P'$ solves $AC_Y = C_{XY}$.

39. Show that Problem 37 is a special case of Problem 30.

8.5: *Estimation of covariance matrices

40. If $X_1, \ldots, X_n$ are i.i.d. zero mean and have variance σ^2, show that

$$\frac{1}{n}\sum_{k=1}^{n} X_k^2$$

is an unbiased estimator of σ^2; i.e., show its expectation is equal to σ^2.

41. If $X_1, \ldots, X_n$ are i.i.d. random vectors with zero mean and covariance matrix C, show that

$$\frac{1}{n}\sum_{k=1}^{n} X_k X_k'$$

is an unbiased estimator of C; i.e., show its expectation is equal to C.

42. **MATLAB.** Suppose $X_1, \ldots, X_n$ are i.i.d. with nonzero mean vector m. The following code uses $X_1, \ldots, X_n$ as the columns of the matrix X. Add code to the end of the script to estimate the mean vector and the covariance matrix.

```
G = [ 1 -2 -2 1 0 ; 0 1 -1 3 -1 ; 1 3 3 -3 4 ; ...
     -1 1 2 1 -4 ; 0 2 0 -4 3 ];
C = G*G'
d = length(G);
n = 1000;
m = [1:d]';        % Use an easy-to-define mean vector
Z = randn(d,n);    % Create d by n array of i.i.d. N(0,1) RVs
X = G*Z;           % Multiply each column by G
X = X + repmat(m,1,n); % Add mean vec to each col of X
```

8.6: *Nonlinear estimation of random vectors

43. Let $X \sim N(0,1)$ and $W \sim \text{Laplace}(\lambda)$ be independent, and put $Y := X + W$. Find the ML estimator of X based on Y.

44. Let $X \sim \exp(\mu)$ and $W \sim \text{Laplace}(\lambda)$ be independent, and put $Y := X + W$. Find the ML estimator of X based on Y. Repeat for $X \sim \text{uniform}[0,1]$.

45. For $X \sim \exp(\mu)$ and Y as in the preceding problem, find the MAP estimator of X based on Y if $\mu < \lambda$. Repeat for $\mu \geq \lambda$.

46. If X and Y are positive random variables with joint density

$$f_{XY}(x,y) = (x/y^2)e^{-(x/y)^2/2} \cdot \lambda e^{-\lambda y}, \quad x, y > 0,$$

find both the MMSE and MAP estimators of X given Y (the estimators should be different).

*47. Let X be a scalar random variable. Show that if $g_1(y)$ and $g_2(y)$ both satisfy

$$\mathsf{E}[(X - g(Y))h(Y)] = 0, \quad \text{for all bounded functions } h,$$

then $g_1 = g_2$ in the sense that

$$\mathsf{E}[|g_2(Y) - g_1(Y)|] = 0.$$

Exam preparation

You may use the following suggestions to prepare a study sheet, including formulas mentioned that you have trouble remembering. You may also want to ask your instructor for additional suggestions.

8.1. Review of matrix operations. Know definitions and properties of matrix multiplication, trace, norm of a vector. In particular, $\text{tr}(AB) = \text{tr}(BA)$.

8.2. Random vectors and random matrices. If X is random and A, B, and G are not, then $\mathsf{E}[AXB + G] = A\mathsf{E}[X]B + G$. Know definitions of the covariance matrix, $\text{cov}(X)$, and the cross-covariance matrix, $\text{cov}(X,Y)$ when X and Y are random vectors. Know definition of correlation and cross-correlation matrices. Covariance and correlation matrices are always symmetric and positive semidefinite. If X has a singular covariance matrix, then there is a component of X that is a linear combination of the remaining components. Know joint characteristic function and the fact that the components of a random vector are independent if and only if their joint characteristic function is equal to the product of their marginal characteristic functions. Know how to compute moments from the joint characteristic function. If X has covariance matrix C and $P'CP = \Lambda$ is diagonal and $P'P = PP' = I$, then $Y := P'X$ has covariance matrix Λ and therefore the components of Y are uncorrelated. Thus P' is a decorrelating transformation. Note also that $PY = PP'X = X$ and that $X = PY$ is the Karhunen–Loève expansion of X.

8.3. Transformations of random vectors. If $Y = G(X)$, then

$$f_Y(y) = f_X(H(y))|\det dH(y)|,$$

where H is the inverse of G; i.e., $X = H(Y)$, and $dH(y)$ is the matrix of partial derivatives in (8.7).

8.4. *Linear estimation of random vectors (Wiener filters).* The linear MMSE estimator of X based on Y is $A(Y - m_Y) + m_X$, where A solves $AC_Y = C_{XY}$. The MSE is given by $\text{tr}(C_X - AC_{YX})$.

8.5. *Estimation of covariance matrices.* Know the unbiased estimators of $\text{cov}(X)$ when X is known to have zero mean and when the mean vector is unknown.

8.6. *Nonlinear estimation of random vectors.* Know formulas for $g_{\text{MMSE}}(y)$, $g_{\text{ML}}(y)$, and $g_{\text{MAP}}(y)$. When X is uniform, the ML and MAP estimators are the same; but in general they are different.

Work any review problems assigned by your instructor. If you finish them, re-work your homework assignments.

9
Gaussian random vectors

9.1 Introduction

Scalar Gaussian or normal random variables were introduced in Chapter 4. Pairs of Gaussian random variables were introduced in Chapter 7. In this chapter, we generalize these notions to random vectors.

The univariate $N(m,\sigma^2)$ density is

$$\frac{\exp[-\frac{1}{2}(x-m)^2/\sigma^2]}{\sqrt{2\pi}\,\sigma}.$$

If $X_1,\ldots,X_n$ are independent $N(m_i,\sigma_i^2)$, then their joint density is the product

$$\begin{aligned}f(x) &= \prod_{i=1}^{n}\frac{\exp[-\frac{1}{2}(x_i-m_i)^2/\sigma_i^2]}{\sqrt{2\pi}\,\sigma_i} \\ &= \frac{\exp\left[-\frac{1}{2}\sum_{i=1}^{n}(x_i-m_i)^2/\sigma_i^2\right]}{(2\pi)^{n/2}\sigma_1\cdots\sigma_n},\end{aligned} \quad (9.1)$$

where $x := [x_1,\ldots,x_n]'$. We now rewrite this joint density using matrix–vector notation. To begin, observe that since the X_i are independent, they are uncorrelated; hence, the covariance matrix of $X := [X_1,\ldots,X_n]'$ is

$$C = \begin{bmatrix} \sigma_1^2 & & 0 \\ & \ddots & \\ 0 & & \sigma_n^2 \end{bmatrix}.$$

Next, put $m := [m_1,\ldots,m_n]'$ and write

$$C^{-1}(x-m) = \begin{bmatrix} 1/\sigma_1^2 & & 0 \\ & \ddots & \\ 0 & & 1/\sigma_n^2 \end{bmatrix}\begin{bmatrix} x_1-m_1 \\ \vdots \\ x_n-m_n \end{bmatrix} = \begin{bmatrix} (x_1-m_1)/\sigma_1^2 \\ \vdots \\ (x_n-m_n)/\sigma_n^2 \end{bmatrix}.$$

It is then easy to see that $(x-m)'C^{-1}(x-m) = \sum_{i=1}^{n}(x_i-m_i)^2/\sigma_i^2$. Since C is diagonal, its determinant is $\det C = \sigma_1^2\cdots\sigma_n^2$. It follows that (9.1) can be written in matrix–vector notation as

$$\boxed{f(x) = \frac{\exp[-\frac{1}{2}(x-m)'C^{-1}(x-m)]}{(2\pi)^{n/2}\sqrt{\det C}}.} \quad (9.2)$$

Even if C is not diagonal, this is the general formula for the density of a Gaussian random vector of length n with mean vector m and covariance matrix C.

One question about (9.2) that immediately comes to mind is whether this formula integrates to one even when C is not diagonal. There are several ways to see that this is indeed the case. For example, it can be shown that the multivariate Fourier transform of (9.2) is

$$\int_{\mathbb{R}^n} e^{jv'x} \frac{\exp[-\frac{1}{2}(x-m)'C^{-1}(x-m)]}{(2\pi)^{n/2}\sqrt{\det C}} \, dx = e^{jv'm - v'Cv/2}. \tag{9.3}$$

Taking $v = 0$ shows that the density integrates to one. Although (9.3) can be derived directly by using a multivariate change of variable,[1] we use a different argument in Section 9.4.

A second question about (9.2) is what to do if C is not invertible. For example, suppose $Z \sim N(0,1)$, and $X_1 := Z$ and $X_2 := 2Z$. Then the covariance matrix of $[X_1 X_2]'$ is

$$\begin{bmatrix} \mathsf{E}[X_1^2] & \mathsf{E}[X_1 X_2] \\ \mathsf{E}[X_2 X_1] & \mathsf{E}[X_2^2] \end{bmatrix} = \begin{bmatrix} 1 & 2 \\ 2 & 4 \end{bmatrix},$$

which is not invertible. Now observe that the right-hand side of (9.3) this involves C but not C^{-1}. This suggests that we define a random vector to be Gaussian if its characteristic function is given by the right-hand side of (9.3). Then when C is invertible, we see that the joint density exists and is given by (9.2).

Instead of defining a random vector to be Gaussian if its characteristic function has the form $e^{jv'm - v'Cv/2}$, in Section 9.2 we define a random vector to be Gaussian if every linear combination of its components is a scalar Gaussian random variable. This definition turns out to be equivalent to the characteristic function definition, but is easier to use in deriving various properties, including the joint density when it exists.

9.2 Definition of the multivariate Gaussian

A random vector $X = [X_1, \ldots, X_n]'$ is said to be **Gaussian** or **normal** if every linear combination of the components of X, e.g.,

$$\sum_{i=1}^{n} c_i X_i, \tag{9.4}$$

is a scalar Gaussian random variable. Equivalent terminology is that $X_1, \ldots, X_n$ are **jointly Gaussian** or **jointly normal**. In order for this definition to make sense when all $c_i = 0$ or when X has a singular covariance matrix (recall the remark following Example 8.5), we agree that any constant random variable is considered to be Gaussian (see Problem 2).

Notation. If X is a Gaussian random vector with mean vector m and covariance matrix C, we write $X \sim N(m, C)$.

Example 9.1 (independent and Gaussian implies jointly Gaussian). If the X_i are independent $N(m_i, \sigma_i^2)$, then it is easy to see using moment generating functions that every linear combination of the X_i is a scalar Gaussian; i.e., X is a Gaussian random vector (Problem 4).

Example 9.2. If X is a Gaussian random vector, then the numerical average of its components,

$$\frac{1}{n}\sum_{i=1}^{n} X_i,$$

is a scalar Gaussian random variable.

An easy consequence of our definition of a Gaussian random vector is that any subvector is also Gaussian. To see this, suppose $X = [X_1, \ldots, X_n]'$ is a Gaussian random vector. Then every linear combination of the components of the subvector $[X_1, X_3, X_5]'$ is of the form (9.4) if we take $c_i = 0$ for i not equal to $1, 3, 5$.

Example 9.3. Let X be a Gaussian random vector of length 5 and covariance matrix

$$C = \begin{bmatrix} 58 & 43 & 65 & 55 & 48 \\ 43 & 53 & 57 & 52 & 45 \\ 65 & 57 & 83 & 70 & 58 \\ 55 & 52 & 70 & 63 & 50 \\ 48 & 45 & 58 & 50 & 48 \end{bmatrix}.$$

Find the covariance matrix of $[X_1, X_3, X_5]'$.

Solution. All we need to do is extract the appropriate 3×3 submatrix of elements C_{ij}, where $i = 1, 3, 5$ and $j = 1, 3, 5$. This yields

$$\begin{bmatrix} 58 & 65 & 48 \\ 65 & 83 & 58 \\ 48 & 58 & 48 \end{bmatrix}.$$

This is easy to do in MATLAB if C is already defined:

```
k = [ 1 3 5 ];
C(k,k)
```

displays the 3×3 matrix above.

Sometimes it is more convenient to express linear combinations as the product of a row vector times the column vector X. For example, if we put $c = [c_1, \ldots, c_n]'$, then

$$\sum_{i=1}^{n} c_i X_i = c'X.$$

Now suppose that $Y = AX$ for some $r \times n$ matrix A. Letting $c = [c_1, \ldots, c_r]'$, every linear combination of the r components of Y has the form

$$\sum_{i=1}^{r} c_i Y_i = c'Y = c'(AX) = (A'c)'X,$$

which is a linear combination of the components of X, and therefore normal.

We can even add a constant vector. If $Y = AX + b$, where A is again $r \times n$, and b is $r \times 1$, then
$$c'Y = c'(AX+b) = (A'c)'X + c'b.$$
Adding the constant $c'b$ to the normal random variable $(A'c)'X$ results in another normal random variable (with a different mean).

In summary, if X is a Gaussian random vector, then so is $AX + b$ for any $r \times n$ matrix A and any r-vector b. Symbolically, we write
$$X \sim N(m,C) \implies AX + b \sim N(Am+b, ACA').$$

In particular, if X is Gaussian, then $Y = AX$ is Gaussian. The converse may or may not be true. In other words, if $Y = AX$ and Y is Gaussian, it is not necessary that X be Gaussian. For example, let X_1 and X_2 be independent with $X_1 \sim N(0,1)$ and X_2 not normal, say Laplace(λ). Put
$$\begin{bmatrix} Y_1 \\ Y_2 \end{bmatrix} = \begin{bmatrix} 1 & 0 \\ 2 & 0 \end{bmatrix} \begin{bmatrix} X_1 \\ X_2 \end{bmatrix}.$$
It is easy to see that Y_1 and Y_2 are jointly Gaussian, while X_1 and X_2 are not jointly Gaussian. On the other hand, if $Y = AX$, where Y is Gaussian and A is invertible, then $X = A^{-1}Y$ must be Gaussian.

9.3 Characteristic function

We now find the joint characteristic function, $\varphi_X(v) := \mathsf{E}[e^{jv'X}]$, when $X \sim N(m,C)$. The key is to observe that since X is normal, so is $Y := v'X$. Furthermore, the mean and variance of the scalar random variable Y are given by $\mu := \mathsf{E}[Y] = v'm$ and $\sigma^2 := \text{var}(Y) = v'Cv$. Now write
$$\varphi_X(v) = \mathsf{E}[e^{jv'X}] = \mathsf{E}[e^{jY}] = \mathsf{E}[e^{j\eta Y}]\big|_{\eta=1} = \varphi_Y(\eta)\big|_{\eta=1}.$$
Since $Y \sim N(\mu, \sigma^2)$, $\varphi_Y(\eta) = e^{j\eta\mu - \eta^2 \sigma^2/2}$. Hence,
$$\varphi_X(v) = \varphi_Y(1) = e^{j\mu - \sigma^2/2} = e^{jv'm - v'Cv/2}.$$

We have shown here that if every linear combination of the X_i is a scalar Gaussian, then the joint characteristic function has the above form. The converse is also true; i.e., if X has the above joint characteristic function, then every linear combination of the X_i is a scalar Gaussian (Problem 11). Hence, many authors use the equivalent definition that a random vector is Gaussian if its joint characteristic function has the above form.

For Gaussian random vectors uncorrelated implies independent

If the components of a random vector are uncorrelated, then the covariance matrix is diagonal. In general, this is not enough to prove that the components of the random vector are independent. However, if X is a Gaussian random vector, then the components are independent. To see this, suppose that X is Gaussian with uncorrelated components. Then C is diagonal, say

$$C = \begin{bmatrix} \sigma_1^2 & & 0 \\ & \ddots & \\ 0 & & \sigma_n^2 \end{bmatrix},$$

where $\sigma_i^2 = C_{ii} = \text{var}(X_i)$. The diagonal form of C implies that

$$v'Cv = \sum_{i=1}^{n} \sigma_i^2 v_i^2,$$

and so

$$\varphi_X(v) = e^{jv'm - v'Cv/2} = \prod_{i=1}^{n} e^{jv_i m_i - \sigma_i^2 v_i^2 / 2}.$$

In other words,

$$\varphi_X(v) = \prod_{i=1}^{n} \varphi_{X_i}(v_i),$$

where $\varphi_{X_i}(v_i)$ is the characteristic function of the $N(m_i, \sigma_i^2)$ density. Multivariate inverse Fourier transformation then yields

$$f_X(x) = \prod_{i=1}^{n} f_{X_i}(x_i),$$

where $f_{X_i} \sim N(m_i, \sigma_i^2)$. This establishes the independence of the X_i.

Example 9.4. If X is a Gaussian random vector and we apply a decorrelating transformation to it, say $Y = P'X$ as in Section 8.2, then Y will be a Gaussian random vector with uncorrelated and therefore *independent* components.

Example 9.5. Let X be an n-dimensional, zero mean Gaussian random vector with a covariance matrix C whose eigenvalues $\lambda_1, \ldots, \lambda_n$ are only zeros and ones. Show that if r of the eigenvalues are one, and $n - r$ of them are zero, then $\|X\|^2$ is a chi-squared random variable with r degrees of freedom.

Solution. Apply the decorrelating transformation $Y = P'X$ as in Section 8.2. Then the Y_i are uncorrelated, Gaussian, and therefore independent. Furthermore, the Y_i corresponding to the zero eigenvalues are zero, and the remaining Y_i have $\mathsf{E}[Y_i^2] = \lambda_i = 1$. With the nonzero Y_i i.i.d. $N(0,1)$,

$$\|Y\|^2 = \sum_{i:\lambda_i=1} Y_i^2,$$

which is a sum of r terms, is chi-squared with r degrees of freedom by Problems 46 and 55 in Chapter 4. It remains to observe that since $PP' = I$,

$$\|Y\|^2 = Y'Y = (P'X)'(P'X) = X'(PP')X = X'X = \|X\|^2.$$

Example 9.6. Let $X_1, \ldots, X_n$ be i.i.d. $N(m, \sigma^2)$ random variables. Let $\overline{X} := \frac{1}{n}\sum_{i=1}^{n} X_i$ denote the average of the X_i. Furthermore, for $j = 1, \ldots, n$, put $Y_j := X_j - \overline{X}$. Show that $\overline{X}$ and $Y := [Y_1, \ldots, Y_n]'$ are jointly normal and independent.

Solution. Let $X := [X_1, \ldots, X_n]'$, and put $a := [\frac{1}{n}, \ldots, \frac{1}{n}]$. Then $\overline{X} = aX$. Next, observe that

$$\begin{bmatrix} Y_1 \\ \vdots \\ Y_n \end{bmatrix} = \begin{bmatrix} X_1 \\ \vdots \\ X_n \end{bmatrix} - \begin{bmatrix} \overline{X} \\ \vdots \\ \overline{X} \end{bmatrix}.$$

Let M denote the $n \times n$ matrix with each row equal to a; i.e., $M_{ij} = 1/n$ for all i, j. Then $Y = X - MX = (I - M)X$, and Y is a jointly normal random vector. Next consider the vector

$$Z := \begin{bmatrix} \overline{X} \\ Y \end{bmatrix} = \begin{bmatrix} a \\ I - M \end{bmatrix} X.$$

Since Z is a linear transformation of the Gaussian random vector X, Z is also a Gaussian random vector. Furthermore, its covariance matrix has the block-diagonal form (see Problem 8)

$$\begin{bmatrix} \text{var}(\overline{X}) & 0 \\ 0 & \mathsf{E}[YY'] \end{bmatrix}.$$

This implies, by Problem 12, that $\overline{X}$ and Y are independent.

9.4 Density function

In this section we give a simple derivation of the fact that if $Y \sim N(m, C)$ and if C is invertible, then

$$f_Y(y) = \frac{\exp[-\frac{1}{2}(y-m)'C^{-1}(y-m)]}{(2\pi)^{n/2}\sqrt{\det C}}.$$

We exploit the Jacobian formulas of Section 8.3, specifically the result of Example 8.9. Put

$$X := C^{-1/2}(Y - m),$$

where the existence of the symmetric matrices $C^{1/2}$ and $C^{-1/2}$ is shown in the Notes.[2] Since Y is Gaussian, the form of X implies that it too is Gaussian. It is also easy to see that X has zero mean and covariance matrix

$$\mathsf{E}[XX'] = C^{-1/2}\mathsf{E}[(Y-m)(Y-m)']C^{-1/2} = C^{-1/2}CC^{-1/2} = I.$$

Hence, the components of X are jointly Gaussian, uncorrelated, and therefore independent. It follows that

$$f_X(x) = \prod_{i=1}^{n} \frac{e^{-x_i^2/2}}{\sqrt{2\pi}} = \frac{e^{-x'x/2}}{(2\pi)^{n/2}}.$$

Since we also have $Y = C^{1/2}X + m$, we can use the result of Example 8.9 with $A = C^{1/2}$ and $b = m$ to obtain

$$f_Y(y) = \frac{f_X(C^{-1/2}(y-m))}{|\det C^{1/2}|}.$$

Using the above formula for $f_X(x)$ with $x = C^{-1/2}(y-m)$, we have

$$f_Y(y) = \frac{e^{-(y-m)'C^{-1}(y-m)/2}}{(2\pi)^{n/2}\sqrt{\det C}},$$

where the fact that $\det C^{1/2} = \sqrt{\det C} > 0$ is shown in Note 2.

Simulation

The foregoing derivation tells us how to simulate an arbitrary Gaussian random vector Y with mean m and covariance matrix C. First generate X with i.i.d. $N(0,1)$ components. The MATLAB command X = randn(n,1) generates such a vector of length n. Then generate Y with the command Y = Chalf*X + m, where Chalf is the square root of the matrix C. From Note 2 at the end of the chapter, Chalf = P*sqrt(Lambda)*P', where the matrices P and Lambda are obtained with [P, Lambda] = eig(C).

Level sets

The **level sets** of a density are sets where the density is constant. The Gaussian density is constant on the **ellipsoids** centered at m,

$$\{x \in \mathbb{R}^n : (x-m)'C^{-1}(x-m) = \text{constant}\}. \quad (9.5)$$

To see why these sets are called ellipsoids, consider the two-dimensional case in which C^{-1} is a diagonal matrix, say $\text{diag}(1/a^2, 1/b^2)$. Then

$$\begin{bmatrix} x & y \end{bmatrix} \begin{bmatrix} 1/a^2 & 0 \\ 0 & 1/b^2 \end{bmatrix} \begin{bmatrix} x \\ y \end{bmatrix} = \frac{x^2}{a^2} + \frac{y^2}{b^2}.$$

The set of (x,y) for which this is constant is an ellipse centered at the origin with principal axes aligned with the coordinate axes.

Returning to the n-dimensional case, let P' be a decorrelating transformation so that $P'CP = \Lambda$ is a diagonal matrix. Then $C^{-1} = P\Lambda^{-1}P'$, and

$$f_X(x) = \frac{\exp[-\frac{1}{2}(P'(x-m))'\Lambda^{-1}(P'(x-m))]}{(2\pi)^{n/2}\sqrt{\det C}}.$$

Since Λ^{-1} is diagonal, the ellipsoid

$$\{y \in \mathbb{R}^n : y'\Lambda^{-1}y = \text{constant}\}$$

is centered at the origin, and its principal axes are aligned with the coordinate axes. Applying the transformation $x = Py + m$ to this centered and aligned ellipsoid yields (9.5). In the two-dimensional case, P is the rotation by the angle θ determined in Problem 18 in Chapter 8, and the level sets in (9.5) are ellipses as shown in Figures 7.9–7.11. In the three-dimensional case, the level sets are ellipsoid surfaces such as the ones in Figure 9.1.

9.5 Conditional expectation and conditional probability

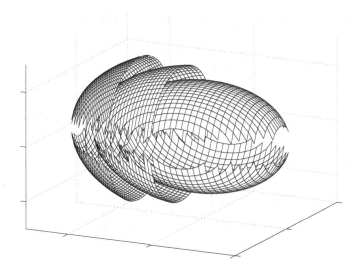

Figure 9.1. Ellipsoid level surfaces of a three-dimensional Gaussian density.

9.5 Conditional expectation and conditional probability

In Section 8.6 we showed that $g(y) := \mathsf{E}[X|Y=y]$ is characterized as the solution of

$$\mathsf{E}[h(Y)'\{X - g(Y)\}] = 0, \quad \text{for all functions } h. \tag{9.6}$$

Here we show that if X and Y are random vectors such that $[X',Y']'$ is a Gaussian random vector, then

$$\mathsf{E}[X|Y=y] = A(y - m_Y) + m_X, \quad \text{where } A \text{ solves } AC_Y = C_{XY}.$$

In other words, when X and Y are jointly Gaussian, the MMSE estimator is equal to the linear MMSE estimator.

To establish this result, we show that if A solves $AC_Y = C_{XY}$ and $g(y) := A(y - m_Y) + m_X$, then (9.6) holds. For simplicity, we assume both X and Y are zero mean. We then observe that

$$\begin{bmatrix} X - AY \\ Y \end{bmatrix} = \begin{bmatrix} I & -A \\ 0 & I \end{bmatrix} \begin{bmatrix} X \\ Y \end{bmatrix}$$

is a linear transformation of $[X',Y']'$ and so the left-hand side is a Gaussian random vector whose top and bottom entries are easily seen to be uncorrelated:

$$\begin{aligned}
\mathsf{E}[(X - AY)Y'] &= C_{XY} - AC_Y \\
&= C_{XY} - C_{XY} \\
&= 0.
\end{aligned}$$

Being jointly Gaussian and uncorrelated, they are independent (cf. Problem 12). Hence, for any function $h(y)$,

$$\mathsf{E}[h(Y)'(X-AY)] = \mathsf{E}[h(Y)]'\mathsf{E}[X-AY]$$
$$= \mathsf{E}[h(Y)]'0 = 0.$$

With a little more work in Problem 17, we can characterize conditional probabilities of X given $Y = y$.

If $[X',Y']'$ is a Gaussian random vector, then given $Y = y$,

$$X \sim N(\mathsf{E}[X|Y=y], C_{X|Y}),$$

where $C_{X|Y} := C_X - AC_{YX}$, $\mathsf{E}[X|Y=y] = A(y - m_Y) + m_X$, and A solves $AC_Y = C_{XY}$. If $C_{X|Y}$ is invertible and X is n-dimensional, then

$$f_{X|Y}(x|y) = \frac{\exp[-\frac{1}{2}(x-g(y))'C_{X|Y}^{-1}(x-g(y))]}{(2\pi)^{n/2}\sqrt{\det C_{X|Y}}},$$

where $g(y) := \mathsf{E}[X|Y=y]$.

Example 9.7 (Gaussian signal in additive Gaussian noise). Suppose that a signal $X \sim N(0,1)$ is transmitted over a noisy channel so that the received measurement is $Y = X + W$, where $W \sim N(0, \sigma^2)$ is independent of X. Find $\mathsf{E}[X|Y=y]$ and $f_{X|Y}(x|y)$.

Solution. Since X and Y are jointly Gaussian, the answers are in terms of the linear MMSE estimator. In other words, we need to write out m_X, m_Y, C_Y, and C_{XY}. In this case, both means are zero. For C_Y, we have $C_Y = C_X + C_W = 1 + \sigma^2$. Since X and W are independent, they are uncorrelated, and we can write

$$C_{XY} = \mathsf{E}[XY'] = \mathsf{E}[X(X+W)'] = C_X = 1.$$

Thus, $AC_Y = C_{XY}$ becomes $A(1+\sigma^2) = 1$. Since $A = 1/(1+\sigma^2)$,

$$\mathsf{E}[X|Y=y] = Ay = \frac{1}{1+\sigma^2}y.$$

Since

$$C_{X|Y} = C_X - AC_{YX} = 1 - \frac{1}{1+\sigma^2} \cdot 1 = \frac{\sigma^2}{1+\sigma^2},$$

we have

$$f_{X|Y}(\cdot|y) \sim N\left(\frac{y}{1+\sigma^2}, \frac{\sigma^2}{1+\sigma^2}\right).$$

These answers agree with those found by direct calculation of $f_{X|Y}(x|y)$ carried out in Example 8.21.

9.6 Complex random variables and vectors[†]

A **complex random variable** is a pair of real random variables, say X and Y, written in the form $Z = X + jY$, where j denotes the square root of -1. The advantage of the complex notation is that it becomes easy to write down certain functions of (X,Y). For example, it is easier to talk about

$$Z^2 = (X+jY)(X+jY) = (X^2 - Y^2) + j(2XY)$$

than the vector-valued mapping

$$g(X,Y) = \begin{bmatrix} X^2 - Y^2 \\ 2XY \end{bmatrix}.$$

Recall that the absolute value of a complex number $z = x + jy$ is

$$|z| := \sqrt{x^2 + y^2}.$$

The **complex conjugate** of z is

$$z^* := x - jy,$$

and so

$$zz^* = (x+jy)(x-jy) = x^2 + y^2 = |z|^2.$$

We also have

$$x = \frac{z+z^*}{2} \quad \text{and} \quad y = \frac{z-z^*}{2j}.$$

The expected value of Z is simply

$$\mathsf{E}[Z] := \mathsf{E}[X] + j\mathsf{E}[Y].$$

The variance of Z is

$$\boxed{\mathsf{var}(Z) := \mathsf{E}[(Z - \mathsf{E}[Z])(Z - \mathsf{E}[Z])^*] = \mathsf{E}\big[|Z - \mathsf{E}[Z]|^2\big].}$$

Note that $\mathsf{var}(Z) = \mathsf{var}(X) + \mathsf{var}(Y)$, while

$$\mathsf{E}[(Z - \mathsf{E}[Z])^2] = [\mathsf{var}(X) - \mathsf{var}(Y)] + j[2\mathsf{cov}(X,Y)],$$

which is zero if and only if X and Y are uncorrelated and have the same variance.

If X and Y are jointly continuous real random variables, then we say that $Z = X + jY$ is a continuous complex random variable with density

$$f_Z(z) = f_Z(x+jy) := f_{XY}(x,y).$$

Sometimes the formula for $f_{XY}(x,y)$ is more easily expressed in terms of the complex variable z. For example, if X and Y are independent $N(0, 1/2)$, then

$$f_{XY}(x,y) = \frac{e^{-x^2}}{\sqrt{2\pi}\sqrt{1/2}} \cdot \frac{e^{-y^2}}{\sqrt{2\pi}\sqrt{1/2}} = \frac{e^{-|z|^2}}{\pi}.$$

[†]Section 9.6 can be skipped without loss of continuity.

Note that $\mathsf{E}[Z] = 0$ and $\mathrm{var}(Z) = 1$. Also, the density is circularly symmetric since $|z|^2 = x^2 + y^2$ depends only on the distance from the origin of the point $(x,y) \in \mathbb{R}^2$.

A **complex random vector** of dimension n, say

$$Z = [Z_1, \ldots, Z_n]',$$

is a vector whose ith component is a complex random variable $Z_i = X_i + jY_i$, where X_i and Y_j are real random variables. If we put

$$X := [X_1, \ldots, X_n]' \quad \text{and} \quad Y := [Y_1, \ldots, Y_n]',$$

then $Z = X + jY$, and the mean vector of Z is $\mathsf{E}[Z] = \mathsf{E}[X] + j\mathsf{E}[Y]$. The covariance matrix of Z is

$$\boxed{\mathrm{cov}(Z) := \mathsf{E}[(Z - \mathsf{E}[Z])(Z - \mathsf{E}[Z])^H],}$$

where the superscript ^{H} denotes the complex conjugate transpose. Letting $K := \mathrm{cov}(Z)$, the ik entry of K is

$$K_{ik} = \mathsf{E}[(Z_i - \mathsf{E}[Z_i])(Z_k - \mathsf{E}[Z_k])^*] =: \mathrm{cov}(Z_i, Z_k).$$

It is also easy to show that

$$K = (C_X + C_Y) + j(C_{YX} - C_{XY}). \tag{9.7}$$

For joint distribution purposes, we identify the n-dimensional complex vector Z with the $2n$-dimensional real random vector

$$[X_1, \ldots, X_n, Y_1, \ldots, Y_n]'. \tag{9.8}$$

If this $2n$-dimensional real random vector has a joint density f_{XY}, then we write

$$f_Z(z) := f_{XY}(x_1, \ldots, x_n, y_1, \ldots, y_n).$$

Sometimes the formula for the right-hand side can be written simply in terms of the complex vector z.

Complex Gaussian random vectors

An n-dimensional complex random vector $Z = X + jY$ is said to be Gaussian if the $2n$-dimensional real random vector in (9.8) is jointly Gaussian; i.e., its characteristic function $\varphi_{XY}(v, \theta) = \mathsf{E}[e^{j(v'X + \theta'Y)}]$ has the form

$$\exp\left\{ j(v'm_X + \theta'm_Y) - \tfrac{1}{2} [\, v' \ \theta' \,] \begin{bmatrix} C_X & C_{XY} \\ C_{YX} & C_Y \end{bmatrix} \begin{bmatrix} v \\ \theta \end{bmatrix} \right\}. \tag{9.9}$$

Now observe that

$$[\, v' \ \theta' \,] \begin{bmatrix} C_X & C_{XY} \\ C_{YX} & C_Y \end{bmatrix} \begin{bmatrix} v \\ \theta \end{bmatrix} \tag{9.10}$$

is equal to
$$v'C_X v + v'C_{XY}\theta + \theta'C_{YX} v + \theta'C_Y \theta,$$
which, upon noting that $v'C_{XY}\theta$ is a scalar and therefore equal to its transpose, simplifies to
$$v'C_X v + 2\theta'C_{YX} v + \theta'C_Y \theta.$$
On the other hand, if we put $w := v + j\theta$, and use (9.7), then (see Problem 22)
$$w^H K w = v'(C_X + C_Y)v + \theta'(C_X + C_Y)\theta + 2\theta'(C_{YX} - C_{XY})v.$$
Clearly, if
$$C_X = C_Y \quad \text{and} \quad C_{XY} = -C_{YX}, \qquad (9.11)$$
then (9.10) is equal to $w^H K w/2$. Conversely, if (9.10) is equal to $w^H K w/2$ for all $w = v + j\theta$, then (9.11) holds (Problem 29). We say that a complex Gaussian random vector $Z = X + jY$ is **circularly symmetric** or **proper** if (9.11) holds. If Z is circularly symmetric and zero mean, then its characteristic function is
$$\mathsf{E}[e^{j(v'X + \theta'Y)}] = e^{-w^H K w/4}, \quad w = v + j\theta. \qquad (9.12)$$

The density corresponding to (9.9) is (assuming zero means)
$$f_{XY}(x,y) = \frac{\exp\left\{-\tfrac{1}{2}[x' \; y']\begin{bmatrix} C_X & C_{XY} \\ C_{YX} & C_Y \end{bmatrix}^{-1}\begin{bmatrix} x \\ y \end{bmatrix}\right\}}{(2\pi)^n \sqrt{\det \Gamma}}, \qquad (9.13)$$
where
$$\Gamma := \begin{bmatrix} C_X & C_{XY} \\ C_{YX} & C_Y \end{bmatrix}.$$
It is shown in Problem 30 that under the assumption of circular symmetry (9.11),
$$f_{XY}(x,y) = \frac{e^{-z^H K^{-1} z}}{\pi^n \det K}, \quad z = x + jy, \qquad (9.14)$$
and that K is invertible if and only if Γ is invertible.

Notes

9.1: Introduction

Note 1. We show that if X has the density in (9.2), then its characteristic function is $e^{jv'm - v'Cv/2}$. Write
$$\mathsf{E}[e^{jv'X}] = \int_{\mathbb{R}^n} e^{jv'x} f(x)\, dx$$
$$= \int_{\mathbb{R}^n} e^{jv'x} \frac{\exp[-\tfrac{1}{2}(x-m)'C^{-1}(x-m)]}{(2\pi)^{n/2}\sqrt{\det C}}\, dx.$$

Now make the multivariate change of variable $y = C^{-1/2}(x-m)$, or equivalently, $x = C^{1/2}y+m$. Then $dx = |\det C^{1/2}|dy = \sqrt{\det C}\,dy$ (see Note 2), and

$$\mathsf{E}[e^{jv'X}] = \int_{\mathbb{R}^n} e^{jv'(C^{1/2}y+m)} \frac{e^{-y'y/2}}{(2\pi)^{n/2}\sqrt{\det C}} \sqrt{\det C}\,dy$$

$$= e^{jv'm} \int_{\mathbb{R}^n} e^{j(C^{1/2}v)'y} \frac{e^{-y'y/2}}{(2\pi)^{n/2}}\,dy$$

$$= e^{jv'm} \int_{\mathbb{R}^n} e^{j(C^{1/2}v)'y} \frac{\exp[-\frac{1}{2}\sum_{i=1}^n y_i^2]}{(2\pi)^{n/2}}\,dy.$$

Put $t := C^{1/2}v$ so that

$$(C^{1/2}v)'y = \sum_{i=1}^n t_i y_i.$$

Then

$$\mathsf{E}[e^{jv'X}] = e^{jv'm} \int_{\mathbb{R}^n} \prod_{i=1}^n \left(e^{jt_i y_i} \frac{e^{-y_i^2/2}}{\sqrt{2\pi}}\right) dy$$

$$= e^{jv'm} \prod_{i=1}^n \left(\int_{-\infty}^\infty e^{jt_i y_i} \frac{e^{-y_i^2/2}}{\sqrt{2\pi}}\,dy_i\right).$$

Since the integral in parentheses is of the form of the characteristic function of a univariate $N(0,1)$ random variable,

$$\mathsf{E}[e^{jv'X}] = e^{jv'm} \prod_{i=1}^n e^{-t_i^2/2}$$

$$= e^{jv'm} e^{-t't/2}$$

$$= e^{jv'm} e^{-v'Cv/2}$$

$$= e^{jv'm - v'Cv/2}.$$

9.4: Density function

Note 2. Recall that an $n \times n$ matrix C is symmetric if it is equal to its transpose; i.e., $C = C'$. It is positive definite if $a'Ca > 0$ for all $a \neq 0$. We show that the determinant of a positive-definite matrix is positive. A trivial modification of the derivation shows that the determinant of a positive-semidefinite matrix is nonnegative. At the end of the note, we also define the square root of a positive-semidefinite matrix.

We start with the well-known fact that a symmetric matrix can be diagonalized [30]; i.e., there is an $n \times n$ matrix P such that $P'P = PP' = I$ and such that $P'CP$ is a diagonal matrix, say

$$P'CP = \Lambda = \begin{bmatrix} \lambda_1 & & 0 \\ & \ddots & \\ 0 & & \lambda_n \end{bmatrix}.$$

Next, from $P'CP = \Lambda$, we can easily obtain $C = P\Lambda P'$. Since the determinant of a product of matrices is the product of their determinants, $\det C = \det P \det \Lambda \det P'$. Since the determinants are numbers, they can be multiplied in any order. Thus,

$$\begin{aligned} \det C &= \det \Lambda \det P' \det P \\ &= \det \Lambda \det(P'P) \\ &= \det \Lambda \det I \\ &= \det \Lambda \\ &= \lambda_1 \cdots \lambda_n. \end{aligned}$$

Rewrite $P'CP = \Lambda$ as $CP = P\Lambda$. Then it is easy to see that the columns of P are eigenvectors of C; i.e., if P has columns $p_1, \ldots, p_n$, then $Cp_i = \lambda_i p_i$. Next, since $P'P = I$, each p_i satisfies $p_i' p_i = 1$. Since C is positive definite,

$$0 < p_i' C p_i = p_i'(\lambda_i p_i) = \lambda_i p_i' p_i = \lambda_i.$$

Thus, each eigenvalue $\lambda_i > 0$, and it follows that $\det C = \lambda_1 \cdots \lambda_n > 0$.

Because positive-semidefinite matrices are diagonalizable with nonnegative eigenvalues, it is easy to define their **square root** by

$$\sqrt{C} := P\sqrt{\Lambda} P',$$

where

$$\sqrt{\Lambda} := \begin{bmatrix} \sqrt{\lambda_1} & & 0 \\ & \ddots & \\ 0 & & \sqrt{\lambda_n} \end{bmatrix}.$$

Thus, $\det \sqrt{C} = \sqrt{\lambda_1} \cdots \sqrt{\lambda_n} = \sqrt{\det C}$. Furthermore, from the definition of $\sqrt{C}$, it is clear that it is positive semidefinite and satisfies $\sqrt{C}\sqrt{C} = C$. We also point out that since $C = P\Lambda P'$, if C is positive definite, then $C^{-1} = P\Lambda^{-1}P'$, where Λ^{-1} is diagonal with diagonal entries $1/\lambda_i$; hence, $\sqrt{C^{-1}} = (\sqrt{C})^{-1}$. Finally, note that

$$\sqrt{C} C^{-1} \sqrt{C} = (P\sqrt{\Lambda}P')(P\Lambda^{-1}P')(P\sqrt{\Lambda}P') = I.$$

Problems

9.1: Introduction

1. Evaluate
$$f(x) = \frac{\exp[-\frac{1}{2}(x-m)'C^{-1}(x-m)]}{(2\pi)^{n/2}\sqrt{\det C}}$$

if $m = 0$ and

$$C = \begin{bmatrix} \sigma_1^2 & \sigma_1 \sigma_2 \rho \\ \sigma_1 \sigma_2 \rho & \sigma_2^2 \end{bmatrix},$$

where $|\rho| < 1$. Show that your result has the same form as the bivariate normal density in (7.25).

9.2: Definition of the multivariate Gaussian

2. MATLAB. Let X be a constant, scalar random variable taking the value m. It is easy to see that $F_X(x) = u(x-m)$, where u is the unit step function. It then follows that $f_X(x) = \delta(x-m)$. Use the following MATLAB code to plot the $N(0, 1/n^2)$ density for $n = 1, 2, 3, 4$ to demonstrate that as the variance of a Gaussian goes to zero, the density approaches an impulse; in other words, a constant random variable can be viewed as the limiting case of the ordinary Gaussian.

```
x=linspace(-3.5,3.5,200);
s = 1;      y1 = exp(-x.*x/(2*s))/sqrt(2*pi*s);
s = 1/4;    y2 = exp(-x.*x/(2*s))/sqrt(2*pi*s);
s = 1/9;    y3 = exp(-x.*x/(2*s))/sqrt(2*pi*s);
s = 1/16;   y4 = exp(-x.*x/(2*s))/sqrt(2*pi*s);
plot(x,y1,x,y2,x,y3,x,y4)
```

3. Let $X \sim N(0,1)$ and put $Y := 3X$.

   (a) Show that X and Y are jointly Gaussian.

   (b) Find their covariance matrix, $\text{cov}([X,Y]')$.

   (c) Show that they are not jointly continuous. *Hint:* Show that the conditional cdf of Y given $X = x$ is a unit-step function, and hence, the conditional density is an impulse.

4. If $X_1, \ldots, X_n$ are independent with $X_i \sim N(m_i, \sigma_i^2)$, show that $X = [X_1, \ldots, X_n]'$ is a Gaussian random vector by showing that for any coefficients c_i, $\sum_{i=1}^n c_i X_i$ is a scalar Gaussian random variable.

5. Let $X = [X_1, \ldots, X_n]' \sim N(m, C)$, and suppose that $Y = AX + b$, where A is a $p \times n$ matrix, and $b \in \mathbb{R}^p$. Find the mean vector and covariance matrix of Y.

6. Let $X_1, \ldots, X_n$ be random variables, and define

$$Y_k := \sum_{i=1}^k X_i, \quad k = 1, \ldots, n.$$

Suppose that $Y_1, \ldots, Y_n$ are jointly Gaussian. Determine whether or not $X_1, \ldots, X_n$ are jointly Gaussian.

7. If X is a zero-mean, multivariate Gaussian with covariance matrix C, show that

$$E[(v'XX'v)^k] = (2k-1)(2k-3)\cdots 5 \cdot 3 \cdot 1 \cdot (v'Cv)^k.$$

Hint: Example 4.11.

8. Let $X_1, \ldots, X_n$ be i.i.d. $N(m, \sigma^2)$ random variables, and denote the average of the X_i by $\overline{X} := \frac{1}{n}\sum_{i=1}^n X_i$. For $j = 1, \ldots, n$, put $Y_j := X_j - \overline{X}$. Show that $E[Y_j] = 0$ and that $E[\overline{X}Y_j] = 0$ for $j = 1, \ldots, n$.

9. **Wick's theorem.** Let $X \sim N(0,C)$ be n-dimensional. Let $(i_1,\ldots,i_{2k})$ be a vector of indices chosen from $\{1,\ldots,n\}$. Repetitions are allowed; e.g., $(1,3,3,4)$. Derive **Wick's theorem**,

$$\mathsf{E}[X_{i_1}\cdots X_{i_{2k}}] = \sum_{j_1,\ldots,j_{2k}} C_{j_1 j_2}\cdots C_{j_{2k-1}j_{2k}},$$

where the sum is over all $j_1,\ldots,j_{2k}$ that are permutations of $i_1,\ldots,i_{2k}$ *and* such that the product $C_{j_1 j_2}\cdots C_{j_{2k-1}j_{2k}}$ is distinct. *Hint:* The idea is to view both sides of the equation derived in Problem 7 as a multivariate polynomial in the n variables $v_1,\ldots,v_n$. After collecting all terms on each side that involve $v_{i_1}\cdots v_{i_{2k}}$, the corresponding coefficients must be equal. In the expression

$$\mathsf{E}[(v'X)^{2k}] = \mathsf{E}\left[\left(\sum_{j_1=1}^{n} v_{j_1} X_{j_1}\right)\cdots\left(\sum_{j_{2k}=1}^{n} v_{j_{2k}} X_{j_{2k}}\right)\right]$$

$$= \sum_{j_1=1}^{n}\cdots\sum_{j_{2k}=1}^{n} v_{j_1}\cdots v_{j_{2k}} \mathsf{E}[X_{j_1}\cdots X_{j_{2k}}],$$

we are only interested in those terms for which $j_1,\ldots,j_{2k}$ is a permutation of $i_1,\ldots,i_{2k}$. There are $(2k)!$ such terms, each equal to

$$v_{i_1}\cdots v_{i_{2k}} \mathsf{E}[X_{i_1}\cdots X_{i_{2k}}].$$

Similarly, from

$$(v'Cv)^k = \left(\sum_{i=1}^{n}\sum_{j=1}^{n} v_i C_{ij} v_j\right)^k$$

we are only interested in terms of the form

$$v_{j_1} v_{j_2}\cdots v_{j_{2k-1}} v_{j_{2k}} C_{j_1 j_2}\cdots C_{j_{2k-1} j_{2k}},$$

where $j_1,\ldots,j_{2k}$ is a permutation of $i_1,\ldots,i_{2k}$. Now many of these permutations involve the same value of the product $C_{j_1 j_2}\cdots C_{j_{2k-1}j_{2k}}$. First, because C is symmetric, each factor C_{ij} also occurs as C_{ji}. This happens in 2^k different ways. Second, the order in which the C_{ij} are multiplied together occurs in $k!$ different ways.

10. Let X be a multivariate normal random vector with covariance matrix C. Use Wick's theorem of the previous problem to evaluate $\mathsf{E}[X_1 X_2 X_3 X_4]$, $\mathsf{E}[X_1 X_3^2 X_4]$, and $\mathsf{E}[X_1^2 X_2^2]$.

9.3: Characteristic function

11. Let X be a random vector with joint characteristic function $\varphi_X(v) = e^{jv'm - v'Cv/2}$. For any coefficients a_i, put $Y := \sum_{i=1}^{n} a_i X_i$. Show that $\varphi_Y(\eta) = \mathsf{E}[e^{j\eta Y}]$ has the form of the characteristic function of a scalar Gaussian random variable.

12. Let $X = [X_1,\ldots,X_n]' \sim N(m,C)$, and suppose C is block diagonal, say

$$C = \begin{bmatrix} S & 0 \\ 0 & T \end{bmatrix},$$

where S and T are square submatrices with S being $s \times s$ and T being $t \times t$ with $s+t = n$. Put $U := [X_1, \ldots, X_s]'$ and $W := [X_{s+1}, \ldots, X_n]'$. Show that U and W are independent. *Hint:* It is enough to show that

$$\varphi_X(v) = \varphi_U(v_1, \ldots, v_s) \varphi_W(v_{s+1}, \ldots, v_n),$$

where φ_U is an s-variate normal characteristic function, and φ_W is a t-variate normal characteristic function. Use the notation $\alpha := [v_1, \ldots, v_s]'$ and $\beta := [v_{s+1}, \ldots, v_n]'$.

9.4: Density function

13. The digital signal processing chip in a wireless communication receiver generates the n-dimensional Gaussian vector X with mean zero and positive-definite covariance matrix C. It then computes the vector $Y = C^{-1/2}X$. (Since $C^{-1/2}$ is invertible, there is no loss of information in applying such a transformation.) Finally, the decision statistic $V = \|Y\|^2 := \sum_{k=1}^n Y_k^2$ is computed.

    (a) Find the multivariate density of Y.
    (b) Find the density of Y_k^2 for $k = 1, \ldots, n$.
    (c) Find the density of V.

14. Let X and Y be independent $N(0,1)$ random variables. Find the density of

$$Z := \det \begin{bmatrix} X & -Y \\ Y & X \end{bmatrix}.$$

15. Review the derivation of (9.3) in Note **1**. Using similar techniques, show directly that

$$\frac{1}{(2\pi)^n} \int_{\mathbb{R}^n} e^{-jv'x} e^{jv'm - v'Cv/2} \, dv = \frac{\exp[-\frac{1}{2}(x-m)'C^{-1}(x-m)]}{(2\pi)^{n/2}\sqrt{\det C}}.$$

9.5: Conditional expectation and conditional probability

16. Let X, Y, U, and V be jointly Gaussian with X and Y independent $N(0,1)$. Put

$$Z := \det \begin{bmatrix} X & Y \\ U & V \end{bmatrix}.$$

If $[X,Y]'$ and $[U,V]'$ are uncorrelated random vectors, find the conditional density $f_{Z|UV}(z|u,v)$.

17. Let X and Y be jointly normal random vectors, and let the matrix A solve $AC_Y = C_{XY}$. Show that given $Y = y$, X is conditionally $N(m_X + A(y - m_Y), C_X - AC_{YX})$. *Hints:* First note that $(X - m_X) - A(Y - m_Y)$ and Y are uncorrelated and therefore independent by Problem 12. Next, observe that $\mathsf{E}[e^{jv'X}|Y = y]$ is equal to

$$\mathsf{E}\left[e^{jv'[(X-m_X)-A(Y-m_Y)]} e^{jv'[m_X+A(Y-m_Y)]} \Big| Y = y\right].$$

Now use substitution on the right-hand exponential, but not the left. Observe that $(X - m_X) - A(Y - m_Y)$ is a zero-mean Gaussian random vector whose covariance matrix you can easily find; then write out its characteristic function.

18. Let X, Y, U, and V be jointly Gaussian with zero means. Assume that X and Y are independent $N(0,1)$. Suppose

$$Z := \det \begin{bmatrix} X & Y \\ U & V \end{bmatrix}.$$

Find the conditional density $f_{Z|UV}(z|u,v)$. Show that if $[X,Y]'$ and $[U,V]'$ are uncorrelated, then your answer reduces to that of Problem 16. *Hint:* Problem 17 may be helpful.

9.6: Complex random variables and vectors

19. Show that for a complex random variable $Z = X + jY$, $\text{cov}(Z) = \text{var}(X) + \text{var}(Y)$.

20. Consider the complex random vector $Z = X + jY$ with covariance matrix K.

    (a) Show that $K = (C_X + C_Y) + j(C_{YX} - C_{XY})$.

    (b) If the circular symmetry conditions $C_X = C_Y$ and $C_{XY} = -C_{YX}$ hold, show that the diagonal elements of C_{XY} are zero; i.e., for each i, the components X_i and Y_i are uncorrelated.

    (c) If the circular symmetry conditions hold, and if K is a real matrix, show that X and Y are uncorrelated.

21. Let X and Y be real, n-dimensional $N(0, \frac{1}{2}I)$ random vectors that are independent of each other. Write out the densities $f_X(x)$, $f_Y(y)$, and $f_{XY}(x,y) = f_X(x)f_Y(y)$. Compare the joint density with

$$\frac{e^{-(x+jy)^H(x+jy)}}{\pi^n}.$$

22. Let Z be a complex random vector with covariance matrix $K = R + jQ$ for real matrices R and Q.

    (a) Show that $R = R'$ and that $Q' = -Q$.

    (b) If $Q' = -Q$, show that $v'Qv = 0$.

    (c) If $w = v + j\theta$, show that

$$w^H K w = v'Rv + \theta'R\theta + 2\theta'Qv.$$

23. Let $Z = X + jY$ be a complex random vector, and let $A = \alpha + j\beta$ be a complex matrix. Show that the transformation $Z \mapsto AZ$ is equivalent to

$$\begin{bmatrix} X \\ Y \end{bmatrix} \mapsto \begin{bmatrix} \alpha & -\beta \\ \beta & \alpha \end{bmatrix} \begin{bmatrix} X \\ Y \end{bmatrix}.$$

Hence, multiplying an n-dimensional complex random vector by an $n \times n$ complex matrix is a linear transformation of the $2n$-dimensional vector $[X', Y']'$. Now show that such a transformation preserves circular symmetry; i.e., if Z is circularly symmetric, then so is AZ.

24. Consider the complex random vector Θ partitioned as

$$\Theta = \begin{bmatrix} Z \\ W \end{bmatrix} = \begin{bmatrix} X+jY \\ U+jV \end{bmatrix},$$

where X, Y, U, and V are appropriately-sized, real random vectors. Since every complex random vector is identified with a real random vector of twice the length, it is convenient to put $\widetilde{Z} := [X', Y']'$ and $\widetilde{W} := [U', V']'$. Since the real and imaginary parts of Θ are $R := [X', U']'$ and $I := [Y', V']'$, we put

$$\widetilde{\Theta} := \begin{bmatrix} R \\ I \end{bmatrix} = \begin{bmatrix} X \\ U \\ Y \\ V \end{bmatrix}.$$

Assume that Θ is Gaussian and circularly symmetric.

(a) Show that $K_{ZW} = 0$ if and only if $C_{\widetilde{Z}\widetilde{W}} = 0$.

(b) Show that the complex matrix $A = \alpha + j\beta$ solves $AK_W = K_{ZW}$ if and only if

$$\widetilde{A} := \begin{bmatrix} \alpha & -\beta \\ \beta & \alpha \end{bmatrix}$$

solves $\widetilde{A}C_{\widetilde{W}} = C_{\widetilde{Z}\widetilde{W}}$.

(c) If A solves $AK_W = K_{ZW}$, show that given $W = w$, Z is conditionally Gaussian and circularly symmetric $N(m_Z + A(w - m_W), K_Z - AK_{WZ})$. *Hint:* Problem 17.

25. Let $Z = X + jY$ have density $f_Z(z) = e^{-|z|^2}/\pi$ as discussed in the text.

(a) Find $\text{cov}(Z)$.

(b) Show that $2|Z|^2$ has a chi-squared density with 2 degrees of freedom.

26. Let $X \sim N(m_r, 1)$ and $Y \sim N(m_i, 1)$ be independent, and define the complex random variable $Z := X + jY$. Use the result of Problem 25 in Chapter 5 to show that $|Z|$ has the Rice density.

27. The base station of a wireless communication system generates an n-dimensional, complex, circularly symmetric, Gaussian random vector Z with mean zero and covariance matrix K. Let $W = K^{-1/2}Z$.

(a) Find the density of W.

(b) Let $W_k = U_k + jV_k$. Find the joint density of the pair of real random variables (U_k, V_k).

(c) If

$$\|W\|^2 := \sum_{k=1}^{n} |W_k|^2 = \sum_{k=1}^{n} U_k^2 + V_k^2,$$

show that $2\|W\|^2$ has a chi-squared density with $2n$ degrees of freedom.

Remark. (*i*) The chi-squared density with $2n$ degrees of freedom is the same as the n-Erlang density, whose cdf has the closed-form expression given in Problem 15(c) in Chapter 4. (*ii*) By Problem 19 in Chapter 5, $\sqrt{2}\,\|W\|$ has a Nakagami-n density with parameter $\lambda = 1$.

28. Let M be a real symmetric matrix such that $u'Mu = 0$ for all real vectors u.

    (a) Show that $v'Mu = 0$ for all real vectors u and v. *Hint:* Consider the quantity $(u+v)'M(u+v)$.

    (b) Show that $M = 0$. *Hint:* Note that $M = 0$ if and only if $Mu = 0$ for all u, and $Mu = 0$ if and only if $\|Mu\| = 0$.

29. Show that if (9.10) is equal to $w^H K w/2$ for all $w = v + j\theta$, then (9.11) holds. *Hint:* Use the result of the preceding problem.

30. Assume that circular symmetry (9.11) holds. In this problem you will show that (9.13) reduces to (9.14).

    (a) Show that $\det \Gamma = (\det K)^2 / 2^{2n}$. *Hint:*

    $$\det(2\Gamma) = \det \begin{bmatrix} 2C_X & -2C_{YX} \\ 2C_{YX} & 2C_X \end{bmatrix}$$

    $$= \det \begin{bmatrix} 2C_X + j2C_{YX} & -2C_{YX} \\ 2C_{YX} - j2C_X & 2C_X \end{bmatrix}$$

    $$= \det \begin{bmatrix} K & -2C_{YX} \\ -jK & 2C_X \end{bmatrix}$$

    $$= \det \begin{bmatrix} K & -2C_{YX} \\ 0 & K^H \end{bmatrix} = (\det K)^2.$$

    Remark. Thus, Γ is invertible if and only if K is invertible.

    (b) **Matrix inverse formula.** For any matrices A, B, C, and D, let $V = A + BCD$. If A and C are invertible, show that

    $$V^{-1} = A^{-1} - A^{-1}B(C^{-1} + DA^{-1}B)^{-1}DA^{-1}$$

    by verifying that the formula for V^{-1} satisfies $VV^{-1} = I$.

    (c) Show that

    $$\Gamma^{-1} = \begin{bmatrix} \Delta^{-1} & C_X^{-1}C_{YX}\Delta^{-1} \\ -\Delta^{-1}C_{YX}C_X^{-1} & \Delta^{-1} \end{bmatrix},$$

    where $\Delta := C_X + C_{YX}C_X^{-1}C_{YX}$, by verifying that $\Gamma\Gamma^{-1} = I$. *Hint:* Note that Δ^{-1} satisfies

    $$\Delta^{-1} = C_X^{-1} - C_X^{-1}C_{YX}\Delta^{-1}C_{YX}C_X^{-1}.$$

    (d) Show that $K^{-1} = (\Delta^{-1} - jC_X^{-1}C_{YX}\Delta^{-1})/2$ by verifying that $KK^{-1} = I$.

    (e) Show that (9.13) and (9.14) are equal. *Hint:* Using the equation for Δ^{-1} given in part (c), it can be shown that $C_X^{-1}C_{YX}\Delta^{-1} = \Delta^{-1}C_{YX}C_X^{-1}$. Selective application of this formula may be helpful.

Exam preparation

You may use the following suggestions to prepare a study sheet, including formulas mentioned that you have trouble remembering. You may also want to ask your instructor for additional suggestions.

9.1. Introduction. Know formula (9.2) for the density of the n-dimensional Gaussian random vector with mean vector m and covariance matrix C. Also know its joint characteristic function is $e^{jv'm - v'Cv/2}$; hence, *a Gaussian random vector is completely determined by its mean vector and covariance matrix*.

9.2. Definition of the multivariate Gaussian. Know key facts about Gaussian random vectors:

1. It is possible for X and Y to be jointly Gaussian, but not jointly continuous (Problem 3).
2. Linear transformations of Gaussian random vectors are Gaussian.
3. In particular, any subvector of a Gaussian vector is Gaussian; i.e., marginals of Gaussian vectors are also Gaussian.
4. In general, just because X is Gaussian and Y is Gaussian, it does not follow that X and Y are jointly Gaussian, even if they are uncorrelated. See Problem 51 in Chapter 7.
5. A vector of independent Gaussians is jointly Gaussian.

9.3. Characteristic function. Know the formula for the Gaussian characteristic function. We used it to show that *if the components of a Gaussian random vector are uncorrelated, they are independent*.

9.4. Density function. Know the formula for the n-dimensional Gaussian density function.

9.5. Conditional expectation and conditional probability. If X and Y are jointly Gaussian then $\mathsf{E}[X|Y=y] = A(Y - m_Y) + m_X$, where A solves $AC_Y = C_{XY}$; more generally, the conditional distribution of X given $Y = y$ is Gaussian with mean $A(y - m_Y) + m_X$ and covariance matrix $C_X - AC_{YX}$ as shown in Problem 17.

9.6. Complex random variables and vectors. An n-dimensional complex random vector $Z = X + jY$ is shorthand for the $2n$-dimensional real vector $[X', Y']'$. The covariance matrix of $[X', Y']'$ has the form

$$\begin{bmatrix} C_X & C_{XY} \\ C_{YX} & C_Y \end{bmatrix}. \quad (9.15)$$

In general, knowledge of the covariance matrix of Z,

$$K = (C_X + C_Y) + j(C_{YX} - C_{XY}),$$

is not sufficient to determine (9.15). However, if circular symmetry holds, i.e., if $C_X = C_Y$ and $C_{XY} = -C_{YX}$, then K and (9.15) are equivalent. If X and Y are jointly Gaussian and circularly symmetric, then the joint characteristic function and joint density can be written easily in complex notation, e.g., (9.12) and (9.14).

Work any review problems assigned by your instructor. If you finish them, re-work your homework assignments.

10
Introduction to random processes[†]

10.1 Definition and examples

A **random process** or **stochastic process** is a family of random variables. In principle this could refer to a finite family of random variables such as $\{X,Y,Z\}$, but in practice the term usually refers to infinite families. The need for working with infinite families of random variables arises when we have an indeterminate amount of data to model. For example, in sending bits over a wireless channel, there is no set number of bits to be transmitted. To model this situation, we use an infinite sequence of random variables. As another example, the signal strength in a cell-phone receiver varies continuously over time in a random manner depending on location. To model this requires that the random signal strength depend on the continuous-time index t. More detailed examples are discussed below.

Discrete-time processes

A **discrete-time random process** is a family of random variables $\{X_n\}$ where n ranges over a specified subset of the integers. For example, we might have

$$\{X_n, n = 1, 2, \ldots\}, \quad \{X_n, n = 0, 1, 2, \ldots\}, \quad \text{or} \quad \{X_n, n = 0, \pm 1, \pm 2, \ldots\}.$$

Recalling that random variables are functions defined on a sample space Ω, we can think of $X_n(\omega)$ in two ways. First, for fixed n, $X_n(\omega)$ is a function of ω and therefore a random variable. Second, for fixed ω we get a sequence of numbers $X_1(\omega), X_2(\omega), X_3(\omega), \ldots$. Such a sequence is called a **realization**, **sample path**, or **sample function** of the random process.

Example 10.1 (sending bits over a noisy channel). In sending a sequence of bits over a noisy channel, bits are flipped independently with probability p. Let $X_n = 1$ if the nth bit is flipped and $X_n = 0$ otherwise. Then $\{X_n, n = 1, 2, \ldots\}$ is an i.i.d. Bernoulli(p) sequence. Three realizations of $X_1, X_2, \ldots$ are shown in Figure 10.1.

As the preceding example shows, a random process can be composed of discrete random variables. The next example shows that a random process can be composed of continuous random variables.

Example 10.2 (sampling thermal noise in an amplifier). Consider the amplifier of a radio receiver. Because all amplifiers internally generate thermal noise, even if the radio is not receiving any signal, the voltage at the output of the amplifier is not zero but is well modeled as a Gaussian random variable each time it is measured. Suppose we measure this voltage once per second and denote the nth measurement by Z_n. Three realizations of $Z_1, Z_2, \ldots$ are shown in Figure 10.2.

[†]The material in this chapter can be covered any time after Chapter 7. No background on random vectors from Chapters 8 or 9 is assumed.

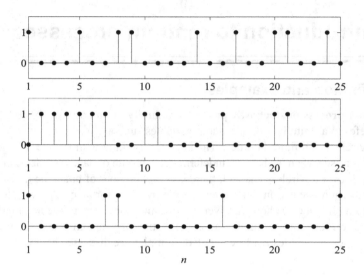

Figure 10.1. Three realizations of an i.i.d. sequence of Bernoulli(p) random variables $\{X_n, n = 1, 2, \ldots\}$.

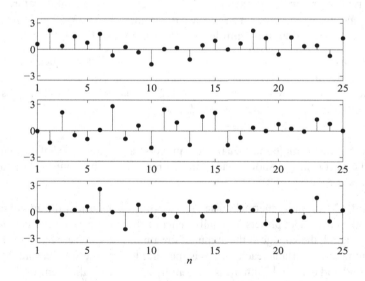

Figure 10.2. Three realizations of an i.i.d. sequence of $N(0,1)$ random variables $\{Z_n, n = 1, 2, \ldots\}$.

10.1 Definition and examples

Example **10.3** (effect of amplifier noise on a signal). Suppose that the amplifier of the preceding example has a gain of 5 and the input signal $\sin(2\pi f t)$ is applied. When we sample the amplifier output once per second, we get $5\sin(2\pi f n) + Z_n$. Three realizations of this process are shown in Figure 10.3.

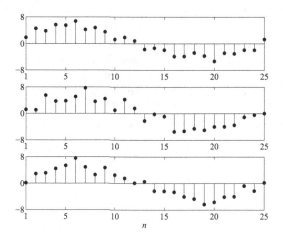

Figure **10.3.** Three realizations of $5\sin(2\pi f n) + Z_n$, where $f = 1/25$. The realizations of Z_n in this figure are the same as those in Figure 10.2.

Example **10.4** (filtering of random signals). Suppose the amplifier noise samples Z_n are applied to a simple digital signal processing chip that computes $Y_n = \frac{1}{2}Y_{n-1} + Z_n$ for $n = 1, 2, \ldots$, where $Y_0 \equiv 0$. Three realizations of $Y_1, Y_2, \ldots$ are shown in Figure 10.4.

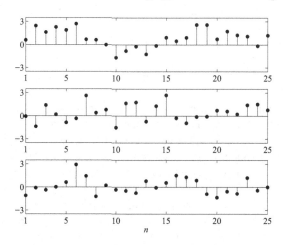

Figure **10.4.** Three realizations of $Y_n = \frac{1}{2}Y_{n-1} + Z_n$, where $Y_0 \equiv 0$. The realizations of Z_n in this figure are the same as those in Figure 10.2.

Continuous-time processes

A **continuous-time random process** is a family of random variables $\{X_t\}$ where t ranges over a specified interval of time. For example, we might have

$$\{X_t, t \geq 0\}, \quad \{X_t, 0 \leq t \leq T\}, \quad \text{or} \quad \{X_t, -\infty < t < \infty\}.$$

Example 10.5 (carrier with random phase). In radio communications, the carrier signal is often modeled as a sinusoid with a random phase. The reason for using a random phase is that the receiver does not know the time when the transmitter was turned on or the distance from the transmitter to the receiver. The mathematical model for this is the continuous-time random process defined by $X_t := \cos(2\pi f t + \Theta)$, where f is the carrier frequency and $\Theta \sim \text{uniform}[-\pi, \pi]$. Three realizations of this process are shown in Figure 10.5.

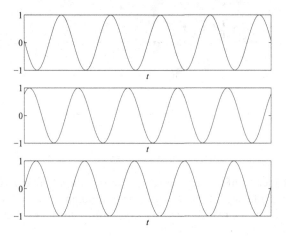

Figure 10.5. Three realizations of the carrier with random phase, $X_t := \cos(2\pi f t + \Theta)$. The three different values of Θ are 1.5, -0.67, and -1.51, top to bottom, respectively.

Example 10.6 (counting processes). In a counting process $\{N_t, t \geq 0\}$, N_t counts the number of occurrences of some quantity that have happened up to time t (including any event happening exactly at time t). We could count the number of hits to a website up to time t, the number of radioactive particles emitted from a mass of uranium, the number of packets arriving at an Internet router, the number of photons detected by a powerful telescope, etc. Three realizations of a counting process are shown in Figure 10.6. The times at which the graph jumps are the times at which something is counted. For the sake of illustration, suppose that N_t counts the number of packets arriving at an Internet router. We see from the figure that in the top realization, the first packet arrives at time $t = 0.8$, the second packet arrives at time $t = 2$, etc. In the middle realization, the first packet arrives at time $t = 0.5$ and the second packet arrives at time $t = 1$. In the bottom realization, the first packet does not arrive until time $t = 2.1$ and the second arrives soon after at time $t = 2.3$.

10.1 Definition and examples

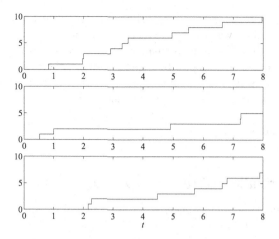

Figure **10.6.** Three realizations of a counting process N_t.

Example **10.7** (Brownian motion or the Wiener process). In 1827, Robert Brown observed that small particles in a liquid were continually in motion and followed erratic paths. A simulation of such a path is shown at the upper left in Figure 10.7. Wiggly paths of this kind are called **Brownian motion**. Let us denote the position of a particle at time t by (X_t, Y_t). A plot of Y_t as a function of time is shown at the right in Figure 10.7. The dashed horizontal lines point out that the maximum vertical position occurs at the final time $t = 1$ and the minimum vertical position occurs at time $t = 0.46$. Similarly, X_t is plotted at the lower left. Note that the vertical axis is time and the horizontal axis is X_t. The dashed vertical lines show that right-most horizontal position occurs at time $t = 0.96$ and the left-most

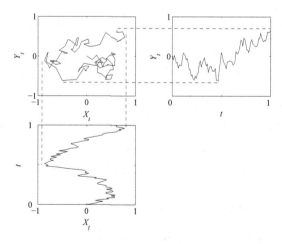

Figure **10.7.** The two-dimensional Brownian motion (X_t, Y_t) is shown in the upper-left plot; the curve starts in the center of the plot at time $t = 0$ and ends at the upper right of the plot at time $t = 1$. The vertical component Y_t as a function of time is shown in the upper-right plot. The horizontal component X_t as a function of time is shown in the lower-left plot; *note here that the vertical axis is time and the horizontal axis is X_t.*

horizontal position occurs at time $t = 0.52$. The random paths observed by Robert Brown are physical phenomena. It was Norbert Wiener who established the existence of random processes X_t and Y_t as well-defined mathematical objects. For this reason, a process such as X_t or Y_t is called a **Wiener process** or a **Brownian motion process**.

Today Wiener processes arise in many different areas. Electrical engineers use them to model integrated white noise in communication and control systems, computer engineers use them to study heavy traffic in Internet routers, and economists use them to model the stock market and options trading.

10.2 Characterization of random processes

For a single random variable X, once we know its pmf or density, we can write down a sum or integral expression for $P(X \in B)$ or $E[g(X)]$ for any set B or function g. Similarly, for a pair of random variables (X,Y), we can write down a sum or integral expression for $P((X,Y) \in A)$ or $E[h(X,Y)]$ for any two-dimensional set A or bivariate function h. More generally, for any finite number of random variables, once we know the joint pmf or density, we can write down expressions for any probability or expectation that arises.

When considering more than finitely many random variables, Kolmogorov showed that a random process X_t is completely characterized once we say how to compute, for every $1 \leq n < \infty$,

$$P((X_{t_1}, \ldots, X_{t_n}) \in B)$$

for arbitrary n-dimensional sets B and distinct times $t_1, \ldots, t_n$. The precise result is discussed in more detail in Chapter 11.

In most real-world problems, we are not told the joint densities or pmfs of all relevant random variables. We have to estimate this information from data. We saw in Chapter 6 how much work it was to estimate $E[X]$ or $f_X(x)$ from data. Imagine trying to estimate an unending sequence of joint densities, $f_{X_1}(x_1)$, $f_{X_1 X_2}(x_1, x_2)$, $f_{X_1 X_2 X_3}(x_1, x_2, x_3)$, ... Hence, in practical problems, we may have to make due with partial characterizations. In the case of a single random variable, we may know only the mean and variance. For a pair of dependent random variables X and Y, we may know only the means, variances, and correlation $E[XY]$. We now present the analogous quantities for random processes.

Mean and correlation functions

If X_t is a random process, then for every value of t, X_t is a random variable with mean $E[X_t]$. We call

$$m_X(t) := E[X_t] \qquad (10.1)$$

the **mean function** of the process. The mean function reflects the average behavior of the process with time.

If X_{t_1} and X_{t_2} are two random variables of a process X_t, their correlation is denoted by

$$R_X(t_1, t_2) := E[X_{t_1} X_{t_2}]. \qquad (10.2)$$

10.2 Characterization of random processes

When regarded as a function of the times t_1 and t_2, we call $R_X(t_1,t_2)$ the **correlation function** of the process. The correlation function reflects how smooth or wiggly a process is.

***Example* 10.8.** In a communication system, the carrier signal at the receiver is modeled by $X_t = \cos(2\pi f t + \Theta)$, where $\Theta \sim$ uniform$[-\pi, \pi]$. Find the mean function and the correlation function of X_t.

Solution. For the mean, write

$$\begin{aligned} \mathsf{E}[X_t] &= \mathsf{E}[\cos(2\pi f t + \Theta)] \\ &= \int_{-\infty}^{\infty} \cos(2\pi f t + \theta) f_\Theta(\theta) \, d\theta \\ &= \int_{-\pi}^{\pi} \cos(2\pi f t + \theta) \frac{d\theta}{2\pi}. \end{aligned}$$

Be careful to observe that this last integral is with respect to θ, *not* t. Hence, this integral evaluates to zero.

For the correlation, first write

$$R_X(t_1,t_2) = \mathsf{E}[X_{t_1} X_{t_2}] = \mathsf{E}\big[\cos(2\pi f t_1 + \Theta)\cos(2\pi f t_2 + \Theta)\big].$$

Then use the trigonometric identity

$$\cos A \cos B = \tfrac{1}{2}[\cos(A+B) + \cos(A-B)] \qquad (10.3)$$

to write

$$R_X(t_1,t_2) = \tfrac{1}{2}\mathsf{E}\big[\cos(2\pi f[t_1+t_2] + 2\Theta) + \cos(2\pi f[t_1-t_2])\big].$$

The first cosine has expected value zero just as the mean did. The second cosine is nonrandom, and therefore equal to its expected value. Thus, $R_X(t_1,t_2) = \cos(2\pi f[t_1 - t_2])/2$.

***Example* 10.9.** Find the correlation function of

$$X_n := Z_1 + \cdots + Z_n, \quad n = 1, 2, \ldots,$$

if the Z_i are zero-mean and uncorrelated with common variance $\sigma^2 := \text{var}(Z_i)$ for all i.

Solution. For $m > n$, observe that

$$X_m = \underbrace{Z_1 + \cdots + Z_n}_{X_n} + Z_{n+1} + \cdots + Z_m.$$

Then write

$$\begin{aligned} \mathsf{E}[X_n X_m] &= \mathsf{E}\bigg[X_n\bigg(X_n + \sum_{i=n+1}^{m} Z_i\bigg)\bigg] \\ &= \mathsf{E}[X_n^2] + \mathsf{E}\bigg[X_n\bigg(\sum_{i=n+1}^{m} Z_i\bigg)\bigg]. \end{aligned}$$

To analyze the first term on the right, observe that since the Z_i are zero mean, so is X_n. Also, X_n is the sum of uncorrelated random variables. Hence,

$$E[X_n^2] = \text{var}(X_n) = \sum_{i=1}^{n} \text{var}(Z_i) = n\sigma^2,$$

since the variance of the sum of uncorrelated random variables is the sum of the variances (recall (2.28)). To analyze the remaining expectation, write

$$E\left[X_n\left(\sum_{i=n+1}^{m} Z_i\right)\right] = E\left[\left(\sum_{j=1}^{n} Z_j\right)\left(\sum_{i=n+1}^{m} Z_i\right)\right]$$

$$= \sum_{j=1}^{n}\sum_{i=n+1}^{m} E[Z_j Z_i]$$

$$= 0$$

since in the double sum $i \neq j$, and since the Z_i are uncorrelated with zero mean. We can now write $E[X_n X_m] = \sigma^2 n$ for $m > n$. Since we can always write $E[X_n X_m] = E[X_m X_n]$, it follows that the general result is

$$R_X(n,m) = E[X_n X_m] = \sigma^2 \min(n,m), \quad n,m \geq 1.$$

Example 10.10. In the preceding example, if the Z_i are i.i.d. $N(0,\sigma^2)$ random variables, then X_n is an $N(0,\sigma^2 n)$ random variable by Problem 55(a) in Chapter 4. For $1 \leq k < l \leq n < m$, the **increments** $X_l - X_k$ and $X_m - X_n$ are independent with

$$X_l - X_k \sim N(0,\sigma^2(l-k)) \quad \text{and} \quad X_m - X_n \sim N(0,\sigma^2(m-n)).$$

After studying the properties of the continuous-time Wiener process in Chapter 11, it will be evident that X_n is the discrete-time analog of the Wiener process.

Example 10.11. Let X_t be a random process with mean function $m_X(t)$. Suppose that X_t is applied to a **linear time-invariant** (LTI) **system** with **impulse response** $h(t)$. Find the mean function of the output process[a]

$$Y_t = \int_{-\infty}^{\infty} h(t-\theta) X_\theta \, d\theta.$$

Solution. To begin, write

$$E[Y_t] = E\left[\int_{-\infty}^{\infty} h(t-\theta) X_\theta \, d\theta\right] = \int_{-\infty}^{\infty} E[h(t-\theta) X_\theta] \, d\theta,$$

---

[a] The precise definition of an integral of a random process is given in Chapter 13.

10.2 Characterization of random processes

where the interchange of expectation and integration is heuristically justified by writing the integral as a Riemann sum and appealing to the linearity of expectation; i.e.,

$$\mathsf{E}\left[\int_{-\infty}^{\infty} h(t-\theta)X_\theta\, d\theta\right] \approx \mathsf{E}\left[\sum_i h(t-\theta_i)X_{\theta_i}\Delta\theta_i\right]$$
$$= \sum_i \mathsf{E}[h(t-\theta_i)X_{\theta_i}\Delta\theta_i]$$
$$= \sum_i \mathsf{E}[h(t-\theta_i)X_{\theta_i}]\Delta\theta_i$$
$$\approx \int_{-\infty}^{\infty} \mathsf{E}[h(t-\theta)X_\theta]\, d\theta.$$

To evaluate this last expectation, note that X_θ is a random variable, while for each fixed t and θ, $h(t-\theta)$ is just a nonrandom constant that can be pulled out of the expectation. Thus,

$$\mathsf{E}[Y_t] = \int_{-\infty}^{\infty} h(t-\theta)\mathsf{E}[X_\theta]\, d\theta,$$

or equivalently,

$$m_Y(t) = \int_{-\infty}^{\infty} h(t-\theta)m_X(\theta)\, d\theta. \quad (10.4)$$

For future reference, make the change of variable $\tau = t - \theta$, $d\tau = -d\theta$, to get

$$m_Y(t) = \int_{-\infty}^{\infty} h(\tau)m_X(t-\tau)\, d\tau. \quad (10.5)$$

The foregoing example has a discrete-time analog in which the integrals are replaced by sums. In this case, a discrete-time process X_n is applied to a discrete-time LTI system with impulse response sequence $h(n)$. The output is

$$Y_n = \sum_{k=-\infty}^{\infty} h(n-k)X_k.$$

The analogs of (10.4) and (10.5) can be derived. See Problem 6.

Correlation functions have special properties. First,

$$R_X(t_1,t_2) = \mathsf{E}[X_{t_1}X_{t_2}] = \mathsf{E}[X_{t_2}X_{t_1}] = R_X(t_2,t_1).$$

In other words, the correlation function is a **symmetric function** of t_1 and t_2. Next, observe that $R_X(t,t) = \mathsf{E}[X_t^2] \geq 0$, and for any t_1 and t_2,

$$|R_X(t_1,t_2)| \leq \sqrt{\mathsf{E}[X_{t_1}^2]\mathsf{E}[X_{t_2}^2]}. \quad (10.6)$$

This is just the Cauchy–Schwarz inequality (2.24), which says that

$$|\mathsf{E}[X_{t_1}X_{t_2}]| \leq \sqrt{\mathsf{E}[X_{t_1}^2]\mathsf{E}[X_{t_2}^2]}.$$

A random process for which $E[X_t^2] < \infty$ for all t is called a **second-order process**. By (10.6), the correlation function of a second-order process is finite for all t_1 and t_2. Such a process also has a finite mean function; again by the Cauchy–Schwarz inequality,

$$|E[X_t]| = |E[X_t \cdot 1]| \le \sqrt{E[X_t^2]E[1^2]} = \sqrt{E[X_t^2]}.$$

Except for the continuous-time white noise processes discussed later, all processes in this chapter are assumed to be second-order processes.

The **covariance function** is

$$C_X(t_1,t_2) := E\big[(X_{t_1} - E[X_{t_1}])(X_{t_2} - E[X_{t_2}])\big].$$

An easy calculation (Problem 3) shows that

$$C_X(t_1,t_2) = R_X(t_1,t_2) - m_X(t_1)m_X(t_2). \tag{10.7}$$

Note that the covariance function is also symmetric; i.e., $C_X(t_1,t_2) = C_X(t_2,t_1)$.

Cross-correlation functions

Let X_t and Y_t be random processes. Their **cross-correlation function** is

$$\boxed{R_{XY}(t_1,t_2) := E[X_{t_1}Y_{t_2}].} \tag{10.8}$$

To distinguish between the terms cross-correlation function and correlation function, the latter is sometimes referred to as the **auto-correlation function**. The **cross-covariance function** is

$$C_{XY}(t_1,t_2) := E[\{X_{t_1} - m_X(t_1)\}\{Y_{t_2} - m_Y(t_2)\}] = R_{XY}(t_1,t_2) - m_X(t_1)m_Y(t_2). \tag{10.9}$$

Since we usually assume that our processes are zero mean; i.e., $m_X(t) \equiv 0$, we focus on correlation functions and their properties.

Example 10.12. Let X_t be a random process with correlation function $R_X(t_1,t_2)$. Suppose that X_t is applied to an LTI system with impulse response $h(t)$. If

$$Y_t = \int_{-\infty}^{\infty} h(\theta)X_{t-\theta}\,d\theta,$$

find the cross-correlation function $R_{XY}(t_1,t_2)$ and the auto-correlation function $R_Y(t_1,t_2)$.

Solution. For the cross-correlation function, write

$$\begin{aligned} R_{XY}(t_1,t_2) &:= E[X_{t_1}Y_{t_2}] \\ &= E\left[X_{t_1}\int_{-\infty}^{\infty} h(\theta)X_{t_2-\theta}\,d\theta\right] \\ &= \int_{-\infty}^{\infty} h(\theta)E[X_{t_1}X_{t_2-\theta}]\,d\theta \\ &= \int_{-\infty}^{\infty} h(\theta)R_X(t_1,t_2-\theta)\,d\theta. \end{aligned}$$

To compute the auto-correlation function, write

$$R_Y(t_1,t_2) := \mathsf{E}[Y_{t_1}Y_{t_2}]$$
$$= \mathsf{E}\left[\left(\int_{-\infty}^{\infty} h(\beta)X_{t_1-\beta}\,d\beta\right)Y_{t_2}\right]$$
$$= \int_{-\infty}^{\infty} h(\beta)\mathsf{E}[X_{t_1-\beta}Y_{t_2}]\,d\beta$$
$$= \int_{-\infty}^{\infty} h(\beta)R_{XY}(t_1-\beta,t_2)\,d\beta.$$

Using the formula that we just derived above for R_{XY}, we have

$$R_Y(t_1,t_2) = \int_{-\infty}^{\infty} h(\beta)\left(\int_{-\infty}^{\infty} h(\theta)R_X(t_1-\beta,t_2-\theta)\,d\theta\right)d\beta.$$

For future reference, we extract from the above example the formulas

$$\mathsf{E}[X_{t_1}Y_{t_2}] = \int_{-\infty}^{\infty} h(\theta)\mathsf{E}[X_{t_1}X_{t_2-\theta}]\,d\theta \qquad (10.10)$$

and

$$\mathsf{E}[Y_{t_1}Y_{t_2}] = \int_{-\infty}^{\infty} h(\beta)\left(\int_{-\infty}^{\infty} h(\theta)\mathsf{E}[X_{t_1-\beta}X_{t_2-\theta}]\,d\theta\right)d\beta. \qquad (10.11)$$

The discrete-time analogs are derived in Problem 6.

10.3 Strict-sense and wide-sense stationary processes

An every-day example of a stationary process is the daily temperature during the summer. During the summer, it is warm every day. The exact temperature varies during the day and from day to day, but we do not check the weather forecast to see if we need a jacket to stay warm.

Similarly, the exact amount of time it takes you to go from home to school or work varies from day to day, but you know when to leave in order not to be late.

In each of these examples, your behavior is the same every day (*time invariant!*) even though the temperature or travel time is not. The reason your behavior is successful is that the statistics of the temperature or travel time do not change. As we shall see in Section 10.4, the interplay between LTI systems and stationary processes yields some elegant and useful results.

Strict-sense stationarity

A random process is **nth order strictly stationary** if for any collection of n times $t_1,\ldots,t_n$, all joint probabilities involving $X_{t_1+\Delta t},\ldots,X_{t_n+\Delta t}$ do not depend on the time shift Δt, whether it be positive or negative. In other words, for every n-dimensional set B,

$$\mathsf{P}\big((X_{t_1+\Delta t},\ldots,X_{t_n+\Delta t}) \in B\big)$$

does not depend on Δt. The corresponding condition for discrete-time processes is that

$$\mathsf{P}\big((X_{1+m},\ldots,X_{n+m}) \in B\big)$$

not depend on the integer time shift m.

If a process is nth order strictly stationary for every positive, finite integer n, then the process is said to be **strictly stationary**.

Example 10.13. Let Z be a random variable, and put $X_t := Z$ for all t. Show that X_t is strictly stationary.

Solution. Given any n-dimensional set B,

$$\mathsf{P}\big((X_{t_1+\Delta t},\ldots,X_{t_n+\Delta t}) \in B\big) = \mathsf{P}\big((Z,\ldots,Z) \in B\big),$$

which does not depend on Δt.

Example 10.14. Show that an i.i.d. sequence of continuous random variables X_n with common density f is strictly stationary.

Solution. Fix any positive integer n and any n-dimensional set B. Let m be any integer, positive or negative. Then $\mathsf{P}\big((X_{1+m},\ldots,X_{n+m}) \in B\big)$ is given by

$$\int\cdots\int_B f(x_{1+m})\cdots f(x_{n+m})\,dx_{1+m}\cdots dx_{n+m}. \tag{10.12}$$

Since $x_{1+m},\ldots,x_{n+m}$ are just dummy variables of integration, we may replace them by $x_1,\ldots,x_n$. Hence, the above integral is equal to

$$\int\cdots\int_B f(x_1)\cdots f(x_n)\,dx_1\cdots dx_n,$$

which does not depend on m.

It is instructive to see how the preceding example breaks down if the X_i are independent but not identically distributed. In this case, (10.12) becomes

$$\int\cdots\int_B f_{X_{1+m}}(x_{1+m})\cdots f_{X_{n+m}}(x_{n+m})\,dx_{1+m}\cdots dx_{n+m}.$$

Changing the dummy variables of integration as before, we obtain

$$\int\cdots\int_B f_{X_{1+m}}(x_1)\cdots f_{X_{n+m}}(x_n)\,dx_1\cdots dx_n,$$

which still depends on m.

Strict stationarity is a strong property with many implications. If a process is first-order strictly stationary, then for any t_1 and $t_1 + \Delta t$, X_{t_1} and $X_{t_1+\Delta t}$ have the same pmf or density.

10.3 Strict-sense and wide-sense stationary processes

It then follows that for any function $g(x)$, $\mathsf{E}[g(X_{t_1})] = \mathsf{E}[g(X_{t_1+\Delta t})]$. Taking $\Delta t = -t_1$ shows that $\mathsf{E}[g(X_{t_1})] = \mathsf{E}[g(X_0)]$, which does not depend on t_1. If a process is second-order strictly stationary, then for any function $g(x_1, x_2)$, we have

$$\mathsf{E}[g(X_{t_1}, X_{t_2})] = \mathsf{E}[g(X_{t_1+\Delta t}, X_{t_2+\Delta t})]$$

for every time shift Δt. Since Δt is arbitrary, let $\Delta t = -t_2$. Then

$$\mathsf{E}[g(X_{t_1}, X_{t_2})] = \mathsf{E}[g(X_{t_1-t_2}, X_0)].$$

It follows that $\mathsf{E}[g(X_{t_1}, X_{t_2})]$ depends on t_1 and t_2 only through the time difference $t_1 - t_2$.

Requiring second-order strict stationarity is a strong requirement. In practice, e.g., analyzing receiver noise in a communication system, it is often enough to require that $\mathsf{E}[X_t]$ not depend on t and that the correlation $R_X(t_1, t_2) = \mathsf{E}[X_{t_1} X_{t_2}]$ depend on t_1 and t_2 only through the time difference, $t_1 - t_2$. This is a much weaker requirement than second-order strict-sense stationarity for two reasons. First, we are not concerned with probabilities, only expectations. Second, we are only concerned with $\mathsf{E}[X_t]$ and $\mathsf{E}[X_{t_1} X_{t_2}]$ rather than $\mathsf{E}[g(X_t)]$ and $\mathsf{E}[g(X_{t_1}, X_{t_2})]$ for arbitrary functions g.

Even if you can justify the assumption of first-order strict-sense stationarity, to fully exploit it, say in the discrete-time case, you would have to estimate the density or pmf of X_i. We saw in Chapter 6 how much work it was for the i.i.d. case to estimate $f_{X_1}(x)$. For a second-order strictly stationary process, you would have to estimate $f_{X_1 X_2}(x_1, x_2)$ as well. For a strictly stationary process, imagine trying to estimate n-dimensional densities for *all* $n = 1, 2, 3, \ldots, 100, \ldots$.

Wide-sense stationarity

We say that a process is **wide-sense stationary (WSS)** if the following two properties *both* hold:

(i) The mean function $\mathsf{E}[X_t]$ does not depend on t.
(ii) The correlation function $\mathsf{E}[X_{t_1} X_{t_2}]$ depends on t_1 and t_2 only through the time difference $t_1 - t_2$.

Notation. For a WSS process, $\mathsf{E}[X_{t+\tau} X_t]$ depends only on the time difference, which is $(t+\tau) - t = \tau$. Hence, for a WSS process, it is convenient to re-use the term **correlation function** to refer to the the univariate function

$$R_X(\tau) := \mathsf{E}[X_{t+\tau} X_t]. \tag{10.13}$$

Observe that since t in (10.13) is arbitrary, taking $t = t_2$ and $\tau = t_1 - t_2$ gives the formula

$$\mathsf{E}[X_{t_1} X_{t_2}] = R_X(t_1 - t_2). \tag{10.14}$$

Example 10.15. In Figure 10.8, three correlation functions $R_X(\tau)$ are shown at the left. At the right is a sample path X_t of a zero-mean process with that correlation function.

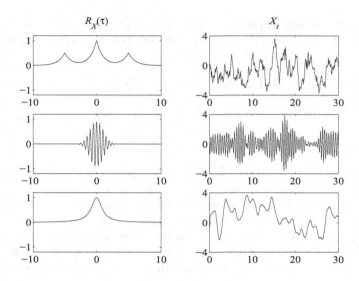

Figure 10.8. Three examples of a correlation function with a sample path of a process with that correlation function.

Example 10.16. Show that univariate correlation functions are always even.

Solution. Write

$$\begin{aligned} R_X(-\tau) &= \mathsf{E}[X_{t-\tau}X_t], &&\text{by (10.13)}, \\ &= \mathsf{E}[X_t X_{t-\tau}], &&\text{since multiplication commutes}, \\ &= R_X(t - [t-\tau]), &&\text{by (10.14)}, \\ &= R_X(\tau). \end{aligned}$$

Example 10.17. The carrier with random phase in Example 10.8 is WSS since we showed that $\mathsf{E}[X_t] = 0$ and $\mathsf{E}[X_{t_1} X_{t_2}] = \cos(2\pi f[t_1 - t_2])/2$. Hence, the (univariate) correlation function of this process is $R_X(\tau) = \cos(2\pi f \tau)/2$.

Example 10.18. Let X_t be WSS with zero mean and correlation function $R_X(\tau)$. If Y_t is a delayed version of X_t, say $Y_t := X_{t-t_0}$, determine whether or not Y_t is WSS.

Solution. We first check the mean value by writing

$$\mathsf{E}[Y_t] = \mathsf{E}[X_{t-t_0}] = 0,$$

since X_t is zero mean. Next we check the correlation function of Y_t. Write

$$\mathsf{E}[Y_{t_1} Y_{t_2}] = \mathsf{E}[X_{t_1-t_0} X_{t_2-t_0}] = R_X([t_1 - t_0] - [t_2 - t_0]) = R_X(t_1 - t_2).$$

Hence, Y_t is WSS, and in fact, $R_Y(\tau) = R_X(\tau)$.

10.3 Strict-sense and wide-sense stationary processes

Example **10.19** (a WSS process that is not strictly stationary). Let X_n be independent with $X_n \sim N(0,1)$ for $n \neq 0$, and $X_0 \sim \text{Laplace}(\lambda)$ with $\lambda = \sqrt{2}$. Show that this process is WSS but not strictly stationary.

Solution. Using the table inside the back cover, it is easy to see that *for all n*, the X_n are zero mean and unit variance. Furthermore, for $n \neq m$, we have by independence that $\mathsf{E}[X_n X_m] = 0$. Hence, for all n and m, $\mathsf{E}[X_n X_m] = \delta(n-m)$, where δ denotes the **Kronecker delta**, $\delta(n) = 1$ for $n=0$ and $\delta(n) = 0$ otherwise. This establishes that the process is WSS.

To show the process is not strictly stationary, it suffices to show that the fourth moments depend on n. For $n \neq 0$, $\mathsf{E}[X_n^4] = 3$ from the table or by Example 4.11. For $n = 0$,

$$\mathsf{E}[X_0^4] = \int_{-\infty}^{\infty} x^4 \frac{\lambda}{2} e^{-\lambda |x|} dx = \int_0^{\infty} x^4 \lambda e^{-\lambda x} dx,$$

which is the fourth moment of an $\exp(\lambda)$ random variable. From the table or by Example 4.17, this is equal to $4!/\lambda^4$. With $\lambda = \sqrt{2}$, $\mathsf{E}[X_0^4] = 6$. Hence, X_n cannot be strictly stationary.

The preceding example shows that in general, a WSS process need not be strictly stationary. However, there is one important exception. If a WSS process is Gaussian, a notion defined in Section 11.4, then the process must in fact be strictly stationary (see Example 11.9).

Estimation of correlation functions

In practical problems, we are not given the correlation function, but must estimate it from the data. Suppose we have discrete-time WSS process X_k. Observe that the expectation of

$$\frac{1}{2N+1} \sum_{k=-N}^{N} X_{k+n} X_k \qquad (10.15)$$

is equal to

$$\frac{1}{2N+1} \sum_{k=-N}^{N} \mathsf{E}[X_{k+n} X_k] = \frac{1}{2N+1} \sum_{k=-N}^{N} R_X(n) = R_X(n).$$

Thus, (10.15) is an **unbiased estimator** of $R_X(n)$ that can be computed from observations of X_k. In fact, under conditions given by **ergodic theorems**, the estimator (10.15) actually converges to $R_X(n)$ as $N \to \infty$. For a continuous-time WSS process X_t, the analogous estimator is

$$\frac{1}{2T} \int_{-T}^{T} X_{t+\tau} X_t \, dt. \qquad (10.16)$$

Its expectation is

$$\frac{1}{2T} \int_{-T}^{T} \mathsf{E}[X_{t+\tau} X_t] \, dt = \frac{1}{2T} \int_{-T}^{T} R_X(\tau) \, dt = R_X(\tau).$$

Under suitable conditions, the estimator (10.16) converges to $R_X(\tau)$ as $T \to \infty$. Ergodic theorems for continuous-time processes are discussed later in Section 10.10 and in the problems for that section.

Transforms of correlation functions

In the next section, when we pass WSS processes through LTI systems, it will be convenient to work with the Fourier transform of the correlation function. The **Fourier transform** of $R_X(\tau)$ is defined by

$$S_X(f) := \int_{-\infty}^{\infty} R_X(\tau) e^{-j2\pi f \tau} d\tau.$$

By the **inversion formula**,

$$R_X(\tau) = \int_{-\infty}^{\infty} S_X(f) e^{j2\pi f \tau} df.$$

***Example* 10.20.** Three correlation functions, $R_X(\tau)$, and their corresponding Fourier transforms, $S_X(f)$, are shown in Figure 10.9. The correlation functions are the same ones shown in Figure 10.8. Notice how the smoothest sample path in Figure 10.8 corresponds to the $S_X(f)$ with the lowest frequency content and the most wiggly sample path corresponds to the $S_X(f)$ with the highest frequency content.

As illustrated in Figure 10.9, $S_X(f)$ is real, even, and nonnegative. These properties can be proved mathematically. We defer the issue of nonnegativity until later. For the moment, we show that $S_X(f)$ is real and even by using the fact that $R_X(\tau)$ is real and even. Write

$$\begin{aligned} S_X(f) &= \int_{-\infty}^{\infty} R_X(\tau) e^{-j2\pi f \tau} d\tau \\ &= \int_{-\infty}^{\infty} R_X(\tau) \cos(2\pi f \tau) d\tau - j \int_{-\infty}^{\infty} R_X(\tau) \sin(2\pi f \tau) d\tau. \end{aligned}$$

Since $R_X(\tau)$ is real and even, and since $\sin(2\pi f \tau)$ is an odd function of τ, the second

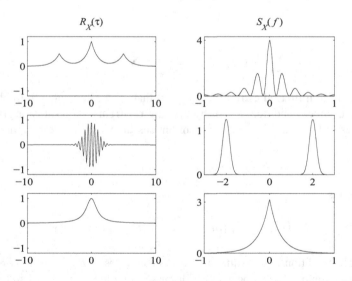

Figure 10.9. Three correlation functions (left) and their corresponding Fourier transforms (right).

10.3 Strict-sense and wide-sense stationary processes

integrand is odd, and therefore integrates to zero. Hence, we can always write

$$S_X(f) = \int_{-\infty}^{\infty} R_X(\tau)\cos(2\pi f \tau)\,d\tau.$$

Thus, $S_X(f)$ is real. Furthermore, since $\cos(2\pi f \tau)$ is an even function of f, so is $S_X(f)$.

Example 10.21. If a carrier with random phase is transmitted at frequency f_0, then from Example 10.17 we know that $R_X(\tau) = \cos(2\pi f_0 \tau)/2$. Verify that its transform is $S_X(f) = [\delta(f-f_0) + \delta(f+f_0)]/4$.

Solution. All we have to do is inverse transform $S_X(f)$. Write

$$\begin{aligned}
\int_{-\infty}^{\infty} S_X(f)e^{j2\pi f \tau}\,df &= \tfrac{1}{4}\int_{-\infty}^{\infty}[\delta(f-f_0)+\delta(f+f_0)]e^{j2\pi f \tau}\,df \\
&= \tfrac{1}{4}[e^{j2\pi f_0 \tau}+e^{-j2\pi f_0 \tau}] \\
&= \tfrac{1}{2}\cdot\frac{e^{j2\pi f_0 \tau}+e^{-j2\pi f_0 \tau}}{2} \\
&= \cos(2\pi f_0 \tau)/2.
\end{aligned}$$

Example 10.22. If $S_X(f)$ has the form shown in Figure 10.10, find the correlation function $R_X(\tau)$.

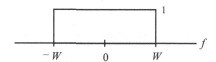

Figure 10.10. Graph of $S_X(f)$ for Example 10.22.

Solution. We must find the inverse Fourier transform of $S_X(f)$. Write

$$\begin{aligned}
R_X(\tau) &= \int_{-\infty}^{\infty} S_X(f)e^{j2\pi f \tau}\,df \\
&= \int_{-W}^{W} e^{j2\pi f \tau}\,df \\
&= \left.\frac{e^{j2\pi f \tau}}{j2\pi \tau}\right|_{f=-W}^{f=W} \\
&= \frac{e^{j2\pi W \tau}-e^{-j2\pi W \tau}}{j2\pi \tau} \\
&= 2W\frac{e^{j2\pi W \tau}-e^{-j2\pi W \tau}}{2j(2\pi W \tau)} \\
&= 2W\frac{\sin(2\pi W \tau)}{2\pi W \tau}.
\end{aligned}$$

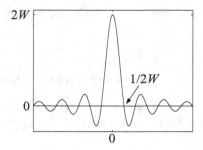

Figure 10.11. Correlation function $R_X(\tau)$ of Example 10.22.

This function is shown in Figure 10.11. The maximum value of $R_X(\tau)$ is $2W$ and occurs at $\tau = 0$. The zeros occur at τ equal to positive integer multiples of $\pm 1/(2W)$.

Remark. Since R_X and S_X are transform pairs, an easy corollary of Example 10.22 is that
$$\int_{-\infty}^{\infty} \frac{\sin t}{t} \, dt = \pi.$$
To see this, first note that
$$1 = S_X(f)|_{f=0} = \left(\int_{-\infty}^{\infty} R_X(\tau) e^{-j2\pi f \tau} \, d\tau\right)\bigg|_{f=0}.$$
This holds for all $W > 0$. Taking $W = 1/2$ we have
$$1 = \int_{-\infty}^{\infty} \frac{\sin(\pi \tau)}{\pi \tau} \, d\tau.$$
Making the change of variable $t = \pi \tau$, $dt = \pi d\tau$, yields the result.

Remark. It is common practice to define $\text{sinc}(\tau) := \sin(\pi \tau)/(\pi \tau)$. The reason for including the factor of π is so that the zero crossings occur on the nonzero integers. Using the sinc function, the correlation function of the preceding example is $R_X(\tau) = 2W \text{sinc}(2W\tau)$.

The foregoing examples are of continuous-time processes. For a discrete-time WSS process X_n with correlation function $R_X(n) = E[X_{k+n} X_k]$,
$$S_X(f) := \sum_{n=-\infty}^{\infty} R_X(n) e^{-j2\pi f n}$$
is the **discrete-time Fourier transform**. Hence, for discrete-time processes, $S_X(f)$ is periodic with period one. Since $S_X(f)$ is a **Fourier series**, the coefficients $R_X(n)$ can be recovered using
$$R_X(n) = \int_{-1/2}^{1/2} S_X(f) e^{j2\pi f n} \, df.$$
Properties of $S_X(f)$ are explored in Problem 20.

10.4 WSS processes through LTI systems

In this section we show that LTI systems preserve wide-sense stationarity. In other words, if a WSS process X_t is applied to an LTI system with impulse response h, as shown in Figure 10.12, then the output

$$Y_t = \int_{-\infty}^{\infty} h(t-\theta) X_\theta \, d\theta$$

is another WSS process. Furthermore, the correlation function of Y_t, and the cross-correlation function of X_t and Y_t can be expressed in terms of convolutions involving h and R_X. By introducing appropriate Fourier transforms the convolution relationships are converted into product formulas in the frequency domain.

The derivation of the analogous results for discrete-time processes and systems is carried out in Problem 31.

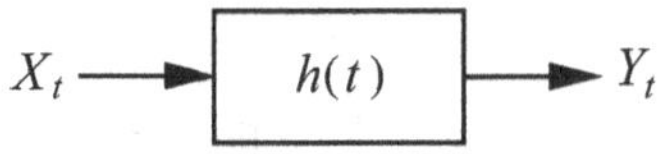

$$X_t \longrightarrow \boxed{h(t)} \longrightarrow Y_t$$

Figure 10.12. Block diagram of an LTI system with impulse response $h(t)$, input random process X_t, and output random process Y_t.

Time-domain analysis

Recall from Example 10.11 that

$$m_Y(t) = \int_{-\infty}^{\infty} h(\tau) m_X(t-\tau) \, d\tau.$$

Since X_t is WSS, its mean function $m_X(t)$ is constant for all t, say $m_X(t) \equiv m$. Then

$$m_Y(t) = m \int_{-\infty}^{\infty} h(\tau) \, d\tau,$$

which does not depend on t. Next, from (10.11),

$$E[Y_{t_1} Y_{t_2}] = \int_{-\infty}^{\infty} h(\beta) \left(\int_{-\infty}^{\infty} h(\theta) E[X_{t_1-\beta} X_{t_2-\theta}] \, d\theta \right) d\beta.$$

Since X_t is WSS, the expectation inside the integral is just

$$R_X([t_1 - \beta] - [t_2 - \theta]) = R_X([t_1 - t_2] - [\beta - \theta]).$$

Hence,

$$E[Y_{t_1} Y_{t_2}] = \int_{-\infty}^{\infty} h(\beta) \left(\int_{-\infty}^{\infty} h(\theta) R_X([t_1 - t_2] - [\beta - \theta]) \, d\theta \right) d\beta,$$

which depends on t_1 and t_2 only through their difference. We have thus shown that **the response of an LTI system to a WSS input is another WSS process** with correlation function

$$R_Y(\tau) = \int_{-\infty}^{\infty} h(\beta) \left(\int_{-\infty}^{\infty} h(\theta) R_X(\tau - \beta + \theta) \, d\theta \right) d\beta. \quad (10.17)$$

Before continuing the analysis of $R_Y(\tau)$, it is convenient to first look at the cross-correlation between X_{t_1} and Y_{t_2}. From (10.10),

$$\mathsf{E}[X_{t_1} Y_{t_2}] = \int_{-\infty}^{\infty} h(\theta) \mathsf{E}[X_{t_1} X_{t_2-\theta}] \, d\theta$$

$$= \int_{-\infty}^{\infty} h(\theta) R_X(t_1 - t_2 + \theta) \, d\theta.$$

If two processes X_t and Y_t are each WSS, and if their cross-correlation $\mathsf{E}[X_{t_1} Y_{t_2}]$ depends on t_1 and t_2 only through their difference, the processes are said to be **jointly wide-sense stationary** (J-WSS). In this case, their univariate **cross-correlation function** is defined by

$$R_{XY}(\tau) := \mathsf{E}[X_{t+\tau} Y_t].$$

The generalization of (10.14) is

$$\mathsf{E}[X_{t_1} Y_{t_2}] = R_{XY}(t_1 - t_2).$$

The foregoing analysis shows that if a WSS process is applied to an LTI system, then the input and output processes are J-WSS with cross-correlation function

$$R_{XY}(\tau) = \int_{-\infty}^{\infty} h(\theta) R_X(\tau + \theta) \, d\theta. \tag{10.18}$$

Comparing (10.18) and the inner integral in (10.17) shows that

$$R_Y(\tau) = \int_{-\infty}^{\infty} h(\beta) R_{XY}(\tau - \beta) \, d\beta. \tag{10.19}$$

Thus, R_Y is the convolution of h and R_{XY}. Furthermore, making the change of variable $\alpha = -\theta$, $d\alpha = -d\theta$ in (10.18) yields

$$R_{XY}(\tau) = \int_{-\infty}^{\infty} h(-\alpha) R_X(\tau - \alpha) \, d\alpha. \tag{10.20}$$

In other words, R_{XY} is the convolution of $h(-\alpha)$ and R_X.

Frequency-domain analysis

The preceding convolutions suggest that by applying the Fourier transform, much simpler formulas can be obtained in the frequency domain. The Fourier transform of the system impulse response h,

$$H(f) := \int_{-\infty}^{\infty} h(\tau) e^{-j2\pi f \tau} \, d\tau,$$

is called the system **transfer function**. The Fourier transforms of $R_X(\tau)$, $R_Y(\tau)$, and $R_{XY}(\tau)$ are denoted by $S_X(f)$, $S_Y(f)$, and $S_{XY}(f)$, respectively. Taking the Fourier transform of (10.19) yields

$$S_Y(f) = H(f) S_{XY}(f). \tag{10.21}$$

Similarly, taking the Fourier transform of (10.20) yields

$$\boxed{S_{XY}(f) = H(f)^* S_X(f),} \tag{10.22}$$

since, as shown in Problem 22, for h real, the Fourier transform of $h(-\tau)$ is $H(f)^*$, where the asterisk * denotes the complex conjugate. Combining (10.21) and (10.22), we have

$$S_Y(f) = H(f)S_{XY}(f) = H(f)H(f)^*S_X(f) = |H(f)|^2 S_X(f).$$

Thus,

$$S_Y(f) = |H(f)|^2 S_X(f). \tag{10.23}$$

Example 10.23. Suppose that the process X_t is WSS with correlation function $R_X(\tau) = e^{-\lambda|\tau|}$. If X_t is applied to an LTI system with transfer function

$$H(f) = \sqrt{\lambda^2 + (2\pi f)^2}\, I_{[-W,W]}(f),$$

find the system output correlation function $R_Y(\tau)$.

Solution. Our approach is to first find $S_Y(f)$ using (10.23) and then take the inverse Fourier transform to obtain $R_Y(\tau)$. To begin, first note that

$$|H(f)|^2 = [\lambda^2 + (2\pi f)^2] I_{[-W,W]}(f),$$

where we use the fact that since an indicator is zero or one, it has the property that it is equal to its square. We obtain $S_X(f)$ from the table of Fourier transforms inside the front cover and write

$$\begin{aligned} S_Y(f) &= S_X(f)|H(f)|^2 \\ &= \frac{2\lambda}{\lambda^2 + (2\pi f)^2}[\lambda^2 + (2\pi f)^2] I_{[-W,W]}(f) \\ &= 2\lambda I_{[-W,W]}(f), \end{aligned}$$

which is proportional to the graph in Figure 10.10. Using the transform table inside the front cover or the result of Example 10.22,

$$R_Y(\tau) = 2\lambda \cdot 2W \frac{\sin(2\pi W \tau)}{2\pi W \tau}.$$

10.5 Power spectral densities for WSS processes

Motivation

Recall that if $v(t)$ is the voltage across a resistance R, then the instantaneous power is $v(t)^2/R$, and the energy dissipated is $\int_{-\infty}^{\infty} v(t)^2/R\, dt$. Similarly, if the current through the resistance is $i(t)$, the instantaneous power is $i(t)^2 R$, and the energy dissipated is $\int_{-\infty}^{\infty} i(t)^2 R\, dt$.

Based on the foregoing observations, the "energy" of any waveform $x(t)$ is defined to be $\int_{-\infty}^{\infty} |x(t)|^2\, dt$. Of course, if $x(t)$ is the voltage across a one-ohm resistor or the current through a one-ohm resistor, then $\int_{-\infty}^{\infty} |x(t)|^2\, dt$ is the physical energy dissipated.

Some signals, such as periodic signals like $\cos(t)$ and $\sin(t)$, do not have finite energy, but they do have finite average power; i.e.,

$$\lim_{T\to\infty} \frac{1}{2T} \int_{-T}^{T} |x(t)|^2 \, dt \; < \; \infty.$$

For periodic signals, this limit is equal to the energy in one period divided by the period (Problem 32).

Power in a process

For a deterministic signal $x(t)$, the energy or average power serves as a single-number characterization. For a random process X_t, the analogous quantities

$$\int_{-\infty}^{\infty} X_t^2 \, dt \quad \text{and} \quad \lim_{T\to\infty} \frac{1}{2T} \int_{-T}^{T} X_t^2 \, dt$$

are *random variables* — they are *not* single-number characterizations (unless extra assumptions such as ergodicity are made; see Section 10.10). However, their expectations *are* single-number characterizations. Since most processes have infinite expected energy (e.g., WSS processes – see Problem 33), we focus on the **expected average power**,

$$P_X := \mathsf{E}\!\left[\lim_{T\to\infty} \frac{1}{2T} \int_{-T}^{T} X_t^2 \, dt \right].$$

For a WSS process, this becomes

$$\lim_{T\to\infty} \frac{1}{2T} \int_{-T}^{T} \mathsf{E}[X_t^2] \, dt \; = \; \lim_{T\to\infty} \frac{1}{2T} \int_{-T}^{T} R_X(0) \, dt \; = \; R_X(0).$$

Since R_X and S_X are Fourier transform pairs,

$$R_X(0) = \left(\int_{-\infty}^{\infty} S_X(f) e^{j2\pi f \tau} \, df \right)\bigg|_{\tau=0}$$
$$= \int_{-\infty}^{\infty} S_X(f) \, df.$$

Since we also have $\mathsf{E}[X_t^2] = R_X(0)$,

$$\boxed{P_X \;=\; \mathsf{E}[X_t^2] \;=\; R_X(0) \;=\; \int_{-\infty}^{\infty} S_X(f) \, df,} \qquad (10.24)$$

and we have three ways to express the power in a WSS process.

Remark. From the definition of P_X, (10.24) says that for a WSS process,

$$\mathsf{E}\!\left[\lim_{T\to\infty} \frac{1}{2T} \int_{-T}^{T} X_t^2 \, dt \right] = \mathsf{E}[X_t^2],$$

which we call the **expected instantaneous power**. Thus, for a WSS process, the expected average power is equal to the expected instantaneous power.

10.5 Power spectral densities for WSS processes

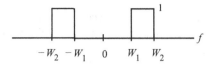

Figure 10.13. Bandpass filter $H(f)$ for extracting the power in the frequency band $W_1 \leq |f| \leq W_2$.

Example 10.24 (power in a frequency band). For a WSS process X_t, find the power in the frequency band $W_1 \leq |f| \leq W_2$.

Solution. We interpret the problem as asking us to apply X_t to the ideal bandpass filter with transfer function $H(f)$ shown in Figure 10.13 and then find the power in the output process. Denoting the filter output by Y_t, we have

$$P_Y = \int_{-\infty}^{\infty} S_Y(f) df$$
$$= \int_{-\infty}^{\infty} |H(f)|^2 S_X(f) df, \quad \text{by (10.23)},$$
$$= \int_{-W_2}^{-W_1} S_X(f) df + \int_{W_1}^{W_2} S_X(f) df,$$

where the last step uses the fact that $H(f)$ has the form in Figure 10.13. Since, as shown at the end of Section 10.4, $S_X(f)$ is even, these last two integrals are equal. Hence,

$$P_Y = 2 \int_{W_1}^{W_2} S_X(f) df.$$

To conclude the example, we use the formula for P_Y to derive the additional result that $S_X(f)$ is a nonnegative function. Suppose that $W_2 = W_1 + \Delta W$, where $\Delta W > 0$ is small. Then

$$P_Y = 2 \int_{W_1}^{W_1 + \Delta W} S_X(f) df \approx 2 S_X(W_1) \Delta W.$$

It follows that

$$S_X(W_1) \approx \frac{P_Y}{2\Delta W} \geq 0,$$

since $P_Y = E[Y_t^2] \geq 0$. Since $W_1 \geq 0$ is arbitrary, and since $S_X(f)$ is even, we conclude that $S_X(f) \geq 0$ for all f.

Example 10.24 shows that $S_X(f)$ is a nonnegative function that, when integrated over a frequency band, yields the process's power in that band. This is analogous to the way a probability density is integrated over an interval to obtain its probability. On account of this similarity, $S_X(f)$ is called the **power spectral density** of the process. The adjective "spectral" means that S_X is a function of frequency. While there are infinitely many nonnegative, even functions of frequency that integrate to P_X, there is only one such function that when integrated over every frequency band gives the power in that band. See Problem 34.

The analogous terminology for $S_{XY}(f)$ is **cross power spectral density**. However, in general, $S_{XY}(f)$ can be complex valued. Even if $S_{XY}(f)$ is real valued, it need not be nonnegative. See Problem 35.

White noise

If a WSS process has constant power across all frequencies, it is called white noise. This is analogous to white light, which contains equal amounts of all the colors found in a rainbow. To be precise, X_t is called **white noise** if its power spectral density is constant for all frequencies. Unless otherwise specified, this constant is usually denoted by $N_0/2$. Taking the Fourier transform of

$$R_X(\tau) = \frac{N_0}{2}\delta(\tau),$$

where δ is the Dirac delta function, yields

$$S_X(f) = \frac{N_0}{2}.$$

Thus, the correlation function of white noise is a delta function. White noise is an idealization of what is observed in physical noise sources. In real noise sources, $S_X(f)$ is approximately constant for frequencies up to about 1000 GHz. For $|f|$ larger than this, $S_X(f)$ decays. However, what real systems see is $|H(f)|^2 S_X(f)$, where the bandwidth of the transfer function is well below 1000 GHz. In other words, any hardware filters the noise so that $S_Y(f)$ is not affected by the exact values of $S_X(f)$ for the large $|f|$ where $S_X(f)$ begins to decay.

Remark. Just as the delta function is not an ordinary function, white noise is not an ordinary random process. For example, since $\delta(0)$ is not defined, and since $\mathsf{E}[X_t^2] = R_X(0) = (N_0/2)\delta(0)$, we cannot speak of the second moment of X_t when X_t is white noise. In particular, white noise is not a second-order process. Also, since $S_X(f) = N_0/2$ for white noise, and since

$$\int_{-\infty}^{\infty} \frac{N_0}{2}\, df = \infty,$$

we often say that white noise has infinite average power.

In Figure 10.10, if we let $W \to \infty$, we get $S_X(f) = 1$ for all f. Similarly, if we let $W \to \infty$ in Figure 10.11, $R_X(\tau)$ begins to look more and more like $\delta(\tau)$. This suggests that a process X_t with correlation function in Figure 10.11 should look more and more like white noise as W increases. For finite W, we call such a process **bandlimited white noise**. In Figure 10.14, we show sample paths X_t with $W = 1/2$ (top), $W = 2$ (middle), and $W = 4$ (bottom). As W increases, the processes become less smooth and more wiggly. In other words, they contain higher and higher frequencies.

Example 10.25. Consider the lowpass *RC* filter shown in Figure 10.15. Suppose that the voltage source is a white-noise process X_t with power spectral density $S_X(f) = N_0/2$. If the filter output is taken to be the capacitor voltage, which we denote by Y_t, find its power spectral density $S_Y(f)$ and the corresponding correlation function $R_Y(\tau)$.

10.5 Power spectral densities for WSS processes

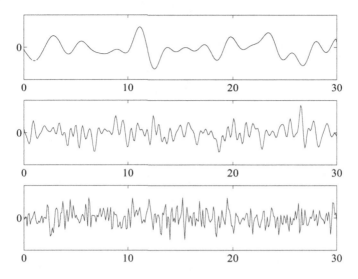

Figure 10.14. Bandlimited white noise processes with the power spectral density in Figure 10.10 and the correlation function in Figure 10.11 for $W = 1/2$ (top), $W = 2$ (middle), and $W = 4$ (bottom).

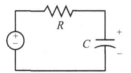

Figure 10.15. Lowpass RC filter.

Solution. From standard circuit-analysis techniques, the system transfer function between the input X_t and output Y_t is

$$H(f) = \frac{1}{1 + j2\pi fRC}.$$

Hence,

$$S_Y(f) = |H(f)|^2 S_X(f) = \frac{N_0/2}{1 + (2\pi fRC)^2}.$$

If we write

$$S_Y(f) = \frac{N_0}{4RC} \cdot \frac{2/RC}{(1/RC)^2 + (2\pi f)^2},$$

then the inverse transform of $S_Y(f)$ can be found by inspection using the table inside the front cover. We find that

$$R_Y(\tau) = \frac{N_0}{4RC} e^{-|\tau|/RC}.$$

Figure 10.16 shows how the sample paths Y_t vary with the filter time constant, RC. In each case, the process wanders between the top and the bottom of the graph. However, the top graph is less wiggly than the bottom one, and the middle one has an intermediate amount

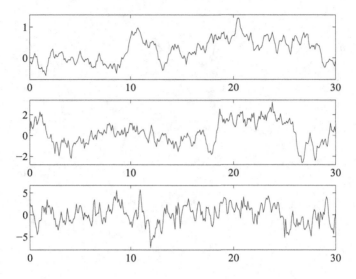

Figure 10.16. Three realizations of a lowpass *RC* filter output driven by white noise. The time constants are $RC = 4$ (top), $RC = 1$ (middle), and $RC = 1/4$ (bottom).

of wiggle. To explain this, recall that the filter time constant is inversely proportional to the filter bandwidth. Hence, when *RC* is large, the filter has a small bandwidth and passes only low-frequency components. When *RC* is small, the filter has a large bandwidth that passes both high and low frequency components. A signal with only low frequency components cannot wiggle as much as a signal with high frequency components.

Example 10.26. A certain communication receiver employs a bandpass filter to reduce white noise generated in the amplifier. Suppose that the white noise X_t has power spectral density $S_X(f) = N_0/2$ and that the filter transfer function $H(f)$ is given in Figure 10.13. Find the expected output power from the filter.

Solution. The expected output power is obtained by integrating the power spectral density of the filter output. Denoting the filter output by Y_t,

$$P_Y = \int_{-\infty}^{\infty} S_Y(f)\,df = \int_{-\infty}^{\infty} |H(f)|^2 S_X(f)\,df.$$

Since $|H(f)|^2 S_X(f) = N_0/2$ for $W_1 \leq |f| \leq W_2$, and is zero otherwise, $P_Y = 2(N_0/2)(W_2 - W_1) = N_0(W_2 - W_1)$. In other words, the expected output power is N_0 times the bandwidth[b] of the filter.

Example 10.27. White noise with power spectral density $N_0/2$ is applied to a lowpass filter with transfer function $H(f) = e^{-2\pi\lambda|f|}$. Find the output noise power from the filter.

---

[b] Bandwidth refers to the range of *positive* frequencies where $|H(f)| > 0$. The reason for this is that in physical systems, the impulse response is real. This implies $H(-f) = H(f)^*$, and then $|H(f)|^2 = H(f)H(f)^* = H(f)H(-f)$ is an even function.

10.5 Power spectral densities for WSS processes

Solution. To begin, write

$$S_Y(f) = |H(f)|^2 S_X(f) = \left|e^{-2\pi\lambda|f|}\right|^2 \frac{N_0}{2} = \frac{N_0}{2} e^{-2\pi(2\lambda)|f|}.$$

Since $P_Y = \int_{-\infty}^{\infty} S_Y(f)\,df$, one approach would be to compute this integral, which can be done in closed form. However, in this case it is easier to use the fact that $P_Y = R_Y(0)$. From the transform table inside the front cover, we see that

$$R_Y(\tau) = \frac{N_0}{2\pi} \cdot \frac{2\lambda}{(2\lambda)^2 + \tau^2},$$

and it follows that $P_Y = R_Y(0) = N_0/(4\pi\lambda)$.

Example 10.28. White noise with power spectral density $S_X(f) = N_0/2$ is applied to a filter with impulse response $h(t) = I_{[0,T]}(t)$ shown in Figure 10.17. Find (a) the cross power spectral density $S_{XY}(f)$; (b) the cross-correlation, $R_{XY}(\tau)$; (c) $\mathsf{E}[X_{t_1} Y_{t_2}]$; (d) the output power spectral density $S_Y(f)$; (e) the output auto-correlation, $R_Y(\tau)$; (f) the output power P_Y.

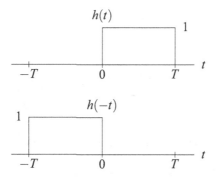

Figure 10.17. Impulse response $h(t) = I_{[0,T]}(t)$ and $h(-t)$ of Example 10.28.

Solution. (a) Since

$$S_{XY}(f) = H(f)^* S_X(f) = H(f)^* \cdot \frac{N_0}{2},$$

we need to compute

$$H(f) = \int_{-\infty}^{\infty} h(t) e^{-j2\pi ft}\,dt = \int_0^T e^{-j2\pi ft}\,dt = \frac{e^{-j2\pi ft}}{-j2\pi f}\bigg|_0^T.$$

We simplify writing

$$H(f) = \frac{1 - e^{-j2\pi fT}}{j2\pi f}$$
$$= e^{-j\pi Tf} T \frac{e^{j\pi Tf} - e^{-j\pi Tf}}{2j\pi Tf}$$
$$= e^{-j\pi Tf} T \frac{\sin(\pi Tf)}{\pi Tf}.$$

It follows that
$$S_{XY}(f) = e^{j\pi Tf} T \frac{\sin(\pi T f)}{\pi T f} \frac{N_0}{2}.$$

(b) Since $h(t)$ is real, the inverse transform of $S_{XY}(f) = H(f)^* N_0/2$ is (recall Problem 22)
$$R_{XY}(\tau) = h(-\tau)N_0/2 = I_{[0,T]}(-\tau)N_0/2 = I_{[-T,0]}(\tau)N_0/2.$$

(c) $E[X_{t_1} Y_{t_2}] = R_{XY}(t_1 - t_2) = I_{[-T,0]}(t_1 - t_2)N_0/2.$

(d) Since we computed $H(f)$ in part (a), we can easily write
$$S_Y(f) = |H(f)|^2 S_X(f) = T^2 \left[\frac{\sin(\pi T f)}{\pi T f}\right]^2 \frac{N_0}{2}.$$

(e) From the transform table inside the front cover,
$$R_Y(\tau) = \frac{TN_0}{2}(1 - |\tau|/T)I_{[-T,T]}(\tau),$$
which is shown in Figure 10.18.

(f) Use part (e) to write $P_Y = R_Y(0) = TN_0/2.$

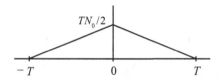

Figure 10.18. Output auto-correlation $R_Y(\tau)$ of Example 10.28.

10.6 Characterization of correlation functions

In the preceding sections we have shown that if $R_X(\tau)$ is the correlation function of a WSS process, then the power spectral density $S_X(f)$ is a real, even, nonnegative function of f.

Conversely, it is shown in Problem 48 of Chapter 11 that given any real, even, nonnegative function of frequency, say $S(f)$, there is a WSS process whose correlation function is the inverse Fourier transform of $S(f)$.

Thus, one can ask many questions of the form, "Is $R(\tau) = \cdots$ a valid correlation function?"

To show that a given $R(\tau)$ is a valid correlation function, you can take its Fourier transform and show that it is real, even, and nonnegative.

On the other hand, if $R(\tau)$ is not a valid correlation function, you can sometimes see this without taking its Fourier transform. For example, if $R(\tau)$ is not even, it cannot be a correlation function since, by Example 10.16, correlation functions are always even.

Another important property of a correlation function $R(\tau)$ is that
$$|R(\tau)| \le R(0), \quad \text{for all } \tau. \tag{10.25}$$

10.6 Characterization of correlation functions

In other words, the maximum absolute value of a correlation function is achieved at $\tau = 0$, and at that point the function is nonnegative. Note that (10.25) does not preclude other maximizing values of τ; it only says that $\tau = 0$ is one of the maximizers. To derive (10.25), we first note that if $R(\tau) = R_X(\tau)$ for some process X_t, then

$$R_X(0) = E[X_t^2] \geq 0.$$

Then use the Cauchy–Schwarz inequality (2.24) to write

$$\begin{aligned} |R_X(\tau)| &= |E[X_{t+\tau} X_t]| \\ &\leq \sqrt{E[X_{t+\tau}^2] E[X_t^2]} \\ &= \sqrt{R_X(0) R_X(0)} \\ &= R_X(0). \end{aligned}$$

Example 10.29. Determine whether or not $R(\tau) := \tau e^{-|\tau|}$ is a valid correlation function.

Solution. Since $R(\tau)$ is odd, it cannot be a valid correlation function. Alternatively, we can observe that $R(0) = 0 < e^{-1} = R(1)$, violating $R(0) \geq |R(\tau)|$ for all τ.

Example 10.30. Determine whether or not $R(\tau) := 1/(1+\tau^2)$ is a valid correlation function.

Solution. It is easy to see that $R(\tau)$ is real, even, and its maximum absolute value occurs at $\tau = 0$. So we cannot rule it out as a valid correlation function. The next step is to check its Fourier transform. From the table inside the front cover, the Fourier transform of $R(\tau)$ is $S(f) = \pi \exp(-2\pi|f|)$. Since $S(f)$ is real, even, and nonnegative, $R(\tau)$ is a valid correlation function.

Correlation functions of deterministic signals

Up to this point, we have discussed correlation functions for WSS random processes. However, there is a connection with correlation functions of a deterministic signals.

The correlation function of a real signal $v(t)$ of finite energy is defined by

$$R_v(\tau) := \int_{-\infty}^{\infty} v(t+\tau) v(t) \, dt.$$

Note that $R_v(0) = \int_{-\infty}^{\infty} v(t)^2 \, dt$ is the signal energy.

Since the formula for $R_v(\tau)$ is similar to a convolution integral, for simple functions $v(t)$ such as the one at the top in Figure 10.19, $R_v(\tau)$ can be computed directly, and is shown at the bottom of the figure.

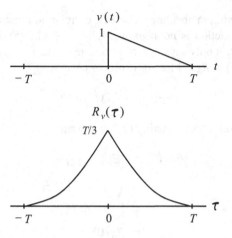

Figure 10.19. Deterministic signal $v(t)$ and its correlation function $R_v(\tau)$.

Further insight into $R_v(\tau)$ can be obtained using Fourier transforms. If $v(t)$ has Fourier transform $V(f)$, then by the inversion formula,

$$v(t) = \int_{-\infty}^{\infty} V(f) e^{j2\pi f t} df.$$

Let us apply this formula to $v(t+\tau)$ in the definition of $R_v(\tau)$. Then

$$R_v(\tau) = \int_{-\infty}^{\infty} \left(\int_{-\infty}^{\infty} V(f) e^{j2\pi f(t+\tau)} df \right) v(t) dt$$

$$= \int_{-\infty}^{\infty} V(f) \left(\int_{-\infty}^{\infty} v(t) e^{-j2\pi f t} dt \right)^* e^{j2\pi f \tau} df,$$

where we have used the fact that $v(t)$ is real. Since the inner integral is just $V(f)$,

$$R_v(\tau) = \int_{-\infty}^{\infty} |V(f)|^2 e^{j2\pi f \tau} df. \tag{10.26}$$

Since $v(t)$ is a real signal, $V(f)^* = V(-f)$. Hence, $|V(f)|^2 = V(f)V(f)^* = V(f)V(-f)$ is real, even, and nonnegative, *just like a power spectral density*. In fact, when a signal has finite energy; i.e., $\int_{-\infty}^{\infty} |v(t)|^2 dt < \infty$, $|V(f)|^2$ is called the **energy spectral density**.

10.7 The matched filter

Consider an air-traffic control system which sends out a known, deterministic radar pulse. If there are no objects in range of the radar, the radar outputs only noise from its amplifiers. We model the noise by a zero-mean WSS process X_t with power spectral density $S_X(f)$. If there is an object in range of the radar, the system returns the reflected radar pulse, say $v(t)$, which is known, plus the noise X_t. We wish to design a system that decides whether the received waveform is noise only, X_t, or signal plus noise, $v(t) + X_t$. As an aid

10.7 The matched filter

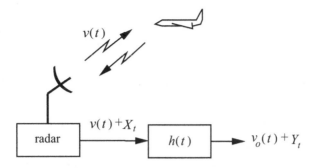

Figure 10.20. Block diagram of radar system and matched filter.

to achieving this goal, we propose to take the received waveform and pass it through an LTI system with impulse response $h(t)$. If the received signal is in fact $v(t) + X_t$, then as shown in Figure 10.20, the output of the linear system is

$$\int_{-\infty}^{\infty} h(t-\tau)[v(\tau) + X_\tau]\,d\tau = v_o(t) + Y_t,$$

where

$$v_o(t) := \int_{-\infty}^{\infty} h(t-\tau)v(\tau)\,d\tau$$

is the output signal, and

$$Y_t := \int_{-\infty}^{\infty} h(t-\tau)X_\tau\,d\tau$$

is the output noise process.

Typically, at the radar output, the signal $v(t)$ is obscured by the noise X_t. For example, at the top in Figure 10.21, a triangular signal $v(t)$ and broadband noise X_t are shown in the same graph. At the bottom is their sum $v(t) + X_t$, in which it is difficult to discern the triangular signal. By passing $v(t) + X_t$ through through the matched filter derived below, the presence of the signal can be made much more obvious at the filter output. The matched filter output is shown later in Figure 10.23.

We now find the impulse response h that maximizes the output **signal-to-noise ratio** (SNR)

$$\text{SNR} := \frac{v_o(t_0)^2}{\mathsf{E}[Y_{t_0}^2]},$$

where $v_o(t_0)^2$ is the instantaneous output signal power at time t_0, and $\mathsf{E}[Y_{t_0}^2]$ is the expected instantaneous output noise power at time t_0. Note that since $\mathsf{E}[Y_{t_0}^2] = R_Y(0) = P_Y$, we can also write

$$\text{SNR} = \frac{v_o(t_0)^2}{P_Y}.$$

Our approach is to obtain an upper bound on the numerator of the form $v_o(t_0)^2 \leq P_Y \cdot B$, where B does not depend on the impulse response h. It will then follow that

$$\text{SNR} = \frac{v_o(t_0)^2}{P_Y} \leq \frac{P_Y \cdot B}{P_Y} = B.$$

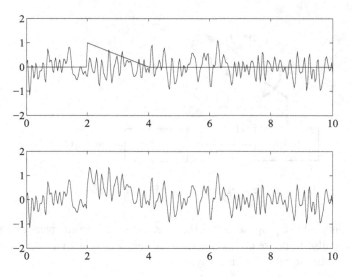

Figure 10.21. A triangular signal $v(t)$ and broadband noise X_t (top). Their sum, $v(t) + X_t$ (bottom), shows that the noise hides the presence of the signal.

We then show how to choose the impulse response so that in fact $v_o(t_0)^2 = P_Y \cdot B$. For this choice of impulse response, we then have SNR $= B$, the maximum possible value.

We begin by analyzing the denominator in the SNR. Observe that

$$P_Y = \int_{-\infty}^{\infty} S_Y(f) \, df = \int_{-\infty}^{\infty} |H(f)|^2 S_X(f) \, df.$$

To analyze the numerator, write

$$v_o(t_0) = \int_{-\infty}^{\infty} V_o(f) e^{j2\pi f t_0} \, df = \int_{-\infty}^{\infty} H(f) V(f) e^{j2\pi f t_0} \, df, \quad (10.27)$$

where $V_o(f)$ is the Fourier transform of $v_o(t)$, and $V(f)$ is the Fourier transform of $v(t)$. Next, write

$$v_o(t_0) = \int_{-\infty}^{\infty} H(f) \sqrt{S_X(f)} \cdot \frac{V(f) e^{j2\pi f t_0}}{\sqrt{S_X(f)}} \, df$$

$$= \int_{-\infty}^{\infty} H(f) \sqrt{S_X(f)} \cdot \left[\frac{V(f)^* e^{-j2\pi f t_0}}{\sqrt{S_X(f)}} \right]^* df,$$

where the asterisk denotes complex conjugation. Applying the Cauchy–Schwarz inequality for time functions (Problem 2), we obtain the upper bound,

$$|v_o(t)|^2 \leq \underbrace{\int_{-\infty}^{\infty} |H(f)|^2 S_X(f) \, df}_{= P_Y} \cdot \underbrace{\int_{-\infty}^{\infty} \frac{|V(f)|^2}{S_X(f)} \, df}_{=: B}.$$

Thus,

$$\text{SNR} = \frac{|v_o(t_0)|^2}{P_Y} \leq \frac{P_Y \cdot B}{P_Y} = B.$$

10.7 The matched filter

Now, the Cauchy–Schwarz inequality holds with equality if and only if $H(f)\sqrt{S_X(f)}$ is a multiple of $V(f)^* e^{-j2\pi f t_0}/\sqrt{S_X(f)}$. Thus, the upper bound on the SNR will be achieved if we take $H(f)$ to solve

$$H(f)\sqrt{S_X(f)} = \alpha \frac{V(f)^* e^{-j2\pi f t_0}}{\sqrt{S_X(f)}},$$

where α is a constant;[c] i.e., we should take

$$\boxed{H(f) = \alpha \frac{V(f)^* e^{-j2\pi f t_0}}{S_X(f)}.} \tag{10.28}$$

Thus, the optimal filter is "matched" to the known signal and known noise power spectral density.

Example 10.31. Consider the special case in which X_t is white noise with power spectral density $S_X(f) = N_0/2$. Taking $\alpha = N_0/2$ as well, we have $H(f) = V(f)^* e^{-j2\pi f t_0}$, which inverse transforms to $h(t) = v(t_0 - t)$, assuming $v(t)$ is real. Thus, the matched filter has an impulse response which is a time-reversed and translated copy of the known signal $v(t)$. An example of $v(t)$ and the corresponding $h(t)$ are shown in Figure 10.22. As the figure illustrates, if $v(t)$ is a finite-duration, causal waveform, as any radar "pulse" would be, then the sampling time t_0 can always be chosen so that $h(t)$ corresponds to a causal system.

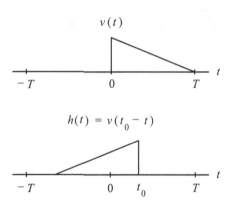

Figure 10.22. Known signal $v(t)$ and corresponding matched filter impulse response $h(t)$ in the case of white noise.

Analysis of the matched filter output

We show that the matched filter forces the components of the output $v_o(t) + Y_t$ to be related by

$$v_o(t) = \tfrac{1}{\alpha} R_Y(t - t_0). \tag{10.29}$$

---

[c] The constant α must be real in order for the matched filter impulse response to be real.

In other words, the matched filter forces the output signal to be proportional to a time-shifted correlation function. Hence, the maximum value of $|v_o(t)|$ occurs at $t = t_0$. Equation (10.29) also implies that the filter output $v_o(t) + Y_t$ has the form of a correlation function plus noise.

We now derive (10.29). Since $S_Y(f) = |H(f)|^2 S_X(f)$, if $H(f)$ is the matched filter, then

$$S_Y(f) = \frac{|\alpha V(f)|^2}{S_X(f)^2} S_X(f) = \frac{|\alpha V(f)|^2}{S_X(f)},$$

and

$$R_Y(\tau) = \int_{-\infty}^{\infty} \frac{|\alpha V(f)|^2}{S_X(f)} e^{j2\pi f \tau} df. \tag{10.30}$$

Now observe that if we put $t_0 = t$ in (10.27), and if $H(f)$ is the matched filter, then

$$v_o(t) = \int_{-\infty}^{\infty} \alpha \frac{|V(f)|^2}{S_X(f)} e^{j2\pi f(t-t_0)} df.$$

Hence, (10.29) holds.

Example 10.32. If the noise is white with power spectral density $S_X(f) = N_0/2$, and if $\alpha = N_0/2$, then comparing (10.30) and (10.26) shows that $R_Y(\tau) = \alpha R_v(\tau)$, where $R_v(\tau)$ is the correlation function of the deterministic signal $v(t)$. We also have $v_o(t) = R_v(t - t_0)$. When $v(t)$ is the triangular waveform shown at the top in Figure 10.21, the signal $v_o(t) = R_v(t - t_0)$ and a sample path of Y_t are shown at the top in Figure 10.23 in the same graph. At the bottom is the sum $v_o(t) + Y_t$.

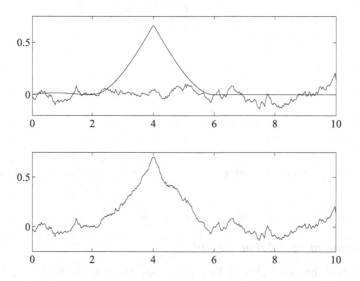

Figure 10.23. Matched filter output terms $v_o(t)$ and Y_t (top) and their sum $v_o(t) + Y_t$ (bottom), when $v(t)$ is the signal at the top in Figure 10.19 and $H(f)$ is the corresponding matched filter.

10.8 The Wiener filter

In the preceding section, the available data was of the form $v(t) + X_t$, where $v(t)$ was a known, nonrandom signal, and X_t was a zero-mean, WSS noise process. In this section, we suppose that V_t is an unknown random process that we would like to estimate based on observing a related random process U_t. For example, we might have $U_t = V_t + X_t$, where X_t is a noise process. However, for generality, we assume only that U_t and V_t are zero-mean, J-WSS with known power spectral densities and known cross power spectral density.

We restrict attention to linear estimators of the form

$$\widehat{V}_t = \int_{-\infty}^{\infty} h(t-\tau) U_\tau \, d\tau = \int_{-\infty}^{\infty} h(\theta) U_{t-\theta} \, d\theta, \qquad (10.31)$$

as shown in Figure 10.24. Note that to estimate V_t at a single time t, we use the entire observed waveform U_τ for $-\infty < \tau < \infty$. Our goal is to find an impulse response h that minimizes the **mean-squared error**, $\mathsf{E}[|V_t - \widehat{V}_t|^2]$. In other words, we are looking for an impulse response h such that if $\widehat{V}_t$ is given by (10.31), and if $\tilde{h}$ is any other impulse response, and we put

$$\tilde{V}_t = \int_{-\infty}^{\infty} \tilde{h}(t-\tau) U_\tau \, d\tau = \int_{-\infty}^{\infty} \tilde{h}(\theta) U_{t-\theta} \, d\theta, \qquad (10.32)$$

then

$$\mathsf{E}[|V_t - \widehat{V}_t|^2] \leq \mathsf{E}[|V_t - \tilde{V}_t|^2].$$

To find the optimal filter h, we apply the **orthogonality principle** (derived below), which says that if

$$\mathsf{E}\left[(V_t - \widehat{V}_t) \int_{-\infty}^{\infty} \tilde{h}(\theta) U_{t-\theta} \, d\theta\right] = 0 \qquad (10.33)$$

for *every* filter $\tilde{h}$, then h is the optimal filter.

Before proceeding any further, we need the following observation. Suppose (10.33) holds for *every* choice of $\tilde{h}$. Then in particular, it holds if we replace $\tilde{h}$ by $h - \tilde{h}$. Making this substitution in (10.33) yields

$$\mathsf{E}\left[(V_t - \widehat{V}_t) \int_{-\infty}^{\infty} [h(\theta) - \tilde{h}(\theta)] U_{t-\theta} \, d\theta\right] = 0.$$

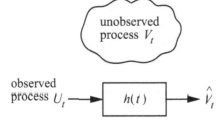

Figure 10.24. Estimation of an unobserved process V_t by passing an observed process U_t through an LTI system with impulse response $h(t)$.

Since the integral in this expression is simply $\widehat{V}_t - \widetilde{V}_t$, we have that

$$E[(V_t - \widehat{V}_t)(\widehat{V}_t - \widetilde{V}_t)] = 0. \tag{10.34}$$

To establish the orthogonality principle, assume (10.33) holds for every choice of $\widetilde{h}$. Then (10.34) holds as well. Now write

$$\begin{aligned}
E[|V_t - \widetilde{V}_t|^2] &= E[|(V_t - \widehat{V}_t) + (\widehat{V}_t - \widetilde{V}_t)|^2] \\
&= E[|V_t - \widehat{V}_t|^2 + 2(V_t - \widehat{V}_t)(\widehat{V}_t - \widetilde{V}_t) + |\widehat{V}_t - \widetilde{V}_t|^2] \\
&= E[|V_t - \widehat{V}_t|^2] + 2E[(V_t - \widehat{V}_t)(\widehat{V}_t - \widetilde{V}_t)] + E[|\widehat{V}_t - \widetilde{V}_t|^2] \\
&= E[|V_t - \widehat{V}_t|^2] + E[|\widehat{V}_t - \widetilde{V}_t|^2] \\
&\geq E[|V_t - \widehat{V}_t|^2],
\end{aligned}$$

and thus, h is the filter that minimizes the mean-squared error.

The next task is to characterize the filter h such that (10.33) holds for every choice of $\widetilde{h}$. Write (10.33) as

$$\begin{aligned}
0 &= E\left[(V_t - \widehat{V}_t)\int_{-\infty}^{\infty} \widetilde{h}(\theta)U_{t-\theta}\,d\theta\right] \\
&= E\left[\int_{-\infty}^{\infty} \widetilde{h}(\theta)(V_t - \widehat{V}_t)U_{t-\theta}\,d\theta\right] \\
&= \int_{-\infty}^{\infty} E[\widetilde{h}(\theta)(V_t - \widehat{V}_t)U_{t-\theta}]\,d\theta \\
&= \int_{-\infty}^{\infty} \widetilde{h}(\theta)E[(V_t - \widehat{V}_t)U_{t-\theta}]\,d\theta \\
&= \int_{-\infty}^{\infty} \widetilde{h}(\theta)[R_{VU}(\theta) - R_{\widehat{V}U}(\theta)]\,d\theta.
\end{aligned}$$

Since this must hold for all $\widetilde{h}$, take $\widetilde{h}(\theta) = R_{VU}(\theta) - R_{\widehat{V}U}(\theta)$ to get

$$\int_{-\infty}^{\infty} |R_{VU}(\theta) - R_{\widehat{V}U}(\theta)|^2\,d\theta = 0. \tag{10.35}$$

Thus, (10.33) holds for all $\widetilde{h}$ if and only if $R_{VU} = R_{\widehat{V}U}$.

The next task is to analyze $R_{\widehat{V}U}$. Recall that $\widehat{V}_t$ in (10.31) is the response of an LTI system to input U_t. Applying (10.18) with X replaced by U and Y replaced by $\widehat{V}$, we have, also using the fact that R_U is even,

$$R_{\widehat{V}U}(\tau) = R_{U\widehat{V}}(-\tau) = \int_{-\infty}^{\infty} h(\theta)R_U(\tau - \theta)\,d\theta.$$

Taking Fourier transforms of

$$R_{VU}(\tau) = R_{\widehat{V}U}(\tau) = \int_{-\infty}^{\infty} h(\theta)R_U(\tau - \theta)\,d\theta \tag{10.36}$$

yields

$$S_{VU}(f) = H(f)S_U(f).$$

10.8 The Wiener filter

Thus,

$$H(f) = \frac{S_{VU}(f)}{S_U(f)} \qquad (10.37)$$

is the optimal filter. This choice of $H(f)$ is called the **Wiener filter**.

*Causal Wiener filters

Typically, the Wiener filter as found above is not causal; i.e., we do not have $h(t) = 0$ for $t < 0$. To find such an h, we need to reformulate the problem by replacing (10.31) with

$$\widehat{V}_t = \int_{-\infty}^{t} h(t - \tau) U_\tau \, d\tau = \int_0^\infty h(\theta) U_{t-\theta} \, d\theta,$$

and replacing (10.32) with

$$\widetilde{V}_t = \int_{-\infty}^{t} \tilde{h}(t - \tau) U_\tau \, d\tau = \int_0^\infty \tilde{h}(\theta) U_{t-\theta} \, d\theta.$$

Everything proceeds as before from (10.33) through (10.35) except that lower limits of integration are changed from $-\infty$ to 0. Thus, instead of concluding $R_{VU}(\tau) = R_{\widehat{V}U}(\tau)$ for all τ, we only have $R_{VU}(\tau) = R_{\widehat{V}U}(\tau)$ for $\tau \geq 0$. Instead of (10.36), we have

$$R_{VU}(\tau) = \int_0^\infty h(\theta) R_U(\tau - \theta) \, d\theta, \quad \tau \geq 0. \qquad (10.38)$$

This is known as the **Wiener–Hopf equation**. Because the equation only holds for $\tau \geq 0$, we run into a problem if we try to take Fourier transforms. To compute $S_{VU}(f)$, we need to integrate $R_{VU}(\tau) e^{-j2\pi f \tau}$ from $\tau = -\infty$ to $\tau = \infty$. But we can use the Wiener–Hopf equation only for $\tau \geq 0$.

In general, the Wiener–Hopf equation is difficult to solve. However, if U is white noise, say $R_U(\theta) = \delta(\theta)$, then (10.38) reduces to

$$R_{VU}(\tau) = h(\tau), \quad \tau \geq 0.$$

Since h is causal, $h(\tau) = 0$ for $\tau < 0$.

The preceding observation suggests the construction of $H(f)$ using a **whitening filter** as shown in Figure 10.25. If U_t is not white noise, suppose we can find a causal filter $K(f)$ such that when U_t is passed through this system, the output is white noise W_t, by which we

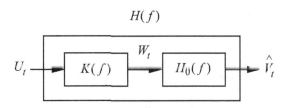

Figure 10.25. Decomposition of the causal Wiener filter using the whitening filter $K(f)$.

mean $S_W(f) = 1$. Letting k denote the impulse response corresponding to K, we can write W_t mathematically as

$$W_t = \int_0^\infty k(\theta) U_{t-\theta} \, d\theta. \quad (10.39)$$

Then

$$1 = S_W(f) = |K(f)|^2 S_U(f). \quad (10.40)$$

Consider the problem of causally estimating V_t based on W_t instead of U_t. The solution is again given by the Wiener–Hopf equation,

$$R_{VW}(\tau) = \int_0^\infty h_0(\theta) R_W(\tau - \theta) \, d\theta, \quad \tau \geq 0.$$

Since K was chosen so that $S_W(f) = 1$, $R_W(\theta) = \delta(\theta)$. Therefore, the Wiener–Hopf equation tells us that $h_0(\tau) = R_{VW}(\tau)$ for $\tau \geq 0$. Using (10.39), it is easy to see that

$$R_{VW}(\tau) = \int_0^\infty k(\theta) R_{VU}(\tau + \theta) \, d\theta, \quad (10.41)$$

and then[d]

$$S_{VW}(f) = K(f)^* S_{VU}(f). \quad (10.42)$$

We now summarize the procedure for designing the causal Wiener filter.

(i) According to (10.40), we must first write $S_U(f)$ in the form

$$S_U(f) = \frac{1}{K(f)} \cdot \frac{1}{K(f)^*},$$

where $K(f)$ is a causal filter (this is known as **spectral factorization**).[e]

(ii) The optimum filter is $H(f) = H_0(f) K(f)$, where

$$H_0(f) = \int_0^\infty R_{VW}(\tau) e^{-j2\pi f \tau} \, d\tau,$$

and $R_{VW}(\tau)$ is given by (10.41) or by the inverse transform of (10.42).

Example 10.33. Let $U_t = V_t + X_t$, where V_t and X_t are zero-mean, WSS processes with $E[V_t X_\tau] = 0$ for all t and τ. Assume that the signal V_t has power spectral density $S_V(f) = 2\lambda/[\lambda^2 + (2\pi f)^2]$ and that the noise X_t is white with power spectral density $S_X(f) = 1$. Find the causal Wiener filter.

---

[d]If $k(\theta)$ is complex valued, so is W_t in (10.39). In this case, as in Problem 46, it is understood that $R_{VW}(\tau) = E[V_{t+\tau} W_t^*]$.

[e]If $S_U(f)$ satisfies the **Paley–Wiener condition**,

$$\int_{-\infty}^\infty \frac{|\ln S_U(f)|}{1 + f^2} \, df < \infty,$$

then $S_U(f)$ can always be factored in this way.

Solution. From your solution of Problem 59, $S_U(f) = S_V(f) + S_X(f)$. Thus,

$$S_U(f) = \frac{2\lambda}{\lambda^2 + (2\pi f)^2} + 1 = \frac{A^2 + (2\pi f)^2}{\lambda^2 + (2\pi f)^2},$$

where $A^2 := \lambda^2 + 2\lambda$. This factors into

$$S_U(f) = \frac{A + j2\pi f}{\lambda + j2\pi f} \cdot \frac{A - j2\pi f}{\lambda - j2\pi f}.$$

Then

$$K(f) = \frac{\lambda + j2\pi f}{A + j2\pi f}$$

is the required causal (by Problem 64) whitening filter. Next, from your solution of Problem 59, $S_{VU}(f) = S_V(f)$. So, by (10.42),

$$S_{VW}(f) = \frac{\lambda - j2\pi f}{A - j2\pi f} \cdot \frac{2\lambda}{\lambda^2 + (2\pi f)^2}$$

$$= \frac{2\lambda}{(A - j2\pi f)(\lambda + j2\pi f)}$$

$$= \frac{B}{A - j2\pi f} + \frac{B}{\lambda + j2\pi f},$$

where $B := 2\lambda/(\lambda + A)$. It follows that

$$R_{VW}(\tau) = Be^{A\tau}u(-\tau) + Be^{-\lambda\tau}u(\tau),$$

where u is the unit-step function. Since $h_0(\tau) = R_{VW}(\tau)$ for $\tau \geq 0$, $h_0(\tau) = Be^{-\lambda\tau}u(\tau)$ and $H_0(f) = B/(\lambda + j2\pi f)$. Next,

$$H(f) = H_0(f)K(f) = \frac{B}{\lambda + j2\pi f} \cdot \frac{\lambda + j2\pi f}{A + j2\pi f} = \frac{B}{A + j2\pi f},$$

and $h(\tau) = Be^{-A\tau}u(\tau)$.

10.9 *The Wiener–Khinchin theorem

The Wiener–Khinchin theorem gives an alternative representation of the power spectral density of a WSS process. A slight modification of the derivation will allow us to derive the mean-square ergodic theorem in the next section.

Recall that the expected average power in a process X_t is

$$P_X := \mathsf{E}\left[\lim_{T \to \infty} \frac{1}{2T} \int_{-T}^{T} X_t^2 \, dt\right].$$

Let

$$X_t^T := \begin{cases} X_t, & |t| \leq T, \\ 0, & |t| > T, \end{cases}$$

so that $\int_{-T}^{T} X_t^2 \, dt = \int_{-\infty}^{\infty} |X_t^T|^2 \, dt$. The Fourier transform of X_t^T is

$$\widetilde{X}_f^T := \int_{-\infty}^{\infty} X_t^T e^{-j2\pi ft} \, dt = \int_{-T}^{T} X_t e^{-j2\pi ft} \, dt, \qquad (10.43)$$

and by Parseval's equation,

$$\int_{-\infty}^{\infty} |X_t^T|^2 \, dt = \int_{-\infty}^{\infty} |\widetilde{X}_f^T|^2 \, df.$$

We can now write

$$\begin{aligned} P_X &= \mathsf{E}\left[\lim_{T\to\infty} \frac{1}{2T} \int_{-T}^{T} X_t^2 \, dt\right] \\ &= \mathsf{E}\left[\lim_{T\to\infty} \frac{1}{2T} \int_{-\infty}^{\infty} |X_t^T|^2 \, dt\right] \\ &= \mathsf{E}\left[\lim_{T\to\infty} \frac{1}{2T} \int_{-\infty}^{\infty} |\widetilde{X}_f^T|^2 \, df\right] \\ &= \int_{-\infty}^{\infty} \left(\lim_{T\to\infty} \frac{\mathsf{E}[|\widetilde{X}_f^T|^2]}{2T}\right) df. \end{aligned}$$

The **Wiener–Khinchin theorem** says that for a WSS process, the above integrand is exactly the power spectral density $S_X(f)$. In particular, since the integrand is nonnegative, the Wiener–Khinchin theorem provides another proof that $S_X(f)$ must be nonnegative.

To derive the Wiener–Khinchin theorem, we begin with the numerator,

$$\mathsf{E}[|\widetilde{X}_f^T|^2] = \mathsf{E}[(\widetilde{X}_f^T)(\widetilde{X}_f^T)^*],$$

where the asterisk denotes complex conjugation. To evaluate the right-hand side, use (10.43) to obtain

$$\mathsf{E}\left[\left(\int_{-T}^{T} X_t e^{-j2\pi ft} \, dt\right)\left(\int_{-T}^{T} X_\theta e^{-j2\pi f\theta} \, d\theta\right)^*\right].$$

We can now write

$$\begin{aligned} \mathsf{E}[|\widetilde{X}_f^T|^2] &= \int_{-T}^{T}\int_{-T}^{T} \mathsf{E}[X_t X_\theta] e^{-j2\pi f(t-\theta)} \, dt \, d\theta \\ &= \int_{-T}^{T}\int_{-T}^{T} R_X(t-\theta) e^{-j2\pi f(t-\theta)} \, dt \, d\theta \qquad (10.44) \\ &= \int_{-T}^{T}\int_{-T}^{T} \left[\int_{-\infty}^{\infty} S_X(\nu) e^{j2\pi \nu(t-\theta)} \, d\nu\right] e^{-j2\pi f(t-\theta)} \, dt \, d\theta \\ &= \int_{-\infty}^{\infty} S_X(\nu) \left[\int_{-T}^{T} e^{j2\pi \theta(f-\nu)} \left(\int_{-T}^{T} e^{j2\pi t(\nu-f)} \, dt\right) d\theta\right] d\nu. \end{aligned}$$

Notice that the inner two integrals decouple so that

$$\begin{aligned} \mathsf{E}[|\widetilde{X}_f^T|^2] &= \int_{-\infty}^{\infty} S_X(\nu) \left[\int_{-T}^{T} e^{j2\pi \theta(f-\nu)} \, d\theta\right]\left(\int_{-T}^{T} e^{j2\pi t(\nu-f)} \, dt\right) d\nu \\ &= \int_{-\infty}^{\infty} S_X(\nu) \cdot \left[2T \frac{\sin(2\pi T(f-\nu))}{2\pi T(f-\nu)}\right]^2 d\nu. \end{aligned}$$

We can then write

$$\frac{\mathsf{E}[|\widetilde{X}_f^T|^2]}{2T} = \int_{-\infty}^{\infty} S_X(v) \cdot 2T \left[\frac{\sin(2\pi T(f-v))}{2\pi T(f-v)}\right]^2 dv. \qquad (10.45)$$

This is a convolution integral. Furthermore, the quantity multiplying $S_X(v)$ converges to the delta function $\delta(f-v)$ as $T \to \infty$.[1] Thus,

$$\lim_{T \to \infty} \frac{\mathsf{E}[|\widetilde{X}_f^T|^2]}{2T} = \int_{-\infty}^{\infty} S_X(v) \delta(f-v) dv = S_X(f),$$

which is exactly the Wiener–Khinchin theorem.

Remark. The preceding derivation shows that $S_X(f)$ is equal to the limit of (10.44) divided by $2T$. Thus,

$$S(f) = \lim_{T \to \infty} \frac{1}{2T} \int_{-T}^{T} \int_{-T}^{T} R_X(t-\theta) e^{-j2\pi f(t-\theta)} dt\, d\theta.$$

As noted in Problem 66, the properties of the correlation function directly imply that this double integral is nonnegative. This is the direct way to prove that power spectral densities are nonnegative.

10.10 *Mean-square ergodic theorem for WSS processes

As an easy corollary of the derivation of the Wiener–Khinchin theorem, we derive the mean-square ergodic theorem for WSS processes. This result shows that $\mathsf{E}[X_t]$ can often be computed by averaging a single sample path over time.

In the process of deriving the weak law of large numbers in Chapter 3, we showed that for an uncorrelated sequence X_n with common mean $m = \mathsf{E}[X_n]$ and common variance $\sigma^2 = \text{var}(X_n)$, the sample mean (or time average)

$$M_n := \frac{1}{n}\sum_{i=1}^{n} X_i$$

converges to m in the sense that $\mathsf{E}[|M_n - m|^2] = \text{var}(M_n) \to 0$ as $n \to \infty$ by (3.7). In this case, we say that M_n converges in mean square to m, and we call this a **mean-square law of large numbers**.

Let Y_t be a WSS process with mean $m = \mathsf{E}[Y_t]$ and covariance function $C_Y(\tau)$. We show below that if the Fourier transform of C_Y is continuous at $f = 0$, then the sample mean (or time average)

$$M_T := \frac{1}{2T}\int_{-T}^{T} Y_t\, dt \to m \qquad (10.46)$$

in the sense that $\mathsf{E}[|M_T - m|^2] \to 0$ as $T \to \infty$.

We can view this result as a mean-square law of large numbers for WSS processes. Laws of large numbers for sequences or processes that are not uncorrelated are often called **ergodic theorems**. The point in all theorems of this type is that the expectation $\mathsf{E}[Y_t]$ can be computed by averaging a single sample path over time.

To prove the above result, put $X_t := Y_t - m$ so that X_t is zero mean and has correlation function $R_X(\tau) = C_Y(\tau)$. If the Fourier transform of C_Y is continuous at $f = 0$, then so is the power spectral density $S_X(f)$. Write

$$M_T - m = \frac{1}{2T}\int_{-T}^{T} X_t\,dt = \frac{\widetilde{X}_f^T\big|_{f=0}}{2T},$$

where $\widetilde{X}_f^T$ was defined in (10.43). Then

$$\mathsf{E}[|M_T - m|^2] = \frac{\mathsf{E}[|\widetilde{X}_0^T|^2]}{4T^2} = \frac{1}{2T} \cdot \frac{\mathsf{E}[|\widetilde{X}_0^T|^2]}{2T}.$$

Now use (10.45) with $f = 0$ to write

$$\mathsf{E}[|M_T - m|^2] = \frac{1}{2T}\int_{-\infty}^{\infty} S_X(\nu) \cdot 2T \left[\frac{\sin(2\pi T \nu)}{2\pi T \nu}\right]^2 d\nu. \tag{10.47}$$

By the argument following (10.45), as $T \to \infty$ the integral in (10.47) is approximately $S_X(0)$ if $S_X(f)$ is continuous at $f = 0$.[2] Thus, if $S_X(f)$ is continuous at $f = 0$,

$$\mathsf{E}[|M_T - m|^2] \approx \frac{S_X(0)}{2T} \to 0$$

as $T \to \infty$.

If the Fourier transform of C_Y is not available, how can we use the above result? Here is a sufficient condition on C_Y that guarantees continuity of its transform without actually computing it. If $R_X(\tau) = C_Y(\tau)$ is absolutely integrable, then $S_X(f)$ is uniformly continuous. To see this write

$$|S_X(f) - S_X(f_0)| = \left|\int_{-\infty}^{\infty} R_X(\tau)e^{-j2\pi f\tau}\,d\tau - \int_{-\infty}^{\infty} R_X(\tau)e^{-j2\pi f_0\tau}\,d\tau\right|$$

$$\leq \int_{-\infty}^{\infty} |R_X(\tau)|\,|e^{-j2\pi f\tau} - e^{-j2\pi f_0\tau}|\,d\tau$$

$$= \int_{-\infty}^{\infty} |R_X(\tau)|\,|e^{-j2\pi f_0\tau}[e^{-j2\pi(f-f_0)\tau} - 1]|\,d\tau$$

$$= \int_{-\infty}^{\infty} |R_X(\tau)|\,|e^{-j2\pi(f-f_0)\tau} - 1|\,d\tau.$$

Now observe that $|e^{-j2\pi(f-f_0)\tau} - 1| \to 0$ as $f \to f_0$. Since R_X is absolutely integrable, Lebesgue's dominated convergence theorem [3, p. 209] implies that the integral goes to zero as well.

We also note here that Parseval's equation shows that (10.47) is equivalent to

$$\mathsf{E}[|M_T - m|^2] = \frac{1}{2T}\int_{-2T}^{2T} R_X(\tau)\left(1 - \frac{|\tau|}{2T}\right)d\tau$$

$$= \frac{1}{2T}\int_{-2T}^{2T} C_Y(\tau)\left(1 - \frac{|\tau|}{2T}\right)d\tau.$$

Thus, M_T in (10.46) converges in mean square to m if and only if

$$\lim_{T\to\infty}\frac{1}{2T}\int_{-2T}^{2T}C_Y(\tau)\left(1-\frac{|\tau|}{2T}\right)d\tau = 0. \tag{10.48}$$

Example 10.34. Let $Z \sim$ Bernoulli(p) for some $0 < p < 1$, and put $Y_t := Z$ for all t. Then Y_t is strictly stationary, but $M_T = Z$ is either 0 or 1 for all T and therefore cannot converge to $\mathsf{E}[Y_t] = p$. It is also easy to see that (10.48) does not hold. Since $C_Y(\tau) = \mathrm{var}(Z) = p(1-p)$,

$$\frac{1}{2T}\int_{-2T}^{2T}C_Y(\tau)\left(1-\frac{|\tau|}{2T}\right)d\tau = p(1-p) \not\to 0.$$

10.11 *Power spectral densities for non-WSS processes

In Section 10.9, we showed that the expected average power in a process X_t can be written in the form

$$P_X = \int_{-\infty}^{\infty}\left(\lim_{T\to\infty}\frac{\mathsf{E}[|\widetilde{X}_f^T|^2]}{2T}\right)df.$$

The derivation of this formula did not assume that X_t is WSS. However, if it is, the Wiener–Khinchin theorem showed that the integrand was the power spectral density. At the end of this section, we show that whether or not X_t is WSS,

$$\lim_{T\to\infty}\frac{\mathsf{E}[|\widetilde{X}_f^T|^2]}{2T} = \int_{-\infty}^{\infty}\overline{R}_X(\tau)e^{-j2\pi f\tau}\,d\tau, \tag{10.49}$$

where[f]

$$\overline{R}_X(\tau) := \lim_{T\to\infty}\frac{1}{2T}\int_{-T}^{T}R_X(\tau+\theta,\theta)\,d\theta. \tag{10.50}$$

Hence, for a non-WSS process, we define its power spectral density to be the Fourier transform of $\overline{R}_X(\tau)$,

$$S_X(f) := \int_{-\infty}^{\infty}\overline{R}_X(\tau)e^{-j2\pi f\tau}\,d\tau.$$

To justify the name power spectral density, we need to show that its integral over every frequency band is the power in that band. This will follow exactly as in the WSS case, provided we can show that the new definition of power spectral density still satisfies $S_Y(f) = |H(f)|^2 S_X(f)$. See Problems 71–73.

An important application the foregoing is to **cyclostationary** processes. A process Y_t is (wide-sense) cyclostationary if its mean function is periodic in t, and if its correlation function has the property that for fixed τ, $R_X(\tau+\theta,\theta)$ is periodic in θ. For a cyclostationary process with period T_0, it is not hard to show that

$$\overline{R}_X(\tau) = \frac{1}{T_0}\int_0^{T_0}R_X(\tau+\theta,\theta)\,d\theta. \tag{10.51}$$

---

[f]Note that if X_t is WSS, then $\overline{R}_X(\tau) = R_X(\tau)$.

Example 10.35. Let X_t be WSS, and put $Y_t := X_t \cos(2\pi f_0 t)$. Show that Y_t is cyclostationary and that

$$S_Y(f) = \frac{1}{4}[S_X(f-f_0) + S_X(f+f_0)].$$

Solution. The mean of Y_t is

$$\mathsf{E}[Y_t] = \mathsf{E}[X_t \cos(2\pi f_0 t)] = \mathsf{E}[X_t] \cos(2\pi f_0 t).$$

Because X_t is WSS, $\mathsf{E}[X_t]$ does not depend on t, and it is then clear that $\mathsf{E}[Y_t]$ has period $1/f_0$. Next consider

$$\begin{aligned} R_Y(t+\theta,\theta) &= \mathsf{E}[Y_{t+\theta}Y_\theta] \\ &= \mathsf{E}[X_{t+\theta}\cos(2\pi f_0\{t+\theta\})X_\theta\cos(2\pi f_0\theta)] \\ &= R_X(t)\cos(2\pi f_0\{t+\theta\})\cos(2\pi f_0\theta), \end{aligned}$$

which is periodic in θ with period $1/f_0$. To compute $S_Y(f)$, first use a trigonometric identity to write

$$R_Y(t+\theta,\theta) = \frac{R_X(t)}{2}[\cos(2\pi f_0 t) + \cos(2\pi f_0\{t+2\theta\})].$$

Applying (10.51) to R_Y with $T_0 = 1/f_0$ yields

$$\overline{R}_Y(t) = \frac{R_X(t)}{2}\cos(2\pi f_0 t).$$

Taking Fourier transforms yields the claimed formula for $S_Y(f)$.

Derivation of (10.49)

We begin as in the derivation of the Wiener–Khinchin theorem, except that instead of (10.44) we have

$$\begin{aligned} \mathsf{E}[|\widetilde{X}_f^T|^2] &= \int_{-T}^{T}\int_{-T}^{T} R_X(t,\theta)e^{-j2\pi f(t-\theta)}\,dt\,d\theta \\ &= \int_{-T}^{T}\int_{-\infty}^{\infty} I_{[-T,T]}(t)R_X(t,\theta)e^{-j2\pi f(t-\theta)}\,dt\,d\theta. \end{aligned}$$

Now make the change of variable $\tau = t - \theta$ in the inner integral. This results in

$$\mathsf{E}[|\widetilde{X}_f^T|^2] = \int_{-T}^{T}\int_{-\infty}^{\infty} I_{[-T,T]}(\tau+\theta)R_X(\tau+\theta,\theta)e^{-j2\pi f\tau}\,d\tau\,d\theta.$$

Change the order of integration to get

$$\mathsf{E}[|\widetilde{X}_f^T|^2] = \int_{-\infty}^{\infty} e^{-j2\pi f\tau}\int_{-T}^{T} I_{[-T,T]}(\tau+\theta)R_X(\tau+\theta,\theta)\,d\theta\,d\tau.$$

To simplify the inner integral, observe that $I_{[-T,T]}(\tau+\theta) = I_{[-T-\tau,T-\tau]}(\theta)$. Now $T-\tau$ is to the left of $-T$ if $2T < \tau$, and $-T-\tau$ is to the right of T if $-2T > \tau$. Thus,

$$\frac{\mathsf{E}[|\widetilde{X}_f^T|^2]}{2T} = \int_{-\infty}^{\infty} e^{-j2\pi f\tau} g_T(\tau)\,d\tau,$$

where

$$g_T(\tau) := \begin{cases} \frac{1}{2T} \int_{-T}^{T-\tau} R_X(\tau+\theta,\theta)\,d\theta, & 0 \leq \tau \leq 2T, \\ \frac{1}{2T} \int_{-T-\tau}^{T} R_X(\tau+\theta,\theta)\,d\theta, & -2T \leq \tau < 0, \\ 0, & |\tau| > 2T. \end{cases}$$

If T much greater than $|\tau|$, then $T - \tau \approx T$ and $-T - \tau \approx -T$ in the above limits of integration. Hence, if R_X is a reasonably-behaved correlation function,

$$\lim_{T \to \infty} g_T(\tau) = \lim_{T \to \infty} \frac{1}{2T} \int_{-T}^{T} R_X(\tau+\theta,\theta)\,d\theta = \overline{R}_X(\tau),$$

and we find that

$$\lim_{T \to \infty} \frac{\mathsf{E}[|\widetilde{X}_f^T|^2]}{2T} = \int_{-\infty}^{\infty} e^{-j2\pi f\tau} \lim_{T \to \infty} g_T(\tau)\,d\tau = \int_{-\infty}^{\infty} e^{-j2\pi f\tau} \overline{R}_X(\tau)\,d\tau.$$

Remark. If X_t is actually WSS, then $R_X(\tau+\theta,\theta) = R_X(\tau)$, and

$$g_T(\tau) = R_X(\tau)\left(1 - \frac{|\tau|}{2T}\right)I_{[-2T,2T]}(\tau).$$

In this case, for each fixed τ, $g_T(\tau) \to R_X(\tau)$. We thus have an alternative derivation of the Wiener–Khinchin theorem.

Notes

10.9: *The Wiener–Khinchin theorem

Note 1. To give a rigorous derivation of the fact that

$$\lim_{T \to \infty} \int_{-\infty}^{\infty} S_X(\nu) \cdot 2T \left[\frac{\sin(2\pi T(f-\nu))}{2\pi T(f-\nu)}\right]^2 d\nu = S_X(f),$$

it is convenient to assume $S_X(f)$ is continuous at f. Letting

$$\delta_T(f) := 2T\left[\frac{\sin(2\pi T f)}{2\pi T f}\right]^2,$$

we must show that

$$\left|\int_{-\infty}^{\infty} S_X(\nu)\,\delta_T(f-\nu)\,d\nu - S_X(f)\right| \to 0.$$

To proceed, we need the following properties of δ_T. First,

$$\int_{-\infty}^{\infty} \delta_T(f)\,df = 1.$$

This can be seen by using the Fourier transform table to evaluate the inverse transform of $\delta_T(f)$ at $t = 0$. Second, for fixed $\Delta f > 0$, as $T \to \infty$,

$$\int_{\{f:|f|>\Delta f\}} \delta_T(f)\,df \to 0.$$

This can be seen by using the fact that $\delta_T(f)$ is even and writing

$$\int_{\Delta f}^{\infty} \delta_T(f)\,df \leq \frac{2T}{(2\pi T)^2} \int_{\Delta f}^{\infty} \frac{1}{f^2}\,df = \frac{1}{2T\pi^2 \Delta f},$$

which goes to zero as $T \to \infty$. Third, for $|f| \geq \Delta f > 0$,

$$|\delta_T(f)| \leq \frac{1}{2T(\pi \Delta f)^2}.$$

Now, using the first property of δ_T, write

$$S_X(f) = S_X(f) \int_{-\infty}^{\infty} \delta_T(f-v)\,dv = \int_{-\infty}^{\infty} S_X(f)\,\delta_T(f-v)\,dv.$$

Then

$$S_X(f) - \int_{-\infty}^{\infty} S_X(v)\,\delta_T(f-v)\,dv = \int_{-\infty}^{\infty} [S_X(f) - S_X(v)]\,\delta_T(f-v)\,dv.$$

For the next step, let $\varepsilon > 0$ be given, and use the continuity of S_X at f to get the existence of a $\Delta f > 0$ such that for $|f - v| < \Delta f$, $|S_X(f) - S_X(v)| < \varepsilon$. Now break up the range of integration into v such that $|f - v| < \Delta f$ and v such that $|f - v| \geq \Delta f$. For the first range, we need the calculation

$$\left| \int_{f-\Delta f}^{f+\Delta f} [S_X(f) - S_X(v)]\,\delta_T(f-v)\,dv \right| \leq \int_{f-\Delta f}^{f+\Delta f} |S_X(f) - S_X(v)|\,\delta_T(f-v)\,dv$$

$$\leq \varepsilon \int_{f-\Delta f}^{f+\Delta f} \delta_T(f-v)\,dv$$

$$\leq \varepsilon \int_{-\infty}^{\infty} \delta_T(f-v)\,dv = \varepsilon.$$

For the second range of integration, consider the integral

$$\left| \int_{f+\Delta f}^{\infty} [S_X(f) - S_X(v)]\,\delta_T(f-v)\,dv \right| \leq \int_{f+\Delta f}^{\infty} |S_X(f) - S_X(v)|\,\delta_T(f-v)\,dv$$

$$\leq \int_{f+\Delta f}^{\infty} (|S_X(f)| + |S_X(v)|)\,\delta_T(f-v)\,dv$$

$$= |S_X(f)| \int_{f+\Delta f}^{\infty} \delta_T(f-v)\,dv$$

$$+ \int_{f+\Delta f}^{\infty} |S_X(v)|\,\delta_T(f-v)\,dv.$$

Observe that

$$\int_{f+\Delta f}^{\infty} \delta_T(f-v)\,dv = \int_{-\infty}^{-\Delta f} \delta_T(\theta)\,d\theta,$$

which goes to zero by the second property of δ_T. Using the third property, we have

$$\int_{f+\Delta f}^{\infty} |S_X(v)|\, \delta_T(f-v)\, dv = \int_{-\infty}^{-\Delta f} |S_X(f-\theta)|\, \delta_T(\theta)\, d\theta$$

$$\leq \frac{1}{2T(\pi\Delta f)^2} \int_{-\infty}^{-\Delta f} |S_X(f-\theta)|\, d\theta,$$

which also goes to zero as $T \to \infty$.

10.10: *Mean-square ergodic theorem for WSS processes

Note 2. In applying the derivation in Note 1 to the special case $f = 0$ in (10.47) we do not need $S_X(f)$ to be continuous for all f, we only need continuity at $f = 0$.

Problems

10.2: Characterization of random processes

1. Given waveforms $a(t)$, $b(t)$, and $c(t)$, let

$$g(t,i) := \begin{cases} a(t), & i=1, \\ b(t), & i=2, \\ c(t), & i=3, \end{cases}$$

and put $X_t := g(t,Z)$, where Z is a discrete random variable with $\mathsf{P}(Z = i) = p_i$ for $i = 1,2,3$. Express the mean function and the correlation function of X_t in terms of the p_i and $a(t)$, $b(t)$, and $c(t)$.

2. Derive the **Cauchy–Schwarz inequality** for complex-valued functions g and h,

$$\left| \int_{-\infty}^{\infty} g(\theta) h(\theta)^* \, d\theta \right|^2 \leq \int_{-\infty}^{\infty} |g(\theta)|^2 \, d\theta \cdot \int_{-\infty}^{\infty} |h(\theta)|^2 \, d\theta,$$

where the asterisk denotes complex conjugation, and for any complex number z, $|z|^2 = z \cdot z^*$. *Hint:* The Cauchy–Schwarz inequality for random variables (2.24) was derived in Chapter 2. Modify the derivation there by replacing expectations of the form $\mathsf{E}[XY]$ with integrals of the form $\int_{-\infty}^{\infty} g(\theta) h(\theta)^* \, d\theta$. Watch those complex conjugates!

3. Derive the formulas

$$C_X(t_1,t_2) = R_X(t_1,t_2) - m_X(t_1) m_X(t_2)$$

and

$$C_{XY}(t_1,t_2) = R_{XY}(t_1,t_2) - m_X(t_1) m_Y(t_2).$$

4. Show that $R_X(t_1,t_2) = \mathsf{E}[X_{t_1} X_{t_2}]$ is a **positive semidefinite function** in the sense that for any real or complex constants $c_1,\ldots,c_n$ and any times $t_1,\ldots,t_n$,

$$\sum_{i=1}^{n} \sum_{k=1}^{n} c_i R_X(t_i,t_k) c_k^* \geq 0.$$

Hint: Observe that
$$E\left[\left|\sum_{i=1}^{n} c_i X_{t_i}\right|^2\right] \geq 0.$$

5. Let X_t for $t > 0$ be a random process with zero mean and correlation function $R_X(t_1, t_2) = \min(t_1, t_2)$. If X_t is Gaussian for each t, write down the probability density function of X_t.

6. Let X_n be a discrete-time random process with mean function $m_X(n) := E[X_n]$ and correlation function $R_X(n,m) := E[X_n X_m]$. Suppose
$$Y_n := \sum_{i=-\infty}^{\infty} h(n-i) X_i.$$

(a) Show that
$$m_Y(n) = \sum_{k=-\infty}^{\infty} h(k) m_X(n-k).$$

(b) Show that
$$E[X_n Y_m] = \sum_{k=-\infty}^{\infty} h(k) R_X(n, m-k).$$

(c) Show that
$$E[Y_n Y_m] = \sum_{l=-\infty}^{\infty} h(l) \left(\sum_{k=-\infty}^{\infty} h(k) R_X(n-l, m-k) \right).$$

10.3: Strict-sense and wide-sense stationary processes

7. Let $X_t = \cos(2\pi f t + \Theta)$ be the carrier signal with random phase as in Example 10.8.

   (a) Are X_{t_1} and X_{t_2} jointly continuous for all choices of $t_1 \neq t_2$? Justify your answer. *Hint:* See discussion at the end of Section 7.2.

   (b) Show that for any function $g(x)$, $E[g(X_t)]$ does not depend on t.

8. Let X_t be a zero-mean, WSS process with correlation function $R_X(\tau)$. Let $Y_t := X_t \cos(2\pi f t + \Theta)$, where $\Theta \sim \text{uniform}[-\pi, \pi]$ and Θ is independent of the process X_t.

   (a) Find the correlation function of Y_t.
   (b) Find the cross-correlation function of X_t and Y_t.
   (c) Is Y_t WSS?

*9. Find the density of X_t in Problem 7. *Hint:* Problem 7 above and Problem 35 in Chapter 5 may be helpful.

*10. If a process is nth order strictly stationary, then for $k = 1, \ldots, n-1$ it is kth order strictly stationary. Show this for $n = 2$; i.e., if X_t is second-order strictly stationary, show that it is first-order strictly stationary.

11. In Problem 1, take $a(t) := e^{-|t|}$, $b(t) := \sin(2\pi t)$, and $c(t) := -1$.

    (a) Give a choice of the p_i and show that X_t is WSS.
    (b) Give a choice of the p_i and show that X_t is not WSS.
    (c) For arbitrary p_i, compute $P(X_0 = 1)$, $P(X_t \leq 0, 0 \leq t \leq 0.5)$, and $P(X_t \leq 0, 0.5 \leq t \leq 1)$.

*12. Let X_k be a strictly stationary process, and let $q(x_1, \ldots, x_L)$ a function of L variables. Put $Y_k := q(X_k, X_{k+1}, \ldots, X_{k+L-1})$. Show that Y_k is also strictly stationary. *Hint:* Show that the joint characteristic function of $Y_{1+m}, \ldots, Y_{n+m}$ does not depend on m.

13. In Example 10.19, we showed that the process X_n was not strictly stationary because $E[X_0^4] \neq E[X_n^4]$ for $n \neq 0$. Now show that for $g(x) := xI_{[0,\infty)}(x)$, $E[g(X_0)] \neq E[g(X_n)]$ for $n \neq 0$, thus giving another proof that the process is not strictly stationary.

14. Let $q(t)$ have period T_0, and let $T \sim \text{uniform}[0, T_0]$. Is $X_t := q(t+T)$ WSS? Justify your answer.

15. Let X_t be as in the preceding problem. Determine whether or not X_t is strictly stationary.

16. A discrete-time random process is WSS if $E[X_n]$ does not depend on n and if the correlation $E[X_n X_m]$ depends on n and m only through their difference. In this case, $E[X_n X_m] = R_X(n-m)$, where R_X is the univariate correlation function. Show that if X_n is WSS, then so is $Y_n := X_n - X_{n-1}$.

17. If a WSS process X_t has correlation function $R_X(\tau) = e^{-\tau^2/2}$, find $S_X(f)$.

18. If a WSS process X_t has correlation function $R_X(\tau) = 1/(1+\tau^2)$, find $S_X(f)$.

19. **MATLAB.** We can use the `fft` command to approximate $S_X(f)$ as follows.

    (a) Show that
    $$S_X(f) = 2\,\text{Re}\int_0^\infty R_X(\tau) e^{-j2\pi f \tau}\,d\tau.$$
    The **Riemann sum** approximation of this integral is
    $$\sum_{n=0}^{N-1} R_X(n\Delta\tau) e^{-j2\pi f n \Delta\tau} \Delta\tau.$$
    Taking $\Delta\tau = 1/\sqrt{N}$ yields the approximation
    $$S_X(f) \approx \frac{2}{\sqrt{N}} \text{Re} \sum_{n=0}^{N-1} R_X(n/\sqrt{N}) e^{-j2\pi f n/\sqrt{N}}.$$
    Specializing to $f = k/\sqrt{N}$ yields
    $$S_X(k/\sqrt{N}) \approx \frac{2}{\sqrt{N}} \text{Re} \sum_{n=0}^{N-1} R_X(n/\sqrt{N}) e^{-j2\pi kn/N}.$$

(b) The above right-hand side can be efficiently computed with the `fft` command. However, as a function of k it is periodic with period N. Hence, the values for $k = N/2$ to $N-1$ are the same as those for $k = -N/2$ to -1. To rearrange the values for plotting about the origin, we use the command `fftshift`. Put the following script into an M-file.

```
N = 128;
rootN = sqrt(N);
nvec = [0:N-1];
Rvec = R(nvec/rootN);  % The function R(.) is defined
                       % in a separate M-file below.
Svec = fftshift((2*real(fft(Rvec))))/rootN;
f = (nvec-N/2)/rootN;
plot(f,Svec)
```

(c) Suppose that $R_X(\tau) = \sin(\pi\tau/2)/(\pi\tau/2)$, where it is understood that $R_X(0) = 1$. Put the following code in an M-file called R.m:

```
function y = R(tau)
y = ones(size(tau));
i = find(tau~=0);
x = pi*tau(i)/2;
y(i) = sin(x)./x;
```

Use your script in part (b) to plot the approximation of $S_X(f)$.

(d) Repeat part (c) if $R_X(\tau) = [\sin(\pi\tau/4)/(\pi\tau/4)]^2$. *Hint:* Remember that to square every element of a vector s, use the command `s.^2`, *not* `s^2`.

20. A discrete-time random process is WSS if $\mathsf{E}[X_n]$ does not depend on n and if the correlation $\mathsf{E}[X_{n+k}X_k]$ does not depend on k. In this case we write $R_X(n) = \mathsf{E}[X_{n+k}X_k]$. For discrete-time WSS processes, the discrete-time Fourier transform of $R_X(n)$ is

$$S_X(f) := \sum_{n=-\infty}^{\infty} R_X(n)e^{-j2\pi fn},$$

which is a periodic function of f with period one. (Hence, we usually plot $S_X(f)$ only for $|f| \leq 1/2$.)

(a) Show that $R_X(n)$ is an even function of n.

(b) Show that $S_X(f)$ is a real and even function of f.

21. **MATLAB.** Let X_n be a discrete-time random process as defined in Problem 20. Then we can use the MATLAB command `fft` to approximate $S_X(f)$ as follows.

(a) Show that

$$S_X(f) = R_X(0) + 2\operatorname{Re}\sum_{n=1}^{\infty} R_X(n)e^{-j2\pi fn}.$$

This leads to the approximation

$$S_X(f) \approx R_X(0) + 2\operatorname{Re}\sum_{n=1}^{N-1} R_X(n)e^{-j2\pi fn}.$$

Specializing to $f = k/N$ yields

$$S_X(k/N) \approx R_X(0) + 2\,\mathrm{Re}\sum_{n=1}^{N-1} R_X(n)e^{-j2\pi kn/N}.$$

(b) The above right-hand side can be efficiently computed with the fft command. However, as a function of k it is periodic with period N. Hence, the values for $k = N/2$ to $N-1$ are the same as those for $k = -N/2$ to -1. To rearrange the values for plotting about the origin, we use the command fftshift. Put the following script into an M-file.

```
N = 128;
nvec = [0:N-1];
Rvec = R(nvec);          % The function R(n) is defined
Rvec(1) = Rvec(1)/2;     % in a separate M-file below.
Svec = fftshift((2*real(fft(Rvec))));
f = (nvec-N/2)/N;
plot(f,Svec)
```

(c) Suppose that $R_X(n) = \sin(\pi n/2)/(\pi n/2)$, where it is understood that $R_X(0) = 1$. Put the following code in an M-file called R.m:

```
function y = R(n)
y = ones(size(n));
i = find(n~=0);
x = pi*n(i)/2;
y(i) = sin(x)./x;
```

Use your script in part (b) to plot the approximation of $S_X(f)$.

(d) Repeat part (c) if $R_X(n) = [\sin(\pi n/4)/(\pi n/4)]^2$. *Hint:* Remember that to square every element of a vector s, use the command s.^2, *not* s^2.

10.4: WSS processes through LTI systems

22. If $h(t)$ is a real-valued function, show that the Fourier transform of $h(-t)$ is $H(f)^*$, where the asterisk * denotes the complex conjugate.

23. If the process in Problem 17 is applied to an ideal differentiator with transfer function $H(f) = j2\pi f$, and the system output is denoted by Y_t, find $R_{XY}(\tau)$ and $R_Y(\tau)$.

24. A WSS process X_t with correlation function $R_X(\tau) = 1/(1+\tau^2)$ is passed through an LTI system with impulse response $h(t) = 3\sin(\pi t)/(\pi t)$. Let Y_t denote the system output. Find $S_Y(f)$.

25. A WSS input signal X_t with correlation function $R_X(\tau) = e^{-\tau^2/2}$ is passed through an LTI system with transfer function $H(f) = e^{-(2\pi f)^2/2}$. Denote the system output by Y_t. Find (a) $S_{XY}(f)$; (b) the cross-correlation, $R_{XY}(\tau)$; (c) $\mathsf{E}[X_{t_1}Y_{t_2}]$; (d) $S_Y(f)$; (e) the output auto-correlation, $R_Y(\tau)$.

26. A zero-mean, WSS process X_t with correlation function $(1-|\tau|)I_{[-1,1]}(\tau)$ is to be processed by a filter with transfer function $H(f)$ designed so that the system output Y_t has correlation function
$$R_Y(\tau) = \frac{\sin(\pi\tau)}{\pi\tau}.$$
Find a formula for the required filter $H(f)$.

27. Let X_t be a WSS random process. Put $Y_t := \int_{-\infty}^{\infty} h(t-\tau)X_\tau \, d\tau$, and $Z_t := \int_{-\infty}^{\infty} g(t-\theta)X_\theta \, d\theta$. Determine whether or not Y_t and Z_t are J-WSS.

28. Let X_t be a zero-mean WSS random process with $S_X(f) = 2/[1+(2\pi f)^2]$. Put $Y_t := X_t - X_{t-1}$.

    (a) Show that X_t and Y_t are J-WSS.
    (b) Find $S_Y(f)$.

29. Let X_t be a WSS random process, and put $Y_t := \int_{t-3}^{t} X_\tau \, d\tau$. Determine whether or not Y_t is WSS.

30. Use the Fourier transform table inside the front cover to show that
$$\int_{-\infty}^{\infty} \left(\frac{\sin t}{t}\right)^2 dt = \pi.$$

31. Suppose that
$$Y_n := \sum_{i=-\infty}^{\infty} h(n-i)X_i,$$
where X_n is a discrete-time WSS process as defined in Problem 20, and $h(n)$ is a real-valued, discrete-time impulse response.

    (a) Use the appropriate formula of Problem 6 to show that
    $$\mathsf{E}[X_n Y_m] = \sum_{k=-\infty}^{\infty} h(k) R_X(n-m+k).$$

    (b) Use the appropriate formula of Problem 6 to show that
    $$\mathsf{E}[Y_n Y_m] = \sum_{l=-\infty}^{\infty} h(l) \left(\sum_{k=-\infty}^{\infty} h(k) R_X([n-m]-[l-k]) \right).$$
    It is now easy to see that X_n and Y_n are discrete-time J-WSS processes.

    (c) If we put $R_{XY}(n) := \mathsf{E}[X_{n+m}Y_m]$, and denote its discrete-time Fourier transform by $S_{XY}(f)$, show that
    $$S_{XY}(f) = H(f)^* S_X(f),$$
    where
    $$H(f) := \sum_{k=-\infty}^{\infty} h(k) e^{-j2\pi fk},$$
    and $S_X(f)$ was defined in Problem 20.

(d) If we put $R_Y(n) := E[Y_{n+m}Y_m]$ and denote its discrete-time Fourier transform by $S_Y(f)$, show that
$$S_Y(f) = |H(f)|^2 S_X(f).$$

10.5: Power spectral densities for WSS processes

32. Let $x(t)$ be a deterministic signal that is periodic with period T_0 and satisfies
$$E_0 := \int_0^{T_0} |x(t)|^2 \, dt < \infty.$$
Show that
$$\lim_{T \to \infty} \frac{1}{2T} \int_{-T}^{T} |x(t)|^2 \, dt = \frac{E_0}{T_0}.$$
Hints: Write T as a multiple of T_0 plus "a little bit," i.e., $T = nT_0 + \tau$, where $0 \le \tau < T_0$. Then write
$$\int_0^T |x(t)|^2 \, dt = \int_0^{nT_0} |x(t)|^2 \, dt + \int_{nT_0}^{nT_0+\tau} |x(t)|^2 \, dt$$
$$= nE_0 + \int_0^{\tau} |x(t)|^2 \, dt,$$
where we have used the fact that $x(t)$ has period T_0. Note that this last integral is less than or equal to E_0.

33. For a WSS process X_t, show that the expected energy
$$E\left[\int_{-\infty}^{\infty} X_t^2 \, dt\right]$$
is infinite.

34. According to Example 10.24, the integral of the power spectral density over every frequency band gives the power in that band. Use the following approach to show that the power spectral density is the unique such function. Show that if
$$\int_0^W S_1(f) \, df = \int_0^W S_2(f) \, df, \quad \text{for all } W > 0,$$
then $S_1(f) = S_2(f)$ for all $f \ge 0$. *Hint:* The function $q(W) := \int_0^W S_1(f) - S_2(f) \, df$ is identically zero for $W \ge 0$.

35. By applying white noise to the LTI system with impulse response $h(t) = I_{[-T,T]}(t)$, show that the cross power spectral density can be real but not nonnegative. By applying white noise to the LTI system with impulse response $h(t) = e^{-t} I_{[0,\infty)}(t)$, show that the cross power spectral density can be complex valued.

36. White noise with power spectral density $S_X(f) = N_0/2$ is applied to a lowpass filter with transfer function
$$H(f) = \begin{cases} 1 - f^2, & |f| \le 1, \\ 0, & |f| > 1. \end{cases}$$
Find the output power of the filter.

37. A WSS process X_t is applied to an LTI system with transfer function $H(f)$. Let Y_t denote the system output. Find the expected instantaneous output power $\mathsf{E}[Y_t^2]$ if

$$R_X(\tau) = e^{-(2\pi\tau)^2/2} \quad \text{and} \quad H(f) = \begin{cases} \sqrt{|f|}, & -1 \leq f \leq 1, \\ 0, & \text{otherwise.} \end{cases}$$

38. White noise with power spectral density $N_0/2$ is applied to a lowpass filter with transfer function $H(f) = \sin(\pi f)/(\pi f)$. Find the output noise power from the filter.

39. White noise with power spectral density $S_X(f) = N_0/2$ is applied to a lowpass RC filter with impulse response $h(t) = \frac{1}{RC} e^{-t/(RC)} I_{[0,\infty)}(t)$. Find (a) the cross power spectral density, $S_{XY}(f)$; (b) the cross-correlation, $R_{XY}(\tau)$; (c) $\mathsf{E}[X_{t_1} Y_{t_2}]$; (d) the output power spectral density, $S_Y(f)$; (e) the output auto-correlation, $R_Y(\tau)$; (f) the output power P_Y.

40. White noise with power spectral density $N_0/2$ is passed through a linear, time-invariant system with impulse response $h(t) = 1/(1+t^2)$. If Y_t denotes the filter output, find $\mathsf{E}[Y_{t+1/2} Y_t]$.

41. White noise with power spectral density $S_X(f) = N_0/2$ is passed though a filter with impulse response $h(t) = I_{[-T/2, T/2]}(t)$. Find the correlation function of the filter output.

42. Consider the system

$$Y_t = e^{-t} \int_{-\infty}^{t} e^{\theta} X_{\theta} \, d\theta.$$

Assume that X_t is zero mean white noise with power spectral density $S_X(f) = N_0/2$. Show that X_t and Y_t are J-WSS, and find $R_{XY}(\tau)$, $S_{XY}(f)$, $S_Y(f)$, and $R_Y(\tau)$.

43. Let $\{X_t\}$ be a zero-mean wide-sense stationary random process with power spectral density $S_X(f)$. Consider the process

$$Y_t := \sum_{n=-\infty}^{\infty} h_n X_{t-n},$$

with h_n real valued.

(a) Show that $\{X_t\}$ and $\{Y_t\}$ are *jointly* wide-sense stationary.

(b) Show that $S_Y(f)$ has the form $S_Y(f) = P(f) S_X(f)$ where P is a real-valued, nonnegative, periodic function of f with period 1. Give a formula for $P(f)$.

44. **System identification.** When white noise $\{W_t\}$ with power spectral density $S_W(f) = 3$ is applied to a certain linear time-invariant system, the output has power spectral density e^{-f^2}. Now let $\{X_t\}$ be a zero-mean, wide-sense stationary random process with power spectral density $S_X(f) = e^{f^2} I_{[-1,1]}(f)$. If $\{Y_t\}$ is the response of the system to $\{X_t\}$, find $R_Y(\tau)$ for all τ.

45. Let W_t be a zero-mean, wide-sense stationary white noise process with power spectral density $S_W(f) = N_0/2$. Suppose that W_t is applied to the ideal lowpass filter of bandwidth $B = 1$ MHz and power gain 120 dB; i.e., $H(f) = GI_{[-B,B]}(f)$, where $G = 10^6$. Denote the filter output by Y_t, and for $i = 1, \ldots, 100$, put $X_i := Y_{i\Delta t}$, where $\Delta t = (2B)^{-1}$. Show that the X_i are zero mean, uncorrelated, with variance $\sigma^2 = G^2 B N_0$.

*46. **Extension to complex random processes.** If X_t is a complex-valued random process, then its auto-correlation function is defined by $R_X(t_1, t_2) := \mathsf{E}[X_{t_1} X_{t_2}^*]$. Similarly, if Y_t is another complex-valued random process, their cross-correlation is defined by $R_{XY}(t_1, t_2) := \mathsf{E}[X_{t_1} Y_{t_2}^*]$. The concepts of WSS, J-WSS, the power spectral density, and the cross power spectral density are defined as in the real case. Now suppose that X_t is a complex WSS process and that $Y_t = \int_{-\infty}^{\infty} h(t - \tau) X_\tau \, d\tau$, where the impulse response h is now possibly complex valued.

(a) Show that $R_X(-\tau) = R_X(\tau)^*$.

(b) Show that $S_X(f)$ must be real valued.

(c) Show that
$$\mathsf{E}[X_{t_1} Y_{t_2}^*] = \int_{-\infty}^{\infty} h(-\beta)^* R_X([t_1 - t_2] - \beta) \, d\beta.$$

(d) Even though the above result is a little different from (10.20), show that (10.22) and (10.23) still hold for complex random processes.

10.6: Characterization of correlation functions

47. Find the correlation function corresponding to each of the following power spectral densities. (a) $\delta(f)$. (b) $\delta(f - f_0) + \delta(f + f_0)$. (c) $e^{-f^2/2}$. (d) $e^{-|f|}$.

48. Let X_t be a WSS random process with power spectral density $S_X(f) = I_{[-W,W]}(f)$. Find $\mathsf{E}[X_t^2]$.

49. Explain why each of the following frequency functions cannot be a power spectral density. (a) $e^{-f} u(f)$, where u is the unit step function. (b) $e^{-f^2} \cos(f)$. (c) $(1 - f^2)/(1 + f^4)$. (d) $1/(1 + jf^2)$.

50. For each of the following functions, determine whether or not it is a valid correlation function.
(a) $\sin(\tau)$. (b) $\cos(\tau)$. (c) $e^{-\tau^2/2}$. (d) $e^{-|\tau|}$. (e) $\tau^2 e^{-|\tau|}$. (f) $I_{[-T,T]}(\tau)$.

51. Let $R_0(\tau)$ be a correlation function, and put $R(\tau) := R_0(\tau) \cos(2\pi f_0 \tau)$ for some $f_0 > 0$. Determine whether or not $R(\tau)$ is a valid correlation function.

*52. Let $R(\tau)$ be a correlation function, and for fixed $\tau_0 > 0$ put
$$\tilde{R}(\tau) := R(\tau - \tau_0) + R(\tau + \tau_0).$$
Select the best answer from the following (justify your choice):

(a) $\tilde{R}(\tau)$ is always a correlation function.

(b) For some choice of $R(\tau)$ and $\tau_0 > 0$, $\overline{R}(\tau)$ is a correlation function.

(c) There is no choice of $R(\tau)$ and $\tau_0 > 0$ for which $\overline{R}(\tau)$ is a correlation function.

*53. Let $S(f)$ be a real-valued, even, nonnegative function, and put

$$R(\tau) := \int_{-\infty}^{\infty} S(f) e^{j2\pi f \tau} d\tau.$$

Show that $R(\tau)$ is real-valued, even, and satisfies $|R(\tau)| \leq R(0)$.

54. Let $R_0(\tau)$ be a real-valued, even function, but not necessarily a correlation function. Let $R(\tau)$ denote the convolution of R_0 with itself, i.e.,

$$R(\tau) := \int_{-\infty}^{\infty} R_0(\theta) R_0(\tau - \theta) d\theta.$$

(a) Show that $R(\tau)$ is a valid correlation function.

(b) Now suppose that $R_0(\tau) = I_{[-T,T]}(\tau)$. In this case, what is $R(\tau)$, and what is its Fourier transform?

10.7: The matched filter

55. Determine the matched filter impulse response $h(t)$ if the known radar pulse is $v(t) = \sin(t) I_{[0,\pi]}(t)$ and X_t is white noise with power spectral density $S_X(f) = N_0/2$. For what values of t_0 is the optimal system causal?

56. Determine the matched filter impulse response $h(t)$ if $v(t) = e^{-(t/\sqrt{2})^2/2}$ and $S_X(f) = e^{-(2\pi f)^2/2}$.

57. Derive the matched filter for a discrete-time received signal $v(n) + X_n$. Hint: Problems 20 and 31 may be helpful.

10.8: The Wiener filter

58. Suppose V_t and X_t are J-WSS. Let $U_t := V_t + X_t$. Show that U_t and V_t are J-WSS.

59. Suppose $U_t = V_t + X_t$, where V_t and X_t are each zero mean and WSS. Also assume that $E[V_t X_\tau] = 0$ for all t and τ. Express the Wiener filter $H(f)$ in terms of $S_V(f)$ and $S_X(f)$.

60. Using the setup of Problem 59, suppose that the signal has correlation function $R_V(\tau) = \left(\frac{\sin \pi \tau}{\pi \tau}\right)^2$ and that the noise has a power spectral density given by $S_X(f) = 1 - I_{[-1,1]}(f)$. Find the Wiener filter $H(f)$ and the corresponding impulse response $h(t)$.

61. Let V_t and U_t be zero-mean, J-WSS. If

$$\widehat{V}_t = \int_{-\infty}^{\infty} h(\theta) U_{t-\theta} d\theta$$

is the estimate of V_t using the Wiener filter, show that the minimum mean-squared error is

$$E[|V_t - \widehat{V}_t|^2] = \int_{-\infty}^{\infty} \left[S_V(f) - \frac{|S_{VU}(f)|^2}{S_U(f)} \right] df.$$

62. Derive the Wiener filter for discrete-time J-WSS signals U_n and V_n with zero means. *Hints:* (i) First derive the analogous orthogonality principle. (ii) Problems 20 and 31 may be helpful.

63. Using the setup of Problem 59, find the Wiener filter $H(f)$ and the corresponding impulse response $h(t)$ if $S_V(f) = 2\lambda/[\lambda^2 + (2\pi f)^2]$ and $S_X(f) = 1$.

    Remark. You may want to compare your answer with the *causal* Wiener filter found in Example 10.33.

64. Find the impulse response of the whitening filter $K(f)$ of Example 10.33. Is it causal?

65. The causal Wiener filter $h(\tau)$ estimates V_t based only on the observation up to time t, $\{U_\tau, -\infty < \tau \leq t\}$. Based on this observation, suppose that instead of estimating V_t, you want to estimate $V_{t+\Delta t}$, where $\Delta t \neq 0$. When $\Delta t > 0$, this is called **prediction**. When $\Delta t < 0$, this is called **smoothing**. (The ordinary Wiener filter can be viewed as the most extreme case of smoothing.) For $\Delta t \neq 0$, let $h_{\Delta t}(\tau)$ denote the optimal filter. Find the analog of the Wiener–Hopf equation (10.38) for $h_{\Delta t}(\tau)$. In the special case that U_t is white noise, express $h_{\Delta t}(\tau)$ as a function of ordinary causal Wiener filter $h(\tau)$.

10.9: *The Wiener–Khinchin theorem

66. Recall that by Problem 4, correlation functions are positive semidefinite. Use this fact to prove that the double integral in (10.44) is nonnegative, assuming that R_X is continuous. *Hint:* Since R_X is continuous, the double integral in (10.44) is a limit of Riemann sums of the form

$$\sum_i \sum_k R_X(t_i - t_k) e^{-j2\pi f(t_i - t_k)} \Delta t_i \Delta t_k.$$

10.10: *Mean-square ergodic theorem for WSS processes

67. Let Y_t be a WSS process. In each of the cases below, determine whether or not $\frac{1}{2T} \int_{-T}^{T} Y_t \, dt \to \mathsf{E}[Y_t]$ in mean square.

    (a) The covariance $C_Y(\tau) = e^{-|\tau|}$.

    (b) The covariance $C_Y(\tau) = \sin(\pi\tau)/(\pi\tau)$.

68. Let $Y_t = \cos(2\pi t + \Theta)$, where $\Theta \sim \text{uniform}[-\pi, \pi]$. As in Example 10.8, $\mathsf{E}[Y_t] = 0$. Determine whether or not

$$\lim_{T \to \infty} \frac{1}{2T} \int_{-T}^{T} Y_t \, dt \to 0.$$

69. Let X_t be a zero-mean, WSS process. For fixed τ, you might expect

$$\frac{1}{2T} \int_{-T}^{T} X_{t+\tau} X_t \, dt$$

to converge in mean square to $E[X_{t+\tau}X_t] = R_X(\tau)$. Give conditions on the process X_t under which this will be true. *Hint:* Define $Y_t := X_{t+\tau}X_t$.

Remark. When $\tau = 0$ this says that $\frac{1}{2T}\int_{-T}^{T} X_t^2\, dt$ converges in mean square to $R_X(0) = P_X$.

70. Let X_t be a zero-mean, WSS process. For a fixed set $B \subset \mathbb{R}$, you might expect[g]

$$\frac{1}{2T}\int_{-T}^{T} I_B(X_t)\, dt$$

to converge in mean square to $E[I_B(X_t)] = P(X_t \in B)$. Give conditions on the process X_t under which this will be true. *Hint:* Define $Y_t := I_B(X_t)$.

10.11: *Power spectral densities for non-WSS processes

71. Give a suitable definition of $\overline{R}_{XY}(\tau)$ and show that the following analog of (10.18) holds,

$$\overline{R}_{XY}(\tau) = \int_{-\infty}^{\infty} h(\alpha)\overline{R}_X(\tau+\alpha)\, d\alpha.$$

Hint: Formula (10.10) may be helpful. You may also use the assumption that for fixed α,

$$\lim_{T \to \infty} \frac{1}{2T}\int_{-T-\alpha}^{T-\alpha} \cdots = \lim_{T \to \infty} \frac{1}{2T}\int_{-T}^{T} \cdots.$$

72. Show that the following analog of (10.19) holds,

$$\overline{R}_Y(\tau) = \int_{-\infty}^{\infty} h(\beta)\overline{R}_{XY}(\tau-\beta)\, d\beta.$$

73. Let $S_{XY}(f)$ denote the Fourier transform of $\overline{R}_{XY}(\tau)$ that you defined in Problem 71. Let $S_X(f)$ denote the Fourier transform of $\overline{R}_X(\tau)$ defined in the text. Show that $S_{XY}(f) = H(f)^* S_X(f)$ and that $S_Y(f) = |H(f)|^2 S_X(f)$.

74. Derive (10.51).

Exam preparation

You may use the following suggestions to prepare a study sheet, including formulas mentioned that you have trouble remembering. You may also want to ask your instructor for additional suggestions.

10.1. Definition and examples. There are two ways to think of random processes $X_t(\omega)$. For each fixed t, we have a random variable (function of ω), and for fixed ω, we have a waveform (function of t).

---

[g] This is the fraction of time during $[-T,T]$ that $X_t \in B$. For example, we might have $B = [v_{\min}, v_{\max}]$ being the acceptable operating range of the voltage of some device. Then we would be interested in the fraction of time during $[-T,T]$ that the device is operating normally.

10.2. Characterization of random processes. For a single random variable, we often do not know the density or pmf, but we may know the mean and variance. Similarly, for a random process, we may know only the mean function (10.1) and the correlation function (10.2). For a pair of processes, the cross-correlation function (10.8) may also be known. Know how the correlation and covariance functions are related (10.7) and how the cross-correlation and cross-covariance functions are related (10.9). The upper bound (10.6) is also important.

10.3. Strict-sense and wide-sense stationary processes. Know properties (i) and (ii) that define a WSS process. Once we know a process is WSS, we write $R_X(\tau) = \mathsf{E}[X_{t+\tau} X_t]$ for any t. This is an even function of τ.

10.4. WSS processes through LTI systems. LTI systems preserve wide-sense stationarity; i.e., if a WSS process is applied to an LTI system, then the input and output are J-WSS. Key formulas include (10.22) and (10.23). Do lots of problems.

10.5. Power spectral densities for WSS processes. Know the three expressions for power (10.24). The power spectral density is a nonnegative function that when integrated over a frequency band yields the power in the band. Do lots of problems.

10.6. Characterization of correlation functions. To guarantee that a function $R(\tau)$ is a correlation function, you must show that its Fourier transform $S(f)$ is real, even, and nonnegative. To show a function $R(\tau)$ is not a correlation function, you can show that its transform fails to have one of these three properties. However, it is sometimes easier to show that $R(\tau)$ fails to be real or even, or fails to have its maximum absolute value at $\tau = 0$ or satisfies $R(0) < 0$.

10.7. The matched filter. This filter is used for detecting the presence of the *known, deterministic signal* $v(t)$ from $v(t) + X_t$, where X_t is a WSS noise process. The transfer function of the matched filter is $H(f)$ in (10.28). Note that the constant α is arbitrary, but should be real to keep the impulse response real. When the noise is white, the optimal impulse response is $h(t)$ is proportional to $v(t_0 - t)$, where t_0 is the filter sampling instant, and may be chosen to make the filter causal.

10.8. The Wiener filter. This filter is used for estimating a *random signal* V_t based on measuring a related process U_t for all time. Typically, U_t and V_t are related by $U_t = V_t + X_t$, but this is not required. All that is required is knowledge of $R_{VU}(\tau)$ and $R_U(\tau)$. The Wiener filter transfer function is given by (10.37). The causal Wiener filter is found by the spectral factorization procedure.

10.9. *The Wiener–Khinchin theorem. This theorem gives an alternative representation of the power spectral density of a WSS process.

10.10. *Mean-square ergodic theorem. This mean-square law of large numbers for WSS processes is given in (10.46). It is equivalent to the condition (10.48). However, for practical purposes, it is important to note that (10.46) holds if $C_Y(\tau)$ is absolutely integrable, or if the Fourier transform of C_Y is continuous at $f = 0$.

10.11. *Power spectral densities for non-WSS processes. For such a process, the power spectral density is defined to be the Fourier transform of $\overline{R}_X(\tau)$ defined in (10.50). For a cyclostationary process, it is important to know that $\overline{R}_X(\tau)$ is more easily expressed by (10.51). And, as expected, for a WSS process, $\overline{R}_X(\tau) = R_X(\tau)$.

Work any review problems assigned by your instructor. If you finish them, re-work your homework assignments.

11
Advanced concepts in random processes

The two most important continuous-time random processes are the Poisson process and the Wiener process, which are introduced in Sections 11.1 and 11.3, respectively. The construction of arbitrary random processes in discrete and continuous time using Kolmogorov's theorem is discussed in Section 11.4.

In addition to the Poisson process, marked Poisson processes and shot noise are introduced in Section 11.1. The extension of the Poisson process to renewal processes is presented briefly in Section 11.2. In Section 11.3, the Wiener process is defined and then interpreted as integrated white noise. The Wiener integral is introduced. The approximation of the Wiener process via a random walk is also outlined. For random walks without finite second moments, it is shown by a simulation example that the limiting process is no longer a Wiener process.

11.1 The Poisson process†

A **counting process** $\{N_t, t \geq 0\}$ is a random process that counts how many times something happens from time zero up to and including time t. A sample path of such a process is shown in Figure 11.1. Such processes always have a staircase form with jumps of height

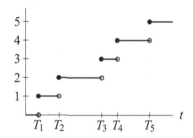

Figure 11.1. Sample path N_t of a counting process.

one. The randomness is in the times T_i at which whatever we are counting happens. Note that counting processes are right continuous.

Here are some examples of things we might count.

- N_t = the number of radioactive particles emitted from a sample of radioactive material up to and including time t.
- N_t = the number of photoelectrons emitted from a photodetector up to and including time t.
- N_t = the number of hits of a website up to and including time t.

---
†Section 11.1 on the Poisson process can be covered any time after Chapter 5.

- N_t = the number of customers passing through a checkout line at a grocery store up to and including time t.
- N_t = the number of vehicles passing through a toll booth on a highway up to and including time t.

Suppose that $0 \leq t_1 < t_2 < \infty$ are given times, and we want to know how many things have happened between t_1 and t_2. Now N_{t_2} is the number of occurrences up to and including time t_2. If we subtract N_{t_1}, the number of occurrences up to and including time t_1, then the difference $N_{t_2} - N_{t_1}$ is simply the number of occurrences that happen after t_1 up to and including t_2. We call differences of the form $N_{t_2} - N_{t_1}$ **increments** of the process.

A counting process $\{N_t, t \geq 0\}$ is called a **Poisson process** if the following three conditions hold.

- $N_0 \equiv 0$; i.e., N_0 is a constant random variable whose value is always zero.
- For any $0 \leq s < t < \infty$, the increment $N_t - N_s$ is a Poisson random variable with parameter $\lambda(t-s)$; i.e.,

$$P(N_t - N_s = k) = \frac{[\lambda(t-s)]^k e^{-\lambda(t-s)}}{k!}, \quad k = 0, 1, 2, \ldots.$$

Also, $\mathsf{E}[N_t - N_s] = \lambda(t-s)$ and $\text{var}(N_t - N_s) = \lambda(t-s)$. The constant λ is called the **rate** or the **intensity** of the process.
- If the time intervals

$$(t_1, t_2], (t_2, t_3], \ldots, (t_n, t_{n+1}]$$

are disjoint, then the increments

$$N_{t_2} - N_{t_1}, N_{t_3} - N_{t_2}, \ldots, N_{t_{n+1}} - N_{t_n}$$

are independent; i.e., the process has **independent increments**. In other words, the numbers of occurrences in disjoint time intervals are independent.

Example 11.1. Photoelectrons are emitted from a photodetector at a rate of λ per minute. Find the probability that during each of two consecutive minutes, more than five photoelectrons are emitted.

Solution. Let N_i denote the number of photoelectrons emitted from time zero up through the ith minute. The probability that during the first minute and during the second minute more than five photoelectrons are emitted is

$$P(\{N_1 - N_0 \geq 6\} \cap \{N_2 - N_1 \geq 6\}).$$

By the independent increments property, this is equal to

$$P(N_1 - N_0 \geq 6) P(N_2 - N_1 \geq 6).$$

Each of these factors is equal to

$$1 - \sum_{k=0}^{5} \frac{\lambda^k e^{-\lambda}}{k!},$$

11.1 The Poisson process

where we have used the fact that the length of the time increments is one. Hence,

$$P(\{N_1 - N_0 \geq 6\} \cap \{N_2 - N_1 \geq 6\}) = \left(1 - \sum_{k=0}^{5} \frac{\lambda^k e^{-\lambda}}{k!}\right)^2.$$

We now compute the mean, correlation, and covariance of a Poisson process. Since $N_0 \equiv 0$, $N_t = N_t - N_0$ is a Poisson random variable with parameter $\lambda(t-0) = \lambda t$. Hence,

$$E[N_t] = \lambda t \quad \text{and} \quad \text{var}(N_t) = \lambda t.$$

This further implies that $E[N_t^2] = \lambda t + (\lambda t)^2$. For $0 \leq s < t$, we can compute the correlation

$$\begin{aligned} E[N_t N_s] &= E[(N_t - N_s)N_s] + E[N_s^2] \\ &= E[(N_t - N_s)(N_s - N_0)] + (\lambda s)^2 + \lambda s. \end{aligned}$$

Since $(0, s]$ and $(s, t]$ are disjoint, the above increments are independent, and so

$$E[(N_t - N_s)(N_s - N_0)] = E[N_t - N_s] \cdot E[N_s - N_0] = \lambda(t - s) \cdot \lambda s.$$

It follows that

$$E[N_t N_s] = (\lambda t)(\lambda s) + \lambda s.$$

We can also compute the covariance,

$$\begin{aligned} \text{cov}(N_t, N_s) &= E[(N_t - \lambda t)(N_s - \lambda s)] \\ &= E[N_t N_s] - (\lambda t)(\lambda s) \\ &= \lambda s. \end{aligned}$$

More generally, given any two times t_1 and t_2,

$$\boxed{\text{cov}(N_{t_1}, N_{t_2}) = \lambda \min(t_1, t_2).}$$

So far, we have focused on the *number* of occurrences between two fixed times. Now we focus on the **jump times**, which are defined by (see Figure 11.1)

$$T_n := \min\{t > 0 : N_t \geq n\}.$$

In other words, T_n is the time of the nth jump in Figure 11.1. In particular, if $T_n > t$, then the nth jump happens after time t; hence, at time t we must have $N_t < n$. Conversely, if at time t, $N_t < n$, then the nth occurrence has not happened yet; it must happen after time t, i.e., $T_n > t$. We can now write

$$P(T_n > t) = P(N_t < n) = \sum_{k=0}^{n-1} \frac{(\lambda t)^k}{k!} e^{-\lambda t}.$$

Since $F_{T_n}(t) = 1 - \mathsf{P}(T_n > t)$, differentiation shows that T_n has the Erlang density with parameters n and λ,

$$f_{T_n}(t) = \lambda \frac{(\lambda t)^{n-1} e^{-\lambda t}}{(n-1)!}, \quad t \geq 0.$$

In particular, T_1 has an exponential density with parameter λ. Depending on the context, the jump times may be called **arrival times** or **occurrence times**.

In the previous paragraph, we defined the occurrence times in terms of counting process $\{N_t, t \geq 0\}$. Observe that we can express N_t in terms of the occurrence times since

$$N_t = \sum_{k=1}^{\infty} I_{(0,t]}(T_k).$$

To see this, note that each term in the sum is either zero or one. A term is one if and only if $T_k \in (0,t]$. Hence, the sum counts the number of occurrences in the interval $(0,t]$, which is exactly the definition of N_t.

We now define the **interarrival times**,

$$\begin{aligned} X_1 &= T_1, \\ X_n &= T_n - T_{n-1}, \quad n = 2, 3, \ldots. \end{aligned}$$

The occurrence times can be recovered from the interarrival times by writing

$$T_n = X_1 + \cdots + X_n.$$

We noted above that T_n is Erlang with parameters n and λ. Recalling Problem 55 in Chapter 4, which shows that a sum of i.i.d. $\exp(\lambda)$ random variables is Erlang with parameters n and λ, we wonder if the X_i are i.i.d. exponential with parameter λ. This is indeed the case, as shown in [3, p. 301]. Thus, for all i,

$$f_{X_i}(x) = \lambda e^{-\lambda x}, \quad x \geq 0.$$

Example 11.2. Micrometeors strike the space shuttle according to a Poisson process. The expected time between strikes is 30 minutes. Find the probability that during at least one hour out of five consecutive hours, three or more micrometeors strike the shuttle.

Solution. The problem statement is telling us that the expected interarrival time is 30 minutes. Since the interarrival times are $\exp(\lambda)$ random variables, their mean is $1/\lambda$. Thus, $1/\lambda = 30$ minutes, or 0.5 hours, and so $\lambda = 2$ strikes per hour. The number of strikes during the ith hour is $N_i - N_{i-1}$. The probability that during at least 1 hour out of five consecutive hours, three or more micrometeors strike the shuttle is

$$\mathsf{P}\left(\bigcup_{i=1}^{5} \{N_i - N_{i-1} \geq 3\}\right) = 1 - \mathsf{P}\left(\bigcap_{i=1}^{5} \{N_i - N_{i-1} < 3\}\right)$$

$$= 1 - \prod_{i=1}^{5} \mathsf{P}(N_i - N_{i-1} \leq 2),$$

11.1 The Poisson process

where the last step follows by the independent increments property of the Poisson process. Since $N_i - N_{i-1} \sim \text{Poisson}(\lambda[i-(i-1)])$, or simply $\text{Poisson}(\lambda)$,

$$P(N_i - N_{i-1} \leq 2) = e^{-\lambda}(1 + \lambda + \lambda^2/2) = 5e^{-2},$$

and we have

$$P\left(\bigcup_{i=1}^{5}\{N_i - N_{i-1} \geq 3\}\right) = 1 - (5e^{-2})^5 \approx 0.86.$$

Example 11.3 (simulation of a Poisson process). Since $T_n = X_1 + \cdots + X_n$, and since the X_i are i.i.d. $\exp(\lambda)$, it is easy to simulate a Poisson process and to plot the result. Suppose we want to plot N_t for $0 \leq t \leq T_{\max}$. Using the MATLAB command X = -log(rand(1))/lambda, we can generate an $\exp(\lambda)$ random variable. We can collect the sequence of arrival times with the command T(n) = T(n-1) + X. We do this until we get an arrival time that exceeds $T_{\max}$. If this happens on the nth arrival, we plot only the first $n-1$ arrivals.

Plotting is a little tricky. In Figure 11.1, the jumps are not connected with vertical lines. However, with MATLAB it is more convenient to include vertical lines at the jump times. Since plot operates by "connecting the dots," to plot N_t on $[0, T_{\max}]$, we connect the dots located at

$$(0,0), (T_1,0), (T_1,1), (T_2,1), (T_2,2), (T_3,2), \ldots, (T_{n-1}, n-1), (T_{\max}, n-1).$$

For the plot command, we need to generate a vector of times in which the T_i are repeated and also a vector in which the values of N_t are repeated. An easy way to do this is with the MATLAB command kron, which implements the **Kronecker product** of matrices. For our purposes, all we need is the observation that kron([4 5 6 7],[1 1]) yields [4 4 5 5 6 6 7 7]. In other words, every entry is repeated. Here is the code to simulate and plot a Poisson process.

```
% Plot Poisson Process on [0,Tmax]
%
Tmax = 10;
lambda = 1;
n = 0;            % Number of points
Tlast = 0;        % Time of last arrival
while Tlast <= Tmax
    n = n + 1;
    X = -log(rand(1))/lambda;   % Generate exp(lambda) RV
    Tlast = Tlast + X;
    T(n) = Tlast;
end
n = n-1;          % Remove last arrival,
T = T(1:n);       % which is after Tmax.
fprintf('There were %g arrivals in [0,%g].\n',n,Tmax)
tt = kron(T,[1 1]);  % Convert [x y z] to [x x y y z z]
tt = [ 0 tt Tmax ];
N = [ 0:n ];            % Values of the Poisson process
```

```
NN = kron(N,[1 1]);
plot(tt,NN)
axis([0 Tmax 0 n+1])
```

Try this yourself! (The command `stairs([0 T Tmax],[0:n n])` can replace the lines between `fprintf` and `axis`.)

*Derivation of the Poisson probabilities

In the definition of the Poisson process, we required that $N_t - N_s$ be a Poisson random variable with parameter $\lambda(t-s)$. In particular, this implies that

$$\frac{\mathsf{P}(N_{t+\Delta t} - N_t = 1)}{\Delta t} = \frac{\lambda \Delta t \, e^{-\lambda \Delta t}}{\Delta t} \to \lambda,$$

as $\Delta t \to 0$. The Poisson assumption also implies that

$$\frac{1 - \mathsf{P}(N_{t+\Delta t} - N_t = 0)}{\Delta t} = \frac{\mathsf{P}(N_{t+\Delta t} - N_t \geq 1)}{\Delta t}$$

$$= \frac{1}{\Delta t} \sum_{k=1}^{\infty} \frac{(\lambda \Delta t)^k}{k!}$$

$$= \lambda \left(1 + \sum_{k=2}^{\infty} \frac{(\lambda \Delta t)^{k-1}}{k!} \right) \to \lambda,$$

as $\Delta t \to 0$.

As we now show, the converse is also true as long as we continue to assume that $N_0 \equiv 0$ and that the process has independent increments. So, instead of assuming that $N_t - N_s$ is a Poisson random variable with parameter $\lambda(t-s)$, we assume the following.

- During a sufficiently short time interval, $\Delta t > 0$, $\mathsf{P}(N_{t+\Delta t} - N_t = 1) \approx \lambda \Delta t$. By this we mean that

$$\lim_{\Delta t \downarrow 0} \frac{\mathsf{P}(N_{t+\Delta t} - N_t = 1)}{\Delta t} = \lambda. \tag{11.1}$$

This property can be interpreted as saying that the probability of having exactly one occurrence during a short time interval of length Δt is approximately $\lambda \Delta t$.

- For sufficiently small $\Delta t > 0$, $\mathsf{P}(N_{t+\Delta t} - N_t = 0) \approx 1 - \lambda \Delta t$. More precisely,

$$\lim_{\Delta t \downarrow 0} \frac{1 - \mathsf{P}(N_{t+\Delta t} - N_t = 0)}{\Delta t} = \lambda. \tag{11.2}$$

By combining this property with the preceding one, we see that during a short time interval of length Δt, we have either exactly one occurrence or no occurrences. In other words, during a short time interval, at most one occurrence is observed.

For $n = 0, 1, \ldots,$ let

$$p_n(t) := \mathsf{P}(N_t - N_s = n), \quad t \geq s. \tag{11.3}$$

11.1 The Poisson process

Note that $p_0(s) = P(N_s - N_s = 0) = P(0 = 0) = P(\Omega) = 1$. Now,

$$\begin{aligned}
p_n(t + \Delta t) &= P(N_{t+\Delta t} - N_s = n) \\
&= P\left(\bigcup_{k=0}^{n}\left[\{N_t - N_s = n - k\} \cap \{N_{t+\Delta t} - N_t = k\}\right]\right) \\
&= \sum_{k=0}^{n} P(N_{t+\Delta t} - N_t = k, N_t - N_s = n - k) \\
&= \sum_{k=0}^{n} P(N_{t+\Delta t} - N_t = k) p_{n-k}(t),
\end{aligned}$$

using independent increments and (11.3). Break the preceding sum into three terms as follows.

$$\begin{aligned}
p_n(t + \Delta t) &= P(N_{t+\Delta t} - N_t = 0) p_n(t) \\
&\quad + P(N_{t+\Delta t} - N_t = 1) p_{n-1}(t) \\
&\quad + \sum_{k=2}^{n} P(N_{t+\Delta t} - N_t = k) p_{n-k}(t).
\end{aligned}$$

This enables us to write

$$\begin{aligned}
p_n(t + \Delta t) - p_n(t) &= -[1 - P(N_{t+\Delta t} - N_t = 0)] p_n(t) \\
&\quad + P(N_{t+\Delta t} - N_t = 1) p_{n-1}(t) \\
&\quad + \sum_{k=2}^{n} P(N_{t+\Delta t} - N_t = k) p_{n-k}(t).
\end{aligned} \quad (11.4)$$

For $n = 0$, only the first term on the right in (11.4) is present, and we can write

$$p_0(t + \Delta t) - p_0(t) = -[1 - P(N_{t+\Delta t} - N_t = 0)] p_0(t). \quad (11.5)$$

It then follows that

$$\lim_{\Delta t \downarrow 0} \frac{p_0(t + \Delta t) - p_0(t)}{\Delta t} = -\lambda p_0(t).$$

In other words, we are left with the first-order differential equation,

$$p_0'(t) = -\lambda p_0(t), \qquad p_0(s) = 1,$$

whose solution is simply

$$p_0(t) = e^{-\lambda(t-s)}, \qquad t \geq s.$$

To handle the case $n \geq 2$, note that since

$$\begin{aligned}
\sum_{k=2}^{n} P(N_{t+\Delta t} - N_t = k) p_{n-k}(t) &\leq \sum_{k=2}^{n} P(N_{t+\Delta t} - N_t = k) \\
&\leq \sum_{k=2}^{\infty} P(N_{t+\Delta t} - N_t = k) \\
&= P(N_{t+\Delta t} - N_t \geq 2) \\
&= 1 - [P(N_{t+\Delta t} - N_t = 0) + P(N_{t+\Delta t} - N_t = 1)],
\end{aligned}$$

it follows that
$$\lim_{\Delta t \downarrow 0} \frac{\sum_{k=2}^{n} P(N_{t+\Delta t} - N_t = k) p_{n-k}(t)}{\Delta t} = \lambda - \lambda = 0.$$

Returning to (11.4), we see that for $n = 1$ and for $n \geq 2$,
$$\lim_{\Delta t \downarrow 0} \frac{p_n(t + \Delta t) - p_n(t)}{\Delta t} = -\lambda p_n(t) + \lambda p_{n-1}(t).$$

This results in the differential-difference equation,
$$p_n'(t) = -\lambda p_n(t) + \lambda p_{n-1}(t), \quad p_0(t) = e^{-\lambda(t-s)}. \tag{11.6}$$

It is easily verified that for $n = 1, 2, \ldots$,
$$p_n(t) = \frac{[\lambda(t-s)]^n e^{-\lambda(t-s)}}{n!},$$
which are the claimed Poisson probabilities, solve (11.6).

Marked Poisson processes

It is frequently the case that in counting arrivals, each arrival is associated with a **mark**. For example, suppose packets arrive at a router according to a Poisson process of rate λ, and that the size of the ith packet is B_i bytes, where B_i is a random variable. The size B_i is the mark. Thus, the ith packet, whose size is B_i, arrives at time T_i, where T_i is the ith occurrence time of the Poisson process. The total number of bytes processed up to time t is
$$M_t := \sum_{i=1}^{N_t} B_i.$$

We usually assume that the mark sequence is i.i.d. and independent of the Poisson process. In this case, the mean of M_t can be computed as in Example 7.24. The characteristic function of M_t can be computed as in Problem 59 in Chapter 7.

Shot noise

Light striking a photodetector generates photoelectrons according to a Poisson process. The rate of the process is proportional to the intensity of the light and the efficiency of the detector. The detector output is then passed through an amplifier of impulse response $h(t)$. We model the input to the amplifier as a train of impulses
$$X_t := \sum_i \delta(t - T_i),$$
where the T_i are the occurrence times of the Poisson process. The amplifier output is
$$Y_t = \int_{-\infty}^{\infty} h(t - \tau) X_\tau \, d\tau$$
$$= \sum_{i=1}^{\infty} \int_{-\infty}^{\infty} h(t - \tau) \delta(\tau - T_i) \, d\tau$$
$$= \sum_{i=1}^{\infty} h(t - T_i). \tag{11.7}$$

11.1 The Poisson process

For any realizable system, $h(t)$ is a causal function; i.e., $h(t) = 0$ for $t < 0$. Then

$$Y_t = \sum_{i:T_i \leq t} h(t - T_i) = \sum_{i=1}^{N_t} h(t - T_i). \qquad (11.8)$$

A process of the form of Y_t is called a **shot-noise** process or a **filtered Poisson process**. If the impulse response $h(t)$ has a jump discontinuity, e.g., $h(t) = e^{-t}u(t)$ as shown at the left in Figure 11.2, then the shot-noise process Y_t has jumps as shown in the middle plot in

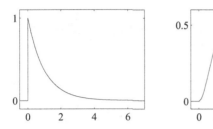

Figure 11.2. Plots of $h(t) = e^{-t}u(t)$ (left) and $h(t) = t^2 e^{-t}u(t)$ (right), where $u(t)$ is the unit step function.

Figure 11.3. On the other hand, if $h(t)$ is continuous; e.g., $h(t) = t^2 e^{-t} u(t)$ as shown at the right in Figure 11.2, then the shot-noise process Y_t is continuous, as shown in the bottom plot in Figure 11.3.

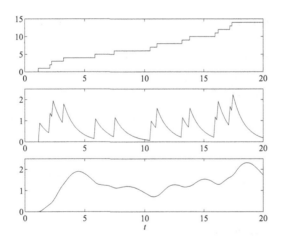

Figure 11.3. Point process N_t (top) and corresponding shot noise Y_t in Eq. (11.8) for $h(t) = e^{-t}u(t)$ (middle) and for $h(t) = t^2 e^{-t}u(t)$ (bottom), where $u(t)$ is the unit step function.

Example 11.4. If

$$Y := \sum_{i=1}^{\infty} g(T_i),$$

where $g(\tau) := cI_{(a,b]}(\tau)$ and c is a constant, find the mean and variance of Y.

Solution. To begin, write

$$Y = c \sum_{i=1}^{\infty} I_{(a,b]}(T_i).$$

Notice that the terms of the sum are either zero or one. A term is one if and only if $T_i \in (a,b]$. Hence, this sum simply counts the number of occurrences in the interval $(a,b]$. But this is just $N_b - N_a$. Thus, $Y = c(N_b - N_a)$. From the properties of Poisson random variables, $E[Y] = c\lambda(b-a)$ and $E[Y^2] = c^2\{\lambda(b-a) + [\lambda(b-a)]^2\}$. We can then write $\text{var}(Y) = c^2\lambda(b-a)$. More important, however, is the fact that we can also write

$$E[Y] = \int_0^\infty g(\tau)\lambda\,d\tau \quad \text{and} \quad \text{var}(Y) = \int_0^\infty g(\tau)^2 \lambda\,d\tau. \quad (11.9)$$

For Y as in the previous example, it is easy to compute its characteristic function. First write

$$\varphi_Y(\nu) = E[e^{j\nu Y}] = E[e^{j\nu c(N_b - N_a)}] = \exp[\lambda(b-a)(z-1)]|_{z=e^{j\nu c}}.$$

Thus,

$$\varphi_Y(\nu) = \exp\left[\lambda(b-a)(e^{j\nu c} - 1)\right]$$
$$= \exp\left[\int_0^\infty (e^{j\nu g(\tau)} - 1)\lambda\,d\tau\right]. \quad (11.10)$$

This last equation follows because when $g(\tau) = 0$, $e^0 - 1 = 0$ too. Equations of the form (11.9) and (11.10) are usually known as **Campbell's theorem**, and they hold for rather general functions g [34, pp. 28–29]. See also Problem 17.

11.2 Renewal processes

Recall that a Poisson process of rate λ can be constructed by writing

$$N_t := \sum_{k=1}^{\infty} I_{[0,t]}(T_k),$$

where the arrival times

$$T_k := X_1 + \cdots + X_k,$$

and the X_k are i.i.d. $\exp(\lambda)$ interarrival times. If we drop the requirement that the interarrival times be exponential and let them have arbitrary density f, then N_t is called a **renewal process**.

Because of the similarity between the Poisson process and renewal processes, it is trivial to modify the MATLAB code of Example 11.3 to simulate a renewal process instead of a Poisson process. All we have to do is change the formula for X. See Problem 19.

Example 11.5. Hits to the Nuclear Engineering Department's website form a renewal process N_t, while hits to the Mechanical Engineering Department's website form a renewal process M_t. Assuming the processes are independent, find the probability that the first hit to the Nuclear Engineering website occurs before the first hit to the Mechanical Engineering website.

Solution. Let X_k denote the kth interarrival time of the N_t process, and let Y_k denote the kth interarrival time of the M_t process. Then we need to compute

$$P(X_1 < Y_1) = \int_0^\infty P(X_1 < Y_1 | Y_1 = y) f_Y(y) \, dy,$$

where we have used the law of total probability, and where f_Y denotes the common density of the Y_k. Since the renewal processes are independent, so are their arrival and interarrival times. Using the law of substitution and independence,

$$\begin{aligned} P(X_1 < Y_1) &= \int_0^\infty P(X_1 < y | Y_1 = y) f_Y(y) \, dy \\ &= \int_0^\infty P(X_1 < y) f_Y(y) \, dy \\ &= \int_0^\infty F_X(y) f_Y(y) \, dy, \end{aligned}$$

where F_X is the common cdf of the X_k. (Since X_k has a density, F_X is continuous.)

If we let F denote the cdf corresponding to the interarrival density f, it is easy to see that the mean of the process is

$$E[N_t] = \sum_{k=1}^\infty F_k(t), \qquad (11.11)$$

where F_k is the cdf of T_k. The corresponding density, denoted by f_k, is the k-fold convolution of f with itself. Hence, in general this formula is difficult to work with. However, there is another way to characterize $E[N_t]$. In the problems you are asked to derive the **renewal equation**,

$$E[N_t] = F(t) + \int_0^t E[N_{t-x}] f(x) \, dx.$$

The mean function $m(t) := E[N_t]$ of a renewal process is called the **renewal function**. Note that $m(0) = E[N_0] = 0$, and that the renewal equation can be written in terms of the renewal function as

$$m(t) = F(t) + \int_0^t m(t-x) f(x) \, dx.$$

11.3 The Wiener process

The theory of wide-sense stationary processes and white noise developed in Chapter 10 provides a satisfactory operational calculus for the analysis and design of linear, time-invariant systems driven by white noise. However, if the output of such a system is passed through a nonlinearity, and if we try to describe the result using a differential equation, we immediately run into trouble.

Example 11.6. Let X_t be a wide-sense stationary, zero-mean process with correlation function $R_X(\tau) = \sigma^2 \delta(\tau)$. If the white noise X_t is applied to an integrator starting at time zero, then the output at time t is

$$V_t := \int_0^t X_\tau \, d\tau,$$

and the derivative of V_t with respect to t is $\dot{V}_t = X_t$. If we now pass V_t through a square-law device, say

$$Y_t := V_t^2,$$

then

$$\dot{Y}_t = 2V_t \dot{V}_t = 2V_t X_t, \quad Y_0 = 0.$$

Hence,

$$\begin{aligned} Y_t &= \int_0^t \dot{Y}_\theta \, d\theta \\ &= 2 \int_0^t V_\theta X_\theta \, d\theta \\ &= 2 \int_0^t \left(\int_0^\theta X_\tau \, d\tau \right) X_\theta \, d\theta. \end{aligned} \quad (11.12)$$

It is now easy to see that

$$\mathsf{E}[Y_t] = 2 \int_0^t \int_0^\theta \mathsf{E}[X_\tau X_\theta] \, d\tau \, d\theta = 2 \int_0^t \int_0^\theta \sigma^2 \delta(\tau - \theta) \, d\tau \, d\theta$$

reduces to

$$\mathsf{E}[Y_t] = 2\sigma^2 \int_0^t d\theta = 2\sigma^2 t.$$

On the other hand,

$$\mathsf{E}[Y_t] = \mathsf{E}[V_t^2] = \mathsf{E}\left[\left(\int_0^t X_\tau \, d\tau \right) \left(\int_0^t X_\theta \, d\theta \right) \right]$$

reduces to

$$\int_0^t \int_0^t R_X(\tau - \theta) \, d\tau \, d\theta = \int_0^t \int_0^t \sigma^2 \delta(\tau - \theta) \, d\tau \, d\theta = \sigma^2 t,$$

which is the correct result.

As mentioned in Section 10.5, white noise does not exist as an ordinary process. This example illustrates the perils of working with objects that are not mathematically well defined. Although white noise does not exist as an ordinary random process, there is a well defined process that can take the place of V_t. The **Wiener process** or **Brownian motion** is a random process that models integrated white noise. The Wiener process, and Wiener integral introduced below, are used extensively in stochastic differential equations. Stochastic differential equations arise in numerous applications. For example, they describe control systems driven by white noise, heavy traffic behavior of communication networks, and economic models of the stock market.

11.3 The Wiener process

We say that $\{W_t, t \geq 0\}$ is a Wiener process if the following four conditions hold.

- $W_0 \equiv 0$; i.e., W_0 is a constant random variable whose value is always zero.
- For any $0 \leq s \leq t < \infty$, the increment $W_t - W_s$ is a Gaussian random variable with zero mean and variance $\sigma^2(t-s)$. In particular, $\mathsf{E}[W_t - W_s] = 0$, and $\mathsf{E}[(W_t - W_s)^2] = \sigma^2(t-s)$.
- If the time intervals
$$(t_1, t_2], (t_2, t_3], \ldots, (t_n, t_{n+1}]$$
are disjoint, then the increments
$$W_{t_2} - W_{t_1}, W_{t_3} - W_{t_2}, \ldots, W_{t_{n+1}} - W_{t_n}$$
are independent; i.e., the process has independent increments.
- For each sample point $\omega \in \Omega$, $W_t(\omega)$ as a function of t is continuous. More briefly, we just say that W_t has **continuous sample paths**.

Remark. (*i*) If the parameter $\sigma^2 = 1$, then the process is called a **standard Wiener process**. Two sample paths of a standard Wiener process are shown in Figure 11.4.

(*ii*) Since the first and third properties of the Wiener process are the same as those of the Poisson process, it is easy to show (Problem 26) that

$$\operatorname{cov}(W_{t_1}, W_{t_2}) = \sigma^2 \min(t_1, t_2).$$

Hence, to justify the claim that W_t is a model for integrated white noise, it suffices to show that the process V_t of Example 11.6 has the same covariance function; see Problem 25.

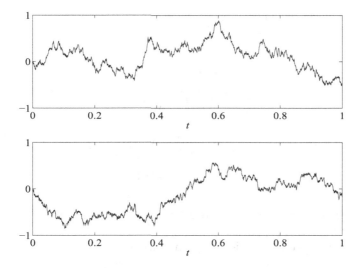

Figure **11.4.** Two sample paths of a standard Wiener process.

(iii) The fourth condition, that as a function of t, W_t should be continuous, is always assumed in practice, and can always be arranged by construction [3, Section 37]. The precise statement of the fourth property is

$$P(\{\omega \in \Omega : W_t(\omega) \text{ is a continuous function of } t\}) = 1,$$

i.e., the realizations of W_t are continuous with probability one.

(iv) As indicated in Figure 11.4, the Wiener process is very "wiggly." In fact, it wiggles so much that it is nowhere differentiable with probability one [3, p. 505, Theorem 37.3].

(v) Although the Wiener process was originally proposed as a model for the path of a small particle suspended in a fluid, the fact that Wiener-process paths are nowhere differentiable implies the particle has infinite velocity! One way around this problem is to use the Wiener process (more precisely, the **Ornstein–Uhlenbeck process**), which is continuous, to model the velocity, and integrate the velocity to model the particle position [23], [40]. See Problems 32 and 33.

The Wiener integral

The Wiener process is a well-defined mathematical object. We argued above that W_t behaves like $V_t := \int_0^t X_\tau \, d\tau$, where X_t is white noise. If such noise is applied to a linear time-invariant system starting at time zero, and if the system has impulse response h, then the output is

$$\int_0^\infty h(t-\tau) X_\tau \, d\tau.$$

If we now suppress t and write $g(\tau)$ instead of $h(t-\tau)$, then we need a well-defined mathematical object to play the role of

$$\int_0^\infty g(\tau) X_\tau \, d\tau.$$

To see what this object should be, suppose that $g(\tau)$ is piecewise constant taking the value g_i on the interval $(t_i, t_{i+1}]$. Then

$$\begin{aligned}
\int_0^\infty g(\tau) X_\tau \, d\tau &= \sum_i \int_{t_i}^{t_{i+1}} g(\tau) X_\tau \, d\tau \\
&= \sum_i g_i \int_{t_i}^{t_{i+1}} X_\tau \, d\tau \\
&= \sum_i g_i \left(\int_0^{t_{i+1}} X_\tau \, d\tau - \int_0^{t_i} X_\tau \, d\tau \right) \\
&= \sum_i g_i (V_{t_{i+1}} - V_{t_i}).
\end{aligned}$$

Thus, for piecewise-constant functions, integrals with white noise should be replaced by sums involving the Wiener process.

The **Wiener integral** of a function $g(\tau)$ is denoted by

$$\int_0^\infty g(\tau) \, dW_\tau,$$

11.3 The Wiener process

and is defined as follows. For piecewise constant functions g of the form

$$g(\tau) = \sum_{i=1}^{n} g_i I_{(t_i, t_{i+1}]}(\tau),$$

where $0 \le t_1 < t_2 < \cdots < t_{n+1} < \infty$, we define

$$\int_0^\infty g(\tau) \, dW_\tau := \sum_{i=1}^{n} g_i (W_{t_{i+1}} - W_{t_i}),$$

where W_t is a Wiener process. Note that the right-hand side is a weighted sum of independent, zero-mean, Gaussian random variables. The sum is therefore Gaussian with zero mean and variance

$$\sum_{i=1}^{n} g_i^2 \operatorname{var}(W_{t_{i+1}} - W_{t_i}) = \sum_{i=1}^{n} g_i^2 \cdot \sigma^2 (t_{i+1} - t_i) = \sigma^2 \int_0^\infty g(\tau)^2 \, d\tau.$$

Because of the zero mean, the variance and second moment are the same. Hence, we also have

$$\mathsf{E}\left[\left(\int_0^\infty g(\tau) \, dW_\tau\right)^2\right] = \sigma^2 \int_0^\infty g(\tau)^2 \, d\tau. \tag{11.13}$$

For functions g that are not piecewise constant, but do satisfy $\int_0^\infty g(\tau)^2 \, d\tau < \infty$, the Wiener integral can be defined by a limiting process, which is discussed in more detail in Chapter 13. Basic properties of the Wiener integral are explored in the problems.

Remark. When the Wiener integral is extended to allow random integrands, it is known as the **Itô integral**. Using the **Itô rule** [67], it can be shown that if $Y_t = W_t^2$, then

$$Y_t = 2 \int_0^t W_\tau \, dW_\tau + \sigma^2 t$$

is the correct version of (11.12) in Example 11.6. Since the expected value of the Itô integral is zero, we find that $\mathsf{E}[Y_t] = \sigma^2 t$, which is the correct result. The extra term $\sigma^2 t$ in the equation for Y_t is called the **Itô correction term**.

Random walk approximation of the Wiener process

We present a three-step construction of a continuous-time, piecewise-constant random process that approximates the Wiener process.

The first step is to construct a symmetric random walk. Let $X_1, X_2, \ldots$ be i.i.d. ± 1-valued random variables with $\mathsf{P}(X_i = \pm 1) = 1/2$. Then each X_i has zero mean and variance one. Let $S_0 \equiv 0$, and for $n \ge 1$, put

$$S_n := \sum_{i=1}^{n} X_i.$$

Then S_n has zero mean and variance n. The process $\{S_n, n \ge 0\}$ is called a symmetric random walk.

The second step is to construct the scaled random walk $S_n/\sqrt{n}$. Note that $S_n/\sqrt{n}$ has zero mean and variance one. By the central limit theorem, which is discussed in detail in Chapter 5, the cdf of $S_n/\sqrt{n}$ converges to the standard normal cdf.

The third step is to construct the continuous-time, piecewise-constant process

$$W_t^{(n)} := \frac{1}{\sqrt{n}} S_{\lfloor nt \rfloor},$$

where $\lfloor \tau \rfloor$ denotes the greatest integer that is less than or equal to τ. For example, if $n = 100$ and $t = 3.1476$, then

$$W_{3.1476}^{(100)} = \frac{1}{\sqrt{100}} S_{\lfloor 100 \cdot 3.1476 \rfloor} = \frac{1}{10} S_{\lfloor 314.76 \rfloor} = \frac{1}{10} S_{314}.$$

For example, a sample path of $S_0, S_1, \ldots, S_{75}$ is shown at the top in Figure 11.5. The corresponding continuous-time, piecewise constant process $W_t^{(75)}$ is shown at the bottom in Figure 11.5. Notice that as the continuous variable t ranges over $[0, 1]$, the values of $\lfloor 75t \rfloor$ range over the integers $0, 1, \ldots, 75$. Thus, the constant levels seen at the bottom in Figure 11.5 are $1/\sqrt{75}$ times those at the top.

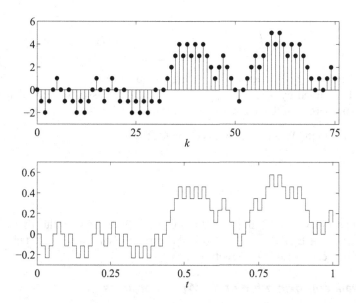

Figure 11.5. Sample path $S_k, k = 0, \ldots, 75$ (top). Sample path $W_t^{(75)}$ (bottom).

Figure 11.6 shows a sample path of $W_t^{(n)}$ for $n = 150$ (top) and for $n = 10\,000$ (bottom). As n increases, the sample paths look more and more like those of the Wiener processes shown in Figure 11.4.

Since the central limit theorem applies to any i.i.d. sequence with finite variance, the preceding convergence to the Wiener process holds if we replace the ± 1-valued X_i by any i.i.d. sequence with finite variance.[a] However, if the X_i only have finite mean but infinite

---

[a] If the mean of the X_i is m and the variance is σ^2, then we must replace S_n by $(S_n - nm)/\sigma$.

11.4 Specification of random processes

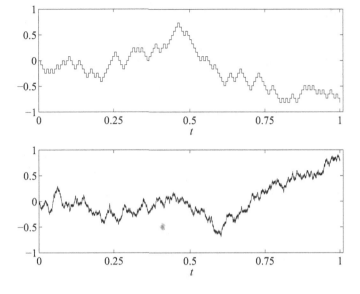

Figure 11.6. Sample path of $W_t^{(n)}$ for $n = 150$ (top) and for $n = 10\,000$ (bottom).

variance, other limit processes can be obtained. For example, suppose the X_i are i.i.d. having Student's t density with $\nu = 3/2$ degrees of freedom. Then the X_i have zero mean and infinite variance (recall Problems 27 and 37 in Chapter 4). As can be seen in Figure 11.7, the limiting process has jumps, which is inconsistent with the Wiener process, which has continuous sample paths.

11.4 Specification of random processes

Finitely many random variables

In this text we have often seen statements of the form, "Let X, Y, and Z be random variables with $P((X,Y,Z) \in B) = \mu(B)$," where $B \subset \mathbb{R}^3$, and $\mu(B)$ is given by some formula. For example, if X, Y, and Z are discrete, we would have

$$\mu(B) = \sum_i \sum_j \sum_k I_B(x_i, y_j, z_k) p_{i,j,k}, \qquad (11.14)$$

where the x_i, y_j, and z_k are the values taken by the random variables, and the $p_{i,j,k}$ are nonnegative numbers that sum to one. If X, Y, and Z are jointly continuous, we would have

$$\mu(B) = \int_{-\infty}^{\infty} \int_{-\infty}^{\infty} \int_{-\infty}^{\infty} I_B(x,y,z) f(x,y,z) \, dx \, dy \, dz, \qquad (11.15)$$

where f is nonnegative and integrates to one. In fact, if X is discrete and Y and Z are jointly continuous, we would have

$$\mu(B) = \sum_i \int_{-\infty}^{\infty} \int_{-\infty}^{\infty} I_B(x_i, y, z) f(x_i, y, z) \, dy \, dz, \qquad (11.16)$$

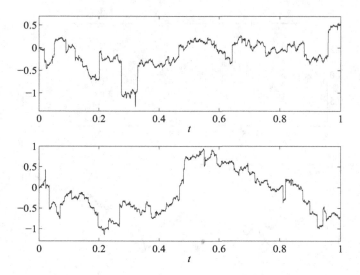

Figure 11.7. Two sample paths of $S_{\lfloor nt \rfloor}/n^{2/3}$ for $n = 10\,000$ when the X_i have Student's t density with 3/2 degrees of freedom.

where f is nonnegative and

$$\sum_i \int_{-\infty}^{\infty} \int_{-\infty}^{\infty} f(x_i, y, z) \, dy \, dz = 1.$$

The big question is, given a formula for computing $\mu(B)$, how do we know that a sample space Ω, a probability measure P, and functions $X(\omega)$, $Y(\omega)$, and $Z(\omega)$ exist such that we indeed have

$$\mathsf{P}\big((X,Y,Z) \in B\big) = \mu(B), \quad B \subset \mathbb{R}^3.$$

As we show in the next paragraph, the answer turns out to be rather simple.

If μ is defined by expressions such as (11.14)–(11.16), it can be shown that μ is a probability measure[b] on $\mathbb{R}^3$; the case of (11.14) is easy; the other two require some background in measure theory, e.g., [3]. More generally, if we are given any probability measure μ on $\mathbb{R}^3$, we take $\Omega = \mathbb{R}^3$ and put $\mathsf{P}(A) := \mu(A)$ for $A \subset \Omega = \mathbb{R}^3$. For $\omega = (\omega_1, \omega_2, \omega_3)$, we define

$$X(\omega) := \omega_1, \quad Y(\omega) := \omega_2, \quad \text{and} \quad Z(\omega) := \omega_3.$$

It then follows that for $B \subset \mathbb{R}^3$,

$$\{\omega \in \Omega : (X(\omega), Y(\omega), Z(\omega)) \in B\}$$

reduces to

$$\{\omega \in \Omega : (\omega_1, \omega_2, \omega_3) \in B\} = B.$$

Hence, $\mathsf{P}(\{(X,Y,Z) \in B\}) = \mathsf{P}(B) = \mu(B)$.

---

[b] See Section 1.4 to review the axioms satisfied by a probability measure.

11.4 Specification of random processes

For fixed $n \geq 1$, the foregoing ideas generalize in the obvious way to show the existence of a sample space Ω, probability measure P, and random variables $X_1,\ldots,X_n$ with

$$P\big((X_1,X_2,\ldots,X_n) \in B\big) = \mu(B), \quad B \subset \mathbb{R}^n,$$

where μ is any given probability measure defined on $\mathbb{R}^n$.

Infinite sequences (discrete time)

Consider an infinite sequence of random variables such as $X_1, X_2, \ldots$. While $(X_1,\ldots,X_n)$ takes values in $\mathbb{R}^n$, the infinite sequence $(X_1, X_2, \ldots)$ takes values in $\mathbb{R}^\infty$. If such an infinite sequence of random variables exists on some sample space Ω equipped with some probability measure P, then[1]

$$P\big((X_1, X_2, \ldots) \in B\big), \quad B \subset \mathbb{R}^\infty,$$

is a probability measure on $\mathbb{R}^\infty$. We denote this probability measure by $\mu(B)$. Similarly, P induces on $\mathbb{R}^n$ the measure

$$\mu_n(B_n) = P\big((X_1,\ldots,X_n) \in B_n\big), \quad B_n \subset \mathbb{R}^n,$$

Of course, P induces on $\mathbb{R}^{n+1}$ the measure

$$\mu_{n+1}(B_{n+1}) = P\big((X_1,\ldots,X_n,X_{n+1}) \in B_{n+1}\big), \quad B_{n+1} \subset \mathbb{R}^{n+1}.$$

If we take $B_{n+1} = B_n \times \mathbb{R}$ for any $B_n \subset \mathbb{R}^n$, then

$$\begin{aligned}
\mu_{n+1}(B_n \times \mathbb{R}) &= P\big((X_1,\ldots,X_n,X_{n+1}) \in B_n \times \mathbb{R}\big) \\
&= P\big((X_1,\ldots,X_n) \in B_n, X_{n+1} \in \mathbb{R}\big) \\
&= P\big(\{(X_1,\ldots,X_n) \in B_n\} \cap \{X_{n+1} \in \mathbb{R}\}\big) \\
&= P\big(\{(X_1,\ldots,X_n) \in B_n\} \cap \Omega\big) \\
&= P\big((X_1,\ldots,X_n) \in B_n\big) \\
&= \mu_n(B_n).
\end{aligned}$$

Thus, we have the **consistency condition**

$$\mu_{n+1}(B_n \times \mathbb{R}) = \mu_n(B_n), \quad B_n \subset \mathbb{R}^n, \quad n = 1,2,\ldots. \tag{11.17}$$

Next, observe that since $(X_1,\ldots,X_n) \in B_n$ if and only if

$$(X_1, X_2, \ldots, X_n, X_{n+1}, \ldots) \in B_n \times \mathbb{R} \times \cdots,$$

it follows that $\mu_n(B_n)$ is equal to

$$P\big((X_1, X_2, \ldots, X_n, X_{n+1}, \ldots) \in B_n \times \mathbb{R} \times \cdots\big),$$

which is simply $\mu(B_n \times \mathbb{R} \times \cdots)$. Thus,

$$\mu(B_n \times \mathbb{R} \times \cdots) = \mu_n(B_n), \quad B_n \in \mathbb{R}^n, \quad n = 1,2,\ldots. \tag{11.18}$$

The big question here is, if we are given a sequence of probability measures μ_n on $\mathbb{R}^n$ for $n = 1,2,\ldots$, does there exist a probability measure μ on $\mathbb{R}^\infty$ such that (11.18) holds?

To appreciate the complexity of this question, consider the simplest possible case of constructing a sequence of i.i.d. Bernoulli(1/2) random variables. In this case, for a finite sequence of zeros and ones, say $(b_1,\ldots,b_n)$, and with B_n being the singleton set $B_n = \{(b_1,\ldots,b_n)\}$, we want

$$\mu_n(B_n) = P(X_1 = b_1,\ldots,X_n = b_n) = \prod_{i=1}^n P(X_i = b_i) = \left(\frac{1}{2}\right)^n.$$

We need a probability measure μ on $\mathbb{R}^\infty$ that concentrates all its probability on the infinite sequences of zeros and ones,

$$S = \{\omega = (\omega_1, \omega_2, \ldots) : \omega_i = 0 \text{ or } \omega_i = 1\} \subset \mathbb{R}^\infty;$$

i.e., we need $\mu(S) = 1$. Unfortunately, as shown in Example 1.9, the set S is not **countable**. Hence, we cannot use a probability mass function to define μ. The general solution to this difficulty was found by **Kolmogorov** and is discussed next.

Conditions under which a probability measure can be constructed on $\mathbb{R}^\infty$ are known as **Kolmogorov's consistency theorem** or as **Kolmogorov's extension theorem**. It says that if the consistency condition (11.17) holds,[c] then a probability measure μ exists on $\mathbb{R}^\infty$ such that (11.18) holds [7, p. 188].

We now specialize the foregoing discussion to the case of integer-valued random variables $X_1, X_2, \ldots$. For each $n = 1, 2, \ldots$, let $p_n(i_1,\ldots,i_n)$ denote a proposed joint probability mass function of $X_1,\ldots,X_n$. In other words, we want a random process for which

$$P\big((X_1,\ldots,X_n) \in B_n\big) = \sum_{i_1=-\infty}^{\infty} \cdots \sum_{i_n=-\infty}^{\infty} I_{B_n}(i_1,\ldots,i_n) p_n(i_1,\ldots,i_n).$$

More precisely, with $\mu_n(B_n)$ given by the above right-hand side, does there exist a measure μ on $\mathbb{R}^\infty$ such that (11.18) holds? By Kolmogorov's theorem, we just need to show that (11.17) holds.

We now show that (11.17) is equivalent to

$$\sum_{j=-\infty}^{\infty} p_{n+1}(i_1,\ldots,i_n,j) = p_n(i_1,\ldots,i_n). \tag{11.19}$$

The left-hand side of (11.17) takes the form

$$\sum_{i_1=-\infty}^{\infty} \cdots \sum_{i_n=-\infty}^{\infty} \sum_{j=-\infty}^{\infty} I_{B_n \times \mathbb{R}}(i_1,\ldots,i_n,j) p_{n+1}(i_1,\ldots,i_n,j). \tag{11.20}$$

---

[c] Knowing the measure μ_n, we can always write the corresponding cdf as

$$F_n(x_1,\ldots,x_n) = \mu_n\big((-\infty, x_1] \times \cdots \times (-\infty, x_n]\big).$$

Conversely, if we know the F_n, there is a unique measure μ_n on $\mathbb{R}^n$ such that the above formula holds [3, Section 12]. Hence, the consistency condition has the equivalent formulation in terms of cdfs [7, p. 189],

$$\lim_{x_{n+1} \to \infty} F_{n+1}(x_1,\ldots,x_n,x_{n+1}) = F_n(x_1,\ldots,x_n).$$

11.4 Specification of random processes

Observe that $I_{B_n \times \mathbb{R}} = I_{B_n} I_{\mathbb{R}} = I_{B_n}$. Hence, the above sum becomes

$$\sum_{i_1=-\infty}^{\infty} \cdots \sum_{i_n=-\infty}^{\infty} \sum_{j=-\infty}^{\infty} I_{B_n}(i_1,\ldots,i_n) p_{n+1}(i_1,\ldots,i_n,j),$$

which, using (11.19), simplifies to

$$\sum_{i_1=-\infty}^{\infty} \cdots \sum_{i_n=-\infty}^{\infty} I_{B_n}(i_1,\ldots,i_n) p_n(i_1,\ldots,i_n), \quad (11.21)$$

which is our definition of $\mu_n(B_n)$. Conversely, if in (11.17), or equivalently in (11.20) and (11.21), we take B_n to be the singleton set

$$B_n = \{(j_1,\ldots,j_n)\},$$

then we obtain (11.19).

The next question is how to construct a sequence of probability mass functions satisfying (11.19). Observe that (11.19) can be rewritten as

$$\sum_{j=-\infty}^{\infty} \frac{p_{n+1}(i_1,\ldots,i_n,j)}{p_n(i_1,\ldots,i_n)} = 1.$$

In other words, if $p_n(i_1,\ldots,i_n)$ is a valid joint pmf, and if we define

$$p_{n+1}(i_1,\ldots,i_n,j) := p_{n+1|1,\ldots,n}(j|i_1,\ldots,i_n) \cdot p_n(i_1,\ldots,i_n),$$

where $p_{n+1|1,\ldots,n}(j|i_1,\ldots,i_n)$ is a valid pmf in the variable j (i.e., is nonnegative and the sum over j is one), then (11.19) will automatically hold!

***Example* 11.7.** Let $q(i)$ be any pmf. Take $p_1(i) := q(i)$, and take $p_{n+1|1,\ldots,n}(j|i_1,\ldots,i_n) := q(j)$. Then, for example,

$$p_2(i,j) = p_{2|1}(j|i) p_1(i) = q(j) q(i),$$

and

$$p_3(i,j,k) = p_{3|1,2}(k|i,j) p_2(i,j) = q(i) q(j) q(k).$$

More generally,

$$p_n(i_1,\ldots,i_n) = q(i_1) \cdots q(i_n).$$

Thus, the X_n are i.i.d. with common pmf q.

***Example* 11.8.** Again let $q(i)$ be any pmf. Suppose that for each i, $r(j|i)$ is a pmf in the variable j; i.e., r is any conditional pmf. Put $p_1(i) := q(i)$, and put

$$p_{n+1|1,\ldots,n}(j|i_1,\ldots,i_n) := r(j|i_n).$$

Then

$$p_2(i,j) = p_{2|1}(j|i) p_1(i) = r(j|i) q(i),$$

and
$$p_3(i,j,k) = p_{3|1,2}(k|i,j)\,p_2(i,j) = q(i)\,r(j|i)\,r(k|j).$$

More generally,
$$p_n(i_1,\ldots,i_n) = q(i_1)\,r(i_2|i_1)\,r(i_3|i_2)\cdots r(i_n|i_{n-1}).$$

As we will see in Chapter 12, p_n is the joint pmf of a Markov chain with stationary transition probabilities $r(j|i)$ and initial pmf $q(i)$.

Continuous-time random processes

The consistency condition for a continuous-time random process is a little more complicated. The reason is that in discrete time, between any two consecutive integers, there are no other integers, while in continuous time, for any $t_1 < t_2$, there are infinitely many times between t_1 and t_2.

Now suppose that for any $t_1 < \cdots < t_{n+1}$, we are given a probability measure $\mu_{t_1,\ldots,t_{n+1}}$ on $\mathbb{R}^{n+1}$. Fix any $B_n \subset \mathbb{R}^n$. For $k = 1,\ldots,n+1$, define $B_{n,k} \subset \mathbb{R}^{n+1}$ by

$$B_{n,k} := \{(x_1,\ldots,x_{n+1}) : (x_1,\ldots,x_{k-1},x_{k+1},\ldots,x_{n+1}) \in B_n \text{ and } x_k \in \mathbb{R}\}.$$

Note the special cases[d]

$$B_{n,1} = \mathbb{R} \times B_n \quad \text{and} \quad B_{n,n+1} = B_n \times \mathbb{R}.$$

The continuous-time consistency condition is that [53, p. 244] for $k = 1,\ldots,n+1$,

$$\mu_{t_1,\ldots,t_{n+1}}(B_{n,k}) = \mu_{t_1,\ldots,t_{k-1},t_{k+1},\ldots,t_{n+1}}(B_n). \tag{11.22}$$

If this condition holds, then there is a sample space Ω, a probability measure P, and random variables X_t such that

$$\mathsf{P}\big((X_{t_1},\ldots,X_{t_n}) \in B_n\big) = \mu_{t_1,\ldots,t_n}(B_n), \quad B_n \subset \mathbb{R}^n,$$

for any $n \geq 1$ and any times $t_1 < \cdots < t_n$.

Gaussian processes

A continuous-time random process X_t is said to be Gaussian if for every sequence of times $t_1,\ldots,t_n$, $[X_{t_1},\ldots,X_{t_n}]'$ is a Gaussian random vector. Since we have defined a random vector to be Gaussian if every linear combination of its components is a scalar Gaussian random variable, we see that a random process is Gaussian if and only if every finite linear combination of samples, say $\sum_{i=1}^{n} c_i X_{t_i}$, is a Gaussian random variable. You can use this fact to show that the Wiener process is a Gaussian process in Problem 46.

---

[d]In the previous subsection, we only needed the case $k = n+1$. If we had wanted to allow two-sided discrete-time processes X_n for n any positive or negative integer, then both $k = 1$ and $k = n+1$ would have been needed (Problem 42).

11.4 Specification of random processes

Example 11.9. Show that if a Gaussian process is wide-sense stationary, then it is strictly stationary.

Solution. Without loss of generality, we assume the process is zero mean. Let $R(\tau) := \mathsf{E}[X_{t+\tau}X_t]$. Let times $t_1,\ldots,t_n$ be given, and consider the vectors

$$X := [X_{t_1},\ldots,X_{t_n}]' \quad \text{and} \quad Y := [X_{t_1+\Delta t},\ldots,X_{t_n+\Delta t}]'.$$

We need to show that $\mathsf{P}(X \in B) = \mathsf{P}(Y \in B)$ for any n-dimensional set B. It suffices to show that X and Y have the same joint characteristic function. Since X and Y are zero-mean Gaussian random vectors, all we need to do is show that they have the same covariance matrix. The ij entry of the covariance matrix of X is

$$\mathsf{E}[X_{t_i}X_{t_j}] = R(t_i - t_j),$$

while for Y it is

$$\mathsf{E}[X_{t_i+\Delta t}X_{t_j+\Delta t}] = R\big((t_i+\Delta t) - (t_j+\Delta t)\big) = R(t_i - t_j).$$

Example 11.10. Show that a real-valued function $R(t,s)$ is the correlation function of a continuous-time random process if and only if for every finite sequence of distinct times, say $t_1,\ldots,t_n$, the $n \times n$ matrix with ij entry $R(t_i,t_j)$ is positive semidefinite.

Solution. If $R(t,s)$ is the correlation function of a process X_t, then the matrix with entries $R(t_i,t_j)$ is the covariance matrix of the random vector $[X_{t_1},\ldots,X_{t_n}]'$. As noted following Example 8.4, the covariance matrix of a random vector must be positive semidefinite.

Conversely, suppose every matrix with entries $R(t_i,t_j)$ is positive semidefinite. Imagine a Gaussian random vector $[Y_1,\ldots,Y_n]'$ having this covariance matrix, and put

$$\mu_{t_1,\ldots,t_n}(B) := \mathsf{P}((Y_1,\ldots,Y_n) \in B).$$

Since any subvector of Y is Gaussian with covariance matrix given by appropriate entries of the covariance matrix of Y, the consistency conditions are satisfied. By Kolmogorov's theorem, the required process exists. In fact, the process constructed in this way is Gaussian.

Another important property of Gaussian processes is that their integrals are Gaussian random variables. If X_t is a Gaussian process, we might consider an integral of the form

$$\int c(t)X_t \, dt = \lim \sum_i c(t_i)X_{t_i}\Delta t_i.$$

Since the process is Gaussian, linear combinations of samples X_{t_i} are scalar Gaussian random variables. We then use the fact that limits of Gaussian random variables are Gaussian. This is just a sketch of how the general argument goes. For details, see the discussion of mean-square integrals at the end of Section 13.2 and also Example 14.9.

Notes

11.4: Specification of random processes

Note **1.** Comments analogous to Note **1** in Chapter 7 apply here. Specifically, the set B must be restricted to a suitable σ-field $\mathscr{B}^\infty$ of subsets of $\mathbb{R}^\infty$. Typically, $\mathscr{B}^\infty$ is taken to be the smallest σ-field containing all sets of the form

$$\{\omega = (\omega_1, \omega_2, \ldots) \in \mathbb{R}^\infty : (\omega_1, \ldots, \omega_n) \in B_n\},$$

where B_n is a Borel subset of $\mathbb{R}^n$, and n ranges over the positive integers [3, p. 485].

Problems

11.1: The Poisson process

1. Hits to a certain website occur according to a Poisson process of rate $\lambda = 3$ per minute. What is the probability that there are no hits in a 10-minute period? Give a formula and then evaluate it to obtain a numerical answer.

2. Cell-phone calls processed by a certain wireless base station arrive according to a Poisson process of rate $\lambda = 12$ per minute. What is the probability that more than three calls arrive in a 20-second interval? Give a formula and then evaluate it to obtain a numerical answer.

3. Let N_t be a Poisson process with rate $\lambda = 2$, and consider a fixed observation interval $(0, 5]$.

    (a) What is the probability that $N_5 = 10$?
    (b) What is the probability that $N_i - N_{i-1} = 2$ for all $i = 1, \ldots, 5$?

4. A sports clothing store sells football jerseys with a certain very popular number on them according to a Poisson process of rate three crates per day. Find the probability that on 5 days in a row, the store sells at least three crates each day.

5. A sporting goods store sells a certain fishing rod according to a Poisson process of rate two per day. Find the probability that on at least 1 day during the week, the store sells at least three rods. (Note: week = 5 days.)

6. A popular music group produces a new hit song every 7 months on average. Assume that songs are produced according to a Poisson process.

    (a) Find the probability that the group produces more than two hit songs in 1 year.
    (b) How long do you expect it to take until the group produces its 10th song?

7. Let N_t be a Poisson process with rate λ, and let $\Delta t > 0$.

    (a) Show that $N_{t+\Delta t} - N_t$ and N_t are independent.
    (b) Show that $\mathsf{P}(N_{t+\Delta t} = k + \ell | N_t = k) = \mathsf{P}(N_{t+\Delta t} - N_t = \ell)$.
    (c) Evaluate $\mathsf{P}(N_t = k | N_{t+\Delta t} = k + \ell)$.

(d) Show that as a function of $k = 0, \ldots, n$, $P(N_t = k | N_{t+\Delta t} = n)$ has the binomial(n, p) probability mass function and identify p.

8. Customers arrive at a store according to a Poisson process of rate λ. What is the expected time until the nth customer arrives? What is the expected time between customers?

9. During the winter, snowstorms occur according to a Poisson process of intensity $\lambda = 2$ per week.

   (a) What is the average time between snowstorms?
   (b) What is the probability that no storms occur during a given 2-week period?
   (c) If winter lasts 12 weeks, what is the expected number of snowstorms?
   (d) Find the probability that during at least one of the 12 weeks of winter, there are at least five snowstorms.

10. Space shuttles are launched according to a Poisson process. The average time between launches is 2 months.

    (a) Find the probability that there are no launches during a 4-month period.
    (b) Find the probability that during at least 1 month out of four consecutive months, there are at least two launches.

11. Internet packets arrive at a router according to a Poisson process of rate λ. Find the variance of the time it takes for the first n packets to arrive.

12. Let U be a uniform$[0, 1]$ random variable that is independent of a Poisson process N_t with rate $\lambda = 1$. Put

    $$Y_t := N_{\ln(1+tU)}.$$

    Find the probability generating function of Y_t, $G(z) := \mathsf{E}[z^{Y_t}]$ for real z, including $z = 0$.

*13. Hits to the websites of the Nuclear and Mechanical Engineering Departments form two independent Poisson processes, N_t and M_t, respectively. Let λ and μ be their respective rates. Find the probability that between two consecutive hits to the Nuclear Engineering website, there are exactly m hits to the Mechanical Engineering website.

14. Diners arrive at popular restaurant according to a Poisson process N_t of rate λ. A confused maitre d' seats the ith diner with probability p, and turns the diner away with probability $1 - p$. Let $Y_i = 1$ if the ith diner is seated, and $Y_i = 0$ otherwise. The number diners seated up to time t is

    $$M_t := \sum_{i=1}^{N_t} Y_i.$$

    Show that M_t is a Poisson random variable and find its parameter. Assume the Y_i are independent of each other and of the Poisson process.

    Remark. M_t is an example of a **thinned Poisson process**.

15. Lightning strikes occur according to a Poisson process of rate λ per minute. The energy of the ith strike is V_i. Assume the energies V_i are i.i.d. random variables that are independent of the occurrence times (independent of the Poisson process). What is the expected energy of a storm that lasts for t minutes? What is the average time between lightning strikes?

16. (This problem uses the methods and notation of Section 6.4.) Let N_t be a Poisson process with unknown intensity λ. For $i = 1, \ldots, 100$, put $X_i = (N_i - N_{i-1})$. Then the X_i are i.i.d. Poisson(λ), and $E[X_i] = \lambda$. If $M_{100} = 5.170$, and $S_{100} = 2.132$, find the 95% confidence interval for λ.

*17. For $0 \le t_0 < \cdots < t_n < \infty$, let
$$g(\tau) = \sum_{k=1}^{n} g_k I_{(t_{k-1}, t_k]}(\tau) \quad \text{and} \quad h(\tau) = \sum_{l=1}^{n} h_l I_{(t_{l-1}, t_l]}(\tau),$$
and put
$$Y = \sum_{i=1}^{\infty} g(T_i) \quad \text{and} \quad Z = \sum_{j=1}^{\infty} h(T_j).$$
Show that formula for $E[Y]$ in (11.9) and the formula for the characteristic function of Y in (11.10) continue to hold. Also show that
$$\text{cov}(Y, Z) = \int_0^{\infty} g(\tau) h(\tau) \lambda \, d\tau.$$

*18. Find the mean and characteristic function of the shot-noise random variable Y_t in equation (11.7). Also find $\text{cov}(Y_t, Y_s)$. *Hint:* Use the results of the previous problem.

11.2: Renewal processes

19. **MATLAB.** Modify the MATLAB code of Example 11.3 and print out a simulation of a renewal process whose interarrival times are i.i.d. chi-squared with one degree of freedom.

20. In Example 11.5, suppose N_t is actually a Poisson process of rate λ. Show that the result of Example 11.5 can be expressed in terms of the moment generating function of the Y_k. Now further simplify your expression in the case that M_t is a Poisson process of rate μ.

21. Internet packets arrive at a router according to a renewal process whose interarrival times are uniform$[0,1]$. Find the variance of the time it takes for the first n packets to arrive.

22. In the case of a Poisson process, show that the right-hand side of (11.11) reduces to λt.

23. Derive the renewal equation
$$E[N_t] = F(t) + \int_0^t E[N_{t-x}] f(x) \, dx$$
as follows.

(a) Show that $E[N_t|X_1 = x] = 0$ for $x > t$.

(b) Show that $E[N_t|X_1 = x] = 1 + E[N_{t-x}]$ for $x \le t$.

(c) Use parts (a) and (b) and the law of total probability to derive the renewal equation.

24. Solve the renewal equation for the renewal function $m(t) := E[N_t]$ if the interarrival density is $f \sim \exp(\lambda)$. *Hint:* Take the one-sided Laplace transform of the renewal equation. It then follows that $m(t) = \lambda t$ for $t \ge 0$, which is what we expect since $f \sim \exp(\lambda)$ implies N_t is a Poisson process of rate λ.

11.3: The Wiener process

25. Let V_t be defined as in Example 11.6. Show that for $0 \le s < t < \infty$, $E[V_t V_s] = \sigma^2 s$.

26. For $0 \le s < t < \infty$, use the definition of the Wiener process to show that $E[W_t W_s] = \sigma^2 s$.

27. Let W_t be a Wiener process with $E[W_t^2] = \sigma^2 t$. Put $Y_t := e^{W_t}$. Find the correlation function $R_Y(t_1, t_2) := E[Y_{t_1} Y_{t_2}]$ for $t_2 > t_1$.

28. Let the random vector $X = [W_{t_1}, \ldots, W_{t_n}]'$, $0 < t_1 < \cdots < t_n < \infty$, consist of samples of a Wiener process. Find the covariance matrix of X, and write it out in detail as

$$\begin{bmatrix} c_{11} & c_{12} & c_{13} & & c_{1n} \\ c_{21} & c_{22} & c_{23} & & c_{2n} \\ c_{31} & c_{32} & c_{33} & & c_{3n} \\ & & & \ddots & \\ c_{n1} & c_{n2} & c_{n3} & & c_{nn} \end{bmatrix},$$

where each c_{ij} is given explicitly in terms of t_i or t_j.

29. For piecewise constant g and h, show that

$$\int_0^\infty g(\tau) \, dW_\tau + \int_0^\infty h(\tau) \, dW_\tau = \int_0^\infty [g(\tau) + h(\tau)] \, dW_\tau.$$

Hint: The problem is easy if g and h are constant *over the same intervals.*

30. Use (11.13) to derive the formula

$$E\left[\left(\int_0^\infty g(\tau) \, dW_\tau\right)\left(\int_0^\infty h(\tau) \, dW_\tau\right)\right] = \sigma^2 \int_0^\infty g(\tau) h(\tau) \, d\tau.$$

Hint: Consider the expectation

$$E\left[\left(\int_0^\infty g(\tau) \, dW_\tau - \int_0^\infty h(\tau) \, dW_\tau\right)^2\right],$$

which can be evaluated in two different ways. The first way is to expand the square and take expectations term by term, applying (11.13) where possible. The second way is to observe that since

$$\int_0^\infty g(\tau)\,dW_\tau - \int_0^\infty h(\tau)\,dW_\tau = \int_0^\infty [g(\tau) - h(\tau)]\,dW_\tau,$$

the above second moment can be computed directly using (11.13).

31. Let
$$Y_t = \int_0^t g(\tau)\,dW_\tau, \quad t \geq 0.$$

(a) Use (11.13) to show that
$$E[Y_t^2] = \sigma^2 \int_0^t g(\tau)^2\,d\tau.$$

Hint: Observe that
$$\int_0^t g(\tau)\,dW_\tau = \int_0^\infty g(\tau) I_{(0,t]}(\tau)\,dW_\tau.$$

(b) Show that Y_t has correlation function
$$R_Y(t_1, t_2) = \sigma^2 \int_0^{\min(t_1,t_2)} g(\tau)^2\,d\tau, \quad t_1, t_2 \geq 0.$$

32. Consider the process
$$Y_t = e^{-\lambda t} V + \int_0^t e^{-\lambda(t-\tau)}\,dW_\tau, \quad t \geq 0,$$

where W_t is a Wiener process independent of V, and V has zero mean and variance q^2. Use Problem 31 to show that Y_t has correlation function
$$R_Y(t_1, t_2) = e^{-\lambda(t_1+t_2)}\left(q^2 - \frac{\sigma^2}{2\lambda}\right) + \frac{\sigma^2}{2\lambda} e^{-\lambda|t_1-t_2|}.$$

Remark. If V is normal, then the process Y_t is Gaussian and is known as an **Ornstein–Uhlenbeck process**.

33. Let W_t be a Wiener process, and put
$$Y_t := \frac{e^{-\lambda t}}{\sqrt{2\lambda}} W_{e^{2\lambda t}}.$$

Show that
$$R_Y(t_1, t_2) = \frac{\sigma^2}{2\lambda} e^{-\lambda|t_1-t_2|}.$$

In light of the remark above, this is another way to define an Ornstein–Uhlenbeck process.

34. Let W_t be a standard Wiener process, and put

$$Y_t := \int_0^t g(\tau)\, dW_\tau$$

for some function $g(t)$.

(a) Evaluate $P(t) := \mathsf{E}[Y_t^2]$. *Hint:* Use Problem 31.
(b) If $g(t) \neq 0$, show that $P(t)$ is strictly increasing.
(c) Assume $P(t) < \infty$ for $t < \infty$ and that $P(t) \to \infty$ as $t \to \infty$. If $g(t) \neq 0$, then by part (b), $P^{-1}(t)$ exists and is defined for all $t \geq 0$. If $X_t := Y_{P^{-1}(t)}$, compute $\mathsf{E}[X_t]$ and $\mathsf{E}[X_t^2]$.

35. So far we have defined the Wiener process W_t only for $t \geq 0$. When defining W_t for all t, we continue to assume that $W_0 \equiv 0$; that for $s < t$, the increment $W_t - W_s$ is a Gaussian random variable with zero mean and variance $\sigma^2(t-s)$; that W_t has independent increments; and that W_t has continuous sample paths. The only difference is that s or both s and t can be negative, and that increments can be located anywhere in time, not just over intervals of positive time. In the following take $\sigma^2 = 1$.

(a) For $t > 0$, show that $\mathsf{E}[W_t^2] = t$.
(b) For $s < 0$, show that $\mathsf{E}[W_s^2] = -s$.
(c) Show that

$$\mathsf{E}[W_t W_s] = \frac{|t| + |s| - |t-s|}{2}.$$

11.4: Specification of random processes

36. Suppose X and Y are random variables with

$$\mathsf{P}((X,Y) \in A) = \sum_i \int_{-\infty}^{\infty} I_A(x_i, y) f_{XY}(x_i, y)\, dy,$$

where the x_i are distinct real numbers, and f_{XY} is a nonnegative function satisfying

$$\sum_i \int_{-\infty}^{\infty} f_{XY}(x_i, y)\, dy = 1.$$

(a) Show that

$$\mathsf{P}(X = x_k) = \int_{-\infty}^{\infty} f_{XY}(x_k, y)\, dy.$$

(b) Show that for $C \subset \mathbb{R}$,

$$\mathsf{P}(Y \in C) = \int_C \left(\sum_i f_{XY}(x_i, y) \right) dy.$$

In other words, Y has marginal density

$$f_Y(y) = \sum_i f_{XY}(x_i, y).$$

(c) Show that
$$P(Y \in C | X = x_k) = \int_C \left[\frac{f_{XY}(x_k, y)}{p_X(x_k)} \right] dy.$$

In other words,
$$f_{Y|X}(y|x_k) = \frac{f_{XY}(x_k, y)}{p_X(x_k)}.$$

(d) For $B \subset \mathbb{R}$, show that if we define
$$P(X \in B | Y = y) := \sum_i I_B(x_i) p_{X|Y}(x_i | y),$$

where
$$p_{X|Y}(x_i | y) := \frac{f_{XY}(x_i, y)}{f_Y(y)},$$

then
$$\int_{-\infty}^{\infty} P(X \in B | Y = y) f_Y(y) \, dy = P(X \in B).$$

In other words, we have the law of total probability.

37. Let F be the standard normal cdf. Then F is a one-to-one mapping from $(-\infty, \infty)$ onto $(0, 1)$. Therefore, F has an inverse, $F^{-1}: (0, 1) \to (-\infty, \infty)$. If $U \sim \text{uniform}(0, 1)$, show that $X := F^{-1}(U)$ has F for its cdf.

38. Consider the cdf
$$F(x) := \begin{cases} 0, & x < 0, \\ x^2, & 0 \le x < 1/2, \\ 1/4, & 1/2 \le x < 1, \\ x/2, & 1 \le x < 2, \\ 1, & x \ge 2. \end{cases}$$

(a) Sketch $F(x)$.

(b) For $0 < u < 1$, sketch
$$G(u) := \min\{x \in \mathbb{R} : F(x) \ge u\}.$$

Hint: First identify the set
$$B_u := \{x \in \mathbb{R} : F(x) \ge u\}.$$

Then find its minimum element.

39. As illustrated in the previous problem, an arbitrary cdf F is usually not invertible, either because the equation $F(x) = u$ has more than one solution, e.g., $F(x) = 1/4$, or because it has no solution, e.g., $F(x) = 3/8$. However, for any cdf F, we can always introduce the function[e]
$$G(u) := \min\{x \in \mathbb{R} : F(x) \ge u\}, \quad 0 < u < 1,$$

---

[e]In the previous problem, it was seen from the graph of $F(x)$ that $\{x \in \mathbb{R} : F(x) \ge u\}$ is a closed semi-infinite interval, whose left-hand end point is its minimum element. This is true for any cdf because cdfs are nondecreasing and right continuous.

which, you will now show, can play the role of F^{-1} in Problem 37. Show that if $0 < u < 1$ and $x \in \mathbb{R}$, then $G(u) \leq x$ if and only if $u \leq F(x)$.

40. Let G be as defined in the preceding problem, and let $U \sim$ uniform$(0,1)$. Put $X := G(U)$ and show that X has cdf F.

41. **MATLAB.** Write a MATLAB function called G to compute the function $G(u)$ that you found in Problem 38. Then use the script

```
n = 10000;
nbins = 20;
U = rand(1,n);
X = G(U);
minX = min(X);
maxX = max(X);
e = linspace(minX,maxX,nbins+1);  % edge sequence
H = histc(X,e);
H(nbins) = H(nbins)+H(nbins+1);   % explained in Section 6.2
H = H(1:nbins);                    % resize H
bw = (maxX-minX)/nbins;            % bin width
a = e(1:nbins);                    % left  edge sequence
b = e(2:nbins+1);                  % right edge sequence
bin_centers = (a+b)/2;             % bin centers
bar(bin_centers,H/(bw*n),'hist')
```

to use your function G to simulate 10 000 realizations of the random variable X with the cdf of Problem 38 and to plot a histogram of the results. Discuss the relationship between the histogram and the density of X.

42. In the text we considered discrete-time processes X_n for $n = 1, 2, \ldots$. The consistency condition (11.17) arose from the requirement that

$$P\big((X_1,\ldots,X_n,X_{n+1}) \in B \times \mathbb{R}\big) = P\big((X_1,\ldots,X_n) \in B\big),$$

where $B \subset \mathbb{R}^n$. For processes X_n with $n = 0, \pm 1, \pm 2, \ldots$, we require not only

$$P\big((X_m,\ldots,X_n,X_{n+1}) \in B \times \mathbb{R}\big) = P\big((X_m,\ldots,X_n) \in B\big),$$

but also

$$P\big((X_{m-1},X_m,\ldots,X_n) \in \mathbb{R} \times B\big) = P\big((X_m,\ldots,X_n) \in B\big),$$

where now $B \subset \mathbb{R}^{n-m+1}$. Let $\mu_{m,n}(B)$ be a proposed formula for the above right-hand side. Then the two consistency conditions are

$$\mu_{m,n+1}(B \times \mathbb{R}) = \mu_{m,n}(B) \quad \text{and} \quad \mu_{m-1,n}(\mathbb{R} \times B) = \mu_{m,n}(B).$$

For integer-valued random processes, show that these are equivalent to

$$\sum_{j=-\infty}^{\infty} p_{m,n+1}(i_m,\ldots,i_n,j) = p_{m,n}(i_m,\ldots,i_n)$$

and

$$\sum_{j=-\infty}^{\infty} p_{m-1,n}(j, i_m, \ldots, i_n) = p_{m,n}(i_m, \ldots, i_n),$$

where $p_{m,n}$ is the proposed joint probability mass function of $X_m, \ldots, X_n$.

43. Let q be any pmf, and let $r(j|i)$ be any conditional pmf. In addition, assume that $\sum_k q(k) r(j|k) = q(j)$. Put

$$p_{m,n}(i_m, \ldots, i_n) := q(i_m) r(i_{m+1}|i_m) r(i_{m+2}|i_{m+1}) \cdots r(i_n|i_{n-1}).$$

Show that both consistency conditions for pmfs in the preceding problem are satisfied.

Remark. This process is strictly stationary as defined in Section 10.3 since the upon writing out the formula for $p_{m+k,n+k}(i_m, \ldots, i_n)$, we see that it does not depend on k.

44. Let μ_n be a probability measure on $\mathbb{R}^n$, and suppose that it is given in terms of a joint density f_n, i.e.,

$$\mu_n(B_n) = \int_{-\infty}^{\infty} \cdots \int_{-\infty}^{\infty} I_{B_n}(x_1, \ldots, x_n) f_n(x_1, \ldots, x_n) \, dx_n \cdots dx_1.$$

Show that the consistency condition (11.17) holds if and only if

$$\int_{-\infty}^{\infty} f_{n+1}(x_1, \ldots, x_n, x_{n+1}) \, dx_{n+1} = f_n(x_1, \ldots, x_n).$$

45. Generalize Problem 44 for the continuous-time consistency condition (11.22).

46. Show that the Wiener process is a Gaussian process. *Hint:* For $0 < t_1 < \cdots < t_n$, write

$$\begin{bmatrix} W_{t_1} \\ W_{t_2} \\ \vdots \\ W_{t_n} \end{bmatrix} = \begin{bmatrix} 1 & 0 & 0 & & 0 \\ 1 & 1 & 0 & & 0 \\ 1 & 1 & 1 & & 0 \\ & & & \ddots & \\ 1 & 1 & 1 & & 1 \end{bmatrix} \begin{bmatrix} W_{t_1} - W_0 \\ W_{t_2} - W_{t_1} \\ \vdots \\ W_{t_n} - W_{t_{n-1}} \end{bmatrix}.$$

47. Let W_t be a standard Wiener process, and let $f_{t_1,\ldots,t_n}$ denote the joint density of $W_{t_1}, \ldots, W_{t_n}$. Find $f_{t_1,\ldots,t_n}$ and show that it satisfies the density version of (11.22) that you derived in Problem 45. *Hint:* Example 8.9 and the preceding problem may be helpful.

48. Let $R(\tau)$ be the inverse Fourier transform of a real, even, nonnegative function $S(f)$. Show that there is a Gaussian random process X_t that has correlation function $R(\tau)$. *Hint:* By the result of Example 11.10, it suffices to show that the matrix with ik entry $R(t_i - t_k)$ is positive semidefinite. In other words, if C is the matrix whose ik entry is $C_{ik} = R(t_i - t_k)$, you must show that for every vector of real numbers $a = [a_1, \ldots, a_n]'$, $a'Ca \geq 0$. Recall from Example 8.4 that $a'Ca = \sum_{i=1}^{n} \sum_{k=1}^{n} a_i a_k C_{ik}$.

Exam preparation

You may use the following suggestions to prepare a study sheet, including formulas mentioned that you have trouble remembering. You may also want to ask your instructor for additional suggestions.

11.1. The Poisson process. Know the three properties that define a Poisson process. The nth arrival time T_n has an Erlang(n,λ) density. The interarrival times X_n are i.i.d. $\exp(\lambda)$. Be able to do simple calculations with a marked Poisson process and a shot-noise process.

11.2. Renewal processes. A renewal process is similar to a Poisson process, except that the i.i.d. interarrival times do not have to be exponential.

11.3. The Wiener process. Know the four properties that define a Wiener process. Its covariance function is $\sigma^2 \min(t_1, t_2)$. The Wiener process is a model for integrated white noise. A Wiener integral is a Gaussian random variable with zero mean and variance given by (11.13). The Wiener process is the limit of a scaled random walk.

11.4. Specification of random processes. Kolmogorov's theorem says that a random process exists with a specified choice for $P((X_{t_1},\ldots,X_{t_n}) \in B)$ if whenever we eliminate one of the variables, we get the specified formula for the remaining variables. The other important result is that if a Gaussian process is wide-sense stationary, then it is strictly stationary.

Work any review problems assigned by your instructor. If you finish them, re-work your homework assignments.

12
Introduction to Markov chains[†]

A Markov chain is a random process with the property that given the values of the process from time zero up through the current time, the conditional probability of the value of the process at any future time depends only on its value at the current time. This is equivalent to saying that the future and the past are **conditionally independent** given the present (cf. Problem 70 in Chapter 1).

Markov chains often have intuitively pleasing interpretations. Some examples discussed in this chapter are random walks (without barriers and with barriers, which may be reflecting, absorbing, or neither), queuing systems (with finite or infinite buffers), birth–death processes (with or without spontaneous generation), life (with states being "healthy," "sick," and "death"), and the gambler's ruin problem.

Section 12.1 briefly highlights some simple properties of conditional probability that are very useful in studying Markov chains. Sections 12.2–12.4 cover basic results about discrete-time Markov chains. Continuous-time chains are discussed in Section 12.5.

12.1 Preliminary results

We present some easily-derived properties of conditional probability. These observations will greatly simplify some of our calculations for Markov chains.[1]

***Example* 12.1.** Given any event A and any two integer-valued random variables X and Y, show that if $P(A|X=i, Y=j)$ depends on i but not j, then in fact $P(A|X=i, Y=j) = P(A|X=i)$.

Solution. We use the law of total probability along with the chain rule of conditional probability (Problem 3), which says that

$$P(A \cap B|C) = P(A|B \cap C)P(B|C). \qquad (12.1)$$

Now, suppose that

$$P(A|X=i, Y=j) = h(i) \qquad (12.2)$$

for some function of i only. We must show that $h(i) = P(A|X=i)$. Write

$$P(A \cap \{X=i\}) = \sum_j P(A \cap \{X=i\}|Y=j)P(Y=j)$$

$$= \sum_j P(A|X=i, Y=j)P(X=i|Y=j)P(Y=j)$$

$$= \sum_j h(i)P(X=i|Y=j)P(Y=j), \quad \text{by (12.2)},$$

[†]Sections 12.1–12.4 can be covered any time after Chapter 3. However, Section 12.5 uses material from Chapter 5 and Chapter 11.

12.2 Discrete-time Markov chains

$$= h(i) \sum_j P(X = i | Y = j) P(Y = j)$$
$$= h(i) P(X = i).$$

Solving for $h(i)$ yields $h(i) = P(A|X=i)$ as required.

Example 12.2. The method used to solve Example 12.1 extends in the obvious way to show that if $P(A|X=i, Y=j, Z=k)$ is a function of i only, then not only does $P(A|X=i, Y=j, Z=k) = P(A|X=i)$, but also

$$P(A|X=i, Y=j) = P(A|X=i)$$

and

$$P(A|X=i, Z=k) = P(A|X=i)$$

as well. See Problem 1.

12.2 Discrete-time Markov chains

A sequence of integer-valued random variables, $X_0, X_1, \ldots$ is called a **Markov chain** if for $n \geq 1$,

$$P(X_{n+1} = i_{n+1} | X_n = i_n, \ldots, X_0 = i_0) = P(X_{n+1} = i_{n+1} | X_n = i_n).$$

In other words, given the sequence of values $i_0, \ldots, i_n$, the conditional probability of what X_{n+1} will be one time unit in the future depends only on the value of i_n. A random sequence whose conditional probabilities satisfy this condition is said to satisfy the **Markov property**.

Consider a person who has had too much to drink and is staggering around. Suppose that with each step, the person randomly moves forward or backward by one step. This is the idea to be captured in the following example.

Example 12.3 (random walk). Let X_0 be an integer-valued random variable that is independent of the i.i.d. sequence $Z_1, Z_2, \ldots$, where $P(Z_n = 1) = a$, $P(Z_n = -1) = b$, and $P(Z_n = 0) = 1 - (a+b)$. Show that if

$$X_n := X_{n-1} + Z_n, \quad n = 1, 2, \ldots,$$

then X_n is a Markov chain.

Solution. It helps to write out

$$X_1 = X_0 + Z_1$$
$$X_2 = X_1 + Z_2 = X_0 + Z_1 + Z_2$$
$$\vdots$$
$$X_n = X_{n-1} + Z_n = X_0 + Z_1 + \cdots + Z_n.$$

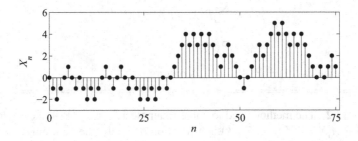

Figure 12.1. Realization of a symmetric random walk X_n.

The point here is that $(X_0,\ldots,X_n)$ is a function of $(X_0, Z_1, \ldots, Z_n)$, and hence, Z_{n+1} and $(X_0,\ldots,X_n)$ are independent. Now observe that

$$P(X_{n+1} = i_{n+1} | X_n = i_n, \ldots, X_0 = i_0) \tag{12.3}$$

is equal to

$$P(X_n + Z_{n+1} = i_{n+1} | X_n = i_n, \ldots, X_0 = i_0).$$

Using the substitution law, this becomes

$$P(Z_{n+1} = i_{n+1} - i_n | X_n = i_n, \ldots, X_0 = i_0).$$

On account of the independence of Z_{n+1} and $(X_0,\ldots,X_n)$, the above conditional probability is equal to

$$P(Z_{n+1} = i_{n+1} - i_n).$$

Putting this all together shows that (12.3) depends on i_n but not on $i_{n-1},\ldots,i_0$. By Example 12.1, (12.3) must be equal to $P(X_{n+1} = i_{n+1} | X_n = i_n)$; i.e., X_n is a Markov chain.

The Markov chain of the preceding example is called a random walk on the integers. The random walk is said to be symmetric if $a = b = 1/2$. A realization of a symmetric random walk is shown in Figure 12.1. Notice that each point differs from the preceding one by ± 1.

To restrict the random walk to the nonnegative integers, we can take $X_n = \max(0, X_{n-1} + Z_n)$ (Problem 2).

Conditional joint PMFs

The Markov property says that

$$P(X_{n+1} = j | X_n = i_n, \ldots, X_0 = i_0) = P(X_{n+1} = j | X_n = i_n).$$

In this subsection, we explore some implications of this equation for conditional joint pmfs.

We show below that

$$P(X_{n+m} = j_m, \ldots, X_{n+1} = j_1 | X_n = i_n, \ldots, X_0 = i_0)$$
$$= P(X_{n+m} = j_m, \ldots, X_{n+1} = j_1 | X_n = i_n). \tag{12.4}$$

12.2 Discrete-time Markov chains

In other words, *the conditional joint pmf of $X_{n+1},\ldots,X_{n+m}$ also satisfies a kind of Markov property.*

Although the conditional probability on the left-hand side of (12.4) involves $i_0,\ldots,i_n$, the right-hand side depends only on i_n. Hence, it follows from the observations in Example 12.2 that we can also write equations like

$$P(X_{n+m} = j_m,\ldots,X_{n+1} = j_1 | X_n = i_n, X_0 = i_0)$$
$$= P(X_{n+m} = j_m,\ldots,X_{n+1} = j_1 | X_n = i_n). \quad (12.5)$$

Furthermore, summing both sides of (12.5) over all values of j_1, all values of j_2, ..., all values of j_{m-1} shows that

$$P(X_{n+m} = j_m | X_n = i_n, X_0 = i_0) = P(X_{n+m} = j_m | X_n = i_n). \quad (12.6)$$

To establish (12.4), first write the left-hand side as

$$P(\underbrace{X_{n+m} = j_m}_{A}, \underbrace{X_{n+m-1} = j_{m-1},\ldots,X_{n+1} = j_1}_{B} | \underbrace{X_n = i_n,\ldots,X_0 = i_0}_{C}).$$

Then use the chain rule of conditional probability (12.1) to write it as

$$P(X_{n+m} = j_m | X_{n+m-1} = j_{m-1},\ldots,X_{n+1} = j_1, X_n = i_n,\ldots,X_0 = i_0)$$
$$\cdot P(X_{n+m-1} = j_{m-1},\ldots,X_{n+1} = j_1 | X_n = i_n,\ldots,X_0 = i_0).$$

Applying the Markov property to the left-hand factor yields

$$P(X_{n+m} = j_m | X_{n+m-1} = j_{m-1})$$
$$\cdot P(X_{n+m-1} = j_{m-1},\ldots,X_{n+1} = j_1 | X_n = i_n,\ldots,X_0 = i_0).$$

Now apply the foregoing two steps to the right-hand factor to get

$$P(X_{n+m} = j_m | X_{n+m-1} = j_{m-1})$$
$$\cdot P(X_{n+m-1} = j_{m-1} | X_{n+m-2} = j_{m-2})$$
$$\cdot P(X_{n+m-2} = j_{m-2},\ldots,X_{n+1} = j_1 | X_n = i_n,\ldots,X_0 = i_0).$$

Continuing in this way, we end up with

$$P(X_{n+m} = j_m,\ldots,X_{n+1} = j_1 | X_n = i_n,\ldots,X_0 = i_0)$$
$$= P(X_{n+m} = j_m | X_{n+m-1} = j_{m-1})$$
$$\cdot P(X_{n+m-1} = j_{m-1} | X_{n+m-2} = j_{m-2}) \quad (12.7)$$
$$\vdots$$
$$\cdot P(X_{n+2} = j_2 | X_{n+1} = j_1)$$
$$\cdot P(X_{n+1} = j_1 | X_n = i_n)$$

Since the right-hand side depends on i_n but not on $i_{n-1},\ldots,i_0$, the result of Example 12.1 tells us that the above left-hand side is equal to

$$P(X_{n+m} = j_m,\ldots,X_{n+1} = j_1 | X_n = i_n).$$

Thus, (12.4) holds. Furthermore, since the left-hand sides of (12.4) and (12.7) are the same, we have the additional formula

$$\begin{aligned}
\mathsf{P}(X_{n+m} &= j_m,\ldots,X_{n+1} = j_1|X_n = i_n) \\
&= \mathsf{P}(X_{n+m} = j_m|X_{n+m-1} = j_{m-1}) \\
&\quad \cdot \mathsf{P}(X_{n+m-1} = j_{m-1}|X_{n+m-2} = j_{m-2}) \\
&\quad \vdots \\
&\quad \cdot \mathsf{P}(X_{n+2} = j_2|X_{n+1} = j_1) \\
&\quad \cdot \mathsf{P}(X_{n+1} = j_1|X_n = i_n).
\end{aligned} \quad (12.8)$$

State space and transition probabilities

The set of possible values that the random variables X_n can take is called the **state space** of the chain. In this chapter, we take the state space to be the set of integers or some specified subset of the integers. The conditional probabilities

$$\mathsf{P}(X_{n+1} = j|X_n = i)$$

are called **transition probabilities**. In this chapter, we assume that the transition probabilities do not depend on time n. Such a Markov chain is said to have **stationary** transition probabilities or to be **time homogeneous**. For a time-homogeneous Markov chain, we use the notation

$$\boxed{p_{ij} := \mathsf{P}(X_{n+1} = j|X_n = i)}$$

for the transition probabilities. The p_{ij} are also called the **one-step transition probabilities** because they are the probabilities of going from state i to state j in one time step. One of the most common ways to specify the transition probabilities is with a **state transition diagram** as in Figure 12.2. This particular diagram says that the state space is the finite set $\{0,1\}$, and that $p_{01} = a$, $p_{10} = b$, $p_{00} = 1-a$, and $p_{11} = 1-b$. Note that the sum of all the probabilities leaving a state must be one. This is because for each state i,

$$\sum_j p_{ij} = \sum_j \mathsf{P}(X_{n+1} = j|X_n = i) = 1.$$

The transition probabilities p_{ij} can be arranged in a matrix P, called the **transition matrix**, whose ij entry is p_{ij}. For the chain in Figure 12.2,

$$P = \begin{bmatrix} 1-a & a \\ b & 1-b \end{bmatrix}.$$

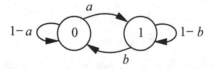

Figure 12.2. A state transition diagram. The diagram says that the state space is the finite set $\{0,1\}$, and that $p_{01} = a$, $p_{10} = b$, $p_{00} = 1-a$, and $p_{11} = 1-b$.

12.2 Discrete-time Markov chains

The top row of P contains the probabilities p_{0j}, which is obtained by noting the probabilities written next to all the arrows leaving state 0. Similarly, the probabilities written next to all the arrows leaving state 1 are found in the bottom row of P.

Examples

The general **random walk** on the integers has the state transition diagram shown in Figure 12.3. Note that the Markov chain constructed in Example 12.3 is a special case in

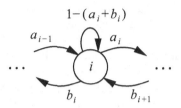

Figure 12.3. State transition diagram for a random walk on the integers.

which $a_i = a$ and $b_i = b$ for all i. The state transition diagram is telling us that

$$p_{ij} = \begin{cases} b_i, & j = i-1, \\ 1-(a_i+b_i), & j = i, \\ a_i, & j = i+1, \\ 0, & \text{otherwise.} \end{cases} \qquad (12.9)$$

Hence, the transition matrix P is infinite, tridiagonal, and its ith row is

$$\begin{bmatrix} \cdots & 0 & b_i & 1-(a_i+b_i) & a_i & 0 & \cdots \end{bmatrix}.$$

Frequently, it is convenient to introduce a barrier at zero, leading to the state transition diagram in Figure 12.4. In this case, we speak of a random walk with a barrier. For $i \geq 1$, the formula for p_{ij} is given by (12.9), while for $i = 0$,

$$p_{0j} = \begin{cases} 1-a_0, & j = 0, \\ a_0, & j = 1, \\ 0, & \text{otherwise.} \end{cases} \qquad (12.10)$$

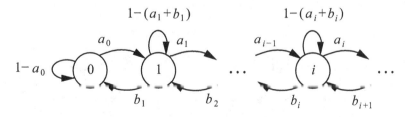

Figure 12.4. State transition diagram for a random walk with a barrier at the origin (also called a birth–death process).

The transition matrix P is the tridiagonal, semi-infinite matrix

$$P = \begin{bmatrix} 1-a_0 & a_0 & 0 & 0 & 0 & \cdots \\ b_1 & 1-(a_1+b_1) & a_1 & 0 & 0 & \cdots \\ 0 & b_2 & 1-(a_2+b_2) & a_2 & 0 & \cdots \\ 0 & 0 & b_3 & 1-(a_3+b_3) & a_3 & \cdots \\ \vdots & \vdots & \vdots & & & \ddots \end{bmatrix}.$$

If $a_0 = 1$, the barrier is said to be **reflecting**. If $a_0 = 0$, the barrier is said to be **absorbing**. Once a chain hits an absorbing state, the chain stays in that state from that time onward.

A random walk with a barrier at the origin has several interpretations. When thinking of a drunken person staggering around, we can view a wall or a fence as a reflecting barrier; if the person backs into the wall, then with the next step the person must move forward away from the wall. Similarly, we can view a curb or step as an absorbing barrier; if the person trips and falls down when stepping over a curb, then the walk is over.

A random walk with a barrier at the origin can be viewed as a model for a queue with an infinite buffer. Consider a queue of packets buffered at an Internet router. The state of the chain is the number of packets in the buffer. This number cannot go below zero. The number of packets can increase by one if a new packet arrives, decrease by one if a packet is forwarded to its next destination, or stay the same if both or neither of these events occurs.

A random walk with a barrier at the origin can also be viewed as a **birth–death process**. With this terminology, the state of the chain is taken to be a population, say of bacteria. In this case, if $a_0 > 0$, there is **spontaneous generation**. If $b_i = 0$ for all i, we have a **pure birth process**.

Sometimes it is useful to consider a random walk with barriers at the origin and at N, as shown in Figure 12.5. The formula for p_{ij} is given by (12.9) above for $1 \leq i \leq N-1$, by (12.10) above for $i = 0$, and, for $i = N$, by

$$p_{Nj} = \begin{cases} b_N, & j = N-1, \\ 1-b_N, & j = N, \\ 0, & \text{otherwise.} \end{cases} \qquad (12.11)$$

This chain can be viewed as a model for a queue with a finite buffer, especially if $a_i = a$ and $b_i = b$ for all i. When $a_0 = 0$ and $b_N = 0$, the barriers at 0 and N are absorbing, and the chain is a model for the **gambler's ruin** problem. In this problem, a gambler starts at time zero with $1 \leq i \leq N-1$ dollars and plays until he either runs out of money, that is, absorption into state zero, or his winnings reach N dollars and he stops playing (absorption

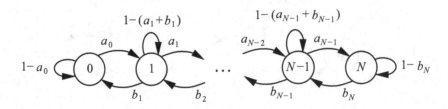

Figure 12.5. State transition diagram for a queue with a finite buffer.

12.2 Discrete-time Markov chains

into state N). If $N=2$ and $b_2 = 0$, the chain can be interpreted as the story of life if we view state $i = 0$ as being the "healthy" state, $i = 1$ as being the "sick" state, and $i = 2$ as being the "death" state. In this model, if you are healthy (in state 0), you remain healthy with probability $1 - a_0$ and become sick (move to state 1) with probability a_0. If you are sick (in state 1), you become healthy (move to state 0) with probability b_1, remain sick (stay in state 1) with probability $1 - (a_1 + b_1)$, or die (move to state 2) with probability a_1. Since state 2 is absorbing ($b_2 = 0$), once you enter this state, you never leave.

Consequences of time homogeneity

Notice that the right-hand side of (12.8) involves only one-step transition probabilities. Hence, for a time-homogeneous chain,

$$\mathsf{P}(X_{n+m} = j_m, \ldots, X_{n+1} = j_1 | X_n = i) = p_{ij_1} p_{j_1 j_2} \cdots p_{j_{m-1} j_m}. \quad (12.12)$$

Taking $n = 0$ yields

$$\mathsf{P}(X_m = j_m, \ldots, X_1 = j_1 | X_0 = i) = p_{ij_1} p_{j_1 j_2} \cdots p_{j_{m-1} j_m}. \quad (12.13)$$

Since the right-hand sides are the same, we conclude that

$$\mathsf{P}(X_{n+m} = j_m, \ldots, X_{n+1} = j_1 | X_n = i) = \mathsf{P}(X_m = j_m, \ldots, X_1 = j_1 | X_0 = i). \quad (12.14)$$

If we sum (12.14) over all values of j_1, all values of $j_2, \ldots,$ all values of j_{m-1}, we find that

$$\mathsf{P}(X_{n+m} = j_m | X_n = i) = \mathsf{P}(X_m = j_m | X_0 = i). \quad (12.15)$$

The ***m*-step transition probabilities** are defined by

$$\boxed{p_{ij}^{(m)} := \mathsf{P}(X_m = j | X_0 = i).} \quad (12.16)$$

This is the probability of going from state i (at time zero) to state j in m steps. In particular,

$$p_{ij}^{(1)} = p_{ij},$$

and

$$p_{ij}^{(0)} = \mathsf{P}(X_0 = j | X_0 = i) = \delta_{ij},$$

where δ_{ij} denotes the **Kronecker delta**, which is one if $i = j$ and is zero otherwise. We also point out that (12.15) says

$$\boxed{\mathsf{P}(X_{n+m} = j | X_n = i) = p_{ij}^{(m)}.} \quad (12.17)$$

In other words, *the m-step transition probabilities are stationary.*

The Chapman–Kolmogorov equation

The m-step transition probabilities satisfy the **Chapman–Kolmogorov equation**,

$$\boxed{p_{ij}^{(n+m)} = \sum_k p_{ik}^{(n)} p_{kj}^{(m)}.} \qquad (12.18)$$

This is easily derived as follows.[2] First write

$$\begin{aligned} p_{ij}^{(n+m)} &= \mathsf{P}(X_{n+m} = j | X_0 = i) \\ &= \sum_k \mathsf{P}(X_{n+m} = j, X_n = k | X_0 = i) \\ &= \sum_k \mathsf{P}(X_{n+m} = j | X_n = k, X_0 = i) \mathsf{P}(X_n = k | X_0 = i). \end{aligned}$$

Now apply the Markov property (cf. (12.6)) and stationarity of the m-step transition probabilities to write

$$\begin{aligned} p_{ij}^{(n+m)} &= \sum_k \mathsf{P}(X_{n+m} = j | X_n = k) \mathsf{P}(X_n = k | X_0 = i) \\ &= \sum_k p_{kj}^{(m)} \mathsf{P}(X_n = k | X_0 = i) \\ &= \sum_k p_{ik}^{(n)} p_{kj}^{(m)}. \end{aligned}$$

If we take $n = m = 1$ in (12.18), we see that

$$p_{ij}^{(2)} = \sum_k p_{ik} p_{kj}.$$

In other words, the matrix with entries $p_{ij}^{(2)}$ is exactly the matrix PP, where P is the transition matrix. Taking $n = 2$ and $m = 1$ in (12.18) shows that the matrix with entries $p_{ij}^{(3)}$ is equal to $P^2 P = P^3$. In general, the matrix with entries $p_{ij}^{(n)}$ is given by P^n. The Chapman–Kolmogorov equation says that

$$P^{n+m} = P^n P^m.$$

Stationary distributions

Until now we have focused on the conditional probabilities p_{ij} and $p_{ij}^{(n)}$. However, we can use the law of total probability to write

$$\mathsf{P}(X_n = j) = \sum_i \mathsf{P}(X_n = j | X_0 = i) \mathsf{P}(X_0 = i).$$

Thus, $\mathsf{P}(X_n = j)$, which is not a conditional probability, depends on the probability mass function of initial state X_0 of the chain. If we put $\rho_j^{(n)} := \mathsf{P}(X_n = j)$ and $v_i := \mathsf{P}(X_0 = i)$, then the above display can be written as

$$\rho_j^{(n)} = \sum_i v_i p_{ij}^{(n)},$$

12.2 Discrete-time Markov chains

or in matrix–vector form as
$$\rho^{(n)} = vP^n,$$
where $\rho^{(n)}$ and v are row vectors. In general, $\rho^{(n)}$ depends on n, and for large n, powers of P are difficult to compute. However, there is one case in which there is great simplification. Suppose that $P(X_0 = i) = \pi_i$, where π is a probability mass function that satisfies the equation[a]
$$\pi = \pi P.$$
Right multiplication by P on both sides of this equation shows that $\pi P = \pi P^2$, and it then follows that $\pi = \pi P^2$. More generally, $\pi = \pi P^n$. Hence,
$$\rho_j^{(n)} = \pi_j,$$
and we see that $P(X_n = j)$ does not depend on n. We make the following general definition. If π_j is a sequence that satisfies

$$\pi_j = \sum_k \pi_k p_{kj}, \quad \pi_j \geq 0, \quad \text{and} \quad \sum_j \pi_j = 1, \qquad (12.19)$$

then π is called a **stationary distribution** or **equilibrium distribution** of the chain.

***Example* 12.4.** Find the stationary distribution of the chain with state transition matrix
$$P = \begin{bmatrix} 0 & 1/4 & 3/4 \\ 0 & 1/2 & 1/2 \\ 2/5 & 2/5 & 1/5 \end{bmatrix}.$$

Solution. We begin by writing out the equations
$$\pi_j = \sum_k \pi_k p_{kj}$$
for each j. Notice that the right-hand side is the inner product of the row vector π and the jth column of P. For $j = 0$, we have
$$\pi_0 = \sum_k \pi_k p_{k0} = 0\pi_0 + 0\pi_1 + \tfrac{2}{5}\pi_2 = \tfrac{2}{5}\pi_2.$$
For $j = 1$, we have
$$\pi_1 = \tfrac{1}{4}\pi_0 + \tfrac{1}{2}\pi_1 + \tfrac{2}{5}\pi_2$$
$$= \tfrac{1}{4}\left(\tfrac{2}{5}\pi_2\right) + \tfrac{1}{2}\pi_1 + \tfrac{2}{5}\pi_2,$$
from which it follows that $\pi_1 = \pi_2$. As it turns out, the equation for the last value of j is always redundant. Instead we use the requirement that $\sum_j \pi_j = 1$. Writing
$$1 = \pi_0 + \pi_1 + \pi_2 = \tfrac{2}{5}\pi_2 + \pi_2 + \pi_2,$$
it follows that $\pi_2 = 5/12$, $\pi_1 = 5/12$, and $\pi_0 = 1/6$.

---

[a]The equation $\pi = \pi P$ says that π is a left **eigenvector** of P with **eigenvalue** 1. To say this another way, $I - P$ is singular; i.e., there are many solutions of $\pi(I - P) = 0$. Since π is a probability mass function, it cannot be the zero vector. Recall that by definition, eigenvectors are precluded from being the zero vector.

The solution in the preceding example suggests the following algorithm to find the stationary distribution a Markov chain with a finite number of states. First, rewrite $\pi P = \pi$ as $\pi(P-I) = 0$, where the right-hand side is a row vector whose length is the number of states. Second, we said above that equation for the last value of j is always redundant, and so we use the requirement $\sum_j \pi_j = 1$ instead. This amounts to solving the equation $\pi A = y$, where A is obtained by replacing the last column of $P - I$ with all ones, and $y = [0, \ldots, 0, 1]$. See Problem 11 for a MATLAB implementation.

The next example involves a Markov chain with an infinite number of states.

Example 12.5. The state transition diagram for a queuing system with an infinite buffer is shown in Figure 12.6. Find the stationary distribution of the chain if $a < b$.

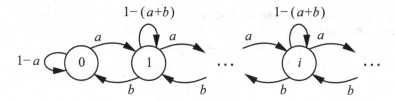

Figure 12.6. State transition diagram for a queuing system with an infinite buffer.

Solution. We begin by writing out

$$\pi_j = \sum_k \pi_k p_{kj} \qquad (12.20)$$

for $j = 0, 1, 2, \ldots$. For each j, the coefficients p_{kj} are obtained by inspection of the state transition diagram and looking at all the arrows that go into state j. For $j = 0$, we must consider

$$\pi_0 = \sum_k \pi_k p_{k0}.$$

We need the values of p_{k0}. From the diagram, the only way to get to state 0 is from state 0 itself (with probability $p_{00} = 1 - a$) or from state 1 (with probability $p_{10} = b$). The other $p_{k0} = 0$. Hence,

$$\pi_0 = \pi_0 (1-a) + \pi_1 b.$$

We can rearrange this to get

$$\pi_1 = \frac{a}{b} \pi_0.$$

Now put $j = 1$ in (12.20). The state transition diagram tells us that the only way to enter state 1 is from states 0, 1, and 2, with probabilities a, $1 - (a+b)$, and b, respectively. Hence,

$$\pi_1 = \pi_0 a + \pi_1 [1 - (a+b)] + \pi_2 b.$$

Substituting $\pi_1 = (a/b)\pi_0$ yields $\pi_2 = (a/b)^2 \pi_0$. In general, if we substitute $\pi_j = (a/b)^j \pi_0$ and $\pi_{j-1} = (a/b)^{j-1} \pi_0$ into

$$\pi_j = \pi_{j-1} a + \pi_j [1 - (a+b)] + \pi_{j+1} b,$$

12.2 Discrete-time Markov chains

then we obtain $\pi_{j+1} = (a/b)^{j+1}\pi_0$. We conclude that

$$\pi_j = \left(\frac{a}{b}\right)^j \pi_0, \quad j = 0, 1, 2, \ldots.$$

To solve for π_0, we use the fact that

$$\sum_{j=0}^{\infty} \pi_j = 1,$$

or

$$\pi_0 \sum_{j=0}^{\infty} \left(\frac{a}{b}\right)^j = 1.$$

The geometric series formula shows that

$$\pi_0 = 1 - a/b,$$

and

$$\pi_j = \left(\frac{a}{b}\right)^j (1 - a/b).$$

In other words, the stationary distribution is a geometric$_0(a/b)$ probability mass function. Note that we needed $a < b$ to apply the geometric series formula. If $a \geq b$, there is no stationary distribution.

In the foregoing example, when $a \geq b$, the chain does not have a stationary distribution. On the other hand, as the next example shows, a chain can have more than one stationary distribution; i.e., there may be more than one solution of (12.19).

Example 12.6. Consider the chain in Figure 12.7. Its transition matrix is

$$P = \begin{bmatrix} 2/3 & 1/3 & 0 & 0 \\ 2/7 & 5/7 & 0 & 0 \\ 0 & 0 & 4/5 & 1/5 \\ 0 & 0 & 3/4 & 1/4 \end{bmatrix}.$$

It is easy to check that

$$\pi = \begin{bmatrix} 6/13 & 7/13 & 0 & 0 \end{bmatrix} \quad \text{and} \quad \pi = \begin{bmatrix} 0 & 0 & 15/19 & 4/19 \end{bmatrix}$$

are both probability mass functions that solve $\pi P = \pi$.

Figure 12.7. A Markov chain with multiple stationary distributions.

We conclude this section with a sufficient condition to guarantee that if a chain has a stationary distribution, it is unique.

> If for every pair of states $i \neq j$ there is a path in the state transition diagram from i to j and a path from j to i, we say that the chain is **irreducible**. It is shown in [23, Section 6.4] that an irreducible chain can have at most one stationary distribution.

Example **12.7.** The chains in Figures 12.2–12.6 are all irreducible (as long as none of the parameters a, a_i, b, or b_i is zero). Hence, the stationary distributions that we found in Examples 12.4 and 12.5 are unique. The chain in Figure 12.7 is not irreducible.

12.3 *Recurrent and transient states

Entrance times and intervisit times

The first time the chain visits state j is given by

$$T_1(j) := \min\{k \geq 1 : X_k = j\}.$$

We call $T_1(j)$ the **first entrance time** or **first passage time** of state j. It may happen that $X_k \neq j$ for any $k \geq 1$. In this case, $T_1(j) = \min \emptyset$, which we take to be ∞. In other words, if the chain never visits state j for any time $k \geq 1$, we put $T_1(j) = \infty$. Given that the chain starts in state j, the conditional probability that the chain returns to state j in finite time is

$$f_{jj} := \mathsf{P}(T_1(j) < \infty | X_0 = j). \tag{12.21}$$

A state j is said to be

$$\begin{cases} \text{recurrent, if } f_{jj} = 1, \\ \text{transient, if } f_{jj} < 1. \end{cases}$$

We describe the condition $f_{jj} = 1$ in words by saying that a **recurrent state** is one that the chain is guaranteed to come back to in finite time. Here "guaranteed" means "happens with conditional probability one." On the other hand, if $f_{jj} < 1$, then

$$\mathsf{P}(T_1(j) = \infty | X_0 = j) = 1 - f_{jj}$$

is positive. Thus, a **transient state** is one for which there is a positive probability of never returning.

Example **12.8.** Show that state 0 in Figure 12.7 is recurrent.

Solution. It suffices to show that the conditional probability of never returning to state 0 is zero. The only way this can happen starting from state zero is to have

$$\{X_1 = 1\} \cap \{X_2 = 1\} \cap \cdots.$$

12.3 Recurrent and transient states

To compute the probability of this, we first use the limit property (1.14) to write

$$P\left(\bigcap_{n=1}^{\infty}\{X_n = 1\}\Big|X_0 = 0\right) = \lim_{N\to\infty} P\left(\bigcap_{n=1}^{N}\{X_n = 1\}\Big|X_0 = 0\right)$$

$$= \lim_{N\to\infty} p_{01}(p_{11})^{N-1}, \quad \text{by (12.13)},$$

$$= \lim_{N\to\infty} \left(\frac{1}{3}\right)\left(\frac{5}{7}\right)^{N-1} = 0.$$

Hence, state 0 is recurrent.

Example 12.9. Show that state 1 in Figure 12.8 is transient.

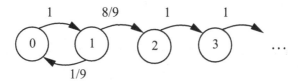

Figure 12.8. State transition diagram for Example 12.9.

Solution. We need to show that there is positive conditional probability of never returning to state 1. From the figure, we see that the chain never returns to state 1 if and only if starting at time 0 in state 1 it then jumps to state 2 at time 1. The probability of this is $P(X_1 = 2|X_0 = 1) = 8/9 > 0$.

Since $T_1(j)$ is a discrete random variable that may take the value ∞, its conditional expectation is given by the formula

$$E[T_1(j)|X_0 = j] := \sum_{k=1}^{\infty} kP(T_1(j) = k|X_0 = j) + \infty \cdot P(T_1(j) = \infty|X_0 = j) \qquad (12.22)$$

$$= \sum_{k=1}^{\infty} kP(T_1(j) = k|X_0 = j) + \infty \cdot (1 - f_{jj}).$$

From this formula we see that the expected time to return to a transient state is infinite. For a recurrent state, the formula reduces to

$$E[T_1(j)|X_0 = j] = \sum_{k=1}^{\infty} kP(T_1(j) = k|X_0 = j),$$

which may be finite or infinite. If the expected time to return to a recurrent state is finite, the state is said to be **positive recurrent**. Otherwise, the expected time to return is infinite, and the state is said to be **null recurrent**.

The ***n*th entrance time** of state j is given by

$$T_n(j) := \min\{k > T_{n-1}(j) : X_k = j\}.$$

The times between visits to state j, called **intervisit times**, are given by

$$D_1(j) := T_1(j) \quad \text{and} \quad D_n(j) := T_n(j) - T_{n-1}(j), \quad n \geq 2.$$

Hence,
$$T_n(j) = D_1(j) + \cdots + D_n(j). \tag{12.23}$$

Notation. When the state j is understood, we sometimes drop the (j) and write T_n or D_n.

Theorem 1. *Given $X_0 = i$, $D_1(j), D_2(j), \ldots$ are independent with $D_2(j), D_3(j), \ldots$ being i.i.d. If $i = j$, then $D_1(j), D_2(j), \ldots$ are i.i.d.*

Proof. We begin with the observation that the first visit to state j occurs at time n if and only if $X_1 \neq j, X_2 \neq j, \ldots, X_{n-1} \neq j$, and $X_n = j$. In terms of events, this is written as

$$\{T_1 = n\} = \{X_n = j, X_{n-1} \neq j, \ldots, X_1 \neq j\}. \tag{12.24}$$

Thus,
$$\begin{aligned} f_{ij}^{(n)} &:= \mathsf{P}(T_1(j) = n | X_0 = i) \\ &= \mathsf{P}(X_n = j, X_{n-1} \neq j, \ldots, X_1 \neq j | X_0 = i) \end{aligned} \tag{12.25}$$

is the conditional probability that given $X_0 = i$, the chain first enters state j at time n. More generally, for $k \geq 2$, if $1 \leq n_1 < n_2 < \cdots < n_k$,

$$\begin{aligned} \{T_1 = n_1, \ldots, T_k = n_k\} = \{&X_1 \neq j, \ldots, X_{n_1-1} \neq j, X_{n_1} = j, \\ &X_{n_1+1} \neq j, \ldots, X_{n_2-1} \neq j, X_{n_2} = j, \\ &\qquad \vdots \\ &X_{n_{k-1}+1} \neq j, \ldots, X_{n_k-1} \neq j, X_{n_k} = j\}. \end{aligned}$$

Using this formula along with the Markov property (12.4) and time homogeneity (12.14), it is not hard to show (Problem 13) that

$$\begin{aligned} \mathsf{P}(T_{k+1} &= n_{k+1} | T_k = n_k, \ldots, T_1 = n_1, X_0 = i) \\ &= \mathsf{P}(X_{n_{k+1}} = j, X_{n_{k+1}-1} \neq j, \ldots, X_{n_k+1} \neq j | X_{n_k} = j) \\ &= \mathsf{P}(X_{n_{k+1}-n_k} = j, X_{n_{k+1}-n_k-1} \neq j, \ldots, X_1 \neq j | X_0 = j) \\ &= f_{jj}^{(n_{k+1}-n_k)}, \quad \text{by (12.25)}. \end{aligned} \tag{12.26}$$

The next step is to let $d_1, \ldots, d_k$ be given positive integers. Then

$$\mathsf{P}(D_1 = d_1, \ldots, D_k = d_k | X_0 = i)$$

12.3 Recurrent and transient states

is equal to

$$P(D_1 = d_1 | X_0 = i) \prod_{m=2}^{k} P(D_m = d_m | D_{m-1} = d_{m-1}, \ldots, D_1 = d_1, X_0 = i). \qquad (12.27)$$

Since $D_1 = T_1$, the left-hand factor is $f_{ij}^{(d_1)}$ by (12.25). Next, on account of (12.23),

$$\{X_0 = i, D_1 = d_1, \ldots, D_{m-1} = d_{m-1}\}$$
$$= \{X_0 = i, T_1 = d_1, T_2 = d_1 + d_2, \ldots, T_{m-1} = d_1 + \cdots + d_{m-1}\}.$$

Hence, the mth factor in (12.27) is equal to

$$P(T_m - T_{m-1} = d_m | T_{m-1} = d_1 + \cdots + d_{m-1}, \ldots, T_1 = d_1, X_0 = i).$$

By the substitution law, this becomes

$$P(T_m = d_1 + \cdots + d_m | T_{m-1} = d_1 + \cdots + d_{m-1}, \ldots, T_1 = d_1, X_0 = i).$$

By (12.26), this is equal to

$$f_{jj}^{([d_1 + \cdots + d_m] - [d_1 + \cdots + d_{m-1}])} = f_{jj}^{(d_m)}.$$

We now have that

$$P(D_1 = d_1, \ldots, D_k = d_k | X_0 = i) = f_{ij}^{(d_1)} f_{jj}^{(d_2)} \cdots f_{jj}^{(d_k)}, \qquad (12.28)$$

which says that the D_k are independent with $D_2, D_3, \ldots$ being i.i.d. □

Number of visits to a state and occupation time

The number of visits to state j up to time m is given by the formula

$$V_m(j) := \sum_{k=1}^{m} I_{\{j\}}(X_k).$$

Since this is equal to the amount of time the chain has spent in state j, $V_m(j)$ is also called the **occupation time** of state j up to time m.

There is an important relationship between the number of visits to a state and the entrance times of that state. To see this relationship, observe that the number of visits to a state up to time m is less than n if and only if the nth visit has not happened yet; i.e., if and only if the nth visit occurs after time m. In terms of events, this is written as

$$\{V_m(j) < n\} = \{T_n(j) > m\}. \qquad (12.29)$$

The *average occupation time* up to time m, denoted by $V_m(j)/m$, is the fraction of time spent in state j up to time m. If an irreducible chain has a stationary distribution π, the ergodic theorem below says that $V_m(j)/m \to \pi_j$ as $m \to \infty$. In other words, if we watch the chain evolve, and we count the fraction of time spent in state j, this fraction is a consistent estimator of the equilibrium probability π_j.

The total number of visits to state j is

$$V(j) := \sum_{k=1}^{\infty} I_{\{j\}}(X_k).$$

Notice that $V(j) = \infty$ if and only if the chain visits state j an infinite number of times; in this case we say that the chain visits state j **infinitely often (i.o.)**. Since $V(j)$ is equal to the total time spent in state j, we call $V(j)$ the **total occupation time** of state j. We show later that $V(j)$ is either a constant random variable equal to ∞ or it has a geometric$_0$ pmf when conditioned on $X_0 = j$. Thus, either the chain visits state j infinitely often with conditional probability one, or it visits state j only finitely many times with conditional probability one, and in this case, the number of visits is a geometric$_0$ random variable.

The key to the derivations in this subsection is the fact that given $X_0 = j$, the intervisit times D_k are i.i.d. by Theorem 1. Hence, by the law of large numbers,[3]

$$\frac{1}{n}\sum_{k=1}^{n} D_k \;\to\; \mathsf{E}[D_k|X_0 = j], \tag{12.30}$$

assuming this expectation is finite. Since the $D_k(j)$ are i.i.d. given $X_0 = j$, and since $T_1(j) = D_1(j)$, we can also write

$$\frac{1}{n}\sum_{k=1}^{n} D_k(j) \;\to\; \mathsf{E}[T_1(j)|X_0 = j] \tag{12.31}$$

if this expectation is finite; i.e., if state j is positive recurrent. On account of (12.23), we can write (12.31) as

$$\frac{T_n(j)}{n} \;\to\; \mathsf{E}[T_1(j)|X_0 = j]. \tag{12.32}$$

The independence of the D_k can be used to give further characterizations of recurrence and transience. The total number of visits to j, $V(j)$, is at least L if and only if the Lth visit occurs in finite time, i.e., $T_L(j) < \infty$, which happens if and only if $D_1, \ldots, D_L$ are all finite. Thus,

$$\mathsf{P}(V(j) \geq L | X_0 = i) = \mathsf{P}(D_1 < \infty, \ldots, D_L < \infty | X_0 = i)$$

$$= \mathsf{P}(D_1 < \infty | X_0 = i) \prod_{k=2}^{L} \mathsf{P}(D_k < \infty | X_0 = i).$$

We now calculate each factor. Since $D_1 = T_1$, we have

$$\boxed{\mathsf{P}(D_1 < \infty | X_0 = i) = \mathsf{P}(T_1(j) < \infty | X_0 = i) =: f_{ij}.} \tag{12.33}$$

12.3 Recurrent and transient states

Note that the definition here of f_{ij} is the obvious generalization of f_{jj} in (12.21). For $k \geq 2$,

$$\begin{aligned}
P(D_k < \infty | X_0 = i) &= \sum_{d=1}^{\infty} P(D_k = d | X_0 = i) \\
&= \sum_{d=1}^{\infty} f_{jj}^{(d)}, \quad \text{by (12.28),} \\
&= \sum_{d=1}^{\infty} P(T_1(j) = d | X_0 = j), \text{ by (12.25),} \\
&= P(T_1(j) < \infty | X_0 = j) \\
&= f_{jj}.
\end{aligned}$$

Thus,
$$P(V(j) \geq L | X_0 = i) = f_{ij}(f_{jj})^{L-1}. \tag{12.34}$$

It then follows that
$$\begin{aligned}
P(V(j) = L | X_0 = i) &= P(V(j) \geq L | X_0 = i) - P(V(j) \geq L+1 | X_0 = i) \\
&= f_{ij}(f_{jj})^{L-1}(1 - f_{jj}).
\end{aligned} \tag{12.35}$$

Theorem 2. *The total number of visits to state j satisfies*

$$P(V(j) = \infty | X_0 = j) = \begin{cases} 1, & \text{if } f_{jj} = 1 \text{ (recurrent)}, \\ 0, & \text{if } f_{jj} < 1 \text{ (transient)}. \end{cases}$$

In the transient case,

$$P(V(j) = L | X_0 = j) = (f_{jj})^L (1 - f_{jj}),$$

which is a geometric$_0(f_{jj})$ pmf; hence,

$$E[V(j)|X_0 = j] = \frac{f_{jj}}{1 - f_{jj}} < \infty.$$

In the recurrent case, $E[V(j)|X_0 = j] = \infty$.

Proof. In Problem 14 you will show that
$$P(V(j) = \infty | X_0 = i) = \lim_{L \to \infty} P(V(j) \geq L | X_0 = i).$$

Now take $i = j$ in (12.34) and observe that $(f_{jj})^L$ converges to one or to zero according to $f_{jj} = 1$ or $f_{jj} < 1$. To obtain the pmf of $V(j)$ in the transient case take $i = j$ in (12.35). The fact that $E[V(j)|X_0 = j] = \infty$ for a recurrent state is immediate since by the first part of the theorem already proved, $V(j) = \infty$ with conditional probability one. □

We next observe that
$$E[V(j)|X_0 = j] = E\left[\sum_{n=1}^{\infty} I_{\{j\}}(X_n) \bigg| X_0 = j\right]$$

$$= \sum_{n=1}^{\infty} \mathsf{E}[I_{\{j\}}(X_n)|X_0 = j]$$

$$= \sum_{n=1}^{\infty} \mathsf{P}(X_n = j|X_0 = j)$$

$$= \sum_{n=1}^{\infty} p_{jj}^{(n)}.$$

Combining this with the foregoing results shows that

$$\boxed{\begin{aligned} f_{jj} = 1 \text{ (recurrence)} &\Leftrightarrow \sum_{n=1}^{\infty} p_{jj}^{(n)} = \infty \\ f_{jj} < 1 \text{ (transience)} &\Leftrightarrow \sum_{n=1}^{\infty} p_{jj}^{(n)} < \infty. \end{aligned}}$$

(12.36)

A slight modification of the preceding analysis (Problem 15) yields[b]

$$\boxed{\begin{aligned} f_{jj} = 1 \text{ (recurrence)} &\Rightarrow \sum_{n=1}^{\infty} p_{ij}^{(n)} = \begin{cases} \infty, & f_{ij} > 0, \\ 0, & f_{ij} = 0, \end{cases} \\ f_{jj} < 1 \text{ (transience)} &\Rightarrow \sum_{n=1}^{\infty} p_{ij}^{(n)} < \infty. \end{aligned}}$$

(12.37)

We next use (12.32) and (12.29) to show that for a positive recurrent state j, $V_m(j)/m \to 1/\mathsf{E}[T_1(j)|X_0 = j]$. To simplify the notation, let $t := \mathsf{E}[T_1(j)|X_0 = j]$ and $v := 1/t$ so that $T_n/n \to t$ and we need to show that $V_m/m \to v$. The first fact we need is that if $\alpha_m \to \alpha$ and $\beta_m \to \beta$ and if $\alpha > \beta$, then for all sufficiently large m, $\alpha_m > \beta_m$. Next, for $\varepsilon > 0$, consider the quantity $\lfloor m(v+\varepsilon) \rfloor$, where $\lfloor x \rfloor$ denotes the greatest integer less than or equal to x. For larger and larger m, $\lfloor m(v+\varepsilon) \rfloor$ takes larger and larger integer values. Since $T_n/n \to t$,

$$\alpha_m := \frac{T_{\lfloor m(v+\varepsilon) \rfloor}}{\lfloor m(v+\varepsilon) \rfloor} - t \to 0 =: \alpha.$$

For β_m, we take

$$\beta_m := \frac{m}{\lfloor m(v+\varepsilon) \rfloor} - t = \frac{m}{\lfloor m(v+\varepsilon) \rfloor} - \frac{1}{v}.$$

Now by Problem 16, for any $\lambda > 0$, $m/\lfloor \lambda m \rfloor \to 1/\lambda$. Hence,

$$\beta_m \to \frac{1}{v+\varepsilon} - \frac{1}{v} = \frac{-\varepsilon}{v(v+\varepsilon)} =: \beta.$$

---

[b]Recall that f_{ij} was defined in (12.33) as the conditional probability that starting from state i, the chain visits state j in finite time. If there is no path in the state transition diagram from i to j, then it is impossible to go from i to j and we must have $f_{ij} = 0$. Conversely, if there is a path, say a path of n transitions, and if this path is taken, then $T_1(j) \leq n$. Hence,

$$0 < \mathsf{P}(\text{particular path taken}|X_0 = i) \leq \mathsf{P}(T_1(j) \leq n|X_0 = i) \leq \mathsf{P}(T_1(j) < \infty|X_0 = i) =: f_{ij}.$$

Hence, $f_{ij} > 0$ if and only if there is a path in the state transition diagram from i to j.

12.3 Recurrent and transient states

Since $\alpha = 0 > -\varepsilon/[v(v+\varepsilon)] = \beta$, for all large m, we have $\alpha_m > \beta_m$. From the definitions of α_m and β_m,

$$\frac{T_{\lfloor m(v+\varepsilon)\rfloor}}{\lfloor m(v+\varepsilon)\rfloor} - t > \frac{m}{\lfloor m(v+\varepsilon)\rfloor} - t,$$

which is equivalent to

$$T_{\lfloor m(v+\varepsilon)\rfloor} > m.$$

From (12.29), this implies

$$V_m < \lfloor m(v+\varepsilon)\rfloor \le m(v+\varepsilon).$$

This can be rearranged to get

$$\frac{V_m}{m} - v < \varepsilon.$$

A similar argument shows that

$$\frac{V_m}{m} - v > -\varepsilon,$$

from which it then follows that $|V_m/m - v| < \varepsilon$. Hence, $V_m/m \to v$. We have thus proved the following result.

Theorem 3. *If state j is positive recurrent; i.e., if $\mathsf{E}[T_1(j)|X_0 = j] < \infty$, then (12.31) and (12.32) hold, and the average occupation time converges:*[4]

$$\frac{V_m(j)}{m} \to \frac{1}{\mathsf{E}[T_1(j)|X_0 = j]}.$$

This raises the question, "When is a state positive recurrent?"

Theorem 4. *If an irreducible chain has a stationary distribution π, then all states are positive recurrent, and*

$$\pi_j = \frac{1}{\mathsf{E}[T_1(j)|X_0 = j]}.$$

Proof. See [23, Section 6.4]. In fact, the results in [23] go further; if an irreducible chain does not have a stationary distribution, then the states of the chain are either all null recurrent or all transient. □

Ergodic theorem for Markov chains. If an irreducible chain has a stationary distribution π, and $h(i)$ is a bounded function of j, then

$$\lim_{m \to \infty} \frac{1}{m} \sum_{k=1}^{m} h(X_k) \to \sum_j h(j)\pi_j. \tag{12.38}$$

Remark. If the initial distribution of the chain is taken to be π, then $\mathsf{P}(X_k = j) = \pi_j$ for all k. In this case, the right-hand side of (12.38) is equal to $\mathsf{E}[h(X_k)]$. Hence, the limiting time average of $h(X_k)$ converges to $\mathsf{E}[h(X_k)]$.

Proof of the ergodic theorem. By Theorem 4, all states are positive recurrent and $\pi_i = 1/\mathsf{E}[T_1(j)|X_0 = j]$. Since all states are positive recurrent, we then have from Theorem 3 that $V_m(j)/m \to 1/\mathsf{E}[T_1(j)|X_0 = j] = \pi_j$ for every state j. Now consider the special case $h(j) = I_{\{s\}}(j)$ for a fixed state s. Then the average on the left-hand side of (12.38) is

$$\frac{1}{m}\sum_{k=1}^{m} I_{\{s\}}(X_k) = V_m(s)/m.$$

The right-hand side of (12.38) is

$$\sum_j I_{\{s\}}(j)\pi_j = \pi_s.$$

Since $V_m(s)/m \to \pi_s$, this establishes (12.38) for the function $h(j) = I_{\{s\}}(j)$. More general cases for $h(j)$ are considered in the problems. □

12.4 *Limiting *n*-step transition probabilities

We showed in Section 12.2 that by the law of total probability,

$$\mathsf{P}(X_n = j) = \sum_i p_{ij}^{(n)} \mathsf{P}(X_0 = i).$$

Now suppose that $\tilde{\pi}_j := \lim_{n\to\infty} p_{ij}^{(n)}$ exists and does not depend on i. Then[5]

$$\lim_{n\to\infty} \mathsf{P}(X_n = j) = \lim_{n\to\infty} \sum_i p_{ij}^{(n)} \mathsf{P}(X_0 = i)$$
$$= \sum_i \lim_{n\to\infty} p_{ij}^{(n)} \mathsf{P}(X_0 = i) \quad (12.39)$$
$$= \sum_i \tilde{\pi}_j \mathsf{P}(X_0 = i)$$
$$= \tilde{\pi}_j \sum_i \mathsf{P}(X_0 = i)$$
$$= \tilde{\pi}_j. \quad (12.40)$$

Notice that the initial probabilities $\mathsf{P}(X_0 = i)$ do not affect the limiting value of $\mathsf{P}(X_n = j)$. Hence, we can approximate $\mathsf{P}(X_n = j)$ by $\tilde{\pi}_j$ when n is large, *no matter what the initial probabilities are.*

Example 12.10. Show that if all states are transient, then

$$\tilde{\pi}_j := \lim_{n\to\infty} p_{ij}^{(n)} = 0 \quad \text{for all } i, j.$$

In particular then $\lim_{n\to\infty} \mathsf{P}(X_n = j) = 0$.

12.4 Limiting n-step transition probabilities

Solution. If all states j are transient, then by (12.37) we have $\sum_{n=1}^{\infty} p_{ij}^{(n)} < \infty$. Since the sum converges to a finite value, the terms must go to zero as $n \to \infty$.[6]

If a stationary distribution π exists and we take $P(X_0 = i) = \pi_i$, then we showed in Section 12.2 that $P(X_n = j) = \pi_j$ for all n. In this case we trivially have

$$\lim_{n \to \infty} P(X_n = j) = \pi_j. \qquad (12.41)$$

Comparing (12.40) and (12.41), we have the following result.

Theorem 5. *If $\tilde{\pi}_j := \lim_{n \to \infty} p_{ij}^{(n)}$ exists and does not depend on i, and if there is a stationary distribution π, then $\pi_j = \tilde{\pi}_j$ for all j. This further implies uniqueness of the stationary distribution, since all stationary distributions have to be equal to $\tilde{\pi}$.*

Since a stationary distribution satisfies $\sum_j \pi_j = 1$, we see from Example 12.10 combined with Theorem 5 that if all states are transient, then a stationary distribution cannot exist.

Theorem 6. *If $\tilde{\pi}_j := \lim_{n \to \infty} p_{ij}^{(n)}$ exists and does not depend on i, and if the chain has a finite number of states, then $\tilde{\pi}_j$ is a stationary distribution. By Theorem 5, $\tilde{\pi}$ is the **unique** stationary distribution.*

Proof. By definition, the $\tilde{\pi}_j$ are nonnegative. It remains to show that $\tilde{\pi} = \tilde{\pi} P$ and that $\sum_j \tilde{\pi}_j = 1$. By the Chapman–Kolmogorov equation,

$$p_{ij}^{(n+1)} = \sum_k p_{ik}^{(n)} p_{kj}.$$

Taking limits on both sides yields

$$\tilde{\pi}_j = \lim_{n \to \infty} p_{ij}^{(n+1)} = \lim_{n \to \infty} \sum_k p_{ik}^{(n)} p_{kj} \qquad (12.42)$$

$$= \sum_k \left[\lim_{n \to \infty} p_{ik}^{(n)} \right] p_{kj} \qquad (12.43)$$

$$= \sum_k \tilde{\pi}_k p_{kj}.$$

Thus, $\tilde{\pi} = \tilde{\pi} P$. Note that since the chain has a finite number of states, the sum in (12.42) has only a finite number of terms. This justifies bringing the limit inside the sum in (12.43). We next show that $\sum_j \tilde{\pi}_j = 1$. Write

$$\sum_j \tilde{\pi}_j = \sum_j \lim_{n \to \infty} p_{ij}^{(n)} = \lim_{n \to \infty} \sum_j p_{ij}^{(n)},$$

where the last step is justified because the sum involves only a finite number of terms. To conclude, recall that as a function of j, $p_{ij}^{(n)}$ is a pmf. Hence, the sum on the right is 1 for all i and all n. □

If a chain has a finite number of states, then they cannot all be transient; i.e., at least one state must be recurrent. For if all states were transient, we would have $\tilde{\pi}_j = 0$ for all j by Example 12.10, but by Theorem 6 the $\tilde{\pi}_j$ would be a stationary distribution summing to one.

Example 12.11. Sometimes $\lim_{n \to \infty} p_{ij}^{(n)}$ does not exist due to periodic behavior. Consider the chain in Figure 12.9. Observe that

$$p_{01}^{(n)} = \begin{cases} 1, & n = \text{odd}, \\ 0, & n = \text{even}. \end{cases}$$

In this case, $\lim_{n \to \infty} p_{01}^{(n)}$ does not exist, although both $\lim_{n \to \infty} p_{01}^{(2n)}$ and $\lim_{n \to \infty} p_{01}^{(2n+1)}$ do exist. Even though $\lim_{n \to \infty} p_{ij}^{(n)}$ does not exist, the chain does have a stationary distribution by Problem 5 with $a = b = 1$.

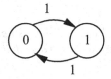

Figure 12.9. A Markov chain with no limit distribution due to periodic behavior. Note that this is a special case of Figure 12.2 with $a = b = 1$. It is also a special case of Figure 12.5 with $N = 1$ and $a_0 = b_1 = 1$.

Example 12.12. Sometimes it can happen that $\lim_{n \to \infty} p_{ij}^{(n)}$ exists, but depends on i. Consider the chain in Figure 12.7 of the previous section. Even though $\lim_{n \to \infty} p_{01}^{(n)}$ exists and is positive, it is clear that $\lim_{n \to \infty} p_{21}^{(n)} = 0$. Thus, $\lim_{n \to \infty} p_{ij}^{(n)}$ depends on i.

Classes of states

When specifying a Markov chain, we usually start either with a state transition diagram or with a specification of the one-step transition probabilities p_{ij}. To compute the n-step transition probabilities $p_{ij}^{(n)}$, we have the Chapman–Kolmogorov equation. However, suppose we just want to know if $p_{ij}^{(n)} > 0$ for some n? If $j = i$, we trivially have $p_{ii}^{(0)} = 1$. What if $j \neq i$? For $j \neq i$, we of course have $p_{ij} > 0$ if and only if there is an arrow in the state transition diagram from state i to state j. However, in many cases, there is no arrow directly from state i to state j because $p_{ij} = 0$. This is the case in the state transition diagrams in Figures 12.3–12.6 except when $j = i \pm 1$. Now consider two states i and j with no arrow

12.4 Limiting n-step transition probabilities

from i to j, but for which there is an intermediate state l such that there is an arrow from i to l and another arrow from l to j. Equivalently, p_{il} and p_{lj} are positive. Now use the Chapman–Kolmogorov equation to write

$$p_{ij}^{(2)} = \sum_k p_{ik} p_{kj} \geq p_{il} p_{lj} > 0.$$

Conversely, if $p_{ij}^{(2)} > 0$, then at least one of the terms in the above sum must be positive. Hence, if $p_{ij}^{(2)} > 0$, there must be some state k with $p_{ik} p_{kj} > 0$; i.e., there is an arrow in the state transition diagram from i to k and another arrow from k to j. In general, to examine $p_{ij}^{(n)}$, we apply the Chapman–Kolmogorov equation $n-1$ times to write

$$p_{ij}^{(n)} = \sum_{k_1} p_{ik_1} p_{k_1 j}^{(n-1)} = \sum_{k_1} p_{ik_1} \left(\sum_{k_2} p_{k_1 k_2} p_{k_2 j}^{(n-2)} \right) = \cdots = \sum_{k_1,\ldots,k_{n-1}} p_{ik_1} p_{k_1 k_2} \cdots p_{k_{n-1} j}.$$

Hence, $p_{ij}^{(n)} > 0$ if and only if there is at least one term on the right with $p_{ik_1} p_{k_1 k_2} \cdots p_{k_{n-1} j} > 0$. But this term is positive if and only if there is an arrow from i to k_1, an arrow from k_1 to $k_2, \ldots$, and an arrow from k_{n-1} to j.

We say that state j is **accessible** or **reachable** from state i if for some $n \geq 0$, $p_{ij}^{(n)} > 0$. In other words, starting from state i, there is a positive conditional probability of going from state i to state j in n steps. We use the notation $i \to j$ to mean that j is accessible from i.

Since $p_{ii}^{(0)} = 1$, state i is always accessible from itself; i.e., $i \to i$. For $j \neq i$, we see from the discussion above that $i \to j$ if and only if there is a path (sequence of arrows) in the state transition diagram from i to j. If $i \to j$ and $j \to i$, we write $i \leftrightarrow j$ and we say that i and j **communicate**. For example, in Figure 12.7, $0 \leftrightarrow 1$ and $2 \leftrightarrow 3$, while $1 \not\leftrightarrow 2$. If a chain satisfies $i \leftrightarrow j$ for all states $i \neq j$, then the chain is **irreducible** as defined at the end of Section 12.2.

***Example* 12.13.** It is intuitively clear that if $i \to j$ and $j \to k$, then $i \to k$. Derive this directly from the Chapman–Kolmogorov equation.

Solution. Since $i \to j$, there is an $n \geq 0$ with $p_{ij}^{(n)} > 0$. Since $j \to k$, there is an $m \geq 0$ with $p_{jk}^{(m)} > 0$. Using the Chapman–Kolmogorov equation, write

$$p_{ik}^{(n+m)} = \sum_l p_{il}^{(n)} p_{lk}^{(m)} \geq p_{ij}^{(n)} p_{jk}^{(m)} > 0.$$

Since, $p_{ik}^{(n+m)} > 0$, we have $i \to k$.

It is easy to see that $\leftrightarrow$ has the three properties:

(i) $i \leftrightarrow i$; it is **reflexive**.
(ii) $i \leftrightarrow j \Leftrightarrow j \leftrightarrow i$; it is **symmetric**.
(iii) $i \leftrightarrow j$ and $j \leftrightarrow k \Rightarrow i \leftrightarrow k$; it is **transitive**.

A relation that is reflexive, symmetric, and transitive is called an **equivalence relation**. As shown in the problems, an equivalence relation partitions a set into disjoint subsets called **equivalence classes**. Each class consists of those elements that are equivalent to each other. In the case of the relation $\leftrightarrow$, two states belong to the same class if and only if they communicate. For an irreducible chain, there is only one class. Otherwise, there are multiple classes.

Example **12.14.** The chain in Figure 12.8 consists of the classes $\{0,1\}$, $\{2\}$, $\{3\}$, and so on. Consider the chain in Figure 12.7. Since $0 \leftrightarrow 1$ and $2 \leftrightarrow 3$, the state space of this chain can be partitioned into the two disjoint classes $\{0,1\}$ and $\{2,3\}$.

Theorem 7. *If $i \leftrightarrow j$, then either both states are transient or both states are recurrent. If both are recurrent, then both are either positive recurrent or null recurrent.*

Proof. Let $p_{ij}^{(n)} > 0$ and $p_{ji}^{(m)} > 0$. Using the Chapman–Kolmogorov equation twice, write

$$\sum_{r=0}^{\infty} p_{ii}^{(n+r+m)} = \sum_{r=0}^{\infty} \sum_{k} \sum_{l} p_{il}^{(n)} p_{lk}^{(r)} p_{ki}^{(m)}$$

$$\geq \sum_{r=0}^{\infty} p_{ij}^{(n)} p_{jj}^{(r)} p_{ji}^{(m)}$$

$$= p_{ij}^{(n)} p_{ji}^{(m)} \sum_{r=0}^{\infty} p_{jj}^{(r)}.$$

Combining this with (12.36), we see that if j is recurrent, so is i, and if i is transient, so is j. To complete the proof of the first part of the theorem, interchange the roles of i and j. For a proof of the second part of the theorem, see [23]. □

Example **12.15.** States 0 and 1 of the chain in Figure 12.7 communicate, and by Example 12.8, state 0 is recurrent. Hence, state 1 is also recurrent.

Similarly, states 0 and 1 of the chain in Figure 12.8 communicate, and by Example 12.9, state 1 is transient. Hence, state 0 is also transient.

In general, the state space of any chain can be partitioned into disjoint sets T, R_1, R_2, ..., where each R_i is a communicating class of recurrent states, and T is the *union* of all classes of transient states. Thus, given any two transient states in T, they may or may not communicate as in Figure 12.8.

The **period** of state i is defined as

$$d(i) := \gcd\{n \geq 1 : p_{ii}^{(n)} > 0\}.$$

where gcd is the **greatest common divisor**. If $d(i) > 1$, i is said to be **periodic** with period $d(i)$. If $d(i) = 1$, i is said to be **aperiodic**.

12.4 Limiting n-step transition probabilities

Lemma. *If $i \leftrightarrow j$, then $d(i) = d(j)$. In other words, if two states communicate, then they have the same period.*

Proof. It suffices to show that if ν divides every element of $\{n \geq 1 : p_{ii}^{(n)} > 0\}$, then ν divides every element of $\{n \geq 1 : p_{jj}^{(n)} > 0\}$, and conversely. Recall that ν divides n if there is an integer λ such that $n = \lambda \nu$. In this case, we write $\nu | n$. Note that

$$\nu | a \text{ and } \nu | b \implies \nu | (a \pm b).$$

Now, since $i \leftrightarrow j$, there exist r and s with $p_{ij}^{(r)} p_{ji}^{(s)} > 0$. Suppose ν divides every element of $\{n \geq 1 : p_{ii}^{(n)} > 0\}$. Then in particular, by the Chapman–Kolmogorov equation,

$$p_{ii}^{(r+s)} \geq p_{ij}^{(r)} p_{ji}^{(s)} > 0,$$

and it follows that $\nu | (r+s)$. Next, if $p_{jj}^{(n)} > 0$, use the Chapman–Kolmogorov equation to write

$$p_{ii}^{(r+n+s)} \geq p_{ij}^{(r)} p_{jj}^{(n)} p_{ji}^{(s)} > 0.$$

Thus, $\nu | (r+n+s)$. It now follows that

$$\nu | [(r+n+s) - (r+s)] \quad \text{or} \quad \nu | n. \qquad \square$$

Example **12.16.** The chain in Figure 12.9 is irreducible, and each state has period 2. The chain in Figure 12.7 is not irreducible. For state 0 in Figure 12.7, $\{n \geq 1 : p_{00}^{(n)} > 0\} \supset \{1\}$. Since the only (positive) divisor of 1 is 1, $d(0) = 1$. Since $0 \leftrightarrow 1$, $d(1) = 1$ too.

Theorem 8. *If a chain is irreducible and aperiodic, then the limits*

$$\tilde{\pi}_j := \lim_{n \to \infty} \mathsf{P}(X_n = j | X_0 = i) = \frac{1}{\mathsf{E}[T_1(j)|X_0 = j]}, \quad \text{for all } i \text{ and } j,$$

exist and do not depend on i.

Proof. See [23, Section 6.4]. $\qquad \square$

Discussion: In a typical application we start with an irreducible chain and try to find a stationary distribution π. If we are successful, then by Theorem 4 it is unique, all states are positive recurrent, and $\pi_j = 1/\mathsf{E}[T_1(j)|X_0 = j]$. If the chain is also aperiodic, then by Theorem 8, $\tilde{\pi}_j = \pi_j$.

On the other hand, if no stationary distribution exists, then as mentioned in the proof of Theorem 4, the states of the chain are either all transient or all null recurrent. In either case, the conditional expectations in Theorem 8 are infinite, and so the $\tilde{\pi}_j$ are all zero.

> In trying to find a stationary distribution, we may be unsuccessful. But is this because no stationary distribution exists or is it because we are not clever enough to find it? For an irreducible, aperiodic chain *with a finite number of states*, a unique stationary distribution always exists. We can see this as follows. First use Theorem 8 to guarantee the existence of the limits $\tilde{\pi}_j$. Then by Theorem 6, $\tilde{\pi}_j$ is the unique stationary distribution.

12.5 Continuous-time Markov chains

A family of integer-valued random variables, $\{X_t, t \geq 0\}$, is called a **Markov chain** if for all $n \geq 1$, and for all $0 \leq s_0 < \cdots < s_{n-1} < s < t$,

$$P(X_t = j | X_s = i, X_{s_{n-1}} = i_{n-1}, \ldots, X_{s_0} = i_0) = P(X_t = j | X_s = i).$$

In other words, given the sequence of values $i_0, \ldots, i_{n-1}, i$, the conditional probability of what X_t will be depends only on the condition $X_s = i$. The quantity $P(X_t = j | X_s = i)$ is called the **transition probability**.

Example 12.17. Show that the Poisson process of rate λ is a Markov chain.

Solution. To begin, observe that

$$P(N_t = j | N_s = i, N_{s_{n-1}} = i_{n-1}, \ldots, N_{s_0} = i_0),$$

is equal to

$$P(N_t - i = j - i | N_s = i, N_{s_{n-1}} = i_{n-1}, \ldots, N_{s_0} = i_0).$$

By the substitution law, this is equal to

$$P(N_t - N_s = j - i | N_s = i, N_{s_{n-1}} = i_{n-1}, \ldots, N_{s_0} = i_0). \tag{12.44}$$

Since

$$(N_s, N_{s_{n-1}}, \ldots, N_{s_0}) \tag{12.45}$$

is a function of

$$(N_s - N_{s_{n-1}}, \ldots, N_{s_1} - N_{s_0}, N_{s_0} - N_0),$$

and since this is independent of $N_t - N_s$ by the independent increments property of the Poisson process, it follows that (12.45) and $N_t - N_s$ are also independent. Thus, (12.44) is equal to $P(N_t - N_s = j - i)$, which depends on i but not on $i_{n-1}, \ldots, i_0$. It then follows that

$$P(N_t = j | N_s = i, N_{s_{n-1}} = i_{n-1}, \ldots, N_{s_0} = i_0) = P(N_t = j | N_s = i),$$

and we see that the Poisson process is a Markov chain.

As shown in the above example,

$$P(N_t = j | N_s = i) = P(N_t - N_s = j - i) = \frac{[\lambda(t-s)]^{j-i} e^{-\lambda(t-s)}}{(j-i)!} \tag{12.46}$$

12.5 Continuous-time Markov chains

depends on t and s only through $t-s$. In general, if a Markov chain has the property that the transition probability $P(X_t = j | X_s = i)$ depends on t and s only through $t-s$, we say that the chain is **time-homogeneous** or that it has **stationary transition probabilities**. In this case, if we put

$$p_{ij}(t) := P(X_t = j | X_0 = i),$$

then $P(X_t = j | X_s = i) = p_{ij}(t-s)$. Note that $p_{ij}(0) = \delta_{ij}$, the Kronecker delta.

In the remainder of the chapter, we assume that X_t is a time-homogeneous Markov chain with transition probability function $p_{ij}(t)$. For such a chain, we can derive the continuous-time **Chapman–Kolmogorov equation**,

$$p_{ij}(t+s) = \sum_k p_{ik}(t) p_{kj}(s).$$

To derive this, we first use the law of total conditional probability (Problem 33) to write

$$\begin{aligned} p_{ij}(t+s) &= P(X_{t+s} = j | X_0 = i) \\ &= \sum_k P(X_{t+s} = j | X_t = k, X_0 = i) P(X_t = k | X_0 = i). \end{aligned}$$

Now use the Markov property and time homogeneity to obtain

$$\begin{aligned} p_{ij}(t+s) &= \sum_k P(X_{t+s} = j | X_t = k) P(X_t = k | X_0 = i) \\ &= \sum_k p_{kj}(s) p_{ik}(t). \end{aligned} \quad (12.47)$$

The reader may wonder why the derivation of the continuous-time Chapman–Kolmogorov equation is so much simpler than the derivation of the discrete-time version. The reason is that in discrete time, the Markov property and time homogeneity are defined in a one-step manner. Hence, induction arguments are first needed to *derive* the discrete-time analogs of the continuous-time *definitions*!

Behavior of continuous-time Markov chains

In the remainder of the chapter, we assume that for small $\Delta t > 0$,

$$p_{ij}(\Delta t) \approx g_{ij} \Delta t, \quad i \neq j, \quad \text{and} \quad p_{ii}(\Delta t) \approx 1 + g_{ii} \Delta t.$$

These approximations tell us the conditional probability of being in state j at time Δt in the near future given that we are in state i at time zero. These assumptions are more precisely written as

$$\lim_{\Delta t \downarrow 0} \frac{p_{ij}(\Delta t)}{\Delta t} = g_{ij} \quad \text{and} \quad \lim_{\Delta t \downarrow 0} \frac{p_{ii}(\Delta t) - 1}{\Delta t} = g_{ii}. \quad (12.48)$$

Note that $g_{ij} \geq 0$, while $g_{ii} \leq 0$. The parameters g_{ij} are called **transition rates**.

As the next example shows, for a Poisson process of rate λ, $g_{i,i+1} = \lambda$.

Example 12.18. Calculate the transition rates g_{ij} for a Poisson process of rate λ.

Solution. Since $p_{i,i+1}(\Delta t) = \mathsf{P}(N_{\Delta t} = i+1|N_0 = i)$, we have from (12.46) that

$$g_{i,i+1} = \lim_{\Delta t \downarrow 0} \frac{p_{i,i+1}(\Delta t)}{\Delta t} = \lim_{\Delta t \downarrow 0} \frac{(\lambda \Delta t)e^{-\lambda \Delta t}}{\Delta t} = \lambda.$$

Similarly, since $p_{ii}(\Delta t) = \mathsf{P}(N_{\Delta t} = i|N_0 = i)$, we have from (12.46) that

$$g_{ii} = \lim_{\Delta t \downarrow 0} \frac{p_{ii}(\Delta t) - 1}{\Delta t} = \lim_{\Delta t \downarrow 0} \frac{e^{-\lambda \Delta t} - 1}{\Delta t} = -\lambda.$$

It is left to Problem 22 to show that $g_{i,i+n} = 0$ for $n \geq 2$.

The length of time a chain spends in one state before jumping to the next state is called the **sojourn time** or **holding time**. It is shown in Problem 31 that the sojourn time in state i is an $\exp(-g_{ii})$ random variable.[c] Hence, the chain operates as follows. Upon arrival in state i, the chain stays a length of time that is an $\exp(-g_{ii})$ random variable. Then the chain jumps to state j with some probability p_{ij}. So, if we look at a continuous-time chain only at the times that it jumps, we get the **embedded chain** or **jump chain** with discrete-time transition probabilities p_{ij}.

The formula for p_{ij} is suggested by the following argument. Suppose the chain is in state i and jumps to a new state at time t. What is the probability that the new state is $j \neq i$? For small $\Delta t > 0$, consider[d]

$$\mathsf{P}(X_t = j | X_t \neq i, X_{t-\Delta t} = i) = \frac{\mathsf{P}(X_t = j, X_t \neq i, X_{t-\Delta t} = i)}{\mathsf{P}(X_t \neq i, X_{t-\Delta t} = i)}.$$

Since $j \neq i$ implies $\{X_t = j\} \subset \{X_t \neq i\}$, the right-hand side simplifies to

$$\frac{\mathsf{P}(X_t = j, X_{t-\Delta t} = i)}{\mathsf{P}(X_t \neq i, X_{t-\Delta t} = i)} = \frac{\mathsf{P}(X_t = j|X_{t-\Delta t} = i)}{\mathsf{P}(X_t \neq i|X_{t-\Delta t} = i)} = \frac{p_{ij}(\Delta t)}{1 - p_{ii}(\Delta t)}.$$

Writing this last quotient as

$$\frac{p_{ij}(\Delta t)/\Delta t}{[1 - p_{ii}(\Delta t)]/\Delta t}$$

and letting $\Delta t \downarrow 0$, we get $-g_{ij}/g_{ii}$.

Intuitively, a continuous-time chain cannot jump from state i directly back into state i; if it did, it really never jumped at all. This suggests that $p_{ii} = 0$, or equivalently, $\sum_{j \neq i} p_{ij} = 1$. Applying this condition to $p_{ij} = -g_{ij}/g_{ii}$ for $j \neq i$ requires that

$$\sum_{j \neq i} g_{ij} = -g_{ii} < \infty. \qquad (12.49)$$

Such a chain is said to be **conservative**.

---

[c] For a Poisson process of rate λ, the sojourn time is just the interarrival time, which is $\exp(\lambda)$.
[d] As in the case of the Poisson process, we assume X_t is a right-continuous function of t.

12.5 Continuous-time Markov chains

***Example* 12.19.** Consider an Internet router with a buffer that can hold N packets. Suppose that in a short time interval Δt, a new packet arrives with probability $\lambda \Delta t$ or a buffered packet departs with probability $\mu \Delta t$. To model the number packets in the buffer at time t, we use a continuous-time Markov chain with rates $g_{i,i+1} = \lambda$ for $i = 0, \ldots, N-1$ and $g_{i,i-1} = \mu$ for $i = 1, \ldots, N$ as shown in the state transition diagram in Figure 12.10.

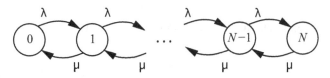

Figure 12.10. State transition diagram for a continuous-time queue with finite buffer.

Notice the diagram follows the convention of not showing g_{ii} since it is tacitly assumed that the chain is conservative. In other words, state transition diagrams for continuous-time Markov chains assume $-g_{ii}$ is equal to the sum of the rates leaving state i. Thus, in Figure 12.10, $g_{00} = -\lambda$, $g_{NN} = -\mu$, and for $i = 1, \ldots, N-1$, $g_{ii} = -(\lambda + \mu)$. The embedded discrete-time chain has the state transition diagram of Figure 12.5 with $a_0 = 1$, $b_N = 1$, $a_i = \lambda/(\lambda+\mu)$, and $b_i = \mu/(\lambda+\mu)$ for $i = 1, \ldots, N-1$. Notice this implies $p_{ii} = 0$ for all i.

Kolmogorov's differential equations

Using the Chapman–Kolmogorov equation, write

$$p_{ij}(t+\Delta t) = \sum_k p_{ik}(t) p_{kj}(\Delta t)$$

$$= p_{ij}(t) p_{jj}(\Delta t) + \sum_{k \neq j} p_{ik}(t) p_{kj}(\Delta t).$$

Now subtract $p_{ij}(t)$ from both sides to get

$$p_{ij}(t+\Delta t) - p_{ij}(t) = p_{ij}(t)[p_{jj}(\Delta t) - 1] + \sum_{k \neq j} p_{ik}(t) p_{kj}(\Delta t). \quad (12.50)$$

Dividing by Δt and applying the limit assumptions (12.48),[7] we obtain

$$p'_{ij}(t) = p_{ij}(t) g_{jj} + \sum_{k \neq j} p_{ik}(t) g_{kj}.$$

This is **Kolmogorov's forward differential equation**, which can be written more compactly as

$$\boxed{p'_{ij}(t) = \sum_k p_{ik}(t) g_{kj}.} \quad (12.51)$$

To derive the backward equation, observe that since $p_{ij}(t+\Delta t) = p_{ij}(\Delta t + t)$, we can write

$$p_{ij}(t+\Delta t) = \sum_k p_{ik}(\Delta t) p_{kj}(t)$$

$$= p_{ii}(\Delta t)p_{ij}(t) + \sum_{k \neq i} p_{ik}(\Delta t)p_{kj}(t).$$

Now subtract $p_{ij}(t)$ from both sides to get

$$p_{ij}(t+\Delta t) - p_{ij}(t) = [p_{ii}(\Delta t) - 1]p_{ij}(t) + \sum_{k \neq i} p_{ik}(\Delta t)p_{kj}(t).$$

Dividing by Δt and applying the limit assumptions (12.48),[7] we obtain

$$p'_{ij}(t) = g_{ii}p_{ij}(t) + \sum_{k \neq i} g_{ik}p_{kj}(t).$$

This is **Kolmogorov's backward differential equation**, which can be written more compactly as

$$\boxed{p'_{ij}(t) = \sum_k g_{ik}p_{kj}(t).} \quad (12.52)$$

Readers familiar with linear system theory may find it insightful to write the forward and backward equations in matrix form. Let $P(t)$ denote the matrix whose ij entry is $p_{ij}(t)$, and let G denote the matrix whose ij entry is g_{ij} (G is called the **generator matrix** or **rate matrix**). Then the forward equation (12.51) becomes

$$P'(t) = P(t)G,$$

and the backward equation (12.52) becomes

$$P'(t) = GP(t),$$

The initial condition in both cases is $P(0) = I$. Under suitable assumptions, the solution of both equations is given by the **matrix exponential**,

$$P(t) = e^{Gt} := \sum_{n=0}^{\infty} \frac{(Gt)^n}{n!}.$$

When the state space is finite, G is a finite-dimensional matrix, and the theory is straightforward. Otherwise, more careful analysis is required.

Stationary distributions

In analogy with the discrete-time case, let us put $p_j(t) := \mathsf{P}(X_t = j)$. By the law of total probability, we have

$$p_j(t) = \sum_i \mathsf{P}(X_0 = i)p_{ij}(t).$$

Can we find a choice for the initial probabilities, say $\mathsf{P}(X_0 = i) = \pi_i$, such that

$$p_j(t) = \sum_i \pi_i p_{ij}(t)$$

does not depend on t; i.e., $p_j(t) = p_j(0) = \pi_j$ for all t? Let us differentiate

$$\pi_j = \sum_i \pi_i p_{ij}(t) \quad (12.53)$$

with respect to t and apply the forward differential equation (12.51). Then

$$0 = \sum_i \pi_i \left(\sum_k p_{ik}(t) g_{kj} \right) = \sum_k \left(\sum_i \pi_i p_{ik}(t) \right) g_{kj}.$$
$$= \sum_k \pi_k g_{kj}, \quad \text{by (12.53)}.$$

Combining

$$\boxed{0 = \sum_k \pi_k g_{kj}}$$

with the normalization condition $\sum_k \pi_k = 1$ allows us to solve for π_k much as in the discrete case.

Example 12.20. Find the stationary distribution of the continuous-time Markov chain with generator matrix

$$G = \begin{bmatrix} -2 & 1 & 1 \\ 2 & -4 & 2 \\ 2 & 4 & -6 \end{bmatrix}.$$

Solution. We begin by writing out the equations

$$0 = \sum_k \pi_k g_{kj}$$

for each j. Notice that the right-hand side is the inner product of the row vector π and the jth column of G. For $j = 0$, we have

$$0 = \sum_k \pi_k g_{k0} = -2\pi_0 + 2\pi_1 + 2\pi_2,$$

which implies $\pi_0 = \pi_1 + \pi_2$. For $j = 1$, we have

$$0 = \pi_0 + -4\pi_1 + 4\pi_2 = (\pi_1 + \pi_2) - 4\pi_1 + 4\pi_2,$$

which implies $\pi_1 = 5\pi_2/3$. As it turns out, the equation for the last value of j is always redundant. Instead we use the requirement that $\sum_j \pi_j = 1$. Writing

$$1 = \pi_0 + \pi_1 + \pi_2 = (\pi_1 + \pi_2) + 5\pi_2/3 + \pi_2,$$

and again using $\pi_1 = 5\pi_2/3$, we find that $\pi_2 = 3/16$, $\pi_1 = 5/16$, and $\pi_0 = 1/2$.

Notes

12.1: Preliminary results

Note 1. The results of Examples 12.1 and 12.2 are easy to derive using the smoothing property (13.30). See Example 13.24.

12.2: Discrete-time Markov chains

Note 2. An alternative derivation of the Chapman–Kolmogorov equation is given in Example 13.25 using the smoothing property.

12.3: *Recurrent and transient states

Note 3. The strong law of large numbers is discussed in Section 14.3. The strong law of large numbers implies that the convergence in (12.30) is almost sure under $\mathsf{P}(\cdot|X_0 = j)$.

Note 4. As mentioned in the Note 3, the convergence in (12.30) is almost sure under $\mathsf{P}(\cdot|X_0 = j)$. Hence, the same is true for the convergence in (12.31)–(12.32) and in Theorem 3.

12.4: *Limiting n-step transition probabilities

Note 5. If the sum in (12.39) contains infinitely many terms, then the interchange of the limit and sum is justified by the dominated convergence theorem.

Note 6. Let x_n be a sequence of real numbers, and let $S_N := \sum_{n=1}^{N} x_n$ denote the sequence of partial sums. To say that the infinite sum $\sum_{n=1}^{\infty} x_n$ converges to some finite real number S means that $S_N \to S$. However, if $S_N \to S$, then we also have $S_{N-1} \to S$. Hence, $S_N - S_{N-1} \to S - S = 0$. However, since

$$S_N - S_{N-1} = \sum_{n=1}^{N} x_n - \sum_{n=1}^{N-1} x_n = x_N,$$

we must have $x_N \to 0$.

12.4: Continuous-time Markov chains

Note 7. The derivations of both the forward and backward differential equations require taking a limit in Δt inside the sum over k. For example, in deriving the backward equation, we tacitly assumed that

$$\lim_{\Delta t \downarrow 0} \sum_{k \neq i} \frac{p_{ik}(\Delta t)}{\Delta t} p_{kj}(t) = \sum_{k \neq i} \lim_{\Delta t \downarrow 0} \frac{p_{ik}(\Delta t)}{\Delta t} p_{kj}(t). \tag{12.54}$$

If the state space of the chain is finite, the above sum is finite and there is no problem. Otherwise, additional technical assumptions are required to justify this step. We now show that a sufficient assumption for deriving the backward equation is that the chain be conservative; i.e., that (12.49) hold. For any finite N, observe that

$$\sum_{k \neq i} \frac{p_{ik}(\Delta t)}{\Delta t} p_{kj}(t) \geq \sum_{\substack{|k| \leq N \\ k \neq i}} \frac{p_{ik}(\Delta t)}{\Delta t} p_{kj}(t).$$

Since the right-hand side is a finite sum,

$$\lim_{\Delta t \downarrow 0} \sum_{k \neq i} \frac{p_{ik}(\Delta t)}{\Delta t} p_{kj}(t) \geq \sum_{\substack{|k| \leq N \\ k \neq i}} g_{ik} p_{kj}(t).$$

Letting $N \to \infty$ shows that

$$\lim_{\Delta t \downarrow 0} \sum_{k \neq i} \frac{p_{ik}(\Delta t)}{\Delta t} p_{kj}(t) \geq \sum_{k \neq i} g_{ik} p_{kj}(t). \quad (12.55)$$

To get an upper bound on the limit, take $N \geq |i|$ and write

$$\sum_{k \neq i} \frac{p_{ik}(\Delta t)}{\Delta t} p_{kj}(t) = \sum_{\substack{|k| \leq N \\ k \neq i}} \frac{p_{ik}(\Delta t)}{\Delta t} p_{kj}(t) + \sum_{|k| > N} \frac{p_{ik}(\Delta t)}{\Delta t} p_{kj}(t).$$

Since $p_{kj}(t) \leq 1$,

$$\sum_{k \neq i} \frac{p_{ik}(\Delta t)}{\Delta t} p_{kj}(t) \leq \sum_{\substack{|k| \leq N \\ k \neq i}} \frac{p_{ik}(\Delta t)}{\Delta t} p_{kj}(t) + \sum_{|k| > N} \frac{p_{ik}(\Delta t)}{\Delta t}$$

$$= \sum_{\substack{|k| \leq N \\ k \neq i}} \frac{p_{ik}(\Delta t)}{\Delta t} p_{kj}(t) + \frac{1}{\Delta t}\left(1 - \sum_{|k| \leq N} p_{ik}(\Delta t)\right)$$

$$= \sum_{\substack{|k| \leq N \\ k \neq i}} \frac{p_{ik}(\Delta t)}{\Delta t} p_{kj}(t) + \frac{1 - p_{ii}(\Delta t)}{\Delta t} - \sum_{\substack{|k| \leq N \\ k \neq i}} \frac{p_{ik}(\Delta t)}{\Delta t}.$$

Since these sums are finite,

$$\lim_{\Delta t \downarrow 0} \sum_{k \neq i} \frac{p_{ik}(\Delta t)}{\Delta t} p_{kj}(t) \leq \sum_{\substack{|k| \leq N \\ k \neq i}} g_{ik} p_{kj}(t) - g_{ii} - \sum_{\substack{|k| \leq N \\ k \neq i}} g_{ik}.$$

Letting $N \to \infty$ shows that

$$\lim_{\Delta t \downarrow 0} \sum_{k \neq i} \frac{p_{ik}(\Delta t)}{\Delta t} p_{kj}(t) \leq \sum_{k \neq i} g_{ik} p_{kj}(t) - g_{ii} - \sum_{k \neq i} g_{ik}.$$

If the chain is conservative, this simplifies to

$$\lim_{\Delta t \downarrow 0} \sum_{k \neq i} \frac{p_{ik}(\Delta t)}{\Delta t} p_{kj}(t) \leq \sum_{k \neq i} g_{ik} p_{kj}(t).$$

Combining this with (12.55) yields (12.54), thus justifying the backward equation.

Problems

12.1: Preliminary results

1. Show that if $P(A|X = i, Y = j, Z = k)$ depends on i only, say $P(A|X = i, Y = j, Z = k) = h(i)$ for some function $h(i)$, then $P(A|X = i, Z = k) = P(A|X = i)$.

12.2: Discrete-time Markov chains

2. Let $X_0, Z_1, Z_2, \ldots$ be a sequence of independent discrete random variables. Put
$$X_n = g(X_{n-1}, Z_n), \quad n = 1, 2, \ldots.$$
Show that X_n is a Markov chain. For example, if $X_n = \max(0, X_{n-1} + Z_n)$, where X_0 and the Z_n are as in Example 12.3, then X_n is a random walk restricted to the nonnegative integers.

3. Derive the chain rule of conditional probability, $P(A \cap B|C) = P(A|B \cap C)P(B|C)$.

4. Let X_n be a time-homogeneous Markov chain with transition probabilities p_{ij}. Put $v_i := P(X_0 = i)$. Express
$$P(X_0 = i, X_1 = j, X_2 = k, X_3 = l)$$
in terms of v_i and entries from the transition probability matrix.

5. Find the stationary distribution of the Markov chain in Figure 12.2.

6. Draw the state transition diagram and find the stationary distribution of the Markov chain whose transition matrix is
$$P = \begin{bmatrix} 1/2 & 1/2 & 0 \\ 1/4 & 0 & 3/4 \\ 1/2 & 1/2 & 0 \end{bmatrix}.$$
Answer: $\pi_0 = 5/12$, $\pi_1 = 1/3$, $\pi_2 = 1/4$.

7. Draw the state transition diagram and find the stationary distribution of the Markov chain whose transition matrix is
$$P = \begin{bmatrix} 0 & 1/2 & 1/2 \\ 1/4 & 3/4 & 0 \\ 1/4 & 3/4 & 0 \end{bmatrix}.$$
Answer: $\pi_0 = 1/5$, $\pi_1 = 7/10$, $\pi_2 = 1/10$.

8. Draw the state transition diagram and find the stationary distribution of the Markov chain whose transition matrix is
$$P = \begin{bmatrix} 1/2 & 1/2 & 0 & 0 \\ 9/10 & 0 & 1/10 & 0 \\ 0 & 1/10 & 0 & 9/10 \\ 0 & 0 & 1/2 & 1/2 \end{bmatrix}.$$
Answer: $\pi_0 = 9/28$, $\pi_1 = 5/28$, $\pi_2 = 5/28$, $\pi_3 = 9/28$.

9. Find the stationary distribution of the queuing system with finite buffer of size N, whose state transition diagram is shown in Figure 12.11.

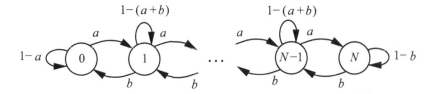

Figure 12.11. State transition diagram for a queue with a finite buffer.

10. Show that the chain in Example 12.6 has an infinite number of stationary distributions.

11. **MATLAB.** Use the following MATLAB code to find the stationary distributions in Problems 6–8. (The algorithm is discussed following Example 12.4.)

```
% Stationary Distribution Solver
%
% Enter transition matrix here:
%
P = [ 1/2 1/2 0 ; 1/4 0 3/4 ; 1/2 1/2 0 ];
%
n = length(P);              % number of states
onecol = ones(n,1);         % col vec of ones
In = diag(onecol);          % n x n identity matrix
y = zeros(1,n);             % Create
y(n) = 1;                   % [ 0 0 0 .... 0 1 ]
A = P - In;
A(:,n) = onecol;
pi = y/A;                   % Solve pi * A = y
fprintf('pi = [ ');         % Print answer in
fprintf(' %g ',pi);         % decimal format
fprintf(' ]\n\n')
[num,den] = rat(pi);        % Print answer using
fprintf('pi = [ ')          % rational numbers
fprintf(' %g/%g ',[num ; den])
fprintf(' ]\n\n')
```

12.3: *Recurrent and transient states

12. Show that
$$E[T_1(j)|X_0 = i] = \sum_{k=1}^{\infty} k f_{ij}^{(k)} + \infty \cdot (1 - f_{ij}).$$

Hint: Equations (12.22), (12.25), and (12.33) may be helpful.

13. Give a detailed derivation of the steps in (12.26) in the following special case:
(a) First show that

$$P(T_2 = 5|T_1 = 2, X_0 = i) = P(X_5 = j, X_4 \neq j, X_3 \neq j | X_2 = j, X_1 \neq j, X_0 = i).$$

(b) Now show that
$$P(X_5 = j, X_4 \neq j, X_3 \neq j | X_2 = j, X_1 \neq j, X_0 = i) = P(X_5 = j, X_4 \neq j, X_3 \neq j | X_2 = j).$$

(c) Conclude by showing that
$$P(X_5 = j, X_4 \neq j, X_3 \neq j | X_2 = j) = P(X_3 = j, X_2 \neq j, X_1 \neq j | X_0 = j),$$
which is $P(T_1(j) = 3 | X_0 = j) = f_{jj}^{(3)} = f_{jj}^{(5-2)}$.

14. Use the identity
$$\{V = \infty\} = \bigcap_{L=1}^{\infty} \{V \geq L\}$$
to show that the total occupation time of state j satisfies
$$P(V(j) = \infty | X_0 = i) = \lim_{L \to \infty} P(V(j) \geq L | X_0 = i).$$

15. Derive (12.37).

16. Given $\lambda > 0$, show that $m/\lfloor m\lambda \rfloor \to 1/\lambda$. *Hint:* Use the identity $x - 1 \leq \lfloor x \rfloor \leq x$.

17. Generalize the proof of the ergodic theorem as follows:

   (a) Show that (12.38) holds if $h(j) = I_S(j)$, where S is a finite subset of states, say $S = \{s_1, \ldots, s_n\}$.

   (b) Show that (12.38) holds if $h(j) = \sum_{l=1}^{n} c_l I_{S_l}(j)$, where each S_l is a finite subset of states.

   (c) Show that (12.38) holds if $h(j) = 0$ for all but finitely many states.

12.4: *Limiting n-step transition probabilities

18. **MATLAB.** Add the following lines to the end of the script of Problem 11:

    ```
    % Now compare with P^m
    %
    m = input('Enter a value of m (0 to quit): ');
    while m > 0
        Pm = P^m
        fprintf('pi = [ ');       % Print pi in decimal
        fprintf(' %g ',pi);       % to compare with P^m
        fprintf(' ]\n\n')
        m = input('Enter a value of m (0 to quit): ');
    end
    ```

    Again using the data in Problems 6–8, in each case find a value of m so that numerically all rows of P^m agree with π.

19. **MATLAB.** Use the script of Problem 18 to investigate the limiting behavior of P^m for large m if P is the transition matrix of Example 12.6.

20. For any state i, put $A_i := \{k : i \leftrightarrow k\}$. The sets A_i are called **equivalence classes**. For any two states i and j, show that $A_i \cap A_j \neq \emptyset$ implies $A_i = A_j$. In other words, two equivalence classes are either disjoint or exactly equal to each other.

21. Consider a chain with a finite number of states. If the chain is irreducible and aperiodic, is the conditional expected time to return to a state finite? Justify your answer.

12.5: Continuous-time Markov chains

22. For a Poisson process of rate λ, show that for $n \geq 2$, $g_{i,i+n} = 0$.

23. Draw the state transition diagram, and find the stationary distribution of the continuous-time Markov chain with generator matrix

$$G = \begin{bmatrix} -1 & 1 & 0 \\ 2 & -5 & 3 \\ 5 & 4 & -9 \end{bmatrix}.$$

Answer: $\pi_0 = 11/15$, $\pi_1 = 1/5$, $\pi_2 = 1/15$.

24. **MATLAB.** Modify the code of Problem 11 to solve for stationary distributions of continuous-time Markov chains with a finite number of states. Check your script with the generator matrix of the previous problem and with the generator matrix of Example 12.20.

25. The general continuous-time random walk is defined by

$$g_{ij} = \begin{cases} \mu_i, & j = i-1, \\ -(\lambda_i + \mu_i), & j = i, \\ \lambda_i, & j = i+1, \\ 0, & \text{otherwise.} \end{cases}$$

Write out the forward and backward equations. Is the chain conservative?

26. The continuous-time queue with infinite buffer can be obtained by modifying the general random walk in the preceding problem to include a barrier at the origin. Put

$$g_{0j} = \begin{cases} -\lambda_0, & j = 0, \\ \lambda_0, & j = 1, \\ 0, & \text{otherwise.} \end{cases}$$

Find the stationary distribution assuming

$$\sum_{j=1}^{\infty} \left(\frac{\lambda_0 \cdots \lambda_{j-1}}{\mu_1 \cdots \mu_j}\right) < \infty.$$

If $\lambda_i = \lambda$ and $\mu_i = \mu$ for all i, simplify the above condition to one involving only the relative values of λ and μ.

27. Modify Problem 26 to include a barrier at some finite N. Find the stationary distribution.

28. For the chain in Problem 26, let

$$\lambda_j = j\lambda + \alpha \quad \text{and} \quad \mu_j = j\mu,$$

where λ, α, and μ are positive. Put $m_i(t) := E[X_t|X_0 = i]$. Derive a differential equation for $m_i(t)$ and solve it. Treat the cases $\lambda = \mu$ and $\lambda \neq \mu$ separately. *Hint:* Use the forward equation (12.51).

29. For the chain in Problem 26, let $\mu_i = 0$ and $\lambda_i = \lambda$. Write down and solve the forward equation (12.51) for $p_{0j}(t)$. *Hint:* Equation (11.6).

30. If a continuous-time Markov chain has conservative transition rates g_{ij}, then the corresponding jump chain has transition probabilities $p_{ij} = -g_{ij}/g_{ii}$ for $j \neq i$, and $p_{ii} = 0$.

   (a) Let $\hat{\pi}_k$ be a pmf that satisfies $0 = \sum_k \hat{\pi}_k g_{kj}$, and put $\hat{D} := \sum_i \hat{\pi}_i g_{ii}$. If $\hat{D}$ is finite, show that $\pi_k := \hat{\pi}_k g_{kk}/\hat{D}$ is a pmf that satisfies $\pi_j = \sum_k \pi_k p_{kj}$,

   (b) Let $\check{\pi}_k$ be a pmf that satisfies $\check{\pi}_j = \sum_k \check{\pi}_k p_{kj}$, and put $\check{D} := \sum_i \check{\pi}_i/g_{ii}$. If $\check{D}$ is finite, show that $\pi_k := (\check{\pi}_k/g_{kk})/\check{D}$ is a pmf that satisfies $0 = \sum_k \pi_k g_{kj}$.

   (c) If g_{ii} does not depend on i, say $g_{ii} = g$, show that in (a) $\pi_k = \hat{\pi}_k$ and in (b) $\pi_k = \check{\pi}_k$. In other words, the stationary distributions of the continuous-time chain and the jump chain are the same when g_{ii} does not depend on i.

31. Let T denote the first time a chain leaves state i,

$$T := \min\{t \geq 0 : X_t \neq i\}.$$

Show that given $X_0 = i$, T is conditionally $\exp(-g_{ii})$. In other words, the time the chain spends in state i, known as the **sojourn time** or **holding time**, has an exponential density with parameter $-g_{ii}$. *Hints:* By Problem 50 in Chapter 5, it suffices to prove that

$$P(T > t + \Delta t | T > t, X_0 = i) = P(T > \Delta t | X_0 = i).$$

To derive this equation, use the fact that if X_t is right-continuous,

$$T > t \quad \text{if and only if} \quad X_s = i \text{ for } 0 \leq s \leq t.$$

Use the Markov property in the form

$$P(X_s = i, t \leq s \leq t + \Delta t | X_s = i, 0 \leq s \leq t)$$
$$= P(X_s = i, t \leq s \leq t + \Delta t | X_t = i),$$

and use time homogeneity in the form

$$P(X_s = i, t \leq s \leq t + \Delta t | X_t = i) = P(X_s = i, 0 \leq s \leq \Delta t | X_0 = i).$$

To identify the parameter of the exponential density, you may use the formula

$$\lim_{\Delta t \downarrow 0} \frac{1 - P(X_s = i, 0 \leq s \leq \Delta t | X_0 = i)}{\Delta t} = \lim_{\Delta t \downarrow 0} \frac{1 - P(X_{\Delta t} = i | X_0 = i)}{\Delta t}.$$

32. The notion of a Markov chain can be generalized to include random variables that are not necessarily discrete. We say that X_t is a continuous-time **Markov process** if for $0 < s_0 < \cdots < s_{n-1} < s < t$,

$$P(X_t \in B | X_s = x, X_{s_{n-1}} = x_{n-1}, \ldots, X_{s_0} = x_0) = P(X_t \in B | X_s = x).$$

Such a process is time homogeneous if $P(X_t \in B | X_s = x)$ depends on t and s only through $t - s$. Show that the Wiener process is a Markov process that is time homogeneous. *Hint:* It is enough to look at conditional cdfs; i.e., show that

$$P(X_t \leq y | X_s = x, X_{s_{n-1}} = x_{n-1}, \ldots, X_{s_0} = x_0) = P(X_t \leq y | X_s = x).$$

33. Let X, Y, and Z be discrete random variables. Show that the following law of total probability for conditional probability holds:

$$P(X = x | Z = z) = \sum_y P(X = x | Y = y, Z = z) P(Y = y | Z = z).$$

34. Let X_t be a time-homogeneous Markov process as defined in Problem 32. Put

$$P_t(x, B) := P(X_t \in B | X_0 = x),$$

and assume that there is a corresponding conditional density, denoted by $f_t(x, y) := f_{X_t | X_0}(y | x)$, such that

$$P_t(x, B) = \int_B f_t(x, y) \, dy.$$

Derive the Chapman–Kolmogorov equation for conditional densities,

$$f_{t+s}(x, y) = \int_{-\infty}^{\infty} f_s(x, z) f_t(z, y) \, dz.$$

Hint: It suffices to show that

$$P_{t+s}(x, B) = \int_{-\infty}^{\infty} f_s(x, z) P_t(z, B) \, dz.$$

To derive this, you may assume that a law of total conditional probability holds for random variables with appropriate conditional densities.

Exam preparation

You may use the following suggestions to prepare a study sheet, including formulas mentioned that you have trouble remembering. You may also want to ask your instructor for additional suggestions.

12.1. Preliminary results. Be familiar with the results of Examples 12.1 and 12.2 and how to apply them as in the rest of the chapter.

12.2. Discrete-time Markov chains. Be able to write down the transition matrix given the state transition diagram, and be able to draw the state transition diagram given the transition matrix. Know the meaning of the m-step transition probability $p_{ij}^{(m)}$ in (12.16). Know that stationarity of the one-step transition probabilities implies stationarity of the m-step transition probabilities as in (12.17). Know the Chapman–Kolmogorov equation (12.18) as well as the matrix formulation $P^{n+m} = P^n P^m$. Be able to find stationary distributions π_j using the conditions (12.19).

12.3. *Recurrent and transient states. Theorem 2 says that the random variable $V(j)$, which is the total number of visits to state j, is infinite with conditional probability one if j is recurrent and is a geometric$_0(f_{jj})$ random variable if j is transient. Formulas (12.36) and (12.37) give alternative characterizations of recurrent and transient states. Theorem 4 says that if an irreducible chain has a stationary distribution π, then all states are positive recurrent, and $\pi_j = 1/\mathsf{E}[T_1(j)|X_0 = j]$. The Ergodic Theorem says that if an irreducible chain has a stationary distribution π, then

$$\lim_{m \to \infty} \frac{1}{m} \sum_{k=1}^{m} h(X_k) = \sum_{j} h(j) \pi_j.$$

If the initial distribution of the chain is taken to be π, then $\mathsf{P}(X_k = j) = \pi_j$ for all k. In this case, right-hand side is equal to $\mathsf{E}[h(X_k)]$. Hence, the limiting time average of $h(X_k)$ converges to $\mathsf{E}[h(X_k)]$.

12.4. *Limiting n-step transition probabilities. In general, the state space of any chain can be partitioned into disjoint sets $T, R_1, R_2, \ldots$, where each R_i is a communicating class of recurrent states, and T is the union of all classes of transient states. When the entire state space belongs to a single class, the chain is irreducible. Know that communicating states are all either transient or recurrent and have the same period. (Hence, transience, recurrence, and periodicity are called class properties.) Be very familiar with the discussion at the end of the section.

12.5. Continuous-time Markov chains. To do derivations, you must know the Chapman–Kolmogorov equation (12.47). The elements of the generator matrix G are related to the $p_{ij}(t)$ by (12.48). Have a qualitative understanding of the behavior of continuous-time Markov chains in terms of the sojourn times and the embedded discrete-time chain. Be able to solve for the stationary distribution π_j.

Work any review problems assigned by your instructor. If you finish them, re-work your homework assignments.

13
Mean convergence and applications

As mentioned at the beginning of Chapter 1, limit theorems are the foundation of the success of Kolmogorov's axiomatic theory of probability. In this chapter and the next, we focus on four different notions of convergence and their implications. The four types of convergence are, in the order to be studied:

(*i*) convergence in mean of order p;
(*ii*) convergence in probability;
(*iii*) convergence in distribution; and
(*iv*) almost sure convergence.

When we say X_n converges to X, we usually understand this intuitively as what is known as almost-sure convergence. However, when we want to talk about moments, say $\mathsf{E}[X_n^2] \to \mathsf{E}[X^2]$, we need to exploit results based on convergence in mean of order 2. When we want to talk about probabilities, say $\mathsf{P}(X_n \in B) \to \mathsf{P}(X \in B)$, we need to exploit results based on convergence in distribution. Examples 14.8 and 14.9 are important applications that require both convergence in mean of order 2 and convergence in distribution. We must also mention that the central limit theorem, which we made extensive use of in Chapter 6 on confidence intervals, is a statement about convergence in distribution. Convergence in probability is a concept we have also been using for quite a while, e.g., the weak law of large numbers in Section 3.3.

The present chapter is devoted to the study of convergence in mean of order p, while the remaining types of convergence are studied in the next chapter.

Section 13.1 introduces the notion of convergence in mean of order p. There is also a discussion of continuity in mean of order p. Section 13.2 introduces the normed L^p spaces. Norms provide a compact notation for establishing results about convergence in mean of order p. We also point out that the L^p spaces are complete. Completeness is used to show that convolution sums like

$$\sum_{k=0}^{\infty} h_k X_{n-k}$$

are well defined. This is an important result because sums like this represent the response of a causal, linear, time-invariant system to a random input X_k. The section concludes with an introduction to mean-square integrals. Section 13.3 introduces the Karhunen–Loève expansion, which is of paramount importance in signal detection problems. Section 13.4 uses completeness to develop the Wiener integral. Section 13.5 introduces the notion of projections. The L^2 setting allows us to introduce a general orthogonality principle that unifies results from earlier chapters on the Wiener filter, linear estimators of random vectors, and minimum mean squared error estimation. The completeness of L^2 is also used to prove the projection theorem. In Section 13.6, the projection theorem is used to establish the existence of conditional expectation and conditional probability for random variables that may not be discrete or jointly continuous. In Section 13.7, completeness is used to establish the spectral representation of wide-sense stationary random sequences.

13.1 Convergence in mean of order p

We say that X_n **converges in mean of order p** to X if

$$\lim_{n \to \infty} \mathsf{E}[|X_n - X|^p] = 0,$$

where $1 \leq p < \infty$. Note that when X is zero, the expression simplifies to $\lim_{n \to \infty} \mathsf{E}[|X_n|^p] = 0$. Mostly we focus on the cases $p = 1$ and $p = 2$. The case $p = 1$ is called **convergence in mean** or **mean convergence**. The case $p = 2$ is called **mean-square convergence** or **quadratic-mean convergence**.

Example 13.1. Let $X_n \sim N(0, 1/n^2)$. Show that $\sqrt{n} X_n$ converges in mean square to zero.

Solution. Write

$$\mathsf{E}[|\sqrt{n} X_n|^2] = n \mathsf{E}[X_n^2] = n \frac{1}{n^2} = \frac{1}{n} \to 0.$$

In the next example, X_n converges in mean square to zero, but not in mean of order 4.

Example 13.2. Let X_n have density

$$f_n(x) = g_n(x)(1 - 1/n^3) + h_n(x)/n^3,$$

where $g_n \sim N(0, 1/n^2)$ and $h_n \sim N(n, 1)$. Show that X_n converges to zero in mean square, but not in mean of order 4.

Solution. For convergence in mean square, write

$$\mathsf{E}[|X_n|^2] = \frac{1}{n^2}(1 - 1/n^3) + (1 + n^2)/n^3 \to 0.$$

However, using Problem 28 in Chapter 4, we have

$$\mathsf{E}[|X_n|^4] = \frac{3}{n^4}(1 - 1/n^3) + (n^4 + 6n^2 + 3)/n^3 \to \infty.$$

The preceding example raises the question of whether X_n might converge in mean of order 4 to something other than zero. However, by Problem 9 at the end of the chapter, if X_n converged in mean of order 4 to some X, then it would also converge in mean square to X. Hence, the only possible limit for X_n in mean of order 4 is zero, and as we saw, X_n does not converge in mean of order 4 to zero.

13.1 Convergence in mean of order p

Example 13.3. Let $X_1, X_2, \ldots$ be uncorrelated random variables with common mean m and common variance σ^2. Show that the sample mean

$$M_n := \frac{1}{n} \sum_{i=1}^{n} X_i$$

converges in mean square to m. We call this the **mean-square law of large numbers** for uncorrelated random variables.

Solution. Since

$$M_n - m = \frac{1}{n} \sum_{i=1}^{n} (X_i - m),$$

we can write

$$\mathsf{E}[|M_n - m|^2] = \frac{1}{n^2} \mathsf{E}\left[\left(\sum_{i=1}^{n}(X_i - m)\right)\left(\sum_{j=1}^{n}(X_j - m)\right)\right]$$

$$= \frac{1}{n^2} \sum_{i=1}^{n} \sum_{j=1}^{n} \mathsf{E}[(X_i - m)(X_j - m)]. \quad (13.1)$$

Since X_i and X_j are uncorrelated, the preceding expectations are zero when $i \neq j$. Hence,

$$\mathsf{E}[|M_n - m|^2] = \frac{1}{n^2} \sum_{i=1}^{n} \mathsf{E}[(X_i - m)^2] = \frac{n\sigma^2}{n^2} = \frac{\sigma^2}{n},$$

which goes to zero as $n \to \infty$.

Example 13.4 (mean-square ergodic theorem). The preceding example gave a mean-square law of large numbers for uncorrelated sequences. This example provides a mean-square law of large numbers for wide-sense stationary sequences. Laws of large numbers for sequences that are not uncorrelated are called **ergodic theorems**. Let $X_1, X_2, \ldots$ be wide-sense stationary; i.e., the X_i have common mean $m = \mathsf{E}[X_i]$, and the covariance $\mathsf{E}[(X_i - m)(X_j - m)]$ depends only on the difference $i - j$. Put

$$C(i) := \mathsf{E}[(X_{j+i} - m)(X_j - m)].$$

Show that

$$M_n := \frac{1}{n} \sum_{i=1}^{n} X_i$$

converges in mean square to m if and only if

$$\lim_{n \to \infty} \frac{1}{n} \sum_{k=0}^{n-1} C(k) = 0. \quad (13.2)$$

Note that a sufficient condition for (13.2) to hold is that $\lim_{k \to \infty} C(k) = 0$ (Problem 3).

Solution. We show that (13.2) implies M_n converges in mean square to m. The converse is left to the reader in Problem 4. From (13.1), we see that

$$n^2 \mathsf{E}[|M_n - m|^2] = \sum_{i=1}^{n}\sum_{j=1}^{n} C(i-j)$$

$$= \sum_{i=j} C(0) + 2\sum_{j<i} C(i-j)$$

$$= nC(0) + 2\sum_{i=2}^{n}\sum_{j=1}^{i-1} C(i-j)$$

$$= nC(0) + 2\sum_{i=2}^{n}\sum_{k=1}^{i-1} C(k).$$

On account of (13.2), given $\varepsilon > 0$, there is an N such that for all $i \geq N$,

$$\left| \frac{1}{i-1} \sum_{k=1}^{i-1} C(k) \right| < \varepsilon.$$

For $n \geq N$, the double sum above can be written as

$$\sum_{i=2}^{n}\sum_{k=1}^{i-1} C(k) = \sum_{i=2}^{N-1}\sum_{k=1}^{i-1} C(k) + \sum_{i=N}^{n} (i-1)\left[\frac{1}{i-1}\sum_{k=1}^{i-1} C(k)\right].$$

The magnitude of the right-most double sum is upper bounded by

$$\left| \sum_{i=N}^{n} (i-1)\left[\frac{1}{i-1}\sum_{k=1}^{i-1} C(k)\right] \right| < \varepsilon \sum_{i=N}^{n} (i-1)$$

$$< \varepsilon \sum_{i=1}^{n} (i-1)$$

$$= \frac{\varepsilon n(n-1)}{2}$$

$$< \frac{\varepsilon n^2}{2}.$$

It now follows that $\lim_{n \to \infty} \mathsf{E}[|M_n - m|^2]$ can be no larger than ε. Since ε is arbitrary, the limit must be zero.

Example 13.5. Let W_t be a Wiener process with $\mathsf{E}[W_t^2] = \sigma^2 t$. Show that W_t/t converges in mean square to zero as $t \to \infty$.

Solution. Write

$$\mathsf{E}\left[\left|\frac{W_t}{t}\right|^2\right] = \frac{\sigma^2 t}{t^2} = \frac{\sigma^2}{t} \to 0.$$

13.1 Convergence in mean of order p

Example 13.6. Let X be a nonnegative random variable with finite mean; i.e., $\mathsf{E}[X] < \infty$. Put
$$X_n := \min(X, n) = \begin{cases} X, & X \le n, \\ n, & X > n. \end{cases}$$
The idea here is that X_n is a bounded random variable that can be used to approximate X. Show that X_n converges in mean to X.

Solution. Since $X \ge X_n$, $\mathsf{E}[|X_n - X|] = \mathsf{E}[X - X_n]$. Since $X - X_n$ is nonnegative, we can write
$$\mathsf{E}[X - X_n] = \int_0^\infty \mathsf{P}(X - X_n > t) \, dt,$$
where we have appealed to (5.16) in Section 5.7. Next, for $t \ge 0$, a little thought shows that
$$\{X - X_n > t\} = \{X > t + n\}.$$
Hence,
$$\mathsf{E}[X - X_n] = \int_0^\infty \mathsf{P}(X > t + n) \, dt = \int_n^\infty \mathsf{P}(X > \theta) \, d\theta,$$
which goes to zero as $n \to \infty$ on account of the fact that
$$\infty > \mathsf{E}[X] = \int_0^\infty \mathsf{P}(X > \theta) \, d\theta.$$

Continuity in mean of order p

A continuous-time process X_t is said to be **continuous in mean of order p** at t_0 if
$$\lim_{t \to t_0} \mathsf{E}[|X_t - X_{t_0}|^p] = 0.$$
If X_t is continuous in mean of order p for all t_0, we just say that X_t is continuous in mean of order p.

Example 13.7. Show that a Wiener process is mean-square continuous.

Solution. For $t > t_0$,
$$\mathsf{E}[(W_t - W_{t_0})^2] = \sigma^2(t - t_0),$$
while for $t < t_0$,
$$\mathsf{E}[(W_t - W_{t_0})^2] = \mathsf{E}[(W_{t_0} - W_t)^2] = \sigma^2(t_0 - t).$$
In either case, $\mathsf{E}[(W_t - W_{t_0})^2] = \sigma^2 |t - t_0|$ and goes to zero as $t \to t_0$.

Example 13.8. Show that a Poisson process of rate λ is continuous in mean.

Solution. For $t > t_0$,
$$\mathsf{E}[|N_t - N_{t_0}|] = \mathsf{E}[N_t - N_{t_0}] = \lambda(t - t_0),$$
while for $t < t_0$,
$$\mathsf{E}[|N_t - N_{t_0}|] = \mathsf{E}[N_{t_0} - N_t] = \lambda(t_0 - t).$$
In either case, $\mathsf{E}[|N_t - N_{t_0}|] = \lambda |t - t_0|$ and goes to zero as $t \to t_0$.

The preceding example is a surprising result since a Poisson process has jump discontinuities. However, it is important to keep in mind that the jump locations are random, and continuity in mean only says that the *expected* or *average* distance between N_t and N_{t_0} goes to zero.

We now focus on the case $p = 2$. If X_t has correlation function $R(t,s) := \mathsf{E}[X_t X_s]$, then X_t is mean-square continuous at t_0 if and only if $R(t,s)$ is continuous at (t_0,t_0). To show this, first suppose that $R(t,s)$ is continuous at (t_0,t_0) and write

$$\begin{aligned}\mathsf{E}[|X_t - X_{t_0}|^2] &= R(t,t) - 2R(t,t_0) + R(t_0,t_0) \\ &= [R(t,t) - R(t_0,t_0)] - 2[R(t,t_0) - R(t_0,t_0)].\end{aligned}$$

Then for t close to t_0, (t,t) is close to (t_0,t_0) and (t,t_0) is also close to (t_0,t_0). By continuity of $R(t,s)$ at (t_0,t_0), it follows that X_t is mean-square continuous at t_0. To prove the converse, suppose X_t is mean-square continuous at t_0 and write

$$\begin{aligned}R(t,s) - R(t_0,t_0) &= R(t,s) - R(t_0,s) + R(t_0,s) - R(t_0,t_0) \\ &= \mathsf{E}[(X_t - X_{t_0})X_s] + \mathsf{E}[X_{t_0}(X_s - X_{t_0})].\end{aligned}$$

Next, by the Cauchy–Schwarz inequality,

$$|\mathsf{E}[(X_t - X_{t_0})X_s]| \leq \sqrt{\mathsf{E}[|X_t - X_{t_0}|^2]\mathsf{E}[|X_s|^2]}$$

and

$$|\mathsf{E}[X_{t_0}(X_s - X_{t_0})]| \leq \sqrt{\mathsf{E}[|X_{t_0}|^2]\mathsf{E}[|X_s - X_{t_0}|^2]}.$$

For (t,s) close to (t_0,t_0), t will be close to t_0 and s will be close to t_0. By mean-square continuity at t_0, both $\mathsf{E}[|X_t - X_{t_0}|^2]$ and $\mathsf{E}[|X_s - X_{t_0}|^2]$ will be small. We also need the fact that $\mathsf{E}[|X_s|^2]$ is bounded for s near t_0 (see Problem 18 and the remark following it). It now follows that $R(t,s)$ is close to $R(t_0,t_0)$.

A similar argument shows that if X_t is mean-square continuous at all t_0, then $R(t,s)$ is continuous at all (τ,θ) (Problem 13).

13.2 Normed vector spaces of random variables

We denote by L^p the set of all random variables X with the property that $\mathsf{E}[|X|^p] < \infty$. We claim that L^p is a vector space. To prove this, we need to show that if $\mathsf{E}[|X|^p] < \infty$ and $\mathsf{E}[|Y|^p] < \infty$, then $\mathsf{E}[|aX + bY|^p] < \infty$ for all scalars a and b. To begin, recall that the **triangle inequality** applied to numbers x and y says that

$$|x + y| \leq |x| + |y|.$$

If $|y| \leq |x|$, then

$$|x + y| \leq 2|x|,$$

and so

$$|x + y|^p \leq 2^p |x|^p.$$

13.2 Normed vector spaces of random variables

A looser bound that has the advantage of being symmetric is

$$|x+y|^p \le 2^p(|x|^p + |y|^p).$$

It is easy to see that this bound also holds if $|y| > |x|$. We can now write

$$\begin{aligned} \mathsf{E}[|aX+bY|^p] &\le \mathsf{E}[2^p(|aX|^p + |bY|^p)] \\ &= 2^p(|a|^p \mathsf{E}[|X|^p] + |b|^p \mathsf{E}[|Y|^p]). \end{aligned}$$

Hence, if $\mathsf{E}[|X|^p]$ and $\mathsf{E}[|Y|^p]$ are both finite, then so is $\mathsf{E}[|aX+bY|^p]$.

For $X \in L^p$, we put

$$\|X\|_p := \mathsf{E}[|X|^p]^{1/p}.$$

We claim that $\|\cdot\|_p$ is a **norm** on L^p, by which we mean the following three properties hold.

(i) $\|X\|_p \ge 0$, and $\|X\|_p = 0$ if and only if X is the zero random variable.
(ii) For scalars a, $\|aX\|_p = |a|\,\|X\|_p$.
(iii) For $X, Y \in L^p$,

$$\|X+Y\|_p \le \|X\|_p + \|Y\|_p.$$

As in the numerical case, this is also known as the **triangle inequality**.

The first two properties are obvious, while the third one is known as **Minkowski's inequality**, which is derived in Problem 10.

Observe now that X_n converges in mean of order p to X if and only if

$$\lim_{n \to \infty} \|X_n - X\|_p = 0.$$

Hence, the three norm properties above can be used to derive results about convergence in mean of order p, as shown next.

***Example* 13.9.** If $X_n \sim N(0, 1/n^2)$ and $Y_n \sim \exp(n)$, show that $X_n - Y_n$ converges in mean of order 2 to zero.

Solution. We show below that X_n and Y_n each converge in mean of order 2 to zero; i.e., $\|X_n\|_2 \to 0$ and $\|Y_n\|_2 \to 0$. By writing

$$\|X_n - Y_n\|_2 \le \|X_n\|_2 + \|Y_n\|_2,$$

it then follows that $X_n - Y_n$ converges in mean of order 2 to zero. It now remains to observe that since

$$\mathsf{E}[X_n^2] = 1/n^2 \quad \text{and} \quad \mathsf{E}[Y_n^2] = 2/n^2,$$

$\|X_n\|_2 = 1/n \to 0$ and $\|Y_n\|_2 = \sqrt{2}/n \to 0$ as claimed.

Recall that a sequence of real numbers x_n is **Cauchy** if for every $\varepsilon > 0$, for all sufficiently large n and m, $|x_n - x_m| < \varepsilon$. A basic fact that can be proved about the set of real numbers is that it is **complete**; i.e., given any Cauchy sequence of real numbers x_n, there is a real number x such that x_n converges to x [51, p. 53, Theorem 3.11]. Similarly, a sequence of random variables $X_n \in L^p$ is said to be **Cauchy** if for every $\varepsilon > 0$, for all sufficiently large n and m,

$$\|X_n - X_m\|_p < \varepsilon.$$

It can be shown that the L^p spaces are complete; i.e., if X_n is a Cauchy sequence of L^p random variables, then there exists an L^p random variable X such that X_n converges in mean of order p to X. This is known as the **Riesz–Fischer theorem** [50, p. 244]. A normed vector space that is complete is called a **Banach space**.

Of special interest is the case $p = 2$ because the norm $\|\cdot\|_2$ can be expressed in terms of the **inner product**[a]

$$\langle X, Y \rangle := \mathsf{E}[XY], \quad X, Y \in L^2.$$

It is easily seen that

$$\langle X, X \rangle^{1/2} = \|X\|_2.$$

Because the norm $\|\cdot\|_2$ can be obtained using the inner product, L^2 is called an **inner-product space**. Since the L^p spaces are complete, L^2 in particular is a complete inner-product space. A complete inner-product space is called a **Hilbert space**.

The space L^2 has several important properties. First, for fixed Y, it is easy to see that $\langle X, Y \rangle$ is linear in X. Second, the simple relationship between the norm and the inner product implies the **parallelogram law** (Problem 23),

$$\|X+Y\|_2^2 + \|X-Y\|_2^2 = 2(\|X\|_2^2 + \|Y\|_2^2). \tag{13.3}$$

Third, there is the **Cauchy–Schwarz inequality**

$$|\langle X, Y \rangle| \leq \|X\|_2 \|Y\|_2,$$

which was derived in Chapter 2.

Example 13.10. Show that

$$\sum_{k=1}^{\infty} h_k X_k$$

is well defined as an element of L^2 assuming that

$$\sum_{k=1}^{\infty} |h_k| < \infty \quad \text{and} \quad \mathsf{E}[|X_k|^2] \leq B, \quad \text{for all } k,$$

---

[a]For complex-valued random variables (defined in Section 9.6), we put $\langle X, Y \rangle := \mathsf{E}[XY^*]$.

13.2 Normed vector spaces of random variables

where B is a finite constant.

Solution. Consider the partial sums,

$$Y_n := \sum_{k=1}^{n} h_k X_k.$$

Observe that each Y_n is an element of L^2, which is complete. If we can show that Y_n is a Cauchy sequence, then there will exist a $Y \in L^2$ with $\|Y_n - Y\|_2 \to 0$. Thus, the infinite-sum expression $\sum_{k=1}^{\infty} h_k X_k$ is understood to be shorthand for "the mean-square limit of $\sum_{k=1}^{n} h_k X_k$ as $n \to \infty$." Next, for $n > m$, write

$$Y_n - Y_m = \sum_{k=m+1}^{n} h_k X_k.$$

Then

$$\begin{aligned}
\|Y_n - Y_m\|_2^2 &= \langle Y_n - Y_m, Y_n - Y_m \rangle \\
&= \left\langle \sum_{k=m+1}^{n} h_k X_k, \sum_{l=m+1}^{n} h_l X_l \right\rangle \\
&\leq \sum_{k=m+1}^{n} \sum_{l=m+1}^{n} |h_k| |h_l| |\langle X_k, X_l \rangle| \\
&\leq \sum_{k=m+1}^{n} \sum_{l=m+1}^{n} |h_k| |h_l| \|X_k\|_2 \|X_l\|_2,
\end{aligned}$$

by the Cauchy–Schwarz inequality. Next, since $\|X_k\|_2 = \mathsf{E}[|X_k|^2]^{1/2} \leq \sqrt{B}$,

$$\|Y_n - Y_m\|_2^2 \leq B \left(\sum_{k=m+1}^{n} |h_k| \right)^2.$$

Since $\sum_{k=1}^{\infty} |h_k| < \infty$ implies[1]

$$\sum_{k=m+1}^{n} |h_k| \to 0 \quad \text{as } n \text{ and } m \to \infty \text{ with } n > m,$$

it follows that Y_n is Cauchy.

Now that we know $\sum_{k=1}^{\infty} h_k X_k$ is well defined (under the assumptions of the example), the next obvious calculation to perform is

$$\mathsf{E}\left[\sum_{k=1}^{\infty} h_k X_k \right] = \sum_{k=1}^{\infty} h_k \mathsf{E}[X_k].$$

Are we justified in pushing the expectation through the infinite sum? It turns out that we are, but to prove it requires the following result.

Example 13.11. Show that if Y_n converges in mean of order p to Y, then

$$\lim_{n \to \infty} \mathsf{E}[Y_n] = \mathsf{E}[Y].$$

Solution. To begin, write

$$\big|\mathsf{E}[Y_n] - \mathsf{E}[Y]\big| = \big|\mathsf{E}[Y_n - Y]\big| \leq \mathsf{E}[|Y_n - Y|],$$

where the inequality follows by Example 4.25. This last expectation goes to zero because, by Problem 9, convergence in mean of order p implies convergence in mean of order one.

In working Example 13.10, we had the partial sum Y_n converging in mean square to the infinite sum Y. Therefore, by Example 13.11, $\mathsf{E}[Y_n] \to \mathsf{E}[Y]$. Substituting the appropriate sums for Y_n and Y, we have

$$\lim_{n \to \infty} \mathsf{E}\left[\sum_{k=1}^{n} h_k X_k\right] = \mathsf{E}\left[\sum_{k=1}^{\infty} h_k X_k\right].$$

Now notice that by linearity,

$$\lim_{n \to \infty} \mathsf{E}\left[\sum_{k=1}^{n} h_k X_k\right] = \lim_{n \to \infty} \sum_{k=1}^{n} \mathsf{E}[h_k X_k] =: \sum_{k=1}^{\infty} \mathsf{E}[h_k X_k].$$

The foregoing example and discussion have the following application. Consider a discrete-time, causal, stable, linear, time-invariant system with impulse response h_k. Now suppose that the random sequence X_k is applied to the input of this system. If $\mathsf{E}[|X_k|^2]$ is bounded as in the example, then[b] the output of the system at time n is

$$\sum_{k=0}^{\infty} h_k X_{n-k},$$

which is a well-defined element of L^2. Furthermore,

$$\mathsf{E}\left[\sum_{k=0}^{\infty} h_k X_{n-k}\right] = \sum_{k=0}^{\infty} h_k \mathsf{E}[X_{n-k}].$$

Mean-square integrals

We sketch the construction of integrals of the form $\int_a^b X_t \, dt$ for a zero-mean, mean-square continuous process X_t with correlation function $R(t,s)$.

First, since $R(t,s)$ is continuous, if $a = t_0 < t_1 < \cdots < t_n = b$ is a partition of $[a,b]$ such that the differences $t_i - t_{i-1}$ are sufficiently small, and if $\tau_i \in [t_{i-1}, t_i]$, then the **Riemann sum**

$$\sum_{i=1}^{n} \sum_{j=1}^{n} R(\tau_i, \tau_j)(t_i - t_{i-1})(t_j - t_{j-1}) \tag{13.4}$$

---

[b] The assumption in the example, $\sum_k |h_k| < \infty$, is equivalent to the assumption that the system is stable.

is close to the double integral

$$\int_a^b \int_a^b R(t,s)\,dt\,ds. \quad (13.5)$$

Next consider the stochastic Riemann sum,

$$Y = \sum_{i=1}^n X_{\tau_i}(t_i - t_{i-1}), \quad (13.6)$$

and note that

$$\mathsf{E}[Y^2] = \mathsf{E}\left[\left(\sum_{i=1}^n X_{\tau_i}(t_i - t_{i-1})\right)\left(\sum_{j=1}^n X_{\tau_j}(t_j - t_{j-1})\right)\right]$$

is exactly (13.4).

Now consider a sequence of partitions of $[a,b]$. For the mth partition, form the corresponding stochastic Riemann sum Y_m. If as $m \to \infty$, $\max_i t_i^{(m)} - t_{i-1}^{(m)} \to 0$, then $\mathsf{E}[Y_m^2]$ converges to (13.5); denote this double integral by I. A slight generalization of the foregoing (Problem 27) shows that as $m, k \to \infty$, $\mathsf{E}[Y_m Y_k] \to I$ as well. Hence,

$$\mathsf{E}[|Y_m - Y_k|^2] = \mathsf{E}[Y_m^2] - 2\mathsf{E}[Y_m Y_k] + \mathsf{E}[Y_k^2] \to I - 2I + I = 0.$$

Thus Y_m is Cauchy in L^2. Hence, there is a limit, which we denote by $\int_a^b X_t\,dt$. Since each Y_m is zero mean, so is the limit (Example 13.11). Furthermore, the second moment of this limit is the limit of the second moments of the Y_m (Problem 22). It follows that the second moment of $\int_a^b X_t\,dt$ is (13.5).

Remark. The reader may wonder why the derivation of the mean-square integral is so short, while the development of the Riemann integral in calculus courses is so long. The answer is that we take the existence of the double Riemann integral for granted; i.e., we exploit the fact that (13.4) gets closer to (13.5) as the partition intervals become small.

13.3 The Karhunen–Loève expansion

The **Karhunen–Loève expansion** says that if a zero-mean process X_t is mean-square continuous for $a \leq t \leq b$, then

$$X_t = \sum_{k=1}^\infty A_k \varphi_k(t), \quad (13.7)$$

where the A_k are uncorrelated random variables, and the φ_k are deterministic, orthonormal time functions. In fact, if X_t is a Gaussian process, the A_k are jointly Gaussian (Problem 19 in Chapter 14), and therefore independent. The reason this expansion is so useful is that all the randomness is collected in the sequence A_k, and all the time dependence is collected in the nonrandom functions $\varphi_k(t)$.

A typical application of the Karhunen–Loève expansion is in the design of receivers for communication systems. In this case, the received waveform is X_t, which is typically fed into a bank of matched filters and sampled at time t_0 to yield a vector of statistics. If we take the mth matched-filter impulse response to be $h_m(t) = \varphi_m(t_0 - t)$, then

$$\left(\int_a^b h_m(t-s)X_s\,ds\right)\bigg|_{t=t_0} = \int_a^b h_m(t_0 - s)X_s\,ds$$
$$= \int_a^b \varphi_m(s)X_s\,ds$$
$$= \int_a^b \varphi_m(s)\left(\sum_{k=1}^\infty A_k\varphi_k(s)\right)ds$$
$$= \sum_{k=1}^\infty A_k \int_a^b \varphi_m(s)\varphi_k(s)\,ds$$
$$= A_m,$$

where the last step follows because the φ_k are orthonormal. In other words, the Karhunen–Loève expansion tells us how to create a bank of matched filters that when sampled yields a vector whose components are uncorrelated, and in the Gaussian case independent.

Here is a sketch of the technical details. The expansion (13.7) is understood in the mean-square sense; i.e., for each $a \le t \le b$,

$$\sum_{k=1}^n A_k\varphi_k(t) \qquad (13.8)$$

converges in mean square to X_t as $n \to \infty$. Next, with $R(t,s) := \mathsf{E}[X_t X_s]$, the φ_k are the eigenfunctions that solve

$$\boxed{\int_a^b R(t,s)\varphi_k(s)\,ds = \lambda_k\varphi_k(t), \quad a \le t \le b,} \qquad (13.9)$$

for the kth **eigenvalue** λ_k. It turns out that the eigenfunctions can be taken to be **orthonormal**; i.e.,

$$\boxed{\int_a^b \varphi_k(t)\varphi_m(t)\,dt = \delta_{km},} \qquad (13.10)$$

Once the eigenfunctions are known, the coefficients A_k are given by the stochastic integral

$$\boxed{A_k := \int_a^b X_s\,\varphi_k(s)\,ds.} \qquad (13.11)$$

We now show that (13.8) converges in mean square to X_t. Denoting (13.8) by Y_n, write

$$\mathsf{E}[|Y_n - X_t|^2] = \mathsf{E}[Y_n^2] - 2\mathsf{E}[Y_n X_t] + \mathsf{E}[X_t^2].$$

The last term is just $R(t,t)$. The cross term is

$$\mathsf{E}[Y_n X_t] = \sum_{k=1}^n \varphi_k(t)\mathsf{E}[A_k X_t],$$

13.3 The Karhunen–Loève expansion

where

$$E[A_k X_t] = E\left[\left(\int_a^b X_s \varphi_k(s)\,ds\right) X_t\right]$$

$$= \int_a^b E[X_s X_t] \varphi_k(s)\,ds$$

$$= \int_a^b R(t,s) \varphi_k(s)\,ds$$

$$= \lambda_k \varphi_k(t), \quad \text{by (13.9)}. \tag{13.12}$$

Hence,

$$E[Y_n X_t] = \sum_{k=1}^n \lambda_k \varphi_k(t) \varphi_k(t) = \sum_{k=1}^n \lambda_k |\varphi_k(t)|^2.$$

To compute $E[Y_n^2]$, write

$$E[Y_n^2] = E\left[\left(\sum_{k=1}^n A_k \varphi_k(t)\right)\left(\sum_{m=1}^n A_m \varphi_m(t)\right)\right]$$

$$= \sum_{k=1}^n \sum_{m=1}^n E[A_k A_m] \varphi_k(t) \varphi_m(t),$$

where

$$E[A_k A_m] = E\left[A_k \left(\int_a^b X_t \varphi_m(t)\,dt\right)\right]$$

$$= \int_a^b E[A_k X_t] \varphi_m(t)\,dt$$

$$= \int_a^b \lambda_k \varphi_k(t) \varphi_m(t)\,dt, \quad \text{by (13.12)},$$

$$= \lambda_k \delta_{km}, \quad \text{by (13.10)}.$$

Thus,[c]

$$E[Y_n^2] = \sum_{k=1}^n \lambda_k \varphi_k(t) \varphi_k(t) = \sum_{k=1}^n \lambda_k |\varphi_k(t)|^2.$$

Putting this all together, we have

$$E[|Y_n - X_t|^2] = R(t,t) - \sum_{k=1}^n \lambda_k \varphi_k(t) \varphi_k(t).$$

As we show next, this goes to zero by **Mercer's theorem** [19, Chapter IV].

It is a result of functional analysis [19, Chapter III] that whenever $R(t,s) = R(s,t)$ and

$$\int_a^b \int_a^b |R(t,s)|^2\,dt\,ds < \infty, \tag{13.13}$$

---

[c] It is shown in Problem 30 that the λ_k are nonnegative.

there exists a sequence of eigenvalues λ_k and corresponding orthonormal eigenfunctions φ_k satisfying (13.9). Mercer's theorem is a deeper result, which requires two additional hypotheses. The first hypothesis is that

$$\int_a^b \int_a^b R(t,s) g(t) g(s) \, dt \, ds \geq 0 \qquad (13.14)$$

for all square-integrable functions g. To verify this, put $Y := \int_a^b g(t) X_t \, dt$, and note that

$$0 \leq \mathsf{E}[Y^2]$$
$$= \mathsf{E}\left[\left(\int_a^b g(t) X_t \, dt\right)\left(\int_a^b g(s) X_s \, ds\right)\right]$$
$$= \int_a^b \int_a^b R(t,s) g(t) g(s) \, dt \, ds.$$

The second hypothesis of Mercer's theorem is that $R(t,s)$ be continuous. This is guaranteed by the assumption that X_t is mean-square continuous. Note that since $R(t,s)$ is continuous, it is bounded for $a \leq t, s \leq b$ [51, p. 89, Theorem 4.15], and therefore (13.13) also holds. Mercer's theorem says that as $n \to \infty$,

$$\sum_{k=1}^n \lambda_k \varphi_k(t) \varphi_k(s)$$

converges absolutely and uniformly to $R(t,s)$ for $a \leq t, s \leq b$.

Remark. The matrix version of Mercer's theorem and the discrete-time version of the Karhunen–Loève expansion are relatively simple. Just use the decorrelating transformation of Section 8.2 and see (8.6).

To conclude the discussion, note that if $R(t,s) = R(t-s)$ is the correlation function of a WSS process with power spectral density $S(f)$ satisfying $0 < \int_{-\infty}^{\infty} S(f) \, df < \infty$, then the φ_k form a **complete orthonormal set** in the sense that every square-integrable function g on $[a,b]$ satisfies

$$g(t) = \sum_{k=1}^\infty g_k \varphi_k(t), \qquad (13.15)$$

where

$$g_k = \int_a^b g(t) \varphi_k(t) \, dt,$$

and the convergence of (13.15) is in the sense that

$$\lim_{n \to \infty} \int_a^b \left| \sum_{k=1}^n g_k \varphi_k(t) - g(t) \right|^2 dt = 0.$$

See [69, Appendix A].

Example 13.12 (signal detection). To transmit a one-bit message, the known, square-integrable signal $g_i(t)$, $0 \leq t \leq T$, $i = 0, 1$, is transmitted over a channel with additive,

13.3 The Karhunen–Loève expansion

zero-mean, WSS noise X_t having power spectral density $S(f)$. Design a receiver to detect the transmitted message.

Solution. The received signal is $Y_t = g_i(t) + X_t$ for $0 \leq t \leq T$. Using the eigenfunctions of the Karhunen–Loève expansion of X_t, Y_t has the representation[2]

$$Y_t = \sum_{k=1}^{\infty} g_{ik}\varphi_k(t) + \sum_{k=1}^{\infty} A_k \varphi_k(t),$$

where $g_{ik} = \int_0^T g_i(t)\varphi_k(t)\,dt$. The receiver passes this signal through a bank of matched filters as discussed at the beginning of the section (take $t_0 = T$). The mth sampled filter output is

$$Z_m := \int_0^T \varphi_m(t) Y_t \, dt = g_{im} + A_m.$$

As a practical matter, the filter bank can have at most a finite number of filters, say M. We then stack $Z_1, \ldots, Z_M$ into an M-dimensional random vector and write down the corresponding likelihood-ratio test (recall Example 5.9 and the discussion following it). If the noise process is Gaussian, the A_m will be jointly Gaussian, and the likelihood ratio takes a simple form, e.g., Problem 17 in Chapter 5.

If the φ_k do not form a complete orthonormal set, then noiseless signal detection is possible, as shown in Problem 32.

Example 13.13 (white noise). Let X_t be zero-mean white noise with correlation function $R(t,s) = \delta(t-s)$. Find the eigenvalues and eigenfunctions for the Karhunen–Loève expansion of X_t for $0 \leq t \leq T$.

Solution. Since the process is not mean-square continuous, we cannot, strictly speaking, apply the expansion. However, let us proceed formally. The eigenvalue problem is

$$\int_0^T R(t,s)\varphi(s)\,ds = \lambda \varphi(t), \quad 0 \leq t \leq T. \tag{13.16}$$

Since $R(t,s) = \delta(t-s)$,

$$\int_0^T R(t,s)\varphi(s)\,ds = \int_0^T \delta(t-s)\varphi(s)\,ds = \varphi(t),$$

and so the eigenvalue problem reduces to $\varphi(t) = \lambda \varphi(t)$. Hence, the only choice for λ is $\lambda = 1$, and every nonzero function is an eigenfunction.

Example 13.14 (Wiener process). Find the eigenvalues and eigenfunctions for the Karhunen–Loève expansion of the standard Wiener process on $[0,T]$.

Solution. Recall that for the standard Wiener process, $R(t,s) = \min(t,s)$. Hence, the eigenvalue problem (13.16) is

$$\lambda \varphi(t) = \int_0^T \min(t,s) \varphi(s)\,ds$$

$$= \int_0^t \min(t,s)\varphi(s)\,ds + \int_t^T \min(t,s)\varphi(s)\,ds$$
$$= \int_0^t s\varphi(s)\,ds + \int_t^T t\varphi(s)\,ds$$
$$= \int_0^t s\varphi(s)\,ds + t\int_t^T \varphi(s)\,ds. \tag{13.17}$$

Differentiating with respect to t yields
$$\lambda \varphi'(t) = t\varphi(t) - t\varphi(t) + \int_t^T \varphi(s)\,ds = \int_t^T \varphi(s)\,ds. \tag{13.18}$$

Differentiating again yields $\lambda \varphi''(t) = -\varphi(t)$, or
$$\varphi''(t) + \frac{1}{\lambda}\varphi(t) = 0.$$

It is easily checked that the solution of this differential equation is
$$\varphi(t) = \alpha \cos(t/\sqrt{\lambda}) + \beta \sin(t/\sqrt{\lambda}),$$
where α and β are constants to be determined. From (13.17), $\varphi(0) = 0$, which implies $\alpha = 0$. From (13.18), $\varphi'(T) = 0$, which implies
$$\frac{\beta}{\sqrt{\lambda}}\cos(T/\sqrt{\lambda}) = 0.$$

Since an eigenfunction cannot be the zero function, $\beta = 0$ is not an option. The only other possibility is that $T/\sqrt{\lambda}$ be an odd multiple of $\pi/2$. Hence, for $n = 1, 2, \ldots$,
$$\lambda_n = \left[\frac{2T}{(2n-1)\pi}\right]^2,$$
and
$$\varphi_n(t) = \sqrt{\frac{2}{T}}\sin\left[\frac{(2n-1)\pi}{2T}t\right],$$
where the coefficient $\sqrt{2/T}$ is chosen so that $\int_0^T |\varphi_n(t)|^2\,dt = 1$.

13.4 The Wiener integral (again)

In Section 11.3, we defined the Wiener integral
$$\int_0^\infty g(\tau)\,dW_\tau := \sum_i g_i(W_{t_{i+1}} - W_{t_i}),$$
for piecewise constant g, say $g(\tau) = g_i$ for $t_i < t \le t_{i+1}$ for a finite number of intervals, and $g(\tau) = 0$ otherwise. In this case, since the integral is the sum of scaled, independent, zero mean, Gaussian increments of variance $\sigma^2(t_{i+1} - t_i)$,
$$\mathsf{E}\left[\left(\int_0^\infty g(\tau)\,dW_\tau\right)^2\right] = \sigma^2 \sum_i g_i^2(t_{i+1} - t_i) = \sigma^2 \int_0^\infty g(\tau)^2\,d\tau.$$

13.4 The Wiener integral (again)

We now define the Wiener integral for arbitrary g satisfying

$$\int_0^\infty g(\tau)^2 \, d\tau < \infty. \tag{13.19}$$

To do this, we use the fact [14, p. 86, Prop. 3.4.2] that for g satisfying (13.19), there always exists a sequence of piecewise-constant functions g_n converging to g in the mean-square sense

$$\lim_{n\to\infty} \int_0^\infty |g_n(\tau) - g(\tau)|^2 \, d\tau = 0. \tag{13.20}$$

The set of g satisfying (13.19) is an inner product space if we use the inner product $\langle g, h \rangle = \int_0^\infty g(\tau) h(\tau) \, d\tau$. The corresponding norm is $\|g\| = \langle g, g \rangle^{1/2}$. Thus, (13.20) implies $\|g_n - g\| \to 0$. In particular, this implies g_n is Cauchy; i.e., $\|g_n - g_m\| \to 0$ as $n, m \to \infty$ (cf. Problem 17). Consider the random variables

$$Y_n := \int_0^\infty g_n(\tau) \, dW_\tau.$$

Since each g_n is piecewise constant, Y_n is well defined and is Gaussian with zero mean and variance

$$\sigma^2 \int_0^\infty g_n(\tau)^2 \, d\tau.$$

Now observe that

$$\begin{aligned}
\|Y_n - Y_m\|_2^2 &= \mathsf{E}[|Y_n - Y_m|^2] \\
&= \mathsf{E}\left[\left|\int_0^\infty g_n(\tau) \, dW_\tau - \int_0^\infty g_m(\tau) \, dW_\tau\right|^2\right] \\
&= \mathsf{E}\left[\left|\int_0^\infty [g_n(\tau) - g_m(\tau)] \, dW_\tau\right|^2\right] \\
&= \sigma^2 \int_0^\infty |g_n(\tau) - g_m(\tau)|^2 \, d\tau,
\end{aligned}$$

since $g_n - g_m$ is piecewise constant. Thus,

$$\|Y_n - Y_m\|_2^2 = \sigma^2 \|g_n - g_m\|^2.$$

Since g_n is Cauchy, we see that Y_n is too. Since L^2 is complete, there exists a random variable $Y \in L^2$ with $\|Y_n - Y\|_2 \to 0$. We denote this random variable by

$$\int_0^\infty g(\tau) \, dW_\tau,$$

and call it the Wiener integral of g.

13.5 Projections, orthogonality principle, projection theorem

Let C be a subset of L^p. Given $X \in L^p$, suppose we want to approximate X by some $\widehat{X} \in C$. We call $\widehat{X}$ a **projection** of X onto C if $\widehat{X} \in C$ and if

$$\|X - \widehat{X}\|_p \leq \|X - Y\|_p, \quad \text{for all } Y \in C.$$

Note that if $X \in C$, then we can take $\widehat{X} = X$.

Example 13.15. Let C be the unit ball,

$$C := \{Y \in L^p : \|Y\|_p \leq 1\}.$$

For $X \notin C$, i.e., $\|X\|_p > 1$, show that

$$\widehat{X} = \frac{X}{\|X\|_p}.$$

Solution. First note that the proposed formula for $\widehat{X}$ satisfies $\|\widehat{X}\|_p = 1$ so that $\widehat{X} \in C$ as required. Now observe that

$$\|X - \widehat{X}\|_p = \left\| X - \frac{X}{\|X\|_p} \right\|_p = \left| 1 - \frac{1}{\|X\|_p} \right| \|X\|_p = \|X\|_p - 1.$$

Next, for any $Y \in C$,

$$\begin{aligned} \|X - Y\|_p &\geq \left| \|X\|_p - \|Y\|_p \right|, \quad \text{by Problem 21,} \\ &= \|X\|_p - \|Y\|_p \\ &\geq \|X\|_p - 1 \\ &= \|X - \widehat{X}\|_p. \end{aligned}$$

Thus, no $Y \in C$ is closer to X than $\widehat{X}$.

Much more can be said about projections when $p = 2$ and when the set we are projecting onto is a subspace rather than an arbitrary subset.

We now present two fundamental results about projections onto subspaces of L^2. The first result is the **orthogonality principle**.

Let M be a subspace of L^2. If $X \in L^2$, then $\widehat{X} \in M$ satisfies

$$\|X - \widehat{X}\|_2 \leq \|X - Y\|_2, \quad \text{for all } Y \in M, \tag{13.21}$$

if and only if

$$\langle X - \widehat{X}, Y \rangle = 0, \quad \text{for all } Y \in M. \tag{13.22}$$

Furthermore, if such an $\widehat{X} \in M$ exists, it is unique.

13.5 Projections, orthogonality principle, projection theorem

Observe that there is no claim that an $\widehat{X} \in M$ exists that satisfies either (13.21) or (13.22). In practice, we try to find an $\widehat{X} \in M$ satisfying (13.22), since such an $\widehat{X}$ then automatically satisfies (13.21). This was the approach used to derive the Wiener filter in Section 10.8, where we implicitly took (for fixed t)

$$M = \{\widehat{V}_t : \widehat{V}_t \text{ is given by (10.31) and } \mathsf{E}[\widehat{V}_t^2] < \infty\}.$$

In Section 8.4, when we discussed linear estimation of random vectors, we implicitly took

$$M = \{AY + b : A \text{ is a matrix and } b \text{ is a column vector }\}.$$

When we discussed minimum mean squared error estimation in Section 8.6, we implicitly took

$$M = \{g(Y) : g \text{ is any function such that } \mathsf{E}[g(Y)^2] < \infty\}.$$

Thus, several estimation problems discussed in earlier chapters are seen to be special cases of finding the projection onto a suitable subspace of L^2. Each of these special cases had its version of the orthogonality principle, and so it should be no trouble for the reader to show that (13.22) implies (13.21). The converse is also true, as we now show. Suppose (13.21) holds, but for some $Y \in M$,

$$\langle X - \widehat{X}, Y \rangle = c \neq 0.$$

Because we can divide this equation by $\|Y\|_2$, there is no loss of generality in assuming $\|Y\|_2 = 1$. Now, since M is a subspace containing both $\widehat{X}$ and Y, $\widehat{X} + cY$ also belongs to M. We show that this new vector is strictly closer to X than $\widehat{X}$, contradicting (13.21). Write

$$\begin{aligned} \|X - (\widehat{X} + cY)\|_2^2 &= \|(X - \widehat{X}) - cY\|_2^2 \\ &= \|X - \widehat{X}\|_2^2 - |c|^2 - |c|^2 + |c|^2 \\ &= \|X - \widehat{X}\|_2^2 - |c|^2 \\ &< \|X - \widehat{X}\|_2^2. \end{aligned}$$

The second fundamental result to be presented is the **projection theorem**. Recall that the orthogonality principle does not guarantee the existence of an $\widehat{X} \in M$ satisfying (13.21). If we are not smart enough to solve (13.22), what can we do? This is where the projection theorem comes in. To state and prove this result, we need the concept of a **closed set**. We say that M is closed if whenever $X_n \in M$ and $\|X_n - X\|_2 \to 0$ for some $X \in L^2$, the limit X must actually be in M. In other words, a set is closed if it contains all the limits of all converging sequences from the set.

***Example* 13.16.** Show that the set of Wiener integrals

$$M := \left\{ \int_0^\infty g(\tau) \, dW_\tau : \int_0^\infty g(\tau)^2 \, d\tau < \infty \right\}$$

is closed.

Solution. A sequence X_n from M has the form

$$X_n = \int_0^\infty g_n(\tau) \, dW_\tau$$

for square-integrable g_n. Suppose X_n converges in mean square to some X. We must show that there exists a square-integrable function g for which

$$X = \int_0^\infty g(\tau)\, dW_\tau.$$

Since X_n converges, it is Cauchy (Problem 17). Now observe that

$$\begin{aligned}
\|X_n - X_m\|_2^2 &= E[|X_n - X_m|^2] \\
&= E\left[\left|\int_0^\infty g_n(\tau)\, dW_\tau - \int_0^\infty g_m(\tau)\, dW_\tau\right|^2\right] \\
&= E\left[\left|\int_0^\infty [g_n(\tau) - g_m(\tau)]\, dW_\tau\right|^2\right] \\
&= \sigma^2 \int_0^\infty |g_n(\tau) - g_m(\tau)|^2\, d\tau \\
&= \sigma^2 \|g_n - g_m\|^2.
\end{aligned}$$

Thus, g_n is Cauchy. Since the set of square-integrable time functions is complete (the Riesz–Fischer theorem again [50, p. 244]), there is a square-integrable g with $\|g_n - g\| \to 0$. For this g, write

$$\begin{aligned}
\left\|X_n - \int_0^\infty g(\tau)\, dW_\tau\right\|_2^2 &= E\left[\left|\int_0^\infty g_n(\tau)\, dW_\tau - \int_0^\infty g(\tau)\, dW_\tau\right|^2\right] \\
&= E\left[\left|\int_0^\infty [g_n(\tau) - g(\tau)]\, dW_\tau\right|^2\right] \\
&= \sigma^2 \int_0^\infty |g_n(\tau) - g(\tau)|^2\, d\tau \\
&= \sigma^2 \|g_n - g\|^2 \to 0.
\end{aligned}$$

Since mean-square limits are unique (Problem 20), $X = \int_0^\infty g(\tau)\, dW_\tau$.

Remark. The argument in the preceding example also shows that the set of Wiener integrals is complete.

Projection theorem. *If M is a closed subspace of L^2, and $X \in L^2$, then there exists a unique $\widehat{X} \in M$ such that (13.21) holds.*

To prove this result, first put $h := \inf_{Y \in M} \|X - Y\|_2$. From the definition of the infimum, there is a sequence $Y_n \in M$ with $\|X - Y_n\|_2 \to h$. We will show that Y_n is a Cauchy sequence. Since L^2 is a Hilbert space, Y_n converges to some limit in L^2. Since M is closed, the limit, say $\widehat{X}$, must be in M.

To show Y_n is Cauchy, we proceed as follows. By the parallelogram law,

$$\begin{aligned}
2(\|X - Y_n\|_2^2 + \|X - Y_m\|_2^2) &= \|2X - (Y_n + Y_m)\|_2^2 + \|Y_m - Y_n\|_2^2 \\
&= 4\left\|X - \frac{Y_n + Y_m}{2}\right\|_2^2 + \|Y_m - Y_n\|_2^2.
\end{aligned}$$

Note that the vector $(Y_n + Y_m)/2 \in M$ since M is a subspace. Hence,

$$2(\|X - Y_n\|_2^2 + \|X - Y_m\|_2^2) \geq 4h^2 + \|Y_m - Y_n\|_2^2.$$

Since $\|X - Y_n\|_2 \to h$, given $\varepsilon > 0$, there exists an N such that for all $n \geq N$, $\|X - Y_n\|_2 < h + \varepsilon$. Hence, for $m, n \geq N$,

$$\begin{aligned}\|Y_m - Y_n\|_2^2 &< 2((h+\varepsilon)^2 + (h+\varepsilon)^2) - 4h^2 \\ &= 4\varepsilon(2h + \varepsilon),\end{aligned}$$

and we see that Y_n is Cauchy.

Since L^2 is a Hilbert space, and since M is closed, $Y_n \to \widehat{X}$ for some $\widehat{X} \in M$. We now have to show that $\|X - \widehat{X}\|_2 \leq \|X - Y\|_2$ for all $Y \in M$. Write

$$\begin{aligned}\|X - \widehat{X}\|_2 &= \|X - Y_n + Y_n - \widehat{X}\|_2 \\ &\leq \|X - Y_n\|_2 + \|Y_n - \widehat{X}\|_2.\end{aligned}$$

Since $\|X - Y_n\|_2 \to h$ and since $\|Y_n - \widehat{X}\|_2 \to 0$,

$$\|X - \widehat{X}\|_2 \leq h.$$

Since $h \leq \|X - Y\|_2$ for all $Y \in M$, $\|X - \widehat{X}\|_2 \leq \|X - Y\|_2$ for all $Y \in M$.
The uniqueness of $\widehat{X}$ is left to Problem 42.

13.6 Conditional expectation and probability

In earlier chapters, we defined conditional expectation and conditional probability separately for discrete and jointly continuous random variables. We are now in a position to introduce a more general definition. The new definition reduces to the old ones in those cases, but can also handle situations where random variables are neither jointly continuous nor discrete. The new definition is closely related to the orthogonality principle and the projection theorem.

We begin with two examples to illustrate the need for a more general definition of conditional expectation and conditional probability.

***Example* 13.17.** Consider a communication channel in which a discrete signal X with pmf $p_X(x_i)$ is subjected to additive noise Z with density $f_Z(z)$. The receiver sees $Y = X + Z$. If the signal X and the noise Z are independent, it is easy to show that Y is a continuous random variable with density $f_Y(y) = \sum_i f_Z(y - x_i) p_X(x_i)$ (see, e.g., Example 5.9) and that $f_{Y|X}(y|x_i) = f_Z(y - x_i)$. How do we define the conditional pmf $\mathsf{P}(X = x_i | Y = y)$?

Solution. Here is a heuristic approach. Consider the "joint density"

$$f_{XY}(x_i, y) := f_{Y|X}(y|x_i) p_X(x_i).$$

Then it is natural to take

$$\mathsf{P}(X = x_i | Y = y) = \frac{f_{XY}(x_i, y)}{f_Y(y)} = \frac{f_{Y|X}(y|x_i) p_X(x_i)}{f_Y(y)},$$

and it follows that for reasonable functions $v(x)$,

$$E[v(X)|Y=y] \;=\; \sum_i v(x_i) \frac{f_{Y|X}(y|x_i)\, p_X(x_i)}{f_Y(y)}.$$

Later we show that these formulas satisfy our new definitions.

Example 13.18. Let $\Theta \sim \text{uniform}[-\pi, \pi]$, and put $X := \cos\Theta$ and $Y := \sin\Theta$. Then X and Y are both arcsine random variables with common density $f_Y(y) = 1/\pi\sqrt{1-y^2}$ for $-1 < y < 1$ (Problem 35 in Chapter 5). However, since

$$X^2 + Y^2 \;=\; \cos^2\Theta + \sin^2\Theta \;=\; 1,$$

the random point (X,Y) lives on the unit circle, which is a set of zero area. As argued at the end of Section 7.2, X and Y are not jointly continuous. Intuitively though, since (X,Y) is uniformly distributed on the unit circle, given $Y = y$, X is equally likely to be $\pm\sqrt{1-y^2}$, and so we expect X to be conditionally a discrete random variable with conditional pmf

$$P(X=x|Y=y) \;=\; \begin{cases} 1/2, & x = \pm\sqrt{1-y^2}, \\ 0, & \text{otherwise,} \end{cases}$$

and it follows that for reasonable functions $v(x)$,

$$E[v(X)|Y=y] \;=\; \tfrac{1}{2}v(\sqrt{1-y^2}) + \tfrac{1}{2}v(-\sqrt{1-y^2}).$$

Later we show that these formulas satisfy our new definitions.

We say that $\widehat{g}(Y)$ is the **conditional expectation of X given Y** if

$$E[Xg(Y)] \;=\; E[\widehat{g}(Y)g(Y)], \quad \text{for all bounded functions } g. \tag{13.23}$$

When X and Y are discrete or jointly continuous it is easy to check that $\widehat{g}(y) = E[X|Y=y]$ solves this equation.[d] However, the importance of this definition is that we can prove the existence and uniqueness of such a function $\widehat{g}$ even if X and Y are not discrete or jointly continuous, as long as $X \in L^1$. Recall that uniqueness was proved in Problem 47 in Chapter 8.

We first consider the case $X \in L^2$. Put

$$M \;:=\; \{g(Y) : E[g(Y)^2] < \infty\}. \tag{13.24}$$

It is a consequence of the Riesz–Fischer theorem [50, p. 244] that M is closed. By the projection theorem combined with the orthogonality principle, there exists a $\widehat{g}(Y) \in M$ such that

$$\langle X - \widehat{g}(Y), g(Y)\rangle \;=\; 0, \quad \text{for all } g(Y) \in M.$$

---

[d] See, for example, the last paragraph of Section 8.6.

13.6 Conditional expectation and probability

Since the above inner product is defined as an expectation, it is equivalent to

$$\mathsf{E}[Xg(Y)] = \mathsf{E}[\widehat{g}(Y)g(Y)], \quad \text{for all } g(Y) \in M.$$

Since boundedness of g implies $g(Y) \in M$, (13.23) holds.

When $X \in L^2$, we have shown that $\widehat{g}(Y)$ is the projection of X onto M. For $X \in L^1$, we proceed as follows. First consider the case of nonnegative X with $\mathsf{E}[X] < \infty$. We can approximate X by the bounded function $X_n = \min(n, X)$ of Example 13.6. Being bounded, $X_n \in L^2$, and the corresponding $\widehat{g}_n(Y)$ exists and satisfies

$$\mathsf{E}[X_n g(Y)] = \mathsf{E}[\widehat{g}_n(Y)g(Y)], \quad \text{for all } g(Y) \in M. \tag{13.25}$$

Since $X_n \leq X_{n+1}$, $\widehat{g}_n(Y) \leq \widehat{g}_{n+1}(Y)$ by Problem 50. Hence, $\widehat{g}(Y) := \lim_{n \to \infty} \widehat{g}_n(Y)$ exists. To verify that $\widehat{g}(Y)$ satisfies (13.23), write[3]

$$\begin{aligned}
\mathsf{E}[Xg(Y)] &= \mathsf{E}\big[\lim_{n \to \infty} X_n g(Y)\big] \\
&= \lim_{n \to \infty} \mathsf{E}[X_n g(Y)] \\
&= \lim_{n \to \infty} \mathsf{E}[\widehat{g}_n(Y)g(Y)], \quad \text{by (13.25)}, \\
&= \mathsf{E}\big[\lim_{n \to \infty} \widehat{g}_n(Y)g(Y)\big] \\
&= \mathsf{E}[\widehat{g}(Y)g(Y)].
\end{aligned}$$

For signed X with $\mathsf{E}[|X|] < \infty$, consider the nonnegative random variables

$$X^+ := \begin{cases} X, & X \geq 0, \\ 0, & X < 0, \end{cases} \quad \text{and} \quad X^- := \begin{cases} -X, & X < 0, \\ 0, & X \geq 0. \end{cases}$$

Since $X^+ + X^- = |X|$, it is clear that X^+ and X^- are L^1 random variables. Since they are nonnegative, their conditional expectations exist. Denote them by $\widehat{g}^+(Y)$ and $\widehat{g}^-(Y)$. Since $X^+ - X^- = X$, it is easy to verify that $\widehat{g}(Y) := \widehat{g}^+(Y) - \widehat{g}^-(Y)$ satisfies (13.23) (Problem 51).

Notation

We have shown that to every $X \in L^1$, there corresponds a unique function $\widehat{g}(y)$ such that (13.23) holds. The standard notation for this function of y is $\mathsf{E}[X|Y=y]$, which, as noted above, is given by the usual formulas when X and Y are discrete or jointly continuous. It is conventional in probability theory to write $\mathsf{E}[X|Y]$ instead of $\widehat{g}(Y)$. We emphasize that $\mathsf{E}[X|Y=y]$ is a deterministic function of y, while $\mathsf{E}[X|Y]$ is a function of Y and is therefore a random variable. We also point out that with the conventional notation, (13.23) becomes

$$\mathsf{E}[Xg(Y)] = \mathsf{E}\big[\mathsf{E}[X|Y]g(Y)\big], \quad \text{for all bounded functions } g. \tag{13.26}$$

To see that (13.26) captures our earlier results about conditional expectation, consider the case in which X and Y are jointly continuous. Then the left-hand side is

$$\int_{-\infty}^{\infty} \int_{-\infty}^{\infty} x g(y) f_{XY}(x, y) \, dx \, dy,$$

and the right-hand side is
$$\int_{-\infty}^{\infty} \mathsf{E}[X|Y=y]g(y)f_Y(y)\,dy.$$

In particular, taking $g(y) \equiv 1$ recovers the law of total probability for expectations,

$$\mathsf{E}[X] = \int_{-\infty}^{\infty} \mathsf{E}[X|Y=y]f_Y(y)\,dy. \tag{13.27}$$

The corresponding result for discrete random variables can be similarly obtained from (13.26). In fact, taking $g(y) \equiv 1$ in (13.26) shows that the law of probability in all cases is expressed by the single unified formula

$$\boxed{\mathsf{E}[X] = \mathsf{E}[\mathsf{E}[X|Y]].}$$

Because $\mathsf{E}[X|Y]$ is unique, the characterization (13.26) is a powerful tool for showing that conjectured formulas for $\mathsf{E}[X|Y]$ are correct and for deriving properties of conditional expectation.

Example 13.19. If $\mathsf{E}[|v(X)|] < \infty$, show that the conditional expectation formula in Example 13.17 satisfies
$$\mathsf{E}[v(X)g(Y)] = \mathsf{E}[\mathsf{E}[v(X)|Y]g(Y)]$$
for all bounded functions g.

Solution. Since
$$\sum_i I_{\{x_i\}}(X) = 1,$$
we can write
$$\mathsf{E}[v(X)g(Y)] = \mathsf{E}\left[v(X)g(Y)\sum_i I_{\{x_i\}}(X)\right]$$
$$= \sum_i \mathsf{E}[v(X)g(Y)I_{\{x_i\}}(X)]$$
$$= \sum_i \mathsf{E}[v(X)g(X+Z)I_{\{x_i\}}(X)], \quad \text{since } Y = X+Z.$$

Since $v(X)g(X+Z)I_{\{x_i\}}(X) = v(x_i)g(x_i+Z)I_{\{x_i\}}(X)$, we continue with
$$\mathsf{E}[v(X)g(Y)] = \sum_i v(x_i)\mathsf{E}[g(x_i+Z)I_{\{x_i\}}(X)]$$
$$= \sum_i v(x_i)\mathsf{E}[g(x_i+Z)]\mathsf{E}[I_{\{x_i\}}(X)], \quad \text{by independence,}$$
$$= \sum_i v(x_i)\int_{-\infty}^{\infty} g(x_i+z)f_Z(z)\,dz\, p_X(x_i)$$
$$= \sum_i v(x_i)\int_{-\infty}^{\infty} g(y)f_Z(y-x_i)\,dy\, p_X(x_i)$$
$$= \int_{-\infty}^{\infty} \left[\sum_i v(x_i)\frac{f_Z(y-x_i)p_X(x_i)}{f_Y(y)}\right]g(y)f_Y(y)\,dy.$$

13.6 Conditional expectation and probability

Denoting the quantity in brackets by $E[v(X)|Y=y]$, we have

$$E[v(X)g(Y)] = \int_{-\infty}^{\infty} E[v(X)|Y=y]g(y)f_Y(y)\,dy$$
$$= E[E[v(X)|Y]g(Y)]$$

as claimed.

Example 13.20. If $E[|v(X)|] < \infty$, show that the conditional expectation formula in Example 13.18 satisfies

$$E[v(X)g(Y)] = E[E[v(X)|Y]g(Y)]$$

for all bounded functions g.

Solution. Write

$$E[v(X)g(Y)] = E[v(\cos\Theta)g(\sin\Theta)]$$
$$= \int_{-\pi}^{\pi} v(\cos\theta)g(\sin\theta)\frac{d\theta}{2\pi}.$$

Break up the range of integration into $[-\pi, -\pi/2]$, $[-\pi/2, 0]$, $[0, \pi/2]$, and $[\pi/2, \pi]$. On each interval, make the change of variable $y = \sin\theta$, $dy = \cos\theta\,d\theta$, and note that $\cos\theta = \pm\sqrt{1-\sin^2\theta} = \pm\sqrt{1-y^2}$. For example,

$$\int_0^{\pi/2} v(\cos\theta)g(\sin\theta)\frac{d\theta}{2\pi} = \int_0^1 v(\sqrt{1-y^2})g(y)\frac{dy}{2\pi\sqrt{1-y^2}},$$

and, since $\cos\theta$ is negative on $[\pi/2, \pi]$,

$$\int_{\pi/2}^{\pi} v(\cos\theta)g(\sin\theta)\frac{d\theta}{2\pi} = \int_1^0 v(-\sqrt{1-y^2})g(y)\frac{dy}{-2\pi\sqrt{1-y^2}}$$
$$= \int_0^1 v(-\sqrt{1-y^2})g(y)\frac{dy}{2\pi\sqrt{1-y^2}}.$$

The other two intervals are similar, and so

$$E[v(X)g(Y)] = \int_{-1}^{1} E[v(X)|Y=y]g(y)f_Y(y)\,dy,$$

where $E[v(X)|Y=y]$ is given as in Example 13.18. Thus,

$$E[v(X)g(Y)] = E[E[v(X)|Y]g(Y)]$$

as claimed.

Example 13.21. If $X \in L^1$ and is independent of Y, show that $E[X|Y] = E[X]$. In other words, $E[X|Y]$ is a constant function of Y equal to the mean of X.

Solution. If $\mathsf{E}[X]$ is to solve (13.26), we must show that for every bounded function g,
$$\mathsf{E}[Xg(Y)] = \mathsf{E}[\mathsf{E}[X]g(Y)].$$
However, by independence,
$$\mathsf{E}[Xg(Y)] = \mathsf{E}[X]\mathsf{E}[g(Y)] = \mathsf{E}[\mathsf{E}[X]g(Y)],$$
where in the last step we have moved the constant $\mathsf{E}[X]$ inside the expectation $\mathsf{E}[g(Y)]$.

Example 13.22 (linearity). If X and Y are in L^1, show that for any constants a and b,
$$\mathsf{E}[aX+bY|Z] = a\mathsf{E}[X|Z]+b\mathsf{E}[Y|Z].$$

Solution. For bounded $g(z)$, $\mathsf{E}[aX+bY|Z]$ is characterized by writing
$$\begin{aligned}\mathsf{E}\big[\mathsf{E}[aX+bY|Z]g(Z)\big] &= \mathsf{E}[(aX+bY)g(Z)] \\ &= a\mathsf{E}[Xg(Z)]+b\mathsf{E}[Yg(Z)],\end{aligned}$$
where we have used the linearity of ordinary expectation. Now the characterization equations for $\mathsf{E}[X|Z]$ and $\mathsf{E}[Y|Z]$ are
$$\mathsf{E}[Xg(Z)] = \mathsf{E}\big[\mathsf{E}[X|Z]g(Z)\big] \quad \text{and} \quad \mathsf{E}[Yg(Z)] = \mathsf{E}\big[\mathsf{E}[Y|Z]g(Z)\big].$$
Thus,
$$\begin{aligned}\mathsf{E}\big[\mathsf{E}[aX+bY|Z]g(Z)\big] &= a\mathsf{E}\big[\mathsf{E}[X|Z]g(Z)\big]+b\mathsf{E}\big[\mathsf{E}[Y|Z]g(Z)\big] \\ &= \mathsf{E}\big[(a\mathsf{E}[X|Z]+b\mathsf{E}[Y|Z])g(Z)\big],\end{aligned}$$
where we have again used the linearity of ordinary expectation.

Example 13.23 (substitution law). If $\mathsf{E}[|w(X,Y)|]<\infty$, show that
$$\mathsf{E}[w(X,Y)|Y=y] = \mathsf{E}[w(X,y)|Y=y].$$

Solution. To begin, fix any y and write
$$1 = I_{\{y\}}(Y)+I_{\{y\}^c}(Y).$$
Then
$$\begin{aligned}w(X,Y) &= w(X,Y)[I_{\{y\}}(Y)+I_{\{y\}^c}(Y)] \\ &= w(X,Y)I_{\{y\}}(Y)+w(X,Y)I_{\{y\}^c}(Y) \\ &= w(X,y)I_{\{y\}}(Y)+w(X,Y)I_{\{y\}^c}(Y).\end{aligned}$$
By linearity of conditional expectation,
$$\mathsf{E}[w(X,Y)|Y] = \mathsf{E}[w(X,y)I_{\{y\}}(Y)|Y]+\mathsf{E}[w(X,Y)I_{\{y\}^c}(Y)|Y].$$

13.6 Conditional expectation and probability

By Problem 53,
$$E[w(X,y)I_{\{y\}}(Y)|Y] = E[w(X,y)|Y]I_{\{y\}}(Y)$$
and
$$E[w(X,Y)I_{\{y\}^c}(Y)|Y] = E[w(X,Y)|Y]I_{\{y\}^c}(Y).$$

Hence,
$$E[w(X,Y)|Y] = E[w(X,y)|Y]I_{\{y\}}(Y) + E[w(X,Y)|Y]I_{\{y\}^c}(Y),$$

and then
$$E[w(X,Y)|Y=y] = E[w(X,y)|Y=y]I_{\{y\}}(y) + E[w(X,Y)|Y=y]I_{\{y\}^c}(y)$$
$$= E[w(X,y)|Y=y]I_{\{y\}}(y)$$
$$= E[w(X,y)|Y=y].$$

Conditional probability

Many times we have used the fact that every probability can be written as an expectation, e.g.,
$$P(X \in B) = E[I_B(X)].$$

This suggests that we define

$$\boxed{P(X \in B|Y=y) := E[I_B(X)|Y=y].}$$

Since $I_B(X)$ is bounded, it is in L^1, and

$$\boxed{P(X \in B|Y) := E[I_B(X)|Y]}$$

exists. For example, replacing X with $I_B(X)$ in (13.27) yields
$$P(X \in B) = \int_{-\infty}^{\infty} P(X \in B|Y=y) f_Y(y)\, dy.$$

If A is a two-dimensional set, say the disk of radius r centered at the origin, then taking $w(x,y) = I_A(x,y)$ in Example 13.23 yields
$$P((X,Y) \in A|Y=y) = P((X,y) \in A|Y=y),$$
or more explicitly,
$$P(X^2+Y^2 \le r^2|Y=y) = P(X^2 \le r^2 - y^2|Y=y).$$

At the beginning of the section, we proposed formulas for $E[v(X)|Y=y]$ in Examples 13.17 and 13.18, and we verified them in Examples 13.19 and 13.20. Hence, to verify the proposed formulas for the conditional pmf of X given Y, it suffices to take $v(x) = I_B(x)$ for suitable choices of B. For Example 13.17, taking $v(x) = I_{\{x_k\}}(X)$ shows that $P(X=x_k|Y=y)$ has the correct form. For Example 13.18, taking $v(X) = I_{\{x\}}(X)$ shows that $P(X=x|Y=y)$ has the proposed form.

The smoothing property

For $X \in L^1$, the **smoothing property** of conditional expectation says that

$$\mathsf{E}[X|q(Y)] = \mathsf{E}[\mathsf{E}[X|Y]|q(Y)]. \tag{13.28}$$

This formula is derived in Problem 54. An important special case arises if $Y = [Y_1, Y_2]'$ and $q(Y) = Y_1$. Then

$$\mathsf{E}[X|Y_1] = \mathsf{E}[\mathsf{E}[X|Y_2, Y_1]|Y_1].$$

This formula is a kind of law of total probability for conditional expectations. By replacing X with $I_B(X)$ in (13.28), we obtain

$$\mathsf{E}[I_B(X)|q(Y)] = \mathsf{E}[\mathsf{E}[I_B(X)|Y]|q(Y)],$$

or, in terms of conditional probability,

$$\mathsf{P}(X \in B|q(Y)) = \mathsf{E}[\mathsf{P}(X \in B|Y)|q(Y)]. \tag{13.29}$$

The special case $Y = [Y_1, Y_2]'$ and $q(Y) = Y_1$ yields

$$\mathsf{P}(X \in B|Y_1) = \mathsf{E}[\mathsf{P}(X \in B|Y_2, Y_1)|Y_1]. \tag{13.30}$$

This is exactly the law of total probability for conditional probability.

Example 13.24. Let U and V be random variables, and let A be an event. Suppose that $\mathsf{P}(A|U,V)$ is a function of V only, say $\mathsf{P}(A|U,V) = h(V)$ for some function h. Show that $\mathsf{P}(A|U,V) = \mathsf{P}(A|V)$.

Solution. By the smoothing property and the definition of h,

$$\mathsf{P}(A|V) = \mathsf{E}[\mathsf{P}(A|U,V)|V] = \mathsf{E}[h(V)|V].$$

By Problem 53, $\mathsf{E}[h(V)|V] = h(V)\mathsf{E}[1|V] = h(V)$.

Example 13.25 (Chapman–Kolmogorov equation). A sequence of discrete random variables $X_0, X_1, \ldots$ is called a **Markov chain** if

$$\mathsf{P}(X_{n+1} = j|X_n, \ldots, X_0) = \mathsf{P}(X_{n+1} = j|X_n). \tag{13.31}$$

The chain is time-homogeneous if

$$p(i,j) := \mathsf{P}(X_{n+1} = j|X_n = i)$$

does not depend on n. Put

$$p^{(n)}(i,j) := \mathsf{P}(X_n = j|X_0 = i).$$

For a time-homogeneous Markov chain, (13.31) can be written as

$$\mathsf{P}(X_{n+1} = j|X_n, \ldots, X_0) = p(X_n, j).$$

The Chapman–Kolmogorov equation says that

$$p^{(n+m)}(i,j) = \sum_k p^{(n)}(i,k)p^{(m)}(k,j),$$

or equivalently,

$$\mathsf{P}(X_{n+m} = j|X_0) = \sum_k p^{(n)}(X_0,k)p^{(m)}(k,j). \tag{13.32}$$

Use the smoothing property to derive (13.32).

Solution. Write

$$\begin{aligned}
\mathsf{P}(X_{n+m} = j|X_0) &= \mathsf{E}[\mathsf{P}(X_{n+m} = j|X_n,\ldots,X_0)|X_0]\\
&= \mathsf{E}[\mathsf{P}(X_{n+m} = j|X_n)|X_0], \qquad \text{by (13.31)},\\
&= \sum_k \mathsf{P}(X_{n+m} = j|X_n = k)\mathsf{P}(X_n = k|X_0)\\
&= \sum_k \mathsf{P}(X_m = j|X_0 = k)\mathsf{P}(X_n = k|X_0), \quad \text{by (12.15)},\\
&= \sum_k p^{(m)}(k,j)p^{(n)}(X_0,k).
\end{aligned}$$

13.7 The spectral representation

Let X_n be a discrete-time, zero-mean, wide-sense stationary process with correlation function

$$R(n) := \mathsf{E}[X_{n+m}X_m]$$

and corresponding power spectral density

$$S(f) = \sum_{n=-\infty}^{\infty} R(n)e^{-j2\pi fn} \tag{13.33}$$

so that[e]

$$R(n) = \int_{-1/2}^{1/2} S(f)e^{j2\pi fn}\,df. \tag{13.34}$$

Below we construct the **spectral process** of X_n. This process is denoted by $\{Z_f, -1/2 \leq f \leq 1/2\}$ and has the following properties. First, $Z_{-1/2} \equiv 0$. Second, $\mathsf{E}[Z_f] = 0$. Third, Z_f has uncorrelated increments with

$$\mathsf{E}[|Z_f|^2] = \int_{-1/2}^{f} S(v)\,dv. \tag{13.35}$$

---

[e] For some correlation functions, the sum in (13.33) may not converge. However, by **Herglotz's theorem**, we can always replace (13.34) by

$$R(n) = \int_{-1/2}^{1/2} e^{j2\pi fn}\,dS_0(f),$$

where $S_0(f)$ is the spectral (cumulative) distribution function, and the rest of the section goes through with the necessary changes. Note that when the power spectral density exists, $S(f) = S_0'(f)$.

Fourth, X_n has the representation

$$X_n = \int_{-1/2}^{1/2} e^{j2\pi f n} \, dZ_f, \qquad (13.36)$$

where the stochastic integral is defined similarly to the Wiener integral. Fifth, more generally, for frequency functions $G(f)$ and $H(f)$,

$$\int_{-1/2}^{1/2} G(f) \, dZ_f \quad \text{and} \quad \int_{-1/2}^{1/2} H(f) \, dZ_f$$

are zero-mean random variables with

$$\mathsf{E}\left[\left(\int_{-1/2}^{1/2} G(f) \, dZ_f\right)\left(\int_{-1/2}^{1/2} H(f) \, dZ_f\right)^*\right] = \int_{-1/2}^{1/2} G(f) H(f)^* S(f) \, df, \qquad (13.37)$$

where $\int_{-1/2}^{1/2} |G(f)|^2 S(f) \, df < \infty$ and $\int_{-1/2}^{1/2} |H(f)|^2 S(f) \, df < \infty$.

Example 13.26. Show that if

$$Y_n := \sum_{k=-\infty}^{\infty} h_k X_{n-k},$$

then Y_n is wide-sense stationary and has power spectral density $|H(f)|^2 S(f)$, where

$$H(f) := \sum_{k=-\infty}^{\infty} h_k e^{-j2\pi f k}.$$

Solution. Using the spectral representation of X_{n-k}, we can write

$$Y_n = \sum_{k=-\infty}^{\infty} h_k \int_{-1/2}^{1/2} e^{j2\pi f(n-k)} \, dZ_f$$

$$= \int_{-1/2}^{1/2} \left(\sum_{k=-\infty}^{\infty} h_k e^{-j2\pi f k}\right) e^{j2\pi f n} \, dZ_f$$

$$= \int_{-1/2}^{1/2} H(f) e^{j2\pi f n} \, dZ_f.$$

Since Y_n is an integral against dZ_f, Y_n has zero mean. Using (13.37), we have

$$\mathsf{E}[Y_n Y_m^*] = \int_{-1/2}^{1/2} |H(f)|^2 e^{j2\pi f(n-m)} S(f) \, df.$$

This shows that Y_n is WSS and that $R_Y(n)$ is the inverse Fourier transform of $|H(f)|^2 S(f)$. It follows that $S_Y(f) = |H(f)|^2 S(f)$.

Construction of the spectral process

We now construct the spectral process and derive (13.37) and (13.36). Consider the space of complex-valued frequency functions,

$$L^2(S) := \left\{ G : \int_{-1/2}^{1/2} |G(f)|^2 S(f) \, df < \infty \right\} \tag{13.38}$$

equipped with the inner product

$$\langle G, H \rangle := \int_{-1/2}^{1/2} G(f) H(f)^* S(f) \, df$$

and corresponding norm $\|G\| := \langle G, G \rangle^{1/2}$. Then $L^2(S)$ is a Hilbert space. Furthermore, every $G \in L^2(S)$ can be approximated by a trigonometric polynomial of the form

$$G_0(f) = \sum_{n=-N}^{N} g_n e^{j2\pi f n}. \tag{13.39}$$

The approximation is in norm; i.e., given any $\varepsilon > 0$, there is a trigonometric polynomial G_0 such that $\|G - G_0\| < \varepsilon$ [4, p. 139].

To each frequency function $G \in L^2(S)$, we now associate an L^2 random variable as follows. For trigonometric polynomials like G_0, put

$$T(G_0) := \sum_{n=-N}^{N} g_n X_n. \tag{13.40}$$

Note that T is well defined (Problem 69). A critical property of these trigonometric polynomials is that (Problem 70)

$$\mathsf{E}[T(G_0) T(H_0)^*] = \int_{-1/2}^{1/2} G_0(f) H_0(f)^* S(f) \, df = \langle G_0, H_0 \rangle,$$

where H_0 is defined similarly to G_0. In particular, T is **norm preserving** on the trigonometric polynomials; i.e.,

$$\|T(G_0)\|_2^2 = \|G_0\|^2.$$

To define T for arbitrary $G \in L^2(S)$, let G_n be a sequence of trigonometric polynomials converging to G in norm; i.e., $\|G_n - G\| \to 0$. Then G_n is Cauchy (cf. Problem 17). Furthermore, the linearity and the norm-preservation properties of T on the trigonometric polynomials tell us that

$$\|G_n - G_m\| = \|T(G_n - G_m)\|_2 = \|T(G_n) - T(G_m)\|_2,$$

and we see that $T(G_n)$ is Cauchy in L^2. Since L^2 is complete, there is a limit random variable, denoted by $T(G)$ and such that $\|T(G_n) - T(G)\|_2 \to 0$. Note that $T(G)$ is well defined, norm preserving, linear, and continuous on $L^2(S)$ (Problem 71).

There is another way approximate elements of $L^2(S)$. Every $G \in L^2(S)$ can also be approximated by a piecewise constant function of the following form. For $-1/2 \le f_0 < \cdots < f_n \le 1/2$, let

$$G_0(f) = \sum_{i=1}^{n} g_i I_{(f_{i-1}, f_i]}(f).$$

Given any $\varepsilon > 0$, there is a piecewise constant function G_0 such that $\|G - G_0\| < \varepsilon$. This is exactly what we did for the Wiener process, but with a different norm. For piecewise constant G_0,

$$T(G_0) = \sum_{i=1}^{n} g_i T(I_{(f_{i-1}, f_i]}).$$

Since

$$I_{(f_{i-1}, f_i]} = I_{[-1/2, f_i]} - I_{[-1/2, f_{i-1}]},$$

if we put

$$Z_f := T(I_{[-1/2, f]}), \quad -1/2 \leq f \leq 1/2, \tag{13.41}$$

then Z_f is well defined by Problem 72, and

$$T(G_0) = \sum_{i=1}^{n} g_i (Z_{f_i} - Z_{f_{i-1}}).$$

The family of random variables $\{Z_f, -1/2 \leq f \leq 1/2\}$ has many similarities to the Wiener process (see Problem 73). We write

$$\int_{-1/2}^{1/2} G_0(f) \, dZ_f := \sum_{i=1}^{n} g_i (Z_{f_i} - Z_{f_{i-1}}).$$

For arbitrary $G \in L^2(S)$, we approximate G with a sequence of piecewise constant functions. Then G_n will be Cauchy. Since

$$\left\| \int_{-1/2}^{1/2} G_n(f) \, dZ_f - \int_{-1/2}^{1/2} G_m(f) \, dZ_f \right\|_2 = \|T(G_n) - T(G_m)\|_2 = \|G_n - G_m\|,$$

there is a limit random variable in L^2, denoted by

$$\int_{-1/2}^{1/2} G(f) \, dZ_f,$$

and

$$\left\| \int_{-1/2}^{1/2} G_n(f) \, dZ_f - \int_{-1/2}^{1/2} G(f) \, dZ_f \right\|_2 \to 0.$$

On the other hand, since G_n is piecewise constant,

$$\int_{-1/2}^{1/2} G_n(f) \, dZ_f = T(G_n),$$

and since $\|G_n - G\| \to 0$, and since T is continuous, $\|T(G_n) - T(G)\|_2 \to 0$. Thus,

$$T(G) = \int_{-1/2}^{1/2} G(f) \, dZ_f. \tag{13.42}$$

In particular, taking $G(f) = e^{j2\pi f n}$ shows that

$$\int_{-1/2}^{1/2} e^{j2\pi f n} \, dZ_f = T(G) = X_n,$$

where the last step follows from the original definition of T. The formula

$$X_n = \int_{-1/2}^{1/2} e^{j2\pi fn} dZ_f$$

is called the **spectral representation** of X_n.

We now summarize the key equations in the construction of the spectral process. First, when $G_0(f)$ has the form in (13.39), $T(G_0)$ is given by (13.40). Once $T(G)$ is defined for all $G \in L^2(S)$, the spectral process Z_f is given by $Z_f := T(I_{(-1/2,f]})$. Finally, for $G \in L^2(S)$, $T(G)$ is given by (13.42). An important step in all of this is that if $\|G_n - G\| \to 0$, then $\|T(G_n) - T(G)\|_2 \to 0$.

Notes

13.2: Normed vector spaces of random variables

Note 1. The assumption

$$\sum_{k=1}^{\infty} |h_k| < \infty$$

should be understood more precisely as saying that the partial sum

$$A_n := \sum_{k=1}^{n} |h_k|$$

is bounded above by some finite constant. Since A_n is also nondecreasing, the **monotonic sequence property** of real numbers [51, p. 55, Theorem 3.14] says that A_n converges to a real number, say A. Now, by the same argument used to solve Problem 17, since A_n converges, it is Cauchy. Hence, given any $\varepsilon > 0$, for large enough n and m, $|A_n - A_m| < \varepsilon$. If $n > m$, we have

$$A_n - A_m = \sum_{k=m+1}^{n} |h_k| < \varepsilon.$$

13.3: The Karhunen–Loève expansion

Note 2. If X_t is WSS with integrable power spectral density, then the correlation function is uniformly continuous. A proof of the dual result, that if the correlation function is integrable, then the power spectral density is uniformly continuous was given in Section 10.9 in the paragraph following (10.47).

13.6: Conditional expectation and probability

Note 3. The interchange of limit and expectation is justified by Lebesgue's monotone and dominated convergence theorems [3, p. 208].

Problems

13.1: Convergence in mean of order p

1. Let $U \sim \text{uniform}[0,1]$, and put
$$X_n := \sqrt{n} I_{[0,1/n]}(U), \quad n = 1, 2, \ldots.$$
For what values of $p \geq 1$ does X_n converge in mean of order p to zero?

2. Let N_t be a Poisson process of rate λ. Show that N_t/t converges in mean square to λ as $t \to \infty$.

3. Show that $\lim_{k \to \infty} C(k) = 0$ implies
$$\lim_{n \to \infty} \frac{1}{n} \sum_{k=0}^{n-1} C(k) = 0,$$
and thus M_n converges in mean square to m by Example 13.4.

4. Show that if M_n converges in mean square to m, then
$$\lim_{n \to \infty} \frac{1}{n} \sum_{k=0}^{n-1} C(k) = 0,$$
where $C(k)$ is defined in Example 13.4. *Hint:* Observe that
$$\sum_{k=0}^{n-1} C(k) = \mathsf{E}\left[(X_1 - m) \sum_{k=1}^{n} (X_k - m)\right]$$
$$= \mathsf{E}[(X_1 - m) \cdot n(M_n - m)].$$

5. Let Z be a nonnegative random variable with $\mathsf{E}[Z] < \infty$. Given any $\varepsilon > 0$, show that there is a $\delta > 0$ such that for any event A,
$$\mathsf{P}(A) < \delta \quad \text{implies} \quad \mathsf{E}[ZI_A] < \varepsilon.$$
Hint: Recalling Example 13.6, put $Z_n := \min(Z, n)$ and write
$$\mathsf{E}[ZI_A] = \mathsf{E}[(Z - Z_n)I_A] + \mathsf{E}[Z_n I_A].$$

6. Use the preceding problem to show that if f is a probability density function, then the corresponding cdf $F(x) = \int_{-\infty}^{x} f(t)\,dt$ is continuous. *Hint:* Let $U \sim \text{uniform}[0,1]$, put $Z := f(x+U)$, and $A := \{U \leq \Delta x\}$.

7. Derive **Hölder's inequality**,
$$\mathsf{E}[|XY|] \leq \mathsf{E}[|X|^p]^{1/p} \mathsf{E}[|Y|^q]^{1/q},$$

if $1 < p, q < \infty$ with $\frac{1}{p} + \frac{1}{q} = 1$. *Hint:* Let α and β denote the factors on the right-hand side; i.e., $\alpha := \mathsf{E}[|X|^p]^{1/p}$ and $\beta := \mathsf{E}[|Y|^q]^{1/q}$. Then it suffices to show that

$$\frac{\mathsf{E}[|XY|]}{\alpha \beta} \leq 1.$$

To this end, observe that by the convexity of the exponential function, for any real numbers u and v, we can always write

$$\exp\left[\tfrac{1}{p}u + \tfrac{1}{q}v\right] \leq \tfrac{1}{p}e^u + \tfrac{1}{q}e^v.$$

Now take $u = \ln(|X|/\alpha)^p$ and $v = \ln(|Y|/\beta)^q$.

8. Derive **Lyapunov's inequality**,

$$\mathsf{E}[|Z|^\alpha]^{1/\alpha} \leq \mathsf{E}[|Z|^\beta]^{1/\beta}, \quad 0 < \alpha \leq \beta < \infty.$$

Hint: Apply Hölder's inequality to $X = |Z|^\alpha$ and $Y = 1$ with $p = \beta/\alpha$.

9. Use Lyapunov's inequality to show that if X_n converges in mean of order $\beta > 1$ to X, then X_n converges in mean of order α to X for all $1 \leq \alpha < \beta$.

10. Derive **Minkowski's inequality**,

$$\mathsf{E}[|X+Y|^p]^{1/p} \leq \mathsf{E}[|X|^p]^{1/p} + \mathsf{E}[|Y|^p]^{1/p},$$

where $1 \leq p < \infty$. *Hint:* Observe that

$$\begin{aligned}\mathsf{E}[|X+Y|^p] &= \mathsf{E}[|X+Y||X+Y|^{p-1}] \\ &\leq \mathsf{E}[|X||X+Y|^{p-1}] + \mathsf{E}[|Y||X+Y|^{p-1}],\end{aligned}$$

and apply Hölder's inequality.

11. Show that a Wiener process is continuous in mean.

12. Show that a Poisson process is mean-square continuous.

13. Show that if X_t is mean-square continuous at all t_0, then $R(t,s)$ is continuous at every point (τ, θ).

    Remark. Since mean-square continuity at t_0 is equivalent to continuity of $R(t,s)$ at (t_0, t_0), we have the important corollary that $R(t,s)$ is continuous everywhere if and only if $R(t,s)$ is continuous at all diagonal points (t_0, t_0).

14. Let X_t be a WSS process with $R(T) = R(0)$ for some $T > 0$.

    (a) Show that X_t is **mean-square periodic** in the sense that

    $$\mathsf{E}[|X_{t+T} - X_t|^2] = 0.$$

    (b) Apply the Cauchy–Schwarz inequality to

    $$R(t+T) - R(t) = \mathsf{E}[(X_{t+T} - X_t)X_0]$$

    to show that $R(t+T) = R(t)$ for all t.

13.2: Normed vector spaces of random variables

15. Show that if X_n converges in mean of order p to X, and if Y_n converges in mean of order p to Y, then $X_n + Y_n$ converges in mean of order p to $X + Y$.

16. If X_t and Y_t are continuous in mean of order p, show that $Z_t := X_t + Y_t$ is also continuous in mean of order p.

17. Let X_n converge in mean of order p to $X \in L^p$. Show that X_n is Cauchy.

18. Let X_n be a Cauchy sequence in L^p. Show that X_n is bounded in the sense that there is a finite constant K such that $\|X_n\|_p \le K$ for all n.

    Remark. Since by the previous problem, a convergent sequence is Cauchy, it follows that a convergent sequence is bounded.

19. Let X_n converge in mean of order p to $X \in L^p$, and let Y_n converge in mean of order q to $Y \in L^q$. Assuming $1 < p, q < \infty$ and $\frac{1}{p} + \frac{1}{q} = 1$, show that $X_n Y_n$ converges in mean to XY.

20. Show that limits in mean of order p are unique; i.e., if X_n converges in mean of order p to both X and Y, show that $\mathsf{E}[|X-Y|^p] = 0$.

21. Show that
$$\big|\|X\|_p - \|Y\|_p\big| \;\le\; \|X-Y\|_p \;\le\; \|X\|_p + \|Y\|_p.$$

22. If X_n converges in mean of order p to X, show that
$$\lim_{n\to\infty} \mathsf{E}[|X_n|^p] \;=\; \mathsf{E}[|X|^p].$$

    Remark. In fact, since convergence in mean of order p implies convergence in mean of order r for $1 \le r \le p$ (Problem 9), we actually have
$$\lim_{n\to\infty} \mathsf{E}[|X_n|^r] \;=\; \mathsf{E}[|X|^r], \quad \text{for } 1 \le r \le p.$$

23. Derive the **parallelogram law**,
$$\|X+Y\|_2^2 + \|X-Y\|_2^2 \;=\; 2(\|X\|_2^2 + \|Y\|_2^2).$$

24. If $\|X_n - X\|_2 \to 0$ and $\|Y_n - Y\|_2 \to 0$, show that
$$\lim_{n\to\infty} \langle X_n, Y_n \rangle \;=\; \langle X, Y \rangle.$$

25. Show that the result of Example 13.10 holds if the assumption $\sum_{k=1}^{\infty} |h_k| < \infty$ is replaced by the two assumptions $\sum_{k=1}^{\infty} |h_k|^2 < \infty$ and $\mathsf{E}[X_k X_l] = 0$ for $k \ne l$.

26. Show that
$$\sum_{k=1}^{\infty} h_k X_k$$
is well defined as an element of L^p assuming that
$$\sum_{k=1}^{\infty} |h_k| < \infty \quad \text{and} \quad E[|X_k|^p] \le B, \quad \text{for all } k,$$
where B is a finite constant.

27. Let $a = s_0 < s_1 < \cdots < s_v = b$ be a partition of $[a,b]$, and let $\theta_j \in [s_{j-1}, s_j]$. If
$$Z = \sum_{j=1}^{v} X_{\theta_j}(s_j - s_{j-1}),$$
and if Y is given by (13.6), show that
$$E[YZ] = \sum_{i=1}^{n} \sum_{j=1}^{v} R(\tau_i, \theta_j)(t_i - t_{i-1})(s_j - s_{j-1}).$$

28. Let X_t be a mean-square continuous process with correlation function $R_X(t,s)$. If $g(t)$ is a continuous function, use the theory of mean-square integrals to show that $\int_a^b g(t) X_t \, dt$ is well defined and has second moment
$$\int_a^b \int_a^b g(t) R_X(t,s) g(s) \, dt \, ds.$$

13.3: The Karhunen–Loève expansion

29. Let X_t be mean-square periodic as in Problem 14. Find the Karhunen–Loève expansion of X_t on $[0,T]$. In other words, find the eigenfunctions $\varphi_k(t)$ and the eigenvalues λ_k of (13.9) when $R(t,s) = R(t-s)$ and R has period T.

30. Show that (13.14) implies the eigenvalues in (13.9) are nonnegative.

31. If (λ_k, φ_k) and (λ_m, φ_m) are eigenpairs satisfying (13.9), show that if $\lambda_k \ne \lambda_m$ then φ_k and φ_m are orthogonal in the sense that $\int_a^b \varphi_k(t) \varphi_m(t) \, dt = 0$.

32. Suppose that in Example 13.12, the process X_t is such that the φ_k do not form a complete orthonormal set. Then there is a function g that is not equal to its expansion; i.e.,
$$\varphi(t) := g(t) - \sum_{k=1}^{\infty} g_k \varphi_k(t)$$
is not the zero function. Here $g_k := \int_0^T g(t) \varphi_k(t) \, dt$.

(a) Use the formula $R(t,s) = \sum_{k=1}^{\infty} \lambda_k \varphi_k(t) \varphi_k(s)$ of Mercer's theorem to show that

$$\int_0^T R(t,s) g(s) \, ds = \sum_{k=1}^{\infty} \lambda_k g_k \varphi_k(t).$$

(b) Show that $\int_0^T R(t,s) \varphi(s) \, ds = 0$.

(c) Use the result of Problem 28 to show that

$$Z := \int_0^T X_t \varphi(t) \, dt = 0$$

in the sense that $\mathsf{E}[Z^2] = 0$.

Remark. It follows that if $Y_t = \varphi(t) + X_t$ is passed through a single matched filter based on φ, the output is

$$\int_0^T Y_t \varphi(t) \, dt = \int_0^T \varphi(t)^2 \, dt + Z.$$

By part (b), $Z = 0$. Hence, we have **noiseless detection**; we can take $g_1(t) = \varphi(t)$, which is a nonzero function, and we can take $g_0(t) = 0$ for the other waveform. This is the continuous-time analog of Example 8.8.

33. Let X_t be the Ornstein–Uhlenbeck process on $[0, T]$ with correlation $R(t,s) = e^{-|t-s|}$. Investigate the eigenvalues and eigenfunctions for the Karhunen–Loève expansion. In particular, show that if $0 < \lambda < 2$ is an eigenvalue, then the corresponding eigenfunction is sinusoidal.

34. Let X_t have the Karhunen–Loève expansion

$$X_t = \sum_{k=1}^{\infty} A_k \varphi_k(t).$$

For fixed t, find the projection of X_t onto the subspace

$$M_L = \left\{ \sum_{i=1}^{L} c_i A_i : \text{the } c_i \text{ are nonrandom} \right\}.$$

13.4: The Wiener integral (again)

35. Recall that the Wiener integral of g was defined to be the mean-square limit of

$$Y_n := \int_0^{\infty} g_n(\tau) \, dW_\tau,$$

where each g_n is piecewise constant, and $\|g_n - g\| \to 0$. Show that

$$\mathsf{E}\left[\int_0^{\infty} g(\tau) \, dW_\tau \right] = 0,$$

and

$$E\left[\left(\int_0^\infty g(\tau)\,dW_\tau\right)^2\right] = \sigma^2 \int_0^\infty g(\tau)^2\,d\tau.$$

Hint: Let Y denote the mean-square limit of Y_n. By Example 13.11, $E[Y_n] \to E[Y]$, and by Problem 22, $E[Y_n^2] \to E[Y^2]$.

36. Use the following approach to show that the limit definition of the Wiener integral is unique. Let Y_n and Y be as in the text. If $\tilde{g}_n$ is another sequence of piecewise constant functions with $\|\tilde{g}_n - g\|^2 \to 0$, put

$$\tilde{Y}_n := \int_0^\infty \tilde{g}_n(\tau)\,dW_\tau.$$

By the argument that Y_n has a limit Y, $\tilde{Y}_n$ has a limit, say $\tilde{Y}$. Show that $\|Y - \tilde{Y}\|_2 = 0$.

37. Show that the Wiener integral is linear even for functions that are not piecewise constant.

38. Let W_t be a standard Wiener process. For $\beta > 0$, put

$$X_t := \int_0^t \tau^\beta \,dW_\tau.$$

Determine all values of β for which X_t/t converges in mean square to zero.

39. Let W_t be a standard Wiener process, and let the random variable $T \sim \exp(\lambda)$ be independent of the Wiener process. If

$$Y_T := \int_0^T \tau^n \,dW_\tau,$$

evaluate $E[Y_T^2]$.

40. Let f be a twice continuously differentiable function, and let W_t be a standard Wiener process.

   (a) Show that the derivative of $g(t) := E[f(W_t)]$ with respect to t is

   $$g'(t) = \tfrac{1}{2}E[f''(W_t)].$$

   Hint: Use the Taylor series approximation

   $$f(W_{t+\Delta t}) - f(W_t) \approx f'(W_t)(W_{t+\Delta t} - W_t) + \tfrac{1}{2}f''(W_t)(W_{t+\Delta t} - W_t)^2$$

   along with the independent increments property of the Wiener process.

   (b) Use the result of part (a) to find a differential equation for $E[e^{W_t}]$. Solve the differential equation.

   (c) To check your answer to part (b), evaluate the moment generating function of W_t at $s = 1$.

13.5: Projections, orthogonality principle, projection theorem

41. Let $C = \{Y \in L^p : \|Y\|_p \leq r\}$. For $\|X\|_p > r$, find a projection of X onto C.

42. Show that if $\widehat{X} \in M$ satisfies (13.22), it is unique.

43. Let $N \subset M$ be subspaces of L^2. Assume that the projection of $X \in L^2$ onto M exists and denote it by $\widehat{X}_M$. Similarly, assume that the projection of X onto N exists and denote it by $\widehat{X}_N$. Show that $\widehat{X}_N$ is the projection of $\widehat{X}_M$ onto N.

44. The preceding problem shows that when $N \subset M$ are subspaces, the projection of X onto N can be computed in two stages: First project X onto M and then project the result onto N. Show that this is not true in general if N and M are not both subspaces. *Hint:* Draw a disk of radius one. Then draw a straight line through the origin. Identify M with the disk and identify N with the line segment obtained by intersecting the line with the disk. If the point X is outside the disk and not on the line, then projecting first onto the ball and then onto the segment does not give the projection.

    Remark. Interestingly, projecting first onto the line (not the line segment) and then onto the disk does give the correct answer.

45. If Y has density f_Y, show that M in (13.24) is closed. *Hints:* (*i*) Observe that

    $$\mathsf{E}[g(Y)^2] = \int_{-\infty}^{\infty} g(y)^2 f_Y(y)\, dy.$$

    (*ii*) Use the fact that the set of functions

    $$G := \left\{g : \int_{-\infty}^{\infty} g(y)^2 f_Y(y)\, dy < \infty\right\}$$

    is complete if G is equipped with the norm

    $$\|g\|_Y^2 := \int_{-\infty}^{\infty} g(y)^2 f_Y(y)\, dy.$$

    (*iii*) Follow the method of Example 13.16.

46. Let f be a given function that satisfies $\int_0^\infty f(t)^2\, dt < \infty$, and let W_t be a Wiener process with $\mathsf{E}[W_t^2] = t$. Find the projection of

    $$\int_0^\infty f(t)\, dW_t$$

    onto the subspace

    $$M := \left\{\int_0^1 g(t)\, dW_t : \int_0^1 g(t)^2\, dt < \infty\right\}.$$

47. Let W_t be a Wiener process. For $X \in L^2$, find g with $\int_0^\infty g(\tau)^2\, d\tau < \infty$ to minimize

    $$\mathsf{E}\left[\left|X - \int_0^\infty g(\tau)\, dW_\tau\right|^2\right].$$

    Hint: State an appropriate orthogonality principle. If you need to consider an arbitrary function $\widetilde{g}$, consider $\widetilde{g}(\tau) = I_{[0,t]}(\tau)$, where t is arbitrary.

13.6: Conditional expectation and probability

48. Let $X \sim \text{uniform}[-1,1]$, and put $Y = X(1-X)$.

    (a) Find the density of Y.
    (b) For $v(x)$ with $\mathsf{E}[|v(X)|] < \infty$, find $\mathsf{E}[v(X)|Y=y]$.

49. Let Θ be any continuous random variable taking values in $[-\pi, \pi]$. Put $X := \cos\Theta$ and $Y := \sin\Theta$.

    (a) Express the density $f_Y(y)$ in terms of f_Θ. *Hint:* Treat $y \geq 0$ and $y < 0$ separately.
    (b) For $v(x)$ with $\mathsf{E}[|v(X)|] < \infty$, find $\mathsf{E}[v(X)|Y=y]$.

50. Fix $X \in L^1$, and suppose $X \geq 0$. Show that $\mathsf{E}[X|Y] \geq 0$ in the sense that $\mathsf{P}(\mathsf{E}[X|Y] < 0) = 0$. *Hints:* To begin, note that

    $$\{\mathsf{E}[X|Y] < 0\} = \bigcup_{n=1}^{\infty} \{\mathsf{E}[X|Y] < -1/n\}.$$

    By limit property (1.15),

    $$\mathsf{P}(\mathsf{E}[X|Y] < 0) = \lim_{n\to\infty} \mathsf{P}(\mathsf{E}[X|Y] < -1/n).$$

    To obtain a proof by contradiction, suppose that $\mathsf{P}(\mathsf{E}[X|Y] < 0) > 0$. Then for sufficiently large n, $\mathsf{P}(\mathsf{E}[X|Y] < -1/n) > 0$ too. Take $g(Y) = I_B(\mathsf{E}[X|Y])$ where $B = (-\infty, -1/n)$ and consider the defining relationship

    $$\mathsf{E}[Xg(Y)] = \mathsf{E}\big[\mathsf{E}[X|Y]g(Y)\big].$$

51. For $X \in L^1$, let X^+ and X^- be as in the text, and denote their corresponding conditional expectations by $\mathsf{E}[X^+|Y]$ and $\mathsf{E}[X^-|Y]$, respectively. Show that for bounded g,

    $$\mathsf{E}[Xg(Y)] = \mathsf{E}\big[(\mathsf{E}[X^+|Y] - \mathsf{E}[X^-|Y])g(Y)\big].$$

52. If $X \in L^1$, show that $\mathsf{E}[X|Y] \in L^1$. *Hint:* You need to show that $\mathsf{E}\big[|\mathsf{E}[X|Y]|\big] < \infty$. Start with

    $$|\mathsf{E}[X|Y]| = |\mathsf{E}[X^+|Y] - \mathsf{E}[X^-|Y]|.$$

53. If $X \in L^1$ and h is a bounded function, show that

    $$\mathsf{E}[h(Y)X|Y] = h(Y)\mathsf{E}[X|Y].$$

54. For $X \in L^1$, derive the **smoothing property** of conditional expectation,

    $$\mathsf{E}\big[\mathsf{E}[X|Y]\big|q(Y)\big] = \mathsf{E}[X|q(Y)].$$

    Hint: The result of Problem 53 may be helpful.
    Remark. If $X \in L^2$, this is an instance of Problem 43.

55. If $X \in L^2$ and $h(Y) \in L^2$, then by Hölder's inequality, $h(Y)X \in L^1$. Use the orthogonality principle to show that $h(Y)\mathsf{E}[X|Y]$ satisfies

$$\mathsf{E}[\{h(Y)X\}g(Y)] = \mathsf{E}[\{h(Y)\mathsf{E}[X|Y]\}g(Y)]$$

for all bounded g. Thus, $\mathsf{E}[h(Y)X|Y]$ exists and is equal to $h(Y)\mathsf{E}[X|Y]$.

56. If $X \in L^1$ and if $h(Y)X \in L^1$, show that

$$\mathsf{E}[h(Y)X|Y] = h(Y)\mathsf{E}[X|Y].$$

Hint: Use a limit argument as in the text to approximate $h(Y)$ by a bounded function of Y. Then apply Problem 53.

Remark. Note the special case $X \equiv 1$, which says that

$$\mathsf{E}[h(Y)|Y] = h(Y),$$

since taking $\mathsf{E}[1|Y] = 1$ clearly satisfies (13.26).

57. Let $X_1, \ldots, X_n$ be i.i.d. L^1 random variables, and put $Y := X_1 + \cdots + X_n$. Find $\mathsf{E}[X_1|Y]$. *Hint:* Consider $\mathsf{E}[Y|Y]$ and use properties of conditional expectation.

58. A sequence of L^1 random variables $\{X_n, n \geq 1\}$ is said to be a **martingale** with respect to another sequence $\{Y_n, n \geq 1\}$ if X_n is a function of $Y_1, \ldots, Y_n$ and if

$$\mathsf{E}[X_{n+1}|Y_n, \ldots, Y_1] = X_n, \quad n \geq 1.$$

If $Y_1, Y_2, \ldots$ are i.i.d. with zero-mean, and if $X_n := Y_1 + \cdots + Y_n$, show that X_n is a martingale with respect to Y_n.

Remark. If the Y_n have positive mean, then $\mathsf{E}[X_{n+1}|Y_n, \ldots, Y_1] \geq X_n$ for $n \geq 1$, in which case X_n is said to be a **submartingale**. If $\mathsf{E}[X_{n+1}|Y_n, \ldots, Y_1] \leq X_n$ for $n \geq 1$, then X_n is called a **supermartingale**. Note that part of the definitions of submartingale and supermartingale are that X_n be a function of $Y_1, \ldots, Y_n$.

59. If X_n is a martingale with respect to Y_n, show that $\mathsf{E}[X_n] = \mathsf{E}[X_1]$ for all n.

60. If X_n is a nonnegative supermartingale with respect to Y_n, show that $0 \leq \mathsf{E}[X_n] \leq \mathsf{E}[X_1]$ for all n.

61. Use the smoothing property of conditional expectation to show that if $Z \in L^1$, then $X_n := \mathsf{E}[Z|Y_n, \ldots, Y_1]$ is a martingale with respect to $\{Y_n\}$.

62. Let $Y_1, Y_2, \ldots$ be i.i.d. with common density $f(y)$, and let $\tilde{f}(y)$ be another density. Put

$$w(y) := \frac{\tilde{f}(y)}{f(y)}.$$

Show that

$$X_n := \prod_{i=1}^{n} w(Y_i)$$

is a martingale with respect to Y_n.

Remark. The function w arises as the likelihood ratio in designing detectors for communication systems, e.g., Problem 17 in Chapter 5.

63. Generalize the preceding problem to the case in which the Y_i are dependent with joint densities $f_{Y_n \cdots Y_1}(y_n, \ldots, y_1)$. Show that

$$X_n := w_n(Y_1, \ldots, Y_n)$$

is a Y_n martingale if

$$w_n(y_1, \ldots, y_n) := \frac{\tilde{f}_{Y_n \cdots Y_1}(y_n, \ldots, y_1)}{f_{Y_n \cdots Y_1}(y_n, \ldots, y_1)},$$

and $\tilde{f}_{Y_n \cdots Y_1}$ is some other joint density.

64. Let $Y_0, Y_1, \ldots$ be a sequence of continuous random variables with conditional densities that satisfy the density version of the Markov property,

$$f_{Y_{n+1}|Y_n \cdots Y_0}(y_{n+1}|y_n, \ldots, y_0) = f_{Y_{n+1}|Y_n}(y_{n+1}|y_n).$$

Further assume that the transition density $f_{Y_{n+1}|Y_n}(z|y) = p(z|y)$ does not depend on n. Put $X_0 \equiv 0$, and for $n \geq 1$, put

$$X_n := \sum_{k=1}^{n} W_k, \quad \text{where} \quad W_k := Y_k - \int_{-\infty}^{\infty} z p(z|Y_{k-1}) \, dz.$$

Determine whether or not X_n is a martingale with respect to $\{Y_n\}$; i.e., determine whether or not $\mathsf{E}[X_{n+1}|Y_n, \ldots, Y_0] = X_n$.

65. Let X_n be a Markov chain with the state transition diagram in Figure 13.1 where

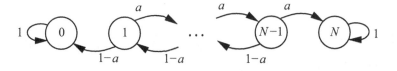

Figure 13.1. State transition diagram for Problem 65.

$0 < a < 1$. Assume that $\mathsf{P}(X_0 = i_0) = 1$ for some $0 < i_0 < N$. Put $\rho := (1-a)/a$, and define $Y_n := \rho^{X_n}$. Determine whether or not Y_n is a martingale with respect to X_n. In other words, determine whether or not $\mathsf{E}[Y_{n+1}|X_n, \ldots, X_0] = Y_n$ for $n \geq 0$.

66. A process A_n is said to be **predictable** with respect to another process Y_n if A_n is a function of $Y_1, \ldots, Y_{n-1}$ for $n \geq 2$, and if A_1 is a constant. The **Doob decomposition** says that every submartingale X_n with respect to Y_n can be written in the form

$$X_n = A_n + M_n,$$

where A_n is nondecreasing, i.e., $A_n \leq A_{n+1}$, and predictable with respect to Y_n, and M_n is a martingale with respect to Y_n. Derive this decomposition. *Hint:* Start with $A_1 = 0$ and $M_1 = X_1$. Then for $n \geq 1$, put

$$A_{n+1} := A_n + (\mathsf{E}[X_{n+1}|Y_n,\ldots,Y_1] - X_n).$$

Verify that A_n and $M_n = X_n - A_n$ satisfy the conditions of the Doob decomposition.

13.7: The spectral representation

67. Use (13.35) and the uncorrelated increments property to show that

$$\mathsf{E}[Z_{f_1}Z_{f_2}^*] = \int_{-1/2}^{\min(f_1,f_2)} S(v)\,dv.$$

Do not derive this result using (13.37).

68. Show that for $f \neq 0$ and $|f| \leq 1/2$, $\frac{1}{n}\sum_{k=1}^{n} e^{j2\pi fn}$ is equal to

$$\frac{e^{j2\pi f}}{n} \cdot \frac{e^{j\pi fn}}{e^{j\pi f}} \cdot \frac{\sin(\pi fn)}{\sin(\pi f)}.$$

69. Show that the mapping T defined by

$$G_0(f) = \sum_{n=-N}^{N} g_n e^{j2\pi fn} \mapsto \sum_{n=-N}^{N} g_n X_n$$

is well defined; i.e., if $G_0(f)$ has another representation, say

$$G_0(f) = \sum_{n=-N}^{N} \tilde{g}_n e^{j2\pi fn},$$

show that

$$\sum_{n=-N}^{N} \tilde{g}_n X_n = \sum_{n=-N}^{N} g_n X_n.$$

Hint: With $d_n := g_n - \tilde{g}_n$, it suffices to show that if $\sum_{n=-N}^{N} d_n e^{j2\pi fn} = 0$, then $Y := \sum_{n=-N}^{N} d_n X_n = 0$, and for this it is enough to show that $\mathsf{E}[|Y|^2] = 0$. Formula (13.34) may be helpful.

70. Show that if $H_0(f)$ is defined similarly to $G_0(f)$ in the preceding problem, then

$$\mathsf{E}[T(G_0)T(H_0)^*] = \int_{-1/2}^{1/2} G_0(f)H_0(f)^* S(f)\,df.$$

71. (a) Let G_n be a sequence of trigonometric polynomials converging in mean square to G. Then $T(G)$ is defined to be the mean-square limit of $T(G_n)$. However, suppose that $\widetilde{G}_n$ is another sequence of trigonometric polynomials converging in mean square to G. Let Y denote the mean-square limit of $T(\widetilde{G}_n)$. Show that $Y = T(G)$. *Hint:* Show that $\|T(G) - Y\|_2 = 0$. Use the fact that T is norm preserving on trigonometric polynomials.

(b) Show that T is norm preserving on $L^2(S)$.

(c) Show that T is linear on $L^2(S)$.

(d) Show that T is continuous on $L^2(S)$.

72. Show that $Z_f := T(I_{[-1/2,f]})$ is well defined. *Hint:* It suffices to show that $I_{[-1/2,f]} \in L^2(S)$.

73. Since
$$\int_{-1/2}^{1/2} G(f) \, dZ_f = T(G),$$
use results about $T(G)$ to prove the following.

(a) Show that
$$\mathsf{E}\left[\int_{-1/2}^{1/2} G(f) \, dZ_f\right] = 0.$$

(b) Show that
$$\mathsf{E}\left[\left(\int_{-1/2}^{1/2} G(f) \, dZ_f\right)\left(\int_{-1/2}^{1/2} H(v) \, dZ_v\right)^*\right] = \int_{-1/2}^{1/2} G(f) H(f)^* S(f) \, df.$$

(c) Show that Z_f has orthogonal (uncorrelated) increments.

74. Let Y_n be as in Example 13.26, and assume $H \in L^2(S)$ defined in (13.38).

(a) Recall that the starting point for defining the spectral process of X_n was the set of frequency functions $L^2(S)$ defined in (13.38). To define the spectral process of Y_n, what is the analog of (13.38)? Denote this analog by $L^2(S_Y)$.

(b) For a trigonometric polynomial G_0 as in the text, put
$$T_Y(G_0) := \sum_{n=-N}^{N} g_n Y_n.$$

Show that
$$T_Y(G_0) = \int_{-1/2}^{1/2} G_0(f) H(f) \, dZ_f.$$

(c) For $G \in L^2(S_Y)$ of part (a), show that if $GH \in L^2(S)$, then
$$T_Y(G) = \int_{-1/2}^{1/2} G(f) H(f) \, dZ_f.$$

(d) Show that the spectral process of Y_n, denoted by $V_f := T_Y(I_{[-1/2,f]})$, is given by
$$V_f = \int_{-1/2}^{f} H(v) \, dZ_v.$$

(e) For $G \in L^2(S_Y)$, show that
$$\int_{-1/2}^{1/2} G(f) \, dV_f = \int_{-1/2}^{1/2} G(f) H(f) \, dZ_f.$$

Exam preparation

You may use the following suggestions to prepare a study sheet, including formulas mentioned that you have trouble remembering. You may also want to ask your instructor for additional suggestions.

13.1. Convergence in mean of order p. Be able to do simple problems based on the definition. As noted in Example 13.4, a sufficient condition for the mean-square law of large numbers for a WSS sequence is that the covariance function $C(k)$ converge to zero. Of course, if the X_i are uncorrelated, $C(k) = 0$ for $k \neq 0$. Know the definition of continuity in mean of order p. A correlation function $R(t,s)$ is jointly continuous at all (t,s) if and only if it is continuous at all diagonal points of the form (t_0, t_0). Also, R is continuous at all diagonal points if and only if X_t is mean-square continuous. A WSS process is mean-square continuous if and only if its correlation function $R(\tau)$ is continuous at $\tau = 0$.

13.2. Normed vector spaces of random variables. The triangle inequality is an important tool for proving results about convergence in mean of order p. The space L^p of random variables with $\mathsf{E}[|X|^p] < \infty$ is complete; i.e., if $X_n \in L^p$ is Cauchy, then there exists an $X \in L^p$ with $\|X_n - X\|_p \to 0$. In the case of $p = 2$, the norm is given by $\|X\|_2 = \langle X, X \rangle^{1/2}$, where the inner product is $\langle X, Y \rangle = \mathsf{E}[XY]$. Thus, for $p = 2$, the Cauchy–Schwarz inequality is an additional tool for proving convergence results. Because $\|\cdot\|_2$ is given by an inner product, the L^2-norm satisfies the parallelogram law (13.3). If X_t is zero-mean and mean-square continuous, then $\int_a^b X_t \, dt$ exists and its variance is given by (13.5).

13.3. The Karhunen–Loève expansion. For a zero-mean, mean-square-continuous process X_t, the relevant formulas are (13.7) and (13.9)–(13.11). When X_t is WSS with integrable power spectral density, the eigenfunctions of the correlation function form a complete orthonormal set.

13.4. The Wiener integral (again). The main thing to know is that if g is square integrable, then the Wiener integral of g exists and is zero mean and has variance $\sigma^2 \int_0^\infty g(\tau)^2 \, d\tau$.

13.5. Projections, orthogonality principle, projection theorem. Understand and be able to use the the orthogonality principle, which says that (13.21) and (13.22) are equivalent; i.e., one holds if and only if the other does. Of course, the orthogonality principle does not guarantee the *existence* of a solution of either (13.21) or (13.22). The projection theorem says that the projection onto a closed subspace of L^2 always exists; i.e., there is always an $\widehat{X}$ satisfying (13.21).

13.6. Conditional expectation and probability. We used the projection theorem to ultimately get the existence of a unique solution $\mathsf{E}[X|Y]$ of characterization (13.26) when $\mathsf{E}[|X|] < \infty$. Remember $\mathsf{E}[X|Y]$ is a function of the random variable Y. Conditional expectation is linear, satisfies the substitution law, the smoothing property (Problem 54), and $\mathsf{E}[h(Y)X|Y] = h(Y)\mathsf{E}[X|Y]$ (Problem 56).

13.7. The spectral representation. Know that $Z_{-1/2} \equiv 0$, Z_f is zero mean with uncorrelated increments, and formulas (13.35)–(13.37). Here are the key equations in the construction of the spectral process. First, when $G_0(f)$ has the form in (13.39),

$T(G_0)$ is given by (13.40). Once $T(G)$ is defined for all $G \in L^2(S)$, the spectral process Z_f is given by $Z_f := T(I_{(-1/2,f]})$. Finally, for $G \in L^2(S)$, $T(G)$ is given by (13.42). An important step in all of this is that if $\|G_n - G\| \to 0$, then $\|T(G_n) - T(G)\|_2 \to 0$.

Work any review problems assigned by your instructor. If you finish them, re-work your homework assignments.

14
Other modes of convergence

The three types of convergence: convergence in probability; convergence in distribution; and almost-sure convergence are related to each other and to convergence in mean of order p as shown in Figure 14.1. In particular, convergence in mean of order p does *not* imply almost-sure convergence, and almost-sure convergence does *not* imply convergence in mean of order p for any p. **The Exam preparation section at the end of the chapter contains a summary of important facts about the different kinds of convergence.**

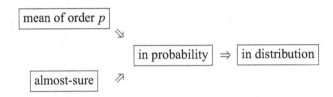

Figure 14.1. Implications of various types of convergence.

Section 14.1 introduces the notion of convergence in probability. Convergence in probability was important in Chapter 6 on parameter estimation and confidence intervals, where it justifies various statistical procedures that are used to estimate unknown parameters.

Section 14.2 introduces the notion of convergence in distribution. Convergence in distribution is often used to approximate probabilities that are hard to calculate exactly. Suppose that X_n is a random variable whose cumulative distribution function $F_{X_n}(x)$ is hard to compute. But suppose that for large n, $F_{X_n}(x) \approx F_X(x)$, where $F_X(x)$ is a cdf that is easy to compute. Loosely speaking, when $F_{X_n}(x) \to F_X(x)$, we say that X_n converges in distribution to X. In this case, we can approximate $F_{X_n}(x)$ by $F_X(x)$ if n is large enough. When the central limit theorem applies, F_X is the normal cdf with mean zero and variance one.

Section 14.3 introduces the notion of almost-sure convergence. This kind of convergence is more technically demanding to analyze, but it allows us to discuss important results such as the strong law of large numbers. Almost-sure convergence also allows us to derive the Skorohod representation, which is a powerful tool for studying convergence in distribution.

14.1 Convergence in probability

We say that X_n **converges in probability** to X if

$$\lim_{n \to \infty} \mathsf{P}(|X_n - X| \geq \varepsilon) = 0 \quad \text{for all } \varepsilon > 0.$$

Note that if X is zero, this reduces to $\lim_{n \to \infty} \mathsf{P}(|X_n| \geq \varepsilon) = 0$.

14.1 Convergence in probability

The first thing to note about convergence in probability is that it is implied by convergence in mean of order p. To see this, use the Markov inequality to write

$$P(|X_n - X| \geq \varepsilon) = P(|X_n - X|^p \geq \varepsilon^p) \leq \frac{E[|X_n - X|^p]}{\varepsilon^p}.$$

Example 14.1 (weak law of large numbers). Since the sample mean M_n defined in Example 13.3 converges in mean square to m, it follows that M_n converges in probability to m. This is known as the weak law of large numbers.

The second thing to note about convergence in probability is that it is possible to have X_n converging in probability to X, while X_n does not converge to X in mean of order p for any $p \geq 1$. For example, if $U \sim \text{uniform}[0,1]$, we show that

$$X_n := n I_{[0,1/n]}(U), \quad n = 1, 2, \ldots$$

converges in probability to zero, but not in mean of order p for any $p \geq 1$. Observe that for $\varepsilon > 0$,

$$\{|X_n| \geq \varepsilon\} = \begin{cases} \{U \leq 1/n\}, & n \geq \varepsilon, \\ \varnothing, & n < \varepsilon. \end{cases}$$

It follows that for all n,

$$P(|X_n| \geq \varepsilon) \leq P(U \leq 1/n) = 1/n \to 0.$$

Thus, X_n converges in probability to zero. However,

$$E[|X_n|^p] = n^p P(U \leq 1/n) = n^{p-1},$$

which does not go to zero as $n \to \infty$.

The third thing to note about convergence in probability is that if $g(x,y)$ is continuous, and if X_n converges in probability to X, and Y_n converges in probability to Y, then $g(X_n, Y_n)$ converges in probability to $g(X, Y)$. This result is derived in Problem 7. In particular, note that since the functions $g(x,y) = x + y$ and $g(x,y) = xy$ are continuous, $X_n + Y_n$ and $X_n Y_n$ converge in probability to $X + Y$ and XY, respectively, whenever X_n and Y_n converge in probability to X and Y, respectively.

Example 14.2. Since Problem 7 is somewhat involved, it is helpful to first see the derivation for the special case in which X_n and Y_n both converge in probability to *constant* random variables, say u and v, respectively.

Solution. Let $\varepsilon > 0$ be given. We show that

$$P(|g(X_n, Y_n) - g(u, v)| \geq \varepsilon) \to 0.$$

By the continuity of g at the point (u, v), there is a $\delta > 0$ such that whenever (u', v') is within δ of (u, v); i.e., whenever

$$\sqrt{|u' - u|^2 + |v' - v|^2} < 2\delta,$$

then
$$|g(u',v') - g(u,v)| < \varepsilon.$$
Since
$$\sqrt{|u'-u|^2 + |v'-v|^2} \le |u'-u| + |v'-v|,$$
we see that
$$|X_n - u| < \delta \text{ and } |Y_n - v| < \delta \implies |g(X_n, Y_n) - g(u,v)| < \varepsilon.$$
Conversely,
$$|g(X_n, Y_n) - g(u,v)| \ge \varepsilon \implies |X_n - u| \ge \delta \text{ or } |Y_n - v| \ge \delta.$$
It follows that
$$\mathsf{P}(|g(X_n, Y_n) - g(u,v)| \ge \varepsilon) \le \mathsf{P}(|X_n - u| \ge \delta) + \mathsf{P}(|Y_n - v| \ge \delta).$$
These last two terms go to zero by hypothesis.

Example 14.3. Let X_n converge in probability to x, and let c_n be a converging sequence of real numbers with limit c. Show that $c_n X_n$ converges in probability to cx.

Solution. Define the sequence of constant random variables $Y_n := c_n$. It is a simple exercise to show that Y_n converges in probability to c. Thus, $X_n Y_n = c_n X_n$ converges in probability to cx.

See the Exam preparation section at the end of the chapter for a summary of important facts about convergence in probability.

14.2 Convergence in distribution

We say that X_n **converges in distribution** to X, or **converges weakly** to X if
$$\lim_{n \to \infty} F_{X_n}(x) = F_X(x), \quad \text{for all } x \in C(F_X),$$
where $C(F_X)$ is the set of points x at which F_X is continuous. If F_X has a jump at a point x_0, then we do not care if
$$\lim_{n \to \infty} F_{X_n}(x_0)$$
exists, and if it does exist, we do not care if the limit is equal to $F_X(x)$.

Example 14.4. Let $X_n \sim \exp(n)$; i.e.,
$$F_{X_n}(x) = \begin{cases} 1 - e^{-nx}, & x \ge 0, \\ 0, & x < 0. \end{cases}$$

As shown in Figure 14.2(a), the pointwise limit of $F_{X_n}(x)$ is the function in Figure 14.2(b). Notice that since this function is left continuous, it cannot be a cdf. However, the function in Figure 14.2(c) is right continuous, and is in fact the cdf of the zero random variable.[1] Since the only discontinuity in Figure 14.2(c) is at $x = 0$, and since $F_{X_n}(x)$ converges to this function for $x \ne 0$, we see that X_n converges in distribution to the zero random variable.

14.2 Convergence in distribution

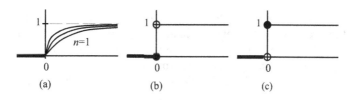

Figure 14.2. (a) Sketch of $F_{X_n}(x)$ for increasing values of n. (b) Pointwise limit of $F_{X_n}(x)$. (c) Limiting cdf $F_X(x)$.

The first thing to note about convergence in distribution is that it is implied by convergence in probability. To derive this result, fix any x at which F_X is continuous. For any $\varepsilon > 0$, write

$$F_{X_n}(x) = \mathsf{P}(X_n \leq x, X \leq x+\varepsilon) + \mathsf{P}(X_n \leq x, X > x+\varepsilon)$$
$$\leq F_X(x+\varepsilon) + \mathsf{P}(X_n - X < -\varepsilon)$$
$$\leq F_X(x+\varepsilon) + \mathsf{P}(|X_n - X| \geq \varepsilon).$$

It follows that

$$\varlimsup_{n \to \infty} F_{X_n}(x) \leq F_X(x+\varepsilon).$$

(The **limit superior** and **limit inferior** are defined in the Notes.[2]) Similarly,

$$F_X(x-\varepsilon) = \mathsf{P}(X \leq x-\varepsilon, X_n \leq x) + \mathsf{P}(X \leq x-\varepsilon, X_n > x)$$
$$\leq F_{X_n}(x) + \mathsf{P}(X - X_n < -\varepsilon)$$
$$\leq F_{X_n}(x) + \mathsf{P}(|X_n - X| \geq \varepsilon),$$

and we obtain

$$F_X(x-\varepsilon) \leq \varliminf_{n \to \infty} F_{X_n}(x).$$

Since the liminf is always less than or equal to the limsup, we have

$$F_X(x-\varepsilon) \leq \varliminf_{n \to \infty} F_{X_n}(x) \leq \varlimsup_{n \to \infty} F_{X_n}(x) \leq F_X(x+\varepsilon).$$

Since F_X is continuous at x, letting ε go to zero shows that the liminf and the limsup are equal to each other and to $F_X(x)$. Hence $\lim_{n \to \infty} F_{X_n}(x)$ exists and equals $F_X(x)$.

The second thing to note about convergence in distribution is that if X_n converges in distribution to X, and if X is a *constant* random variable, say $X \equiv c$, then X_n converges in probability to c. To derive this result, first observe that

$$\{|X_n - c| < \varepsilon\} = \{-\varepsilon < X_n - c < \varepsilon\}$$
$$= \{c - \varepsilon < X_n\} \cap \{X_n < c + \varepsilon\}.$$

It is then easy to see that

$$\mathsf{P}(|X_n - c| \geq \varepsilon) \leq \mathsf{P}(X_n \leq c - \varepsilon) + \mathsf{P}(X_n \geq c + \varepsilon)$$
$$\leq F_{X_n}(c - \varepsilon) + \mathsf{P}(X_n > c + \varepsilon/2)$$
$$= F_{X_n}(c - \varepsilon) + 1 - F_{X_n}(c + \varepsilon/2).$$

Since $X \equiv c$, $F_X(x) = I_{[c,\infty)}(x)$. Therefore, $F_{X_n}(c-\varepsilon) \to 0$ and $F_{X_n}(c+\varepsilon/2) \to 1$.

The third thing to note about convergence in distribution is that it is possible to have X_n converging in distribution to X, while X_n does not converge in probability to X. (Of course, such an X cannot be a constant random variable.) For example, let $U \sim \text{uniform}[0,1]$, and put $X_n := U$ and $X := 1-U$. It is easy to show that X is also uniform$[0,1]$. Thus, $F_{X_n}(x) = F_X(x)$ for all n and all x, and thus X_n converges in distribution to X. However, given any $0 < \varepsilon < 1$, observe that $|X_n - X| < \varepsilon$ if and only if

$$-\varepsilon < X_n - X < \varepsilon.$$

Using the definitions of X_n and X, we have

$$-\varepsilon < U - (1-U) < \varepsilon,$$

or

$$\frac{1-\varepsilon}{2} < U < \frac{1+\varepsilon}{2}.$$

Thus,

$$P(|X_n - X| < \varepsilon) = P\left(\frac{1-\varepsilon}{2} < U < \frac{1+\varepsilon}{2}\right)$$
$$= \frac{1+\varepsilon}{2} - \frac{1-\varepsilon}{2} = \varepsilon.$$

This is equivalent to

$$P(|X_n - X| \geq \varepsilon) = 1 - \varepsilon \not\to 0 \quad \text{as} \quad n \to \infty.$$

The fourth thing to note about convergence in distribution is that it is equivalent to the condition

$$\lim_{n \to \infty} \mathsf{E}[g(X_n)] = \mathsf{E}[g(X)] \quad \text{for every bounded continuous function } g. \tag{14.1}$$

Example 14.5. If X_n converges in distribution to X, show that for every continuous function $c(x)$, $Y_n := c(X_n)$ converges in distribution to $Y := c(X)$.

Solution. It suffices to show that for every bounded continuous function $g(y)$, $\mathsf{E}[g(Y_n)] \to \mathsf{E}[g(Y)]$. Since $g(c(x))$ is a bounded and continuous function of x, we have $\mathsf{E}[g(c(X_n))] \to \mathsf{E}[g(c(X))]$.

A proof that (14.1) implies convergence in distribution is sketched in Problem 24. Conversely, suppose that X_n converges in distribution to X. We sketch a derivation of (14.1). For any $x_0 < x_1 < \cdots < x_m$, write

$$\mathsf{E}[g(X)] = \mathsf{E}[g(X)I_{(-\infty,x_0]}(X)] + \sum_{i=1}^m \mathsf{E}[g(X)I_{(x_{i-1},x_i]}(X)] + \mathsf{E}[g(X)I_{(x_m,\infty)}(X)].$$

Since g is bounded, say by B, the first and last terms on the right can be made as small as we like by taking x_0 small and x_m large. For example, to bound the last term, write

$$\left|\mathsf{E}[g(X)I_{(x_m,\infty)}(X)]\right| \leq B\mathsf{P}(X > x_m) \to 0$$

14.2 Convergence in distribution

as the position of x_m becomes large. The first term can be similarly bounded by letting $x_0 \to -\infty$. Once the *positions* of the first and last points x_0 and x_m have been fixed, assume that m is large enough that the spacing $x_i - x_{i-1}$ is small enough to exploit the uniform continuity of g on $[x_0, x_m]$ to write $g(x) \approx g(x_i)$ for $x \in (x_{i-1}, x_i]$. Then

$$E[g(X)] \approx \sum_{i=1}^{m} E[g(X)I_{(x_{i-1},x_i]}(X)]$$

$$\approx \sum_{i=1}^{m} E[g(x_i)I_{(x_{i-1},x_i]}(X)]$$

$$= \sum_{i=1}^{m} g(x_i) E[I_{(x_{i-1},x_i]}(X)]$$

$$= \sum_{i=1}^{m} g(x_i)[F_X(x_i) - F_X(x_{i-1})]$$

$$\approx \sum_{i=1}^{m} g(x_i)[F_{X_n}(x_i) - F_{X_n}(x_{i-1})]$$

$$= \sum_{i=1}^{m} E[g(x_i)I_{(x_{i-1},x_i]}(X_n)]$$

$$\approx \sum_{i=1}^{m} E[g(X_n)I_{(x_{i-1},x_i]}(X_n)]$$

$$\approx E[g(X_n)],$$

where in the fifth equation we have assumed that n is large and that the x_i are continuity points of F_X. (This can always be arranged since the set of discontinuities is at most countable by Problem 52.)

Example 14.6. Show that if X_n converges in distribution to X, then the characteristic function of X_n converges to the characteristic function of X.

Solution. Fix any v, and take $g(x) = e^{jvx}$. Then

$$\varphi_{X_n}(v) = E[e^{jvX_n}] = E[g(X_n)] \to E[g(X)] = E[e^{jvX}] = \varphi_X(v).$$

The fifth thing to note about convergence in distribution is that it is equivalent to convergence of the corresponding characteristic functions. The fact that $\varphi_{X_n}(v) \to \varphi_X(v)$ for all v implies X_n converges in distribution to X is proved in [3, p. 349, Theorem 26.3].

Example 14.7. Let $X_n \sim N(m_n, \sigma_n^2)$, where $m_n \to m$ and $\sigma_n^2 \to \sigma^2$. If $X \sim N(m, \sigma^2)$, show that X_n converges in distribution to X.

Solution. It suffices to show that the characteristic function of X_n converges to the characteristic function of X. Write

$$\varphi_{X_n}(v) = e^{jm_n v - \sigma_n^2 v^2/2} \to e^{jmv - \sigma^2 v^2/2} = \varphi_X(v).$$

Example 14.8. Let $X_n \sim N(m_n, \sigma_n^2)$. If X_n converges in mean square to some $X \in L^2$, show that $X \sim N(m, \sigma^2)$, where

$$m := \lim_{n \to \infty} m_n \quad \text{and} \quad \sigma^2 := \lim_{n \to \infty} \sigma_n^2.$$

Solution. Recall that if X_n converges in mean square to X, then $\mathsf{E}[X_n] \to \mathsf{E}[X]$ (Example 13.11), and $\mathsf{E}[X_n^2] \to \mathsf{E}[X^2]$ (Problem 22 in Chapter 13). It follows that

$$\sigma_n^2 = \mathsf{E}[X_n^2] - m_n^2 \to \mathsf{E}[X^2] - (\mathsf{E}[X])^2 =: \sigma^2.$$

To conclude, we use the fact that convergence in mean square implies convergence in probability, which implies convergence in distribution, which is equivalent to convergence of the characteristic functions. Then since

$$\varphi_{X_n}(v) = e^{jvm_n - \sigma_n^2 v^2/2} \to e^{jvm - \sigma^2 v^2/2},$$

it follows that $X \sim N(m, \sigma^2)$.

We summarize the preceding example by saying that **the mean-square limit of a sequence of Gaussians is Gaussian with mean and variance equal to the limit of the means and variances**. Note that in these last two examples, it is not required that the X_n be jointly Gaussian, only that they be individually Gaussian.

Example 14.9. If X_t is a mean-square-continuous Gaussian process, then the mean-square integral $\int_a^b X_t \, dt$ is a Gaussian random variable.

Solution. The mean-square integral was developed in Section 13.2 as the mean-square limit of partial sums of the form

$$\sum_{i=1}^n X_{\tau_i}(t_i - t_{i-1}).$$

Since X_t is a Gaussian process, these partial sums are Gaussian, and so their mean-square limit must be a Gaussian random variable.

An important instance of convergence in distribution is the **central limit theorem**. This result says that if $X_1, X_2, \ldots$ are independent, identically distributed random variables with finite mean m and finite variance σ^2, and if

$$M_n := \frac{1}{n} \sum_{i=1}^n X_i \quad \text{and} \quad Y_n := \frac{M_n - m}{\sigma/\sqrt{n}},$$

then

$$\lim_{n \to \infty} F_{Y_n}(y) = \Phi(y) := \frac{1}{\sqrt{2\pi}} \int_{-\infty}^y e^{-t^2/2} \, dt.$$

14.2 Convergence in distribution

In other words, for all y, $F_{Y_n}(y)$ converges to the standard normal cdf $\Phi(y)$ (which is continuous at all points y). To better see the difference between the weak law of large numbers and the central limit theorem, we specialize to the case $m = 0$ and $\sigma^2 = 1$. In this case,

$$Y_n = \frac{1}{\sqrt{n}} \sum_{i=1}^{n} X_i.$$

In other words, the sample mean M_n divides the sum $X_1 + \cdots + X_n$ by n while Y_n divides the sum only by $\sqrt{n}$. The difference is that M_n converges in probability (and in distribution) to the *constant* zero, while Y_n converges in distribution to an $N(0,1)$ *random variable*. Notice also that the weak law requires only uncorrelated random variables, while the central limit theorem requires i.i.d. random variables. The central limit theorem was derived in Section 5.6, where examples and problems can also be found. The central limit theorem was also used to determine confidence intervals in Chapter 6.

Example 14.10. Let N_t be a Poisson process of rate λ. Show that

$$Y_n := \frac{\frac{N_n}{n} - \lambda}{\sqrt{\lambda/n}}$$

converges in distribution to an $N(0,1)$ random variable.

Solution. Since $N_0 \equiv 0$,

$$\frac{N_n}{n} = \frac{1}{n} \sum_{k=1}^{n} (N_k - N_{k-1}).$$

By the independent increments property of the Poisson process, the terms of the sum are i.i.d. Poisson(λ) random variables, with mean λ and variance λ. Hence, Y_n has the structure to apply the central limit theorem, and so $F_{Y_n}(y) \to \Phi(y)$ for all y.

The next example is a version of **Slutsky's theorem**. We used it in our analysis of confidence intervals in Chapter 6.

Example 14.11. Let Y_n be a sequence of random variables with corresponding cdfs F_n. Suppose that F_n converges to a continuous cdf F. Suppose also that U_n converges in probability to 1. Show that

$$\lim_{n \to \infty} \mathsf{P}(Y_n \leq yU_n) = F(y).$$

Solution. The result is intuitive. For large n, $U_n \approx 1$ and $F_n(y) \approx F(y)$ suggest that

$$\mathsf{P}(Y_n \leq yU_n) \approx \mathsf{P}(Y_n \leq y) = F_n(y) \approx F(y).$$

The precise details are more involved. Fix any $y > 0$ and $0 < \delta < 1$. Then $\mathsf{P}(Y_n \leq yU_n)$ is equal to

$$\mathsf{P}(Y_n \leq yU_n, |U_n - 1| < \delta) + \mathsf{P}(Y_n \leq yU_n, |U_n - 1| \geq \delta).$$

The second term is upper bounded by $\mathsf{P}(|U_n - 1| \geq \delta)$, which goes to zero. Rewrite the first term as

$$\mathsf{P}(Y_n \leq yU_n, 1 - \delta < U_n < 1 + \delta),$$

which is equal to
$$P(Y_n \leq yU_n, y(1-\delta) < yU_n < y(1+\delta)). \quad (14.2)$$
Now this is upper bounded by
$$P(Y_n \leq y(1+\delta)) = F_n(y(1+\delta)).$$
Thus,
$$\overline{\lim_{n \to \infty}} P(Y_n \leq yU_n) \leq F(y(1+\delta)).$$
Next, (14.2) is lower bounded by
$$P(Y_n \leq y(1-\delta), |U_n - 1| < \delta),$$
which is equal to
$$P(Y_n \leq y(1-\delta)) - P(Y_n \leq y(1-\delta), |U_n - 1| \geq \delta).$$
Now the second term satisfies
$$P(Y_n \leq y(1-\delta), |U_n - 1| \geq \delta) \leq P(|U_n - 1| \geq \delta) \to 0.$$
In light of these observations,
$$\varliminf_{n \to \infty} P(Y_n \leq yU_n) \geq F(y(1-\delta)).$$
Since δ was arbitrary, and since F is continuous,
$$\lim_{n \to \infty} P(Y_n \leq yU_n) = F(y).$$
The case $y < 0$ is similar.

See the Exam preparation section at the end of the chapter for a summary of important facts about convergence in distribution.

14.3 Almost-sure convergence

Let X_n be any sequence of random variables, and let X be any other random variable. Put
$$G := \{\omega \in \Omega : \lim_{n \to \infty} X_n(\omega) = X(\omega)\}.$$
In other words, G is the set of sample points $\omega \in \Omega$ for which the sequence of real numbers $X_n(\omega)$ converges to the real number $X(\omega)$. We think of G as the set of "good" ω's for which $X_n(\omega) \to X(\omega)$. Similarly, we think of the complement of G, G^c, as the set of "bad" ω's for which $X_n(\omega) \not\to X(\omega)$.

We say that X_n **converges almost surely (a.s.)** to X if[3]
$$P(\{\omega \in \Omega : \lim_{n \to \infty} X_n(\omega) \neq X(\omega)\}) = 0. \quad (14.3)$$

14.3 Almost-sure convergence

In other words, X_n converges almost surely to X if the "bad" set G^c has probability zero. We write $X_n \to X$ a.s. to indicate that X_n converges almost surely to X.

If it should happen that the bad set $G^c = \emptyset$, then $X_n(\omega) \to X(\omega)$ for every $\omega \in \Omega$. This is called **sure convergence**, and is a special case of almost-sure convergence.

Because almost-sure convergence is so closely linked to the convergence of ordinary sequences of real numbers, many results are easy to derive.

***Example* 14.12.** Show that if $X_n \to X$ a.s. and $Y_n \to Y$ a.s., then $X_n + Y_n \to X + Y$ a.s.

Solution. Let $G_X := \{X_n \to X\}$ and $G_Y := \{Y_n \to Y\}$. In other words, G_X and G_Y are the "good" sets for the sequences X_n and Y_n respectively. Now consider any $\omega \in G_X \cap G_Y$. For such ω, the sequence of real numbers $X_n(\omega)$ converges to the real number $X(\omega)$, and the sequence of real numbers $Y_n(\omega)$ converges to the real number $Y(\omega)$. Hence, from convergence theory for sequences of real numbers,

$$X_n(\omega) + Y_n(\omega) \to X(\omega) + Y(\omega). \tag{14.4}$$

At this point, we have shown that

$$G_X \cap G_Y \subset G, \tag{14.5}$$

where G denotes the set of all ω for which (14.4) holds. To prove that $X_n + Y_n \to X + Y$ a.s., we must show that $\mathsf{P}(G^c) = 0$. On account of (14.5), $G^c \subset G_X^c \cup G_Y^c$. Hence,

$$\mathsf{P}(G^c) \leq \mathsf{P}(G_X^c) + \mathsf{P}(G_Y^c),$$

and the two terms on the right are zero because X_n and Y_n both converge almost surely.

In order to discuss almost-sure convergence in more detail, it is helpful to characterize when (14.3) holds. Recall that a sequence of real numbers x_n is said to converge to a real number x if for every $\varepsilon > 0$, there is a positive integer N such that for all $n \geq N$, $|x_n - x| < \varepsilon$. Hence,

$$\{X_n \to X\} = \bigcap_{\varepsilon > 0} \bigcup_{N=1}^{\infty} \bigcap_{n=N}^{\infty} \{|X_n - X| < \varepsilon\}.$$

Equivalently,

$$\{X_n \not\to X\} = \bigcup_{\varepsilon > 0} \bigcap_{N=1}^{\infty} \bigcup_{n=N}^{\infty} \{|X_n - X| \geq \varepsilon\}.$$

It is convenient to put

$$A_n(\varepsilon) := \{|X_n - X| \geq \varepsilon\}, \quad B_N(\varepsilon) := \bigcup_{n=N}^{\infty} A_n(\varepsilon),$$

and

$$A(\varepsilon) := \bigcap_{N=1}^{\infty} B_N(\varepsilon). \tag{14.6}$$

Then
$$P(\{X_n \not\to X\}) = P\left(\bigcup_{\varepsilon>0} A(\varepsilon)\right).$$

If $X_n \to X$ a.s., then
$$0 = P(\{X_n \not\to X\}) = P\left(\bigcup_{\varepsilon>0} A(\varepsilon)\right) \geq P(A(\varepsilon_0))$$

for any choice of $\varepsilon_0 > 0$. Conversely, suppose $P(A(\varepsilon)) = 0$ for every positive ε. We claim that $X_n \to X$ a.s. To see this, observe that in the earlier characterization of the convergence of a sequence of real numbers, we could have restricted attention to values of ε of the form $\varepsilon = 1/k$ for positive integers k. In other words, a sequence of real numbers x_n converges to a real number x if and only if for every positive integer k, there is a positive integer N such that for all $n \geq N$, $|x_n - x| < 1/k$. Hence,

$$P(\{X_n \not\to X\}) = P\left(\bigcup_{k=1}^{\infty} A(1/k)\right) \leq \sum_{k=1}^{\infty} P(A(1/k)).$$

From this we see that if $P(A(\varepsilon)) = 0$ for all $\varepsilon > 0$, then $P(\{X_n \not\to X\}) = 0$.

To say more about almost-sure convergence, we need to examine (14.6) more closely. Observe that
$$B_N(\varepsilon) = \bigcup_{n=N}^{\infty} A_n(\varepsilon) \supset \bigcup_{n=N+1}^{\infty} A_n(\varepsilon) = B_{N+1}(\varepsilon).$$

By limit property (1.16),
$$P(A(\varepsilon)) = P\left(\bigcap_{N=1}^{\infty} B_N(\varepsilon)\right) = \lim_{N \to \infty} P(B_N(\varepsilon)).$$

The next two examples use this equation to derive important results about almost-sure convergence.

Example 14.13. Show that if $X_n \to X$ a.s., then X_n converges in probability to X.

Solution. Recall that, by definition, X_n converges in probability to X if and only if $P(A_N(\varepsilon)) \to 0$ for every $\varepsilon > 0$. If $X_n \to X$ a.s., then for every $\varepsilon > 0$,
$$0 = P(A(\varepsilon)) = \lim_{N \to \infty} P(B_N(\varepsilon)).$$

Next, since
$$P(B_N(\varepsilon)) = P\left(\bigcup_{n=N}^{\infty} A_n(\varepsilon)\right) \geq P(A_N(\varepsilon)),$$

it follows that $P(A_N(\varepsilon)) \to 0$ too.

14.3 Almost-sure convergence

Example 14.14. Show that if

$$\sum_{n=1}^{\infty} \mathsf{P}(A_n(\varepsilon)) < \infty, \qquad (14.7)$$

holds for all $\varepsilon > 0$, then $X_n \to X$ a.s.

Solution. For any $\varepsilon > 0$,

$$\mathsf{P}(A(\varepsilon)) = \lim_{N \to \infty} \mathsf{P}(B_N(\varepsilon))$$
$$= \lim_{N \to \infty} \mathsf{P}\left(\bigcup_{n=N}^{\infty} A_n(\varepsilon)\right)$$
$$\leq \lim_{N \to \infty} \sum_{n=N}^{\infty} \mathsf{P}(A_n(\varepsilon))$$
$$= 0, \quad \text{on account of (14.7)}.$$

What we have done here is derive a particular instance of the first **Borel–Cantelli lemma** (cf. Problem 38 in Chapter 1).

Example 14.15. Let $X_1, X_2, \ldots$ be i.i.d. zero-mean random variables with finite fourth moment. Show that

$$M_n := \frac{1}{n} \sum_{i=1}^{n} X_i \to 0 \text{ a.s.}$$

Solution. We already know from Example 13.3 that M_n converges in mean square to zero, and hence, M_n converges to zero in probability and in distribution as well. Unfortunately, as shown in the next example, convergence in mean does not imply almost-sure convergence. However, by the previous example, the almost-sure convergence of M_n to zero will be established if we can show that for every $\varepsilon > 0$,

$$\sum_{n=1}^{\infty} \mathsf{P}(|M_n| \geq \varepsilon) < \infty.$$

By the Markov inequality,

$$\mathsf{P}(|M_n| \geq \varepsilon) = \mathsf{P}(|M_n|^4 \geq \varepsilon^4) \leq \frac{\mathsf{E}[|M_n|^4]}{\varepsilon^4}.$$

By Problem 43, there are finite, nonnegative constants α (depending on $\mathsf{E}[X_i^4]$) and β (depending on $\mathsf{E}[X_i^2]$) such that

$$\mathsf{E}[|M_n|^4] \leq \frac{\alpha}{n^3} + \frac{\beta}{n^2}.$$

We can now write

$$\sum_{n=1}^{\infty} \mathsf{P}(|M_n| \geq \varepsilon) \leq \frac{\alpha}{\varepsilon^4} \sum_{n=1}^{\infty} \frac{1}{n^3} + \frac{\beta}{\varepsilon^4} \sum_{n=1}^{\infty} \frac{1}{n^2} < \infty$$

by Problem 42.

The preceding example is an instance of the **strong law of large numbers**. The derivation in the example is quite simple because of the assumption of finite fourth moments (which implies finiteness of the third, second, and first moments by Lyapunov's inequality). A derivation assuming only finite second moments can be found in [23, pp. 326–327], and assuming only finite first moments in [23, pp. 329–331].

Strong law of large numbers (SLLN). *Let $X_1, X_2, \ldots$ be independent, identically distributed random variables with finite mean m. Then*

$$M_n := \frac{1}{n} \sum_{i=1}^{n} X_i \to m \text{ a.s.}$$

Since almost-sure convergence implies convergence in probability, the following form of the weak law of large numbers holds.[a]

Weak law of large numbers (WLLN). *Let $X_1, X_2, \ldots$ be independent, identically distributed random variables with finite mean m. Then*

$$M_n := \frac{1}{n} \sum_{i=1}^{n} X_i$$

converges in probability to m.

The weak law of large numbers in Example 14.1, which relied on the mean-square version in Example 13.3, required finite second moments and uncorrelated random variables. The above form does not require finite second moments, but does require independent random variables.

Example 14.16. Let W_t be a Wiener process with $\mathsf{E}[W_t^2] = \sigma^2 t$. Use the strong law to show that W_n/n converges almost surely to zero.

Solution. Since $W_0 \equiv 0$, we can write

$$\frac{W_n}{n} = \frac{1}{n} \sum_{k=1}^{n} (W_k - W_{k-1}).$$

By the independent increments property of the Wiener process, the terms of the sum are i.i.d. $N(0, \sigma^2)$ random variables. By the strong law, this sum converges almost surely to zero.

Example 14.17. We construct a sequence of random variables that converges in mean to zero, but does not converge almost surely to zero. Fix any positive integer n. Then n

---

[a] A proof that does not rely on the strong law can be given [7, pp. 128–130, Theorem 4].

14.3 Almost-sure convergence

can be uniquely represented as $n = 2^m + k$, where m and k are integers satisfying $m \geq 0$ and $0 \leq k \leq 2^m - 1$. Define

$$g_n(x) = g_{2^m + k}(x) = I_{\left[\frac{k}{2^m}, \frac{k+1}{2^m}\right)}(x).$$

For example, taking $m = 2$ and $k = 0, 1, 2, 3$, which corresponds to $n = 4, 5, 6, 7$, we find

$$g_4(x) = I_{\left[0, \frac{1}{4}\right)}(x), \quad g_5(x) = I_{\left[\frac{1}{4}, \frac{1}{2}\right)}(x),$$

and

$$g_6(x) = I_{\left[\frac{1}{2}, \frac{3}{4}\right)}(x), \quad g_7(x) = I_{\left[\frac{3}{4}, 1\right)}(x).$$

For fixed m, as k goes from 0 to $2^m - 1$, $g_{2^m + k}$ is a sequence of pulses moving from left to right. This is repeated for $m + 1$ with twice as many pulses that are half as wide. The two key ideas are that the pulses get narrower and that for any fixed $x \in [0, 1)$, $g_n(x) = 1$ for infinitely many n.

Now let $U \sim \text{uniform}[0, 1)$. Then

$$\mathsf{E}[g_n(U)] = \mathsf{P}\left(\frac{k}{2^m} \leq X < \frac{k+1}{2^m}\right) = \frac{1}{2^m}.$$

Since $m \to \infty$ as $n \to \infty$, we see that $g_n(U)$ converges in mean to zero. It then follows that $g_n(U)$ converges in probability to zero. Since almost-sure convergence also implies convergence in probability, the only possible almost-sure limit is zero.[b] However, we now show that $g_n(U)$ does not converge almost surely to zero. Fix any $x \in [0, 1)$. Then for each $m = 0, 1, 2, \ldots$,

$$\frac{k}{2^m} \leq x < \frac{k+1}{2^m}$$

for some k satisfying $0 \leq k \leq 2^m - 1$. For these values of m and k, $g_{2^m + k}(x) = 1$. In other words, there are infinitely many values of $n = 2^m + k$ for which $g_n(x) = 1$. Hence, for $0 \leq x < 1$, $g_n(x)$ does not converge to zero. Therefore,

$$\{U \in [0, 1)\} \subset \{g_n(U) \not\to 0\},$$

and it follows that

$$\mathsf{P}(\{g_n(U) \not\to 0\}) \geq \mathsf{P}(\{U \in [0, 1)\}) = 1.$$

The Skorohod representation

The **Skorohod representation** says that if X_n converges in distribution to X, then we can construct random variables Y_n and Y with $F_{Y_n} = F_{X_n}$, $F_Y = F_X$, and such that Y_n converges almost surely to Y. This can often simplify proofs concerning convergence in distribution.

---

[b] Limits in probability are unique by Problem 5.

***Example* 14.18.** Let X_n converge in distribution to X, and let $c(x)$ be a continuous function. Show that $c(X_n)$ converges in distribution to $c(X)$.

Solution. Let Y_n and Y be as given by the Skorohod representation. Since Y_n converges almost surely to Y, the set

$$G := \{\omega \in \Omega : Y_n(\omega) \to Y(\omega)\}$$

has the property that $\mathsf{P}(G^c) = 0$. Fix any $\omega \in G$. Then $Y_n(\omega) \to Y(\omega)$. Since c is continuous,

$$c(Y_n(\omega)) \to c(Y(\omega)).$$

Thus, $c(Y_n)$ converges almost surely to $c(Y)$. Now recall that almost-sure convergence implies convergence in probability, which implies convergence in distribution. Hence, $c(Y_n)$ converges in distribution to $c(Y)$. To conclude, observe that since Y_n and X_n have the same cumulative distribution function, so do $c(Y_n)$ and $c(X_n)$. Similarly, $c(Y)$ and $c(X)$ have the same cumulative distribution function. Thus, $c(X_n)$ converges in distribution to $c(X)$.

We now derive the Skorohod representation. For $0 < u < 1$, let

$$G_n(u) := \min\{x \in \mathbb{R} : F_{X_n}(x) \geq u\},$$

and

$$G(u) := \min\{x \in \mathbb{R} : F_X(x) \geq u\}.$$

By Problem 39(a) in Chapter 11,

$$G(u) \leq x \Leftrightarrow u \leq F_X(x),$$

or, equivalently,

$$G(u) > x \Leftrightarrow u > F_X(x),$$

and similarly for G_n and F_{X_n}. Now let $U \sim \text{uniform}(0,1)$. By Problem 39(b) in Chapter 11, $Y_n := G_n(U)$ and $Y := G(U)$ satisfy $F_{Y_n} = F_{X_n}$ and $F_Y = F_X$, respectively.

From the definition of G, it is easy to see that G is nondecreasing. Hence, its set of discontinuities, call it D, is at most countable (Problem 52), and so $\mathsf{P}(U \in D) = 0$ by Problem 53. We show below that for $u \notin D$, $G_n(u) \to G(u)$. It then follows that

$$Y_n(\omega) := G_n(U(\omega)) \to G(U(\omega)) =: Y(\omega),$$

except for $\omega \in \{\omega : U(\omega) \in D\}$, which has probability zero.

Fix any $u \notin D$, and let $\varepsilon > 0$ be given. Then between $G(u) - \varepsilon$ and $G(u)$ we can select a point x,

$$G(u) - \varepsilon < x < G(u),$$

that is a continuity point of F_X. Since $x < G(u)$,

$$F_X(x) < u.$$

Since x is a continuity point of F_X, $F_{X_n}(x)$ must be close to $F_X(x)$ for large n. Thus, for large n, $F_{X_n}(x) < u$. But this implies $G_n(u) > x$. Thus,

$$G(u) - \varepsilon < x < G_n(u),$$

and it follows that

$$G(u) \leq \varliminf_{n \to \infty} G_n(u).$$

To obtain the reverse inequality involving the $\varlimsup$, fix any u' with $u < u' < 1$. Fix any $\varepsilon > 0$, and select another continuity point of F_X, again called x, such that

$$G(u') < x < G(u') + \varepsilon.$$

Then $G(u') \leq x$, and so $u' \leq F_X(x)$. But then $u < F_X(x)$. Since $F_{X_n}(x) \to F_X(x)$, for large n, $u < F_{X_n}(x)$, which implies $G_n(u) \leq x$. It then follows that

$$\varlimsup_{n \to \infty} G_n(u) \leq G(u').$$

Since u is a continuity point of G, we can let $u' \to u$ to get

$$\varlimsup_{n \to \infty} G_n(u) \leq G(u).$$

It now follows that $G_n(u) \to G(u)$ as claimed.

See the Exam preparation section at the end of the chapter for a summary of important facts about almost-sure convergence.

Notes

14.2: Convergence in distribution

Note 1. In the text up to now, we have usually said that if $\mathsf{E}[X^2] = 0$, then X is the zero random variable. However, it is customary to define X to be the **zero random variable** if $\mathsf{P}(X = 0) = 1$; with this definition, it is easy to see that the cdf of X is the unit-step function, $u(x) = I_{[0,\infty)}(x)$. Fortunately, these two definitions are equivalent. If $\mathsf{P}(X = 0) = 1$, then X is a discrete random variable that takes only the value zero with positive probability. Hence, $\mathsf{E}[X^2] = 0$. Conversely, observe that

$$\mathsf{P}(X \neq 0) = \mathsf{P}(X^2 > 0) = \mathsf{P}\left(\bigcup_{n=1}^{\infty} \{X^2 > 1/n\}\right) = \lim_{N \to \infty} \mathsf{P}(X^2 > 1/N),$$

where the last step follows by the limit property (1.15). If $\mathsf{E}[X^2] = 0$, we can apply the Markov inequality to obtain $\mathsf{P}(X^2 > 1/N) \leq N\mathsf{E}[X^2] = 0$. Hence, $\mathsf{P}(X \neq 0) = 0$ and then $\mathsf{P}(X = 0) = 1$.

Note 2. The **limit inferior**, denoted by $\liminf$ or $\varliminf$, and the **limit superior**, denoted by $\limsup$ or $\varlimsup$, of a sequence of real numbers z_n are defined as follows. First put

$$\underline{z}_n := \inf_{k \geq n} z_k \quad \text{and} \quad \bar{z}_n := \sup_{k \geq n} z_k.$$

Then $\underline{z}_n \leq \bar{z}_n$. Furthermore, a little thought shows that

$$\underline{z}_n \leq \underline{z}_{n+1} \quad \text{and} \quad \bar{z}_n \geq \bar{z}_{n+1}.$$

In other words, the $\underline{z}_n$ are getting larger, and the $\bar{z}_n$ are getting smaller. Since bounded monotonic sequences always have limits [51, p. 56], we have[c]

$$\lim_n \underline{z}_n \leq \lim_n \bar{z}_n.$$

Defining

$$\varliminf_n z_n := \lim_n \underline{z}_n \quad \text{and} \quad \varlimsup_n z_n := \lim_n \bar{z}_n,$$

we have

$$\varliminf_n z_n \leq \varlimsup_n z_n.$$

Furthermore, it is not too hard to show that the lim inf equals the lim sup if and only if $\lim_n z_n$ exists, in which case, the limit is the common value of the lim inf and the lim sup.

Remark. The reader should note that since $\underline{z}_n$ is increasing, and since $\bar{z}_n$ is decreasing,

$$\lim_n \underline{z}_n = \sup_n \underline{z}_n \quad \text{and} \quad \lim_n \bar{z}_n = \inf_n \bar{z}_n.$$

Hence, in the literature, one frequently finds the formulas

$$\varliminf_n z_n = \sup_n \inf_{k \geq n} z_k \quad \text{and} \quad \varlimsup_n z_n = \inf_n \sup_{k \geq n} z_k.$$

14.3: Almost-sure convergence

Note 3. In order that (14.3) be well defined, it is necessary that the set

$$\{\omega \in \Omega : \lim_{n \to \infty} X_n(\omega) \neq X(\omega)\}$$

be an **event** in the technical sense of Note **1** in Chapter 1. This is assured by the assumption that each X_n is a random variable (the term "random variable" is used in the technical sense of Note **1** in Chapter 2). The fact that this assumption is sufficient is demonstrated in more advanced texts, e.g., [3, pp. 183–184].

Problems

14.1: Convergence in probability

1. Let $X_n \sim \text{Cauchy}(1/n)$. Show that X_n converges in probability to zero.

---

[c]The sequences treated in this chapter are always bounded. What happens if z_n is not bounded? Consider the following examples: (*i*) $z_n = n$; (*ii*) $z_n = (-1)^n n$; (*iii*) $z_n = [1 - (-1)^n]n$.

2. Let c_n be a converging sequence of real numbers with limit c. Define the constant random variables $Y_n \equiv c_n$ and $Y \equiv c$. Show by direct analysis of $\mathsf{P}(|Y_n - Y| \geq \varepsilon)$ that Y_n converges in probability to Y.

   Remark. Here is an easier approach (that you are *not* to use for your solution of this problem). Since c_n and c are deterministic,
   $$\mathsf{E}[|c_n - c|] = |c_n - c| \to 0.$$
   Since convergence in mean implies convergence in probability, Y_n converges in probability to Y.

3. Let $U \sim \text{uniform}[0,1]$, and put
   $$X_n := nI_{[0,1/\sqrt{n}]}(U), \quad n = 1, 2, \ldots.$$
   Does X_n converge in probability to zero?

4. Let V be any continuous random variable with an even density, and let c_n be any positive sequence with $c_n \to \infty$. Show that $X_n := V/c_n$ converges in probability to zero by direct analysis of $\mathsf{P}(|X_n| \geq \varepsilon)$.

   Remark. Here are some easier approaches (that you are *not* to use for your solution of this problem). If $\mathsf{E}[|V|] < \infty$, write
   $$\mathsf{E}[|X_n|] = \frac{\mathsf{E}[|V|]}{c_n} \to 0.$$
   Thus, X_n converges in mean, and therefore in probability, to zero. If $\mathsf{E}[|V|] = \infty$, put $U_n \equiv 1/c_n$ and $Y_n = V$. Then by Problem 2, U_n converges in probability to zero, and it is easy to see that Y_n converges in probability to V. Then, as pointed out in the text, $U_n Y_n$ converges in probability to $0 \cdot V = 0$.

5. Show that limits in probability are unique; i.e., show that if X_n converges in probability to X, and X_n converges in probability to Y, then $\mathsf{P}(X \neq Y) = 0$. *Hint:* Write
   $$\{X \neq Y\} = \bigcup_{k=1}^{\infty} \{|X - Y| \geq 1/k\},$$
   and use limit property (1.15).

6. Suppose you have shown that given any $\varepsilon > 0$, for sufficiently large n,
   $$\mathsf{P}(|X_n - X| \geq \varepsilon) < \varepsilon.$$
   Show that
   $$\lim_{n \to \infty} \mathsf{P}(|X_n - X| \geq \varepsilon) = 0 \quad \text{for every } \varepsilon > 0.$$

7. Let $g(x,y)$ be continuous, and suppose that X_n converges in probability to X, and that Y_n converges in probability to Y. In this problem you will show that $g(X_n, Y_n)$ converges in probability to $g(X, Y)$.

(a) Fix any $\varepsilon > 0$. Show that for sufficiently large α and β,
$$P(|X| > \alpha) < \varepsilon/4 \quad \text{and} \quad P(|Y| > \beta) < \varepsilon/4.$$

(b) Once α and β have been fixed, we can use the fact that $g(x,y)$ is uniformly continuous on the rectangle $|x| \leq 2\alpha$ and $|y| \leq 2\beta$. In other words, there is a $\delta > 0$ such that for all (x',y') and (x,y) in the rectangle and satisfying
$$|x' - x| \leq \delta \quad \text{and} \quad |y' - y| \leq \delta,$$
we have
$$|g(x',y') - g(x,y)| < \varepsilon.$$
There is no loss of generality if we assume that $\delta \leq \alpha$ and $\delta \leq \beta$. Show that if the four conditions
$$|X_n - X| < \delta, \quad |Y_n - Y| < \delta, \quad |X| \leq \alpha, \quad \text{and} \quad |Y| \leq \beta$$
hold, then
$$|g(X_n, Y_n) - g(X, Y)| < \varepsilon.$$

(c) Show that if n is large enough that
$$P(|X_n - X| \geq \delta) < \varepsilon/4 \quad \text{and} \quad P(|Y_n - Y| \geq \delta) < \varepsilon/4,$$
then
$$P(|g(X_n, Y_n) - g(X, Y)| \geq \varepsilon) < \varepsilon.$$

8. Let $X_1, X_2, \ldots$ be i.i.d. with common finite mean m and common finite variance σ^2. Also assume that $\gamma^4 := \mathsf{E}[X_i^4] < \infty$. Put
$$M_n := \frac{1}{n} \sum_{i=1}^{n} X_i \quad \text{and} \quad V_n := \frac{1}{n} \sum_{i=1}^{n} X_i^2.$$

(a) Explain (briefly) why V_n converges in probability to $\sigma^2 + m^2$.

(b) Explain (briefly) why
$$S_n^2 := \frac{n}{n-1}\left[\left(\frac{1}{n}\sum_{i=1}^{n} X_i^2\right) - M_n^2\right]$$
converges in probability to σ^2.

9. Let X_n converge in probability to X. Assume that there is a nonnegative random variable Y with $\mathsf{E}[Y] < \infty$ and such that $|X_n| \leq Y$ for all n.

(a) Show that $P(|X| \geq Y + 1) = 0$. (In other words, $|X| < Y + 1$ with probability one, from which it follows that $\mathsf{E}[|X|] \leq \mathsf{E}[Y] + 1 < \infty$.)

(b) Show that X_n converges in mean to X. Hints: Write

$$E[|X_n - X|] = E[|X_n - X|I_{A_n}] + E[|X_n - X|I_{A_n^c}],$$

where $A_n := \{|X_n - X| \geq \varepsilon\}$. Then use Problem 5 in Chapter 13 with $Z = Y + |X|$.

10. (a) Let g be a bounded, nonnegative function satisfying $\lim_{x \to 0} g(x) = 0$. Show that $\lim_{n \to \infty} E[g(X_n)] = 0$ if X_n converges in probability to zero.

 (b) Show that

$$\lim_{n \to \infty} E\left[\frac{|X_n|}{1 + |X_n|}\right] = 0$$

if and only if X_n converges in probability to zero. *Hint:* The function $x/(1+x)$ is increasing for $x \geq 0$.

14.2: Convergence in distribution

11. Let c_n be a converging sequence of real numbers with limit c. Define the constant random variables $Y_n \equiv c_n$ and $Y \equiv c$. Show that Y_n converges in distribution to Y by direct analysis of F_{Y_n} and F_Y.

 Remark. Here are some easier approaches (that you are *not* to use for your solution of this problem). Since convergence in probability implies convergence in distribution, this problem is an easy consequence of Problem 2. Alternatively, since convergence of characteristic functions implies convergence in distribution, observe that

$$\varphi_{Y_n}(v) = E[e^{jvc_n}] = e^{jvc_n} \to e^{jvc} = \varphi_Y(v).$$

12. Let X be a random variable, and let c_n be a positive sequence converging to limit c. Show that $Y_n := c_n X$ converges in distribution to $Y := cX$ by direct analysis of F_{Y_n} and F_Y. Consider separately the cases $c = 0$ and $0 < c < \infty$.

 Remark. Here are some easier approaches (that you are *not* to use for your solution of this problem). By Problem 2, $U_n \equiv c_n$ converges in probability to $U := c$. It is also easy to see that $V_n := X$ converges in probability to $V := X$. Then $U_n V_n$ converges in probability to $UV = cX$. Now use the fact that convergence in probability implies convergence in distribution. A simple proof using characteristic functions as in the previous remark is also possible.

13. For $t \geq 0$, let $X_t \leq Y_t \leq Z_t$ be three continuous-time random processes such that

$$\lim_{t \to \infty} F_{X_t}(y) = \lim_{t \to \infty} F_{Z_t}(y) = F(y)$$

for some continuous cdf F. Show that $F_{Y_t}(y) \to F(y)$ for all y.

14. Show that if X_n is sequence of exponential random variables that converges in mean to some X with $0 < E[X] < \infty$, then X is an exponential random variable.

15. Let X_n converge in distribution to a constant x, and let Y_n converge in distribution to a constant y. Determine whether or not $X_n + Y_n$ converges in distribution to $x + y$.

16. Show that the Wiener integral $Y := \int_0^\infty g(\tau) \, dW_\tau$ is Gaussian with zero mean and variance $\sigma^2 \int_0^\infty g(\tau)^2 \, d\tau$. *Hints:* The desired integral Y is the mean-square limit of the sequence Y_n defined in Problem 35 in Chapter 13. Use Example 14.8.

17. Let $g(t, \tau)$ be such that for each t, $\int_0^\infty g(t, \tau)^2 \, d\tau < \infty$. Define the process
$$X_t = \int_0^\infty g(t, \tau) \, dW_\tau.$$
Use the result of the preceding problem to show that for any $0 \le t_1 < \cdots < t_n < \infty$, the random vector of samples $X := [X_{t_1}, \ldots, X_{t_n}]'$ is Gaussian. *Hint:* Read the first paragraph of Section 9.2.

18. Let W_t be a Wiener process, and let $\int_0^\infty g(\tau)^2 \, d\tau < \infty$. Put
$$X_t := \int_0^t g(\tau) \, dW_\tau.$$
Determine whether or not X_t has independent increments.

19. Let X_t be a mean-square-continuous process for $a \le t \le b$. If X_t is a Gaussian process, show that the coefficients A_k in (13.11) of the Karhunen–Loève expansion are jointly Gaussian and independent.

20. If the moment generating functions $M_{X_n}(s)$ converge to the moment generating function $M_X(s)$, show that X_n converges in distribution to X. Also show that for nonnegative, integer-valued random variables, if the probability generating functions $G_{X_n}(z)$ converge to the probability generating function $G_X(z)$, then X_n converges in distribution to X. *Hint:* The paragraph following Example 14.6 may be useful.

21. Let X_n and X be integer-valued random variables with probability mass functions $p_n(k) := \mathsf{P}(X_n = k)$ and $p(k) := \mathsf{P}(X = k)$, respectively.

    (a) If X_n converges in distribution to X, show that for each k, $p_n(k) \to p(k)$.

    (b) If X_n and X are nonnegative, and if for each $k \ge 0$, $p_n(k) \to p(k)$, show that X_n converges in distribution to X.

22. Let p_n be a sequence of numbers lying between 0 and 1 and such that $np_n \to \lambda > 0$ as $n \to \infty$. Let $X_n \sim$ binomial(n, p_n), and let $X \sim$ Poisson(λ). Show that X_n converges in distribution to X. *Hints:* By the previous problem, it suffices to prove that the probability mass functions converge. **Stirling's formula,**
$$n! \sim \sqrt{2\pi} \, n^{n+1/2} e^{-n},$$
by which we mean
$$\lim_{n \to \infty} \frac{n!}{\sqrt{2\pi} \, n^{n+1/2} e^{-n}} = 1,$$
and the formula
$$\lim_{n \to \infty} \left(1 - \frac{q_n}{n}\right)^n = e^{-q}, \quad \text{if } q_n \to q,$$
may be helpful.

23. Let $X_n \sim$ binomial(n, p_n) and $X \sim$ Poisson(λ), where p_n and λ are as in the previous problem. Show that the probability generating function $G_{X_n}(z)$ converges to $G_X(z)$.

24. Suppose that X_n and X are such that for every bounded continuous function $g(x)$,
$$\lim_{n\to\infty} \mathsf{E}[g(X_n)] = \mathsf{E}[g(X)].$$
Show that X_n converges in distribution to X as follows:

(a) For $a < b$, sketch the three functions $I_{(-\infty,a]}(t), I_{(-\infty,b]}(t)$, and
$$g_{a,b}(t) := \begin{cases} 1, & t < a, \\ \dfrac{b-t}{b-a}, & a \le t \le b, \\ 0, & t > b. \end{cases}$$

(b) Your sketch in part (a) shows that
$$I_{(-\infty,a]}(t) \le g_{a,b}(t) \le I_{(-\infty,b]}(t).$$
Use these inequalities to show that for any random variable Y,
$$F_Y(a) \le \mathsf{E}[g_{a,b}(Y)] \le F_Y(b).$$

(c) For $\Delta x > 0$, use part (b) with $a = x$ and $b = x + \Delta x$ to show that
$$\overline{\lim_{n\to\infty}} F_{X_n}(x) \le F_X(x + \Delta x).$$

(d) For $\Delta x > 0$, use part (b) with $a = x - \Delta x$ and $b = x$ to show that
$$F_X(x - \Delta x) \le \underline{\lim_{n\to\infty}} F_{X_n}(x).$$

(e) If x is a continuity point of F_X, show that
$$\lim_{n\to\infty} F_{X_n}(x) = F_X(x).$$

25. Show that X_n converges in distribution to zero if and only if
$$\lim_{n\to\infty} \mathsf{E}\left[\frac{|X_n|}{1+|X_n|}\right] = 0.$$

Hint: Recall Problem 10.

26. Let $f(x)$ be a probability density function. Let X_n have density $f_n(x) = nf(nx)$. Determine whether or not X_n converges in probability to zero.

27. Let X_n converge in mean of order 2 to X. Determine whether or not
$$\lim_{n\to\infty} \mathsf{E}\left[X_n^2 e^{-X_n^2}\right] = \mathsf{E}\left[X^2 e^{-X^2}\right].$$

28. For $t \geq 0$, let Z_t be a continuous-time random process. Suppose that as $t \to \infty$, $F_{Z_t}(z)$ converges to a continuous cdf $F(z)$. Let $u(t)$ be a positive function of t such that $u(t) \to 1$ as $t \to \infty$. Show that

$$\lim_{t \to \infty} P(Z_t \leq z u(t)) = F(z).$$

Hint: Your answer should be simpler than the derivation in Example 14.11.

29. Let Z_t be as in the preceding problem. Show that if $c(t) \to c > 0$, then

$$\lim_{t \to \infty} P(c(t) Z_t \leq z) = F(z/c).$$

30. Let Z_t be as in Problem 28. Let $s(t) \to 0$ as $t \to \infty$. Show that if $X_t = Z_t + s(t)$, then $F_{X_t}(x) \to F(x)$.

31. Let N_t be a Poisson process of rate λ. Show that

$$Y_t := \frac{\frac{N_t}{t} - \lambda}{\sqrt{\lambda/t}}$$

converges in distribution to an $N(0,1)$ random variable. *Hints:* By Example 14.10, Y_n converges in distribution to an $N(0,1)$ random variable. Next, since N_t is a nondecreasing function of t,

$$N_{\lfloor t \rfloor} \leq N_t \leq N_{\lceil t \rceil},$$

where $\lfloor t \rfloor$ denotes the greatest integer less than or equal to t, and $\lceil t \rceil$ denotes the smallest integer greater than or equal to t. The preceding two problems and Problem 13 may be useful.

14.3: Almost-sure convergence

32. Let X_n converge almost surely to X.

   (a) Show that

   $$\frac{1}{1+X_n^2} \text{ converges almost surely to } \frac{1}{1+X^2}.$$

   (b) Determine whether or not

   $$\lim_{n \to \infty} \mathsf{E}\left[\frac{1}{1+X_n^2}\right] = \mathsf{E}\left[\frac{1}{1+X^2}\right].$$

   Justify your answer.

33. Let $X_n \to X$ a.s. and let $Y_n \to Y$ a.s. If $g(x,y)$ is a continuous function, show that $g(X_n, Y_n) \to g(X, Y)$ a.s.

34. Let $X_n \to X$ a.s., and suppose that $X = Y$ a.s. Show that $X_n \to Y$ a.s. (The statement $X = Y$ a.s. means $P(X \neq Y) = 0$.)

35. Show that almost-sure limits are unique; i.e., if $X_n \to X$ a.s. and $X_n \to Y$ a.s., then $X = Y$ a.s. (The statement $X = Y$ a.s. means $P(X \neq Y) = 0$.)

36. Suppose $X_n \to X$ a.s. and $Y_n \to Y$ a.s. Show that if $X_n \leq Y_n$ a.s. for all n, then $X \leq Y$ a.s. (The statement $X_n \leq Y_n$ a.s. means $P(X_n > Y_n) = 0$.)

37. If X_n converges almost surely and in mean, show that the two limits are equal almost surely. *Hint:* Problem 5 may be helpful.

38. In Problem 12, suppose that $\lim_{n \to \infty} c_n = c = \infty$. For each ω, compute the value of $Y(\omega) := \lim_{n \to \infty} c_n X(\omega)$.

39. Suppose state j of a discrete-time Markov chain is "transient" in the sense that

$$\sum_{n=1}^{\infty} p_{ij}^{(n)} < \infty,$$

where $p_{ij}^{(n)} := P(X_n = j | X_0 = i)$. Show that the chain visits state j infinitely often with probability zero; i.e., show that

$$P\left(\bigcap_{N=1}^{\infty} \bigcup_{n=N}^{\infty} \{X_n = j\} \,\Big|\, X_0 = i\right) = 0.$$

40. Let S be a nonnegative random variable with $E[S] < \infty$. Show that $S < \infty$ a.s. *Hints:* It is enough to show that $P(S = \infty) = 0$. Observe that

$$\{S = \infty\} = \bigcap_{n=1}^{\infty} \{S > n\}.$$

Now appeal to the limit property (1.16) and use the Markov inequality.

41. Under the assumptions of Problem 26 in Chapter 13, show that

$$\sum_{k=1}^{\infty} h_k X_k$$

is well defined as an almost-sure limit. *Hints:* It is enough to prove that

$$S := \sum_{k=1}^{\infty} |h_k X_k| < \infty \quad \text{a.s.}$$

Hence, the result of the preceding problem can be applied if it can be shown that $E[S] < \infty$. To this end, put

$$S_n := \sum_{k=1}^{n} |h_k| |X_k|.$$

By Problem 26 in Chapter 13, S_n converges in mean to $S \in L^1$. Use the nonnegativity of S_n and S along with Problem 22 in Chapter 13 to show that

$$E[S] = \lim_{n \to \infty} \sum_{k=1}^{n} |h_k| E[|X_k|] < \infty.$$

42. For $p > 1$, show that $\sum_{n=1}^{\infty} 1/n^p < \infty$.

43. Let $X_1, X_2, \ldots$ be i.i.d. with $\gamma := \mathsf{E}[X_i^4]$, $\sigma^2 := \mathsf{E}[X_i^2]$, and $\mathsf{E}[X_i] = 0$. Show that

$$M_n := \frac{1}{n}\sum_{i=1}^{n} X_i$$

satisfies $\mathsf{E}[M_n^4] = [n\gamma + 3n(n-1)\sigma^4]/n^4$.

44. If $X_n \sim \text{Laplace}(n)$, show that X_n converges almost surely to zero.

45. If $X_n \sim \text{Rayleigh}(1/n)$, show that X_n converges almost surely to zero.

46. Let X_n have the Pareto density

$$f_n(x) = \frac{(p-1)}{n^{p-1}} x^{-p}, \quad x \geq 1/n,$$

for some fixed $p > 2$. Show that X_n converges almost surely to zero.

47. Let $Y_n \sim \text{Bernoulli}(p_n)$, and put $X_n := X + n^2(-1)^n Y_n$, where $X \sim N(0,1)$.

    (a) Determine whether or not there is a sequence p_n such that X_n converges almost surely to X but not in mean.

    (b) Determine whether or not there is a sequence p_n such that X_n converges almost surely and in mean to X.

48. In Problem 8, explain why the assumption $\mathsf{E}[X_i^4] < \infty$ can be omitted.

49. Let N_t be a Poisson process of rate λ. Show that N_t/t converges almost surely to λ. *Hint:* First show that N_n/n converges almost surely to λ. Second, since N_t is a nondecreasing function of t, observe that

$$N_{\lfloor t \rfloor} \leq N_t \leq N_{\lceil t \rceil},$$

where $\lfloor t \rfloor$ denotes the greatest integer less than or equal to t, and $\lceil t \rceil$ denotes the smallest integer greater than or equal to t.

50. Let N_t be a renewal process as defined in Section 11.2. Next, let $X_1, X_2, \ldots$ denote the i.i.d. interarrival times, and let $T_n := X_1 + \cdots + X_n$ denote the nth occurrence time. Assume that the interarrival times have finite, positive mean μ.

    (a) Show that almost surely, for any $\varepsilon > 0$ and for all sufficiently large n,

    $$\sum_{k=1}^{n} X_k < n(\mu + \varepsilon).$$

    (b) Show that as $t \to \infty$, $N_t \to \infty$ a.s.; i.e., show that for any M, $N_t \geq M$ for all sufficiently large t.

    (c) Show that $n/T_n \to 1/\mu$ a.s.

(d) Show that $N_t/t \to 1/\mu$ a.s. *Hints:* On account of (c), if we put $Y_n := n/T_n$, then $Y_{N_t} \to 1/\mu$ a.s. since $N_t \to \infty$ by part (b). Also note that

$$T_{N_t} \le t < T_{N_t+1}.$$

51. Give an example of a sequence of random variables that converges almost surely to zero but not in mean.

52. Let G be a nondecreasing function defined on the closed interval $[a,b]$. Let D_ε denote the set of discontinuities of size greater than ε on $[a,b]$,

$$D_\varepsilon := \{u \in [a,b] : G(u+) - G(u-) > \varepsilon\},$$

with the understanding that $G(b+)$ means $G(b)$ and $G(a-)$ means $G(a)$. Show that if there are n points in D_ε, then

$$n < 2[G(b) - G(a)]/\varepsilon.$$

Remark. The set of all discontinuities of G on $[a,b]$, denoted by $D[a,b]$, is simply $\bigcup_{k=1}^\infty D_{1/k}$. Since this is a countable union of finite sets, $D[a,b]$ is at most countably infinite. If G is defined on the open interval $(0,1)$, we can write

$$D(0,1) = \bigcup_{n=3}^\infty D[1/n, 1-1/n].$$

Since this is a countable union of countably infinite sets, $D(0,1)$ is also countably infinite by Problem 17 in Chapter 1.

53. Let D be a countably infinite subset of $(0,1)$. Let $U \sim \text{uniform}(0,1)$. Show that $P(U \in D) = 0$. *Hint:* Since D is countably infinite, we can enumerate its elements as a sequence u_n. Fix any $\varepsilon > 0$ and put $K_n := (u_n - \varepsilon/2^n, u_n + \varepsilon/2^n)$. Observe that

$$D \subset \bigcup_{n=1}^\infty K_n.$$

Now bound

$$P\left(U \in \bigcup_{n=1}^\infty K_n\right).$$

Exam preparation

You may use the following suggestions to prepare a study sheet, including formulas mentioned that you have trouble remembering. You may also want to ask your instructor for additional suggestions.

14.1. Convergence in probability. There are three important facts to know.

(i) Convergence in mean of order p for any $1 \le p < \infty$ implies convergence in probability.

(ii) It is possible to have X_n converge in probability to X but not in mean of order p for any $1 \leq p < \infty$.

(iii) If X_n converges in probability to X and Y_n converges in probability to Y, and if $g(x,y)$ is continuous, then $g(X_n, Y_n)$ converges in probability to $g(X,Y)$.

14.2. Convergence in distribution. There are seven important facts to know.

(i) If X_n converges in probability to X, then X_n converges in distribution to X.

(ii) If X_n converges in distribution to a constant, then X_n converges in probability to that constant.

(iii) It is possible to have X_n converge in distribution to X but not in probability.

(iv) X_n converges in distribution to X if and only if for every bounded continuous function $g(x)$, $\mathsf{E}[g(X_n)] \to \mathsf{E}[g(X)]$. This can be used to give a simple proof that if X_n converges in distribution to X, and if $c(x)$ is a continuous function, then $c(X_n)$ converges in distribution to $c(X)$.

(v) X_n converges in distribution to X if and only if the characteristic functions $\varphi_{X_n}(\nu)$ converge to $\varphi_X(\nu)$. Examples 14.7–14.9 are important applications of this fact.

(vi) The central limit theorem.

(vii) Slutsky's theorem.

14.3. Almost-sure convergence. There are seven important facts to know.

(i) If X_n converges almost surely to X, then X_n converges in probability to X.

(ii) It is possible to have X_n converge in mean to X but not almost surely.

(iii) It is possible to have X_n converge almost surely to X but not in mean (Problem 51).

(iv) A sufficient condition to guarantee that X_n converges almost surely to X is that for every $\varepsilon > 0$,

$$\sum_{n=1}^{\infty} \mathsf{P}(|X_n - X| \geq \varepsilon) < \infty.$$

(v) The strong law of large numbers (SLLN).

(vi) Know two forms of the weak law of large numbers (WLLN).

(vii) The Skorohod representation.

Work any review problems assigned by your instructor. If you finish them, re-work your homework assignments.

15
Self similarity and long-range dependence

Prior to the 1990s, network analysis and design was carried out using long-established Markovian models such as the Poisson process [41, p. 1]. As self similarity was observed in the traffic of local-area networks [35], wide-area networks [43], and in World Wide Web traffic [13], a great research effort began to examine the impact of self similarity on network analysis and design. This research has yielded some surprising insights into questions about buffer size versus bandwidth, multiple-time-scale congestion control, connection duration prediction, and other issues [41, pp. 9–11].

The purpose of this chapter is to introduce the notion of self similarity and related concepts so that the student can be conversant with the kinds of stochastic processes being used to model network traffic. For more information, the student may consult the text by Beran [2], which includes numerous physical models and a historical overview of self similarity and long-range dependence.

Section 15.1 introduces the Hurst parameter and the notion of distributional self similarity for continuous-time processes. The concept of stationary increments is also presented. As an example of such processes, fractional Brownian motion is developed using the Wiener integral. In Section 15.2, we show that if one samples the increments of a continuous-time self-similar process with stationary increments, then the samples have a covariance function with a specific formula. It is shown that this formula is equivalent to specifying the variance of the sample mean for all values of n. Also, the power spectral density is found up to a multiplicative constant. Section 15.3 introduces the concept of asymptotic second-order self similarity and shows that it is equivalent to specifying the limiting form of the variance of the sample mean. The main result here is a sufficient condition on the power spectral density that guarantees asymptotic second-order self similarity. Section 15.4 defines long-range dependence. It is shown that every long-range-dependent process is asymptotically second-order self similar. Section 15.5 introduces ARMA processes, and Section 15.6 extends this to fractional ARIMA processes. Fractional ARIMA process provide a large class of models that are asymptotically second-order self similar.

15.1 Self similarity in continuous time

Loosely speaking, a continuous-time random process W_t is said to be **self similar** with **Hurst parameter** $H > 0$ if the process $W_{\lambda t}$ "looks like" the process $\lambda^H W_t$. If $\lambda > 1$, then time is speeded up for $W_{\lambda t}$ compared to the original process W_t. If $\lambda < 1$, then time is slowed down. The factor λ^H in $\lambda^H W_t$ either increases or decreases the magnitude (but not the time scale) compared with the original process. Thus, for a self-similar process, when time is speeded up, the apparent effect is the same as changing the magnitude of the original process, rather than its time scale.

The precise definition of "looks like" will be in the sense of finite-dimensional distributions. That is, for W_t to be self similar, we require that for every $\lambda > 0$, for every finite collection of times $t_1,\ldots,t_n$, all joint probabilities involving $W_{\lambda t_1},\ldots,W_{\lambda t_n}$ are the same as those involving $\lambda^H W_{t_1},\ldots,\lambda^H W_{t_n}$. The best example of a self-similar process is the Wiener process. This is easy to verify by comparing the joint characteristic functions of $W_{\lambda t_1},\ldots,W_{\lambda t_n}$ and $\lambda^H W_{t_1},\ldots,\lambda^H W_{t_n}$ for the correct value of H (Problem 2).

Implications of self similarity

Let us focus first on a single time point t. For a self-similar process, we must have

$$P(W_{\lambda t} \leq x) = P(\lambda^H W_t \leq x).$$

Taking $t = 1$, results in

$$P(W_\lambda \leq x) = P(\lambda^H W_1 \leq x).$$

Since $\lambda > 0$ is a dummy variable, we can call it t instead. Thus,

$$P(W_t \leq x) = P(t^H W_1 \leq x), \quad t > 0.$$

Now rewrite this as

$$P(W_t \leq x) = P(W_1 \leq t^{-H} x), \quad t > 0,$$

or, in terms of cumulative distribution functions,

$$F_{W_t}(x) = F_{W_1}(t^{-H} x), \quad t > 0.$$

It can now be shown (Problem 1) that W_t converges in distribution to the zero random variable as $t \to 0$. Similarly, as $t \to \infty$, W_t converges in distribution to a discrete random variable taking the values 0 and $\pm\infty$.

We next look at expectations of self-similar processes. We can write

$$E[W_{\lambda t}] = E[\lambda^H W_t] = \lambda^H E[W_t].$$

Setting $t = 1$ and replacing λ by t results in

$$E[W_t] = t^H E[W_1], \quad t > 0. \tag{15.1}$$

Hence, for a self-similar process, its mean function has the form of a constant times t^H for $t > 0$.

As another example, consider

$$E[W_{\lambda t}^2] = E[(\lambda^H W_t)^2] = \lambda^{2H} E[W_t^2]. \tag{15.2}$$

Arguing as above, we find that

$$E[W_t^2] = t^{2H} E[W_1^2], \quad t > 0. \tag{15.3}$$

We can also take $t = 0$ in (15.2) to get

$$E[W_0^2] = \lambda^{2H} E[W_0^2].$$

Since the left-hand side does not depend on λ, $E[W_0^2] = 0$, which implies $W_0 = 0$ a.s. Hence, (15.1) and (15.3) both continue to hold even when $t = 0$.[a]

---

[a] Using the formula $a^b = e^{b \ln a}$, $0^H = e^{H \ln 0} = e^{-\infty}$, since $H > 0$. Thus, $0^H = 0$.

15.1 Self similarity in continuous time

Example 15.1. Assuming that the Wiener process is self similar, show that the Hurst parameter must be $H = 1/2$.

Solution. Recall that for the Wiener process, we have $\mathsf{E}[W_t^2] = \sigma^2 t$. Thus, (15.2) implies that

$$\sigma^2(\lambda t) = \lambda^{2H} \sigma^2 t.$$

Hence, $H = 1/2$.

Stationary increments

A process W_t is said to have **stationary increments** if for every increment $\tau > 0$, the **increment process**

$$Z_t := W_t - W_{t-\tau}$$

is a stationary process in t. If W_t is self similar with Hurst parameter H, and has stationary increments, we say that W_t is **H-sssi**.

If Z_t is H-sssi, then $\mathsf{E}[Z_t]$ cannot depend on t; but by (15.1),

$$\mathsf{E}[Z_t] = \mathsf{E}[W_t - W_{t-\tau}] = [t^H - (t-\tau)^H]\mathsf{E}[W_1].$$

If $H \neq 1$, then we must have $\mathsf{E}[W_1] = 0$, which by (15.1), implies $\mathsf{E}[W_t] = 0$. As we see later, the case $H = 1$ is not of interest if W_t has finite second moments, and so we always take $\mathsf{E}[W_t] = 0$.

If Z_t is H-sssi, then the stationarity of the increments and (15.3) imply

$$\mathsf{E}[Z_t^2] = \mathsf{E}[Z_\tau^2] = \mathsf{E}[(W_\tau - W_0)^2] = \mathsf{E}[W_\tau^2] = \tau^{2H}\mathsf{E}[W_1^2].$$

Similarly, for $t > s$,

$$\mathsf{E}[(W_t - W_s)^2] = \mathsf{E}[(W_{t-s} - W_0)^2] = \mathsf{E}[W_{t-s}^2] = (t-s)^{2H}\mathsf{E}[W_1^2].$$

For $t < s$,

$$\mathsf{E}[(W_t - W_s)^2] = \mathsf{E}[(W_s - W_t)^2] = (s-t)^{2H}\mathsf{E}[W_1^2].$$

Thus, for arbitrary t and s,

$$\mathsf{E}[(W_t - W_s)^2] = |t-s|^{2H}\mathsf{E}[W_1^2].$$

Note in particular that

$$\mathsf{E}[W_t^2] = |t|^{2H}\mathsf{E}[W_1^2].$$

Now, we also have

$$\mathsf{E}[(W_t - W_s)^2] = \mathsf{E}[W_t^2] - 2\mathsf{E}[W_t W_s] + \mathsf{E}[W_s^2],$$

and it follows that

$$\mathsf{E}[W_t W_s] = \frac{\mathsf{E}[W_1^2]}{2}[|t|^{2H} - |t-s|^{2H} + |s|^{2H}]. \tag{15.4}$$

Fractional Brownian motion

Let W_t denote the standard Wiener process on $-\infty < t < \infty$ as defined in Problem 35 in Chapter 11. The standard **fractional Brownian motion** is the process $B_H(t)$ defined by the Wiener integral

$$B_H(t) := \int_{-\infty}^{\infty} g_{H,t}(\tau) \, dW_\tau,$$

where $g_{H,t}$ is defined below. Then $\mathsf{E}[B_H(t)] = 0$, and

$$\mathsf{E}[B_H(t)^2] = \int_{-\infty}^{\infty} g_{H,t}(\tau)^2 \, d\tau.$$

To evaluate this expression as well as the correlation $\mathsf{E}[B_H(t)B_H(s)]$, we must now define $g_{H,t}(\tau)$. To this end, let

$$q_H(\theta) := \begin{cases} \theta^{H-1/2}, & \theta > 0, \\ 0, & \theta \leq 0, \end{cases}$$

and put

$$g_{H,t}(\tau) := \frac{1}{C_H} [q_H(t-\tau) - q_H(-\tau)],$$

where

$$C_H^2 := \int_0^{\infty} \left[(1+\theta)^{H-1/2} - \theta^{H-1/2}\right]^2 d\theta + \frac{1}{2H}.$$

First note that since $g_{H,0}(\tau) = 0$, $B_H(0) = 0$. Next,

$$B_H(t) - B_H(s) = \int_{-\infty}^{\infty} [g_{H,t}(\tau) - g_{H,s}(\tau)] \, dW_\tau$$

$$= C_H^{-1} \int_{-\infty}^{\infty} [q_H(t-\tau) - q_H(s-\tau)] \, dW_\tau,$$

and so

$$\mathsf{E}[|B_H(t) - B_H(s)|^2] = C_H^{-2} \int_{-\infty}^{\infty} |q_H(t-\tau) - q_H(s-\tau)|^2 \, d\tau.$$

If we now assume $s < t$, then this integral is equal to the sum of

$$\int_{-\infty}^{s} [(t-\tau)^{H-1/2} - (s-\tau)^{H-1/2}]^2 \, d\tau$$

and

$$\int_s^t (t-\tau)^{2H-1} \, d\tau = \int_0^{t-s} \theta^{2H-1} \, d\theta = \frac{(t-s)^{2H}}{2H}.$$

To evaluate the integral from $-\infty$ to s, let $\xi = s - \tau$ to get

$$\int_0^{\infty} [(t-s+\xi)^{H-1/2} - \xi^{H-1/2}]^2 \, d\xi,$$

which is equal to

$$(t-s)^{2H-1} \int_0^{\infty} \left[(1+\xi/(t-s))^{H-1/2} - (\xi/(t-s))^{H-1/2}\right]^2 d\xi.$$

Making the change of variable $\theta = \xi/(t-s)$ yields

$$(t-s)^{2H} \int_0^\infty [(1+\theta)^{H-1/2} - \theta^{H-1/2}]^2 \, d\theta.$$

It is now clear that

$$\mathsf{E}[|B_H(t) - B_H(s)|^2] = (t-s)^{2H}, \quad t > s.$$

Since interchanging the positions of t and s on the left-hand side has no effect, we can write for arbitrary t and s,

$$\mathsf{E}[|B_H(t) - B_H(s)|^2] = |t-s|^{2H}.$$

Taking $s = 0$ yields

$$\mathsf{E}[B_H(t)^2] = |t|^{2H}.$$

Furthermore, expanding $\mathsf{E}[|B_H(t) - B_H(s)|^2]$, we find that

$$|t-s|^{2H} = |t|^{2H} - 2\mathsf{E}[B_H(t)B_H(s)] + |s|^{2H},$$

or,

$$\mathsf{E}[B_H(t)B_H(s)] = \frac{|t|^{2H} - |t-s|^{2H} + |s|^{2H}}{2}. \tag{15.5}$$

Observe that $B_H(t)$ is a **Gaussian** process in the sense that if we select any sampling times, $t_1 < \cdots < t_n$, then the random vector $[B_H(t_1), \ldots, B_H(t_n)]'$ is Gaussian; this is a consequence of the fact that $B_H(t)$ is defined as a Wiener integral (Problem 17 in Chapter 14). Furthermore, the covariance matrix of the random vector is completely determined by (15.5). On the other hand, by (15.4), we see that *any* H-sssi process has the same covariance function (up to a scale factor). If that H-sssi process is Gaussian, then as far as the joint probabilities involving any finite number of sampling times, we may as well assume that the H-sssi process is fractional Brownian motion. In this sense, there is only one H-sssi process with finite second moments that is *Gaussian*: fractional Brownian motion. Sample paths of fractional Brownian motion are shown in Figure 15.1.

15.2 Self similarity in discrete time

Let W_t be an H-sssi process. By choosing an appropriate time scale for W_t, we can focus on the unit increment $\tau = 1$. Furthermore, the advent of digital signal processing suggests that we sample the increment process. This leads us to consider the discrete-time increment process

$$X_n := W_n - W_{n-1}.$$

Since W_t is assumed to have zero mean, the covariance of X_n is easily found using (15.4). For $n > m$,

$$\mathsf{E}[X_n X_m] = \frac{\mathsf{E}[W_1^2]}{2}[(n-m+1)^{2H} - 2(n-m)^{2H} + (n-m-1)^{2H}]$$

Since this depends only on the time difference, the covariance function of X_n is

$$C(n) = \frac{\sigma^2}{2}[|n+1|^{2H} - 2|n|^{2H} + |n-1|^{2H}], \tag{15.6}$$

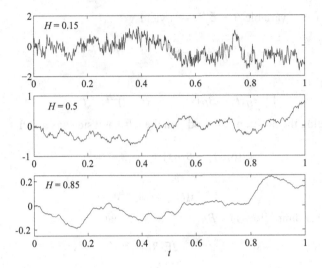

Figure 15.1. Fractional Brownian motions with $H = 0.15$, $H = 0.5$, and $H = 0.85$.

where $\sigma^2 := \mathsf{E}[W_1^2]$.

The foregoing analysis assumed that X_n was obtained by sampling the increments of an H-sssi process. More generally, a discrete-time, wide-sense stationary (WSS) process is said to be **second-order self similar** if its covariance function has the form in (15.6). In this context it is not assumed that X_n is obtained from an underlying continuous-time process or that X_n has zero mean. A second-order self-similar process that is Gaussian is called **fractional Gaussian noise**, since one way of obtaining it is by sampling the increments of fractional Brownian motion.

Convergence rates for the mean-square ergodic theorem

Suppose that X_n is a discrete-time, WSS process with mean $\mu := \mathsf{E}[X_n]$. It is shown later in Section 15.4 that if X_n is second-order self similar; i.e., if (15.6) holds, then $C(n) \to 0$. On account of the mean-square ergodic theorem of Example 13.4, the sample mean

$$\frac{1}{n}\sum_{i=1}^{n} X_i$$

converges in mean square to μ. But how fast does the sample mean converge? We show that (15.6) holds if and only if

$$\mathsf{E}\left[\left|\frac{1}{n}\sum_{i=1}^{n} X_i - \mu\right|^2\right] = \sigma^2 n^{2H-2} = \frac{\sigma^2}{n} n^{2H-1}. \tag{15.7}$$

In other words, X_n is second-order self similar if and only if (15.7) holds. To put (15.7) into perspective, first consider the case $H = 1/2$. Then (15.6) reduces to zero for $n \neq 0$. In other words, the X_n are uncorrelated. Also, when $H = 1/2$, the factor n^{2H-1} in (15.7) is not present. Thus, (15.7) reduces to the mean-square law of large numbers for uncorrelated

15.2 Self similarity in discrete time

random variables derived in Example 13.3. If $2H-1<0$, or equivalently $H<1/2$, then the convergence is faster than in the uncorrelated case. If $1/2<H<1$, then the convergence is slower than in the uncorrelated case. This has important consequences for determining confidence intervals, as shown in Problem 6.

To show the equivalence of (15.6) and (15.7), the first step is to define

$$Y_n := \sum_{i=1}^{n}(X_i - \mu),$$

and observe that (15.7) is equivalent to $\mathsf{E}[Y_n^2] = \sigma^2 n^{2H}$.

The second step is to express $\mathsf{E}[Y_n^2]$ in terms of $C(n)$. Write

$$\mathsf{E}[Y_n^2] = \mathsf{E}\left[\left(\sum_{i=1}^{n}(X_i - \mu)\right)\left(\sum_{k=1}^{n}(X_k - \mu)\right)\right]$$

$$= \sum_{i=1}^{n}\sum_{k=1}^{n}\mathsf{E}[(X_i - \mu)(X_k - \mu)]$$

$$= \sum_{i=1}^{n}\sum_{k=1}^{n}C(i-k). \tag{15.8}$$

The above sum amounts to summing all the entries of the $n \times n$ matrix with ik entry $C(i-k)$. This matrix is symmetric and is constant along each diagonal. Thus,

$$\mathsf{E}[Y_n^2] = nC(0) + 2\sum_{\nu=1}^{n-1}C(\nu)(n-\nu). \tag{15.9}$$

Now that we have a formula for $\mathsf{E}[Y_n^2]$ in terms of $C(n)$, we follow Likhanov [37, p. 195] and write

$$\mathsf{E}[Y_{n+1}^2] - \mathsf{E}[Y_n^2] = C(0) + 2\sum_{\nu=1}^{n}C(\nu),$$

and then

$$\left(\mathsf{E}[Y_{n+1}^2] - \mathsf{E}[Y_n^2]\right) - \left(\mathsf{E}[Y_n^2] - \mathsf{E}[Y_{n-1}^2]\right) = 2C(n). \tag{15.10}$$

Applying the formula $\mathsf{E}[Y_n^2] = \sigma^2 n^{2H}$ shows that for $n \geq 1$,

$$C(n) = \frac{\sigma^2}{2}[(n+1)^{2H} - 2n^{2H} + (n-1)^{2H}].$$

Finally, it is a simple exercise (Problem 7) using induction on n to show that (15.6) implies $\mathsf{E}[Y_n^2] = \sigma^2 n^{2H}$ for $n \geq 1$.

Aggregation

Consider the partitioning of the sequence X_n into blocks of size m:

$$\underbrace{X_1,\ldots,X_m}_{\text{1st block}}\underbrace{X_{m+1},\ldots,X_{2m}}_{\text{2nd block}}\cdots\underbrace{X_{(n-1)m+1},\ldots,X_{nm}}_{n\text{th block}}\cdots.$$

The average of the first block is $\frac{1}{m}\sum_{k=1}^{m} X_k$. The average of the second block is $\frac{1}{m}\sum_{k=m+1}^{2m} X_k$. The average of the nth block is

$$X_n^{(m)} := \frac{1}{m} \sum_{k=(n-1)m+1}^{nm} X_k. \quad (15.11)$$

The superscript (m) indicates the block size, which is the number of terms used to compute the average. The subscript n indicates the block number. We call $\{X_n^{(m)}\}_{n=-\infty}^{\infty}$ the **aggregated process**. We now show that if X_n is second-order self similar, then the covariance function of $X_n^{(m)}$, denoted by $C^{(m)}(n)$, satisfies

$$C^{(m)}(n) = m^{2H-2} C(n). \quad (15.12)$$

In other words, if the original sequence is replaced by the sequence of averages of blocks of size m, then the new sequence has a covariance function that is the same as the original one except that the magnitude is scaled by m^{2H-2}.

The derivation of (15.12) is similar to the derivation of (15.7). Put

$$\widetilde{X}_\nu^{(m)} := \sum_{k=(\nu-1)m+1}^{\nu m} (X_k - \mu). \quad (15.13)$$

Since X_k is WSS, so is $\widetilde{X}_\nu^{(m)}$. Let its covariance function be denoted by $\widetilde{C}^{(m)}(\nu)$. Next define

$$\widetilde{Y}_n := \sum_{\nu=1}^{n} \widetilde{X}_\nu^{(m)}.$$

Just as in (15.10),

$$2\widetilde{C}^{(m)}(n) = \left(\mathsf{E}[\widetilde{Y}_{n+1}^2] - \mathsf{E}[\widetilde{Y}_n^2]\right) - \left(\mathsf{E}[\widetilde{Y}_n^2] - \mathsf{E}[\widetilde{Y}_{n-1}^2]\right).$$

Now observe that

$$\widetilde{Y}_n = \sum_{\nu=1}^{n} \widetilde{X}_\nu^{(m)}$$
$$= \sum_{\nu=1}^{n} \sum_{k=(\nu-1)m+1}^{\nu m} (X_k - \mu)$$
$$= \sum_{\nu=1}^{nm} (X_\nu - \mu)$$
$$= Y_{nm},$$

where this Y is the same as the one defined in the preceding subsection. Hence,

$$2\widetilde{C}^{(m)}(n) = \left(\mathsf{E}[Y_{(n+1)m}^2] - \mathsf{E}[Y_{nm}^2]\right) - \left(\mathsf{E}[Y_{nm}^2] - \mathsf{E}[Y_{(n-1)m}^2]\right). \quad (15.14)$$

Now we use the fact that since X_n is second-order self similar, $\mathsf{E}[Y_n^2] = \sigma^2 n^{2H}$. Thus,

$$\widetilde{C}^{(m)}(n) = \frac{\sigma^2}{2}\left[\left((n+1)m\right)^{2H} - 2(nm)^{2H} + \left((n-1)m\right)^{2H}\right].$$

Since $C^{(m)}(n) = \widetilde{C}^{(m)}(n)/m^2$, (15.12) follows.

15.2 Self similarity in discrete time

The power spectral density

We show that the power spectral density[b] of a second-order self-similar process is proportional to[c]

$$\sin^2(\pi f) \sum_{i=-\infty}^{\infty} \frac{1}{|i+f|^{2H+1}}. \tag{15.15}$$

Since (15.15) is real, even, nonnegative, and has period one, it is a valid power spectral density of a wide-sense stationary process. We show that the corresponding covariance function satisfies (15.6), where, of course, σ^2 is the integral of (15.15) over $[-1/2, 1/2]$.

The proof rests on several observations, all but one of which are easy to see. We showed above that second-order self similarity in the form of $\mathsf{E}[Y_n^2] = \sigma^2 n^{2H}$ implies (15.12). Conversely, given a wide-sense stationary process X_n and aggregated process $X_n^{(m)}$, if (15.12) holds, then X_n is second-order self similar. To see this, put $n=0$ in (15.12) and obtain

$$\frac{\mathsf{E}[Y_m^2]}{m^{2H}} = \frac{C^{(m)}(0)}{m^{2H-2}} = C(0) = \sigma^2, \quad m=1,2,\ldots,$$

i.e., $\mathsf{E}[Y_m^2] = \sigma^2 m^{2H}$, which is equivalent to the second-order self similarity of X_n. Thus, (15.12) implies second-order self similarity, which means that (15.6) holds. But if (15.6) holds, then the corresponding power spectral density is determined up to the constant σ^2.

The remainder of the proof rests on the fact (derived below) that for any wide-sense stationary process with power spectral density $S(f)$ and aggregated process $X_n^{(m)}$,

$$\frac{C^{(m)}(n)}{m^{2H-2}} = \int_0^1 e^{j2\pi fn} \left(\sum_{k=0}^{m-1} \frac{S([f+k]/m)}{m^{2H+1}} \left[\frac{\sin(\pi f)}{\sin(\pi[f+k]/m)} \right]^2 \right) df, \quad m=1,2,\ldots.$$

Since for any wide-sense stationary process we also have

$$C(n) = \int_{-1/2}^{1/2} S(f) e^{j2\pi fn} df = \int_0^1 S(f) e^{j2\pi fn} df,$$

if $S(f)$ satisfies

$$\sum_{k=0}^{m-1} \frac{S([f+k]/m)}{m^{2H+1}} \left[\frac{\sin(\pi f)}{\sin(\pi[f+k]/m)} \right]^2 = S(f), \quad m=1,2,\ldots, \tag{15.16}$$

then (15.12) holds. Now observe that if $S(f)$ is proportional to (15.15), then (15.16) holds. The integral formula for $C^{(m)}(n)/m^{2H-2}$ is derived following Sinai [57, p. 66]. Write

$$\frac{C^{(m)}(n)}{m^{2H-2}} = \frac{1}{m^{2H-2}} \mathsf{E}[(X_{n+1}^{(m)} - \mu)(X_1^{(m)} - \mu)]$$

$$- \frac{1}{m^{2H}} \sum_{i=nm+1}^{(n+1)m} \sum_{k=1}^{m} \mathsf{E}[(X_i - \mu)(X_k - \mu)]$$

---

[b] In Chapter 10 we defined the power spectral density to be the Fourier transform of the correlation function. In this chapter, we define the power spectral density to be the Fourier transform of the covariance function.
[c] The constant of proportionality can be found using results from Section 15.3; see Problem 14.

$$= \frac{1}{m^{2H}} \sum_{i=nm+1}^{(n+1)m} \sum_{k=1}^{m} C(i-k)$$

$$= \frac{1}{m^{2H}} \sum_{i=nm+1}^{(n+1)m} \sum_{k=1}^{m} \int_{-1/2}^{1/2} S(f) e^{j2\pi f(i-k)} df$$

$$= \frac{1}{m^{2H}} \int_{-1/2}^{1/2} S(f) \sum_{i=nm+1}^{(n+1)m} \sum_{k=1}^{m} e^{j2\pi f(i-k)} df.$$

Now write

$$\sum_{i=nm+1}^{(n+1)m} \sum_{k=1}^{m} e^{j2\pi f(i-k)} = \sum_{v=1}^{m} \sum_{k=1}^{m} e^{j2\pi f(nm+v-k)}$$

$$= e^{j2\pi nm} \sum_{v=1}^{m} \sum_{k=1}^{m} e^{j2\pi f(v-k)}$$

$$= e^{j2\pi nm} \left| \sum_{k=1}^{m} e^{-j2\pi fk} \right|^2.$$

Using the finite geometric series,

$$\sum_{k=1}^{m} e^{-j2\pi fk} = e^{-j2\pi f} \frac{1 - e^{-j2\pi fm}}{1 - e^{-j2\pi f}} = e^{-j\pi f(m+1)} \frac{\sin(m\pi f)}{\sin(\pi f)}.$$

Thus,

$$\frac{C^{(m)}(n)}{m^{2H-2}} = \frac{1}{m^{2H}} \int_{-1/2}^{1/2} S(f) e^{j2\pi fnm} \left[\frac{\sin(m\pi f)}{\sin(\pi f)} \right]^2 df.$$

Since the integrand has period one, we can shift the range of integration to $[0,1]$ and then make the change of variable $\theta = mf$. Thus,

$$\frac{C^{(m)}(n)}{m^{2H-2}} = \frac{1}{m^{2H}} \int_0^1 S(f) e^{j2\pi fnm} \left[\frac{\sin(m\pi f)}{\sin(\pi f)} \right]^2 df$$

$$= \frac{1}{m^{2H+1}} \int_0^m S(\theta/m) e^{j2\pi \theta n} \left[\frac{\sin(\pi \theta)}{\sin(\pi \theta/m)} \right]^2 d\theta$$

$$= \frac{1}{m^{2H+1}} \sum_{k=0}^{m-1} \int_k^{k+1} S(\theta/m) e^{j2\pi \theta n} \left[\frac{\sin(\pi \theta)}{\sin(\pi \theta/m)} \right]^2 d\theta.$$

Now make the change of variable $f = \theta - k$ to get

$$\frac{C^{(m)}(n)}{m^{2H-2}} = \frac{1}{m^{2H+1}} \sum_{k=0}^{m-1} \int_0^1 S([f+k]/m) e^{j2\pi[f+k]n} \left[\frac{\sin(\pi[f+k])}{\sin(\pi[f+k]/m)} \right]^2 df$$

$$= \frac{1}{m^{2H+1}} \sum_{k=0}^{m-1} \int_0^1 S([f+k]/m) e^{j2\pi fn} \left[\frac{\sin(\pi f)}{\sin(\pi[f+k]/m)} \right]^2 df$$

$$= \int_0^1 e^{j2\pi fn} \left(\sum_{k=0}^{m-1} \frac{S([f+k]/m)}{m^{2H+1}} \left[\frac{\sin(\pi f)}{\sin(\pi[f+k]/m)} \right]^2 \right) df.$$

15.3 Asymptotic second-order self similarity

Engineering versus statistics/networking notation

We have been using the term **correlation function** to refer to the quantity $E[X_n X_m]$. This is the usual practice in engineering. However, engineers studying network traffic follow the practice of statisticians and use the term **correlation function** to refer to

$$\frac{\text{cov}(X_n, X_m)}{\sqrt{\text{var}(X_n)\text{var}(X_m)}}.$$

In other words, in networking, the term correlation function refers to our covariance function $C(n)$ divided by $C(0)$. We use the notation

$$\rho(n) := \frac{C(n)}{C(0)}.$$

Now assume that X_n is second-order self similar. We have by (15.6) that $C(0) = \sigma^2$, and so

$$\rho(n) = \frac{1}{2}[|n+1|^{2H} - 2|n|^{2H} + |n-1|^{2H}].$$

Let $\rho^{(m)}$ denote the correlation function of $X_n^{(m)}$. Then (15.12) tells us that

$$\rho^{(m)}(n) := \frac{C^{(m)}(n)}{C^{(m)}(0)} = \frac{m^{2H-2}C(n)}{m^{2H-2}C(0)} = \rho(n). \tag{15.17}$$

15.3 Asymptotic second-order self similarity

We showed in the previous section that second-order self similarity (Eq. (15.6)) is equivalent to (15.7), which specifies

$$E\left[\left|\frac{1}{n}\sum_{i=1}^{n} X_i - \mu\right|^2\right] \tag{15.18}$$

exactly for all n. While this is a nice result, it applies only when the covariance function has exactly the form in (15.6). However, if we only need to know the behavior of (15.18) for large n, say

$$\lim_{n\to\infty} \frac{E\left[\left|\frac{1}{n}\sum_{i=1}^{n} X_i - \mu\right|^2\right]}{n^{2H-2}} = \sigma_\infty^2 \tag{15.19}$$

for some finite, positive σ_∞^2, then we can allow more freedom in the behavior of the covariance function. The key to obtaining such a result is suggested by (15.12), which says that for a second-order self similar process,

$$\frac{C^{(m)}(n)}{m^{2H-2}} = \frac{\sigma^2}{2}[|n+1|^{2H} - 2|n|^{2H} + |n-1|^{2H}].$$

You are asked to show in Problems 9 and 10 that (15.19) holds if and only if

$$\lim_{m\to\infty} \frac{C^{(m)}(n)}{m^{2H-2}} = \frac{\sigma_\infty^2}{2}[|n+1|^{2H} - 2|n|^{2H} + |n-1|^{2H}]. \tag{15.20}$$

A wide-sense stationary process that satisfies (15.20) is said to be **asymptotically** second-order self similar. In the literature, (15.20) is usually written in terms of the correlation function $\rho^{(m)}(n)$; see Problem 11.

If a wide-sense stationary process has a covariance function $C(n)$, how can we check if (15.19) or (15.20) holds? Below we answer this question in the frequency domain with a sufficient condition on the power spectral density. In Section 15.4, we answer this question in the time domain with a sufficient condition on $C(n)$ known as long-range dependence.

Let us look into the frequency domain. Suppose that $C(n)$ has a power spectral density $S(f)$ so that[d]

$$C(n) = \int_{-1/2}^{1/2} S(f) e^{j2\pi f n} \, df.$$

The following result is proved in [24, Appendix C].

Theorem. *If*

$$\lim_{f \to 0} \frac{S(f)}{|f|^{\alpha-1}} = s \tag{15.21}$$

for some finite, positive s, and if for every $0 < \delta < 1/2$, $S(f)$ is bounded on $[\delta, 1/2]$, then the process is asymptotically second-order self similar with $H = 1 - \alpha/2$ and

$$\sigma_\infty^2 = s \cdot \frac{4\cos(\alpha\pi/2)\Gamma(\alpha)}{(2\pi)^\alpha (1-\alpha)(2-\alpha)}. \tag{15.22}$$

Notice that $0 < \alpha < 1$ implies $H = 1 - \alpha/2 \in (1/2, 1)$.

Below we give a specific power spectral density that satisfies the above conditions.

Example 15.2 (Hosking [31, Theorem 1(c)]). Fix $0 < d < 1/2$, and let[e]

$$S(f) = |1 - e^{-j2\pi f}|^{-2d}.$$

Since

$$1 - e^{-j2\pi f} = 2je^{-j\pi f} \frac{e^{j\pi f} - e^{-j\pi f}}{2j},$$

we can write

$$S(f) = [4\sin^2(\pi f)]^{-d}.$$

Since

$$\lim_{f \to 0} \frac{[4\sin^2(\pi f)]^{-d}}{[4(\pi f)^2]^{-d}} = 1,$$

---

[d]The power spectral density of a discrete-time process is periodic with period 1, and is real, even, and nonnegative. It is integrable since

$$\int_{-1/2}^{1/2} S(f) \, df = C(0) < \infty.$$

[e]As shown in Section 15.6, a process with this power spectral density is an ARIMA$(0,d,0)$ process. The covariance function that corresponds to $S(f)$ is derived in Problem 12.

15.3 Asymptotic second-order self similarity

if we put $\alpha = 1 - 2d$, then

$$\lim_{f \to 0} \frac{S(f)}{|f|^{\alpha - 1}} = (2\pi)^{-2d}.$$

Notice that to keep $0 < \alpha < 1$, we needed $0 < d < 1/2$.

The power spectral density $S(f)$ in the above example factors into

$$S(f) = [1 - e^{-j2\pi f}]^{-d}[1 - e^{j2\pi f}]^{-d}.$$

More generally, let $S(f)$ be any power spectral density satisfying (15.21), boundedness away from the origin, and having a factorization of the form $S(f) = G(f)G(f)^*$. Then pass any wide-sense stationary, uncorrelated sequence though the discrete-time filter $G(f)$. Then the output power spectral density is proportional to $S(f)$,[f] and therefore asymptotically second-order self similar.

Example 15.3 (Hosking [31, Theorem 1(a)]). Find the impulse response g_n of the filter $G(f) = [1 - e^{-j2\pi f}]^{-d}$.

Solution. Observe that $G(f)$ can be obtained by evaluating the z transform $(1 - z^{-1})^{-d}$ on the unit circle, $z = e^{j2\pi f}$. Hence, the desired impulse response can be found by inspection of the series for $(1 - z^{-1})^{-d}$. To this end, it is easy to show that the Taylor series for $(1+z)^d$ is[g]

$$(1+z)^d = 1 + \sum_{n=1}^{\infty} \frac{d(d-1)\cdots(d-[n-1])}{n!} z^n.$$

Hence,

$$(1-z^{-1})^{-d} = 1 + \sum_{n=1}^{\infty} \frac{(-d)(-d-1)\cdots(-d-[n-1])}{n!}(-z^{-1})^n$$

$$= 1 + \sum_{n=1}^{\infty} \frac{d(d+1)\cdots(d+[n-1])}{n!} z^{-n}.$$

By inspection,

$$g_n = \begin{cases} d(d+1)\cdots(d+[n-1])/n!, & n > 1, \\ 1, & n = 0, \\ 0, & n < 0. \end{cases}$$

Note that the impulse response is causal.

---

[f] See Problem 31 in Chapter 10 or Example 13.26.
[g] Notice that if $d \geq 0$ is an integer, the product

$$d(d-1)\cdots(d-[n-1])$$

contains zero as a factor for $n \geq d+1$; in this case, the sum contains only $d+1$ terms and converges for all complex z. In fact, the formula reduces to the binomial theorem.

Once we have a process whose power spectral density satisfies (15.21) and boundedness away from the origin, it remains so after further filtering by *stable* linear time-invariant systems. For if $\sum_n |h_n| < \infty$, then

$$H(f) = \sum_{n=-\infty}^{\infty} h_n e^{-j2\pi fn}$$

is an absolutely convergent series and therefore continuous. If $S(f)$ satisfies (15.21), then

$$\lim_{f \to 0} \frac{|H(f)|^2 S(f)}{|f|^{\alpha-1}} = |H(0)|^2 s.$$

A wide class of stable filters is provided by autoregressive moving average (ARMA) systems discussed in Section 15.5.

15.4 Long-range dependence

Loosely speaking, a wide-sense stationary process is said to be **long-range dependent** (**LRD**) if its covariance function $C(n)$ decays slowly as $n \to \infty$. The precise definition of slow decay is the requirement that for some $0 < \alpha < 1$,

$$\lim_{n \to \infty} \frac{C(n)}{n^{-\alpha}} = c, \tag{15.23}$$

for some finite, positive constant c. In other words, for large n, $C(n)$ looks like c/n^{α}.

In this section, we prove two important results. The first result is that a second-order self-similar process is long-range dependent. The second result is that long-range dependence implies asymptotic second-order self similarity.

To prove that second-order self similarity implies long-range dependence, we proceed as follows. Write (15.6) for $n \geq 1$ as

$$C(n) = \frac{\sigma^2}{2} n^{2H} [(1+1/n)^{2H} - 2 + (1-1/n)^{2H}] = \frac{\sigma^2}{2} n^{2H} q(1/n),$$

where

$$q(t) := (1+t)^{2H} - 2 + (1-t)^{2H}.$$

For n large, $1/n$ is small. This suggests that we examine the Taylor expansion of $q(t)$ for t near zero. Since $q(0) = q'(0) = 0$, we expand to second order to get

$$q(t) \approx \frac{q''(0)}{2} t^2 = 2H(2H-1)t^2.$$

So, for large n,

$$C(n) = \frac{\sigma^2}{2} n^{2H} q(1/n) \approx \sigma^2 H(2H-1) n^{2H-2}. \tag{15.24}$$

It appears that $\alpha = 2 - 2H$ and $c = \sigma^2 H(2H-1)$. Note that $0 < \alpha < 1$ corresponds to $1/2 < H < 1$. Also, $H > 1/2$ corresponds to $\sigma^2 H(2H-1) > 0$. To prove that these values of α and c work, write

$$\lim_{n \to \infty} \frac{C(n)}{n^{2H-2}} = \frac{\sigma^2}{2} \lim_{n \to \infty} \frac{q(1/n)}{n^{-2}} = \frac{\sigma^2}{2} \lim_{t \downarrow 0} \frac{q(t)}{t^2},$$

15.4 Long-range dependence

and apply l'Hôpital's rule twice to obtain

$$\lim_{n\to\infty} \frac{C(n)}{n^{2H-2}} = \sigma^2 H(2H-1). \tag{15.25}$$

This formula implies the following two facts. First, if $H > 1$, then $C(n) \to \infty$ as $n \to \infty$. This contradicts the fact that covariance functions are bounded (recall that $|C(n)| \leq C(0)$ by the Cauchy–Schwarz inequality; cf. Section 10.3). Thus, a second-order self-similar process cannot have $H > 1$. Second, if $H = 1$, then $C(n) \to \sigma^2$. In other words, the covariance does not decay to zero as n increases. Since this situation does not arise in applications, we do not consider the case $H = 1$.

We now show that long-range dependence (15.23) implies[h] asymptotic second-order self similarity with $H = 1 - \alpha/2$ and $\sigma_\infty^2 = 2c/[(1-\alpha)(2-\alpha)]$. From (15.9),

$$\frac{\mathsf{E}[Y_n^2]}{n^{2-\alpha}} = \frac{C(0)}{n^{1-\alpha}} + 2\frac{\sum_{v=1}^{n-1} C(v)}{n^{1-\alpha}} - 2\frac{\sum_{v=1}^{n-1} vC(v)}{n^{2-\alpha}}.$$

We claim that if (15.23) holds, then

$$\lim_{n\to\infty} \frac{\sum_{v=1}^{n-1} C(v)}{n^{1-\alpha}} = \frac{c}{1-\alpha}, \tag{15.26}$$

and

$$\lim_{n\to\infty} \frac{\sum_{v=1}^{n-1} vC(v)}{n^{2-\alpha}} = \frac{c}{2-\alpha}. \tag{15.27}$$

Since $n^{1-\alpha} \to \infty$, it follows that

$$\lim_{n\to\infty} \frac{\mathsf{E}[Y_n^2]}{n^{2-\alpha}} = \frac{2c}{1-\alpha} - \frac{2c}{2-\alpha} = \frac{2c}{(1-\alpha)(2-\alpha)}.$$

Since $n^{1-\alpha} \to \infty$, to prove (15.26), it is enough to show that for some k,

$$\lim_{n\to\infty} \frac{\sum_{v=k}^{n-1} C(v)}{n^{1-\alpha}} = \frac{c}{1-\alpha}.$$

Fix any $0 < \varepsilon < c$. By (15.23), there is a k such that for all $v \geq k$,

$$\left| \frac{C(v)}{v^{-\alpha}} - c \right| < \varepsilon.$$

---

[h] Actually, the weaker condition,

$$\lim_{n\to\infty} \frac{C(n)}{\ell(n)n^{-\alpha}} = c,$$

where ℓ is a **slowly varying function**, is enough to imply asymptotic second-order self similarity [64]. The derivation we present results from taking the proof in [64, Appendix A] and setting $\ell(n) \equiv 1$ so that no theory of slowly varying functions is required.

Then
$$(c-\varepsilon)\sum_{v=k}^{n-1} v^{-\alpha} \le \sum_{v=k}^{n-1} C(v) \le (c+\varepsilon)\sum_{v=k}^{n-1} v^{-\alpha}.$$

Hence, we only need to prove that
$$\lim_{n\to\infty} \frac{\sum_{v=k}^{n-1} v^{-\alpha}}{n^{1-\alpha}} = \frac{1}{1-\alpha}.$$

This is done in Problem 16 by exploiting the inequality
$$\sum_{v=k}^{n-1}(v+1)^{-\alpha} \le \int_k^n t^{-\alpha}\,dt \le \sum_{v=k}^{n-1} v^{-\alpha}. \tag{15.28}$$

Note that
$$I_n := \int_k^n t^{-\alpha}\,dt = \frac{n^{1-\alpha} - k^{1-\alpha}}{1-\alpha} \to \infty \quad \text{as } n \to \infty.$$

A similar approach is used in Problem 17 to derive (15.27).

15.5 ARMA processes

We say that X_n is an **autoregressive moving average** (ARMA) process if X_n satisfies the equation
$$X_n + a_1 X_{n-1} + \cdots + a_p X_{n-p} = Z_n + b_1 Z_{n-1} + \cdots + b_q Z_{n-q}, \tag{15.29}$$

where Z_n is an uncorrelated sequence of zero-mean random variables with common variance $\sigma^2 = \mathsf{E}[Z_n^2]$. In this case, we say that X_n is ARMA(p,q). If $a_1 = \cdots = a_p = 0$, then
$$X_n = Z_n + b_1 Z_{n-1} + \cdots + b_q Z_{n-q},$$

and we say that X_n is a **moving average** process, denoted by MA(q). If instead $b_1 = \cdots = b_q = 0$, then
$$X_n = -(a_1 X_{n-1} + \cdots + a_p X_{n-p}) + Z_n,$$

and we say that X_n is an **autoregressive** process, denoted by AR(p).

To gain some insight into (15.29), rewrite it using convolution sums as
$$\sum_{k=-\infty}^{\infty} a_k X_{n-k} = \sum_{k=-\infty}^{\infty} b_k Z_{n-k}, \tag{15.30}$$

where
$$a_0 := 1, \quad a_k := 0, \quad k < 0 \text{ and } k > p,$$

and
$$b_0 := 1, \quad b_k := 0, \quad k < 0 \text{ and } k > q.$$

Taking z transforms of (15.30) yields
$$A(z)X(z) = B(z)Z(z),$$

15.5 ARMA processes

or

$$X(z) = \frac{B(z)}{A(z)} Z(z),$$

where

$$A(z) := 1 + a_1 z^{-1} + \cdots + a_p z^{-p} \quad \text{and} \quad B(z) := 1 + b_1 z^{-1} + \cdots + b_q z^{-q}.$$

This suggests that if h_n has z transform $H(z) := B(z)/A(z)$, and if

$$X_n := \sum_k h_k Z_{n-k} = \sum_k h_{n-k} Z_k, \qquad (15.31)$$

then (15.30) holds. This is indeed the case, as can be seen by writing

$$\sum_i a_i X_{n-i} = \sum_i a_i \sum_k h_{n-i-k} Z_k$$
$$= \sum_k \left(\sum_i a_i h_{n-k-i} \right) Z_k$$
$$= \sum_k b_{n-k} Z_k,$$

since $A(z)H(z) = B(z)$.

The "catch" in the preceding argument is to make sure that the infinite sums in (15.31) are well defined. If h_n is causal ($h_n = 0$ for $n < 0$), and if h_n is stable ($\sum_n |h_n| < \infty$), then (15.31) holds in L^2, L^1, and almost surely (recall Example 13.10, Problem 26 in Chapter 13, and Problem 41 in Chapter 14). Hence, it remains to prove the key result of this section, that if $A(z)$ has all roots strictly inside the unit circle, then h_n is causal and stable.

To begin the proof, observe that since $A(z)$ has all its roots inside the unit circle, the polynomial $\alpha(z) := A(1/z)$ has all its roots strictly outside the unit circle. Hence, for small enough $\delta > 0$, $1/\alpha(z)$ has the power series expansion

$$\frac{1}{\alpha(z)} = \sum_{n=0}^{\infty} \alpha_n z^n, \quad |z| < 1 + \delta,$$

for unique coefficients α_n. In particular, this series converges for $z = 1 + \delta/2$. Since the terms of a convergent series go to zero, we must have $\alpha_n (1 + \delta/2)^n \to 0$. Since a convergent sequence is bounded, there is some finite M for which $|\alpha_n (1 + \delta/2)^n| \leq M$, or $|\alpha_n| \leq M(1 + \delta/2)^{-n}$, which is summable by the geometric series. Thus, $\sum_n |\alpha_n| < \infty$. Now write

$$H(z) = \frac{B(z)}{A(z)} = \frac{B(z)}{\alpha(1/z)} = B(z) \frac{1}{\alpha(1/z)},$$

or

$$H(z) = B(z) \sum_{n=0}^{\infty} \alpha_n z^{-n} = \sum_{n=-\infty}^{\infty} h_n z^{-n},$$

where h_n is given by the convolution

$$h_n = \sum_{k=-\infty}^{\infty} \alpha_k b_{n-k}.$$

Since α_n and b_n are causal, so is their convolution h_n. Furthermore, for $n \geq 0$,

$$h_n = \sum_{k=\max(0,n-q)}^{n} \alpha_k b_{n-k}. \qquad (15.32)$$

In Problem 18, you are asked to show that $\sum_n |h_n| < \infty$. In Problem 19, you are asked to show that (15.31) is the *unique* solution of (15.30).

15.6 ARIMA processes

Before defining ARIMA processes, we introduce the **differencing filter**, whose z transform is $1 - z^{-1}$. If the input to this filter is X_n, then the output is $X_n - X_{n-1}$.

A process X_n is said to be an **autoregressive integrated moving average** (ARIMA) process if instead of $A(z)X(z) = B(z)Z(z)$, we have

$$A(z)(1-z^{-1})^d X(z) = B(z)Z(z), \qquad (15.33)$$

where $A(z)$ and $B(z)$ are defined as in the previous section. In this case, we say that X_n is an ARIMA(p,d,q) process. If we let $\widetilde{A}(z) = A(z)(1-z^{-1})^d$, it would seem that ARIMA(p,d,q) is just a fancy name for ARMA$(p+d,q)$. While this is true when d is a nonnegative integer, there are two problems. First, recall that the results of the previous section assume $A(z)$ has all roots strictly inside the unit circle, while $\widetilde{A}(z)$ has a root at $z=1$ repeated d times. The second problem is that we will be focusing on fractional values of d, in which case $\widetilde{A}(1/z)$ is no longer a polynomial, but an infinite power series in z.

Let us rewrite (15.33) as

$$X(z) = (1-z^{-1})^{-d} \frac{B(z)}{A(z)} Z(z) = H(z) G_d(z) Z(z),$$

where $H(z) := B(z)/A(z)$ as in the previous section, and

$$G_d(z) := (1-z^{-1})^{-d}.$$

From the calculations following Example 15.2,[i]

$$G_d(z) = \sum_{n=0}^{\infty} g_n z^{-n},$$

where $g_0 = 1$, and for $n \geq 1$,

$$g_n = \frac{d(d+1)\cdots(d+[n-1])}{n!}.$$

The plan then is to set

$$Y_n := \sum_{k=0}^{\infty} g_k Z_{n-k} \qquad (15.34)$$

---

[i] Since $1-z^{-1}$ is a differencing filter, $(1-z^{-1})^{-1}$ is a summing or **integrating filter**. For noninteger values of d, $(1-z^{-1})^{-d}$ is called a *fractional* integrating filter. The corresponding process is sometimes called a **fractional ARIMA process** (FARIMA).

15.6 ARIMA processes

and then[j]

$$X_n := \sum_{k=0}^{\infty} h_k Y_{n-k}. \qquad (15.35)$$

Note that the power spectral density of Y_n is[k]

$$S_Y(f) = |G_d(e^{j2\pi f})|^2 \sigma^2 = |1-e^{-j2\pi f}|^{-2d}\sigma^2 = [4\sin^2(\pi f)]^{-d}\sigma^2,$$

using the result of Example 15.2. If $p = q = 0$, then $A(z) = B(z) = H(z) = 1$, $X_n = Y_n$, and we see that the process of Example 15.2 is ARIMA$(0,d,0)$.

Now, the problem with the above plan is that we have to make sure that Y_n is well defined. To analyze the situation, we need to know how fast the g_n decay. To this end, observe that

$$\Gamma(d+n) = (d+[n-1])\Gamma(d+[n-1])$$
$$\vdots$$
$$= (d+[n-1])\cdots(d+1)\Gamma(d+1).$$

Hence,

$$g_n = \frac{d \cdot \Gamma(d+n)}{\Gamma(d+1)\Gamma(n+1)}.$$

Now apply Stirling's formula,[l]

$$\Gamma(x) \sim \sqrt{2\pi} x^{x-1/2} e^{-x},$$

to the gamma functions that involve n. This yields

$$g_n \sim \frac{de^{1-d}}{\Gamma(d+1)} \left(1 + \frac{d-1}{n+1}\right)^{n+1/2} (n+d)^{d-1}.$$

Since

$$\left(1 + \frac{d-1}{n+1}\right)^{n+1/2} = \left(1 + \frac{d-1}{n+1}\right)^{n+1} \left(1 + \frac{d-1}{n+1}\right)^{-1/2} \to e^{d-1},$$

and since $(n+d)^{d-1} \sim n^{d-1}$, we see that

$$g_n \sim \frac{d}{\Gamma(d+1)} n^{d-1}$$

as in Hosking [31, Theorem 1(a)]. For $0 < d < 1/2$, $-1 < d-1 < -1/2$, and we see that the g_n are not absolutely summable. However, since $-2 < 2d-2 < -1$, they are square summable. Hence, Y_n is well defined as a limit in mean square by Problem 25 in Chapter 13. The sum defining X_n is well defined in L^2, L^1, and almost surely by Example 13.10, Problem 26 in Chapter 13, and Problem 41 in Chapter 14. Since X_n is the result of filtering the long-range dependent process Y_n with the stable impulse response h_n, X_n is still long range dependent as pointed out in Section 15.4.

---

[j] Recall that h_n is given by (15.32).
[k] See Problem 31 in Chapter 10 or Example 13.26.
[l] We derived Stirling's formula for $\Gamma(n) = (n-1)!$ in Example 5.18. A proof for noninteger x can be found in [5, pp. 300–301].

Problems

15.1: Self similarity in continuous time

1. Show that for a self-similar process, W_t converges in distribution to the zero random variable as $t \to 0$. Next, identify $X(\omega) := \lim_{t \to \infty} t^H W_1(\omega)$ as a function of ω, and find the probability mass function of the limit in terms of $F_{W_1}(x)$.

2. Use joint characteristic functions to show that the Wiener process is self similar with Hurst parameter $H = 1/2$.

3. Use joint characteristic functions to show that the Wiener process has stationary increments.

4. Show that for $H = 1/2$,
$$B_H(t) - B_H(s) = \int_s^t dW_\tau = W_t - W_s.$$
Taking $t > s = 0$ shows that $B_H(t) = W_t$, while taking $s < t = 0$ shows that $B_H(s) = W_s$. Thus, $B_{1/2}(t) = W_t$ for all t.

5. Show that for $0 < H < 1$,
$$\int_0^\infty [(1+\theta)^{H-1/2} - \theta^{H-1/2}]^2 \, d\theta < \infty.$$

15.2: Self similarity in discrete time

6. Let X_n be a second-order self-similar process with mean $\mu = \mathsf{E}[X_n]$, variance $\sigma^2 = \mathsf{E}[(X_n - \mu)^2]$, and Hurst parameter H. Then the sample mean
$$M_n := \frac{1}{n} \sum_{i=1}^n X_i$$
has expectation μ and, by (15.7), variance σ^2/n^{2-2H}. If X_n is a Gaussian sequence,
$$\frac{M_n - \mu}{\sigma/n^{1-H}} \sim N(0,1),$$
and so given a confidence level $1 - \alpha$, we can choose y (e.g., by Table 6.2) such that
$$\mathsf{P}\left(\left|\frac{M_n - \mu}{\sigma/n^{1-H}}\right| \le y\right) = 1 - \alpha.$$
For $1/2 < H < 1$, show that the width of the corresponding confidence interval is wider by a factor of $n^{H-1/2}$ than the confidence interval obtained if the X_n had been independent as in Section 6.3.

7. Use (15.10) and induction on n to show that (15.6) implies $\mathsf{E}[Y_n^2] = \sigma^2 n^{2H}$ for $n \ge 1$.

8. Suppose that X_k is wide-sense stationary.

    (a) Show that the process $\widetilde{X}_\nu^{(m)}$ defined in (15.13) is also wide-sense stationary.
    (b) If X_k is second-order self similar, prove (15.12) for the case $n = 0$.

15.3: Asymptotic second-order self similarity

9. Show that asymptotic second-order self similarity (15.20) implies (15.19). *Hint:* Observe that $C^{(n)}(0) = E[(X_1^{(n)} - \mu)^2]$.

10. Show that (15.19) implies asymptotic second-order self similarity (15.20). *Hint:* Use (15.14), and note that (15.19) is equivalent to $E[Y_n^2]/n^{2H} \to \sigma_\infty^2$.

11. Show that a process is asymptotically second-order self similar; i.e., (15.20) holds, if and only if the conditions

$$\lim_{m \to \infty} \rho^{(m)}(n) = \frac{1}{2}[|n+1|^{2H} - 2|n|^{2H} + |n-1|^{2H}],$$

and

$$\lim_{m \to \infty} \frac{C^{(m)}(0)}{m^{2H-2}} = \sigma_\infty^2$$

both hold.

12. Show that the covariance function corresponding to the power spectral density of Example 15.2 is

$$C(n) = \frac{(-1)^n \Gamma(1-2d)}{\Gamma(n+1-d)\Gamma(1-d-n)}.$$

This result is due to Hosking [31, Theorem 1(d)]. *Hints:* First show that

$$C(n) = \frac{1}{\pi} \int_0^\pi [4\sin^2(v/2)]^{-d} \cos(nv) \, dv.$$

Second, use the change of variable $\theta = 2\pi - v$ to show that

$$\frac{1}{\pi} \int_\pi^{2\pi} [4\sin^2(v/2)]^{-d} \cos(nv) \, dv = C(n).$$

Third, use the formula [21, p. 372]

$$\int_0^\pi \sin^{p-1}(t) \cos(at) \, dt = \frac{\pi \cos(a\pi/2)\Gamma(p+1)2^{1-p}}{p\Gamma\left(\frac{p+a+1}{2}\right)\Gamma\left(\frac{p-a+1}{2}\right)}.$$

13. Show that for $0 < \alpha < 1$,

$$\int_0^\infty \theta^{\alpha-3} \sin^2 \theta \, d\theta = \frac{2^{1-\alpha} \cos(\alpha\pi/2)\Gamma(\alpha)}{(1-\alpha)(2-\alpha)}.$$

Remark. The formula actually holds for complex α with $0 < \text{Re}\,\alpha < 2$ [21, p. 447]. The formula is used to obtain (15.22) [24, Appendix B].

Hints: (i) Fix $0 < \varepsilon < r < \infty$, and apply integration by parts to

$$\int_\varepsilon^r \theta^{\alpha-3} \sin^2 \theta \, d\theta$$

with $u = \sin^2\theta$ and $dv = \theta^{\alpha-3}\, d\theta$.

(ii) Apply integration by parts to the integral

$$\int t^{\alpha-2} \sin t\, dt$$

with $u = \sin t$ and $dv = t^{\alpha-2}\, dt$.

(iii) Use the fact that for $0 < \alpha < 1$,[m]

$$\lim_{\substack{r\to\infty \\ \varepsilon\to 0}} \int_\varepsilon^r t^{\alpha-1} e^{-jt}\, dt = e^{-j\alpha\pi/2}\Gamma(\alpha).$$

14. Let $S(f)$ be given by (15.15). For $1/2 < H < 1$, put $\alpha = 2 - 2H$.

 (a) Evaluate the limit in (15.21). *Hint:* You may use the fact that

$$Q(f) := \sum_{i=1}^\infty \frac{1}{|i+f|^{2H+1}}$$

converges uniformly for $|f| \le 1/2$ and is therefore a continuous and bounded function on $[-1/2, 1/2]$.

 (b) Evaluate

$$\int_{-1/2}^{1/2} S(f)\, df.$$

Hint: The above integral is equal to $C(0) = \sigma^2$. Since (15.15) corresponds to a second-order self-similar process, not just an *asymptotically* second-order self-similar process, $\sigma^2 = \sigma_\infty^2$. Now apply (15.22).

15.4: Long-range dependence

15. Show directly that if a wide-sense stationary sequence has the covariance function $C(n)$ given in Problem 12, then the process is long-range dependent; i.e., (15.23) holds with appropriate values of α and c [31, Theorem 1(d)]. *Hints:* Use the Remark following Problem 14 in Chapter 4, Stirling's formula,

$$\Gamma(x) \sim \sqrt{2\pi} x^{x-1/2} e^{-x},$$

and the formula $(1 + d/n)^n \to e^d$.

---

[m] For $s > 0$, a change of variable shows that

$$\lim_{\substack{r\to\infty \\ \varepsilon\to 0}} \int_\varepsilon^r t^{\alpha-1} e^{-st}\, dt = \frac{\Gamma(\alpha)}{s^\alpha}.$$

As in Notes **7** and **8** in Chapter 4, a permanence of form argument allows us to set $s = j = e^{j\pi/2}$.

16. For $0 < \alpha < 1$, show that
$$\lim_{n \to \infty} \frac{\sum_{v=k}^{n-1} v^{-\alpha}}{n^{1-\alpha}} = \frac{1}{1-\alpha}.$$

Hints: Rewrite (15.28) in the form
$$B_n + n^{-\alpha} - k^{-\alpha} \le I_n \le B_n.$$

Then
$$1 \le \frac{B_n}{I_n} \le 1 + \frac{k^{-\alpha} - n^{-\alpha}}{I_n}.$$

Show that $I_n/n^{1-\alpha} \to 1/(1-\alpha)$, and note that this implies $I_n/n^{-\alpha} \to \infty$.

17. For $0 < \alpha < 1$, show that
$$\lim_{n \to \infty} \frac{\sum_{v=k}^{n-1} v^{1-\alpha}}{n^{2-\alpha}} = \frac{1}{2-\alpha}.$$

15.5: ARMA processes

18. Use the bound $|\alpha_n| \le M(1 + \delta/2)^{-n}$ to show that $\sum_n |h_n| < \infty$, where
$$h_n = \sum_{k=\max(0, n-q)}^{n} \alpha_k b_{n-k}, \quad n \ge 0,$$
and $h_n = 0$ for $n < 0$.

19. Assume (15.30) holds and that $A(z)$ has all roots strictly inside the unit circle. Show that (15.31) must hold. *Hint:* Compute the convolution
$$\sum_n \alpha_{m-n} Y_n$$
first for Y_n replaced by the left-hand side of (15.30) and again for Y_n replaced by the right-hand side of (15.30).

20. Let $\sum_{k=0}^{\infty} |h_k| < \infty$. Show that if X_n is WSS, then $Y_n = \sum_{k=0}^{\infty} h_k X_{n-k}$ and X_n are J-WSS. Be sure to justify the interchange of any expectations and infinite sums.

Exam preparation

You may use the following suggestions to prepare a study sheet, including formulas mentioned that you have trouble remembering. You may also want to ask your instructor for additional suggestions.

15.1. Self similarity in continuous time. Know the definition and some of the implications of self similarity. If the process also has stationary increments and $H \neq 1$, then it is zero mean and its covariance function is given by (15.4). If an H-sssi process is Gaussian with finite second moments, then the process can be represented by fractional Brownian motion.

15.2. Self similarity in discrete time. This notion is obtained by sampling a continuous-time H-sssi process on the integers. More generally, a discrete-time WSS process whose covariance function has the form (15.6) is said to be second-order self similar. It is important to know that (15.6) holds if and only if (15.7) holds. Know what the aggregated process is. Know the relationship between formulas (15.12) and (15.17). The power spectral density of a second-order self-similar process is proportional to (15.15).

15.3. Asymptotic second-order self similarity. A process is asymptotically second-order self similar if instead of (15.12), we have only (15.20) Know that (15.20) holds if and only if (15.19) holds. The theorem containing (15.21) gives sufficient conditions on the power spectral density to guarantee that the process is asymptotically second-order self similar.

15.4. Long-range dependence. In the time-domain, if a process is long-range dependent as in (15.23), then the process is asymptotically second-order self similar.

15.5. ARMA processes. An ARMA process X_n satisfying (15.29) exists and is given by (15.31) if the impulse response h_n is causal and stable. Under these conditions, the sum in (15.31) converges in L^2, L^1, and almost surely. If $A(z)$ has all roots strictly inside the unit circle, then h_n is causal and stable.

15.6. ARIMA processes. An ARIMA process determined by (15.33) with $0 < d < 1/2$ exists and is given by (15.34) and (15.35). The sum in (15.34) converges in mean square, and the sum in (15.35) converges in L^2, L^1, and almost surely.

Work any review problems assigned by your instructor. If you finish them, re-work your homework assignments.

Bibliography

[1] Abramowitz M. and I. A. Stegun, eds. *Handbook of Mathematical Functions, with Formulas, Graphs, and Mathematical Tables.* New York: Dover, 1970.

[2] Beran J. *Statistics for Long-Memory Processes.* New York: Chapman & Hall, 1994.

[3] Billingsley P. *Probability and Measure*, 3rd ed. New York: Wiley, 1995.

[4] Brockwell P. J. and R. A. Davis. *Time Series: Theory and Methods.* New York: Springer-Verlag, 1987.

[5] Buck R. C. *Advanced Calculus*, 3rd ed. New York: McGraw-Hill, 1978.

[6] Chernoff H. "A measure of asymptotic efficiency for tests of a hypothesis based on the sum of observations," *Ann. Math. Statist.*, **23**, 493–507, 1952.

[7] Chow Y. S. and H. Teicher. *Probability Theory: Independence, Interchangeability, Martingales*, 2nd ed. New York: Springer, 1988.

[8] Chung K. L. *Markov Chains with Stationary Transition Probabilities*, 2nd ed. Berlin: Springer, 1967.

[9] Churchill R. V., J. W. Brown and R. F. Verhey. *Complex Variables and Applications*, 3rd ed. New York: McGraw-Hill, 1976.

[10] Craig J. W. "A new, simple and exact result for calculating the probability of error for two-dimensional signal constellations," in *Proc. IEEE Milit. Commun. Conf. MILCOM '91*, McLean, VA, Oct. 1991, pp. 571–5.

[11] Cramér H. "Sur un nouveaux théorème-limite de la théorie des probabilités," *Actualités Scientifiques et Industrielles 736*, pp. 5–23. *Colloque consacré à la théorie des probabilités*, Vol. 3. Paris: Hermann, Oct. 1937.

[12] Cramér H. *Mathematical Methods of Statistics.* Princeton: Princeton University Press, 1946.

[13] Crovella M. and A. Bestavros. "Self-similarity in World Wide Web traffic: Evidence and possible causes," *Perf. Eval. Rev.*, **24**, 160–9, 1996.

[14] Davis M. H. A. *Linear Estimation and Stochastic Control.* London: Chapman and Hall, 1977.

[15] Devroye L. *Non-Uniform Random Variate Generation.* New York: Springer, 1986.

[16] Feller W. *An Introduction to Probability Theory and its Applications*, Vol. 1, 3rd ed. New York: Wiley, 1968.

[17] Fine T. L. *Probability and Probabilistic Reasoning for Electrical Engineering.* Upper Saddle River, NJ: Pearson, 2006.

[18] Gersho A. and R. M. Gray. *Vector Quantization and Signal Compression.* Boston: Kluwer, 1992.

[19] Gohberg I. and S. Goldberg. *Basic Operator Theory.* Boston: Birkhäuser, 1980.

[20] Gonick L. and W. Smith. *The Cartoon Guide to Statistics.* New York: HarperPerennial, 1993.

[21] Gradshteyn I. S. and I. M. Ryzhik. *Table of Integrals, Series, and Products.* Orlando, FL: Academic Press, 1980.

[22] Gray R. M. and L. D. Davisson. *Introduction to Statistical Signal Processing.* Cambridge, UK: Cambridge University Press, 2005.

[23] Grimmett G. R. and D. R. Stirzaker. *Probability and Random Processes*, 3rd ed. Oxford, UK: Oxford University Press, 2001.

[24] Gubner J. A. "Theorems and fallacies in the theory of long-range-dependent processes," *IEEE Trans. Inform. Theory*, **51** (3), 1234–9, 2005.

[25] Haykin S. and B. Van Veen. *Signals and Systems*, 2nd ed. New York: Wiley, 2003.

[26] Heidelberger P. "Fast simulation of rare events in queuing and reliability models," *ACM Trans. Modeling*

Comput. Simul., **5** (1), 43–85, 1995.

[27] Helstrom C. W. *Statistical Theory of Signal Detection*, 2nd ed. Oxford, UK: Pergamon, 1968.

[28] Helstrom C. W. *Probability and Stochastic Processes for Engineers*, 2nd ed. New York: Macmillan, 1991.

[29] Hoel P. G., S. C. Port and C. J. Stone. *Introduction to Probability Theory*. Boston: Houghton Mifflin, 1971.

[30] Hoffman K. and R. Kunze. *Linear Algebra*, 2nd ed. Englewood Cliffs, NJ: Prentice-Hall, 1971.

[31] Hosking J. R. M. "Fractional differencing," *Biometrika*, **68** (1), 165–76, 1981.

[32] Karlin S. and H. M. Taylor. *A First Course in Stochastic Processes*, 2nd ed. New York: Academic Press, 1975.

[33] Karlin S. and H. M. Taylor. *A Second Course in Stochastic Processes*. New York: Academic Press, 1981.

[34] Kingman J. F. C. *Poisson Processes*. Oxford, U.K.: Clarendon, 1993.

[35] Leland W. E., M. S. Taqqu, W. Willinger and D. V. Wilson. "On the self-similar nature of Ethernet traffic (extended version)," *IEEE/ACM Trans. Networking*, **2** (1), 1–15, 1994.

[36] Leon-Garcia A. *Probability and Random Processes for Electrical Engineering*, 2nd ed. Reading, MA: Addison-Wesley, 1994.

[37] Likhanov N. "Bounds on the buffer occupancy probability with self-similar traffic," pp. 193–213, in *Self-Similar Network Traffic and Performance Evaluation*, Eds. K. Park and W. Willinger. New York: Wiley, 2000.

[38] O'Hagen A. *Probability: Methods and Measurement*. London: Chapman and Hall, 1988.

[39] Oppenheim A. V. and A. S. Willsky, with S. H. Nawab. *Signals & Systems*, 2nd ed. Upper Saddle River, NJ: Prentice Hall, 1997.

[40] Papoulis A. and S. U. Pillai. *Probability, Random Variables, and Stochastic Processes*, 4th ed. New York: McGraw-Hill, 2002.

[41] Park K. and W. Willinger. "Self-similar network traffic: An overview," pp. 1–38, in *Self-Similar Network Traffic and Performance Evaluation*, Eds. K. Park and W. Willinger. New York: Wiley, 2000.

[42] Parzen E. *Stochastic Processes*. San Francisco: Holden-Day, 1962.

[43] Paxson V. and S. Floyd. "Wide-area traffic: The failure of Poisson modeling," *IEEE/ACM Trans. Networking*, **3**, 226–44, 1995.

[44] Picinbono B. *Random Signals and Systems*. Englewood Cliffs, NJ: Prentice Hall, 1993.

[45] Ripley B. D. *Stochastic Simulation*. New York: Wiley, 1987.

[46] Ross S. M. *A First Course in Probability*, 6th ed. Upper Saddle River, NJ: Prentice-Hall, 2002.

[47] Ross S. M. *Simulation*, 3rd ed. San Diego: Academic Press, 2002.

[48] Rothenberg R. I. *Probability and Statistics*. San Diego: Harcourt Brace Jovanovich, 1991.

[49] Roussas G. *A First Course in Mathematical Statistics*. Reading, MA: Addison-Wesley, 1973.

[50] Royden H. L. *Real Analysis*, 2nd ed. New York: MacMillan, 1968.

[51] Rudin W. *Principles of Mathematical Analysis*, 3rd ed. New York: McGraw-Hill, 1976.

[52] Samorodnitsky G. and M. S. Taqqu. *Stable non-Gaussian Random Processes: Stochastic Models with Infinite Variance*. New York: Chapman & Hall, 1994.

[53] Shiryayev A. N. *Probability*. New York: Springer, 1984.

[54] Simon M. K. "A new twist on the Marcum Q function and its application," *IEEE Commun. Lett.*, **2** (2), 39–41, 1998.

[55] Simon M. K. and D. Divsalar. "Some new twists to problems involving the Gaussian probability integral," *IEEE Trans. Commun.*, **46** (2), 200–10, 1998.

[56] Simon M. K. and M.-S. Alouini. "A unified approach to the performance analysis of digital communication over generalized fading channels," *Proc. IEEE*, **86** (9), 1860–77, 1998.

[57] Sinai Ya. G. "Self-similar probability distributions," *Theory of Probability and its Applications*, **XXI** (1), 64–80, 1976.

[58] Smith P. J., M. Shafi and H. Gao. "Quick simulation: A review of importance sampling techniques in communication systems," *IEEE J. Select. Areas Commun.*, **15** (4), 597–613, 1997.

[59] Snell J. L. *Introduction to Probability*. New York: Random House, 1988.

[60] Stacy E. W. "A generalization of the gamma distribution," *Ann. Math. Stat.*, **33** (3), 1187–92, 1962.

[61] Stark H. and J. W. Woods. *Probability and Random Processes with Applications to Signal Processing*, 3rd ed. Upper Saddle River, NJ: Prentice Hall, 2002.

[62] Stein E. M. and R. Shakarchi. *Fourier Analysis: An Introduction (Princeton Lectures in Analysis I)*. Princeton, NJ: Princeton University Press, 2003.

[63] Student. "The probable error of a mean," *Biometrika*, **VI** (1), 1–25, 1908.

[64] Tsybakov B. and N. D. Georganas. "Self-similar processes in communication networks," *IEEE Trans. Inform. Theory*, **44** (5), 1713–25, 1998.

[65] Verdú S. *Multiuser Detection*. Cambridge, UK: Cambridge University Press, 1998.

[66] Viniotis Y. *Probability and Random Processes for Electrical Engineers*. New York: McGraw-Hill, 1998.

[67] Wong E. and B. Hajek. *Stochastic Processes in Engineering Systems*. New York: Springer, 1985.

[68] Yates R. D. and D. J. Goodman. *Probability and Stochastic Processes: A Friendly Introduction for Electrical and Computer Engineers*, 2nd ed. New York: Wiley, 2005.

[69] Youla D. C. "The use of the method of maximum likelihood in estimating continuous-modulated intelligence which has been corrupted by noise," *Trans. IRE Prof. Group on Inform. Theory*, **PGIT-3**, 90–106, 1954.

[70] Ziemer R. Z. *Elements of Engineering Probability and Statistics*. Upper Saddle River, NJ: Prentice Hall, 1997.

Index

A
Abel's theorem, 130
absolutely continuous random variables, 221, 318
absorbing state, 482
acceptance region, 264
accessible state, 499
affine function, 186, 344
aggregated process, 598
almost-sure convergence, 572
almost-sure event, 23
alternative hypothesis, 263
analog-to-digital converter, 150
Anderson–Darling test, 248
angle
    of a point in the plane, 354
AR process, *see* autoregressive process
arcsine random variable, 233, 302, 538
    relation to beta, 233
ARIMA, *see* autoregressive integrated moving average process
ARMA process, *see* autoregressive moving average process
arrival times, 446
associative laws, 10
asymptotic second-order self similarity, 602, 611
asymptotically unbiased estimator, 243
atomic weight, 115
auto-correlation function, 392
autoregressive integrated moving average, 608
autoregressive integrated moving average process, 602
    fractional, 608
autoregressive moving average process, 606
autoregressive process, 606
Avogadro's number, 115

B
Banach space, 524
bandlimited white noise, 406
bandwidth, 408
Bayes' rule, 28, 29
Bernoulli random variable, 71
    mean, 81
    second moment and variance, 86
    simulation, 196
Bernoulli trials, 3
Bernoulli, Jacob, 117
Bessel function, 227
    properties, 229
beta function, 175, 176
beta random variable, 175
    relation to arcsine random variable, 233
    relation to gamma and chi-squared, 325
betting on fair games, 104
biased estimator, 243
binary symmetric channel, 58
binomial approximation
    by normal, 213
    by Poisson, 115, 584
binomial coefficient, 38, 114
binomial random variable, 113
    mean, variance, and pgf, 133
    simulation, 197
binomial theorem, 38, 113, 133, 603
birth process, *see* Markov chain
birth–death process, *see* Markov chain
birthday problem, 36
bivariate characteristic function, 301
bivariate Gaussian random variables, 309
block matrices, 332
Borel–Cantelli lemma
    first, 54, 575
    second, 60
Borel set, 56, 96
Borel sets of $\mathbb{R}^2$, 317
Borel σ-field, 56, 96
Brown, Robert, 387
Brownian motion, 387
    fractional, *see* fractional Brownian motion
    ordinary, *see* Wiener process

C
Campbell's theorem, 452
Cantor set, 55
cardinality, 15, 18
Cartesian product, 289
Cauchy random variable, 144
    as quotient of Gaussians, 323
    as tangent of uniform, 194
    cdf, 186
    characteristic function, 180
    nonexistence of mean, 154
    simulation, 194
    special case of Student's t, 176
Cauchy–Schwarz inequality
    for column vectors, 331, 355
    for random variables, 92, 524
    for time functions, 429
Cauchy sequence
    of L^p random variables, 524
    of real numbers, 524
causal Wiener filter, 419
    prediction, 439
    smoothing, 439

Index

cdf, *see* cumulative distribution function
central chi-squared random variable, *see* chi-squared random variable
central limit theorem, 6, 185, 208, 252, 458, 570
    compared with weak law of large numbers, 571
central moment, 86
certain event, 23
chain rule, 190
    of calculus, 319
    of conditional probability, 58, 510
change of variable (multivariate), 341
Chapman–Kolmogorov equation
    continuous time, 503
    derivation via smoothing property, 544
    discrete time, 484
    for Markov processes, 515
characteristic function
    bivariate, 301
    compared with pgf and mgf, 161
    multivariate (joint), 337
    univariate, 159
Chebyshev inequality, 89, 164, 165, 182
    used to derive the weak law, 116
Chernoff bound, 164, 165, 182
Chevalier de Mere, 3
chi-squared random variable, 148, 174
    as squared zero-mean Gaussian, 179, 192, 222
    cdf – special case of gamma, 225
    characteristic function, 179
    moment generating function, 179
    parameter estimation, 276
    relation to F random variable, 325
    relation to beta random variable, 325
    relation to generalized gamma, 224
    see also noncentral chi-squared, 180
    simulation, 276
    square root of = Rayleigh, 223
chi-squared test, 248
circularly symmetric complex Gaussian, 373
closed set, 535
CLT, *see* central limit theorem
co-domain of a function, 13
combinatorics, 34
communicating states, 499
commutative laws, 10
complement of a set, 8
complementary cdf
    Gaussian, 187, 225
complementary error function, 219
complete orthonormal set, 530
completeness
    of the L^p spaces, 524
    of the real numbers, 521
complex conjugate, 371
complex Gaussian random vector, 372
complex random variable, 371
complex random vector, 372
conditional cdf, 192, 303

conditional density, 192, 303
conditional expectation
    abstract definition, 538
    for discrete random variables, 127
    for jointly continuous random variables, 302
    linearity, 542
    smoothing property, 544, 557
conditional independence, 60, 476
conditional probability, 27
    for jointly continuous random variables, 303
conditional probability mass functions, 118
confidence interval, 250
confidence level, 250
conservative Markov chain, 504, 508
consistency condition
    continuous-time processes, 464
    discrete-time processes, 461
continuity in mean of order p, 521
continuous random variable, 139
    arcsine, 233
    beta, 175
    Cauchy, 144
    chi-squared, 174
    Erlang, 174
    exponential, 141
    F, 325
    gamma, 173
    Gaussian = normal, 145
    generalized gamma, 224
    Laplace, 143
    lognormal, 190
    Maxwell, 222
    multivariate Gaussian, 363
    Nakagami, 224
    noncentral chi-squared, 182
    noncentral Rayleigh, 227
    Pareto, 237
    Rayleigh, 177
    Rice, 227
    Student's t, 176
    uniform, 140
    Weibull, 171
continuous sample paths, 455
convergence
    almost-sure (a.s.), 572
    in distribution, 566
    in mean of order p, 518
    in mean square, 518
    in probability, 564
    in quadratic mean, 518
    of real numbers, 573
    sure, 573
    weak, 566
convex function, 105
convolution
    of densities, 163
    of probability mass functions, 125
correlation, 91

correlation coefficient, 92, 104, 311
correlation function, 392
  engineering definition, 601
  of a deterministic signal, 411
  of a random process, 389
  properties, 391
  statistics/networking definition, 601
  unbiased estimator of, 397
  univariate, for WSS processes, 395
correlation matrix, 334
countable additivity, 23
countable set, 15, 462
countable subadditivity, 26
countably infinite set, 15
counting process, 443
covariance, 94
  distinction between scalar and matrix, 335
  function, 392
  matrix, 335
covering of intervals, 221
Craig's formula, 322
critical region, 264
critical value, 249, 264
cross power spectral density, 406
cross-correlation
  function, 392
    univariate, for WSS processes, 402
  matrix, 337
cross-covariance
  function, 392
  matrix, 336
crossover probabilities, 58, 121
cumulative distribution function (cdf), 184
  continuous random variable, 185
  discrete random variable, 194
  joint, 291
  multivariate, 351
  properties, 205
curve fitting, *see* regression
cyclostationary process, 425

D

dB, *see* decibel
de Moivre, Abraham, 208
de Moivre–Laplace theorem, 255
De Morgan's laws, 10
  generalized, 12
decibel, 188, 437
decorrelating transformation, 338
  applied to a Gaussian vector, 366
delta function, 406
  Dirac, 199, 406
  Kronecker, 397, 483
diagonal argument, 17
difference of sets, 9
differencing filter, 608
differential entropy, 178
Dirac delta function, 199

discrete random variable, 66
  Bernoulli, 71
  binomial, 113
  geometric, 74
  hypergeometric, 256
  negative binomial = Pascal, 133
  Poisson, 69
  uniform, 68
  zeta = Zipf, 105
discrete-time Fourier transform, 400
disjoint sets, 9
distribution, 97
distributive laws, 10
  generalized, 12
domain of a function, 13
dominated convergence theorem, 424, 508, 549
Doob decomposition, 559
dot product, *see* inner product
double factorial, 153
double-sided exponential = Laplace, 143

E

eigenvalue, 485, 528
eigenvector, 485
ellipsoids, 368
embedded chain, 504
empty set, 8
energy spectral density, 412
ensemble mean, 241
ensemble variance, 241
entropy, 105
  differential, 178
equilibrium distribution, 485
equivalence classes, 500, 513
equivalence relation, 500
ergodic theorem, 397
  for Markov chains, 495
  mean-square
    for WSS processes, 424
    for WSS sequences, 519
Erlang random variable, 148, 174
  as nth arrival time of Poisson process, 446
  as sum of i.i.d. exponentials, 181
  cdf – special case of gamma, 225
  cumulative distribution function, 174
  moment generating function, 179
  relation to generalized gamma, 224
  simulation, 277
error function, 188, 219
  complementary, 219
estimation of nonrandom parameters
  covariance matrices, 348
estimation of random vectors
  linear MMSE, 344
  maximum likelihood (ML), 350
  MMSE, 350
estimator
  asymptotically unbiased, 243

biased, 243
unbiased, 241
event, 7, 43, 580
expectation
   additive operator, 84
   homogeneous operator, 83
   linearity for arbitrary random variables, 163
   linearity for discrete random variables, 84
   monotonicity for arbitrary random variables, 163
   monotonicity for discrete random variables, 106
   of a discrete random variable, 80
   of an arbitrary random variable, 155
   when it is undefined, 82, 154
expected average power, 404
   and the Wiener–Khinchin theorem, 421
expected instantaneous power, 404
exponential random variable, 141
   difference of = Laplace, 180
   double sided, *see* Laplace random variable
   memoryless property, 171
   moment generating function, 158
   moments, 158
   relation to generalized gamma, 224

F

F random variable, 325
   relation to chi-squared, 325
factorial
   double, 153
factorial function, 173
factorial moment, 111
factorization property, 109
fading
   channel, 223
   Rayleigh, 324
failure rate, 216
   constant, 218
   Erlang, 237
   Pareto, 237
   Weibull, 237
FARIMA, *see* fractional ARIMA
filtered Poisson process, 451
first entrance time, 488
first passage time, 488
Fourier series, 400
   as characteristic function, 161
Fourier transform, 398
   as bivariate characteristic function, 301
   as multivariate characteristic function, 337
   as univariate characteristic function, 160
   discrete time, 400, 432
   inversion formula, 398
fractional ARIMA process, 608
fractional Brownian motion, 594
fractional Gaussian noise, 596
fractional integrating filter, 608
function
   co-domain, 13

definition, 13
domain, 13
inverse image, 14
invertible, 14
one-to-one, 14
onto, 14
probability measure as a function, 23
range, 13

G

gambler's ruin, 482
gamma function, 79, 148, 173
   incomplete, 225
gamma random variable, 147, 173
   cdf, 225
   characteristic function, 160, 179, 180
   generalized, 224, 325
   moment generating function, 179
   moments, 177
   parameter estimation, 276, 277
   relation to beta random variable, 325
   with scale parameter, 174
Gaussian pulse, 160
Gaussian random process, 464
   fractional, 595
   Karhunen–Loève expansion, 584
Gaussian random variable, 145
   ccdf
      approximation, 225
      Craig's formula, 322
      definition, 187
      table, 189
   cdf, 187
      related to error function, 219
      table, 189
   characteristic function, 160, 180
   complex, 372
   complex circularly symmetric, 373
   moment generating function, 157, 159
   moments, 152
   quotient of = Cauchy, 323
   simulation, 194, 278
Gaussian random vector, 363
   characteristic function, 365
   complex circularly symmetric, 373
   joint density, 367
   multivariate moments, Wick's theorem, 377
   proper, 373
   simulation, 368
generalized density, 199
generalized gamma random variable, 224, 325
   relation to Rayleigh, Maxwell, Weibull, 224
generator matrix, 506
geometric random variable, 74
   mean, variance, and pgf, 132
   memoryless property, 101
geometric series, 52
goodness-of-fit tests, 248

greatest common divisor, 500

H
H-sssi, 593
Herglotz's theorem, 545
Hilbert space, 524
histogram, 244
Hölder's inequality, 550
holding time, 504, 514
Hurst parameter, 591
hypergeometric random variable, 256
　derivation, 274
hypothesis, 248
hypothesis testing, 262, 263

I
i.i.d., *see* independent identically distributed
i.o., *see* infinitely often
ideal gas, 224
identically distributed random variables, 72
importance sampling, 272
impossible event, 22
impulse function, 199
impulse response, 390
impulsive, 199
inclusion–exclusion formula, 24
incomplete gamma function, 225
increment process, 593
increments, 390
increments of a random process, 444
independent events
　more than two events, 32
　pairwise, 32, 46
　two events, 30
independent identically distributed (i.i.d.), 72
independent increments, 444
independent random variables, 71
　cdf characterization, 295
　ch. fcn. characterization, 302, 338
　jointly continuous, 300
　multiple, 72
　pdf characterization, 301
　pmf characterization, 76
　uncorrelated does not imply independent, 104, 322, 327
indicator function, 87
infinitely often (i.o.), 492
infinity conventions, 15
　and expectation, 82, 154
inner product
　of column vectors, 331
　of matrices, 355
　of random variables, 524
inner-product space, 524
integrating filter, 608
integration by parts formula, 168
intensity of a Poisson process, 444
interarrival times, 446

intersection of sets, 9
intervisit times, 490
inverse image, 14
inverse tangent
　principal, 354
irreducible Markov chain, 488, 499
Itô correction term, 457
Itô integral, 457
Itô rule, 457

J
J-WSS, *see* jointly wide-sense stationary
Jacobian, 341
　formulas, 341
Jensen's inequality, 105
joint characteristic function, 337
joint cumulative distribution function, 291
joint density, 295
joint probability mass function, 75
joint wide-sense stationarity, 402
　for discrete-time processes, 434
jointly continuous random variables
　bivariate, 295
jointly Gaussian random variables, 363
jointly normal random variables, 363
jump chain, 504
jump times
　of a Poisson process, 445

K
Karhunen–Loève expansion, 527
　finite-dimensional, 338
　Gaussian process, 584
　Ornstein–Uhlenbeck process, 554
　signal detection, 530
　white noise, 531
　Wiener process, 531
Kolmogorov
　and axiomatic theory of probability, 5, 517
　backward equation, 506
　characterization of random processes, 388
　consistency/extension theorem, 462
　forward equation, 505
Kolmogorov–Smirnov test, 248
Kronecker delta, 397, 483
Kronecker product, 103, 447
kurtosis, 86

L
Laplace random variable, 143
　as difference of exponentials, 180
　parameter estimation, 277
　quotient of, 324
　simulation, 277
　variance and moment generating function, 179
Laplace transform, 158
Laplace, Pierre-Simon, 208
law of large numbers

Index

convergence rates, 596
  mean square, for second-order self-similar sequences, 596
  mean square, uncorrelated, 519
  mean square, WSS sequences, 519
  strong, 273, 576
  weak, for independent random variables, 576
  weak, for uncorrelated random variables, 116, 565
law of the unconscious statistician, 83, 149
law of total probability, 27, 29
  discrete conditioned on continuous, 472
  for conditional expectation, 544
  for conditional probability, 503, 515, 544
  for continuous random variables, 304
  for expectation (continuous random variables), 308, 315
  for expectation (discrete random variables), 129
  unified formula, 540
Lebesgue
  dominated convergence theorem, 424, 549
  measure, 45, 57
  monotone convergence theorem, 169, 549
Leibniz' rule, 191, 307
  derivation, 318
level curves, 310
level sets, 368
likelihood, 127, 192
likelihood ratio
  continuous random variables, 193
  discrete random variables, 127
  martingale, 559
likelihood-ratio test, 127, 136, 193, 223
limit inferior, 567, 579
limit properties of P, 25
limit superior, 567, 579
Lindeberg–Lévy theorem, 208
linear estimators, 535
linear MMSE estimator, 344
linear time-invariant system, 390
location parameter, 146
lognormal random variable
  definition, 190
  moments, 222
long-range dependence, 604
LOTUS, *see* law of the unconscious statistician
LRD, *see* long-range dependence
LTI, *see* linear time-invariant (system)
Lyapunov's inequality, 576
  derived from Hölder's inequality, 551
  derived from Jensen's inequality, 105

M

MA process, *see* moving average process
MAP, *see* maximum a posteriori probability
Marcum Q function, 228, 322
marginal cumulative distributions, 292
marginal density, 299
marginal probability, 290

marginal probability mass functions, 75
Markov chain, 544
  absorbing barrier, 482
  accessible state, 499
  aperiodic state, 500
  birth–death process, 482
  Chapman–Kolmogorov equation, 484
  communicating states, 499
  conservative, 504, 508
  continuous time, 502
  discrete time, 477
  embedded chain, 504
  equilibrium distribution, 485
  ergodic theorem, 495
  first entrance time, 488
  first passage time, 488
  gambler's ruin, 482
  generator matrix, 506
  holding time, 504, 514
  intervisit times, 490
  irreducible, 488, 499
  jump chain, 504
  Kolmogorov's backward equation, 506
  Kolmogorov's forward equation, 505
  m-step transition probabilities, 483
  model for queue
    with finite buffer, 482
    with infinite buffer, 482, 513
  nth entrance time, 489
  null recurrent, 489
  occupation time, 491
    average, 491
    convergence, 495
    total, 492, 512
  period of a state, 500
  periodic state, 500
  positive recurrent, 489
  pure birth process, 482
  random walk
    construction, 477
    continuous time, 513
    definition, 481
    symmetric, 478
  rate matrix, 506
  reachable state, 499
  recurrent state, 488
  reflecting barrier, 482
  sojourn time, 504, 514
  state space, 480
  state transition diagram, 480
  stationary distribution, 485
  time homogeneous
    continuous time, 503
    discrete time, 480
  transient state, 488
  transition probabilities
    continuous time, 502
    discrete time, 480

transition probability matrix, 480
transition rates, 503
Markov inequality, 88, 164, 182
Markov process, 515
Markov property, 477
martingale, 558
    likelihood ratio, 559
Matlab commands
    ./, 79
    .^, 78
    axis, 282
    bar, 245
    besseli, 227
    chi2cdf, 227
    chi2inv, 259
    diag, 340
    eig, 340
    erfc, 219
    erfinv, 252
    eye, 346
    factorial, 134
    fft, 432
    fftshift, 433
    find, 79, 197
    for, 78
    format rat, 80
    gamcdf, 225
    gamma, 173
    gammainc, 225
    gammaln, 231
    geopdf, 79
    histc, 244
    hold off, 246
    hold on, 245
    kron, 447
    linspace, 247
    max, 244
    mean, 241
    mean (to compute mean vectors), 350
    min, 244
    nchoosek, 38
    ncx2cdf, 227
    normcdf, 187
    norminv, 252
    ones, 197
    plot, 247
    poisspdf, 79
    polyfit, 270
    polyval, 270
    rand, 194
    randn, 194
    repmat, 359
    semilogy, 183
    size, 197
    sqrt, 276
    stairs, 448
    std, 241
    stem, 231
    subplot, 282
    sum, 78
    sum (of matrix), 80
    tan, 194
    tinv, 257
    trace, 331
    var, 241
    zeros, 197
Matlab M-files
    allpairs, 102
    bernrnd, 197
    binpmf, 231
matrix exponential, 506
matrix inverse formula, 358, 381
maximum a posteriori probability
    estimator, 350, 360
maximum a posteriori probability rule
    continuous observations, 193
    derivation, 131
    discrete observations, 126
maximum-likelihood estimator, 350
maximum-likelihood rule, 127, 193
Maxwell random variable, 343
    as square root of chi-squared, 223
    cdf, 222
    relation to generalized gamma, 225
    speed of particle in ideal gas, 224
mean, *see* expectation
mean function, 388
mean matrix, 333
mean time to failure, 216
mean vector, 333
mean-square convergence, 518
mean-square ergodic theorem
    for WSS processes, 423
    for WSS sequences, 519
mean-square law of large numbers
    for uncorrelated random variables, 519
    for WSS processes, 424
mean-square periodicity, 551
mean-squared error, 103, 104, 344, 417
measure, 45
median, 170
memoryless property
    exponential random variable, 171
    geometric random variable, 101
Mercer's theorem, 529
mgf, *see* moment generating function
minimum mean squared error, 535
Minkowski's inequality, 523, 551
mixed random variable, 199
mixture density, 172
    noncentral chi-squared, 182
ML, *see* maximum likelihood
MMSE, *see* minimum mean-squared error
modified Bessel function of the first kind, 227
    properties, 229
moment, 84

central, 86
factorial, 111
moment generating function (mgf), 156
  compared with pgf and char. fcn., 162
monotone convergence theorem, 169, 549
monotonic sequence property, 549
monotonicity
  of E, 106, 163
  of P, 24
Monte Carlo estimation, 271
Mother Nature, 23
moving average process, 606
MSE, *see* mean-squared error
MTTF, *see* mean time to failure
multinomial coefficient, 42
multivariate change of variable, 374
mutually exclusive sets, 9
mutually independent events, 32

N

Nakagami random variable, 224, 381
  as square root of chi-squared, 224
negative binomial random variable, 133
noiseless detection, 554
  discrete time, 339
noncentral chi-squared random variable
  as squared non-zero-mean Gaussian, 180, 192, 223
  cdf (series form), 227
  density (closed form using Bessel function), 228
  density (mixture form), 182
  moment generating function, 180, 182
  noncentrality parameter, 180
  parameter estimation, 276
  simulation, 277
  square root of = Rice, 227
noncentral Rayleigh random variable, 227
  square of = noncentral chi-squared, 227
noncentrality parameter, 180, 182
norm
  L^p random variables, 523
  matrix, 355
  vector, 331
norm preserving, 547
normal approximation of the binomial, 213
normal random variable, *see* Gaussian
nth entrance time, 489
null hypothesis, 263
null recurrent, 489
null set, 8

O

occupation time, 491
  average, 491
  convergence, 495
  total, 492, 512
occurrence times, 446
odds, 104

one-sided test, 266
one-tailed test, 266
one-to-one, 14
onto, 14
open set, 57
Ornstein–Uhlenbeck process, 456, 470
  Karhunen–Loève expansion, 554
orthogonal increments, 561
orthogonality principle
  for regression, 269
  general statement, 534
  in the derivation of linear estimators, 347
  in the derivation of the Wiener filter, 417
orthonormal, 528
outcomes, 7
outer product, 331
overshoot, 234

P

pairwise disjoint sets, 12
pairwise independent events, 32
Paley–Wiener condition, 420
paradox of continuous random variables, 149
parallelogram law, 524, 552
Pareto failure rate, 237
Pareto random variable, 154, 170, 177, 179, 182, 237, 588
partition, 12
Pascal, 3
Pascal random variable = negative binomial, 133
Pascal's triangle, 114
pdf, *see* probability density function
period, 500
periodic state, 500
permanence of form argument, 169, 612
permutation, 37
pgf, *see* probability generating function
π–λ theorem, 221
pmf, *see* probability mass function
Poisson approximation of binomial, 115, 584
Poisson process, 444
  arrival times, 446
  as a Markov chain, 502
  filtered, 451
  independent increments, 444
  intensity, 444
  interarrival times, 446
  marked, 450
  occurrence times, 446
  rate, 444
  shot noise, 451
  thinned, 467
Poisson random variable, 69
  mean, 81
  mean, variance, and pgf, 111
  probability generating function, 108
  second moment and variance, 86
population mean, 241

population variance, 241
positive definite matrix, 336
positive recurrent, 489
positive semidefinite
    function, 429
    matrix, 336
posterior probability, 30, 126
power
    expected average, 404
    expected instantaneous, 404
power set, 44
power spectral density, 403, 405
    nonnegativity, 422
predictable process, 559
prediction
    using the Wiener filter, 439
principal
    angle, 354
    inverse tangent, 354
principal inverse tangent, 354
prior probabilities, 30, 127
probability
    written as an expectation, 87
probability density function (pdf), 139
probability generating function (pgf), 108
    compared with mgf and char. fcn., 161
    related to z transform, 108
probability mass function (pmf), 67
probability measure, 22, 460
probability space, 43
projection, 534
    in linear estimation, 347
    onto the unit ball, 534
    theorem, 535, 536
proper subset, 8

Q

Q function
    Gaussian, 225, 226
    Marcum, 228
quadratic-mean convergence, 518
quantizer, 150
queue, *see* Markov chain

R

$\mathbb{R} := (-\infty, \infty)$, the real numbers, 11
random matrix, 333
random points on the unit sphere, 325
random process, 383
    continuous-time, 386
    discrete-time, 383
random sum, 316
random variable
    absolutely continuous, 221
    complex-valued, 371
    continuous, 139
    definition, 63
    discrete, 66
    integer-valued, 67
    precise definition, 96
    singular, 221
    traditional interpretation, 63
random variables
    identically distributed, 72
    independent, 71
random vector, 333
random walk
    approximation of the Wiener process, 457
    construction, 477
    definition, 481
    symmetric, 478
    with a barrier at the origin, 481
range of a function, 13
rate matrix, 506
rate of a Poisson process, 444
Rayleigh random variable
    as square root of chi-squared, 223
    cdf, 222
    distance from origin, 141, 224
    generalized, 223
    moments, 177
    parameter estimation, 276
    quotient of, 324
    relation to generalized gamma, 225
    simulation, 276
    square of = chi-squared, 223
reachable state, 499
real numbers, $\mathbb{R} := (-\infty, \infty)$, 11
realization, 383
rectangle formula, 291
recurrent state, 488
reflecting state, 482
reflexive
    property of an equivalence relation, 499
regression, 267
    curve, 267
    relation to conditional expectation, 282
rejection region, 264
relative frequency, 3
reliability function, 215
renewal equation, 453
    derivation, 468
renewal function, 453
renewal process, 452, 588
resonant frequency, 233
Rice random variable, 227, 380
    square of = noncentral chi-squared, 227
Riemann sum, 391, 431, 439, 526
Riesz–Fischer theorem, 524, 536, 538

S

sample, 240
    mean, 115, 240
    standard deviation, 241
    variance, 241
sample function, 383

sample path, 383
sample space, 6, 22
sampling
    with replacement, 255
    without replacement, 255
sampling without replacement, 274
scale parameter, 146, 174, 224
scatter plot, 268
second-order process, 392
second-order self similarity, 596
self similarity, 591
sequential continuity, 26
set difference, 9
shot noise, 451
σ-algebra, 43
σ-field, 43, 96, 317, 466
signal-to-noise ratio, 94, 188, 413
significance level, 248, 264
Simon's formula, 327
simulation, 271
    confidence intervals, 271
    continuous random variables, 193
    discrete random variables, 196
    Gaussian random vectors, 368
    importance sampling, 272
sinc function, 400
singular random variable, 221
skewness, 86
Skorohod representation, 577
    derivation, 578
SLLN, *see* strong law of large numbers
slowly varying function, 605
Slutsky's theorem, 274, 571
smoothing
    using the Wiener filter, 439
smoothing property, 544, 557
SNR, *see* signal-to-noise ratio
sojourn time, 504, 514
spectral distribution, 545
spectral factorization, 420
spectral process, 545
spectral representation, 549
spontaneous generation, 482
square root of a nonnegative definite matrix, 375
standard deviation, 85
standard normal density, 145
state space of a Markov chain, 480
state transition diagram, *see* Markov chain
stationary distribution, 485
stationary increments, 593
stationary process, 394
    i.i.d. example, 394
    of order n, 393
stationary random process
    Markov chain example, 474
statistic, 240
statistical independence, 30
statistical regularity, 4

Stirling's formula, 176, 584, 609, 612
    derivation using exponential, 212
    derivation using Poisson, 236
    more precise version, 212
stochastic process, 383
strictly stationary process, 394
    Markov chain example, 474
    of order n, 393
strong law of large numbers, 6, 273, 576
Student's t, 176, 325
    cdf converges to normal cdf, 258
    density converges to normal density, 176
    generalization of Cauchy, 176
    moments, 177, 178
submartingale, 558
subset, 8
    proper, 8
substitution law, 304
    continuous random variables, 308, 315
    discrete random variables, 124, 129
    general case, 542
sum of squared errors, 268
supermartingale, 558
sure event, 23
symmetric
    function, 391
    matrix, 334, 335, 374
    property of an equivalence relation, 499
    random walk, 478

T

t, *see* Student's t
thinned Poisson process, 467
tilted density, 273
time constant, 407
time-homogeneity, *see* Markov chain
trace, 331
transfer function, 402
transient state, 488
transition matrix, *see* Markov chain
transition probability, *see* Markov chain
transition rates, 503
transitive
    property of an equivalence relation, 499
transpose of a matrix, 330
trial, 3
triangle inequality
    for L^p random variables, 523
    for numbers, 522
trigonometric identity, 389
twisted density, 273
two-sided test, 265
two-tailed test, 265
Type I error, 264
Type II error, 264

U

unbiased estimator, 241

of a correlation function, 397
uncorrelated random variables, 93
  example that are not independent, 104, 322, 327
uncountable set, 16
uniform random variable (continuous), 140
  cdf, 186
  simulation, 194
  tangent of = Cauchy, 194
uniform random variable (discrete), 68
union bound, 26
  derivation, 54
union of sets, 9
unit impulse, 199
unit-step function, 87, 421

V

variance, 84
variance formula, 85
Venn diagrams, 8

W

weak law of large numbers, 6, 116, 423, 565, 576
  compared with the central limit theorem, 571
Weibull failure rate, 237
Weibull random variable, 171, 222
  moments, 178
  relation to generalized gamma, 225
white noise, 406
  bandlimited, 406
  infinite average power, 406
  Karhunen–Loève expansion, 531
whitening filter, 419
Wick's theorem, 377
wide-sense stationarity
  continuous time, 395
  discrete time, 431, 432
Wiener filter, 419, 535
  causal, 419
  for random vectors, 344
  prediction, 439
  smoothing, 439
Wiener integral, 456, 532
  normality, 584
Wiener process, 388, 454
  approximation using random walk, 457
  as a Markov process, 515
  defined for negative and positive time, 471
  independent increments, 455
  Karhunen–Loève expansion, 531
  normality, 474
  relation to Ornstein–Uhlenbeck process, 470
  self similarity, 592, 610
  standard, 455
  stationarity of its increments, 610
Wiener, Norbert, 388
Wiener–Hopf equation, 419
Wiener–Khinchin theorem, 422
  alternative derivation, 427

WLLN, *see* weak law of large numbers
WSS, *see* wide-sense stationary

Z

z transform, 606
  related to pgf, 108
Zener diode, 266
zero random variable, 579
zeta random variable, 105
Zipf random variable = zeta, 82, 105

Continuous random variables

uniform[a, b]

$$f_X(x) = \frac{1}{b-a} \quad \text{and} \quad F_X(x) = \frac{x-a}{b-a}, \quad a \leq x \leq b.$$

$$E[X] = \frac{a+b}{2}, \quad \text{var}(X) = \frac{(b-a)^2}{12}, \quad M_X(s) = \frac{e^{sb} - e^{sa}}{s(b-a)}.$$

exponential, exp(λ)

$$f_X(x) = \lambda e^{-\lambda x} \text{ and } F_X(x) = 1 - e^{-\lambda x}, \quad x \geq 0.$$

$$E[X] = 1/\lambda, \quad \text{var}(X) = 1/\lambda^2, \quad E[X^n] = n!/\lambda^n.$$

$$M_X(s) = \lambda/(\lambda - s), \quad \text{Re}\, s < \lambda.$$

Laplace(λ)

$$f_X(x) = \tfrac{\lambda}{2} e^{-\lambda |x|}.$$

$$E[X] = 0, \quad \text{var}(X) = 2/\lambda^2. \quad M_X(s) = \lambda^2/(\lambda^2 - s^2), \quad -\lambda < \text{Re}\, s < \lambda.$$

Cauchy(λ)

$$f_X(x) = \frac{\lambda/\pi}{\lambda^2 + x^2}, \quad F_X(x) = \frac{1}{\pi} \tan^{-1}\left(\frac{x}{\lambda}\right) + \frac{1}{2}.$$

$$E[X] = \text{undefined}, \quad E[X^2] = \infty, \quad \varphi_X(v) = e^{-\lambda |v|}.$$

Odd moments are not defined; even moments are infinite. Since the first moment is not defined, central moments, including the variance, are not defined.

Gaussian or normal, $N(m, \sigma^2)$

$$f_X(x) = \frac{1}{\sqrt{2\pi}\,\sigma} \exp\left[-\frac{1}{2}\left(\frac{x-m}{\sigma}\right)^2\right]. \quad F_X(x) = \texttt{normcdf(x,m,sigma)}.$$

$$E[X] = m, \quad \text{var}(X) = \sigma^2, \quad E[(X-m)^{2n}] = 1 \cdot 3 \cdots (2n-3)(2n-1)\sigma^{2n},$$

$$M_X(s) = e^{sm + s^2\sigma^2/2}.$$

Rayleigh(λ)

$$f_X(x) = \frac{x}{\lambda^2} e^{-(x/\lambda)^2/2} \text{ and } F_X(x) = 1 - e^{-(x/\lambda)^2/2}, \quad x \geq 0.$$

$$E[X] = \lambda\sqrt{\pi/2}, \quad E[X^2] = 2\lambda^2, \quad \text{var}(X) = \lambda^2(2 - \pi/2).$$

$$E[X^n] = 2^{n/2} \lambda^n \Gamma(1 + n/2).$$

Note. To obtain the characteristic function from $M_X(s)$, let $s = jv$. In other words, $\varphi_X(v) = M_X(jv)$.

Continuous random variables

gamma(p, λ)

$$f_X(x) = \lambda \frac{(\lambda x)^{p-1} e^{-\lambda x}}{\Gamma(p)}, \quad x > 0, \quad \text{where } \Gamma(p) := \int_0^\infty x^{p-1} e^{-x} dx, \quad p > 0.$$

Recall that $\Gamma(p) = (p-1) \cdot \Gamma(p-1), \quad p > 1$.

$F_X(x) = \mathtt{gamcdf(x,p,1/lambda)}$.

$$E[X^n] = \frac{\Gamma(n+p)}{\lambda^n \Gamma(p)}. \quad M_X(s) = \left(\frac{\lambda}{\lambda - s}\right)^p, \quad \operatorname{Re} s < \lambda.$$

Note that gamma($1, \lambda$) is the same as exp(λ).

Erlang(m, λ) := gamma(m, λ), m = integer

Since $\Gamma(m) = (m-1)!$

$$f_X(x) = \lambda \frac{(\lambda x)^{m-1} e^{-\lambda x}}{(m-1)!} \quad \text{and} \quad F_X(x) = 1 - \sum_{k=0}^{m-1} \frac{(\lambda x)^k}{k!} e^{-\lambda x}, \quad x > 0.$$

Note that Erlang($1, \lambda$) is the same as exp(λ).

---

chi-squared(k) := gamma($k/2, 1/2$)

If k is an even integer, then chi-squared(k) is the same as Erlang($k/2, 1/2$).
Since $\Gamma(1/2) = \sqrt{\pi}$,

for $k = 1$, $f_X(x) = \dfrac{e^{-x/2}}{\sqrt{2\pi x}}, \quad x > 0.$

Since $\Gamma\left(\frac{2m+1}{2}\right) = \frac{(2m-1)\cdots 5 \cdot 3 \cdot 1}{2^m} \sqrt{\pi}$,

for $k = 2m+1$, $f_X(x) = \dfrac{x^{m-1/2} e^{-x/2}}{(2m-1)\cdots 5 \cdot 3 \cdot 1 \sqrt{2\pi}}, \quad x > 0.$

$F_X(x) = \mathtt{chi2cdf(x,k)}$.

Note that chi-squared(2) is the same as exp($1/2$).

Weibull(p, λ)

$f_X(x) = \lambda p x^{p-1} e^{-\lambda x^p} \quad \text{and} \quad F_X(x) = 1 - e^{-\lambda x^p}, \quad x > 0.$

$$E[X^n] = \frac{\Gamma(1+n/p)}{\lambda^{n/p}}.$$

Note that Weibull($2, \lambda$) is the same as Rayleigh($1/\sqrt{2\lambda}$) and that Weibull($1, \lambda$) is the same as exp(λ).

Printed in the United States
By Bookmasters